Lecture Notes in Electrical Engineering

Volume 854

The book series *Lecture Notes in Electrical Engineering* (LNEE) publishes the latest developments in Electrical Engineering - quickly, informally and in high quality. While original research reported in proceedings and monographs has traditionally formed the core of LNEE, we also encourage authors to submit books devoted to supporting student education and professional training in the various fields and applications areas of electrical engineering. The series cover classical and emerging topics concerning:

- Communication Engineering, Information Theory and Networks
- Electronics Engineering and Microelectronics
- Signal, Image and Speech Processing
- Wireless and Mobile Communication
- Circuits and Systems
- Energy Systems, Power Electronics and Electrical Machines
- Electro-optical Engineering
- Instrumentation Engineering
- Avionics Engineering
- Control Systems
- Internet-of-Things and Cybersecurity
- Biomedical Devices, MEMS and NEMS

For general information about this book series, comments or suggestions, please contact leontina.dicecco@springer.com.

To submit a proposal or request further information, please contact the Publishing Editor in your country:

China

Jasmine Dou, Editor (jasmine.dou@springer.com)

India, Japan, Rest of Asia

Swati Meherishi, Editorial Director (Swati.Meherishi@springer.com)

Southeast Asia, Australia, New Zealand

Ramesh Nath Premnath, Editor (ramesh.premnath@springernature.com)

USA, Canada:

Michael Luby, Senior Editor (michael.luby@springer.com)

All other Countries:

Leontina Di Cecco, Senior Editor (leontina.dicecco@springer.com)

**** This series is indexed by EI Compendex and Scopus databases. ****

More information about this series at https://link.springer.com/bookseries/7818

Qilian Liang · Wei Wang ·
Jiasong Mu · Xin Liu · Zhenyu Na
Editors

Artificial Intelligence in China

Proceedings of the 3rd International Conference on Artificial Intelligence in China

Editors
Qilian Liang
Department of Electrical Engineering
University of Texas at Arlington
Arlington, TX, USA

Jiasong Mu
Tianjin Normal University
Tianjin, China

Zhenyu Na
School of Information Science
and Technology
Dalian Maritime University
Dalian, China

Wei Wang
Tianjin Normal University
Tianjin, China

Xin Liu
Dalian University of Technology
Dalian, China

ISSN 1876-1100 ISSN 1876-1119 (electronic)
Lecture Notes in Electrical Engineering
ISBN 978-981-16-9425-7 ISBN 978-981-16-9423-3 (eBook)
https://doi.org/10.1007/978-981-16-9423-3

This Springer imprint is published by the registered company Springer Nature Singapore Pte Ltd.
The registered company address is: 152 Beach Road, #21-01/04 Gateway East, Singapore 189721, Singapore

Contents

Inception Based Medical Image Registration

Wenrui Yan(✉), Baoju Zhang, Cuiping Zhang, Jin Zhang, and Chuyi Chen

Tianjin Key Laboratory of Wireless Mobile Communication and Power Transmission, Tianjin Normal University, Tianjin, China
ywr_tj@163.com

Abstract. Biological tissue has strong absorbability and strong scattering, which may lead to edge blur, low signal-to-noise ratio and so on, which affects the registration of medical images. This chapter simulates and makes the biological tissue imitation image data set under the above conditions, cleverly designs a regression structure embedded in the U-net Inception module, and forms an unsupervised deep learning medical image registration method. It is applied to the biological tissue imitation data set for the first time. The experimental results show that the image registration method based on SIFT, ORB, BRISK, AKAZE and so on can not successfully complete the registration work because it can not find a sufficient number of effective key points on the data set, SURF although it can complete the registration work. But the effect is weaker than the proposed method. Based on the medical image registration method proposed by Inception, this chapter can effectively solve the problem of biological tissue image registration with strong absorption and strong scattering.

Keywords: Medical image registration · Unsupervised · Inception · Unet

1 Introduction

Image registration technology is a high-dimensional optimization process. The aim is to find an optimal spatial transformation to map the corresponding points of two or more images obtained under different image devices and different time conditions to a specified space, that is, to register two or more images obtained by different imaging devices at different time [1] space. At present, image registration technology is widely used in medical image processing, word recognition, material mechanics, remote sensing data analysis and computer vision. Medical image registration technology is an indispensable key step in medical image analysis. It is a prerequisite for medical image fusion, segmentation, contrast and reconstruction.

Medical image registration is an important research direction in medical image processing. It is often necessary to analyze the image data of the same patient with different shooting time and different scales in clinic, and medical image registration technology is needed at this time. Image registration algorithm finds the corresponding relationship between the search data, transforms the image space mapping, rearranges the pixel position in the image, so that the patient area or tissue and organ with diagnostic significance in the target image has the consistency of space and gray diagnosis standard, so as to

Q. Liang et al. (Eds.): Artificial Intelligence in China, LNEE 854, pp. 1–7, 2022.
https://doi.org/10.1007/978-981-16-9423-3_1

provide more accurate auxiliary diagnosis results for doctors. Therefore, the selection of accurate registration algorithm needs to be associated with the object of study and image characteristics in order to obtain good diagnostic results.

At present, image registration methods emerge in endlessly. If classified, they can be divided into the following three types: (1) registration method based on gray information; (2) registration method based on transform domain; (3) registration method based on feature point.

The registration method based on gray information adopts the content of image gray information, so it is called gray registration method. The algorithm mainly uses the optimization algorithm to search the transformation parameters when the registration function reaches the extreme value. It mainly includes the error square sum algorithm (SSD) [2], the normalized correlation (NCC) [3], the maximum likelihood (ML) [4], the mutual information (MI) [5] method and so on the registration method based on transform domain is usually based on Fourier transform [6] and wavelet transform [7], and the image transformation is registered in frequency domain. The method has low computational complexity and good anti-noise ability, and can meet the real-time requirements of registration in some fields.

The registration method based on feature points can solve the image registration of different scales, different rotation angles and even different modes. The scope of application is the most extensive. Scale invariant feature transformation (SIFT) [8] is a classical algorithms in registration method based on feature points. SIFT doesn't change as the image scales and rotates. Part of the situation changes in light and camera view, successfully applied in the registration of visible images. Some other improvements have been made in the feature-based methods such as SURF [9] and BRISK [10], which are mainly to improve the computational efficiency. Compared with SIFT and SURF, they can't retain the edges due to the Gaussian filtering, which will lose the position accuracy and uniqueness. Based on KAZE algorithm [11], the fast explicit diffusion algorithm for accelerating features in nonlinear scale space (AKAZE) [12] introduces an efficient improved local differential binary descriptor (M-LDB) [13], which improves the repeatability and uniqueness compared with SIFT and SURF algorithm.

Deep learning and the development of neural network provide a new direction for image registration. Because neural network can imitate human visual mechanism and obtain feature expression at different scales in image, the feature information of image can be generalized better. This makes the algorithm based on deep learning and neural network have unparalleled prospect in image registration research [14]. P. Weinzaepfel [15] combined deep learning with optical flow for image registration. In recent years, there are [16, 17] registration methods based on deep learning. These methods belong to supervised image registration methods, and the training process of network has a strong dependence on annotated data. Medical image data sets are often small and difficult to label. So it is difficult to meet the learning of neural networks. The spatial conversion function (STN) method [18] allows the network to explicitly process transformations by exploiting the invariance of convolution. It can realize the registration of unsupervised learning in the training stage of neural network.

2 Inception-Based Image Registration Networks

2.1 Inception Structure

The Inception structure [19] first proposed by Christian Szegedy et al. The deep convolution neural network built by this structure has achieved the best detection and classification performance in the ILSVRC14. Its structure is shown below (Fig. 1).

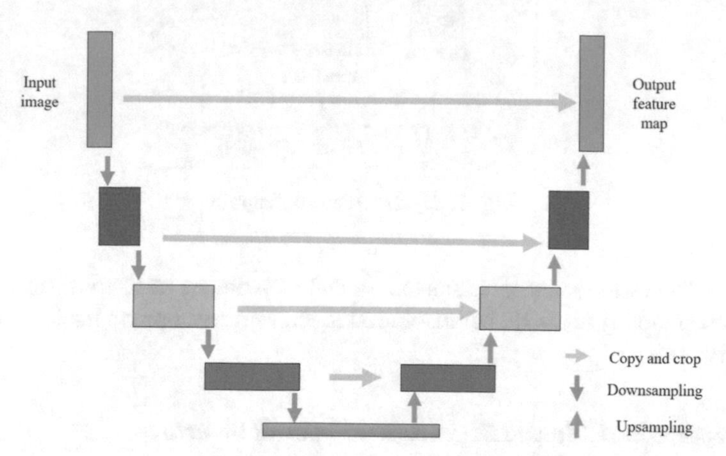

Fig. 1. Inception structured chart

This structure carries on the convolution operation with the convolution kernel size of 1,3,5 and the maximum pooling operation for the incoming input from the upper layer, and then centralizes the results of each branch and continues to pass backward as the output of the next layer. Based on the split-conversion-merger strategy, it makes full use of multiple filters of different scales. The fusion of different features learned by multi-scale receptive field is more beneficial to the network to learn objects of different sizes in the image. The 1×1 convolution added before different scale convolution kernels in Inception structure can reduce the computation and introduce more nonlinear transformations into the network to enhance the learning ability of the whole network to the features.

2.2 U-net Structure

A U-shaped network structure [20] has been proposed by the method which has won many firsts in the ISBI cell tracking competition. It performs well in the field of biomedical image segmentation such as fundus retinal segmentation and lung image segmentation. The detailed structure is shown below (Fig. 2).

U-net structure obtains different size feature layers by convolution and pooling operation, and adds feature channels through upsampling process. This design enables the network to be regarded as a combination of encoder and decoder composed of neural network. It is beneficial for the network to mine higher value features through less data sets. At the same time, the feature fusion method of the U structure splicing links each

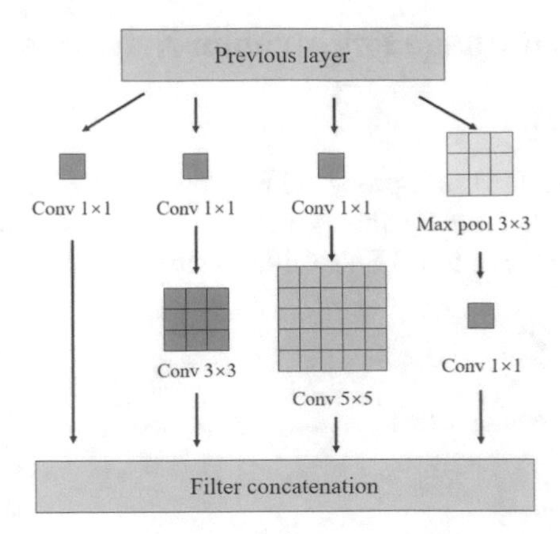

Fig. 2. U-net structure diagram

feature together on the channel dimension, so that more image texture information can be propagated smoothly in the higher resolution layer. Further improve the learning ability of the network.

2.3 Inception-Based Image Registration Network Structure

As a common processing technique in medical image field, registration is more complex than classification, segmentation and detection. It has a very important application in the field of medical image such as detection of lesions, navigation surgery, diagnosis of diseases and so on. This method designs an image registration network structure based on Inception by skillfully combining Inception and U-net structure, and further explores the unsupervised deep learning technology in the field of medical image registration (Fig. 3).

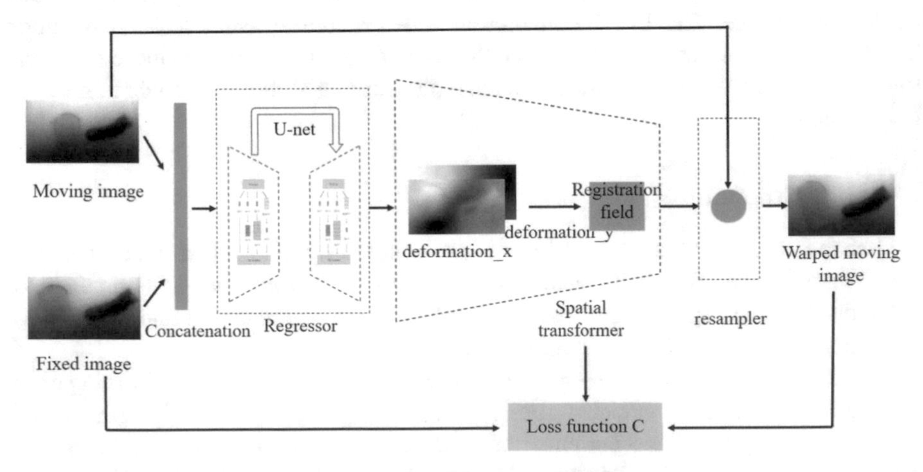

Fig. 3. Inception - based image registration network

This registration network structure is mainly composed of three modules: regression module, spatial conversion module and resampling module, in which regression module is composed of U-net embedded in Inception module. The whole network takes fixed image and floating image as input, outputs the registered mapping image after learning and transformation, and calculates the similarity measure and the value of positive smoothing constraint between the fixed image and the mapped image, which is used as loss function, and updates the parameters by gradient descent method. The loss function formula is as follows:

$$C(f,m,\phi) = C_{sim}(f, m^{\circ}\phi) + \lambda C_{smooth}(\phi) \tag{1}$$

Among them, f and m represent fixed image and floating image respectively, ϕ represents transformation field, $m^{\circ}\phi$ represents floating image after transformation, C_{sim} is used to measure the similarity between the result image and the transformed image, C_{smooth} is regularization term, and constraint space is smooth deformation.

Based on the Inception cleverly designed image registration network structure, it has a multi-scale receptive field self-coding decoder, and allows the original low-level features to participate in the high-level operation, which greatly enriches the diversity of available features. In addition, as an unsupervised deep learning network structure, this network is more suitable for medical image field data set small, calibration data less.

3 Experimental Results and Analysis

The 700 images and fixed images in the dataset are trained in pairs of input image registration network based on Inception. The learning rate is set to 4e-4, and the trained registration network is tested. The test comparison diagram is shown below (Fig. 4).

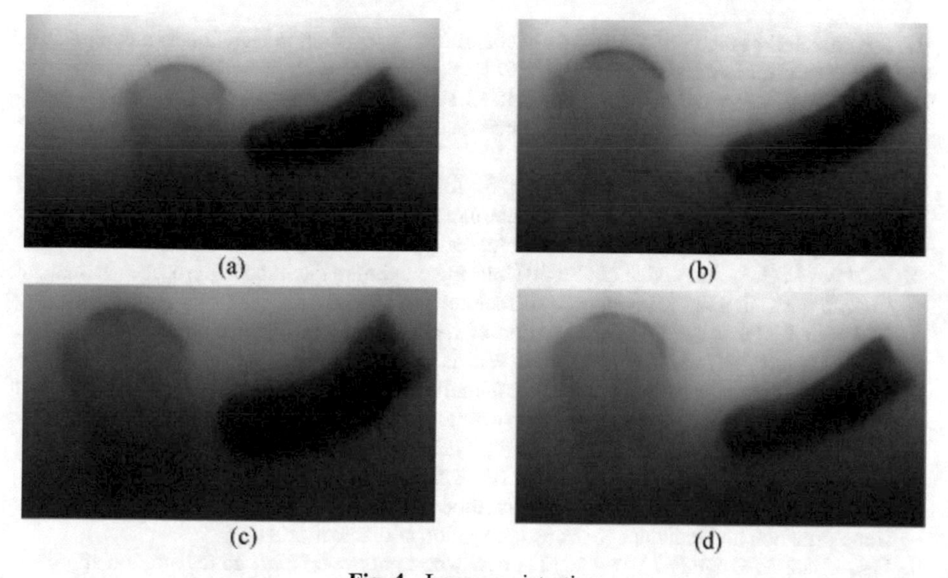

Fig. 4. Image registration

As shown above, (a) is a floating image, (b) is a fixed image, (c) represents the image obtained by registration after detecting key points by SURF method, and (d) represents the image obtained after processing by the method proposed in this chapter. The floating image and the fixed image have great fuzziness, the effective key point is difficult to find, the ORB, BRISK, AKAZE method can not get the effective result, the SURF method obtains the transformed image by the limited key point registration. The image of the SURF method retains the texture of the floating image, but the vertical stretch of the image makes the shape of the simulated heterogeneous body inconsistent with the fixed image. Compared with the fixed image, the result image obtained by this method is closer to the fixed image in shape, and the registration effect is better.

4 Conclusion

This chapter introduces the Inception structure and U-net structure. An Inception-based image registration network is designed by using the ingenious chimerism of Inception and U-net, and it is first applied to the biological tissue imitation dataset with sparse effective feature points. Then this chapter introduces the preparation of experimental data sets, and compares the key points of different methods for the detection of experimental data sets and ordinary natural images, and analyzes the reasons why some classical registration schemes can not effectively complete the registration task. Lastly, through the experimental study, the method can accomplish the task of medical image registration which can not be completed by SIFT, ORB, BRISK, AKAZE and other methods because of blurred image and sparse feature points, and achieve the image registration effect better than the classical SURF method in the whole and details.

References

1. S Karthick S Maniraj 2019 Different medical image registration techniques: a comparative analysis Curr. Med. Imaging 15 10 911 921
2. Chen, W.-J.: Standard deviation normalized summed squared difference for image registration. In: 2017 International Conference on Digital Image Computing: Techniques and Applications (DICTA), pp. 1–8 (2017).
3. Sarvaiya, J.N., Patnaik, S., Bombaywala, S.: Image registration by template matching using normalized cross-correlation. In: 2009 International Conference on Advances in Computing, Control, and Telecommunication Technologies, pp. 819–822 (2009)
4. R Alexis M Grégoire A Nichlolas 2000 Unifying maximum likelihood approaches in medical image registration Int. J. Imaging Syst. Technol. 11 1 71 80
5. M Unser P Thevenaz 2000 Optimization of mutual information for multiresolution image registration IEEE Trans. Image Process. 9 12 2083 2099
6. W Pan K Qin Y Chen 2009 An adaptable-multilayer fractional fourier transform approach for image registration IEEE Trans. Pattern Anal. Mach. Intell. 31 3 400 414
7. MP Sampat Z Wang S Gupta AC Bovik MK Markey 2009 Complex wavelet structural similarity: a new image similarity index IEEE Trans. Image Process. 18 11 2385 2401
8. Zhou, X., et al.: Image registration method integrating image scale invariant feature transformation and individual entropy correlation coefficient (2016).
9. G-Q Zhang W-Z Wu H-J Wang 2012 A new wood microscopic image registration approach based on speeded up robust features (SURF) J. Zhejiang A & F Univ. 29 4 600 605

10. Leutenegger, S., Chli, M., Siegwart, R.Y.: BRISK: Binary Robust Invariant Scalable Keypoints. In: 2011 International Conference on Computer Vision, pp. 2548–2555 (2011).
11. Alcantarill, P.F., Bartoli, A., Davison, A.J.: KAZE features. In: ECCV'12 Proceedings of the 12th European Conference on Computer Vision, vol. Part VI, pp. 214–227 (2012).
12. Bartoli, A., Nuevo, J., Alcantarilla, P.F.: Fast explicit diffusion for accelerated features in nonlinear scale spaces. In: British Machine Vision Conference (BMVC), Bristol, UK (2013).
13. X Liu Y Qiu 2020 Local feature point matching algorithm with anti-affine property J. Comput. Appl. 40 4 1133 1137
14. Viergever, M.A., Antoine Maintz, J.B., Klein, S., Murphy, K., Staring, M., Pluim, J.P.W.: A survey of medical image registration–under review. Med. Image Anal. **33**, 140–144 (2016)
15. Weinzaepfel, P., Revaud, J., Harchaoui, Z., Schmid, C.: Deepflow: Large displacement optical flow with deep matching. In: IEEE International Conference on Computer Vision (ICCV), pp.1385–1392 (2013).
16. M-M Rohé M Datar T Heimann M Sermesant X Pennec 2017 SVF-Net: learning deformable image registration using shape matching M Descoteaux L Maier-Hein A Franz D Pierre Jannin L Collins S Duchesne Eds Medical Image Computing and Computer Assisted Intervention – MICCAI 2017: 20th International Conference, Quebec City, QC, Canada, September 11-13, 2017, Proceedings, Part I Springer International Publishing Cham 266 274 https://doi.org/10.1007/978-3-319-66182-7_31
17. J Krebs T Mansi H Delingette L Zhang FC Ghesu S Miao AK Maier N Ayache R Liao A Kamen 2017 Robust non-rigid registration through agent-based action learning M Descoteaux L Maier-Hein A Franz D Pierre Jannin L Collins S Duchesne Eds Medical Image Computing and Computer Assisted Intervention – MICCAI 2017: 20th International Conference, Quebec City, QC, Canada, September 11-13, 2017, Proceedings, Part I Springer International Publishing Cham 344 352 https://doi.org/10.1007/978-3-319-66182-7_40
18. Jaderberg, M., Simonyan, K., Zisserman, A., Kavukcuoglu, K.: Spatial Transformer Networks. In: Advances in neural information processing systems, pp. 2017–2025 (2015)
19. Szegedy, C., et al.: Going deeper with convolutions. In: Proceedings of the IEEE conference on computer vision and pattern recognition, pp.1–9 (2015)
20. O Ronneberger P Fischer T Brox 2015 U-Net: convolutional networks for biomedical image segmentation N Navab J Hornegger WM Wells AF Frangi Eds Medical Image Computing and Computer-Assisted Intervention – MICCAI 2015: 18th International Conference, Munich, Germany, October 5–9, 2015, Proceedings, Part III Springer International Publishing Cham 234 241 https://doi.org/10.1007/978-3-319-24574-4_28
21. ST Flock SL Jacques BC Wilson WM Star MJC Gemert van 1992 Optical properties of intralipid: a phantom medium for light propagation studies Lasers Surg. Med. 12 5 510 519 https://doi.org/10.1002/lsm.1900120510

Research on Smart Home System Based on Internet of Things

Hai Wang[1], Peng Sun[2], Guiling Sun[1], Zhihong Wang[1], Xiaomei Jiang[1], Chaoran Bi[3], and Ying Zhang[1,4(✉)]

[1] Teaching Center for Experimental Electronic Information, College of Electronic Information and Optical Engineering, Nankai University, Tianjin 300350, China
caroline_zy@nankai.edu.cn

[2] Tianjin Key Laboratory of Optical Thin Film, Tianjin Jinhang Technical Physics Institute, Tianjin 300308, China

[3] College of Media Design, Tianjin Modern Vocational Technology College, Tianjin 300350, China

[4] Tianjin Key Laboratory of Optoelectronic Sensor and Sensing Network Technology, Tianjin 300071, China

Abstract. In the era of Internet of Things (IOT), a more convenient and efficient life concept has created by the smart home system. In this integrated home solution, in order to realize the intelligent control, information technique is frequently utilized to interconnect ordinary household appliances to form an internal family network. With the help of sand table, the real environment is simulated in this design. Different types of sensors are integrated on ZigBee module which controller is CC2531. The intelligent gateway is responsible for analyzing and processing the environmental information obtained by the sensors. The proprietor can control one or more household electrical appliance and set personalized and intelligent home control scenarios according to the actual needs, such as home mode, out mode, sleep mode, etc.

Keywords: Smart home system · Internet of Things (IOT) · ZigBee · Sensor

1 Introduction

IOT is an information carrier based on the Internet and traditional telecommunication network [1, 2]. It enables all the ordinary physical objects, which can be independently addressed, to form an interconnected network [3, 4]. The information of sound, light, heat, electricity, mechanics, chemistry, biology and position of these objects are detected and collected in real time through all kinds of possible network access [5–7]. Thus, there is a ubiquitous connection among the things and controller which realizes the intelligent perception, identification and management [8, 9].

Q. Liang et al. (Eds.): Artificial Intelligence in China, LNEE 854, pp. 8–14, 2022.
https://doi.org/10.1007/978-981-16-9423-3_2

Smart home system is a typical embodiment of the Internet of Things [10–12]. Based on this information platform, various sensors, RFID technology, global positioning system, laser scanners and other devices or technologies are utilized for overall management of the intelligent home appliance controlling, indoor environment detecting, security monitoring and other systems [13–15]. It provides a safe, comfortable, energy-efficient and eco-friendly living environment for the proprietors [16–18]. According to the personalized demand, they can manage the operation of the intelligent system through the remote control at home. Outdoors, the smartphones or computers are the proprietors' preferred controller.

2 Design of Smart Home System

The networking of smart home based on Internet of Things adopts hierarchical architecture, which is composed of sensor layer, transport layer and application layer. The sensor layer includes various types of sensors, controllers and actuators related to domestic appliances, as well as the physical connection of wired network combined with wireless ubiquitous network. The transport layer, which is responsible for the interoperability of different application protocol specifications, mainly contains two parts: home internal network and backbone network access. The former supports lots of network protocols such as IEEE 802.3, IEEE1394, CEBus and USB3.0. The latter includes RF technology, Bluetooth, ZigBee, etc. The application layer is a consumer-centered integrated business layer, which provides human–computer interaction interface and other technical support services of smart home system. It can adequately guarantee the media, security and other business services through the multi-layer cooperative QoS, as well as the adaptive matching of heterogeneous network and terminal equipment.

2.1 Hardware Components

The IOT gateway of smart home system, which processor is S3C6410 chip, supports ARM11 and Cortex-A8 core boards. It enables the applications to run stably in Linux, Wince, Android and other operating systems. ZigBee2007/PRO is applied as the sensor network protocol. The controller of ZigBee wireless communication module is CC2531 with low power consumption. Some wireless modules are integrated with detection equipment, such as high-precision temperature and humidity sensor, light sensor, pressure sensor, 3-axis acceleration sensor, smoke sensor, alcohol sensor, Hall sensor. Some are integrated with the electrical relay, stepping motor, infrared remote controller and other control units. Lots of hardware expansion interfaces, which are Ethernet, USB, serial port, voice, JATG, SD card and VGA, are supported by wireless sensor module.

The structure diagram of smart home system is illustrated in Fig. 1, including three major parts: environmental monitoring, power control and intelligent security, which are responsible for detecting and collecting environmental data, as well as controlling the working status of household appliances and other equipment. Temperature, humidity, light, pressure and other information are real-time monitored by various types of sensors. Smart home gateway is the core component in charge of the management of the whole ZigBee wireless sensor network, and it is also the bridge between the server and the three parts as mention before. It records the type, physical address and network address of each sensor node in the network, and sends the information to the server. A personal computer or smart phone is selected to be the control terminal for establishing communication with the intelligent gateway. As a result, the information of sensor nodes in the whole network can be displayed on the software platform, which is convenient for users to process the data.

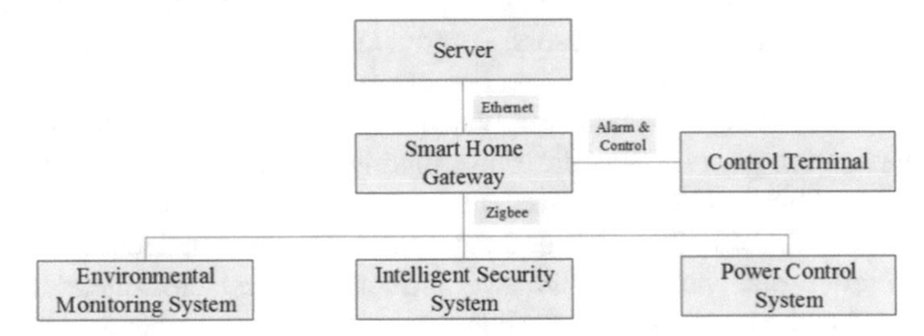

Fig. 1. Structure diagram of smart home system

2.2 Software Architecture

The software platform, including wireless sensor network software, intelligent gateway software and PC server software, is mainly utilized for the configuration of system operation parameters and platform management. It is developed based on .NET Framework, and Visual C# is the development language. IOT server and application software is an application graphical development platform, which can analyze the message and process of the whole system. It is responsible for the topology management of ZigBee network nodes, real-time acquisition and graphical display of sensor network data, smart home system implementation, etc. In addition, the management software of ARM gateway can automatically identify the access type of devices, provide selection and setting of remote communication mode, collect and process the data of sensor network, report the data with PC server software.

Consumers can not only browse details of the node, but also alter the configuration, for example the collection cycle of data. Besides, they can install Python for secondary development, building a personalized IOT application environment. According to the characteristics of different sensors and the form of various networks, more and more practical application modes of Internet of Things have been developed.

3 System Function

Sand table of smart home system is shown in Fig. 2. Connect the ARM gateway and the computer to ensure the normal operation of the hardware platform, then build the software platform. Install the necessary software development environment such as .NET Framework and Python. Establish the ZigBee wireless sensor network protocol and contact with the sensor nodes to get the network topology. The consumer uses the computer to complete the related operations afterward, for example, detecting the environmental data and simulating the relevant scenarios.

Fig. 2. Sand table of smart home system

3.1 Sensor Node Control

Smart home system contains various types of sensor node modules, which can real-time monitor temperature, humidity, light intensity, acceleration, pressure and other information. According to the requirements of the configuration of "acquisition cycle" and "range" and other basic parameters, the live data can be obtained. Figure 3 is the curve of light intensity with time. The vertical axis is light intensity which unit is lux, and the horizontal axis is time.

After mastering the working principle, major function and control method of all sensors in the system platform, users can carry out the comprehensive application of two or more devices. For example, they can set the detection threshold before the light sensor is turned on for real-time monitoring. When the ambient light intensity is greater than the threshold, turn off the light bulb and open the curtain in the living room through the relay, otherwise the opposite operation will be carried out. It implies that the corresponding devices can be controlled by the smart home system according to different results of presupposed conditions.

3.2 Personalized Scenario Design

The smart home system can meet the personalized requirements of different consumers in actual use, which is the same as "outdoor mode", "conference mode" and "ordinary

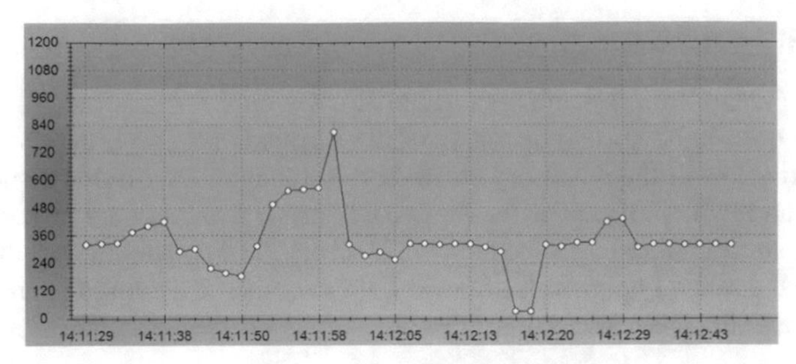

Fig. 3. Light intensity curve

mode" in mobile phones. As shown in Fig. 4, the temperature, humidity, smoke, acceleration, pressure and other sensors are reasonably configured in the system platform. These various sensors perform their duties through the gateway and coordinator network. The collected data is transmitted to the upper computer and compared with the preset value, so as to control the light bulb, air conditioner, electric fan, curtain and other household appliances.

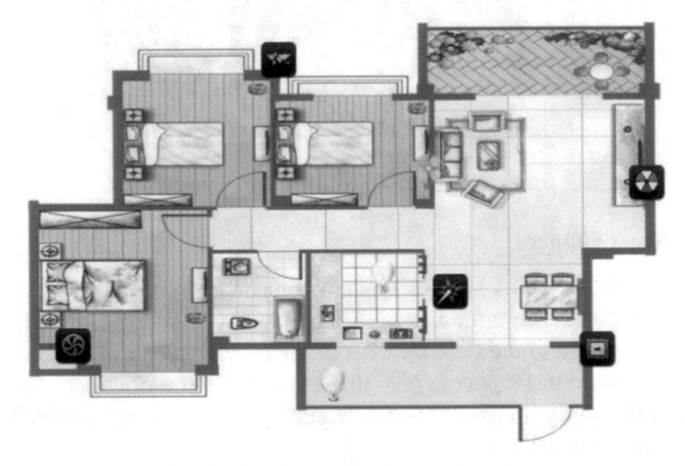

Fig. 4. Deployment scheme of sensors

The correct program, written in Python, is obtained after repeated compilation and verification. It can be run on the upper computer in order to meet the designed requirement. The fan, air conditioner and other electrical appliances are turned off when the proprietor is not at home,, but the access control system works normally. The resident swipes the RFID card to open the door while he or she returns. Meanwhile, the smart home system switches to "home mode" (as shown in Fig. 5), turning on the corresponding electrical appliances to provide a comfortable living environment. Similarly, there are "out mode", "sleep mode", "get up mode" and other personalized and customized scenarios.

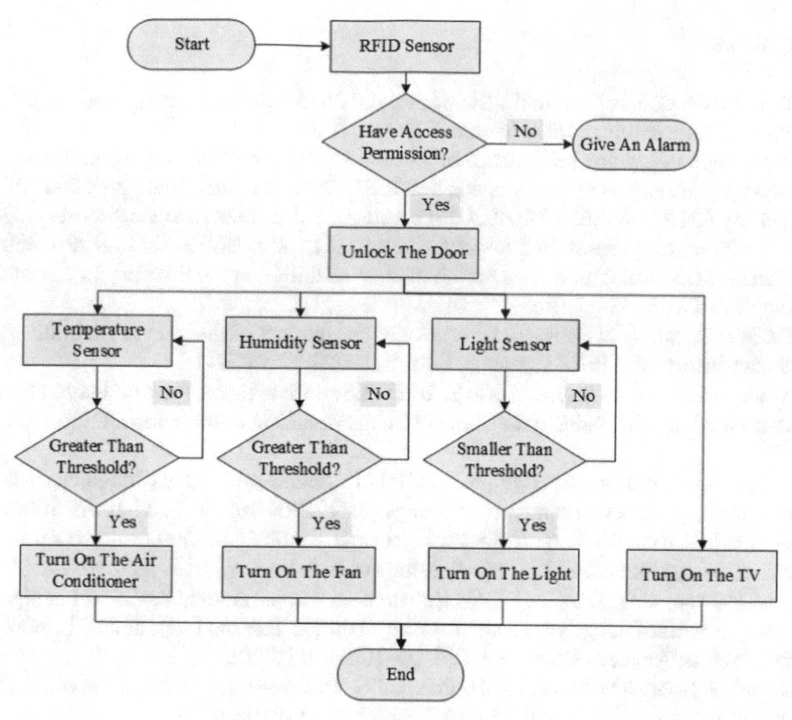

Fig. 5. Flow chart of home mode

4 Conclusion

A smart home system based on Internet of Things is designed in this paper. It adopts the hierarchical architecture and integrates various types of sensors on the ZigBee module which microcontroller is CC2531. The personalized requirements in different scenarios are realized by networking technology. The improvement of smart home application technology is of great significance to the realization of smart city. Besides, it also provides more possibilities for the application of artificial intelligence technology in different scenarios.

Acknowledgements. This work was supported by the 2021 Self-made Experimental Teaching Instrument and Equipment Project Fund of Nankai University (21NKZZYQ01); the 1st batch of industry university cooperative education projects in 2021 (202101186002 and 202101186014); the 2nd batch of industry university cooperative education projects in 2021 (202102296002); Tianjin Science and Technology Project (20YDTPJC00760); the 2020 Rolling Cultivation Project Fund of Self-made Experimental Teaching Instrument and Equipment of Nankai University; the 2020 Undergraduate Education Reform Project Fund (NKJG2020326); Transformation and Extension of Agricultural Scientific and Technological Achievements in Tianjin (201901090) and Tianjin Key Laboratory of Optoelectronic Sensor and Sensing Network Technology.

References

1. MD Sanctis E Cianca G Araniti I Bisio R Prasad 2015 Satellite communications supporting internet of remote things IEEE Internet Things J. 3 1 113 123
2. P Porambage J Okwuibe M Liyanage M Ylianttila T Taleb 2018 Survey on multi-access edge computing for internet of things realization IEEE Commun. Surv.Tutor. 20 4 2961 2991
3. W Osamy AM Khedr A Salim 2019 ADSDA: adaptive distributed service discovery algorithm for internet of things based mobile wireless sensor networks IEEE Sens. J. 19 22 10869 10880
4. C Buratti 2015 Testing protocols for the internet of things on the EuWIn platform IEEE Int. Things J. 3 1 124 133
5. I Butun P Osterberg H Song 2019 Security of the internet of things: vulnerabilities, attacks and countermeasures IEEE Commun. Surv. Tutor. 22 1 616 644
6. T Wang H Luo W Jia A Liu M Xie 2020 MTES: an intelligent trust evaluation scheme in sensor-cloud-enabled industrial internet of things IEEE Trans. on Industr. Inform. 16 3 2054 2062
7. H Jamali-Rad VV Beveren X Campman D Hohl J Brand van den 2020 Continuous Subsurface Tomography over Cellular Internet of Things (IoT) IEEE Sens. J. 20 17 10079 10091
8. MA Bhatti R Riaz SS Rizvi S Shokat F Riaz SJ Kwon 2020 Outlier detection in indoor localization and internet of things (IoT) using machine learning J. Commn. Net. 22 3 236 243
9. Wang, H., Sun, G.L., Gao, Y., Li, X.: Research on escape strategy based on intelligent fire-fighting internet of things virtual simulation system. In: The 2nd International Conference on Artificial Intelligence in China, vol. 653, pp. 102–110 (2020).
10. BD Davis JC Mason M Anwar 2020 Vulnerability studies and security postures of IoT devices: a smart home case study IEEE Internet Things J. 7 10 10102 10110
11. M Yamauchi Y Ohsita M Murata K Ueda Y Kato 2020 Anomaly detection in smart home operation from user behaviors and home conditions IEEE Trans. on Consum. Electr. 66 2 183 192
12. K Cao G Xu J Zhou T Wei M Chen S Hu 2018 QoS-adaptive approximate real-time computation for mobility-aware IoT lifetime optimization IEEE Trans. Comput.-Aided Des. Integr. Circuits Syst. 38 10 1799 1810
13. BC Chifor I Bica VV Patriciu I Bica 2017 A security authorization scheme for smart home internet of things devices Future Gener. Comput. Syst. 86 740 749
14. A Zielonka M Wozniak S Garg G Kaddoum M Jalil Piran G MuHammad 2021 Smart homes: how much will they support us? a research on recent trends and advances IEEE Access 9 26388 26419
15. MA Paredes-Valverde G Alor-Hernández JL García-Alcaráz M Pilar Salas del LO Colombo-Mendoza JL Sánchez-Cervantes 2019 IntelliHome: An Internet of Things-Based System for Electrical Energy Saving in Smart Home Environment Comput. Intell. 36 1 203 224
16. P Kumar A Braeken A Gurtov J Iinatti P Hoai Ha 2017 Anonymous secure framework in connected smart home environments IEEE Trans. Inf. Foren. Sec. 12 4 968 979
17. M Tatan 2019 Internet of things based smart energy management for smart home KSII Trans. Internet Inf. Syst. 13 6 2781 2798
18. B Aksanli TS Rosing 2017 Human behavior aware energy management in residential cyber-physical systems IEEE Trans. Emerg. Top. Com. 8 1 45 57

Autoencoder-Based Baseline Parameterized by Central Limit Theorem for ICS Cybersecurity

Gang Yue[1](✉), Zhuo Sun[1], Jianwei Tian[2], Hongyu Zhu[2], and Bo Zhang[3]

[1] Beijing University of Posts and Telecommunications, Beijing 100876, China
{yuegang,zhuosun}@bupt.edu.cn

[2] State Grid Hunan Electric Power Company, Changsha 410007, Hunan, China
{tianjw,zhuhy}@sgcc.com.cn

[3] Global Energy Internet Research Institute, Beijing 102209, China
zhangbo@geiri.sgcc.com.cn

Abstract. Industrial control system as the core of the industry is concerned about the cybersecurity problem and vulnerable to be threatened by the cyber-attacker. However, the conventional IDS aims to mine intrusion features and realizes intrusion detection by matching the abstract features of intrusion, so it could not recognize unknown and zero-day intrusion. In fact, the ICS as the closed-loop control system is different from the commercial internet and has stable interactive features. In the paper, we analyze the ICS network interaction and construct a parameterized baseline by an autoencoder to detect the intrusion. The experiment with an open ICS dataset shows that this baseline could achieve intrusion detection accuracy above 90% and the false alarm rate below 5%.

Keywords: Industrial control system · Intrusion detection · Autoencoder · Baseline

1 Introduction

Industrial control systems (ICS) are widely used in the national critical infrastructure and control the manufacturing process automatically. Combined with the Internet technology, the ICS is exposed to public networks and suffers from cyber-attacks which lead the facility malfunction, performance reduction and even catastrophe to the nation. For examples, the Iran nuclear station ICS is infected by the virus called "Stuxnet" and operates abnormally in 2010; the railway's SCADA in Japan is threated by the backer and out of control in 2011 ; the BlackEnergy attacks power grid of Ukraine by advanced persistent threat in 2015

G. Yue—Supported by 2020 industrial Internet innovation and development project - smart energy Internet security situation awareness platform project.

Q. Liang et al. (Eds.): Artificial Intelligence in China, LNEE 854, pp. 15–23, 2022.
https://doi.org/10.1007/978-981-16-9423-3_3

and leads the power cutting in the massive area. [1] Therefore, the protection of ICS as the cornerstone for developing is important to all industry.

Intrusion detection as the principal method to protect ICS consists of the positive and passive mode. The signification-based detection as the primary passive method has the defect of low detection accuracy and difficulty in updating the signification database. So, we propose the baseline as a positive way to monitor the ICS operating status and protect it from intrusion. Baseline could outline the profile of flow interaction, represent the in-depth stable relations and detect the malicious data. As for ICS network, the facilities transmit data with each other by request-respond mode. The interaction could represent the ICS network intrinsic attribute and be reflected in the flow sequence. [2] To construct the baseline, an autoencoder model is proposed to learn and fit the distribution of benign data, then the latent data of autoencoder is calculated to output the parameters of baseline by the central limit theorem and tested for intrusion detection.

The rest of this paper is structured as follows. In Sect. 2, we briefly present the state-of-the-art model in intrusion detection of ICS. In Sect. 3, we introduce the ICS network feature and construct autoencoder-based parameterized baseline of an open ICS dataset. In Sect. 4, we utilize the baseline to detect five types of intrusion in the open dataset. Finally, in Sect. 5, we present our conclusion and future work.

2 Relative Works

In this section, we briefly review the relative works about ICS intrusion detection. Intrusion detection is to distinguish the malicious data from the integral dataset and can be recognized as the classification problem. The conventional network security tools such as firewall, IDS etc. are implemented in ICS and abolish the data which matches to the malicious signification rules. As the passive protection mode, it does not support the real-time recognition and could not detect the APT attack. [3] Meanwhile, no matter protocol-based, signification-based or packet-based detection methods rely on the prior knowledge or experts' experience to enact the examination rules. It costs much expense and has no ability to detect zero-day intrusion.

Several ML-based classifiers have been explored over the last years for network intrusion detection. In [4], a CNN based on the transformer structure with attention mechanism was proposed to predict the flow. [5] combined the spark distribution technology and CNN-LSTM model to deal with the immense data in real-time. [6] the performance of different ML-based classification over internet traffic datasets and showed that different flow features affect the classification accuracy and deep learning-based neural network performed better. However, ICS network which is rather different from commercial network has unique interactive features and rare malicious data in reality.

Inspired by the approaches in the [7], they focused on how to distinguish the in-distribution and out-of-distribution data precisely by improving the neural network to fit with the benign data only. Intrusion detection is also the process to distinguish the out-of-distribution data if the model has learned and fitted the in-distribution data accurately. If the date is out of distribution, the fitted model will provide low-confidence prediction to it. So how to construct the baseline of in-distribution data and then use the baseline to detect intrusion is our purpose.

3 Background and Structure

ICS network is different from the commercial internet. In the section, we will briefly introduce the ICS network unique features and how to utilize autoencoder to construct a parameterized baseline in detail.

3.1 ICS Network Introduction

Although the advances in ICS network such as the incorporation of Ethernet technology have started to blur the line between industrial and commercial networks, at their cores they each have fundamentally different requirements. The most essential difference is that ICS network is connect to physical facilities in from form and used to control and monitor real-world actions and conditions. As Fig. 1 shows, ICS network consists of 3 layers commonly and is composed of the sensors, actuators in field layer, PLC, DCS in control layer and SCADA in management layer interacted mutually by fieldbus and industrial ethernet.

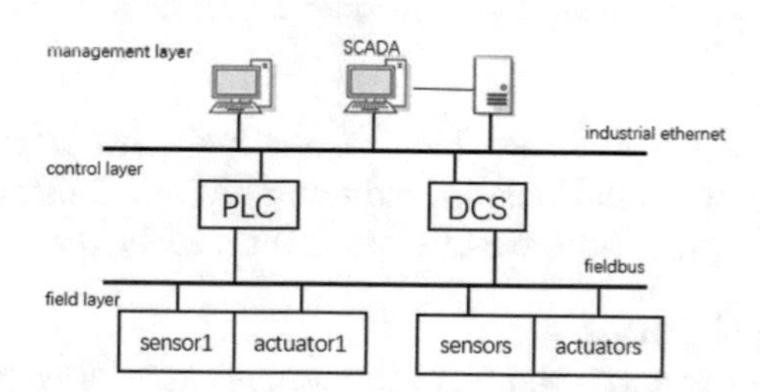

Fig. 1. ICS network abstract common structure

The data transmitted in ICS network is treated to complete the process tasks and mainly provoked by aperiodic and periodic events. The topology of ICS network is stable and the logic of interaction is defined clearly. Due to the real-time determinism, the packet size is usually quite small and transmitted highly frequently in industry, especially in the filed layer. Thus, the ICS network has unique features of stability, periodicity and sparsity which could be reflected in the data interaction. In this paper, we focus on the periodic interactive data which is in a large proportion of the entire data.

3.2 Autoencoder-Baseline Structure and Processes

Autoencoder has the prominent capability of feature extraction and noise reduction and cascades encoder with decoder in sequence. [8] constructs an autoencoder by LSTM to recover the clear signal from the interrupted signal and concludes that the autoencoder could learn the clear signal feature and withdraw the noise influence. We also implement an autoencoder to learn the ICS network feature by rebuilding the interaction mode which is similar to signal process. Compared with the struct in [8], a CNN-based autoencoder instead of LSTM is proposed due to the different input implication. CNN owns the local receptive fields feature to extract internal relationship between interaction more adaptively. The autoencoder-baseline structure is illustrated in Fig. 2.

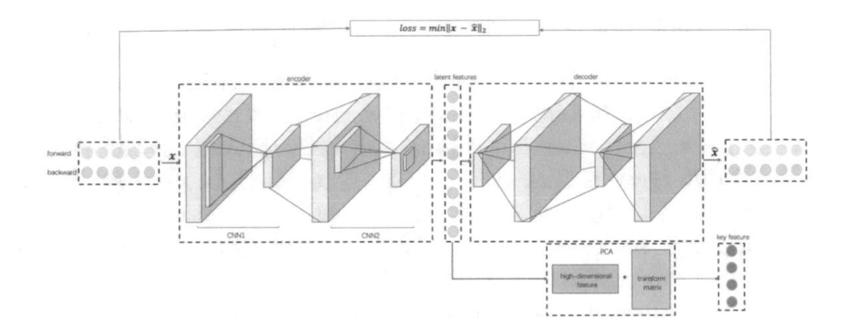

Fig. 2. Integral autoencoder struct of baseline

The struct consists of 4 parts which respectively are extracting original feature, establishing feature relationship, reducing high-dimension features and generating network baseline. More details are discussed below:

1) extracting original feature
   Original features are extracted from the unpacked interactive packets between different IP address periodically and divided into meta features and superior features. Meta features infer to the raw packet information such as source IP, source port, destination IP, destination port, protocol, timestamps, packet size etc. and superior features are calculated by the meta features, such as packet rate, interval arrival time (IAT) mean, packet max size etc. We aggregate the original features of the same interactive IP address which compose a feature matrix which represents the relationship between forward and backward features and contains the meta and superior feature of packet size, packet number, packet IAT, packet content flag and TCP/IP configuration that represent ICS network interactive information quantity, mode, frequency, application and initial information respectively.
2) establishing feature relationship
   The CNN-base autoencoder could establish relationship between forward and backward features in the matrix by convolution with kernel. The latent feature

dimension is relative with kernel number and usually multiple. Autoencoder is trained iteratively by setting loss function which minimize the deviation between input and output as the Eq. (1). The vector $\mathbf{x}$ represents input and $\hat{\mathbf{x}}$ represents output.

$$loss = \min(|\mathbf{x} - \hat{\mathbf{x}}|) = \min(\frac{1}{m}\sum_{k=1}^{m}|x_i - \hat{x}_i|) \tag{1}$$

3) reducing high-dimension features
High-dimensional latent features complicate the final baseline and have reluctance in each dimension. We adopt PCA algorithm to reduce the reluctance and obtain the key features. PCA is the principal component analysis method. By mapping high-dimensional features to low-dimensional new orthogonal basis, the key features are obtained. The main process of PCA is to calculate the eigenvalues and eigenvectors of latent feature covariance matrix, generate transition matrix by eigenvectors and then multiply the latent feature vector with transition matrix to output key feature vector.
4) calculating baseline parameters
As the same interactive IP address data has the same probability distribution, the central limit theorem is used to transform the unknown key features distribution into a normal distribution, and the key feature distribution parameters are calculated through maximum likelihood estimation. The maximum likelihood estimation expression of the normal distribution is as Eqs. (2) and (3), where k_{ij} represents the j-th dimension key feature value of the i-th interactive data, μ_{ij} represents the j-th dimension key feature mean value of the i-th interactive data, σ_{ij}^2 represents the variance of the j-th dimension key feature of the i-th interactive data:

$$\mu_{ij} = \frac{1}{m}\sum_{1}^{m} k_{ij} \tag{2}$$

$$\sigma_{ij}^2 = \frac{1}{m}\sum_{1}^{m}(k_{ij} - \mu_{ij})^2 \tag{3}$$

3.3 Training Autoencoder and Parameterizing Baseline

We use an open ICS dataset to train the autoencoder and parameterize the baseline. The open ICS dataset was generated on a small-scale process automation scenario using MODBUS/TCP equipment [9]. This dataset contains a raw clean data and five types of intrusion data which are mitm-change, mitm-read, pingflood, tcpflood and modbusqueryflood respectively. Only the clean data is adopted to train the autoencoder.

The original features are extracted from the raw clean data with a period of 1 min. By analysis, there are six types of interaction mode in the entire data grouped by limited IP address and protocol and the three of them account for 96% of the total data. As for the autoencoder parameters, both encoder and decoder have two convolution layers constructed by 20 kernels with size (2,5) and 8 kernels with size (1,1). The optimizer is adam and the loss function is MSE. Finally, we train the autoencoder 20 batches by the input feature matrix with size (2,22) and the loss tendency is showed in Fig. 3.

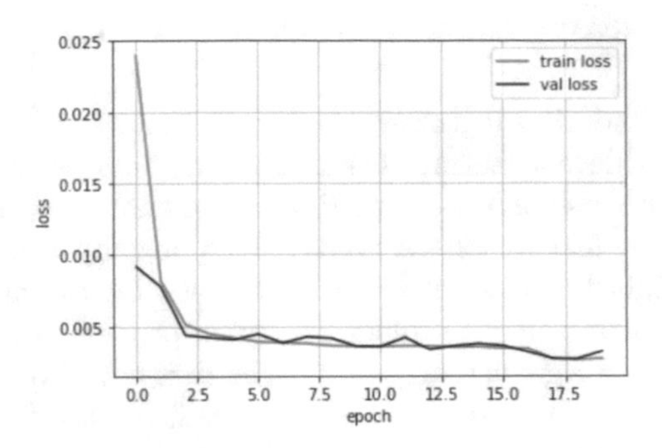

Fig. 3. Training and validating loss tendency of autoencoder

As we can see from Fig. 3, the autoencoder could learn and fit the original feature distribution as both training and validating loss value decrease simultaneously. The latent features and key features of one main interaction mode are illustrated in Fig. 4 (a) and Fig. 4 (b). As the PCA is linear transition algorithm, the key features have similar attributions to the latent features.

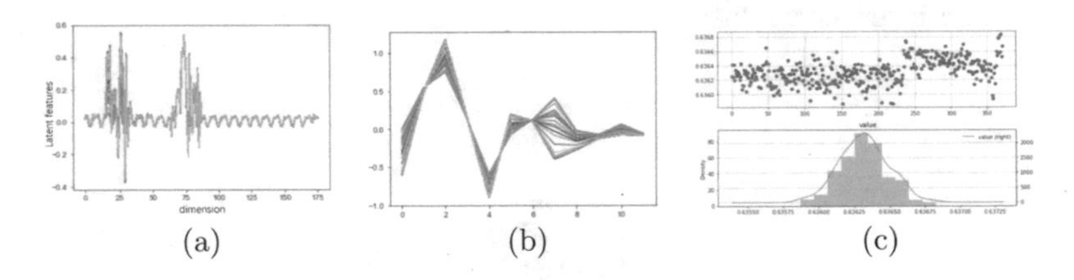

Fig. 4. Fig (a) represents the latent features of one interaction by autoencoder; Fig (b) represents the key features of the same interaction with fig (a) by PCA; Fig (c) represents distribution of the key features' one dimension and the parameters is N(0.632,0.000178)

The parameter of baseline's each dimension is calculated by central limit theorem and whether the distribution of dimension fits normalization is verified by K-S algorithm. The experiment result shows that all dimension of the key feature conform to normalization and Fig. 4 (c) illustrates the distribution of one dimension in key features. As a whole, the ICS network baselines of all type interactions could be parameterized and each dimension accords with Gaussian distribution.

4 Result

In this sector, we would utilise an open ICS dataset to verify the efficiency of intrusion detection. The open ICS dataset above mentioned synthesizes 5 types of intrusion and every type of intrusion contains abnormal duration of 1 min, 5 min, 15 min and 30 min separately. We use the clean data to train the autoencoder and obtain the ICS network baseline, then verify whether the baseline could detect the intrusion successfully.

Since there are multiple types of interaction every minute, we adapt Eq. (4) to calculate the mean value of all interaction by minute as the ICS network baseline. Figure 5 shows the difference between benign data's key features and baseline. By Eq. (4), the ICS network baseline ranges from 0.027 to 0.06 which represents the lower and upper threshold for intrusion detection respectively.

$$\bar{D} = \frac{1}{m}\sum_{i=1}^{m} D_i, D_i \in \{same\ timestamps\ interval\} \tag{4}$$

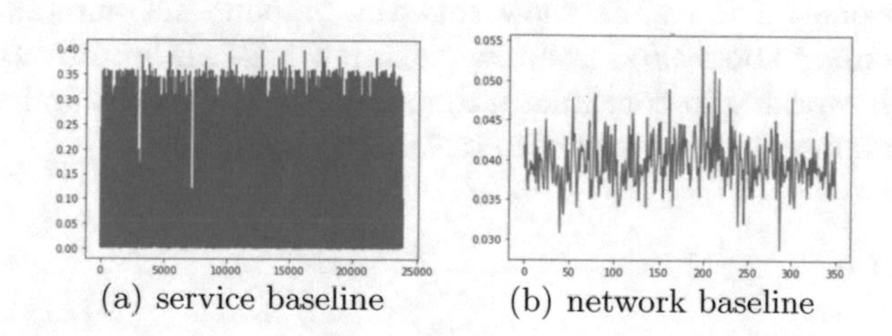

(a) service baseline (b) network baseline

Fig. 5. Fig (a) illustrates every service deviation from service baseline; Fig (b) illustrates network baseline by minute.

As for mitm intrusion, no matter mitm-change or mitm-read intrusion, the interactive sequence is disturbed and the network deviation is lower than the threshold. However, for flood intrusion, the malicious key feature's deviation is greater than the benign which results in the network deviation greater than upper threshold. All detection results is showed in Fig. 6.

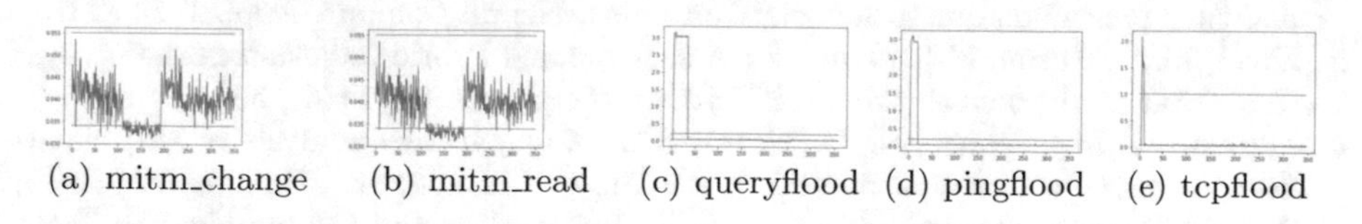

(a) mitm_change (b) mitm_read (c) queryflood (d) pingflood (e) tcpflood

Fig. 6. Fig (a)(b) illustrates the mitm intrusion detection result; Fig (c)(d)(e) illustrates the flood intrusion detection result.

Details of the detection result are listed in Table 1 and measured by the metric of accuracy and false alert rate (FAR). As a whole, the mean accuracy

Table 1. Cross-validate result of datasets with periodic of 30 s, 60 s, 90 s

Intrusion type	Intrusion duration	Total data	Correct detection	Total intrusion	False alert number	Detection accuracy	FAR
mitm_change	5 min	351	333	80	5	94.5%	6.25%
mitm_read	30 min	354	321	155	14	90.1%	9.03%
queryflood	30 min	359	359	30	0	100%	0%
pingflood	15 min	356	356	15	0	100%	0%
tcpflood	5 min	354	354	5	0	100%	0%

is above 90% and the mean FAR is below 5%. The intrusion detection for open dataset can achieve outstanding performance.

5 Conclusion

In this paper, we propose a new baseline model for ICS intrusion detection by using autoencoder. The results show that the baseline has outstanding performance and achieve about 90% accuracy with low false alarm rate. In the future, we are already working in construct a dynamic baseline model and the adaptive threshold for improving the intrusion detection capability.

References

1. Wei, C.: Dynamic baseline detection method for power data network service. In: Green Energy and Sustainable Development I: Proceedings of the International Conference on Green Energy and Sustainable Development (GESD 2017). American Institute of Physics Conference Series (2017)
2. Xu, J., et al.: Identification of ICS security risks toward the analysis of packet interaction characteristics using state sequence matching based on SF-FSM. Secur. Commun. Netw. **2017**, 1–17 (2017)
3. Kwon, D., Kim, H., Kim, J., et al.: A survey of deep learning-based network anomaly detection. Cluster Comput. **22**, 949–961 (2017). https://doi.org/10.1007/s10586-017-1117-8
4. Verma, A.K., Nagpal, S., Desai, A., et al.: An efficient neural-network model for real-time fault detection in industrial machine. Neural Comput. Appl. 1–14 (2020)
5. Khan, M.A., Karim, M.R., Kim, Y.: A scalable and hybrid intrusion detection system based on the convolutional-LSTM network. Symmetry **11**(4), 583 (2019)
6. Mahfouz, A.M., Venugopal, D., Shiva, S.G.: Comparative analysis of ML classifiers for network intrusion detection. In: Yang, X.S., Sherratt, S., Dey, N., Joshi, A. (eds.) Fourth International Congress on Information and Communication Technology. Advances in Intelligent Systems and Computing, vol. 1027, pp. 193–207. Springer, Singapore (2020). https://doi.org/10.1007/978-981-32-9343-4_16
7. Hendrycks, D., Gimpel, K.: A baseline for detecting misclassified and out-of-distribution examples in neural networks (2016)

8. Wu, Q., Sun, Z., Zhou, X.: Interference detection and recognition based on signal reconstruction using recurrent neural network. In: 2019 IEEE Globecom Workshops (GC Wkshps), Waikoloa, HI, USA, pp. 1–6 (2019). https://doi.org/10.1109/GCWkshps45667.2019.9024542
9. Frazão, I., Abreu, P., Cruz, T., Araújo, H., Simões, P.: Cyber-security modbus ICS dataset. IEEE Dataport (2019). https://doi.org/10.21227/pjff-1a03

Secrecy Capacity-Approaching Neural Communications for Gaussian Wiretap Channel

Jingjing Li[1(✉)], Zhuo Sun[1], Jinpo Fan[1], and Hongyu Zhu[2]

[1] Beijing University of Posts and Telecommunications, Beijing 100876, China
{jingjing,zhuosun}@bupt.edu.cn, fanjinpo@yeah.net

[2] State Grid Hunan Electric Power Company, Changsha 410007, Hunan, China
zhuhy@sgcc.com.cn

Abstract. Recently, some researches are devoted to the topic of end-to-end learning a physical layer secure communication system based on autoencoder under Gaussian wiretap channel. However, in those works, the reliability and security of the encoder model were learned through necessary decoding outputs of not only legitimate receiver but also the eavesdropper. In fact, the assumption of known eavesdropper's decoder or its output is not practical. To address this issue, in this paper we propose a dual mutual information neural estimation (MINE) based neural secure communications model. The security constraints of this method is constructed only with the input and output signal samples of the legal and eavesdropper channels and benefit that training the encoder is completely independent of the decoder. Moreover, since the design of secure coding does not rely on the eavesdropper's decoding results, the security performance would not be affected by the eavesdropper's decoding means. Numerical results show that the performance of our model is guaranteed whether the eavesdropper learns the decoder himself or uses the legal decoder.

Keywords: Wiretap channel · Autoencoder · Mutual information neural estimation · Security capacity

1 Introduction

Compared with traditional wired networks, wireless communication is vulnerable to eavesdropping, attacks, and interference due to the openness of the transmission medium and the characteristics of broadcasting. Most of the traditional security mechanisms are based on cryptography. However, with the increasing performance of processors, the rapid improvement of computer computing capacity, and the maturity of distributed computing theory, the classical encryption

J. Li—Supported by 2020 industrial Internet innovation and development project - smart energy Internet security situation awareness platform project.

Q. Liang et al. (Eds.): Artificial Intelligence in China, LNEE 854, pp. 24–31, 2022.
https://doi.org/10.1007/978-981-16-9423-3_4

algorithm system based on complexity is becoming increasingly unstable. Therefore, physical layer security technology is triggering wide interest in the academic field, which process transmitted signals or implement corresponding security policies using channel information with the characteristics of lightweight, high security, and can communicate safely without using the pre-shared key.

When it comes to using Deep neural networks (DNNs) to learn the physical layer security communication method, the autoencoder is the first attempt. In [1], the author first proposed to apply deep learning to the communication system, and an autoencoder structure was used to construct a communication system model. The model greatly improved the reliability of communication through end-to-end training. An application example of autoencoder in the wiretap channel can be found in [2] which considers the encoding and decoding scheme of the Gaussian wiretap channel with a single antenna. In our previous work [3], we also used end-to-end learning to solve two communication security problems: confidential transmission and user authentication. However, learning the end-to-end communication system is a challenging task. It is necessary to establish a differentiable channel model to meet the reverse transfer of parameters in the process of training the neural networks, which is impossible in real-world channel conditions.

One way to dissolve the above issue is to utilize mutual information estimation [4]. Using this method, it was shown in [5]. However, the decoding results of the eavesdropper needed in their mixing function are difficult to obtain in the actual environment, especially in passive eavesdropping scenario. Besides, they only use one eavesdropping structure to verify the security of the system, which led to an ideal result.

In this paper, we present a dual MINE-based neural secure communications model, two MINE modules are used to estimate the channel capacity of the legitimate channel and eavesdropper channel respectively, the estimated mutual information are subsequently used to construct the security capacity constraint as the objective function for training the encoder. Our main contributions and novelties are summarized as follows: (1) the encoder and decoder are completely independently training, avoiding the problem of differentiable channel model needed. (2) We can train the encoder reliably just with the signal samples of the main channel and eavesdropper channel, enabling deep learning for encoder without explicit knowledge of decoders. (3) The model achieves presentable security regardless of the eavesdropping means. (4) The method in this paper can realize a security capacity-approaching encoding scheme, which is a very important task in wireless communication.

The rest of this paper is organized as follows. We present the basis of the Gaussian wiretap channel used in our work in Sect. 2. The implementation details of the proposed dual MINE-based neural secure communications model and the training methods of each block are detailed in Sect. 3. We provide the numerical results and correlation analysis in Sect. 4. Finally, the paper is concluded in Sect. 5.

Notations—Random variables of all dimensions are denoted in boldface upper case letters, e.g., M. The sets of codebooks are denoted with calligraphic letters, i.e. $\mathcal{M}$. The dimension n is marked in the upper right corner of the letter. e.g., X^n. The element i is marked in the lower right corner of the letter (e.g., X_i). The probability mass or density function of event x is denoted as p(x). $\mathbb{R}(\mathbb{C})$ is the set of real numbers (complex numbers). The expectation is denoted by $E\{\cdot\}$.

2 Basis of Gaussian Wiretap Channel

In this paper, we consider the degraded Gaussian wiretap channel extended from Wyner's wiretap channel [6] as depicted in Fig. 1. It is a degraded broadcast channel with memoryless additive Gaussian white noise [7]. It is believed that this will lay a foundation for us to understand and study more complex communication scenarios.

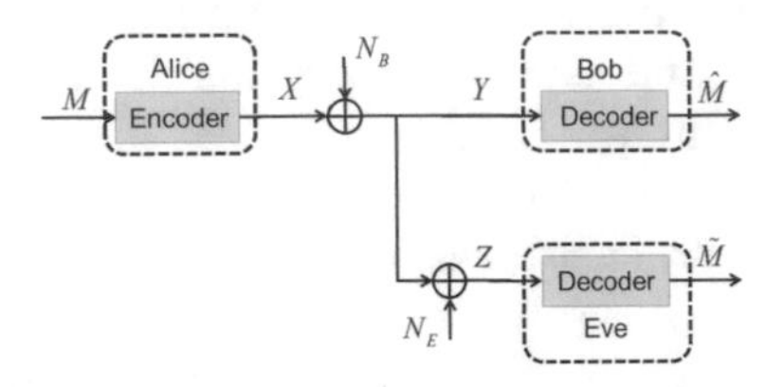

Fig. 1. Degraded Gaussian wiretap channel. The sender (Alice) sends confidential information M to the legitimate receiver (Bob) while keeping the external eavesdropper (Eve) unaware of it.

First, Alice transmits a message M from the set of $\mathcal{M} = \{1, 2, \ldots, 2^{nR}\}$ at a rate R to the legitimate user Bob. Alice encodes it into a codeword of block length n using an encoding function $f(m) = x^n(m) \in \mathbb{C}^n$. Besides, we assume that the variance of the transmitted codewords is not allowed to exceed a certain value P. Second, the main channel between Alice and Bob and the eavesdropper channel between Alice and Eve are defined by

$$Y_i = X_i + N_{B,i} \qquad for\, i \in \{1, \ldots, n\}, N_{B,i} \sim \mathcal{CN}\left(0, \sigma_B^2\right) \tag{1}$$

$$Z_i = Y_i + N_{E,i} \qquad for\, i \in \{1, \ldots, n\}, N_{E,i} \sim \mathcal{CN}\left(0, \sigma_E^2\right) \tag{2}$$

where X_i, Y_i, Z_i denote the channel input, Bob's observation, and Eve's observation, respectively. In the following, we consider the degraded case such that $\sigma_B^2 < \sigma_E^2$.

Finally, The symbol error rate (SER) of the two receivers is then defined as $P_B = E\{\hat{M} \neq M \mid Y\}$, $P_E = E\{\tilde{M} \neq M \mid Z\}$, respectively.

3 Dual MINE-Based Neural Secure Communications Model

Aiming at provide a secure communication over the degraded Gaussian wiretap channel, we build a fully learning based communication system shown in Fig. 2,

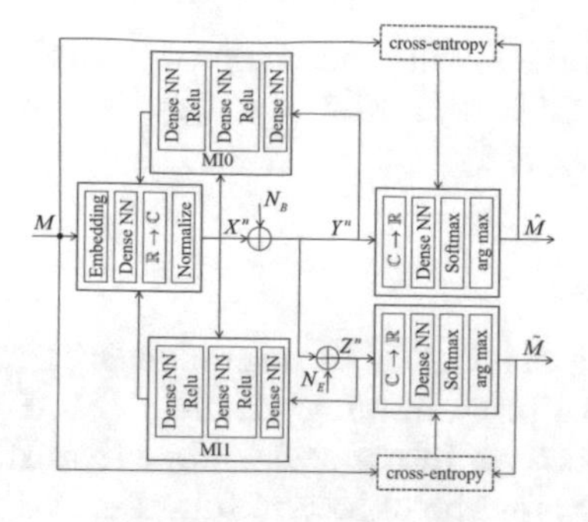

Fig. 2. The architecture of dual MINE-based neural secure communications model.

of which all the modules are implemented with DNNs, including an encoder block, two mutual information estimators, and two decoder blocks.

Wiretap channel security capacity is the basis of physical layer security technology research, which plays an important role in guiding the physical layer security mechanism. Therefore, we consider to use the security capacity as a metric to learn the optimal encoding function of the Gaussian wiretap channel.

The secrecy capacity of a Gaussian wiretap channel was derived in [8] and was found to be the capacity difference between the main and the eavesdropper channels, i.e., $C_s = \max_{p(x)} \{C_m - C_w\}$. Where C_m is the main channel capacity and C_w is the eavesdropper channel capacity. And the capacity C for a Gaussian channel is known to be $C = \max_{p(x)} I(X;Y)$. Therefore, the secrecy capacity calculation further simplifies to just the difference in the mutual information of two channels, $C_s = \max_{p(x)}[I(X;Y) - I(X;Z)]$.

The two MINE modules of our proposed model are used to achieve the estimation of the mutual information. Based on this, the encoder loss function can satisfy the reliability and security constraints.

3.1 MINE Block

As described above, we use the mutual information between input-output measurements of the wireless channel to optimize the encoder weight. In this work, we choose to utilize MINE to achieve this goal, which using SGD and Donsker-Varadhan representation of the Kullback-Leibler (KL) divergence.

The specific operation of calculating mutual information is explained by taking MI0 as an example. We first sample the joint distribution $p(x^n, y^n)$ and the marginal distributions $p(x^n)$ and $p(y^n)$, and then send the results to MI0 to get the T_θ. Finally, the mean value of T_θ is calculated to approximate the expectations. Given that T_θ is expressive enough, the above lower bound converges to the true mutual information. Hence the mutual information for X^n and Y^n is calculated as

$$\tilde{I}_\theta\left(X^n;Y^n\right) := \frac{1}{k}\sum_{i=1}^{k}\left[T_\theta\left(x^n_{(i)}, y^n_{(i)}\right)\right] - \log\frac{1}{k}\sum_{i=1}^{k}\left[e^{T_\theta\left(x^n_{(i)}, \bar{y}^n_{(i)}\right)}\right] \tag{3}$$

where the $k = 64$ represents the number of samples. The objective functions for optimization MI0 and MI1 are then set as $\max_{\theta} \tilde{I}_{\theta}(X^n; Y^n)$ and $\max_{\vartheta} \tilde{I}_{\vartheta}(X^n; Z^n)$, respectively.

3.2 Encoder Block

The purpose of training encoder is to minimize SER and information leakage. This is a multi-objective programming problem (MOP) that can be solved by combining multiple objectives into a scalar objective. Here we use the weighted sum and scalar objective for the objective function, expressed as

$$\max_{\varphi} \alpha I\left(X_{\varphi}^{n}(M); Y^{n}\right) - (1-\alpha) I\left(X_{\varphi}^{n}(M); Z^{n}\right) \tag{4}$$

where the value of $I\left(X_{\varphi}^{n}(M); Y^{n}\right)$ will affect the reliability of the system, while the value of $I\left(X_{\varphi}^{n}(M); Z^{n}\right)$ will affect the security of the system. The α in (4) is a specific weight that can be used to balance two objectives.

3.3 Decoder Block

The structure of the decoder is almost symmetrical to the encoder, except that a softmax activation function is used in the final dense layer. Its output $p_{\hat{M}}$ is a probability vector that assigns a probability to each of the possible messages. Then we use an arg max layer to select the index of the maximum value in the probability vector, which represents the information value $\hat{M}$ of the decoding result. In this case, the cross-entropy loss function with variants of stochastic gradient descent is a satisfactory choice, $L := H\left(M; \hat{M}\right) = -\frac{1}{k}\sum_{i=1}^{k} \log p_{\hat{M}}$. The result describe the distance between the actual output and the expected output, that is, the smaller the cross-entropy, the smaller the SER.

Our training procedure is divided into three phases. Firstly, we train the MI estimation network T_{θ} for 5 epochs with 500 iterations. At this point, the parameters of the encoder network are only randomly initialized and the estimated mutual information might not reflect the final estimated value. Secondly, the security of the system is learned by training the encoder and two estimators iteratively. The parameters of one model will be fixed while training the other so that the optimal point can be found.

Finally, we train the legal decoder to achieve reliable communication with the transmitter and train the eavesdropper's decoder to verify the security of the system.

Note that, all neural networks in the system are optimized by Nadam optimizer [9] with a learning rate of 0.001. The signal-to-noise ratio (SNR) of both channels is fixed to 7 dB. And the output of the encoder network has a unit average power normalization.

4 Numerical Results and Analysis

4.1 Reliability and Security

In this part, we test the system under different $SNRs \in (0\,\text{dB}, 21\,\text{dB})$ of the legal channel. Firstly, we set the SNR of the eavesdropper channel to 7 dB. The SER curves can be seen in Fig. 3a. The 16QAM is taken as the benchmark for comparison. One can see that the reliability performance of our method is close to that of 16QAM at low SNR, and the reliability performance gap is within 1 dB when SNR gets higher. This is due to the trade-off between reliability and security we considered, the system sacrifices part of reliability for security. Moreover, the SER of Eve is limited to a large value even at high SNR. Subsequently, we conducted a test to see if Eve could achieve the purpose of eavesdropping using Bob's decoding network. The result is shown in Fig. 3b. As expected, the decoding result of Eve has a high symbol error rate, which proves that the model achieves presentable security regardless of the eavesdropping means. Furthermore, it can be seen from that the coefficient in (4) is an important factor in determining the performance of the system. In order to balance the security and reliability of the system, we choose $\alpha = 0.7$. The resulting constellations can be seen in Fig. 4.

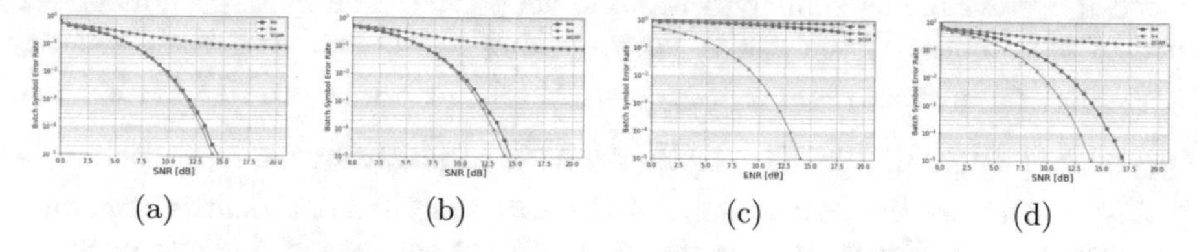

Fig. 3. SER performance of the proposed method for a 16-dimensional codeword constellation. A theoretical estimation of 16QAM as a reference. (a) Eve uses its own trained neural network as the decoder; (b) Eve uses Bob's neural network as the decoder. The weight α in (4) is 0.7. (c) the encoder training weight $\alpha = 0.4$; (d) the encoder training weight $\alpha = 0.6$.

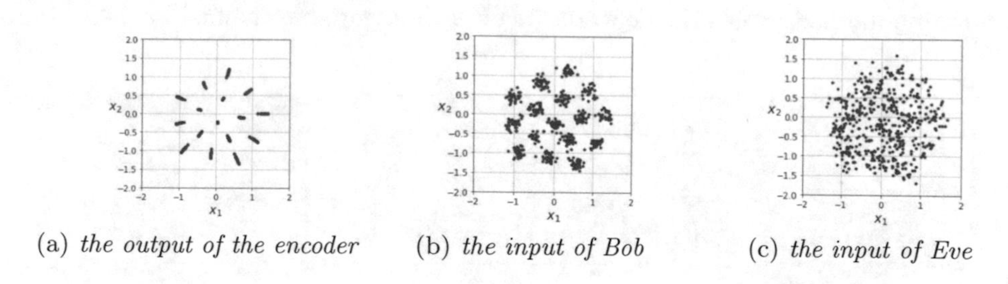

(a) *the output of the encoder* (b) *the input of Bob* (c) *the input of Eve*

Fig. 4. The resulting constellations are shown for 16 symbols.

The secrecy capacity of the Gaussian wiretap channel is known [8] and is given by $C_s = \log(1 + \frac{P}{\sigma_B^2}) - \log(1 + \frac{P}{\sigma_B^2 + \sigma_E^2})$. We take it as the theoretical value and compare it with the capacity measurement of this model. The result is shown in Fig. 5, which proves that our encoding scheme can achieve the performance close to the security capacity.

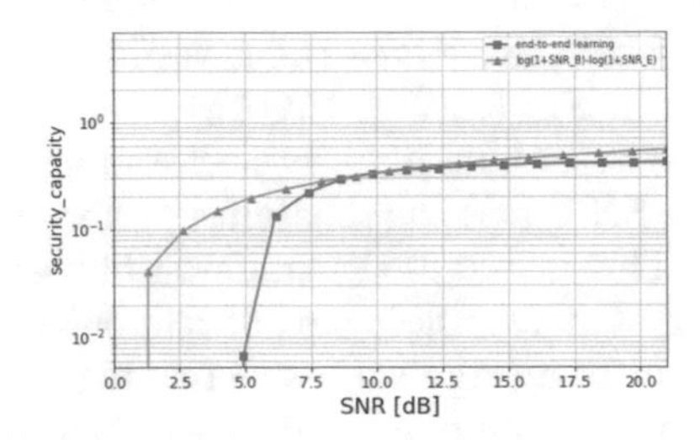

Fig. 5. The resulting security capacity is shown for 32 symbols under different SNR of the main channel. $\log(1+SNR\_B)-\log(1+SNR\_E)$ represents the theoretical value.

4.2 Comparison to Existing Methods

To compare our approach with others presented in [2] and [5], we reproduced a system model similar to the one they used. The most prominent feature of this model is the use of coset coding algorithm. And their security loss functions are expressed as $L_{sec} := \alpha H\left(M;\hat{M}\right) + (1-\alpha)H\left(\bar{M};\tilde{M}\right)$ and $L_{sec} := \alpha I\left(X^n(M);Y^n\right) - (1-\alpha)H\left(M;\tilde{M}\right)$, respectively.

Figure 6 shows the performance of the three methods in security encoding. One can see that, Fig. 6c has a better reliability in the case of similar security as Fig. 6b. This shows that our method can achieve good performance even without prior knowledge of the decoder. By comparing Fig. 6a and Fig. 6b, it can be found that the former has better security in the case of similar reliability. This is because the latter two methods train the encoder and decoder separately which cannot ascertain the global optimality of the system, even if they all achieve their respective optimal performance. But as described in the first section, this training method avoids the dependence on a differentiable channel model.

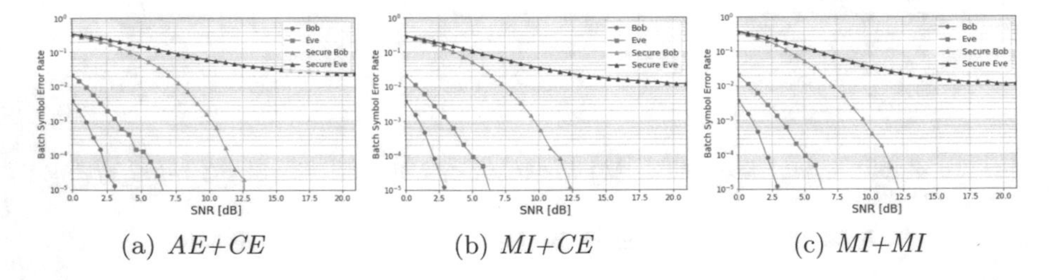

(a) *AE+CE* (b) *MI+CE* (c) *MI+MI*

Fig. 6. Secure Bob and Secure Eve indicate the error rate after the security encoding, reflecting the reliability and security of the system respectively.

5 Conclusions

In this paper, we contribute to give a model based on two mutual information estimators for secure communication in the degraded Gaussian wiretap channel where each party is equipped with DNNs. The encoder is trained by security capacity in a manner to provide reliability for legal decoder while ensuring security against wiretapper. The presented results are compared with the other two security constraint methods based on deep learning which highlights the superiority of our proposed method in robustness to different eavesdropping environments without prior knowledge of the decoder. However, future research needs to investigate the performance under unknown eavesdropping channel conditions. Moreover, we can continue to explore the direction corresponds to the case where the broadcast channel is no longer degraded.

References

1. O'Shea, T., Hoydis, J.: An introduction to deep learning for the physical layer. IEEE Trans. Cogn. Commun. Netw. **3**(4), 563–575 (2017). https://doi.org/10.1109/TCCN.2017.2758370
2. Fritschek, R., Schaefer, R.F., Wunder, G.: Deep learning for the Gaussian wiretap channel. In: ICC 2019–2019 IEEE International Conference on Communications (ICC), Shanghai, China, pp. 1–6 (2019). https://doi.org/10.1109/ICC.2019.8761681
3. Sun, Z., Wu, H., Zhao, C., Yue, G.: End-to-end learning of secure wireless communications: confidential transmission and authentication. IEEE Wirel. Commun. **27**(5), 88–95 (2020). https://doi.org/10.1109/MWC.001.2000005
4. Belghazi, I., Rajeswar, S., Baratin, A., Hjelm, R.D., Courville, A.: MINE: mutual information neural estimation arXiv preprint arXiv:1801.04062 (2018)
5. Fritschek, R., Schaefer, R.F., Wunder, G.: Deep learning based wiretap coding via mutual information estimation. In: Proceedings of the 2nd ACM Workshop on Wireless Security and Machine Learning (2020)
6. Wyner, A.: The wiretap channel. Bell Syst Tech. J. **54**, 1355–1387 (1975)
7. Bergmans, P.: Random coding theorem for broadcast channels with degraded components. IEEE Trans. Inf. Theor. **19**(2), 197–207 (1973). https://doi.org/10.1109/TIT.1973.1054980
8. Leung-Yan-Cheong, S., Hellman, M.: The Gaussian wire-tap channel. IEEE Trans. Inf. Theor. **24**(4), 451–456 (1978). https://doi.org/10.1109/TIT.1978.1055917
9. Dozat, T.: Incorporating Nesterov momentum into Adam. Stanford University Technical Report (2015)

Image Compression Based on Mixed Matrix Decomposition of NMF and SVD

Zhiyang Zhao, Baoju Zhang(✉), and Cuiping Zhang

Tianjin Normal University, TianJin 300087, China
wdxyzbj@163.com

Abstract. Image compression has always been a key research hotspots in image processing. An efficient image compression approach will not only save considerable storage resources, but also exceedingly ease the communication pressure of the network transmission, which has great research significance and practical value. Since the essence of the image is matrix, so mixed matrix decomposition of NMF and SVD is introduced to perform two-level compression framework on images. The experimental results demenstrated that this approach based on mixed matrix decomposition had a CR with larger dynamic range through flexible parameter adjustment and the PSNR of the restored image is 29 dB–36 dB. It verifiy that this method is effective.

Keywords: Image compression · NMF · SVD · Two-level compression

1 Introduction

Nowadays, with the development of computer technology and multimedia technology, the network has become an extremely common form of collecting and transmitting information [1]. The digital image occupies a pivotal position in multimedia images. With the widespread application of 5G, the growth rate of image data far exceeds the development speed of storage devices and transmission technology. Image transmission and storage of images require a large amount of data space, which seriously affects the transmission rate and real-time processing capacity, greatly restricted the development of image communication. So, it is quite necessary to compress the image.

2 Related Work

Image compression methods are generally categorized into lossless and lossy approaches [2]. In lossless compression, all information originally is preserved after uncompressed. In lossy compression, only part of the original information is preserved when the images were uncompressed. Some lossless compression methods were Huffman coding [3], arithmetic coding [4] and dictionary-based

Q. Liang et al. (Eds.): Artificial Intelligence in China, LNEE 854, pp. 32–39, 2022.
https://doi.org/10.1007/978-981-16-9423-3_5

encoding such as LZ78 and LZW [5]. Another was the lossy approaches. A common transformation employed in image compression was SVD. A hybrid system with KL vectors and SVD for compression was designed [6]. A scheme of SVD after preprocessing was proposed to compress images [7]. Combined the SVD with WDR, original image was first compressed using the SVD and then compressed again with the WDR [8]. This paper is organized as follows: some relevant concepts and approaches are briefly described in Sect. 2. The proposed method is presented in Sect. 3. Image quality measures are introduced in Sect. 4. Final experiment results are presented in Sect. 5.

3 Mixed Matrix Decomposition

Compression framework of this paper is shown in Fig. 1.

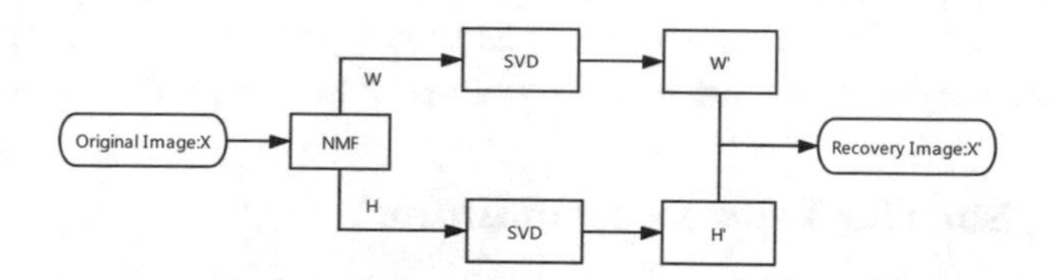

Fig. 1. Compression framework.

3.1 NMF

Factor analysis and principal component analysis are two of the many classical methods used to accomplish the goal of reducing the number of variables [9,10].

Often the data to be analyzed is non-negative, and the low-rank data are further required to be comprised of non-negative values in order to avoid contradicting physical realities [11]. The approach of finding reduced rank non-negative factors to approximate a given non-negative data matrix thus becomes a natural choice. This is the so-called NMF problem [12,13]. NMF is given as follows:

Given $X \in R_+^{m\times n}$ and $k \ll \frac{mn}{m+n}$, find $W \in R_+^{m\times k}$ and $H \in R_+^{k\times n}$ to make the functional:

$$X \approx WH = \sum_k W_{mk}H_{kn} \tag{1}$$

among them, W is a set of basis vectors; h_i is the projection of X_i on W. For the error $E \in R^{m\times n}$:

$$E = X - WH \tag{2}$$

assume that the PDF of random noise obey the Gaussian distribution and the noise variance satisfies $\sigma_{ij}^2 = \sigma^2$, then Maximum Likelihood Estimation is:

$$L(W,H) = \prod_{(i,j)} \frac{1}{\sqrt{2\pi}\sigma} e^{-\frac{E_{ij}^2}{2\sigma^2}} = \prod_{(i,j)} \frac{1}{\sqrt{2\pi}\sigma} e^{-\frac{(X_{ij}-W_{ik}H_{ij})^2}{2\sigma^2}} \tag{3}$$

$$\ln L(W, H) = \sum_{(i,j)} \left[\ln \frac{1}{\sqrt{2\pi}\sigma} - \frac{1}{2\sigma^2} \left(X_{ij} - W_{ik} H_{kj} \right)^2 \right] \tag{4}$$

where monotonicity of $L(W;H)$ and $lnL(W;H)$ is the same. To maximize Eq. 4, it is only necessary to minimize $J(W;H)$.

$$J(W, H) = \frac{1}{2} \sum_i \left[X_{ij} - W_{ik} H_{kj} \right]^2 \tag{5}$$

$$\frac{\partial J(W,h)}{\partial W_{ik}} = \left(XH^T \right)_{ik} - \left(WHH^T \right)_{ik} \quad \frac{\partial J(W,h)}{\partial H_{kj}} = \left(W^T X \right)_{kj} - \left(W^T WH \right)_{kj} \tag{6}$$

$$W_{ik} = W_{ik} \frac{\left(XH^T \right)_{ik}}{\left(WHH^T \right)_{ik}} \quad H_{kj} = H_{kj} \frac{\left(W^T X \right)_{kj}}{\left(W^T WH \right)_{kj}} \tag{7}$$

The W and H were found by multiplying the current value by W_{ik} and H_{kj}. It is a multiplicative iterative, which means that repeated iteration of the update rules is guaranteed to converge to a locally optimal matrix factorization [14].

3.2 SVD, Singular Value Decomposition

SVD is a PCA (Principal Component Analysis) [15] that uses several PCs to represent most of the information of X. In this section, we use SVD to retain a few PCs in W to compress W. SVD of W can be accomplished by as follows:

$$W_{m \times k} = U_{m \times m} S_{m \times k} V_{k \times k}^T \tag{8}$$

$$S = \begin{bmatrix} \sum_p & 0 \\ 0 & 0 \end{bmatrix} \tag{9}$$

W is decomposed to form U, S and V by SVD. U is the eigen-vector of WW^T. S is a unique diagonal matrix, diagonal element is $\sigma_1 > \sigma_2 > \sigma_3 \cdots > \sigma_p$. V is the eigen-vector of $W^T W$. Assume $\hat{p}$-PCs were selected to reconstruct W, then:

$$\hat{W} \approx \sigma_1 u_1 v_1^T + \sigma_2 u_2 v_2^T + \ldots \sigma_{\hat{p}} u_{\hat{p}} v_{\hat{p}}^T \tag{10}$$

SVD of *H* Same as above, SVD of H as follows:

$$H_{k \times n} = U_{k \times k} S_{k \times n} V_{n \times n}^T \tag{11}$$

$$S = \begin{bmatrix} \sum_q & 0 \\ 0 & 0 \end{bmatrix} \tag{12}$$

H is also decomposed to form U, S and V. U is the eigen-vector of HH^T. S is a unique diagonal matrix, diagonal element is $\sigma_1 > \sigma_2 > \sigma_3 \cdots > \sigma_q$. V is the eigen-vector of $H^T H$. Assume $\hat{q}$-PCs were selected to reconstruct H, then:

$$\hat{H} \approx \sigma_1 u_1 v_1^T + \sigma_2 u_2 v_2^T + \ldots \sigma_{\hat{q}} u_{\hat{q}} v_{\hat{q}}^T \tag{13}$$

4 Image Quality Measures

The quality of the restored image is one of the most important criteria for evaluating image compression [16].

4.1 MSE, Mean Square Error

MSE [17] is to calculate the sum of squared variances between $I(x, y)$ and $I'(x, y)$, as shown in Eq. 14. The image sizes are M and N.

$$MSE = \frac{1}{MN}\sum_{x=1}^{M}\sum_{y=1}^{N}\left[I(x,y) - I'(x,y)\right]^2 \tag{14}$$

4.2 PSNR, Peak Signal to Noise Ratio

$PSNR$ [18] is a measure of the peak error of two images, as shown in Eq. 15, R represents the number of pixels required for the maximum value of X. $PSNR$ is a full-reference image evaluation index.

$$PSNR = 10\log_{10}\frac{\left(2^R - 1\right)^2}{MSE} \tag{15}$$

4.3 CR, Compression Ratio

CR is an important index of image compression. This metric measures the performance of compression, given by:

$$CR\left[I(X,Y), I(X,Y)'\right] = \frac{I(X,Y)}{I(X,Y)'} = \frac{m \times n \times r(bits)}{m_1 \times n_1 \times r(bits)} \tag{16}$$

5 Experiment Result

This algorithm in Fig. 2 was tested. This image is the most common test image in image processing. It is a grayscale image and its size is $(512, 512)$.

5.1 Performance Analysis of NMF

In NMF, K has the most important impact on the compression performance of NMF. The experimental results in Fig. 3 prove MSE, CR and $PSNR$ change as K increases. Although the image becomes clearer, the main performance of the compression method is greatly reduced. Therefore, it is not only NMF is used to fulfill the purpose of compression (Table 1 and Fig. 4).

Fig. 2. Test image: Lena.

Table 1. Comparison of compression performance of NMF under various K

	CR	PSNR (dB)	MSE
$K=8$	32	30.1579	62.7079
$K=16$	16	31.2380	48.8966
$K=32$	8	32.5910	35.8079
$K=64$	4	33.6870	27.8213
$K=128$	2	34.8837	21.1207
$K=256$	1	36.4986	14.5619

Fig. 3. (a) $K=8$, Restored image. (b) $K=16$, Restored image. (c) $K=32$, Restored image. (d) $K=64$, Restored image. (e) $K=128$, Restored image. (f) $K=256$, Restored image.

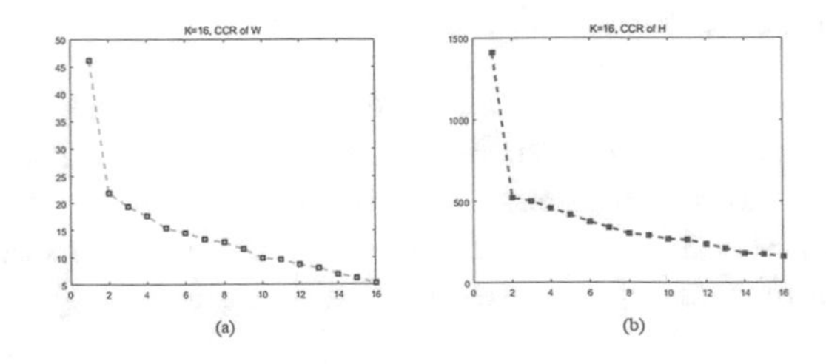

Fig. 4. CCR of W and H.

Table 2. Comparison of compression performance $k = 16$ and various CCR

$K = 16$	CCR	CR	PSNR (dB)	MSE
6-PCs	59.70%	41.2955	28.6788	88.1450
8-PCs	71.51%	30.9716	29.4101	74.4885
10-PCs	81.00%	24.7773	29.8401	67.4631
13-PCs	92.22%	19.0595	30.4586	58.5095

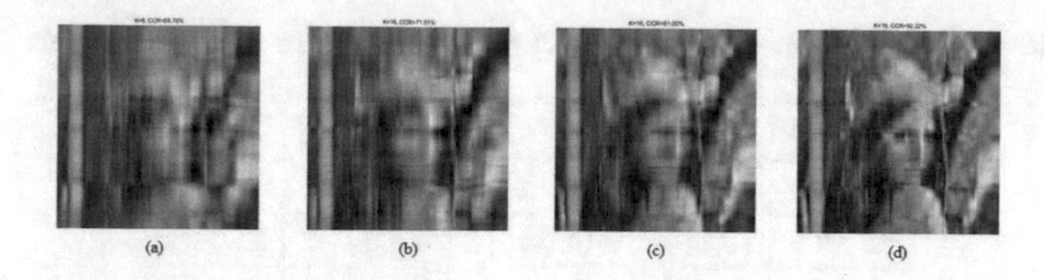

Fig. 5. Restored image under $k = 16$ and various CCR.

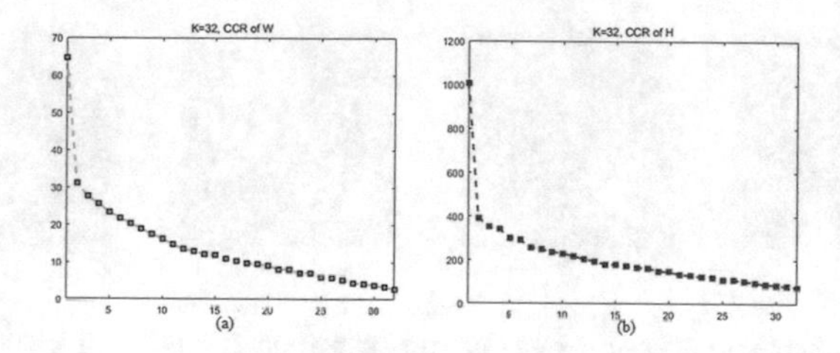

Fig. 6. CCR of W and H.

Table 3. Comparison of compression performance $k = 32$ and various CCR

$K = 32$	CCR	CR	PSNR (dB)	MSE
11-PCs	60.46%	21.8636	29.6141	71.0678
14-PCs	69.30%	17.1785	29.9605	66.2932
18-PCs	79.60%	13.3611	30.5582	57.1816
23-PCs	89.53%	10.4565	31.1717	49.6487

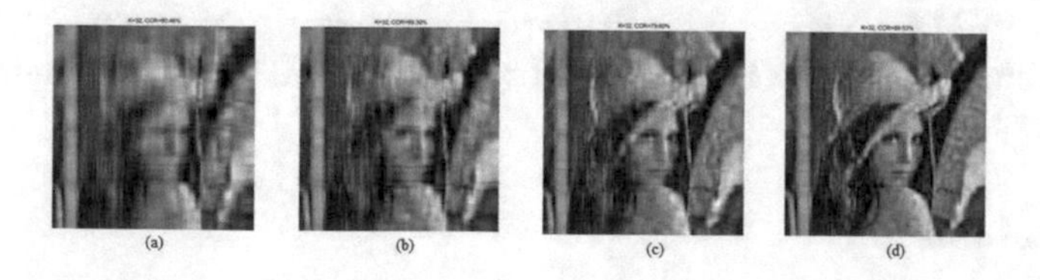

Fig. 7. Comparison of compression performance $k = 32$ and various CCR.

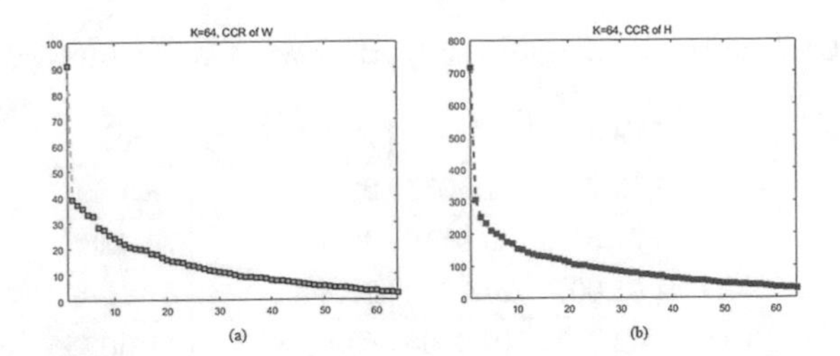

Fig. 8. Comparison of compression performance $k = 64$ and various CCR.

Table 4. Comparison of compression performance $k = 64$ and various CCR

$K = 64$	CCR	CR	PSNR (dB)	MSE
19-PCs	60.71%	11.9559	31.4978	50.8748
25-PCs	70.29%	9.0864	32.0658	42.6485
33-PCs	80.15%	6.8837	32.8318	36.1450
44-PCs	90.00%	5.1628	33.5503	28.2927

Fig. 9. Comparison of compression performance $k = 64$ and various CCR.

5.2 Performance Analysis of NMF and SVD

In SVD of W and H, the CCR (Cumulative Contribution Rate) of PCs was defined, which represents the contribution rate of selected PCs to all components.

$$CCR(i) = \frac{\sum_{e=1}^{i} \sigma_e}{\sum_{e=1}^{r} \sigma_e} \quad (i = 1, 2, \cdots, r) \tag{17}$$

Abundant experiments on the $NMF{+}SVD$ were conducted and divided the CCR into 4 thresholds: 0.6, 0.7, 0.8, 0.9. The PCs corresponding to these four thresholds were restored. From the results in Figs. 5, 6, 7, 8 and 9 and Tables 2,3 and 4, it can be seen that image compression based on mixed matrix decomposition of $NMF{+}SVD$ had a better compression performance. CR with a larger dynamic range can be obtained by adjusting different parameters. PSNR is 26 dB–36 dB.

6 Conclusion

Although there have been very mature algorithms in image compression, its performance is poor in the face of massive images in the 5G era. Therefore, this paper analyzes the existing problems of traditional methods, and introduces the mixed matrix factorization on this basis. The experimental results show that the mixed matrix decomposition has good CR and PSNR.

References

1. Mou, J., Yang, F., Chu, R., Cao, Y.: Image compression and encryption algorithm based on hyper-chaotic map. Mob. Netw. Appl. **3**, 1–13 (2019). https://doi.org/10.1007/s11036-019-01293-9
2. Sayood, K.: Introduction to Data Compression. Morgan Kaufmann, Burlington (2017)
3. Sharma, M.: Compression using Huffman coding. Int. J. Comput. Sci. Netw. Secur. **10**, 133–141 (2010)
4. Gonzales, C.A., Anderson, K.L., Pennebaker, W.B.: DCT-based video compression using arithmetic coding. In: Electronic Imaging: Advanced Devices and Systems, Santa Clara, CA, pp. 305–312 (1990)
5. Morita, H., Kobayashi, K.: An extension of LZW coding algorithm to source coding subject to a fidelity criterion. In: 4th Joint Swedish-Soviet International Workshop on Information Theory, Gotland, Sweden, pp. 105–109 (1989)
6. Waldemar, P., Ramstad, T.: Hybrid KLT-SVD image compression. In: IEEE International Conference on Acoustics, Speech, and Signal Processing, Munich, 21–24 April 1997, pp. 2713–2716 (1997)
7. Rufai, A.M., Anbarjafari, G., Demirel, H.: Lossy image compression using singular value decomposition and wavelet difference reduction. Digit. Signal Process. **24**, 117–123 (2014)
8. Wang, Z., Bovik, A.C., Sheikh, H.R., Simoncelli, E.P.: Image quality assessment: from error visibility to structural similarity. IEEE Trans. Image Process. **13**, 600–612 (2004)
9. Berry, M.W., Browne, M., Langville, A.N., Pauca, V.P., Plemmons, R.J.: Algorithms and applications for approximate nonnegative matrix factorization. Comput. Stat. Data Anal. **52**(1), 155–173 (2007)
10. Wang, Y.: Fisher non-negative matrix factorization for learning local features. In: ACCV, 27–30 January 2004 (2004)
11. Wang, Y., Jia, Y., Hu, C., Turk, M.: Non-negative matrix factorization framework for face recognition. Int. J. Pattern Recogn. Artif. Intell. **19**(4), 495 511 (2005)
12. Cichocki, A., Zdunek, R., Phan, A.H., Amari, S.I.: Quasi-Newton Algorithms for Nonnegative Matrix Factorization. Wiley, Hoboken (2009)
13. Ding, C.H.Q., Li, T., Jordan, M.I.: Convex and semi-nonnegative matrix factorizations. IEEE Trans. Pattern Anal. Mach. Intell. **32**(1), 45–55 (2010)
14. Zhao, Z., et al.: A novel optimization method for WSN based on mixed matrix decomposition of NMF and 2-SVD-QR. Ad Hoc Netw. **115**(6), 102454 (2021)
15. Thanushkodi, K.G., Bhavani, S.: Comparison of fractal coding methods for medical image compression. IET Image Process. **7**(7), 686–693 (2013)
16. Amirjanov, A., Dimililer, K.: Image compression system with an optimisation of compression ratio. IET Image Process. **13**, 1960–1969 (2019)
17. Sankur, B.: Statistical evaluation of image quality measures. J. Electron. Imaging 11(2) (2002)
18. Horé, A., Ziou, D.: Image quality metrics: PSNR vs. SSIM. In: 20th International Conference on Pattern Recognition, ICPR 2010, Istanbul, Turkey, 23–26 August 2010. IEEE Computer Society (2010)

MocNet: Less Motion Artifacts, More Clean MRI

Bin Zhao[1], Shuxue Ding[1,2], Mengran Wu[1], Guohua Liu[1], Chen Cao[3], Song Jin[3], Zhiyang Liu[1(✉)], and Hong Wu[1(✉)]

[1] Tianjin Key Laboratory of Optoelectronic Sensor and Sensing Network Technology, College of Electronic Information and Optical Engineering, Nankai University, Tianjin 300350, China
{liuzhiyang,wuhong}@nankai.edu.cn

[2] School of Artificial Intelligence, Guilin University of Electronic Technology, Guilin 541004, Guangxi, China

[3] Key Laboratory for Cerebral Artery and Neural Degeneration of Tianjin, Department of Medical Imaging, Tianjin Huanhu Hospital, Tianjin 300350, China

Abstract. Magnetic Resonance Imaging is a common way of diagnosing related diseases. However, the magnetic resonance images are easily defected by motion artifacts in their acquisition process, which severely affects the clinicians' diagnosis. To resolve the problem, we propose a motion correction network (MocNet) to correct motion artifacts. The experiments of motion artifacts simulation demonstrate that our MocNet outperforms the comparison methods with a mean PSNR of 34.397 ± 3.155 dB and a mean SSIM of 0.971 ± 0.015.

Keywords: Magnetic Resonance Imaging · Motion artifacts · MocNet · PSNR · SSIM

1 Introduction

Magnetic Resonance Imaging (MRI) is sensitive to the motion due to prolonged data acquisition time and the strategies of filling the k-space. The reasons leading to motion artifacts during brain MRI acquisition includes the sudden involuntary movements and conscious motions of body parts. [1]. As reported in [2], about 10%–42% of brain examinations bring motion artifacts, which interfere with clinicians' diagnosis. Therefore, it is necessary to develop a method to correct motion artifacts.

To obtain clean images during their acquisition process, efforts have been made by improving the acquisition sequence and the imaging techniques to prevent or compensate the motion artifacts, such as fast single-shot pulse sequence, non-Cartesian k-space acquisition strategy, and measuring the head pose in a real-time manner either in image space or k-space [3–7]. On the other hand, when the corrupted images have been collected, post-processing methods should be used to correct the raw images after acquisition by optimizing image entropy

Q. Liang et al. (Eds.): Artificial Intelligence in China, LNEE 854, pp. 40–46, 2022.
https://doi.org/10.1007/978-981-16-9423-3_6

or other measures of artifacts to reconstruct the MR image [8–10]. Generally, improving the quality of signals in the k-space of magnetic resonance (MR) image is the directly way to correct the motion artifacts [11]. However, such methods are hindered in clinical use by their availability varies among MRI manufacturers.

Recently, deep learning has presented great potential in medical imaging tasks, such as lesion classification and segmentation [12]. Thanks to its strong ability in learning representations, it can also be applied in motion artifact correction. For instance, a recurrent neural network based method is proposed to reduce cardiac MRI motion artifacts and the multi-scale structures are used in the method to extract local and global features [13]. Tamada et al. develop a method based on DnCNN [14] for motion artifact reduction of liver, which significantly reduces the magnitude of the artifacts and blurring induced by respiratory motion [15]. For brain motion artifact reduction, a DRN-DCMB [16] model has recently been proposed, and achieved higher performance compared with the comparison methods. However, the DRN-DCMB is trained on the small image patches that are randomly selected from the full size image, which leads to decrease conspicuity of small anatomic structures of the original full image. In the meanwhile, DRN-DCMB adopted mean squared error as the loss function to train DRN-DCMB, which defects in reconstructing the high frequency components and therefore omits the underlying anatomical structure. Therefore, in this paper, we propose the MocNet to correct motion artifacts, which is trained on the original full MR images and focuses on the whole correction. Experimental result reveals that the proposed method is able to achieve better reconstruction performance on the motion-corrupted brain MRIs.

2 Materials and Method

2.1 Research Subjects

The experimental data includes 45 brain T1-weighted (T1W) MRI scans with matrix sizes of $336 \times 448 \times 56$, which were collected from a retrospective database of Tianjin Huanhu Hospital and anonymized prior to the use of researchers. The research has been approved by the institutional ethic committee.

Two experienced clinicians (Dr. Song Jin and Dr. Chen Cao) from Tianjin Huanhu Hospital confirmed that these T1W images have no motion artifacts. The images are randomly divided into training set, validation set and test set, with 27, 9, and 9 subjects, respectively.

2.2 Motion Artifacts Simulation

Since the motion artifacts are caused unintentionally during MRI acquisition, it is impossible to deliberately make patients cooperate to obtain MR image pairs with and without motion artifacts. Therefore, it is necessary to simulate the motion artifacts on clean MR image.

Note that the head motion is rigid, and can be modeled as a combination of translational and rotational motions with six degrees of freedom. By defining a voxel coordinate of MR image as $V = (v_x, v_y, v_z)$ and its coordinate after rigid motion can be modeled as

$$\hat{V} = \begin{bmatrix} r_{xx} & r_{yx} & r_{zx} & \Delta X \\ r_{xy} & r_{yy} & r_{zy} & \Delta Y \\ r_{xz} & r_{yz} & r_{zz} & \Delta Z \\ 0 & 0 & 0 & 1 \end{bmatrix} \times \begin{bmatrix} v_x \\ v_y \\ v_z \\ 1 \end{bmatrix}, \tag{1}$$

where the left matrix on the right hand side represents the rigid motion, which is recorded as T for convenience. The upper-left 3×3 sub-matrix and the last column of T indicate the rotation and translation along the three axes, respectively.

Fourier theorem indicates that translation in the spatial domain results in phase errors in the k-space domain along the phase-encoding direction, while rotation in spatial domain results in identical rotation of the k-space data with the rotation axis through the origin. Given a clean MR image P, its corresponding MR image Q with motion artifacts can be calculated as

$$Q = \mathcal{F}^{-1}(\mathcal{F}(T * P)_{<\alpha,\gamma>}), \tag{2}$$

where $\mathcal{F}$ and $\mathcal{F}^{-1}$ indicate the Fourier transform and inverse Fourier transform, respectively. α and γ are the parameters of motion artifacts simulation. The center $\alpha\%$ of the k-space lines of P are kept intact to reserve the low frequency data that determined image contrast. $\gamma\%$ of the remaining peripheral k-space lines are randomly selected for rigid motion transformation.

2.3 MocNet

The architecture of the MocNet is illustrated in Fig. 1. As shown in Fig. 1, there are input module, two feature-extraction modules (FEM) and output module in the MocNet, which are connected through the attention dense connections to reuse the former features. The feature-extraction module is proposed based on the U-Net [17], while convolutional block and copy connection are replaced with attention paralleling block (APB) and attention residual block (ARB), respectively. The APB and the ARB can help the MocNet to extract more features and further improve the correction results. The attention mechanism used in this paper is coodinate attention (CA) [18], which inherits the advantage of channel attention methods that model inter-channel relationships while capturing the long-range dependencies with precise position information. The input and the output of the MocNet are the motion-corrupted and the clean MR images, respectively.

Since the T1W image reflect the anatomical structure, motion artifact correction should focus on the luminance, contrast, and texture. Inspired by the work in image restoration [19], we propose to use the sum of multi-scale structural similarity index (MS-SSIM) loss and ℓ_1 loss to correct motion artifacts, where

MS-SSIM loss preserves the texture and contrast in high-frequency regions, and ℓ_1 loss preserves the luminance. The loss used in the proposed method is defined as

$$\mathcal{L}(x, \hat{x}) = \lambda \cdot (1 - MS\_SSIM(x, \hat{x})) + (1 - \lambda) \cdot G_{\sigma_G^M} \cdot ||x - \hat{x}||_1, \tag{3}$$

where x and $\hat{x}$ are the clean MR image and the reconstructed MR images, respectively. MS_SSIM denotes MS-SSIM loss. $G_{\sigma_G^M}$ is the Gaussian coefficient. λ is a tradeoff coefficient between the MS-SSIM and the ℓ_1 loss. In our experiment, λ is set to be 0.84.

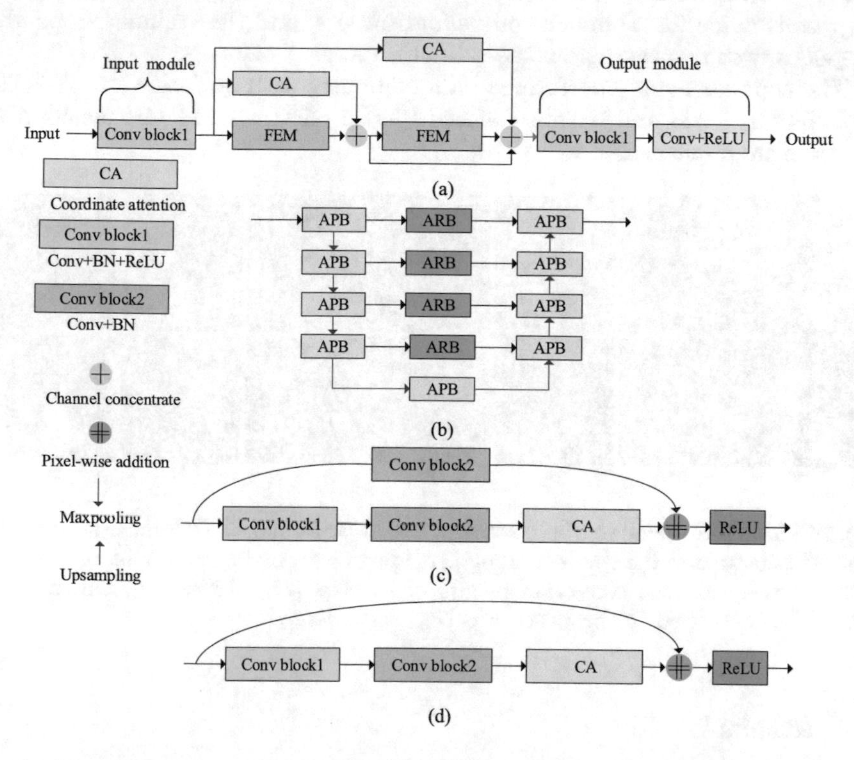

Fig. 1. The architecture of MocNet. (a) MocNet. (b) Feature-extraction module (FEM). (c) Attention paralleling block (APB). (d) Attention residual block (ARB). Best viewed in color.

3 Experiment

3.1 Data Preprocessing and Augmentation

Each MR image is resampled to $192 \times 192 \times 56$ using linear interpolation before simulating motion artifacts. The intensities of the clean and corrupted images are normalized to a range of [0,1] before entering the networks. Data augmentation techniques including randomly rotated by a degree ranging from 1 to 360°, flipped horizontally and vertically are used in this research.

3.2 Implementation Details

The parameters used in simulating motion artifacts are randomly generated with (i) maximum translation of voxel along the x-axis, y-axis and z-axis are 8 mm, 8 mm and 4 mm, respectively, (ii) the maximum rotation angle is 9°, and (iii) α is set to 10. γ is drawn from a uniform distribution within [20, 60].

The networks are initialized using Kaiming's method [20] and the optimizer is the Adam method [21] with $\beta_1 = 0.9$ and $\beta_2 = 0.999$. The initial learning rate is 10^{-3}. During training, the learning rate is scaled down by a factor of 0.1 if no progress is made for 15 epochs on validation loss, and the training stops after 30 epochs with no progress on the validation loss.

The experiments are performed on a computer with an Intel Core i7-6800K CPU, 64 GB RAM and Nvidia Geforce 1080Ti GPU with 11 GB memory. All networks are implemented in PyTorch.

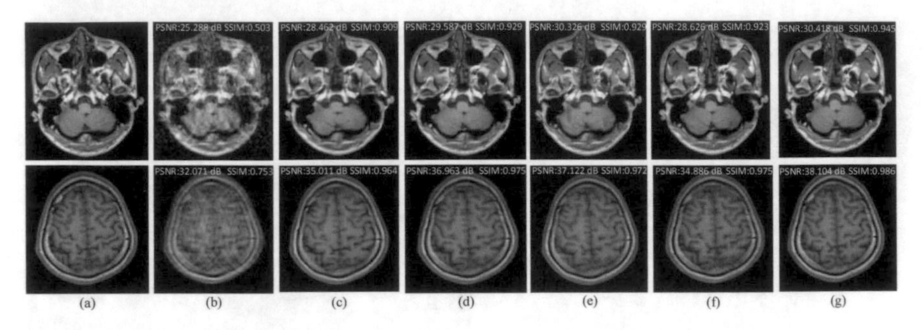

Fig. 2. Visualized examples of the reconstruction performance. Columns (a–b) are the clean MR image and the corresponding MR image with motion artifacts, respectively. Columns (c–g) are the correction results of FCN-8s [22], U-Net [17], DnCNN [14], DRN-DCMB [16] and the proposed MocNet, respectively.

3.3 Results

In this research, structural similarity (SSIM) and peak-to-noise ratio (PSNR) are used to evaluate the performance of MocNet. For the sake of comparison, we also train and evaluate FCN-8s [22], U-Net [17], DnCNN [14] and DRN-DCMB [16] using the same settings and loss function on our dataset. Figure 2 visualizes some slices of motion artifacts and their correction results. As the Fig. 2 shows, the input MR images contain severe motion artifacts after the rigid motion transformation. Although FCN-8s and U-Net are generally used for segmentation task, they present high performance on motion artifacts correction and obtain clean MR images. In particular, U-Net achieves better correction results than DnCNN, which is used to image denoising, despite the appearance of some artifacts remained obvious. However, as a specified model for correcting motion artifacts, DRN-DCMB ignores some structure contrast. In comparison, the proposed MocNet substantially corrects the motion artifacts and produces

Table 1. Quantitative evaluation results on the test set. MA denotes the MR image with motion artifacts. In particular, the mean PSNR and the mean SSIM have presented in the way of mean ± standard deviation. The best results has been highlighted in bold.

Method	PSNR (dB)	SSIM
MA	27.507 ± 2.347	0.635 ± 0.070
FCN-8s [22]	31.526 ± 2.607	0.942 ± 0.023
U-Net [17]	33.391 ± 2.817	0.961 ± 0.019
DnCNN [14]	33.580±3.032	0.954 ± 0.022
DRN-DCMB [16]	29.852 ± 2.912	0.955 ± 0.022
MocNet	**34.397 ± 3.155**	**0.971 ± 0.015**

cleaner MR images, and meanwhile the image contrast is maintained and subtle image details are preserved.

The quantitative evaluation results are summarized in Table 1. As Table 1 shows, benefit from multi-scale feature fusion, U-Net achieves a mean PSNR of 33.391 ± 2.817 dB and a mean SSIM of 0.961 ± 0.019, which exceeds the comparison methods except the MocNet. Our proposed MocNet, however, achieves the best correction results with a mean PSNR of 34.397 ± 3.155 dB and a mean SSIM of 0.971 ± 0.015 thanks to the utilization of the attention dense connection to reuse the former features and attention residual block to extract more features.

4 Conclusion

In this paper, we propose the MocNet to correct motion artifacts of MR images. The feature-extraction modules extract rich semantic information and the attention dense connection used in MocNet is responsible for reusing more features. Experiment results demonstrate that our MocNet outperforms the comparison methods.

Acknowledgments. This work is supported in part by the National Natural Science Foundation of China (61871239, 62076077) and the Natural Science Foundation of Tianjin (20JCQNJC0125).

References

1. Godenschweger, F., et al.: Motion correction in MRI of the brain. Phys. Med. Biol. **61**(5), R32 (2016)
2. Andre, J.B.: Toward quantifying the prevalence, severity, and cost associated with patient motion during clinical MR examinations. J. Am. Coll. Radiol. **12**(7), 689–695 (2015)
3. Pipe, J.G.: Motion correction with propeller MRI: application to head motion and free-breathing cardiac imaging. Magn. Reson. Med.: Off. J. Int. Soc. Magn. Reson. Med. **42**(5), 963–969 (1999)

4. Thesen, S., Heid, O., Mueller, E., Schad, L.R.: Prospective acquisition correction for head motion with image-based tracking for real-time FMRI. Magn. Reson. Med.: Off. J. Int. Soc. Magn. Reson. Med. **44**(3), 457–465 (2000)
5. Van Der Kouwe, A.J.W., Benner, T., Dale, A.M.: Real-time rigid body motion correction and shimming using cloverleaf navigators. Magn. Reson. Med.: Off. J. Int. Soc. Magn. Reson. Med. **56**(5), 1019–1032 (2006)
6. Qin, L., Gelderen, P., Zwart, J., Jin, F., Tao, Y., Duyn, J.: Head movement correction for MRI with a single camera. In: Proceedings of the 16th Scientific Meeting, International Society for Magnetic Resonance in Medicine, Toronto, Canada, p. 1467 (2008)
7. Zaitsev, M., Maclaren, J., Herbst, M.: Motion artifacts in MRI: a complex problem with many partial solutions. J. Magn. Reson. Imaging **42**(4), 887–901 (2015)
8. Atkinson, D., et al.: Automatic compensation of motion artifacts in MRI. Magn. Reson. Med.: Off. J. Int. Soc. Magn. Reson. Med. **41**(1), 163–170 (1999)
9. McGee, K.P., Felmlee, J.P., Jack, C.R., Jr., Manduca, A., Riederer, S.J., Ehman, R.L.: Autocorrection of three-dimensional time-of-flight MR angiography of the Circle of Willis. Am. J. Roentgenol. **176**(2), 513–518 (2001)
10. Loktyushin, A., Nickisch, H., Pohmann, R., Schölkopf, B.: Blind retrospective motion correction of MR images. Magn. Reson. Med. **70**(6), 1608–1618 (2013)
11. Bydder, M., Larkman, D.J., Hajnal, J.V.: Detection and elimination of motion artifacts by regeneration of k-space. Magn. Reson. Med.: Off. J. Int. Soc. Magn. Reson. Med. **47**(4), 677–686 (2002)
12. Shen, D., Guorong, W., Suk, H.-I.: Deep learning in medical image analysis. Ann. Rev. Biomed. Eng. **19**, 221–248 (2017)
13. Lyu, Q., et al.: Cine cardiac MRI motion artifact reduction using a recurrent neural network. IEEE Trans. Med. Imaging **40**, 2170–2181 (2021)
14. Zhang, K., Zuo, W., Chen, Y., Meng, D., Zhang, L.: Beyond a Gaussian denoiser: residual learning of deep CNN for image denoising. IEEE Trans. Image Process. **26**(7), 3142–3155 (2017)
15. Tamada, D., Kromrey, M.-L., Ichikawa, S., Onishi, H., Motosugi, U.: Motion artifact reduction using a convolutional neural network for dynamic contrast enhanced MR imaging of the liver. Magn. Reson. Med. Sci. **19**(1), 64 (2020)
16. Liu, J., Kocak, M., Supanich, M., Deng, J.: Motion artifacts reduction in brain MRI by means of a deep residual network with densely connected multi-resolution blocks (DRN-DCMB). Magn. Reson. Imaging **71**, 69–79 (2020)
17. Ronneberger, O., Fischer, P., Brox, T.: U-net: convolutional networks for biomedical image segmentation. In: International Conference on Medical Image Computing and Computer-assisted Intervention, pp. 234–241 (2015)
18. Hou, Q., Zhou, D., Feng, J.: Coordinate attention for efficient mobile network design. In: Proceedings of the IEEE Conference on Computer Vision and Pattern Recognition (2021)
19. Zhao, H., Gallo, O., Frosio, I., Kautz, J.: Loss functions for image restoration with neural networks. IEEE Trans. Comput. Imaging **3**(1), 47–57 (2016)
20. He, K., Zhang, X., Ren, S., Sun, J.: Delving deep into rectifiers: surpassing human-level performance on imagenet classification. In: Proceedings of the IEEE International Conference on Computer Vision, pp. 1026–1034 (2015)
21. Kingma, D.P., Ba, J.: Adam: a method for stochastic optimization. arXiv preprint arXiv:1412.6980 (2014)
22. Long, L., Shelhamer, E., Darrell, T.: Fully convolutional networks for semantic segmentation. In: Proceedings of the IEEE Conference on Computer Vision and Pattern Recognition, pp. 3431–3440 (2015)

The Application Exploration of Digital Twin in the Space Launch Site

Cai Hongwei[1(✉)], Zhou Bo[1], Yang Hui[2], and Li Xu[1]

[1] Xichang Satellite Launch Center, Group 5, box 16, Xichang 615000, Sichuan Province, China
11899944@qq.com

[2] China Astronaut Research and Training Center, Beijing 100094, China

Abstract. In view of the future development of space launch site, combined with the present situation and development of digital twin technology, the paper puts forward the idea of the construction of digital twin in the space launch site. The system analyzes the application prospect of digital twin technology in the space launch site, and designs the architecture of the digital twin of the space launch site, which is concluded that the digital twin body of the space launch site will completely change the operation management mode, the equipment guarantee mode, the task process, the organization command, and so on.

Keyword: Digital twin · Space launch site · Heterogeneous model · Three-dimensional vision · Virtual interaction · Visual control · Health management

1 Digital Twin

With the explosion of the new generation of technology represented by the Internet of things, large Numbers, cloud computing and artificial intelligence, the development of the intelligent industry has been opened, and the space launch site in China is also moving in a smart and intelligent direction. Digital twin, a frontier technology, are listed as the top six major technology in the future of defense and aerospace industry in 2017, being widespread attention by industry and academia [1]. In July and 2019, the world's most authoritative it research and consulting firm Gartner has listed digital twin as one of the top 10 strategic technology trends in recent three years [2], and has become an important cornerstone of the Internet of things and industry 4.0.

Digital twin is the object of the pointer to the physical world, and by digital means to build an entity that is identical to the digital world to realize the understanding, analysis, and optimization of the physical entity [3]. Digital twin is a virtual model of physical entities in a digital way, using data to simulate the behavior of physical entities in the real environment, and increase or expand new capabilities for physical entities by means of virtual interaction feedback, data fusion analysis, decision iteration optimization, etc.

Digital twin is a multi-dimensional, multi-scale, multi-probability simulation model for highly integrated multi-physical fields. It can use physical models, sensor data and

Q. Liang et al. (Eds.): Artificial Intelligence in China, LNEE 854, pp. 47–54, 2022.
https://doi.org/10.1007/978-981-16-9423-3_7

historical data to reflect the function, real-time state and evolution trend of the corresponding entity of the model, and be applied in several areas such as aviation, aerospace, production, design, manufacturing, and intelligent city [4]. In the space field of NASA in 2010, NASA introduced the concept of digital twin in the release of the NASA space technology roadmap, which is intended to achieve a comprehensive diagnostic and predictive function of the flight system. The route map to achieve NASA's digital twin goals by 2027. NASA estimates that by 2035, the use of digital twin technologies will halve the cost of aircraft maintenance and extend the service lifespan to 10 times the current level. In recent years, the application of digital twin in aerospace has been widely watched by scholars both at home and abroad, and the digital twin technology has been widely used in the health maintenance of domestic and foreign aerospace vehicles. Li Shu and other people carry rocket launch site digital practice, demonstrate and develop the new rocket in the launch field full process and practice test, provide technical support for model development [5].

2 Artificial Intelligence

Artificial intelligence is a cross-subject that involves informatics, logic, cognitive learning, thinking, systems and biology, through a lot of research, using computers to simulate certain thinking processes and intelligent behavior, and to produce a new type of intelligent machine that can respond in a way that is similar to human intelligence [6]. It has been used in the field of knowledge processing, game theory, automatic programming, automatic programming, expert system, knowledge base, intelligent machine, etc., especially in the areas of unmanned driving, intelligent machine, image recognition and language recognition, which is changing the way people make life, which is driving the improvement of technological innovation, economic development and people's lives. In industries such as finance, healthcare, automobile and retail, artificial intelligence is relatively mature in these industries, such as companies such as finance, healthcare, automobile and retail, and the companies such as Google, Baidu, Tesla and Audi are joining in the case of self-driving cars [7–10].

3 Digital Twin Applications And Analysis In Space

Space always has a lot of risks. The space launch site is a prerequisite for space launch. The non-standard, maintenance, use and storage of environmental bad, short-term and reliable characteristics, and the continuous high density task of the development of the launch field for a long period of time, must seek more active and more scientific methods, to find out, to find out, to meet the problem of the resolution of the resolution and the reliability of the launch site, and to implement preventive maintenance and maintenance, to improve the overall reliability and safety of the launch site.

With the explosion of the new generation of technology represented by the Internet of things, large Numbers, cloud computing and artificial intelligence, the development of the intelligent industry has been opened, and the global space launch site is gradually transforming in the direction of intelligence and wisdom. Based on the technology of the Internet of things, artificial intelligence and BIM&CIM model, the space launch site digital twin body is built to realize the high integration of the reality launch site and digital launch field, integrating personnel, process, data, technology and business system, realize the whole process of the launch site, the digital, network and intelligence of the whole elements, and promote the transformation and development of the space launch site.

All in all, it is possible to build the space launch site with high information fusion, which can provide strong support for the maintenance, equipment maintenance, equipment maintenance, failure prediction, health management, simulation training, task implementation, process optimization, evaluation, command decision, etc., and will enhance the comprehensive ability of the space launch site.

4 The Design of the Digital Twin System of the Space Launch Site

The space launch site digital twin system has large volumes and complex data. Through the combination of artificial intelligence algorithm and data, the consistency of the digital twin and physical ontology can be further improved. At the same time, artificial intelligence and large Numbers are analyzed and analyzed, which can predict the changing trends of the data of the ontology, including fault, life expectancy, etc.

The launch site digital twin system is a key function of the aided digital twin system by building a sound data storage structure, using cutting-edge data mining technology to search for key data in the full state of the twin body information.

The space launch site digital twin system consists mainly of four parts, including physical objects, virtual models, twin data and service systems. Physical objects are the basis of digital twin. Based on this, the virtual model of physical objects is constructed, and the virtual model is gradually consistent with the physical object through iterative optimization. Through the drive of the twin data, the virtual model and the service system are more coordinated by the continuous iteration optimization, making the service system more coordinated with the virtual model (Fig. 1).

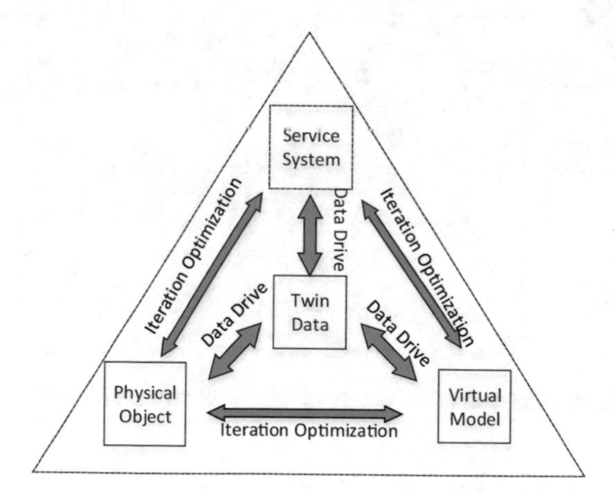

Fig. 1. Architecture of Space launch site digital twin

The key of the digital twin of the launch site is the information physics fusion, the multi-dimensional virtual model, the twin data, the dynamic real-time interaction connection, the service application and the whole element physical entity, etc.

Information physical integration is mainly refers to the realization of the intelligence perception and interconnection of physical elements, the construction of virtual model, the fusion of twin data, the realization of the connection interaction, the formation of the application service, and so on. The multi-dimensional virtual model is the core component of the overall design and planning, mission implementation, failure prediction, and health management of the launch site, and is the function engine of digital twin. The twin data is the core element of the digital twin of the launch site, which is derived from the launch field physical entity, virtual model, and service system, and is integrated into the each group of the digital twin system after fusion processing. Dynamic real-time interaction is the artery of the digital twin system of the launch site, and the dynamic real-time interaction connects the physical entity, virtual model, and service system to an organic whole.

The digital twin system of the space launch site involves many factors such as mathematical model, system integration, platform software and business. In order to realize the actual business requirements of the launch task, this paper reflects the dynamic energy of the unit of the launch field, the operation of the equipment status monitoring, the process control of the launch process, and the requirements of the multi-system expansion and access in the future, and the structure design of the platform, as shown in Fig. 2.

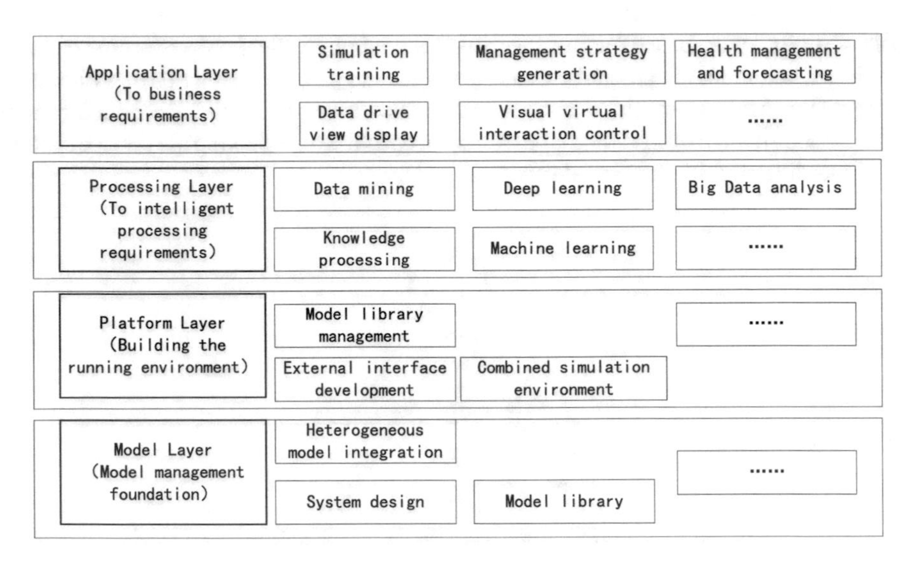

Fig. 2. Hierarchical structure of Space launch site digital twin.

1) Model layer: provides multi-source heterogeneous model integration and system design function, compatible with existing models and supports the subsequent model

extension, and supports the framework of system design based on the model integration base for the overall perspective of the launch site. The virtual object building of the launch site is realized by means of mixed particle degree and multi-level model.
2) Platform layer: provide the operating environment for virtual objects, form the experimental environment based on the model simulation of the model, and consider the external extension interface, support the external object of the external group 3d view, control device, and control signal, and realize the dynamic coupling of multi-type elements.
3) Processing layer: the data processing of a number of twin systems is supported by the intelligent processing of the twin data, which meets the requirements of the different business log processing application, which mainly implements the real data, the data mining, the machine learning, the knowledge processing, the large number analysis and so on, and the ability to improve the data application through the intelligent processing algorithm.
4) Application layer: provides auxiliary function of docking with launch field business, supports the management of space data, data driven visual view, algorithm based device health management, model based equipment control, launch field digital model library management, simulation task management and execution, etc.

The space launch site digital twin system is based on platform software, and it is a functional module such as model integration, system design, simulation test environment and external element extension. The model integration module solves the standardization description of different professional model elements, and provides prerequisites for multi-professional models and platform applications. The system design module solves the problem of the digital launch field, which can be designed by the model of the graphical system based on the model, and builds the simulation object of the application scene quickly by using the model library form management model and the data interface. The simulation test environment solves the problem of digital engineering management and model operation, and supports the performance of typical launch field workflow and task execution. The external element extension module solves the problem of the digital launch site and the external data interaction, and supports the dynamic display of three dimensional viewing based on real-time data.

With the development of artificial intelligence technology, artificial intelligence technology will gradually spread into all walks of life, and it will change the work and lifestyle of people. With the maturity of the digital twin system of the space launch site, artificial intelligence technology will be gradually introduced, and the dual-wheel drive of digital twin and artificial intelligence will be achieved, and the launch site will be continuously driven and improved (Fig. 3).

Through the full integration of artificial intelligence and digital twin, based on the three-dimensional visual human–computer interaction and artificial intelligent data processing platform, the space launch site digital twin body, the carrier digital twin, the spacecraft digital twin, and the full simulation of the space launch mission, the prediction of the emergency in the mission. For the space launch site important facilities, the digital twin is constructed, which can predict the structure of the important facilities of the space launch site, the fatigue damage of the material, the failure of the system caused by the prevention structure and material problems. By building the digital twin

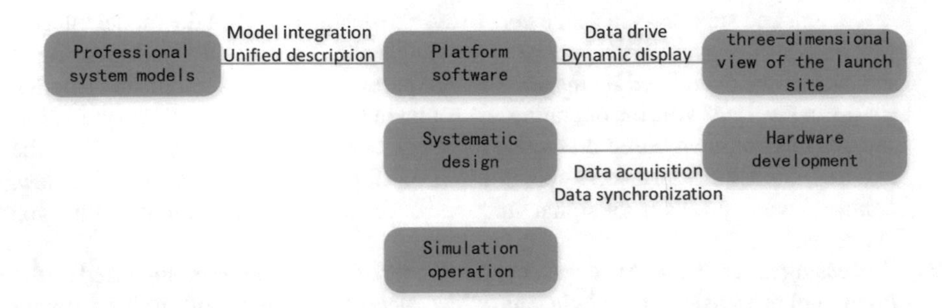

Fig. 3. Multi-level system design framework schematic diagram

of the launch site, the real-time data drive can be used to observe the real-time operation state of the space launch site more fully, and detect the situation in real time. Using the space launch site digital power, it can carry out fault prediction, fault localization and promotion, and through artificial intelligence processing, it can provide optimal control strategy and method for the real-time operation management of the space launch site, realize the effective and the system's launch field health management, and effectively avoid the risk of the space launch site.

Typical application of the digital twin body of the space launch site.

1) Application in visual human–computer interaction.

The launch field digital twin is preset to the external interface at the model level, which can receive external data or pass the model operation to the outside. This method can be used to make a link to the group broadcast data of the cloud platform using the cloud platform, and realize the control of the external equipment of the digital launch site and the state feedback of the external equipment, which is the system of the launch field of the fusion (Fig. 4).

2) Application in visual simulation training.

The digital twin model is based on the control instruction of external equipment or control input device, and the corresponding mathematical model component of the launch field digital twin is simulated by the simulation control instruction, and the work data of each group in the system is passed to the two-dimensional display interface or three-dimensional view, and the performance of the digital launch site is shown in the digital launch field. The staff can simulate the workflow path of the real space launch mission in this virtual scenario to provide the role of the simulation training environment for the staff without the mission (Fig. 5).

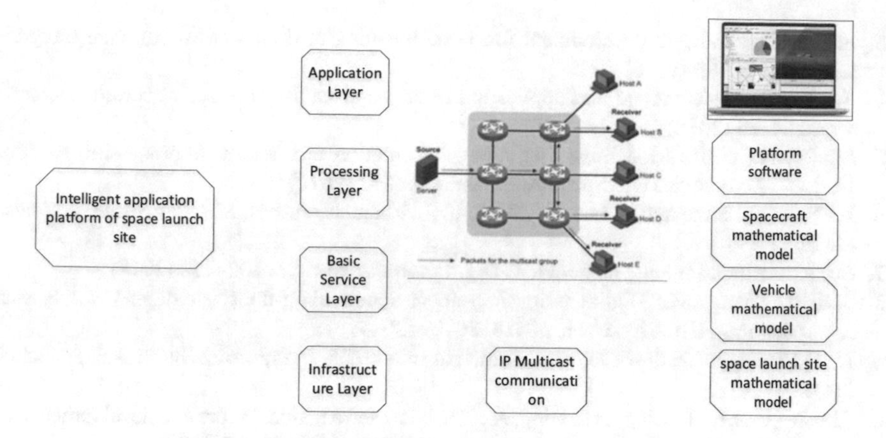

Fig. 4. Equipment data visual interaction diagram

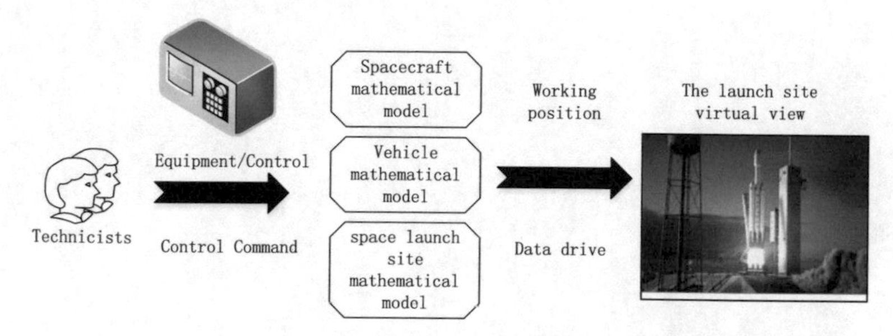

Fig. 5. Simulation training application schematic diagram

5 Conclusion

The digital twin technology has been applied in the space and aviation industry at home and abroad, and has played an important role in design, manufacturing and application. The application of the space launch site must be the result of the deep fusion of digital dual technology and artificial intelligence technology, which will bring revolutionary changes in the operation management mode, equipment guarantee mode, task process and organization command, which will provide strong support for the daily operation, equipment maintenance, failure prediction, health management, simulation training, task implementation, process optimization, evaluation, command decision, etc., and the application prospect is very broad.

References

1. Tao, F., et al.: Digital twin and its potential application exploration. Comput. Integr. Manuf. Syst. **24**(1), 1–18 (2018)
2. Meng, S.H., et al.: Digital twin and its aerospace applications. Acta Aeronautica et Astronautica Sinica **41** (2020). (in Chinese). https://doi.org/10.7527/S1000-6893.2020.23615

3. Shi, P.: The concept, development form and meaning of digital twin. Software Integrated Circuits **9**, 28–33 (2018)
4. Su, R.: The development and application of artificial intelligence are reviewed. Electronic world 84–86 (2018)
5. Shafto, M., et al.: Modeling, simulation, information technology & processing roadmap. National Aeronautics and Space Administration 5–7 (2010)
6. Li, S., et al.: Research of launch vehicle digital launch process. Missiles Space Vehicles **2**, 22–26 (2019)
7. Su, X.: Artificial intelligence review. Dig. Commun. World **1**, 105–112 (2018)
8. Tao, F., Zhang, M.: Digital twin shop-floor: a new shop-floor paradigm towards smart manufacturing. IEEE Access **5**, 20418–20427 (2017)
9. Li, H.: Thoughts on the development route of smart rocket. Astronaut. Systems Eng. Technol. **1**(1), 1–7 (2017)
10. Zhuang, C., Liu, J., Xiong, H., Ding, X.: The connotation, architecture and development trend of product digital twin. Comput. Integr. Manuf. Syst. **23**(4), 753–768 (2017)

Reverse Attention U-Net for Brain Grey Matter Nuclei Segmentation

Mengran Wu[1], Bin Zhao[1], Chao Chai[2], Guohua Liu[1], and Zhiyang Liu[1](✉)

[1] Tianjin Key Laboratory of Optoelectronic Sensor and Sensing Network Technology, College of Electronic Information and Optical Engineering, Nankai University, Tianjin 300350, China
liuzhiyang@nankai.edu.cn

[2] School of Medicine, Tianjin First Central Hospital, Nankai University, Tianjin 300192, China

Abstract. It has been reported that abnormal high iron deposition in brain grey matter nuclei is closely related to neurodegenerative diseases. Therefore, precise nuclei segmentation is beneficial to further explore the pathological mechanism of neurodegenerative diseases. However, segmenting nuclei manually is an extremely time-consuming work. To this end, we proposed a Reverse Attention U-Net (RAU-Net) for segmenting nuclei automatically, where multiple reverse attention (RA) modules are added between the encoder and decoder to aggregate different features. In the meanwhile, we use the estimations of segmentation maps at multiple levels as the guide information for the RA module. The model with RA module implicitly erases the predicted nuclei regions while highlighting background, which guides the network to explore the missing nuclei parts sequentially. Experimental results on our nuclei dataset imply that the RAU-Net performs favorably against the state-of-the-art methods.

Keywords: Deep learning · Nuclei segmentation · Reverse attention

1 Introduction

Quantitative susceptibility mapping (QSM) has been widely adopted to measure the tissue magnetic susceptibility which reflects the degree of iron deposition. It has been reported that a variety of neurodegenerative diseases, such as Alzheimer's disease and Parkinson's disease, are highly related with the abnormally high iron accumulation in brain grey matter nuclei shown in Fig. 1, including caudate nucleus (CN), globus pallidum (GP), putamen (Put), thalamus (Thal), red nucleus (RN), substantia nigra (SN), and dentate nucleus (DN) [2, 5, 7–9]. To quantitively evaluate the magnetic susceptibility level, the nuclei should be first segmented from the images by jointly considering the QSM and the T1-weighted imaging (T1WI) with high-resolution anatomical structures. The segmentation task, however, was usually done by manual delineation in the literature [11, 16, 18], which is a time-consuming task when the number of subjects is large. Moreover, the accuracy of segmentation is also highly dependent on the clinician's experiences. Therefore, it is necessary to develop an automatic segmentation method for brain grey matter nuclei segmentation.

Q. Liang et al. (Eds.): Artificial Intelligence in China, LNEE 854, pp. 55–62, 2022.
https://doi.org/10.1007/978-981-16-9423-3_8

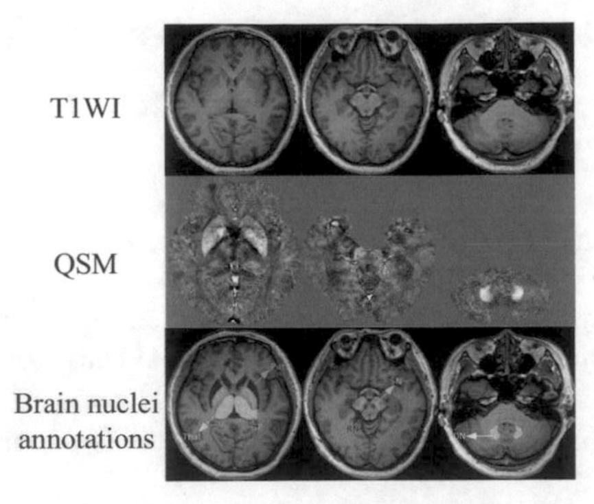

Fig. 1. Visualized examples of T1WI, QSM, and the corresponding brain grey matter nuclei annotations

Conventionally, the atlas-based methods were adopted for nuclei segmentation [3, 14, 19]. An atlas was established by manually annotating the regions of interest (ROIs) on a standard brain. During segmentation, a transform was first obtained by registering the subject-in-question to the standard brain, and the segmentation results could be obtained by performing the inverse transform on the ROI templates. It is clear that the segmentation accuracy of an atlas-based method is highly dependent on the registration accuracy. The deformation of anatomical structures that originates from aging and tumor may also reduce the segmentation accuracy. To overcome the drawbacks of the atlas-based methods, deep learning methods have recently been adopted in the medical image segmentation tasks.

One of the most popular methods is U-Net [12]. Originated from the full convolutional network (FCN) [13], U-Net adopted a U-shaped symmetric structure to obtain multi-level contextual and semantic information, and the extracted features were fused through the skip connections between the encoder and decoder. However, as the useful information was usually accompanied by many redundant information, the redundant information may mislead the decoder and reduce the segmentation accuracy [17]. Therefore, attention mechanism was adopted to improve the performance by adding an attention gate between the encoder and the decoder to filter misleading information [10]. Qiangguo Jin et al. proposed a residual attention-aware mechanism, which divided the attention module into a trunk branch and a soft mask branch to achieve multi-layer feature fusion [4].

Note that although U-Net-like structures perform well on many segmentation tasks such as organ and tumor segmentations, they may not be well-performed in segmenting nuclei. The organs and tumor have very prominent textual differences when compared with the other tissues. The nuclei, however, present similar textual appearances on both QSM and T1WI, as shown in Fig. 1, while the most prominent feature that distinguishes them is the position and the shape. This motivates us to adopt reverse attention mechanism to improve the segmentation accuracy.

In particular, we propose a Reverse Attention U-Net (RAU-Net) that incorporate attention mechanism into conventional U-Net for the nuclei segmentation task. The RAU-Net first obtains an initial guidance area by generating a rough global prediction map, then uses the reverse attention module to guide the network to gradually refine the nuclei segmentation results. Experiment results reveal that the proposed RAU-Net outperforms the U-Net and the Attention U-Net in both symmetric and surface distance metrics.

2 Methods

2.1 Overview

As shown in Fig. 2, the proposed Reverse Attention U-Net (RAU-Net) employs a U-Net like structure in general, but makes several important modifications. First, the encoder part generates an initial estimate of the segmentation map, which, although not accurate enough, roughly locates the approximate positions of the nuclei. Second, several reverse attention (RA) modules with structures shown in Fig. 3 are added on the skip connections between the encoder and decoder. Finally, multi-level supervision losses are adopted to improve network segmentation accuracy.

The RA modules and the estimations of segmentation maps at multiple levels are helpful to guide the network to hierarchically and gradually refine the segmentation results and improve the segmentation accuracy.

2.2 Reverse Attention Module

The RA module is the key component of our proposed RAU-Net. The motivation comes from the fact that high-level features have rich semantic information, which can be used to guide low-level features extraction, while low-level features contain fine spatial information, which can improve the edges of high-level features. By integrating high-level and low-level features, the contrast between foreground and background regions can be more obvious. However, simply concatenating or uniform weighting different levels of feature maps may introduce misleading information. In this paper, we propose a reverse guidance framework that guides the network to gradually discover supplementary areas and detailed information in each train-stage [1]. The output feature of the i-th layer can be obtained as

$$D_i = cat(up(D_{i+1}), (E_i \cdot A_i)) \tag{1}$$

for $i = 1, 2, 3, 4$. where E_i and D_i represent the i-th encoder and decoder feature. A_i denotes the reverse attention, and it is defined as:

$$A_i = softm\_0(S_{i+1}) \tag{2}$$

where S_i denotes the side-output of the i-th layer. $cat(\cdot)$ denotes the cascade operation, $up(\cdot)$ denotes the up-sampling operation. $softm\_0(\cdot)$ is the softmax operation followed by extracting only the background channel, i.e., channel 0 (Fig. 2).

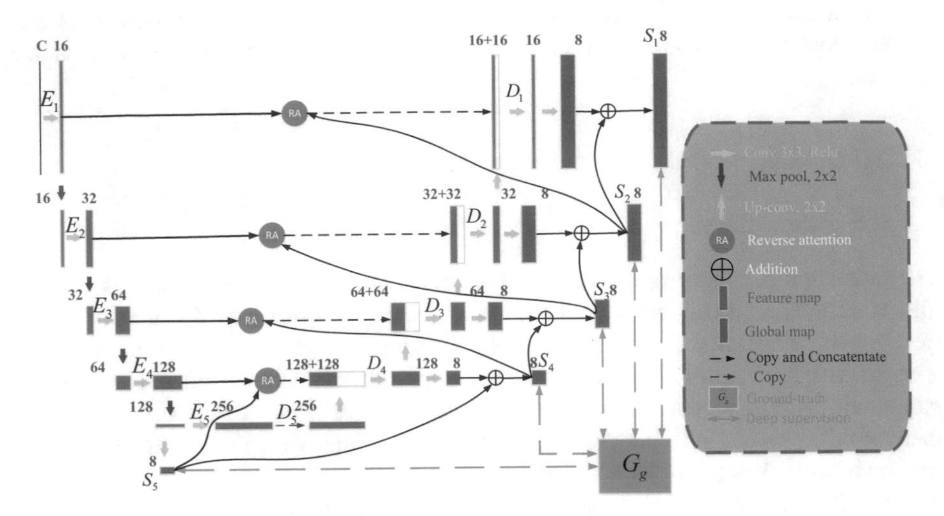

Fig. 2. The architecture of the proposed RAU-Net model

2.3 Loss Function

In the brain nuclei segmentation, the positive and negative samples are extremely unbalanced, so we choose focal loss [6] as the loss function, which is given as

$$L_{fl}(y_t, \hat{y}_t) = -(1 - \hat{y}_t * y_t)^{\gamma} \log \hat{y}_t \tag{3}$$

where $\hat{y}_t$ and y_t denote the prediction map and the ground-truth map, respectively. γ is the tunable focusing parameter, set $\gamma = 2$. At the same time, in order to obtain better experimental results, we also add four auxiliary losses for depth supervision.

We use the conv3 × 3 to obtain the side-output of each layer of the decoder, and use trilinear interpolation to up-sample it to the same size as the ground-truth map, so the overall loss can be expressed as

$$L_{total} = L_{fl}(G, S_g) + \sum_{i=2}^{i=5} L_{fl}(G, S_i) \tag{4}$$

where G and S_g are the ground-truth map and final prediction map. In addition, S_i is the side-output of the i-th layer (Fig. 3).

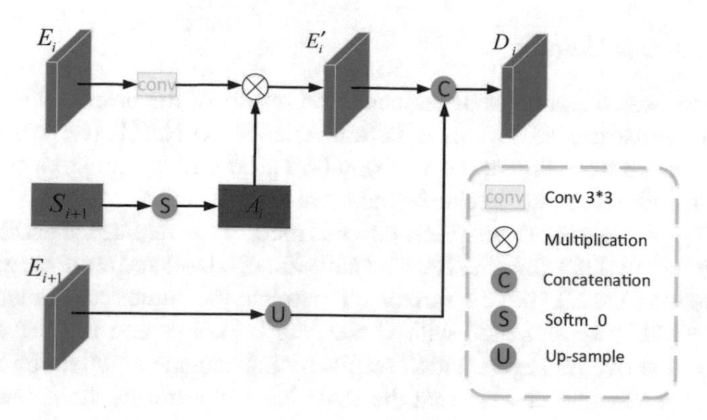

Fig. 3. The architecture of the reverse attention (RA) module

3 Experiment Results

3.1 Experiment Data and Data Processing

We evaluate the segmentation performance on an institutional dataset with 43 subjects, which are collected from the Tianjin First Central Hospital (Tianjin, China). The dataset consists of two kinds of MR sequences, namely Gradient Recalled Echo (GRE) and T1-weighted imaging (T1WI). The ground truth segmentations were annotated by experienced medical imaging clinicians. In this work, we use T1WI and QSM for nuclei segmentation. The QSM can be obtain by comprehensively processing the phase and amplitude maps of GRE following the method in [15]. The whole dataset is randomly divided into training set, validation set and test set, with 19, 3 and 21 subjects, respectively.

Before training, we accurately registered QSM and T1WI, and resampled the T1WI to ensure that they have the same voxel spacing (2 mm × 0.5134 mm × 0.5134 mm) and matrix size (56 × 336 × 448). Meanwhile, we normalized the intensity value of QSM and T1WI to [0, 1][0, 1], respectively. In the evaluation phase, we used test time augmentation for post-processing.

3.2 Implementation Settings

In our experiments, we train our model on a workstation with Intel Core i7-8700 K CPU, 32 GB RAM and Nvidia Geforce GTX 1080Ti GPU. The inputs of our network are cropped to 48 × 112 × 112 before training, and the batch size is set to be 2. In order to prevent over-fitting, we use data augmentation including random rotating and random flipping during the training process. We adopt the momentum stochastic gradient descent (SGD) optimizer with a weight decay of 3×10^{-5} and momentum of 0.9. The initial learning rate is set to 0.01. If the validation loss does not improve for 20 training epochs, the learning rate is reduced by a factor of 0.1. The training terminates if no progress is made for 40 epochs.

3.3 Segmentation Results

Figure 4 shows some examples of segmentation results of the proposed RAU-Net. For comparison, we also trained the U-Net [12] and Attention U-Net [10] on the same training set. As we can see from Fig. 4, the segmentation produced by the proposed method is more smooth and with more precise boundaries.

Following [20], we use three widely adopted metrics to evaluate the performance, i.e. the Dice coefficient (DC), the 95% Hausdorff distance (HD_{95}) and the average symmetric surface distance (ASSD). Table 1 further summarized the numerical evaluation results. As shown in Table 1, compared with U-Net, the RAU-Net and the Attention U-Net both improve the overall segmentation results by introducing an attention mechanism. In addition, we can also observe that the RAU-Net outperforms the Attention U-Net. This is due to the texture variance between the nuclei is very small. The Attention U-Net directly uses high-level features that lack of spatial information to filter low-level features, which will produce unsatisfactory results. Nevertheless, the RAU-Net uses the global prediction map of each layer as the guide information to provide rough position information of the nuclei. Reverse attention module is then used to further explore target clues and boundary clues, so that the model can better distinguish different nuclei. Specifically, we can conclude that the RA module can effectively improve the segmentation accuracy.

Table 1. Evaluation results on the test set. The most prominent result has been highlighted in bold

Metric	Model	CN	GP	PUT	THA	RN	SN	DN
DC	U-Net	0.716	0.823	0.819	0.765	0.704	0.754	**0.790**
	Attention U-Net	0.745	0.822	**0.824**	0.810	0.673	0.754	0.747
	RAU-Net	**0.802**	**0.834**	0.822	**0.848**	**0.722**	**0.772**	0.782
HD_{95} (mm)	U-Net	4.717	2.496	1.654	6.649	2.128	1.491	2.305
	Attention U-Net	3.548	**1.667**	**1.414**	2.732	2.056	1.145	2.799
	RAU-Net	**2.633**	1.784	1.693	**1.300**	**1.879**	**0.912**	**1.987**
ASSD (mm)	U-Net	0.822	0.367	0.449	1.349	0.365	0.276	0.339
	Attention U-Net	0.676	0.335	**0.395**	0.616	0.408	0.261	0.429
	RAU-Net	**0.465**	**0.315**	0.412	**0.375**	**0.309**	**0.222**	**0.317**

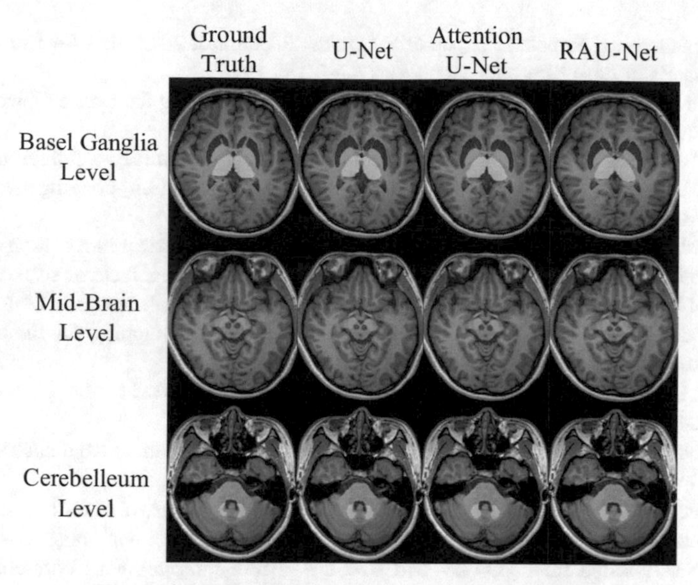

Fig. 4. Visualized examples of segmentation results

4 Conclusion

In this paper, we proposed a deep learning model, named RAU-Net, for automatically segmenting nuclei. The core module of the model is the RA module, which uses the side-outputs as the guide information to erase the predicted foreground regions, thereby forcing the network to focus on the background and gradually mine complementary areas and boundary information, so as to implicitly establish a relationship between the area and the boundary. Experiments demonstrated that the performance of the proposed model is better than the U-Net and Attention U-Net, indicating that the proposed model has great potential to be applied in brain tissue segmentation.

Acknowledgements. This work is supported in part by the National Natural Science Foundation of China (61871239, 81901728) and the Science and Technology talent cultivation Project of Tianjin Health Commission (grant number RC20185 to Chao Chai).

References

1. Chen, S., et al.: Reverse Attention for Salient Object Detection. ArXiv, abs/1807.09940 (2018)
2. C Finke 2015 Altered basal ganglia functional connectivity in multiple sclerosis patients with fatigue Mult. Scler. J. 21 925 934
3. L Igual 2011 A fully-automatic caudate nucleus segmentation of brain MRI: application in volumetric analysis of pediatric attention-deficit/hyperactivity disorder BioMed. Eng. Online 10 105 105
4. Q Jin Z Meng C Sun H Cui R Su 2020 RA-UNet: A Hybrid deep attention-aware network to extract liver and tumor in CT scans Front. Bioeng. Biotechnol. 23 8 605132

5. C Langkammer S Ropele L Pirpamer F Fazekas R Schmidt 2013 MRI for iron mapping in Alzheimer's disease Neurodegener. Dis. 13 2–3 189 191
6. T-Y Lin P Goyal RB Girshick K He P Dollár 2020 Focal Loss for Dense Object Detection IEEE Trans. Pattern Anal. Mach. Intell. 42 318 327
7. C Liu W Li KA Tong KW Yeom S Kuzminski 2015 Susceptibility-weighted imaging and quantitative susceptibility mapping in the brain: brain susceptibility imaging and mapping J. Magn. Reson. Imaging 42 1 23 41
8. R Morecraft J Louie J Herrick KS Stilwell-Morecraft 2001 Cortical innervation of the facial nucleus in the non-human primate: a new interpretation of the effects of stroke and related subtotal brain trauma on the muscles of facial expression Brain J. Neurol. 124 1 176 208
9. Y Murakami 2015 Usefulness of quantitative susceptibility mapping for the diagnosis of Parkinson disease Am. J. Neuroradiol. 36 1102 1108
10. Oktay, O., et al.: Attention U-Net: learning where to look for the pancreas. ArXiv, abs/1804.03999 (2018)
11. KM Rodrigues de 2015 A FreeSurfer-compliant consistent manual segmentation of infant brains spanning the P 2 year age range Front. Human Neurosci. 9 21
12. O Ronneberger P Fischer T Brox 2015 U-Net: convolutional networks for biomedical image segmentation N Navab J Hornegger WM Wells AF Frangi Eds Medical Image Computing and Computer-Assisted Intervention – MICCAI 2015: 18th International Conference, Munich, Germany, October 5-9, 2015, Proceedings, Part III Springer International Publishing Cham 234 241 https://doi.org/10.1007/978-3-319-24574-4_28
13. E Shelhamer J Long T Darrell 2017 Fully convolutional networks for semantic segmentation IEEE Trans. Pattern Anal. Mach. Intell. 39 640 651
14. J Su 2019 Thalamus optimized multi atlas segmentation (THOMAS): fast, fully automated segmentation of thalamic nuclei from structural MRI NeuroImage 194 272 282
15. J Tang S Liu J Neelavalli YCN Cheng S Buch EM Haacke 2013 Improving susceptibility mapping using a threshold-based K-space/image domain iterative reconstruction approach Magn. Reson. Med. 69 5 1396 1407
16. JD Theiss C Ridgewell M McHugo S Heckers JU Blackford 2017 Manual segmentation of the human bed nucleus of the stria terminalis using 3T MRI NeuroImage 146 288 292
17. Wang, J., Zhao, X., Ning, Q., Qian, D.: AEC-Net: attention and edge constraint network for medical image segmentation. In: 2020 42nd Annual International Conference of the IEEE Engineering in Medicine & Biology Society (EMBC), pp. 1616–1619 (2020)
18. Wild, D., Weber, M., Egger, J.: Client/Server Based Online Environment for Manual Segmentation of Medical Images. ArXiv, abs/1904.08610 (2019)
19. Y Xia K Bettinger L Shen AL Reiss 2007 Automatic segmentation of the caudate nucleus from human brain MR images IEEE Trans. Med. Imaging 26 509 517
20. R Zhang 2018 Automatic segmentation of acute ischemic stroke from DWI using 3-D fully convolutional DenseNets IEEE Trans. Med. Imaging 37 9 2149 2160

Transmit RRH Selection in User-Centric Cell-Free Massive MIMO Using Discrete Particle Swarm Optimization

Jiaxiang Li, Yingxin Zhao, Hong Wu, and Zhiyang Liu(✉)

Tianjin Key Laboratory of Optoelectronic Sensor and Sensing Network Technology, College of Electronic Information and Optical Engineering, Nankai University, Tianjin 300350, China
liuzhiyang@nankai.edu.cn

Abstract. Cell-free massive multiple input multiple output (MIMO) has been expected to improve the spectrum efficiency in future mobile communication systems. To reduce the fronthaul link burden, a user-centric structure is expected, where a subset of remote radio heads (RRHs) are expected to form a virtual cell and serve the user. However, how to select serving RRHs for each user so as to maximize the rate performance remains to be an open question. In this paper, by assuming that each RRH can only be selected by at most one user, an improved discrete particle swarm optimization (IDPSO) has been proposed to accelerate its convergency. To further reduce the fronthaul link burden, a large-scale-fading-based fitness function is also adopted. Numerical results reveal that the proposed algorithm is able to achieve a good tradeoff between sum rate performance and the computational complexity.

Keywords: User-centric cell-free massive MIMO · RRH selection · Discrete particle swarm optimization

1 Introduction

Millimeter wave (mmWave) band has been widely adopted in the 5th generation (5G) mobile communication system thanks to its abundant spectrum resources and the resulting high data rate. The mmWave, however, suffers from much severer propagation attenuation due to its short wavelength. Therefore, a cell-free massive MIMO system has been proposed, where a large number of RRHs are connected to a central processing unit (CPU) and simultaneously serve a much smaller number of users [1]. Cell-free massive MIMO can exploit macroscopic diversity against the shadow fading and offer much higher probability of coverage than co-located massive MIMO, but at the cost of increased fronthaul requirements [2]. As the RRHs are geographically distributed, each user mainly benefits most from its surrounding RRHs, and therefore a user-centric structure, which is known as virtual cell, becomes more necessary to reduce fronthaul link burden [3]. Figure 1 shows basic structure of user-centric cell-free massive MIMO with RRH selection, where each user chooses one or several RRHs to request the downlink signals before communication. The user-centric structure significantly reduces the

Q. Liang et al. (Eds.): Artificial Intelligence in China, LNEE 854, pp. 63–72, 2022.
https://doi.org/10.1007/978-981-16-9423-3_9

fronthaul link burden, as it only transmits the symbols of the serving users to each RRH. As the serving RRHs also produce inter-cell interference to the other users, how to select RRHs to form virtual cells for each user becomes the main challenge to be addressed.

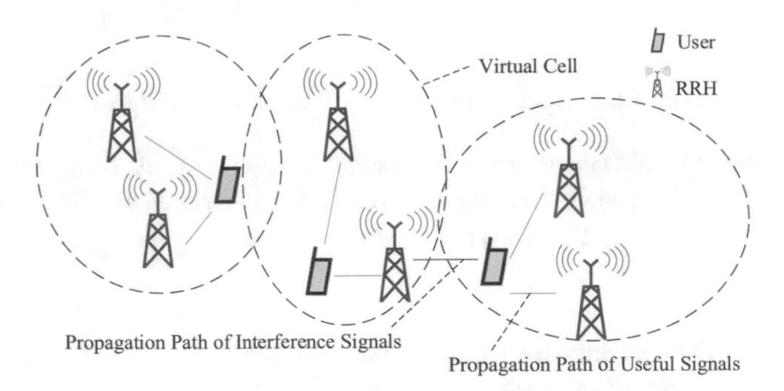

Fig. 1. User-centric cell-free massive MIMO

The matching design between users and RRHs is a discrete combinatorial optimization problem and the choices of different users are coupled. The discrete particle swarm optimization (DPSO), an intelligent optimization algorithm developed from particle swarm optimization (PSO) designed by Kennedy et al. [4], has good performance to solve discrete nonconvex problems with low complexity and some paper use DPSO and its variants to solve RRH or antenna selection problems. In single multi-antenna user model, Hei et al. [5] use the DPSO algorithm for the receive antenna selection problem to maximize capacity in co-located massive MIMO system with single multiple-antennas receiver and Shang et al. [6] propose that RRHs can be selected based on maximum path fading in single user distributed MIMO systems. For multiple single-antenna users model, Liu et al. [7] develop two algorithms based on PSO to achieve the Pareto optimal between energy efficiency and spectral efficiency tradeoff in multiuser MIMO system, but the small-scale fading information is necessary for algorithms. Ari et al. [8] propose an efficient resource allocation scheme based on PSO for distributed user-centric system, to design a logical joint mapping between user and RRHs. However, the algorithm proposed in [8] cannot converge efficiently if one RRH can serve limited number of users.

Additionally, the fronthaul link burden is still heavy in user-centric structure, because the small-scale fading information is necessary in [5–8]. To utilize the small-scale fading information, the CPU has to estimate the channels between all users and all RRHs, which significantly increases the computational burden and the system latency, making it unscalable when the numbers of users and RRHs are large. To solve those problems, we consider that each RRH serves users by utilizing the local channel state information (CSI), and only feedbacks the large-scale fading information to CPU. In this paper, we propose an IDPSO of RRH selection inspired by [9, 10] and we design a new fitness function without small-scale fading information. Numerical result shows the algorithm proposed in this paper can converge in a few iterations. Besides, the algorithm also can provide a good tradeoff between performance and execution complexity if the small-scale fading information is not available.

2 System Model

We assume that M RRHs with N_t antennas and K single-antenna users are uniformly distributed in the service area, with $M > K$. Each user will select s RRHs from M RRHs and serving RRHs would send signals requested by that user. In addition, each RRH serves at most one user. The received signal at user k can be modeled as

$$y_k = \sum_{m=1}^{M} p_{mk}\mathbf{h}_{m,k}^H\mathbf{x}_{m,k} + \sum_{m=1}^{M} \sum_{k'\neq k} p_{mk'}\mathbf{h}_{m,k}^H\mathbf{x}_{m,k'} + n_k, \tag{1}$$

where $\mathbf{x}_{m,k}$ denotes the transmitted signal from RRH m to user k, and n_k denotes the additive white Gaussian noise, which can be modeled as a circular Gaussian random variable with zero mean and variance σ_0^2. $\mathbf{h}_{m,k}$ denotes the channel fading vector between user k and RRH m, which defined as [3, 11]

$$\mathbf{h}_{m,k} = \gamma \sum_{i=1}^{N_{cl}} \sum_{l=1}^{N_{ray}} \alpha_{i,l}\sqrt{L(r_{i,l})}\mathbf{a}_{AP}\left(\theta_{i,l,m,k}^{AP}\right) + \mathbf{h}_{LOS}, \tag{2}$$

where the first item represents the non-line-of-sight (nLOS) component, and the second term accounts for the line-of-sight (LOS) [12]. N_{cl} is the number of clusters in channel, N_{ray} is the number of the rays that we consider for each cluster. $L(r_{i,l})$ and $\alpha_{i,l}$ are the complex path gain and the attenuation associated to the $(i, l)^{th}$ propagation path with $r_{i,l}$ length, respectively. $\mathbf{a}_{AP}$ is the array responses of Uniform Linear Array (ULA) to the mth RRH which depends on the angles of departure θ^{AP}. γ is a normalization factor defined as $\sqrt{\frac{N_t N_r}{N_{cl}N_{ray}}}$ with received antenna $N_r = 1$ of users. The LOS component $\mathbf{h}_{LOS}$ is given as

$$\mathbf{h}_{LOS} = I(d)\sqrt{N_t N_r}e^{j\alpha}\sqrt{L(d)}\mathbf{a}_{AP}\left(\theta_{LOS}^{AP}\right), \tag{3}$$

where $I(d) \in \{0, 1\}$ denotes the existence of direct path between user k and RRH m, which follows the binomial distribution with the probability of $I(d) = 1$ as [12]

$$P\{I(d) = 1\} = \min\left(\tfrac{20}{d}, 1\right)\left(1 - e^{-\frac{d}{39}}\right) + e^{-\frac{d}{39}}, \tag{4}$$

and $L(d)$ denotes the path loss, which is modeled in dB form as

$$L(d) = -20\log_{10}\left(\tfrac{4\pi}{\lambda}\right) - 10n\log_{10}(d) - X_\sigma, \tag{5}$$

where X_σ denotes the shadowing effect, which follows normal distribution with standard deviation σ. n is the path-loss factor, and λ is the carrier wavelength.

In this paper, we assume that all serving RRHs transmit its full power. To reduce fronthaul link burden, we assume that each RRH serves user by utilizing the local channel state information. In this case, maximum ratio transmission is adopted, and the transmitted signal is given as

$$\mathbf{x}_{m,k} = \mathbf{w}_{m,k}b_k, \tag{6}$$

where $\mathbf{w}_{m,k} = \frac{\mathbf{h}_{m,k}}{||\mathbf{h}_{m,k}||}$ is the beamforming vector. b_k is the transmitted symbol, which can be modeled as a circular Gaussian random variable with zero mean and variance η.

The RRH selection is defined as a $K \times M$ dimensional binary matrix $\mathbf{P}$. p_{mk} is the element at row m and column k in $\mathbf{P}$ and $p_{mk} = 1$ if the kth user selects the mth RRH, otherwise $p_{mk} = 0$. The downlink rate of user k in user-centric cell-free massive MIMO with RRH selection can be defined as

$$R_k = \log_2\left(1 + \frac{\sum_{m=1}^{M} \eta_{m,k} p_{mk} \mathbf{h}_{m,k}^H \mathbf{w}_{m,k} \mathbf{w}_{m,k}^H \mathbf{h}_{m,k}}{\sigma_0^2 + \sum_{m=1}^{M} \sum_{k' \neq k} \eta_{m,k'} p_{mk'} \mathbf{h}_{m,k}^H \mathbf{w}_{m,k'} \mathbf{w}_{m,k'}^H \mathbf{h}_{m,k}}\right). \tag{7}$$

In (7), we regard the interference signal as a zero mean Gaussian random variable, and the variance is equal to $\sum_{m=1}^{M} \sum_{k' \neq k} \eta p_{mk'} \beta_{m,k}$, where $\beta_{m,k}$ is the large-scale fading coefficient which can be calculated from (5), and every RRH has same transmit power so $\eta_{m,k}$ can be replaced by η. Then, the downlink rate of user k can be written as

$$R_k = \log_2\left(1 + \frac{\sum_{m=1}^{M} \eta p_{mk} \mathbf{h}_{m,k}^H \mathbf{w}_{m,k} \mathbf{w}_{m,k}^H \mathbf{h}_{m,k}}{\sigma_0^2 + \sum_{m=1}^{M} \sum_{k' \neq k} \eta p_{mk'} \beta_{m,k}}\right). \tag{8}$$

In this paper, we focus on the transmit RRH selection scheme that maximizes the sum rate, and the problem can be formulated as

$$\begin{aligned} \underset{\mathbf{P}}{\text{maximize}} \ & \sum_{k=1}^{K} R_k \\ \text{subject to} \ & \sum_{k=1}^{K} p_{mk} \leq 1, \quad \forall m \\ & \sum_{m=1}^{M} p_{mk} = s, \quad \forall k \\ & p_{mk} \in \{0, 1\}, \quad \forall m, n \end{aligned} \tag{9}$$

The first constraint means that each RRH serves at most one user and the second one stands for that each user selects s RRHs.

In (9), $\mathbf{h}_{m,k}^H$ and $\mathbf{w}_{m,k}$ are used so the small-scale fading information is needed, and that means it costs a lot. In order to reduce the cost, the sum rate should be estimated without small-scale fading information, so we calculate the expectation of R_k and use Jensen's inequality to find the upper bound as

$$\begin{aligned} \mathbf{E}[R_k] &= \mathbf{E}\left[\log_2\left(1 + \frac{\sum_{m=1}^{M} \eta p_{mk} \mathbf{h}_{m,k}^H \mathbf{w}_{m,k} \mathbf{w}_{m,k}^H \mathbf{h}_{m,k}}{\sigma_0^2 + \sum_{m=1}^{M} \sum_{k' \neq k} \eta p_{mk'} \beta_{m,k}}\right)\right] \\ &\leq \log_2\left(1 + \mathbf{E}\left[\frac{\sum_{m=1}^{M} \eta p_{mk} \mathbf{h}_{m,k}^H \mathbf{w}_{m,k} \mathbf{w}_{m,k}^H \mathbf{h}_{m,k}}{\sigma_0^2 + \sum_{m=1}^{M} \sum_{k' \neq k} \eta p_{mk'} \beta_{m,k}}\right]\right) \\ &= \log_2\left(1 + \frac{\mathbf{E}\left[\sum_{m=1}^{M} \eta p_{mk} \mathbf{h}_{m,k}^H \mathbf{w}_{m,k} \mathbf{w}_{m,k}^H \mathbf{h}_{m,k}\right]}{\sigma_0^2 + \sum_{m=1}^{M} \sum_{k' \neq k} \eta p_{mk'} \beta_{m,k}}\right) \\ &= \log_2\left(1 + \frac{\sum_{m=1}^{M} p_{mk} \beta_{m,k}}{\frac{\sigma_0^2}{\eta} + \sum_{m=1}^{M} \sum_{k' \neq k} p_{mk'} \beta_{m,k}}\right) \end{aligned} \tag{10}$$

Then, $\sum_{k=1}^{K} \mathbf{E}[R_k]$ is used as the objective function in (9) to solve the problem which is difficult to obtain small-scale fading information. Though small-scale fading information can be avoided by using the expectation of R_k, the problem in (9) is still a combinatorial optimization problem, and therefore is NP hard. To obtain a low-complexity algorithm, we propose an intelligent RRH selection algorithm based on PSO method.

3 Intelligent RRH Selection Algorithm

To solve the combinatorial problem in (9), an efficient RRH selection algorithm is proposed based on the DPSO. The basic motivation of original DPSO is to find a good position of a particle of population through the interaction among individuals, where the position of a particle is defined as a binary vector, whose i-th entry is set to be "1" if the i-th item is chosen. Fitness function can evaluate the position of particles. The historical optimal position of every particle and the global best position also be used to update the position of particles in iterations.

Note that the original DPSO algorithm that proposed in [5–8] cannot be directly applied to solve the problem in (9), as the update rule will generate a large number of particles that violates the constraints, making it very slow or even impossible to converge. To efficiently solve the RRH selection problem in user-centric cell-free massive MIMO, an IDPSO algorithm is proposed and summarized in Algorithm 1. In Algorithm 1, $\mathbf{I}_n$ and $\mathbf{G}$ are individual historical best of particles and global best of population, respectively. $\mathbf{C}$ marks the difference between $\mathbf{I}_n$ and $\mathbf{P}_n$ and the difference between $\mathbf{G}$ and $\mathbf{P}_n$ with non-zero elements. c_{mk} is the element at row m and column k in $\mathbf{C}$ and p_{mk} has a trend to change if $c_{mk} \neq 0$. $\mathbf{R}_1$ and $\mathbf{R}_2$ are $K \times M$ dimensional random matrices, which denote random disturbance for $\mathbf{C}$, to increase diversity of particles in iterations. The elements in $\mathbf{R}_1$ and $\mathbf{R}_2$ follow the uniform distribution from 0 to 1 and "rand(1)" also stands for the variable following uniform distribution from 0 to 1 in step 14. $\overline{\mathbf{C}^+}$ and $\overline{\mathbf{C}^-}$ can quantify that trend as probability after normalization in step 10 and 11, where $\overline{c_{ij}^+}$ and $\overline{c_{ij}^-}$ stand for the element at row i and column j of $\overline{\mathbf{C}^+}$ and $\overline{\mathbf{C}^-}$. In detail, the non-zero elements in $\overline{\mathbf{C}^+}$ represents the changing probability from "0" to "1" at corresponding positions of $\mathbf{P}_n$, while that in $\overline{\mathbf{C}^-}$ stand for the changing probability from "1" to "0". The next particle position is generated in iterations by changing p_{mk} according to that probability in $\overline{\mathbf{C}^+}$ and $\overline{\mathbf{C}^-}$. $\odot$ means Hadamard product. $\mathbf{c}_i^+$ is the row i of $\mathbf{C}^+$. Function $\mathcal{F}$ is the fitness function. N and T_{max} stand for the numbers of particles and the maximum number of iterations, respectively.

Algorithm 1 Improved Discrete Particle Swarm Optimization

Input: N, K, M, T_{max}, s, distance of users to RRHs, max service range.
Output: $\mathbf{G}$.
1: Generate N particles and the nth particle's position named $\mathbf{P}_n$.
2: Give feasible initial position randomly to every $\mathbf{P}_n$.
3: Initialize $\mathbf{I}_n$ and $\mathbf{G}$, $\mathbf{I}_n = \mathbf{P}_n$ and $\mathbf{G} = argmax_{\{\mathbf{I}_n\}}(\mathcal{F}(\mathbf{I}_n))$.
4: **for** t=1 to T_{max} **do**
5: **for** n=1 to N **do**
6: $\mathbf{P}_n^{t+1} = \mathbf{P}_n^t$
7: $\mathbf{C} = \mathbf{R_1} \odot (\mathbf{I}_n - \mathbf{P}_n^t) + \mathbf{R_2} \odot (\mathbf{G} - \mathbf{P}_n^t)$
8: $\mathbf{C}^+ = \mathbf{C}$ but replace negative elements with 0. $\overline{\mathbf{C}^+} = \mathbf{C}^+$
9: $\mathbf{C}^- = \mathbf{C}$ but replace positive elements with 0. $\overline{\mathbf{C}^-} = \mathbf{C}^-$
10: $\overline{c_{ij}^+} = \frac{c_{ij}^+}{\|\mathbf{c}_i^+\|_1}$ **if** $\|\mathbf{c}_i^+\|_1 \neq \mathbf{0}$ **for every** row i **and** column j
11: $\overline{c_{ij}^-} = -\frac{c_{ij}^-}{\|\mathbf{c}_i^-\|_1}$ **if** $\|\mathbf{c}_i^-\|_1 \neq \mathbf{0}$ **for every** row i **and** column j
12: **for** i=1 to K **do**
13: **for** j=1 to M **do**
14: **if** $\overline{c_{ij}^+} >$ rand(1)
and distance between user i and RRH j < range
and the column j of $\mathbf{P}_n^t$ do not contains "1" **do**
15: $p_{ij} = 1,\ p_{ij} \in \mathbf{P}_n^{t+1}$
16: **and** change a "1" to "0" in this row as probability in $\overline{\mathbf{c}_i^-}$
17: **end if**
18: **end for**
19: $\mathbf{I}_n = \mathbf{P}_n^{t+1}$ **if** $\mathcal{F}(\mathbf{P}_n^{t+1}) > \mathcal{F}(\mathbf{I}_n)$
20: **end for**
21: $\mathbf{G} = argmax_{\{\mathbf{I}_n\}}(\mathcal{F}(\mathbf{I}_n)),\ \forall n$
22: **end for**
23: **break if** meet the stop condition
24: **end for**

4 Numerical Result

Suppose that K users and M RRHs are uniform distributed in a 100 m × 100 m area. By the way, if the location of user is close to the edge, users will not be surrounded by RRHs because of the impact of the borders. Hence, users are only distributed in the central 40 m × 40 m areas of the 100 m × 100 m squares to reduce the border error in our simulations.

Without small-scale fading information, as derived in (10), the fitness function of improved discrete particle swarm optimization based on distance (IDPSO-D) is defined as

$$\mathcal{F}_{IDPSO-D}(\mathbf{P}) = \sum_{k=1}^{K} \log_2 \left(1 + \frac{\sum_{m=1}^{M} p_{mk} \beta_{m,k}}{\frac{\sigma_0^2}{\eta} + \sum_{m=1}^{M} \sum_{k' \neq k} p_{mk'} \beta_{m,k}} \right) \tag{11}$$

In order to show the effectiveness of the algorithm, the result of Nearest Neighborhood (NN) is also presented for comparison, where the user selects the nearest s RRH to form the virtual cells. To study the influence of the loss of small-scale fading information on RRH selection, improved discrete particle swarm optimization based on capacity (IDPSO-C), which can obtain small-scale fading information, uses objective function in (9) as fitness function directly. The fitness function of IDPSO-C is defined as

$$\mathcal{F}_{IDPSO-C}(\mathbf{P}) = \sum_{k=1}^{K} R_k, \tag{12}$$

where R_k is given in (7).

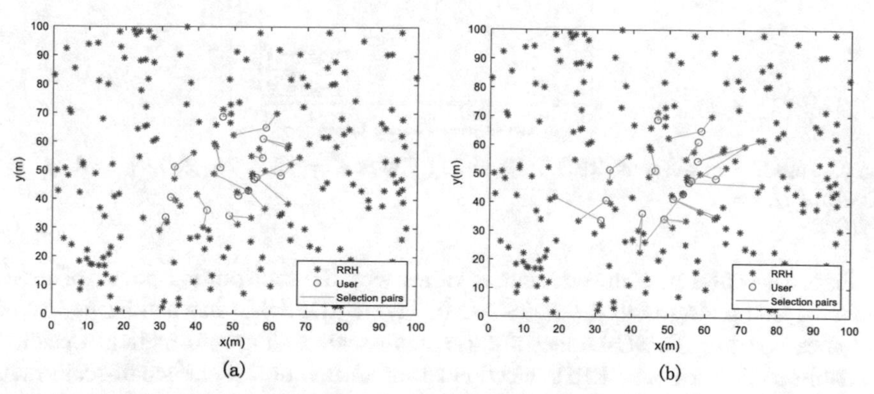

Fig. 2. RRH selection results under different methods $M = 200$, $N_t = 4$, $\sigma_z^2 = -143$ dBm/Hz, $\eta = -90$ dBm, (a) NN, $\sum_{k=1}^{K} R_k = 11.05$ bit/s/Hz, (b) IDPSO-D, $\sum_{k=1}^{K} R_k = 16.36$ bit/s/Hz.

Figure 2 shows the selection results in different selection methods, and each user selects 2 RRHs. The sum rate can only reach 11.51 bit/s/Hz with NN, and it is obvious in Fig. 2 (a) that some users suffer from very severe interference. With IDPSO-D, however, some users selected RRHs with farther distances to reduce interference, leading to a significant improvement in sum rate.

Figure 3 shows the convergence performance of IDPSO algorithm with different fitness functions. The values of two fitness functions are improved in the iteration and IDPSO algorithm can converge in a small number of iterations. That means IDPSO converges and obtains the RRH selection result in a few iterations.

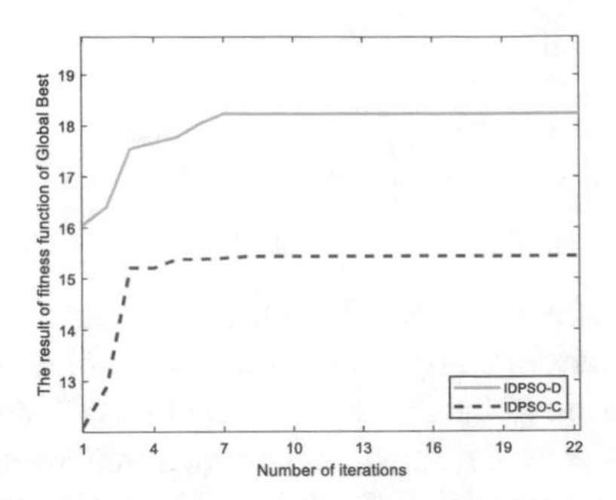

Fig. 3. Convergence performance of IDPSO

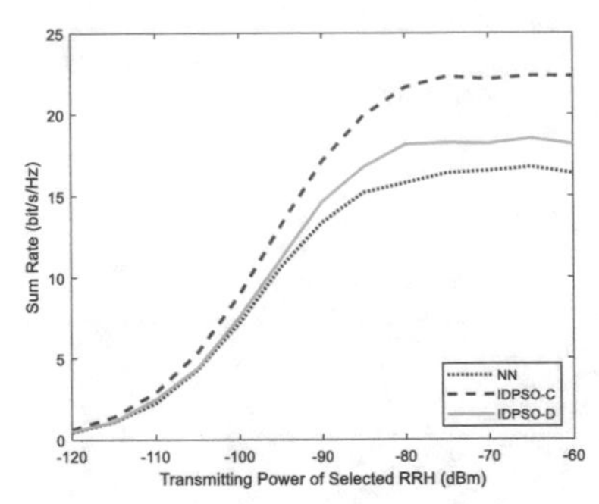

Fig. 4. Sum rate in different RRH Transmitting Power $K = 15$, $M = 200$, $N_t = 4$, $\sigma_z^2 = -143$ dBm/Hz, $s = 2$

Figure 4 shows how the sum rate R varies with the transmitting power of serving RRHs η. Comparing with the IDPSO-C in Fig. 4, IDPSO-D has a little poorer performance because IDPSO-D does not use small-scale fading information. Besides, it is unable to find a better RRH selection result in iteration if the small-scale fading information is missing. However, IDPSO-D requires less time and lower fronthaul overhead because transmitter does not need to obtain small-scale fading information. In other words, IDPSO-D provides a good tradeoff between performance and execution complexity.

Figure 5 shows how the sum rate R varies with the number of serving RRHs of each user s in different user density. IDPSO-C reaches the peak when s equals to 2 in the case of low user density but with the continuous increase of s, the sum rate decreases gradually. However, IDPSO-D and NN decrease continuously with the increase of s. That illustrates enlarge s can get a slight improvement of sum rate under low user density, but the improvement cannot be detected without small-scale fading information. When $K = 15$, the sum rate decreases with the increase of s. This phenomenon shows that increase serving RRHs of each user will reduce the sum rate, except that the case of low user density and available small-scale fading information, and IDPSO-D cannot find a better RRH selection result without small-scale fading information so the sum rate of IDPSO-D is worse than that of IDPSO-C in any user density.

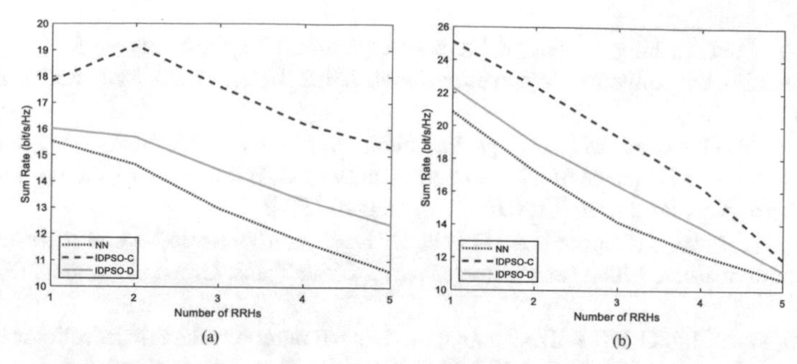

Fig. 5. Sum rate in different Number of serving RRHs $M = 200$, $N_t = 4$, $\eta = -80$ dBm, $\sigma_z^2 = -143$ dBm/Hz, (a) $K = 8$, (b) $K = 15$

5 Conclusion

This paper proposes IDPSO-D algorithm based PSO to implement the RRH selection in user-centric cell-free massive MIMO at downlink transmission only by using large-scale fading information. Simulation results show that IDPSO-D has a good convergence performance and it can obtain an acceptable RRH selection result. As a result, IDPSO-D can provide a good tradeoff between performance of sum rate and execution complexity in user-centric cell-free massive MIMO system.

Acknowledgment. This work is supported by the National Natural Science Foundation of China (61871239).

References

1. Ngo, H.Q., Ashikhmin, A., Yang, H., Larsson, E.G., Marzetta, T.L.: Cell-free massive MIMO versus small cells. IEEE Trans. Wirel. Commun. **16**(3), 1834–1850 (2017)
2. Buzzi, S., D'Andrea, C., Zappone, A., D'Elia, C.: User-centric 5G cellular networks: resource allocation and comparison with the cell-free massive MIMO approach. IEEE Trans. Wirel. Commun. **19**(2), 1250–1264 (2020)
3. Buzzi, S., D'Andrea, C.: Cell-free massive MIMO: user-centric approach. IEEE Wirel. Commun. Lett. **6**(6), 706–709 (2017)
4. Eberhart, R., Kennedy, J.: A new optimizer using particle swarm theory. In: Proceedings of the Sixth International Symposium on Micro Machine and Human Science, MHS 1995, pp. 4–6 (1995)
5. Hei, Y., Li, W., Li, X.: Particle swarm optimization for antenna selection in MIMO system. Wirel. Pers. Commun. **68**(3), 1013–1029 (2013)
6. Shang, P., Zhu, G., Tan, L., Su, G., Li, T.: Transmit antenna selection for the distributed MIMO systems. In: 2009 International Conference on Networks Security, Wireless Communications and Trusted Computing (2009)
7. Liu, Z., Du, W., Sun, D.: Energy and spectral efficiency tradeoff for massive MIMO systems with transmit antenna selection. IEEE Trans. Veh. Technol. **66**(5), 4453–4457 (2017)
8. Ari, A.A.A., Gueroui, A., Titouna, C., Thiare, O., Aliouat, Z.: Resource allocation scheme for 5G C-RAN: a Swarm Intelligence based approach. Comput. Netw. **165**, 106957 (2019)

9. Jeong, Y., Park, J., Jang, S., Lee, K.Y.: A new quantum-inspired binary PSO: application to unit commitment problems for power systems. IEEE Trans. Power Syst. **25**(3), 1486–1495 (2010)
10. Kaushik, A., Goswami, M., Manuja, M., Indu, S., Gupta, D.: A binary PSO approach for improving the performance of wireless sensor networks. Wirel. Pers. Commun. **113**(1), 263–297 (2020). https://doi.org/10.1007/s11277-020-07188-3
11. Alonzo, M., Buzzi, S., Zappone, A., D'Elia, C.: Energy-efficient power control in cell-free and user-centric massive MIMO at millimeter wave. IEEE Trans. Green Commun. **3**(3), 651–663 (2019)
12. Haneda, K., et al.: 5G 3GPP-like channel models for outdoor urban microcellular and macrocellular environments. In: 2016 IEEE 83rd Vehicular Technology Conference (VTC Spring) (2016)

Research on Emotion Recognition Based on GA-BP-Adaboost Algorithm

Ruijuan Chen[1,2], Xiaofei Diao[1], Zhihui Sun[1], Guanghua Deng[1], Zhe Zhao[1,2], and Zhuanping Qin[3](✉)

[1] School of Life Sciences, Tiangong University, Tianjin, China
[2] Tianjin Key Laboratory of Optoelectronic Detection Technology and System, Tianjin, China
[3] School of Automation and Electrical Engineering, Tianjin University of Technology and Education, Tianjin, China
qinzhuanping@126.com

Abstract. With the development of wearable portable electrocardio (ECG) acquisition devices, real-time emotion recognition based on ECG signals becomes possible. In order to study the difference in the characteristics of human heart rate variability in different emotional states, and classify the positive and negative valence emotions, the emotional induction paradigm was designed, and a wearable ECG collection device was used to collect the ECG signals in the corresponding emotional state. Using heart rate variability as a characteristic parameter, an emotion recognition model based on GA-BP-adaboost algorithm is proposed. This model uses genetic algorithm to optimize the threshold and bias of BP neural network, and uses adaboost ensemble learning method to construct multiple BP neural networks as strong classifier. Results show that the GA-BP-adaboost algorithm has a correct recognition rate of 84.27% for positive and negative emotion samples. The algorithm proposed in this paper can significantly improve the accuracy of classification, which is of great significance in the research of emotion recognition based on heart rate variability.

Keywords: Genetic algorithm · Neural network · Ensemble learning · Heart rate variability

1 Introduction

Emotion is a state of coexistence of psychology and physiology that accompanies consciousness and cognitive processes, and it plays an important role in human social interaction [1]. Negative emotions have an important impact on the human and may cause mental illness or physical harm, so it is of great significance to effectively identify and distinguish between positive and negative emotions [2]. In the research of emotion recognition, there are methods based on physiological signals, gesture, expression and voice. The emotion recognition method based on physiological signals is not easy to disguise and can obtain more effective results [3]. Heart Rate Variability (HRV) is a non-invasive technique to assess the balance between the two branches of the sympathetic and parasympathetic nerves in the autonomic nervous system, using the time

Q. Liang et al. (Eds.): Artificial Intelligence in China, LNEE 854, pp. 73–81, 2022.
https://doi.org/10.1007/978-981-16-9423-3_10

domain, frequency domain, nonlinear and time–frequency domain characteristics of the R wave interval of the ECG signal [4]. Compared with other physiological signals, it has unique advantages such as convenient operation and easy collection, low equipment cost, and good recognition effect. Mirmohamadsadeghi et al. used Support Vector Machine (SVM) algorithm to identify ECG data in DEAP emotional data, and the recognition rate of positive and negative valence emotions was 72% [5]. GUO et al. used the PCA-SVM algorithm to classify positive and negative emotions with an accuracy of 71.4% by extracting HRV feature parameters [6]. CHENG et al. extracted HRV feature parameters and used SVM algorithm to classify positive and negative emotions with an accuracy of 79.51% [7]. However, the current emotion recognition effect based on ECG signals is not significant, and the classification accuracy is low. Most of the classifiers used are traditional machine learning algorithms. Among them, support vector machines, random forests, K nearest neighbor algorithms, and decision trees are most commonly used in physiology signal emotion recognition. In order to improve the accuracy of emotion classification, this paper proposes the GA-BP-adaboost algorithm for emotion recognition of ECG signals. The algorithm uses BP neural network as the base classifier, and integrates several BP networks based on the adaboost algorithm to form a strong classifier. Compared with the Bagging algorithm, the adaboost algorithm has an upper error rate, strong generalization ability, is not easy to overfit, and can maximize the classification performance of the model [8]. In addition, the Weights and thresholds of the BP base classifier is optimized by genetic algorithm to prevent falling into the local minimum [9].

2 Emotional Data Acquisition

2.1 Experimental Paradigm

The subjects recruited in this experiment are from college students, aged between 18–25 years old. None of the subjects had mental or cardiovascular disease. Four emotion videos were selected to induce happy and sad emotions, and each emotion contained two induced videos.

The induction experiment was carried out in a quiet laboratory, and the room temperature was controlled at about 24 °C. The collection device used in the experiment is a portable and wearable ECG collection device designed and manufactured by this research group, with a sampling rate of 500 Hz, which can be attached to the lower left edge of the heart through ECG, and will not cause discomfort when worn. Figure 1 shows the ECG acquisition device, and the wearing mode of the ECG acquisition device. Considering that different subjects have different levels of understanding of emotional video content, the experiment introduced an emotional feedback scale to evaluate the intensity of the subject's emotional stimulation. The scale is divided into four levels: very strong, strong, normal, and non-sense. This article selects samples with two levels of strong and strong for analysis.

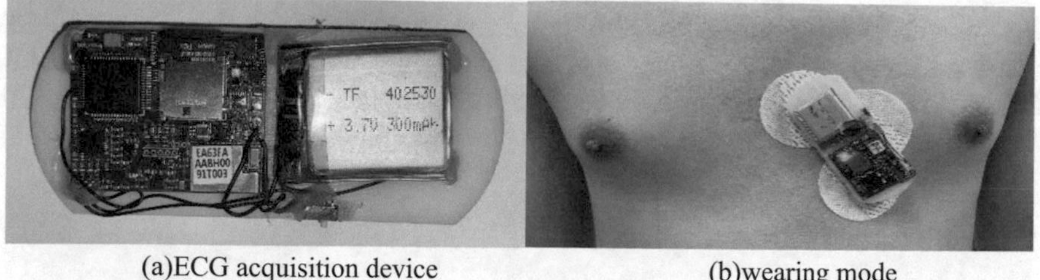

(a)ECG acquisition device (b)wearing mode

Fig. 1. a is the ECG acquisition device and b is wearing mode of ECG device

2.2 Feature Extraction

The characteristics of HRV can be counted and analyzed by a variety of methods. The most commonly used methods are time-domain statistics and frequency-domain analysis, which are convenient for calculation and intuitive. With the development of chaotic nonlinear dynamics, nonlinear features are used for HRV analysis [10]. This paper extracts 17 features in time domain, frequency domain and nonlinearity based on HRV. The eight time domain features are: Standard deviation of heart rate (SDhr), Standard deviation of the RR interval (SDNN), standard deviation of the difference between adjacent RR intervals (SDSD) and root mean square (RMSSD), The number of RR intervals greater than 50 ms and 20 ms as a percentage of the total number of intervals (PNN50, PNN20), (HRVti)(TINN). Based on the Fast Fourier Transform (FFT) method, the RR interval is resampled and interpolated to calculate the Power Spectral Density (PSD). Divided according to frequency bands of extremely low frequency (0.003–0.04 Hz, VLF), low frequency (0.04–0.15 Hz, LF), and high frequency (0.15–0.4 Hz, HF), Calculate the proportions of the power in the three frequency bands as pVLF, pLF, and pHF, Normalized low frequency power LFn, normalized high frequency power HFn, ratio of low frequency to high frequency power LFHF. The formula for calculating the power of the three frequency bands is as follows:

$$pVLF = \left(\frac{VLF}{TP}\right) * 100 \tag{1}$$

$$pLF = \left(\frac{LF}{TP}\right) * 100 \tag{2}$$

$$pHF = \left(\frac{HF}{TP}\right) * 100 \tag{3}$$

Where TP = VLF + LF + HF, which means the total power of the whole frequency band.

The poincare scatter is a geometric model constructed by drawing continuous RR interval scatter points in a two-dimensional coordinate system. Each scattered point takes the nth RR interval as the X axis and the n + 1th RR interval as the Y axis. This paper extracts three features, namely the short axis (SD1), long axis (SD2), and the ratio of the long axis to the short axis (SD2/SD1) of the ellipse in the scatter graph. The SD1 reflects the variability of the RR interval in a short time, the SD2 reflects the variability of the RR interval in a long time, and SD2/SD1 can reflect the overall variability of the RR interval [11].

3 Research Method

3.1 Optimization of BP Neural Network Based on GA

BP neural network (BPNN) is a classical artificial neural network algorithm with good nonlinear fitting ability. BP neural network adjusts the weights and thresholds of the network by error back propagation and minimizes the mean square error between the expected output and the actual output. The traditional BP algorithm has poor global search ability, and the initial network weights and thresholds are set randomly, so it is easy to fall into the local minimum [12]. Genetic algorithm (GA) is a kind of intelligent algorithm with strong global optimization ability, which can automatically obtain and accumulate the characteristics of the search space in the search process, and adaptively control the search process to obtain the optimal solution [13]. Aiming at the defects of BP neural network, this paper applies genetic algorithm to optimize the weights and thresholds of each node.

3.2 GA-BP-Adaboost Algorithm

By optimizing the weights and thresholds of BP network with genetic algorithm, the defect that BP network is easy to fall into local minimum can be effectively solved. However, a single training network can not get good classification results. In this paper, AdaBoost algorithm is used to construct several strong classification models based on GA-BP network. Figure 2 is the flow chart of GA-BP-adaboost algorithm. AdaBoost is an adaptive enhanced iterative algorithm different from other integration methods [14]. With the increase of the number of iterations, the weight of the wrong samples will increase, the weight of the correct samples will decrease, and the classifier will pay more attention to the samples with large weight. The implementation steps of GA-BP-adaboost algorithm are as follows:

a. Initialize sample weights. Randomly select N training samples from the data, The weight of each training sample is:

$$D_1(i) = (w_1, w_2, \ldots w_n) = \left(\frac{1}{n}, \frac{1}{n}, \ldots, \frac{1}{n}\right) \tag{4}$$

b. Iterative training is based on BP weak classifier optimized by GA. The error e between the predicted output g and the expected output y is calculated. When iteratively training to the t-th weak classifier, the error et is:

$$e_t = \sum\nolimits_{i=1}^{n} D_t(i);\ g(t) \neq y \tag{5}$$

G (T) is the prediction output of the t-th weak classifier.

Calculate the weight of weak classifier. The weight of the t-th weak classifier is:

$$\alpha_t = \frac{1}{2} ln\left(\frac{1-e_t}{e_t}\right) \tag{6}$$

c. Update training sample weights. According to the weight of the current weak classifier, the weight of the next training sample is updated. The calculation formula is:

$$D_{t+1}(i) = \frac{D_t(i)}{B_t} * exp(-\alpha_t y_i g_t(x_i)) \tag{7}$$

BT is the normalization factor.

d. Strong classifier prediction. After t-round training, t if classifiers and corresponding weights are obtained. A strong classifier is constructed:

$$h(x) = sign\left[\sum\nolimits_{t=1}^{T} \alpha_t * f(g_t, \alpha_t)\right] \tag{8}$$

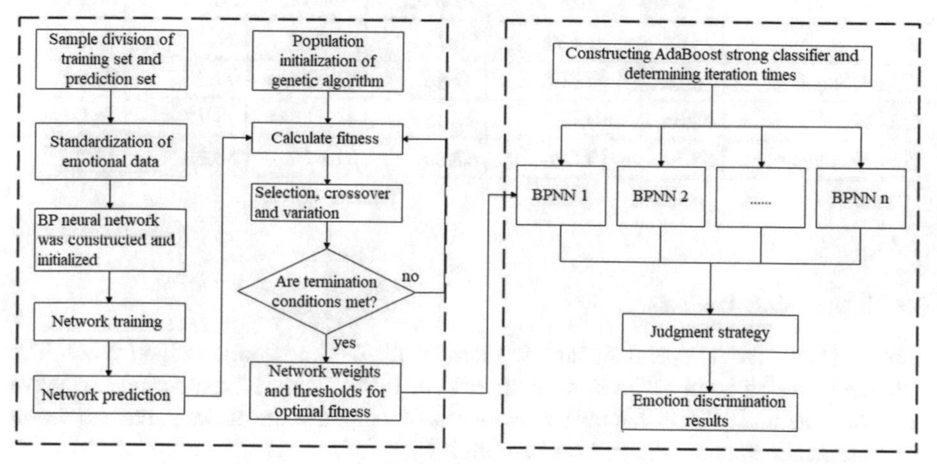

Fig. 2. Flow chart of GA-BP-adaboost algorithm

4 Results and Discussion

4.1 Results

In this paper, 17 HRV features are extracted. The input layer has 17 nodes and the output layer has 1 node. In order to prevent over fitting of network model, the number of hidden layer nodes should not be too large. The final number of selected nodes is 18 and The topological structure of BP network is 17-18-1. In order to enhance the accuracy and authenticity of the classification results, this paper uses a 50% cross validation strategy to evaluate the model. Based on the above extracted eigenvalues, BP, BP-adaboost, GA-BP and GA-BP-adaboost algorithm models are constructed. As shown in Table 1, the sample correct recognition rate of cross validation of the four algorithms is shown. Genetic algorithm is used to optimize the initial weights and thresholds of BP network, which can improve the recognition performance of BP network. By using AdaBoost adaptive algorithm to collect multiple BP based classifiers, the recognition rate can be improved. The GA-BP-adaboost algorithm proposed in this paper, on the one hand, optimizes the initial value of the network by genetic algorithm, on the other hand, strengthens the classification model by ensemble learning method. In the classification of positive and negative emotions, the average recognition rate is 84.27%. Compared with the other three algorithms, GA-BP-adaboost algorithm shows more accurate classification performance.

Table 1. Comparison of recognition rate of four classification models in positive and negative emotion classification

	Correct recognition rate (CCR)					
	90% off	80% off	70% off	60% off	50% off	average
BP	75.43	74.72	75.71	81.94	74.72	76.51
BP-adaboost	80.00	75.00	82.86	91.67	77.78	81.46
GA-BP	84.85	75.56	77.43	81.11	74.17	78.62
GA-BP-adaboost	**88.57**	**77.78**	**82.86**	**91.67**	**80.55**	**84.27**

4.2 Eigenvalue Analysis

Table 2 shows the independent sample t-test results of time domain eigenvalues. The extracted time domain features have significant differences in positive and negative emotional states. It shows that the two emotions of happiness and sadness have different physiological responses in heart rate variability.

Frequency domain analysis of heart rate variability can reflect the whole activity of autonomic nervous system. In order to explore the specific changes of positive and negative emotions in the autonomic nervous system, each emotion sample was randomly selected and divided into extremely low frequency, low frequency and high frequency

Table 2. Time domain characteristic difference analysis

Characteristic value	Joy	Sadness	Sig
SDhr	7.17 ± 2.40	4.87 ± 1.34	0.000^{***}
SDNN	60.77 ± 18.59	49.40 ± 14.31	0.000^{***}
SDSD	31.17 ± 10.14	39.53 ± 16.62	0.000^{***}
RMSSD	31.30 ± 10.15	39.55 ± 16.62	0.000^{***}
PNN50	11.41 ± 8.65	21.67 ± 18.18	0.000^{***}
PNN20	42.98 ± 15.02	57.61 ± 18.99	0.000^{***}
HRVti	11.72 ± 2.00	10.90 ± 1.89	0.005^{**}
TINN	193.93 ± 62.04	163.03 ± 56.99	0.001^{**}

bands, Based on Welch's classical power spectrum estimation method, the PSD distribution maps of happy and sad emotions are drawn as shown in Fig. 3. It can be seen from the figure that the PSD of the two kinds of positive and negative emotions have obvious differences in different frequency bands. Among them, the PSD of happy emotion is mainly distributed in extremely low frequency and low frequency band. The proportion of high frequency PSD is only 13.2%, and the ratio of low frequency to high frequency power is 4.268. However, the PSD distribution of sad emotion in high frequency band increased significantly. The proportion of high frequency is 34.8%, and the ratio of low frequency to high frequency power is 1.375.

The results showed that the activity of parasympathetic nerve increased in sad state. Independent sample t-test analysis of all data showed that pvlf, PLF and PHF showed significant differences between positive and negative emotions ($P < 0.05$).

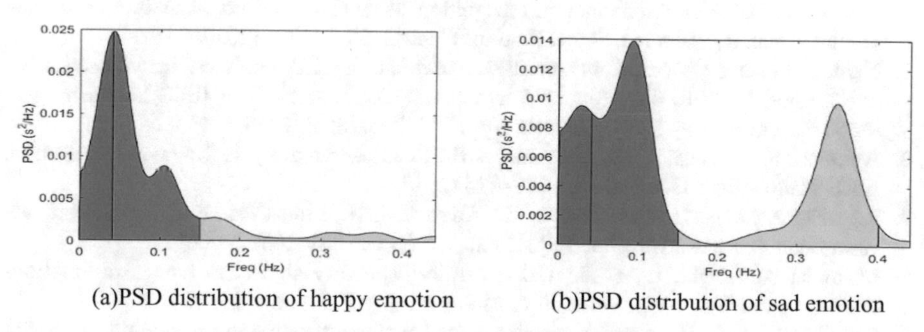

(a)PSD distribution of happy emotion (b)PSD distribution of sad emotion

Fig. 3. PSD distribution of two kinds of positive and negative emotions

In Fig. 4, one emotion sample is randomly selected to draw Poincare scatter diagram of two emotions. It can be seen that the value of happy emotion is larger than sad emotion SD2. In the scatter diagram, the ellipse of happy emotion SD1 and SD2 is narrower and longer. Independent sample t-test analysis showed that SD1, SD2 and their ratio SD2 / SD2 showed significant differences between positive and negative emotions ($P < 0.05$).

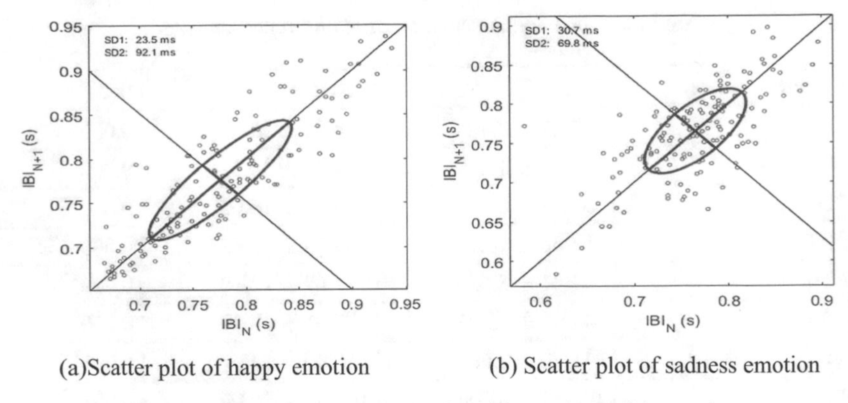

(a)Scatter plot of happy emotion (b) Scatter plot of sadness emotion

Fig. 4. Scatter plot of two kinds of positive and negative emotions

5 Conclusion

In this paper, the ECG data of happy and sad states are collected through experiments, Based on heart rate variability, 17 characteristic parameters were extracted from time domain, frequency domain and Poincare scatter plot. Moreover, ga-bp-adaboost algorithm is proposed to improve the correct recognition rate of samples. The results show that the correct recognition rate of GA-BP-adaboost algorithm is 84.27%. The extracted feature parameters show significant differences in both positive and negative emotions. It is of great significance to realize emotion recognition using ECG signal.

References

1. Wei, C., et al.: EEG-based emotion recognition using simple recurrent units network and ensemble learning. Biomed. Signal Process. Control. **58**, 101756 (2020)
2. Nardelli, M., et al.: A multiclass arousal recognition using HRV nonlinear analysis and affective images. In: 2018 40th Annual International Conference of the IEEE Engineering in Medicine and Biology Society (EMBC), pp. 392–395. IEEE (2018
3. Agrafioti, F., Hatzinakos, D., Anderson, A.K.: ECG pattern analysis for emotion detection. IEEE Trans. Affect. Comput. **3**(1), 102–115 (2012)
4. Rajendra Acharya, U., Paul Joseph, K., Kannathal, N., Lim, C.M., Suri, J.S.: Heart rate variability: a review. Med. Biol. Eng. Comput. **44**(12), 1031–1051 (2006)
5. Mirmohamadsadeghi, L., et al.: Using cardio-respiratory signals to recognize emotions elicited by watching music video clips. Multimed. Signal Process. 1–5 (2016)
6. Guo, H.W., et al.: Heart rate variability signal features for emotion recognition by using principal component analysis and support vectors machine. In: IEEE International Conference on Bioinformatics & Bioengineering. IEEE (2016)
7. Cheng, Z., et al.: A novel ECG-based real-time detection method of negative emotions in wearable applications. In: 2017 International Conference on Security, Pattern Analysis, and Cybernetics (SPAC) (2017)
8. Thongkam, J., Xu, G., Zhang, Y.: AdaBoost algorithm with random forests for predicting breast cancer survivability. In: IEEE International Joint Conference on Neural Networks. IEEE (2008)

9. Hu, Q., Hao, X., Lei, L.: Classification algorithm of solitary pulmonary nodules based on genetic algorithm and BP neural network. Computer Sci. 43(S1), 37–39 + 54 (2016)
10. Germán-Salló, Z., Germán-Salló, M.: Non-linear methods in HRV analysis. Procedia Technol. **22**, 645–651 (2016)
11. Charlet, A., Rodeau, J.L., Poisbeau, P.: Poincaré plot descriptors of heart rate variability as markers of persistent pain expression in freely moving rats. Physiol. Behav. **104**(5), 694–701 (2011)
12. Ding, S., Su, C., Yu, J.: An optimizing BP neural network algorithm based on genetic algorithm. Artif. Intell. Rev. **36**(2), 153–162 (2011)
13. Pasolli, E., Melgani, F.: Genetic algorithm-based method for mitigating label noise issue in ECG signal classification. Biomed. Signal Process. Control **19**, 130–136 (2015)
14. Cao, Y., Miao, Q., Liu, J., Gao, L.: Advance and prospects of AdaBoost algorithm. Acta Autom. Sin. **39**(6), 745–758 (2013). https://doi.org/10.1016/S1874-1029(13)60052-X

Information Extraction of Air-Traffic Control Instructions via Pre-trained Models

Xuan Wang(✉), Hui Ding, Qiucheng Xu, and Zeyuan Liu

State Key Laboratory of Air Traffic Management System and Technology, The 28th Research Institute of China Electronics Technology Group Corporation, Nanjing 210007, China
wangxuan10@cetc.com.cn

Abstract. The air traffic controllers always use air-traffic control instructions (ATC instructions) to command aircrafts. The ATC instruction consists of some situation mentions such as flight number, status, and target location etc. The deep-learning based approach can extract such information for situation awareness. In practice, it is difficult to prepare huge amount of labelled ATC instructions for training the deep-learning model due to expensive costs of handcraft annotations. The large scale pre-trained model (PTMs) can solve this problem by "pre-training" and "fine-tuning". This paper proposes: 1) pre-trained models to extract information from few scale ATC instructions; 2) the probing task to find which layer of model achieves the best performance of information extraction task.

Keywords: ATC instructions · Named entity recognition · Pre-trained model · Fine-tuning · Probing task

1 Introduction

The radiotelephony communication is a kind of voice communication mode between air-traffic controller and aircraft. The air-traffic controllers always use ATC instructions in radiotelephony communication to command aircrafts. It is important to use the correct ATC instructions during communication because incorrect or ambiguous ATC instructions would cause aviation unsafe incidents. The ATC instruction always consists of some situation mentions such as flight number, status, location etc. Analysing these mentions can help ATM systems to avoid the potential accidents. However, the practical ATC instruction occurs in text style and ATM systems cannot apply directly. Named Entity Recognition (NER) is kind of method to extract the situation information in ATC instructions. Then the ATM systems can predict some potential accidents with such situation information and other air-traffic management systems' information such as ADS-B, A-CDM, ASMGCS, etc.

NER [1] aims to recognize mentions of rigid designators from text belonging to predefined semantic types such as person, location, organization etc. The input of NER system is sequence of tokens, and its output is sequence of corresponding labels. NER

Supported by China National Key R&D Program (No. 2020YFB1600100).

Q. Liang et al. (Eds.): Artificial Intelligence in China, LNEE 854, pp. 82–91, 2022.
https://doi.org/10.1007/978-981-16-9423-3_11

plays an essential role in aviation text understanding. There are four main streams of techniques applied in NER: 1) rule-based approaches, 2) feature-based supervised machine learning based approaches, 3) supervised deep-learning based approaches, 4) pre-trained model based approaches [2].

Applying supervised machine learning approaches, NER is regard as sequence labelling task. Some machine learning algorithms are proposed such as hidden Markov model (HMM) [3] and conditional random field (CRF) [4]. These approaches train the model by a large annotated corpus, memorize lists of entities, and creates disambiguation rules based on transition probability. Feature engineering is critical in supervised machine learning approaches. HMM and CRF needs feature vector representation to be inputs.

Deep learning composed of multiple processing layers to learn useful representations of data automatically. Compared with shallow models of machine learning, the deep-learning models learn complex and intricate features from data. The traditional deep-learning models such as Long Short-Term Model (LSTM) is able to capture the non-linear mappings between input and output. Hammerton et al. [5] attempted NER with LSTM, which can capture long distance dependence of input text. Lample et al. [6] proposed LSTM-CRF for NER, where BiLSTM layer computes the deep contextual word embedding and CRF layer outputs tagging sequence via viterbi decoding. Although applying whole sentence information to infer the global optimum sequence of tagging, CRF layer cannot explicitly capture dependency between any two labels due to Markov assumptions. In order to break the limitation, Cui et al. [7] proposed the hierarchically-refined label attention network to replace CRF in inference layer. In Cui's model, the multi-headed self-attention [8] can help NER models learn dependency between any labels. The convolutional neural networks (CNN) are also applied in NER models due to its ability of modelling character-level information, such as LSTM-CNNs [9] and BiLSTM-CNNs-CRF [10]. Deep-learning based approach is still a kind of supervised learning approach, which needs large scale labelled data for training. However, it is difficult to satisfy such requirement of training data due to expensive costs of handcraft annotations.

The pre-trained model based pre-trained models are effective for low-resource NER via fine-tuning. The PTMs learn language knowledge from large amount of unlabeled data in pre-trained stage, then use few scales labelled data of downstream tasks to train the models in fine-tuning stage [11] and let model adapt to any downstream tasks. Radford et al. [12] proposed Generative Pre-trained Transformer (GPT) for language generating task. Devlin et al. [13] proposed Bidirectional Encoder Representations from Transformers (BERT), which encodes tokens by joint conditioning on both left and right context in all layers.

In this paper, the PTMs-CRF model is presented to extract the situation information of ATC instructions. The pre-trained model is defined to be embedding layer and CRF model to be inference layer. There are two contributions in the paper: 1) choose the pre-trained models to solve low-resource problem; 2) apply probing task to search which layer of pre-trained models can obtain the best performance. Finally, we verify PTMs-CRF on practical ATC instructions datasets and discuss the experiment results.

2 Related Work

Pre-trained Models. In the past decade, the technology of pre-trained models has developed rapidly. The technology development is divided into two generations. First generation is to learn pre-trained word embedding, PTMs learn words representation from large scale unlabeled data. Second generation is pre-trained contextual encoder, PTMs learn contextual word embedding by building language model.

In order to capture syntactic and semantic information of words, word embedding via neural network language model is proposed [14]. Mikolov et al. [15] proposed Continuous Bags-of-Words (CBOW) and Ship-Gram (SG) models to represent words. Word2vec can learn high-quality word embedding to capture the latent sematic similarities among words. Glove et al. [16] obtains the global vectors for word representation by word-word cooccurrence matrix. Although improving the performance in different NLP tasks, the pre-trained word embeddings are context-independent and do not apply information of the entire input sentence for representing words.

The pre-trained contextual encoder can represent the word semantic depending on its context. Peters et al. [17] proposed a type of deep contextualized word representation by using stacked BiLSTM and pre-trained on a large text corpus. ULMFiT [18] attempted to fine-tune pre-trained language model for text classification. After Transformer was proposed, the vast majority of PTMs are made up of its encoder or decoder, such as GPT and BERT. Moreover, some advanced models were proposed to improve performance such as Roberta [19], ALBERT [20], XLNET [21]. Compared with pre-trained word embedding, these PTMs can learn more knowledge from large scale text corpora by self-supervised training.

Probing Task. The PTMs are hierarchical structure, which are stacked by multi-layer transformer. The probing task is an approach to find what knowledge do each layer of PTMs learn. Jawahar et al. [22] proposed that BERT's lower layers capture the phrase-level information, intermediate layers learn the linguistic information and higher layers encode the sematic information of. Wallace et al. [23] apply probing task to search BERT's representing ability for numbers, which shows that BERT only uses sub-word units and is less exact.

3 PTMs-CRF Model

This section will describe the PTMs-CRF model. The architecture of PTMs-CRF model includes embedding layer and inference layer, which is similar to the one presented by Lample et al. [6]. In this paper, BERT and its advanced versions are applied in model's embedding layer due to their excellent natural language encoding performance.

3.1 BERT Family

BERT has some advanced versions: RoBERTa trains longer with bigger batches and over more data; and ALBERT, a lite version of BERT.

3.1.1 Original BERT

BERT is a stack of Transformer encoder layers which consist of multi-headed self-attention, fully-connected, layer normalization and residual connection. Every token of input is defined as query, key and value vectors, each head self-attention creates the weighted representation by them. Then fully-connected layer is used to combine the outputs of all heads in the same layer. The residual connection makes model deeper, and layer normalization can improve the performance by reducing gradient disappearance and explosion (Fig. 1).

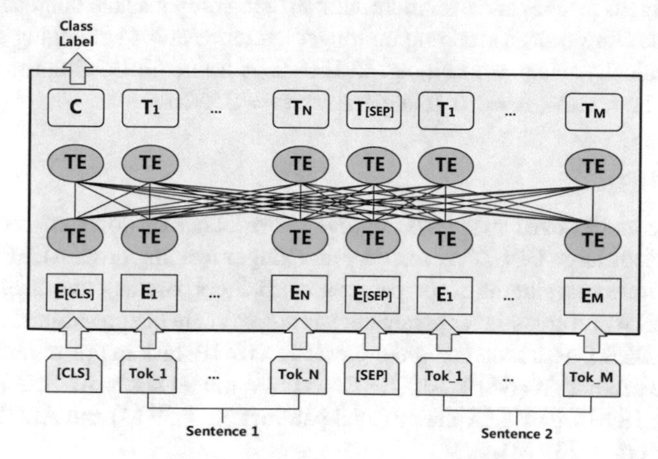

Fig. 1. The architecture of BERT

The traditional workflow for BERT includes two stages, first stage is pre-training by large scale text corpora with two semi-supervised tasks: masked language modeling (MLM) and next sentence prediction (NSP); the second stage is fine-tuning for downstream applications by supervised training via few scales labelled data.

For every word of input sequence, BERT first tokenize it into word-pieces. The final input embeddings are combined by token embedding, position embedding and segment embedding. The special token [CLS] represents the whole sentence and [SEP] is used for separating input segments.

Define the input sequence of ATC instruction as $x_1, x_2,..., x_n$ and the output sequence as $y_1, y_2,..., y_n$, where N is the instruction's length, x_i means ith token of input sequence and y_i means ith label. Denote L to be number of layers, and H to be hidden size of model. Then input embeddings of BERT is computed as:

$$E_i = W_e x_i + W_p + W_s \tag{3.1}$$

where $E_i \in R^{H\times 1}$ represents the ith token, equals to the sum of token embedding $W_e x_i$, position embedding W_p and segment embedding W_s.

The output of BERT can be computed by following equations:

$$h_i^{l+1} = transformer\_encoder(h_i^l), \quad \text{where } h_i^0 = E_i, y_i = h_i^L \tag{3.2}$$

The *transformer_encoder* denotes the encoder of Transformer and A is the number of self-attention heads in one layer of encoder. The h_i^l is hidden state of the lth layer. BERT has two model sizes: BERT-base (L = 12, H = 768, A = 12, total parameters = 110 M) and BERT-large (L = 24, H = 1024, A = 16, total parameters = 340 M).

3.1.2 RoBERTa

The performance of BERT is not excellent because of significantly undertrained when pre-training. In order to develop it, four modifications were applied: 1) training the model longer with bigger batches and over more data; 2) using only masked language modelling (MLM) pre-training task; 3) training on longer sentence and 4) changing the masking pattern dynamically when pre-training. RoBERTa is trained following the BERT-large with L = 24, H = 1024, A = 16, total parameters = 255 M.

3.1.3 ALBERT

Increasing the model size can always improve performance on downstream tasks. However, it also needs large GPU/TPU memory and longer training time. ALBERT is a kind of lite BERT due to reducing some parameters. It incorporates factorized embedding parameterization and cross-layer parameter sharing to scale the pre-trained models. Furthermore, ALBERT uses sentence-order prediction (SOP) task in pre-training instead of next sentence prediction (NSP). ALBERT has three model sizes: ALBERT-large (total parameters = 18 M), ALBERT-xlarge (total parameters = 59 M) and ALBERT-xxlarge (total parameters = 233 M).

3.2 CRF Tagging Model

In NER task, the dependencies across output labels are strong. The NER model can joint use CRF to capture dependency between any two labels. The input of CRF is the hidden state sequence of BERT's last layer $h^L = (h_1{}^L, h_2{}^L, \ldots, h_n{}^L)$, the corresponding prediction output sequence of is $y = (y_1, y_2, \ldots, y_n)$.

Consider $P \in R^{n \times k}$ to be the matrix of scores output from BERT networks, where k is the number of labels, and $P_{i,j}$ means the corresponding score of the jth label of the ith token of input sequence.

Define the score of CRF as:

$$S(h, y) = \sum_{i=0}^{N} A_{y_i, y_{i+1}} + \sum_{i=1}^{N} P_{i, y_i} \tag{3.3}$$

where A is a matrix of transition scores, $A_{i,j}$ represents the score of the transition from label i to label j. The start and end of label sequence y are "*start*" label and "*end*" label, the number of labels should be $k + 2$ and the size of square matrix A becomes $k + 2$.

A softmax over all possible tag sequences yields a probability for the sequence y:

$$p(y|h) = \frac{e^{S(h,y)}}{\sum_{\tilde{y} \in Y_h} e^{S(h,\tilde{y})}} \tag{3.4}$$

where Y_h denotes all possible labels of input h. Maximizing the log-probability of the correct label sequence during training. While decoding, the model predicts the correct output sequence by obtaining the maximize score as:

$$y^* = \arg\max_{\tilde{y} \in Y_h} s(h, \tilde{y}) \tag{3.5}$$

The viterbi algorithm is used to compute the optimal output sequence y^* in (3.5) during decoding.

4 Experiments

In this section, we deal two works: 1) comparing the performance of BERT-CRF, ALBERT-CRF and RoBERTa-CRF; 2) applying probing task to find which layer obtains the best performance of three models in entity extraction task.

4.1 Datasets

In practical, the whole procedure of ATC operation is: 1) if any flight wants to leave the stand, the pilot needs to apply for push-back, after agreeing, 2) the flight enters the taxiway and then taxing to waiting point and 3) holds for take-off clearance. When ATC controller giving clearance, 4) flight enters runway and 5) takes off, and then ATC controller will command the flight to 6) depart from the airport. During these stages, ATC controller may adjust the 7) height and 8) speed of the fight. The flight 9) contacts the approach when it arrives the target airport, then it 10) lands the runway and taxis into the stand.

Therefore, the practical data consists of 12 parts: 1) push-back (short for p-b), 2) taxi, 3) holding, 4) line up, 5) take-off (short for t-off), 6) departure (short for dept), 7) height adjust (short for h-adj), 8) speed adjust (short for s-adj), 9) approach (short for

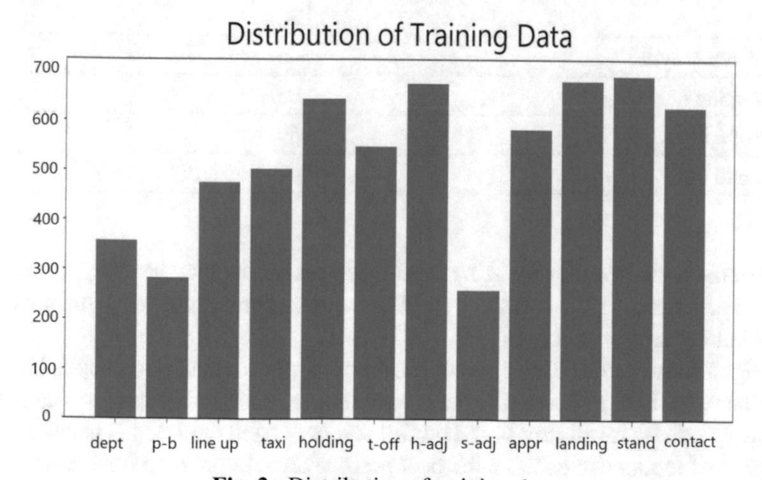

Fig. 2. Distribution of training data

appr), 10) landing, 11) into stand (short for stand), and 12) contact between controller and pilot (short for contact). The distribution of data is shown in Fig. 2.

The practical data includes more than 5000 ATC instructions with average length of 15 words. The longest instruction has 41 words and the shortest one has only 6 words. Experiments choose 90% to be training data and 10% to be testing data. There are 10 types of pre-defined entity: flight number, organization, action, status, location, height, speed, weather, time and other.

4.2 Performance

In this section, we use BERT-CRF, ALBERT-CRF and RoBERTa-CRF to extract the entities of ATC instructions. Table shows their parameters.

Table 1. Model parameters

Models	Total layers	Hidden size	Attn heads	Total parameters
BERT	12	768	12	110 M
ALBERT	12	768	12	59 M
RoBERTa	24	1024	16	255 M

As Table 1 shown, experiments choose BERT-base, ALBERT-base and RoBERTa-large in NER models and Adam to be optimizer. The experiment is done by applying one NVIDIA GeForce GTX 1060 GPU to train the models and the major parameters are shown following (Table 2).

Table 2. Parameter settings

Batch size	32
Max sentence length	45
Adam learning rate	0.01
Number of epochs	30
Dropout rate	0.1

Define the label of entity by using "BOI" format for data, where "B_" means beginning word of phrase, "I_" means the word belongs to phrase and "O_" means the word does not belong to any entity (Fig. 3).

As Fig. 3 shown, after 30th epochs, the error of three models converges below 3% . Among them, ALBERT obtains the best performance, which achieves less than 2% error rate. However, these results are based on full layers of each model, it is necessary to find which layer of model can achieve the best performance in order to further optimize the model.

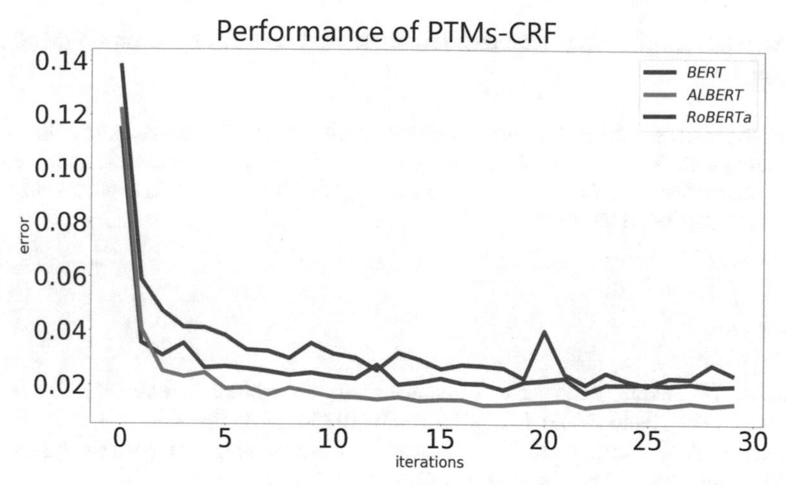

Fig. 3. Performance of BERTs-CRF models for entity extraction

4.3 Probing Task

In this section, probing task will investigate the effectiveness of features from different layers for extracting entities.

In entity extraction task of ATC instructions, it shows that the middle layers of models obtain the best performance. The 6th layer of BERT achieves 0.72% testing error, the 5th layer of ALBERT achieves 1.32% testing error and the 7th layer of RoBERTa achieves 1.15% testing error (Table 3).

Table 3. Error rate of models

Models	Num of layer	Testing error
BERT	6	0.72%
ALBERT	5	1.32%
RoBERTa	7	1.15%

There are two findings: 1) entity extraction task is more depend on syntactic feature of ATC instructions, and 2) ATC instruction is a kind of short text and it has simple semantic feature, so the models depend less on semantics feature.

5 Conclusions

This paper proposes PTMs-CRF model for extracting the information of ATC instructions and conducts experiments to investigate different versions and layers of BERT for entity extraction task. There are some experiments findings: 1) BERT achieves excellent performance, 2) the middle layers of BERT outperform other layers, and 3) The ATC instruction has simple semantic feature so the model depends more on syntactic feature of instructions.

Acknowledgements. The authors are grateful to CAAC North China Regional Administration for providing data. They also thank the anonymous reviewers for their critical and constructive review of the manuscript. This study was supported by the Fund of the China National Key R&D Program (No. 2020YFB1600100).

References

1. Nadeau, D., Sekine, S.: A survey of named entity recognition and classification. Lingvist. Investig. **30**(1), 3–26 (2007). https://doi.org/10.1075/li.30.1.03nad
2. Li, J., Sun, A.X., Han, J.L., Li, C.L.: A survey on deep learning for named entity recognition. IEEE Trans. Knowl. Data Eng. **29**, 1–1 (2020)
3. Bikel, D., Miller, S., Schwartz, R., Weischedel, R.: Nymble: a high-performance learning name-finder. In: Proc. Conference on Applied Natural Language Processing (1997)
4. McCallum, A., Li, W.: Early results for named entity recognition with conditional random fields, features induction and web-enhanced lexicons. In: Proc. Conference on Computational Natural Language Learning (2003)
5. Hammerton, J.: Named entity recognition with long short-term memory. In: HLT-NAACL 2003, vol.: 4, pp. 172–175 (2003)
6. Lample, G., Ballesteros, M., Subramanian, S., Kawakami, K., Dyer, C.: Natural architectures for named entity recognition. In: arXiv preprint arXiv: 1603.01360 (2016)
7. Cui, L., Zhang, Y.: Hierarchically-refined label attention network for sequence labelling. In: arXiv preprint arXiv: 1908.08676 (2019)
8. Vaswani, A., et al.: Attention is all you need. In: Advances in Neural Information Processing Systems, pp. 5998–6008 (2017)
9. Chiu, J., Nichols, E.: Named entity recognition with bidirectional lstm-cnns. In: arXiv preprint arXiv: 1511.08308 (2015)
10. Ma, X., Hovy, E.: End-to-end sequence labelling via bi-directional lastm-cnns-crf. In: arXiv preprint arXiv: 1603.01354 (2016)
11. Qiu, S.P., Sun, T.X., Xu, Y.G., Shao, Y.F., Dai, N., Huang, X.J.: Pre-trained models for natural language processing: a survey. Sci. China Technol. Sci. **63**(10), 1872–1897 (2020). https://doi.org/10.1007/s11431-020-1647-3
12. Radford, A., Narasimhan, K., Salimans, T., Sutskever, I.: Improving language understanding by generative pre-training. In: URL https://s3-us-west-2.amazonaws.com/openai-assets/researchcovers/languageunsupervised/languageunderstandingpaper.pdf (2018)
13. Devlin, J., Chang, M.W., Lee, K., Toutanova, K.: BERT: pre-training of deep bidirectional transformers for language understanding. In: NAACL-HLT (2019)
14. Bengio, Y., Ducharme, R., Vincent, P., Jauvin, C.: A neural probabilistic language model. J. Mach. Learn. Res. **3**(Feb), 1137–1155 (2003)
15. Mikolov, T., Sutskever, I., Chen, K., Corrado, G.S., Dean, J.: Distributed representations of words and phrases and their compositionality. In: NeurIPS (2013)
16. Pennington, J., Socher, R., Manning, C.D.: GloVe: global vectors for word representation. In: EMNLP (2014)
17. Peters, M.E., et al.: Deep contextualized word representations. In: NAACL-HLT (2018)
18. Howard, J., Ruder, S.: Universal language model fine-tuning for text classification. In: ACL, pp. 328–339 (2018)

19. Liu, Y.H., et al.: RoBERTa: a robustly optimized BERT pretraining approach. In: arXiv preprint arXiv:1907.11692 (2019)
20. Lan, Z.Z., Chen, M.D., Goodman, S., Gimpel, K., Sharma, P., Soricut, R.: ALBERT: a lite-BERT for self-supervised learning of language representations. In: arXiv preprint arXiv: 1909.11942 (2019)
21. Yang, Z.L., Dai, Z.H., Yang, Y.M., Carbonell, J., Salakhutdinov, R.R., Le, Q.V.: XLNet: generalize autoregressive pretraining for language understanding. In: NeurIPS, pp. 5754–5764 (2019)
22. Jawahar, G., Sagot, B., Seddah, D.: What does BERT learn about the structure of language? In: ACL, pp. 3651–3657 (2019)
23. Wallace, E., Wang, Y.Z., Li, S.J., Singh, S., Gardner, M.: Do NLP models know numbers? probing numeracy in embeddings. In: Proc of the 2019 Conference on Empirical Methods in Natural Language Processing and the 9th International Joint Conference on Natural Language Processing, pp. 5307–5315 (2019)

Medical Image Segmentation Using Transformer

Qian Wang[1(✉)], Longyan Li[1], Bo Ni[2], Yu Li[1], Dejin Kong[1], Chen Wang[1], and Zan Li[3]

[1] School of Electronic and Electrical Engineering, Wuhan Textile University, Wuhan 430074, China
{wqian,leewoo,djkou}@wtu.edu.cn, 1915053006@mail.wtu.edu.cn

[2] Computer School of Hubei Polytechnic University, Huangshi 435003, China
nb@hbpu.edu.cn

[3] International College Beijing, China Agricultural University, Beijing 100089, China

Abstract. For the past few years, the U-Net structure shows strong performance in the field of medical image segmentation. However, due to the inherent locality of convolution operations, U-shaped structures are often limited in modeling long-range dependencies. Transformer, a global self-attention mechanism designed for sequence-to-sequence prediction, has been successfully used in the field of computer vision. In this paper, we propose a novel network, named TransHarDNet. HarDNet, which is a low memory traffic CNN. We combine it as backbone with Transformer. Our network enables the global semantic context information and low-level spatial details of the input image to be captured more effectively. We evaluate the effectiveness of the proposed network on five medical image datasets.

Keywords: Transformer · Medical image segmentation · Convolutional neural networks

1 Introduction

Convolutional Neural Networks (CNNs) have achieved excellent performance in lots of medical image tasks, such as cardiac segmentation, polyp segmentation and brain tumor segmentation.

In different CNN variants, U-Net [1] uses symmetric encoder and decoder structure with skip connection, which has become the mainstream structure of medical image segmentation. Many variants of U-Net, such as U-Net++ [2], ResUNet++ [3], Attention U-Net [4] have further improved the performance of medical image segmentation. Although the CNN-based approach has special presentation capabilities, it is usually limited in modeling remote relationships due to the inherent locality of the convolution operation.

Q. Liang et al. (Eds.): Artificial Intelligence in China, LNEE 854, pp. 92–99, 2022.
https://doi.org/10.1007/978-981-16-9423-3_12

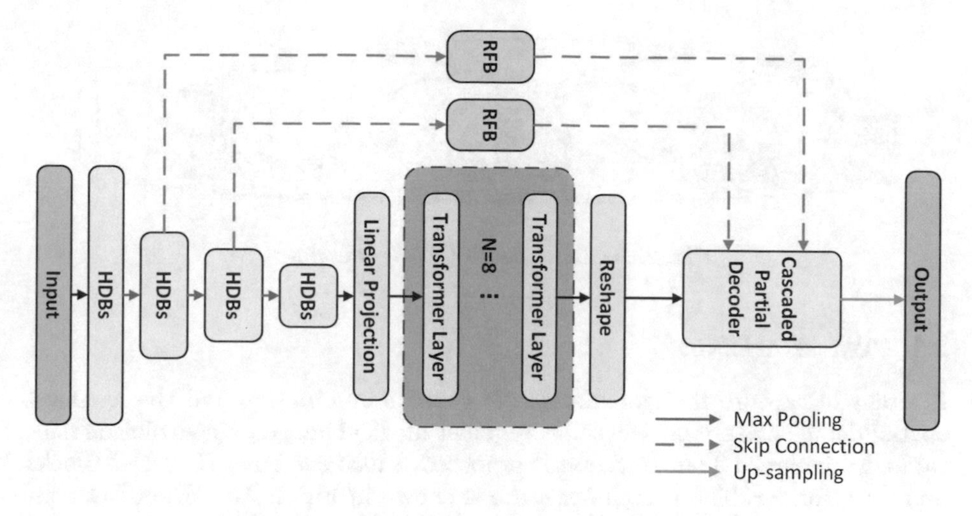

Fig. 1. Overall architecture of the proposed TransHarDNet.

Recently, Transformer, a sequence-to-sequence prediction model for Natural Language Processing (NLP) tasks, has produced promising results in the fields of image classification, object detection, and image segmentation. Vision Transformer(ViT) [5], an image classification network using a purely self-attention mechanism, achieves excellent results on ImageNet [6] while requiring less computing resources for training. DETR [7] shows impressive performance on object detection, with comparable accuracy and speed with the popular and well-established Faster R-CNN [8] baseline on COCO bench-mark. SETR [9] replaces the encoders with transformers in the conventional encoder-decoder based networks to successfully achieve state-of-the-arts results on the natural image segmentation task. TransUNet [10] employs ViT for medical image segmentation and sets new records in the CT multi-organ segmentation task. PraNet [11] is a parallel reverse attention network for accurate polyp segmentation. HarDNet-MSEG [12] is a simple encoder-decoder architecture without any attention module for polyp segmentation.

In this paper, we propose a novel convolution neural network, named TransHarDNet. The model we proposed uses HarDNet [13] as the backbone and Transformer to obtain global semantic information, so as to improve the accuracy of medical image segmentation.

2 Method

In this section, we first give an overview of TransHarDNet, and then discuss each module of the network in detail.

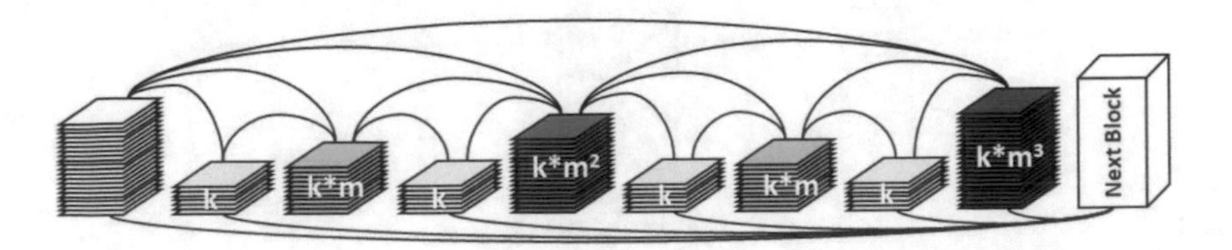

Fig. 2. HarDNet Block (HDB) overview.

2.1 TransHarDNet

In order to capture the global semantic context information and the low-level spatial details better and accurately segment medical images, a convolution neural network named TransHarDNet is proposed, which combines HarDNet Blocks and Transformer. The overall structure is shown in Fig. 1. TransHarDNet consists of encoder part and decoder part. We use HarDNet Blocks to capture spatial and depth information, and then extract global semantic context information by Transformer. Then the feature maps are fused by cascaded partial decoder and up-sampled to the same resolution as the input image to get the segmentation result. The details of the model are given in the rest of this section.

2.2 HarDNet Block (HDB)

HarDNet is an improvement of DenseNet [14], as shown in Fig. 2. Shortcuts are reduced to improve the inference speed and alleviate the problem of overfitting, and channel width of key layer is increased to make up for the loss of accuracy. HarDNet-68 remove the global dense connections. Max pooling is used in the down-sampling process. In DenseNet, the sequence of convolution operations is batch normalization layer, activation layer and convolution layer, but in HarDNet-68, the convolution layer is used first. And the inference time of the HarDNet-68 is reduced by 30% compared to the ResNet-50 [15]. Therefore, we choose HarDNet-68 as the backbone of our model.

2.3 Transformer

Input image after HDBS encoding, the feature map $F \in \mathbb{R}^{C\times\frac{H}{32}\times\frac{W}{32}} (C = 512)$ is obtained. Note that H and W are the size of the input image. F contains rich local semantic information, and we compress it to one dimension to get the feature map $f \in \mathbb{R}^{C\times N} \left(N = \frac{H}{32} \times \frac{W}{32}\right)$, which can be input into Transformer to further learn global context semantic information. We add position embedding to retain the location information of features in f, as shown below:

$$z_0 = f + PE. \tag{1}$$

where $PE \in \mathbb{R}^{C\times N}$ denotes the position embedding, and $z_0 \in \mathbb{R}^{C\times N}$ refers to the feature embedding.

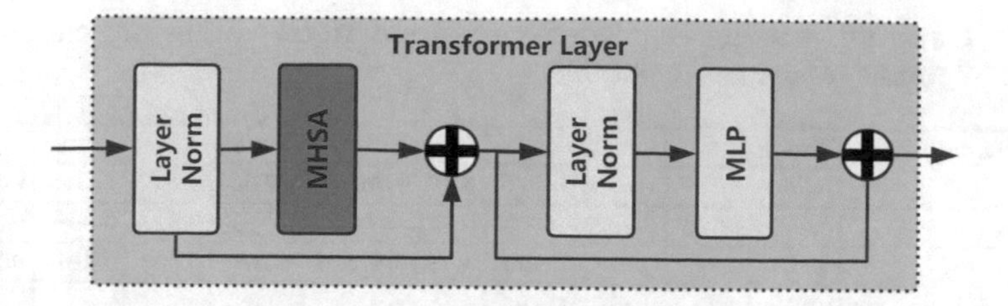

Fig. 3. Transformer Layer overview.

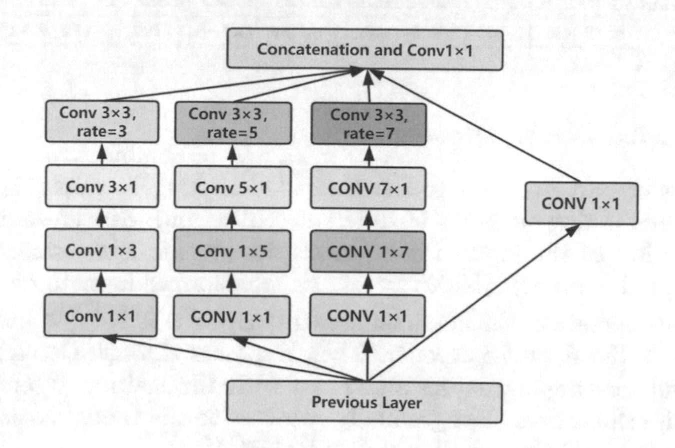

Fig. 4. RFB overview.

The Transformer encoder consists of L transformer layers, each of them has a Multi-Head Self-Attention (MHSA) block and a Multi-Layer Perceptron (MLP) block. Therefore the output of the l-th layer can be written as follows:

$$\mathrm{z}_l' = \mathrm{MHSA}\left(\mathrm{LN}\left(\mathrm{z}_{l-1}\right)\right) + \mathrm{z}_{l-1}. \tag{2}$$

$$\mathrm{z}_l = \mathrm{MLP}\left(\mathrm{LN}\left(\mathrm{z}_l'\right) + \mathrm{z}_l'\right. . \tag{3}$$

where $LN(\cdot)$ denotes the layer normalization and z_l is the encoded image representation of l-th layer. Figure 3 shows the structure of a transformer layer.

2.4 RFB

To improve the discriminability and robustness of features, we add Receptive Field Block(RFB) [16] into the skip connection, and use multiple convolution kernels of different sizes and dilated convolution kernels to enhance feature extraction. Figure 4 shows the structure of a RFB.

Table 1. Quantitative results on polyp segmentation datasets comparing with previous models. '–' means results not applicable.

Methods	Kvasir		ClinicDB		ColonDB		EndoScene		ETIS	
	mDice	mIoU	mDice	mIoU	mDice	mIoU	mDice	mIoU	mDice	mIoU
U-Net	0.818	0.746	0.823	0.750	0.512	0.444	0.710	0.627	0.398	0.335
U-Net++	0.821	0.743	0.794	0.729	0.483	0.410	0.707	0.624	0.401	0.344
ResUNet++	0.813	0.793	0.796	0.796	–	–	–	–	–	–
PraNet	0.898	0.840	0.899	0.849	0.709	0.640	0.871	0.797	0.628	0.567
HarDNet-MSEG	**0.912**	**0.857**	**0.932**	**0.882**	0.731	0.660	0.887	0.821	0.677	0.613
TransHarDNet	0.899	0.844	0.892	0.835	**0.782**	**0.698**	**0.891**	**0.822**	**0.717**	**0.634**

2.5 Cascaded Partial Decoder

In the decoder part, we refer to the Cascaded Partial Decoder [17]. Through the fusion of the deep features of different scales, and then up-sample to the same resolution as the input, the segmentation results are obtained. First of all, we shape the output $z_l \in \mathbb{R}^{C \times N}$ of the transformer layer to $C \times \frac{H}{32} \times \frac{W}{32}$. Through this operation, the obtained feature map $Z \in \mathbb{R}^{C \times \frac{H}{32} \times \frac{W}{32}}$ has the same dimensions as the F of the encoder. Then Z passes through Cascaded Partial Decoder and up-sampling layers and fuses with the features of encoder part through skip-connection, then gradually recovers to the same size as the input to get the pixel-level segmentation result $R \in \mathbb{R}^{C \times H \times W}$.

2.6 Loss Function

We use the weighted IoU loss and binary cross-entropy loss $L = L^w_{IoU} + L^w_{bce}$ for end-to-end training of the entire network. To obtain the final segmentation result, we directly up-sample the feature map of the last part of the decoder to the same resolution as the input image and adjust it to P channels through a 1 $\times$ 1 convolution operation, where P is the number of classes.

3 Experiments

We evaluate the validity of our model in Kvasir-SEG [18], CVC-ColonDB [19], ETIS-Larib Polyp DB [20], EndoScene [21] and CVC-ClinicDB [22] polyp segmentation datasets and compare our TransHarDNet with other medical image segmentation methods.

3.1 Training and Inference Environment Setting

Our model is carried on a NVIDIA TITAN RTX GPU with 24 G memory. We use Adam optimizer to train our model end-to-end for 100 epochs. The initial learning rate is set to 0.0001.

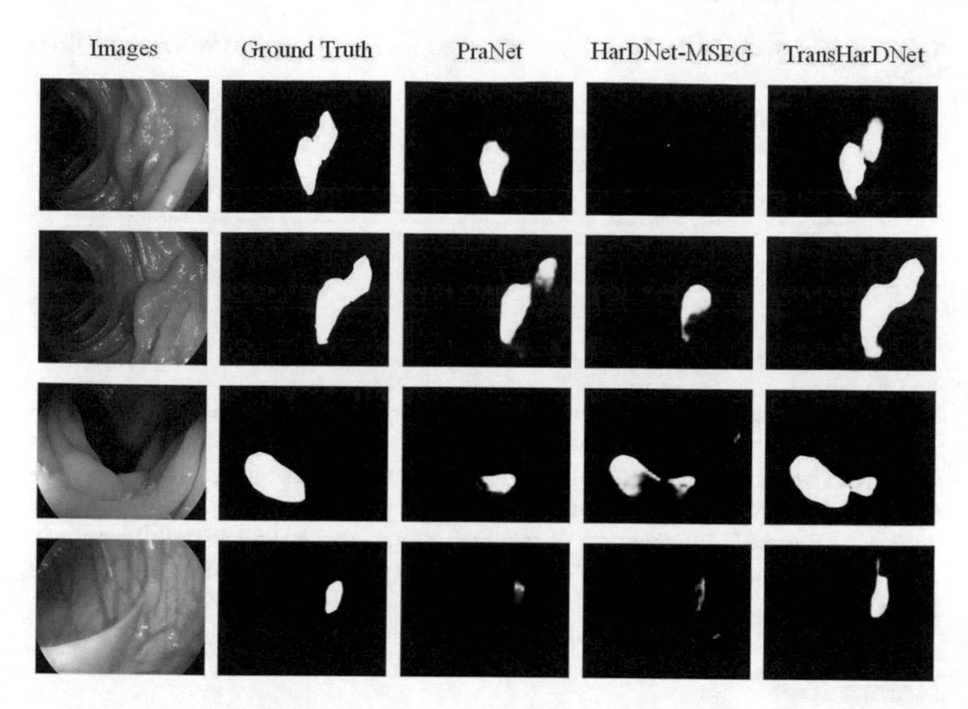

Fig. 5. Inference results on ETIS-Larib Polyp DB comparing with other SOTA methods.

3.2 Polyp Segmentation

The train sets include 900 images in Kvasir-SEG and 550 images in CVC-ClinicDB. And the test sets include 100 images from Kvasir-SEG, 62 images from CVC-ClinicDB, 380 images from CVC-ColonDB, 60 images from EndoScene and 196 images from ETIS-Larib Polyp DB. The training input size is set to 352×352. We employ mean IoU and mean Dice for quantitative evaluation.

The results comparing with other SOTA methods are in Table 1. TransHarDNet shows the greatest accuracy on most metrics. Figure 5 shows some inference results on ETIS-Larib Polyp DB comparing with other SOTA methods.

The number of parameters comparing with previous models is in Table 2. As you can see from Table 2, the number of parameters in our model is 10% fewer than the latest method. Through comparison, it can be found that the mean IoU score and mean Dice score of our model on ColonDB, EndoScene and ETIS-Larib Polyp DB are higher than the previous methods. This proves the performance of transformer in medical image segmentation task.

Table 2. Comparison of the number of model parameters with PraNet, HarDNet-MSEG.

Methods	Backbones	Parameters	ColonDB	EndoScene	ETIS
PraNet	Res2Net-50	32.5M	0.709	0.871	0.628
HarDNet-MSEG	HarDNet-68	33.3M	0.731	0.887	0.677
TransHarDNet	HarDNet-68	**29.6M**	**0.782**	**0.891**	**0.717**

4 Conclusion

In this paper, we designed a novel network combining Transformer and HarDNet for medical image segmentation. TransHarDNet enables more efficient access to low-level spatial details and global semantic context information. The experimental results showed that TransHarDNet has a better ability to segment the boundary semantic information of medical images and it has fewer parameters. This again demonstrated the strong ability of Transformer to capture global semantic information.

References

1. Ronneberger, O., Fischer, P., Brox, T.: U-Net: convolutional networks for biomedical image segmentation. In: Navab, N., Hornegger, J., Wells, W., Frangi, A. (eds.) MICCAI 2015. LNCS, vol. 9351, pp. 234–241. Springer, Cham (2015). https://doi.org/10.1007/978-3-319-24574-4_28
2. Zhou, Z., Siddiquee, M.M.R., Tajbakhsh, N., Liang, J.: UNet++: a nested U-Net architecture for medical image segmentation. In: IEEE TMI, pp. 3–11 (2019)
3. Jha, D., et al.: ResUNet++: an advanced architecture for medical image segmentation. In: IEEE ISM, pp. 225–230 (2019)
4. Oktay, O., et al.: Attention U-Net: Learning Where to Look for the Pancreas. ArXiv Preprint ArXiv:1804.03999 (2018)
5. Dosovitskiy, A., et al.: An image is worth 16×16 words: transformers for image recognition at scale. In: ICLR (2021)
6. ImageNet. http://image-net.org
7. Carion, N., Massa, F., Synnaeve, G., Usunier, N., Kirillov, A., Zagoruyko, S.: End-to-end object detection with transformers. In: Vedaldi, A., Bischof, H., Brox, T., Frahm, J.M. (eds.) ECCV 2020. LNCS, vol. 12346, pp. 213–229. Springer, Cham (2020). https://doi.org/10.1007/978-3-030-58452-8_13
8. Ren, S., He, K., Girshick, R., Sun, J.: Faster R-CNN: towards real-time object detection with region proposal networks. In: IEEE TPAMI, pp. 1137–1149 (2017)
9. Zheng, S., et al.: Rethinking Semantic Segmentation from a Sequence-to-Sequence Perspective with Transformers. ArXiv Preprint ArXiv:2012.15840 (2020)
10. Chen, J., et al.: TransUNet: Transformers Make Strong Encoders for Medical Image Segmentation. ArXiv Preprint ArXiv:2102.04306 (2021)
11. Fan, D.P., et al.: PraNet: parallel reverse attention network for polyp segmentation. In: Martel, A.L. et al. (eds.) MICCAI 2020. LNCS, vol. 12266, pp. 263–273 . Springer, Cham (2020). https://doi.org/10.1007/978-3-030-59725-2_26

12. Huang, C.H., Wu, H.Y., Lin, Y.L.: HarDNet-MSEG: a simple encoder-decoder polyp segmentation neural network that achieves over 0.9 mean dice and 86 FPS. ArXiv Preprint ArXiv:2101.07172 (2021)
13. Chao, P., Kao, C.-Y., Ruan, Y., Huang, C.-H., Lin, Y.-L.: HarDNet: a low memory traffic network. In: IEEE/CVF ICCV, pp. 3552–3561 (2019)
14. Huang, G., Liu, Z., van der Maaten, L., Weinberger, K.Q.: Densely connected convolutional networks. In: IEEE CVPR, pp. 2261–2269 (2017)
15. He, K., Zhang, X., Ren, S., Sun, J.: Deep residual learning for image recognition. In: IEEE CVPR, pp. 770–778 (2016)
16. Liu, S., Huang, D., Wang, Y.: Receptive field block net for accurate and fast object detection. In: Ferrari, V., Hebert, M., Sminchisescu, C., Weiss, Y. (eds.) ECCV 2018. LNCS, vol. 11215, pp. 385–400. Springer, Cham (2018). https://doi.org/10.1007/978-3-030-01252-6_24
17. Wu, Z., Su, L., Huang, Q.: Cascaded partial decoder for fast and accurate salient object detection. In: IEEE CVPR, pp. 3907–3916 (2019)
18. Jha, D., et al.: Kvasir-SEG: a segmented polyp dataset. In: Ro, Y., et al. (eds.) MMM 2020. LNCS, vol. 11962, pp. 451–462. Springer, Cham (2020). https://doi.org/10.1007/978-3-030-37734-2_37
19. Tajbakhsh, N., Gurudu, S.R., Liang, J.: Automated polyp detection in colonoscopy videos using shape and context information. IEEE TMI **35**(2), 630–644 (2015)
20. Silva, J., Histace, A., Romain, O., Dray, X., Granado, B.: Toward embedded detection of polyps in WCE images for early diagnosis of colorectal cancer. Int. J. Comput. Assist. Radiol. Surg. **9**(2), 283–293 (2013). https://doi.org/10.1007/s11548-013-0926-3
21. Vázquez, D., et al.: A benchmark for endoluminal scene segmentation of colonoscopy images. J. Healthc. Eng. **2017**, 1–10 (2017)
22. Bernal, J., Sánchez, F.J., Fernández-Esparrach, G., Gil, D., Rodríguez, C., Vilariño, F.: WM-DOVA maps for accurate polyp highlighting in colonoscopy: validation vs. saliency maps from physicians. In: CMIG, vol. 43, pp. 99–111 (2015)

Satellite Online Scheduling Algorithm Based on Proximal Policy

XueFei Li(✉), Jia Chen, XianTao Cai, and NingBo Cui

School of Computer Science, Wuhan University, Wuhan 300350, China
snowfly_li@163.com

Abstract. Aiming at the problems of high labor cost, low task execution efficiency, and unable to adapt to the normalized dynamic observation requirements caused by satellite scheduling under off-line control of satellite and ground, this paper considers each element of the satellite online scheduling process, and according to the Markov nature of satellite online scheduling process, combined with the proximal policy optimization algorithm, proposes a satellite online scheduling algorithm, which effectively solves the problems of satellite resource conflict and time window conflict in the scheduling process, and realizes the real-time response to the task.

Keywords: Satellite online scheduling · Markov decision process · Proximal policy optimization

1 Introduction

With the continuous launch and operation of civil and commercial military earth observation satellites in recent years, user observation requirements have also shown a complex and diverse trend. At present, ground control is still used for multi-satellite scheduling problems. But it is found that the satellite scheduling ground control cannot meet the timely response of emergency tasks.

Wang [1] firstly classified the arrival of large-scale emergency tasks as a satellite online scheduling problem. According to the characteristics of satellite task online scheduling problem, a centralized satellite task online scheduling model and algorithm were proposed. The reinforcement learning A3C algorithm is used for the solution. However, in the study of the centralized task online scheduling model, the actual impact of the periodic schedule on scheduling decision-making strategy is not considered. Moreover, the A3C algorithm is sensitive to the setting of super parameters, and its performance is unstable and the solution effect is general. Based on his research, this paper will improve the centralized satellite online scheduling algorithm and propose a satellite online scheduling algorithm combined with the proximal policy optimization algorithm.

2 Problem Formulation

In this paper, the satellite online scheduling problem is represented as follows:

Q. Liang et al. (Eds.): Artificial Intelligence in China, LNEE 854, pp. 100–108, 2022.
https://doi.org/10.1007/978-981-16-9423-3_13

The arrival time of each observation task is independent of the other. When an observation task arrives, the decision-maker can obtain the income of the observation task, the time window, and the execution time needed. Decision-maker may choose either to accept an observation mission or to reject it. Its decision-making is only related to the current satellite resource state and the information of the demand itself, therefore, the satellite's online scheduling is a Markov decision process [2], so it is suitable to use a reinforcement learning algorithm for processing. Acceptance is filtered according to the available time window and execution time of the current observation task. If the satellite constraints are not met, no return is made. If a satellite that meets the conditions is found, the observation mission is assigned to the satellite and the return is made. Once the observation task is implemented, the satellite resources are always occupied and cannot be released in this scheduling cycle. Once the observation demand is rejected, the observation demand cannot be recalled again. At the end of the use of satellite resources or the end of the observation cycle. The goal of satellite decision-making is to maximize the total rewards of the scheduling process (Fig. 1).

3 Algorithm

3.1 Overview of Algorithm

Step 1: When a task arrives, the task data is preprocessed to extract information about its execution income, available time window, and execution time.
Step 2: According to the current task and related information to obtain the current environmental state, see the environmental state quad-tuple of the decision model in the constraint description.
Step 3: Input environmental state into the actor-new to obtain decision action.
Step 4: According to the decision action to choose whether to carry out the task.
Step 5: If executed, the execution order of satellites is sorted according to a load of executable satellites, and each satellite with sufficient resources is processed by time window conflict to find the earliest time window to meet the task execution time.
Step 6: Update the total rewards and the environmental state.
Step 7: According to the data obtained from the execution process to collect and update into the actor-new network, see Fig. 2.
Step 8: Repeat Step 1–7 for output (r, s _, done, info) after each observation mission is performed until a certain amount of [(s, a, r) …] is collected, entered into the Critic network, and calculate the discount rewards, each t represents a batch.

$$Advantage(\pi_\theta) = E\left\{\sum_{t=1}^{\infty} \gamma^{t-1} r_t | s_0, \pi_\theta\right\} \tag{1}$$

Step 9: According to the loss function of the critic:

$$critic\_loss = mean(square(Advantage)) \tag{2}$$

Renewal of Criticism Network by Back Propagation.

Step 10: Input all collected [(s, a, r) …] into the Actor (old) and Actor (new) networks for operation, normal distribution Normal1 and Normal2 are obtained. Enter all the stored actions as actions into the Normal1 and Normal2 normal distribution, get prob1 and prob2 for each action and divide prob2 by prob1 to get an important weight, that's ratio. Then updating the Actor (new) network by reverse propagation following formula (8) (9) (10).

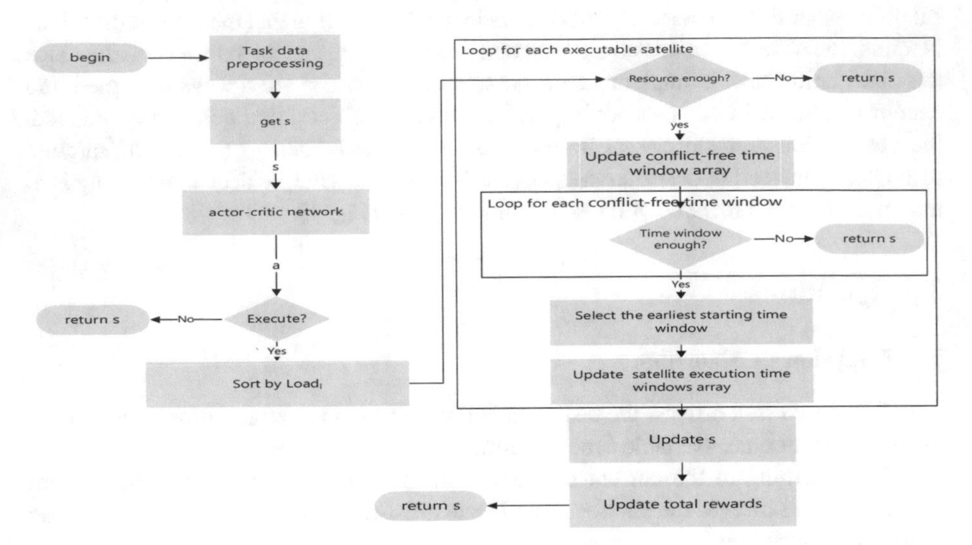

Fig. 1. Algorithm Process for per task

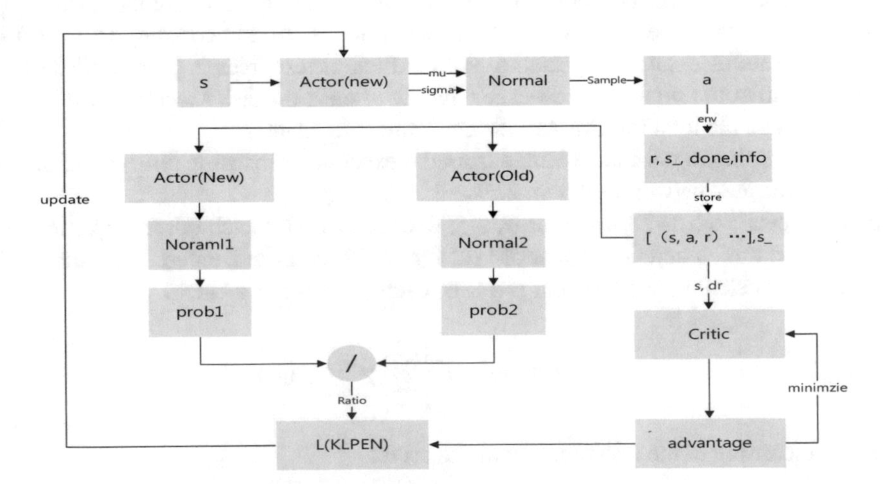

Fig. 2. Network updating Algorithm

3.2 Parameter Description

(See Table 1).

Table 1. Parameter description.

Parameter	Description
$Task_i$	observation task i
A_i	time of arrival of task i
S_l	satellite l
E_t	execute Time
$StoreR_l$	remaining resource of satellite l
$Store_l$	total resource of satellite l
R_i	required resource for task i
$Load_l$	load of satellite l
t_free_l	available time window strip length of satellite l
T_l	total time window strip length of satellite l
S_n	total num of satellite
D_i	action for task i

3.3 Constraint of Problem

Time Window. When the satellite is moving, it will leave a trajectory on the ground. The observation area of each point in the trajectory varies with the attitude and performance of the satellite, see Fig. 3:

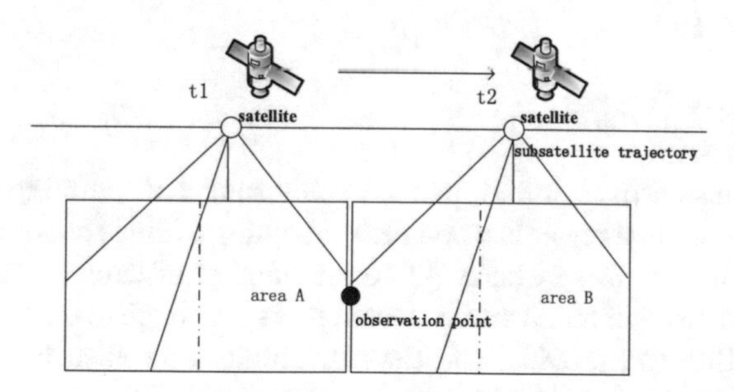

Fig. 3. Satellite observation of point

It is assumed that the satellite observation area is area A at time t1, and the observation point can be observed at this time. As the satellite moves, the observation area moves to area B. At this time, the satellite is the last position that can be observed. Therefore, [t1, t2] is called the time window of the observation point on the satellite. When the observation mission arrives, it can only be executed on the satellite when it has an available time window larger than its execution time.

Resource of Satellite. When accepting a task, it is assumed that the number of resources required for the task is R_i, and the total number of satellite remaining resources is:

$$StoreR_l = Store_l - R_i \quad (R_i < Store_l) \tag{3}$$

Action. For each incoming observation task, the decision-maker makes decisions according to the state of the environment:

$$D_i = \begin{cases} 1\ Accept \\ 0\ Reject \end{cases} \tag{4}$$

Once an observation task is accepted, the task is assigned to the executable satellite with the largest current load, load of satellite l is defined as:

$$Load_l = \frac{t\_free_l}{T_l} * \frac{StoreR_l}{Store_l} \tag{5}$$

Reward. For each observation task, if the observation task is successfully executed, the reward is obtained, and if not, the reward is 0:

$$R_i(s_i, D_i) = \begin{cases} w_i\ D_i = 1 \\ 0\quad D_i = 0 \end{cases} \tag{6}$$

End State. There are two ending states, one is the complete resource consumption and the other is the end of the scheduling cycle.

$$\sum\nolimits_l^{S_n} \frac{StoreR_l}{Store_l} = 0\, or\, A_i > T_i \tag{7}$$

Environmental State.

1) Satellite state: $\sum\nolimits_l^{S_n} \frac{StoreR_l}{Store_l}$
2) Reward of Task i:R_i
3) The ratio of task arrival time A_i to scheduling time T_i:$\frac{A_i}{T_i}$, aiming to execute more tasks to increase total rewards when the scheduling cycle is about to end.
4) The ratio of the number of tasks AT to the number of decision-making tasks PT: AT/PT. When the total number of executed tasks is enough to reach the total number that the satellite can provide, and the scheduling time is sufficient, the focus of decision-making is to find higher-value observation tasks.

3.4 Time Window Conflict Handling

As shown in Fig. 4, suppose a satellite strip has an allocated time window [S1_start, S1_end], [S2_start, S2_end], …

When an observation task arrives, one of its observation time windows of the satellite is [ST_start, ST_end]. Then, there are five states. If it is at State_1, the observation task time window cannot be allocated. At State_2, the observation time window intercepts the free zone on the satellite strip, and the new observation time window is [ST_start, S1_start]. At State_3, the new observation time window becomes [S1_end, ST_end]. At State_4, the new observation time window becomes [S1_start, S1_start] and [S1_end, ST_end]. At State_5, the observation time window remains unchanged. Of course, there may still be conflicts with the newly generated observation time window, so it is necessary to continue to solve the subsequent conflicts. But the first half of the observation window will be no subsequent conflict.

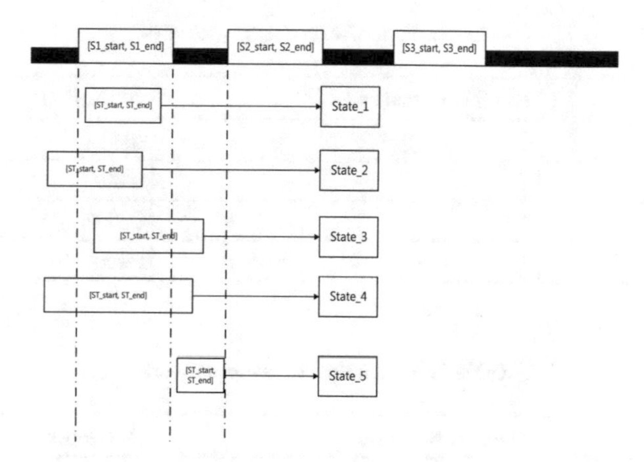

Fig. 4. Time window conflict

3.5 Proximal Policy Optimization

The Idea of Reinforcement Learning originates from Behavioral Psychology [3]. The Satellite online scheduling process has been proved to be a Markov decision process, so a reinforcement learning algorithm is used to deal with it.

John [4] et al. found that the step was not easy to select and easy to fall into a locally optimal solution through the learning policy gradient algorithm [5], so they proposed the proximal policy optimization algorithm to achieve further learning effect. Proximal policy optimization algorithm uses the importance sampling method. The data samples generated by the same strategy are updated to evaluate the policy. PPO paper presents two methods to limit the step size of each update, this model uses its first method:

$$L^{KLPEN}(\theta) = \hat{E}_t\left[\frac{\tilde{\pi}_\theta(a_t|s_t)}{\pi_{old}(a_t|s_t)}\hat{A}_t - \beta KL\left[\pi_{\theta_{old}}(\cdot|s), \tilde{\pi}(\cdot|s)\right]\right] \quad (8)$$

To obtain similar actions, KL divergence is used as a constraint, but it is difficult to determine the penalty hyperparameter β. Therefore, PPO adaptively selects β by the following rules:

$$d = \hat{E}_t\left[KL\left[\pi_{\theta_{old}}(\cdot|s), \tilde{\pi}(\cdot|s)\right]\right] \tag{9}$$

$$\beta = \begin{cases} \beta/2 \; d < d_{targ}/1.5 \\ 2\beta \;\; d > d_{targ} * 1.5 \end{cases} \tag{10}$$

3.6 Network Parameters

(See Tables 2 and 3).

Table 2. Parameters of actor-network.

Layer	Number of neurons	Activation
FC_1	40	tanh
FC_2	20	tanh
FC_3	10	tanh
FC_4	1	tanh

Table 3. Parameters of critic-network.

Layer	Number of neurons	Activation
FC_1	40	tanh
FC_2	14	tanh
FC_3	5	tanh
FC_4	1	Ø

Note: obs_dim is the dimension of the input state, act _dim is the dimension of the input action. Strategy network learning rate set to 0.000045, evaluation network learning rate set to 1/1400. Activation function tanh:tan h$x = \frac{e^x - e^{-x}}{e^x + e^{-x}}$

4 Simulation and Result Analysis

4.1 Parameter Setting

(See Table 4).

Table 4. Simulation parameter

Parameter	Value
S_n	5
Observation area	Longitude: [20°,40°] latitude: [90°,120°]
Task number	100
R_i	5G
$Store_l$	40G
Simulation period	Start: 28 Dec 2020 00:00:00.000 end: 29 Dec 2020 00:00:00.000
Reward for per task	Random between 10–20

4.2 Case Study

This experiment simulates 5000 scheduling cycles. Each scheduling cycle contains 100 random tasks. The following is the result of evaluation in the training process:

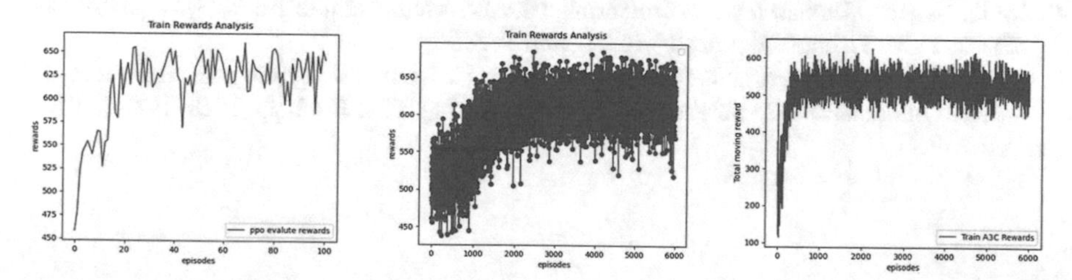

Fig. 5. Left is evaluate result of PPO Algorithm (3 episode), cost 5 min 48 s. Center is evaluate result of DQN [6] Algorithm, cost 20 min 19 s. Right is evaluate result of A3C [7] Algorithm, cost 4 min 46 s.

In Fig. 5 PPO training figure, the ordinate represents the total benefit of training, and the abscissa represents the number of assessments. Each assessment is conducted after 1000 steps of training. The sampling amount is set as 3 episodes. As can be seen from the left figure, with the gradual training, the total revenue of the model showed an upward trend and reached a relatively stable state around the 23rd. Compared with DQN and A3C algorithm, when the 2000th episode, DQN training gains converge, that is, in the 6–7 min training convergence, while the PPO algorithm reaches the convergence state in the 20th assessment, so the learning efficiency of PPO is better than DQN. The profit of the A3C algorithm is far less than DQN and PPO algorithm.

5 Conclusion

Firstly, the online scheduling problem of a satellite is characterized. According to the constraint conditions, the environmental state tuple of the satellite online scheduling process is proposed. According to the Markov property of the scheduling process and combining the proximal policy optimization algorithm, a satellite online scheduling algorithm is proposed. Through the simulation experiment, the online scheduling algorithm in this paper is evaluated to prove its self-learning ability. By comparing the online scheduling algorithms combined with other reinforcement learning algorithms, it is confirmed that the learning efficiency and learning effect of the online scheduling algorithm combined with the proximal policy optimization algorithm is better.

References

1. Haijiao, W., Zhen, Y., Wugen, Z., Dalin, L.I.: Online scheduling of image satellites based on neural networks and deep reinforcement learning. Chinese J. Aeronaut. **32**(4), 1011–1019 (2019)
2. Zhang, Z., et al.: Flow shop scheduling with reinforcement learning. Asia Pac. J. Oper. Res. **30**(5), 1350011–1350014 (2013)
3. Thorndike, E.L.: The law of effect. Am. J. Psychol. **39**(1–4), 212–222 (1927)
4. Schulman, J., Wolski, F., Dhariwal, P., Radford, A., Klimov, O.: Proximal policy optimization algorithms. https://arxiv.org/abs/1707.06347. Last Revised 28 Aug 2017
5. Sutton, R.S., McAllester, D.A., Singh, S.P., Mansour, Y.: Policy gradient methods for reinforcement learning with function approximation. Adv. Neural Inf. Process. Syst. **12**, 1057–1063 (2000)
6. Mnih, V., et al.: Human-level control through deep reinforcement learning. Nature **518**(7540), 529–533 (2015). https://doi.org/10.1038/nature14236
7. Mnih, V., et al.: Asynchronous methods for deep reinforcement learning. In: Proceedings of the 33rd International Conference on Machine Learning, PMLR 48, pp. 1928–1937 (2016)

Generation Method of Control Strategy for Aircrafts Based on Hierarchical Reinforcement Learning

Zeyuan Liu(✉), Qiucheng Xu, Yanyang Shi, Ke Xu, and Qingqing Tan

State Key Laboratory of Air Traffic Management System and Technology, Nanjing Research Institute of Electronic Engineering, Nanjing 210007, China
liuzeyuan@cetc.com.cn

Abstract. With the increasing density of air traffic and the complexity of the terminal sector, air traffic controllers will face more challenges and pressures in ensuring the safe and efficient operation of air traffic. In this work, an artificial intelligence (AI) agent is built to handle dense, complex and dynamic air traffic in the future. In this work, an artificial intelligence (AI) agent based on deep reinforcement learning is built to mimic air traffic controllers, such that the dense, complex and dynamic air traffic flows in terminal airspace can be handled sequentially and separated. To solve the problem, hierarchical reinforcement learning method is proposed, the flights choose agent and the flights action agent are achieved by DDQN. Results show that the built AI agent can guide 16 aircrafts safely and efficiently through Sector 01 of Nanjing Terminal, simultaneously.

Keywords: Terminal sector · Artificial intelligence agent · Aircraft control strategy · Hierarchical reinforcement learning

1 Introduction

With the rapid development of China's air transport industry, the number of flights is increasing at an average annual rate of over 10% in recent years [1]. What is more, by 2030, there will be more than 450 civil transport airports, and the volume of passenger traffic will reach 1.8 billion [2]. However, the rapid development of air transport would lead to the increasing challenges and pressures [3] for air traffic controllers to ensure the safe and efficient operation of air transport.

To deal with the challenges of current and future air traffic demands, Civil Aviation Administration of China (CAAC) has proposed the idea of Four Enhanced Air Traffic Management (ATM) Solutions in 2018 [4], including enhanced security, enhanced efficiency, enhanced intelligence and enhanced collaboration, where the enhanced intelligence is to suggest that the ATM should take new technologies, such as big data, blockchain, artificial intelligence (AI) and so on, to promote the operational effectiveness significantly. Therefore, some researchers are working on how to apply these new technologies to the aviation.

Supported by China National Key R&D Program (No. 2018YFE0208700).

Q. Liang et al. (Eds.): Artificial Intelligence in China, LNEE 854, pp. 109–116, 2022.
https://doi.org/10.1007/978-981-16-9423-3_14

Deep Reinforcement Learning (DRL) framework and algorithm is one of the most famous AI technologies, which has a great ability of dealing with continuous sequential decision-making problems [5, 6], and have been demonstrated to perform high level tasks and learn complex strategies, such as play the games of AlphaGo [7] and StarCraft-II [8].

Inspired by this, many researchers try to solve many difficult decision-making problems in ATM by using the deep-reinforcement Learning Algorithm.

In [9], the authors adopted a reinforcement learning method to predict the taxi-out time of the flight, and the predicted taxi-out time result is then compared with the actual taxi-out time to reduce the taxi-out time error. In [10], a Multi-agent system using Reinforcement Learning is developed for both simulation and daily operations to support human decisions, where two types of reward functions are proposed for air traffic flow management (ATFM) decision making to control safety separation of Ground Holding Problem (GHP) and Air Holding Problem (AHP).

In [12], K. Tumer and et al. proposed a multi-agent algorithm based on reinforcement learning for traffic flow management, where each agent is associated with a fix location and its goal is to set separation and speed up or slow down traffic flows to manage congestion. At last, the proposed method is tested on an air traffic flow simulator, FACET. In their following work [11], the authors proposed a distributed agent based solution where agents provide suggestions to human controllers, and an agent reward structure is designed well to allow agents to learn good actions in the indirect environment, such that the "Human-in-the-Loop" solution can be achieved.

M. Brittain and P. Wei [13] proposed a hierarchical deep reinforcement learning algorithm to build an AI agent, which takes the NASA Sector 33 app as the simulator. The well-trained AI agent can guide aircraft safely and efficiently through "Sector 33" and achieve required separation at the metering fix. And then, the authors [14] also proposed a deep multi-agent reinforcement learning framework to identify and resolve conflicts between aircrafts in a high-density, stochastic, and dynamic en route sector with multiple intersections. However, these works only considered the horizontal space separation and ignored the vertical separation of airspace.

In this work, an artificial intelligence (AI) agent based on deep reinforcement learning is built to mimic air traffic controllers, such that the dense, complex and dynamic air traffic flows in terminal airspace can be handled sequentially and separated. To simplify the problem, the complex three-dimensional terminal airspace is projected onto the vertical plane by dispersing state space and action space. And then, the typical reinforcement learning algorithm, double deep Q-network, is taken to realize the AI agent. Results show that the built AI agent can guide 6 aircrafts safely and efficiently through Sector 01 of Nanjing Terminal, simultaneously.

The remainder of the paper is organized as follows. Section 2 gives a simple example of ATC tasks in terminal sector and describes the problem definition. Section 3 shows the proposed algorithm for hierarchical reinforcement learning. Experimental results and conclusions are discussed in Section 4 and Section 5, respectively.

2 Problem Description

Air traffic control (ATC) is a service provided by ground-based air traffic controllers who direct aircraft on the ground and through controlled airspace, where the goal is to guide aircrafts to their runway for landing. The job of an air traffic controller is to prevent collisions of aircraft, safely and efficiently organize the flow of traffic and to provide support to pilots. Although traffic flow and efficiency are important factors, their primary goal is to guarantee safety of the aircraft. To accomplish this, air traffic controllers use traffic separation rules which ensure the distance between each pair of aircraft is above a minimum value all the time.

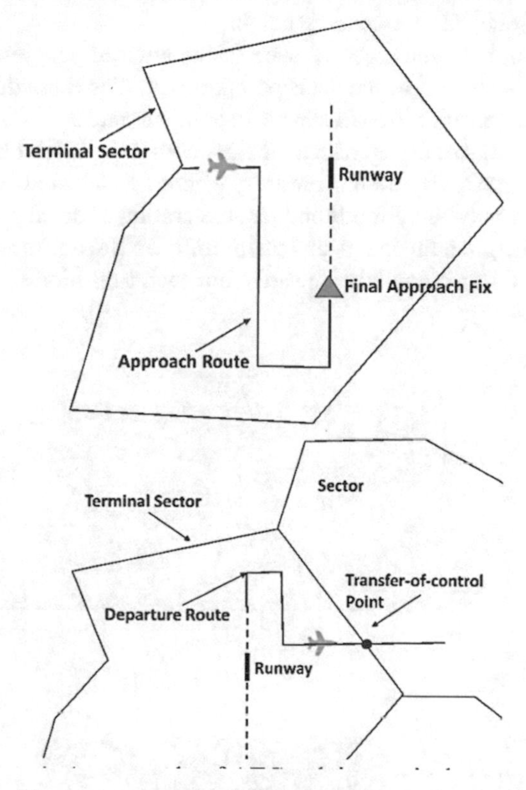

Fig. 1. An example of ATC task in terminal sector

In terminal sector, the ATC task includes two types, one is the approach control and the other is the departure control. Figure 1 gives an example of the ATC task in terminal sector. One ATC task is approach control, which is the job of directing aircrafts which are approaching an airport onto the final approach course at the correct altitude.

Our goal is to train an AI agent to perform basic Air Traffic Control tasks through hierarchical reinforcement learning. In this work, we use two deep reinforcement learning agent to accomplish the task, the flights choose agent and the flights action agent, respectively. Fig. 2 show the relationship between the two agents and the simulation environment. Their state space and action space are as follows.

(1) State space S, which contains all information about the environment and each element $s_t \in S$ can be considered a snapshot of the environment at time t. The state space of the flights choose agent is the latitude, longitude and altitude of all aircraft in the sector. The state space of flights action agent is the combination of any two flight positions within the current sector.
(2) Action space A, which is the set of all actions that AI agent could select in the environment. In this work, the flights choose agent selects a pair of flights in the current sector, and the flights action agent decides whether the two selected flights are descending height or ascending height.
(3) State transition, the agents get the state of the environment every 4 s and choose decision from a set of feasible decision options A. Corresponding to the decision, we can get the transition from a state s_i to another state s_j.
(4) Objective: The goal of the agent is to interact with the emulator by selecting actions in a way that maximises future rewards, where the selected action can maintain safe separation between aircraft and resolve conflicts for all aircraft in the sector by providing height adjustment, and all aircrafts arrival at the target positions with maximization of the cumulative reward from each transition.

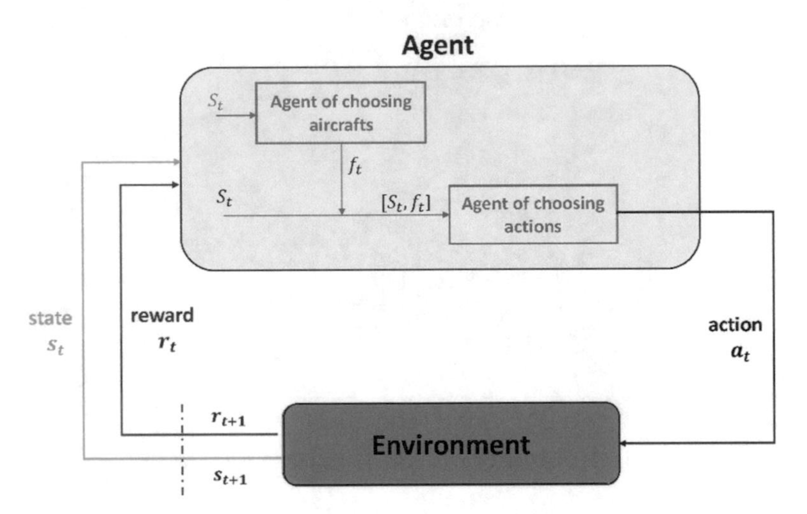

Fig. 2. The relationship between the two agents and the simulation environment

3 The Proposed Algorithm

In our research, we used two double deep Q-network (DDQN) to generate two different strategies, which are flights choose strategy and flights action strategy. The flights choose agent contains two fully-connected hidden layers, and the number of nodes in each layer are 256, 64, the flights action agent contains three fully connected layers, the number of nodes in each layer are 256, 128 and 128. **Algorithm 1** show the overall flow of our method.

During the training progress, ϵ-greedy search strategy is taken, and ϵ is decayed from 1.0 to 0.01, in the experiment, the max buffer length is set to 2000. During the agent's decision-making process, the flights choose agent first judges a pair of flights with potential conflicts according to the state of the environment. The environment selects the states of potential conflicts flights based on the action of flights choose agent and uses them as part of the input of flights action agent, then flights action agent gives the control strategies of these two flights.

The design of the reward function for the flights choose agent and flights action agent should be consistent with the goal of this paper, which are defined as follows:

$$r_{fca} = 1000/dis_{flights} + 10000/dis_{airport} \tag{3.1}$$

$$r_{faa} = \begin{cases} -\alpha * 0.1 & if\ action\ is\ raise\ height \\ 0.05 & if\ action\ is\ maintain\ height \\ \alpha * 0.1 & if\ action\ is\ descent\ height \end{cases} \tag{3.2}$$

where $dis_{flights}$ presents the distance between two flights with potential conflicts, $dis_{airport}$ presents the minimum distance of these two flights from the airport. α is a flag that represents whether the aircraft is approaching or departing. If the aircraft is approaching, we set $\alpha = -1$, otherwise, $\alpha = 1$. Reward function r_{fca} and r_{faa} is calculated at each time-step. And, once safe separation is not satisfied, or aircraft overpasses the terminal sector boundary, or sector handover condition is not satisfied, reward r_{fca} and r_{faa} will minus 5. If all aircrafts reached their corresponding target positions, reward r_{fca} and r_{faa} will add 5.

In this work, the learning process in one episode is terminated when one of the following four situation is satisfied:

(1) All aircrafts reached their target positions $(x^i_{target}, y^i_{target}, h^i_{target})$ without collision, that is,

$$\sqrt{\left(x_i - x^i_{target}\right)^2 + \left(y_i - y^i_{target}\right)^2 + \left(h_i - h^i_{target}\right)^2} = 0, \forall i \tag{3.3}$$

(2) An aircraft overpasses the terminal sector boundary;
(3) Sector handover condition is not satisfied, that is,

$$\sqrt{\left(x_i - x^i_{target}\right)^2 + \left(y_i - y^i_{target}\right)^2} = 0\ and\ h_i - h^i_{target} \neq 0, \forall i \tag{3.4}$$

(4) Collision is occurred between aircrafts, that is,

$$\sqrt{\left(x_i - x_j\right)^2 + \left(y_i - y_j\right)^2 + \left(h_i - h_j\right)^2} < \delta, \forall i \neq j \tag{3.5}$$

At last, by training the proposed DDQN model until it converges, we can obtain the optimal control strategy for aircrafts in terminal sectors.

4 Experimental Results

The proposed AI agent construction method have been implemented in Python language on a 64-bit workstation (Intel 2.4 GHz, 256 GB RAM).

In this work, the simulator based on Sector 01 of Nanjing terminal is constructed as our air traffic control environment. Figure 3 gives an example of the constructed simulator environment, where there are 10 approach routes and 15 departure routes.

In the experiments, we considered 6 aircrafts with control to evaluate the performance of our reinforcement learning framework. By training the AI agent on around 1200 episodes and choosing a time-step of four seconds, we can obtain the optimal solution for this problem and Fig. 4 shows the experiment results. Figure 4(a) shows the height profile under agent control, and we can see that, these aircrafts can successfully reach their target positions using the control strategy generated by AI agent. Figure 4(b) shows the average loss during training, Fig. 4(c) shows the agents' average scores during training, which shows that the score increases with the number of training episodes. Figure 4(d) shows the number of conflicts during the training, as we can see in this fig, the number of conflicts is decrease to zero at the end of training, which demonstrates the effectiveness of the proposed method.

Algorithm 1 Hierarchical RL Agent

1: Initialize Flights Choose Agent FCA and Flights action Agent FAA;
2: Initialize flights choose score $score1 = 0$ and flights action score $score2 = 0$, Initialize state s_t;
3: Initialize queue Flights Choose Replay Buffer $FCRB$, queue Flights action Replay Buffer $FARB$;
4: **for** $i = 1$ to n **do**:
5:     **while** not is_end **do**:
6:         $a_{fca} = FCA.ChooseAction(s_t)$;
7:         $s_{FlightAction} = ChooseState(s_t, a_{fca})$;
8:         $a_{faa} = FAA.ChooseAction(s_{FlightAction}, a_{fca})$;
9:         $s_{t+1}, r_{fca}, r_{faa}, is\_end = Environment.ExecuteAction(a_{faa})$;
10:         $score1 += r_{fca}, score2 += r_{faa}$;
11:         **if** $FCRB.size() > MaxBuffer$ and $FARB.size() > MaxBuffer$:
12:             update $FCRB, FARB$ parameters
13:         **end if**
14:     **end while**
15: **end for**

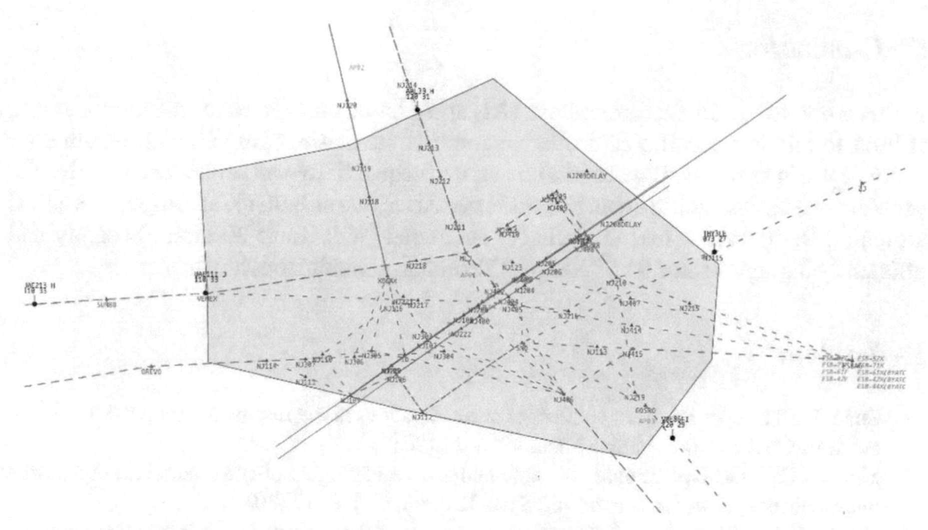

Fig. 3. The simulator environment based on Sector 01 of Nanjing terminal

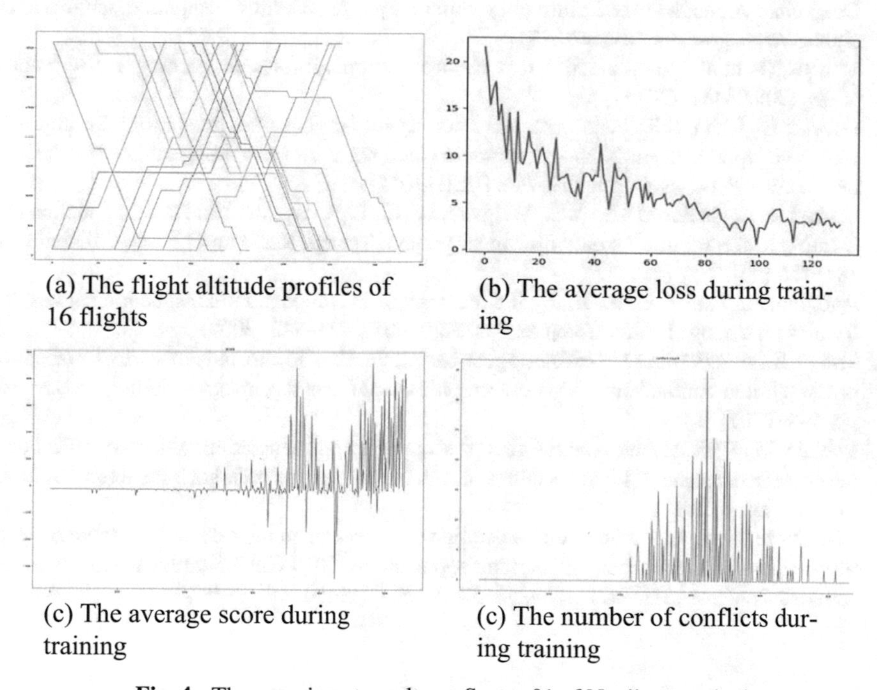

(a) The flight altitude profiles of 16 flights

(b) The average loss during training

(c) The average score during training

(c) The number of conflicts during training

Fig. 4. The experiment results on Sector 01 of Nanjing terminal

5 Conclusion

In this work, two artificial intelligence (AI) agent based on deep reinforcement learning is built to mimic air traffic controllers, such that the dense, complex and dynamic air traffic flows in terminal airspace can be handled sequentially and separated. To solve the problem, the flights choose agent and flight action agent was built to formed a hierarchical structure. Results show that the built AI agent can guide 16 to 20 aircrafts safely and efficiently through Sector 01 of Nanjing Terminal, simultaneously.

References

1. Zhao, W.: The opportunities, challenges and obligations in internationalization of China civil aviation. Civil Aviation Management **09**, 6–11 (2017)
2. Yan, Y., Cao, G.: Operational concepts and key technologies of next generation air traffic management system. Command Inf. Syst. Technol. **9**(3), 8–17 (2018)
3. Ma, X., Xu, X., Yan, Y., et al.: Correlation analysis on delay propagation in aviation network. Command Inf. Syst. Technol. **9**(4), 23–28 (2018)
4. http://www.caac.gov.cn/XWZX/MHYW/201803/t20180315_55771.html
5. Sutton, R.S., Barto, A.G.: Reinforcement Learning: An Introduction (2011)
6. Mnih, V., et al.: Human-level control through deep reinforcement learning. Nature **518**(7540), 529–533 (2015)
7. Deepmind: Alphago at the Future of go Summit, pp. 23–27. http://deepmind.com/research/alphago/alphago-china/ (May 2017)
8. Vinyals, O., et al.: StarCraft ii: A new challenge for reinforcement learning. arXiv preprint arXiv:1708.04782. (2017)
9. George, E., Khan, S.S.: Reinforcement learning for taxi-out time prediction: An improved q-learning approach. In: 2015 International Conference on Computing and Network Communications (CoCoNet), pp. 757–764. IEEE (2015)
10. Cruciol, L.L., de Arruda Jr, A.C., Weigang, L., Li, L., Crespo, A.M.: Reward functions for learning to control in air traffic flow management. Transp. Res. Part C Emerg. Technol. **35**, 141–155 (2013)
11. Agogino, A., Tumer, K.: Learning indirect actions in complex domains: action suggestions for air traffic control. Adv. Complex Sys. **12**(04n05), 493–512 (2009)
12. Tumer, K., Agogino, A.: Distributed agent-based air traffic flow management. In: Proceedings of the 6th International Joint Conference on Autonomous Agents and Multiagent Systems, pp. 1–8 (2007)
13. Brittain, M., Wei, P.: Autonomous aircraft sequencing and separation with hierarchical deep reinforcement learning. In: Proceedings of the International Conference for Research in Air Transportation (2018)
14. Brittain, M., Wei, P.: Autonomous separation assurance in an high-density en route sector: a deep multi-agent reinforcement learning approach. In: 2019 IEEE Intelligent Transportation Systems Conference (ITSC), pp. 3256–3262. IEEE (2019)

Finding Significant Influencing Factors of Core Quality and Ability Development of Teachers Based on Improved Genetic Algorithm

Jian Dang[1], Yueyuan Kang[2], Xiu Zhang[1](✉), and Xin Zhang[1]

[1] Tianjin Key Laboratory of Wireless Mobile Communications and Power Transmission, Tianjin Normal University, Tianjin 300387, China
zhang210@126.com

[2] Faculty of Education, Tianjin Normal University, Tianjin 300387, China

Abstract. The professional development of teachers is crucial for the education development of a country. For understanding the core qualities and ability development of teachers, it is very important to identify the associated influencing factors. In this paper, an improved genetic algorithm is used to find out the significant influencing factors among the many influencing factors of teachers' core qualities and ability development, the idea of the improvement is to introduce elite solution retention strategy and adaptive mutation probability on the basis of traditional genetic algorithm. At the same time, the collected samples are processed by back propagation neural network, and the accuracy of its prediction is taken as the criterion to measure the significant degree of the influencing factors. The experimental results show that the improved genetic algorithm is more effective than traditional genetic algorithm in finding the significant influencing factors.

Keywords: Genetic algorithm · Elite strategy · Adaptive mutation probability · Back propagation neural network

1 Introduction

Education is very important to the development of a country and a nation, and the development of education is a very important standard to measure the development level of a country.

The improvement of individual and overall comprehensive quality is the foundation of a strong country. At present, China's economy is faced with the task of economic transformation. A sufficient and high-quality talent pool can provide a strong driving force for the optimization and transformation of national economic development. There are many scenarios and forms of education, among which school education is an important form of education. The main form of school education is collective learning under the guidance of teachers' organizations, which puts forward high requirements on teachers' professional knowledge level, classroom discipline, education methods, as well as their own moral and cultural accomplishment. An excellent teacher has a great impact on the

Q. Liang et al. (Eds.): Artificial Intelligence in China, LNEE 854, pp. 117–124, 2022.
https://doi.org/10.1007/978-981-16-9423-3_15

comprehensive quality of students. Therefore, if a country wants to improve its education level, it is very important to cultivate the comprehensive quality and ability of teachers. Therefore, it is very necessary to conduct scientific and systematic research on the influencing factors of excellent teachers' core accomplishment and ability development.

As for the influencing factors of excellent teachers' core quality and ability development, many scholars have put forward various views on the factors affecting teachers' professional development from both internal and external aspects. Although many scholars find out a lot of influence factors influencing teachers' professional development, however, we can't intuitively extract the information with practical guiding significance from these complicated and numerous factors, so it is necessary to further analyze the most significant influential factors through technical means, in this way, we can clearly and effectively know the key points of an excellent teacher's core quality and ability, which has important guiding significance for the self-improvement of teachers and the formulation of teacher training programs, and can provide a scientific and effective reference for the development of China's education cause. As an excellent machine learning algorithm, back propagation (BP) neural network can learn samples by establishing a BP neural network model, and then we can test the prediction accuracy of the trained neural network, which can be used as the basis for judging the influence of each group's input factors. As a commonly used optimization algorithm, we can use the excellent large-range optimization ability of genetic algorithm and combine with the prediction results obtained from the neural network model to find out the influential factors with greater influence among many influencing factors. This paper presents an improved genetic algorithm based on the traditional genetic algorithm by introducing the optimal solution retention strategy and the method of nonlinear adaptive adjustment of mutation probability, and applied the improved algorithm to find the most influential factors of teachers' professional development, by comparing the simulation results with the original genetic algorithm, the improved genetic algorithm has more excellent searching ability.

2 Preliminary Basic Works

First, based on the comprehensive consideration of the factors that influence teacher professional development proposed by various scholars at home and abroad, the corresponding theoretical model is constructed and the framework of influencing factors is further constructed. Finally, a total of 43 influencing factors are obtained after strict selection.

Then, questionnaires were prepared according to the framework of influencing factors and the guidance of relevant experts, Likert five-point scale was used in the questionnaire, and each influence factor is divided into five level 1, 2, 3, 4, 5, respectively corresponding to "very small", "small", "general", "large", "very powerful". The evaluation criteria of excellent teachers are: (1) excellent learning and teaching ability; (2) good teacher ethics; (3) play a leading and exemplary role; (4) outstanding performance and contribution; (5) win wide praise and recognition.

Finally, sampling survey was carried out in 31 provinces and cities through questionnaire survey. A total of 1960 valid questionnaires were collected, including 548 excellent teachers and 1421 ordinary teachers, and the proportion of outstanding teachers is 28% [1].

The next work is to use the collected data samples to select the 10 influencing factors with the most significant effect from the 43 influencing factors.

3 Study Method of the Paper

Genetic algorithm is an efficient optimization algorithm with excellent overall search and global optimization capabilities, but the original genetic algorithm also has its inherent limitations, In some optimization problems, it can't achieve very good optimization results, so for specific problems, it is necessary to improve and adjust the original genetic algorithm according to the actual situation of the specific problem so as to achieve the ideal optimization effect. At present, many scholars have proposed various improved algorithms for the traditional genetic algorithm, for example, Shan Shi proposed the method of using cosine function to adjust crossover and mutation operators, which effectively avoided the algorithm falling into the local optimal situation and improved the searching ability of the algorithm [2], Zhou [3], Sang [4], Chen [5], Ma [6], Wang [7], Bai [8], Bai [9] et al. proposed various improved genetic algorithms that adaptively adjusts crossover and mutation probabilities according to population fitness. In this paper, according to the optimization results of the traditional genetic algorithm, the unsatisfactory parts are improved, and the improved genetic algorithm is used to find the relatively significant influencing factors among the numerous influencing factors.

The basic process of genetic algorithm generally includes the design of coding method, the initialization of population, the determination of fitness function, selection, crossover, mutation.

(1) Coding method. The traditional genetic algorithm generally adopt the binary coding method to encode the variables, the problem variable is analogous to the chromosome of an organism, and a set of binary digits is used to represent a problem variable, every 0 or 1 on the binary string is called a gene, although such encoding is very easy to understand, however, this encoding method needs repeated encoding and decoding in the solving process of the problem, which wastes the running time of the algorithm. Meanwhile, this encoding method needs a very long binary string in some problems, which takes up a relatively large memory space, moreover, this encoding method is not accurate enough for some problems. In view of some defects of binary coding, some scholars put forward a real coding method. This encoding method uses a real string within a certain range to represent a problem variable, and its encoding length is much smaller than that of binary coding. Moreover, its gene is one-to-one corresponding to the actual problem, which is easier to understand, and there is no need to carry out repeated encoding and decoding operations in the process of algorithm running, so this paper adopts the method of real number encoding for encoding.

(2) Determination of fitness function. In the genetic algorithm, the function of fitness function is to evaluate the strength and weakness of each individual in the population. For different practical problems, the fitness function is different, which also reflects the universality of genetic algorithm. What this paper will study is the optimization of the factors that influence excellent teachers, among which it is a key issue to judge whether a teacher is an excellent teacher according to some specific factors, for this problem, this paper uses BP neural network to analyze the influencing factors of teachers and then evaluates whether a teacher is an excellent teacher. BP neural network is a machine learning algorithm of supervised learning, the learning process of the algorithm includes the forward propagation process of the signal and the back propagation process of the error, the two processes are repeated alternately, and the weights and thresholds of the neural network are constantly adjusted, so as to fully train the neural network. In this paper, 10 of the 43 impact factors were selected as an individual in the population, and 1500 of the total 1960 samples were selected as the training set to train the neural network, while the remaining 460 samples were used as the test set to test the results of neural network training. Using the trained neural network to evaluate the accuracy of the test sample as the fitness function of the problem, the higher the value of the fitness function is, the better the individual is.

(3) Population initialization. In this paper, 10 influencing factors were selected as an individual from 43 influencing factors, and the population size of each generation population was set as 10.

(4) Select operation. Roulette wheel selection is a common method of selection operation in genetic algorithms, the basic idea is to construct a roulette wheel according to the fitness of individuals in each generation, in which each individual corresponds to a small sector in the roulette wheel, the larger the fitness of the individual, the larger the area of the corresponding sector, and then the individual is selected according to the sector the random number between 0 and 1 falls into. This paper makes some improvements to the traditional selection operation, and added the elite solution reservation strategy [10] on the basis of the traditional selection operation. The optimal individual in each generation of the population will directly enter the next generation without selection operation and will not undergo subsequent crossover and mutation operations until it is replaced by a better individual. In this way, the optimal individuals generated in each generation of the population can be well preserved, which improves the convergence of the algorithm.

(5) Crossover operation. Crossover is to select two individuals from the population and make their chromosomes partially interchange with a certain probability. There are many ways of crossover, and this paper adopts the method of selecting one gene at a time for crossover. This paper set up the population in each generation is 10, the population size is relatively small, in order to allow sufficient interchange of genes between individuals, crossover probabilities are no longer set. The 10 individuals are interchanged in pairs, so that each generation of individuals needs to be interchanged five times. At the same time, in order to avoid the duplication of genes produced by individuals in the crossover process, duplicate checking operation is used to detect

whether the genes to be exchanged produce duplication before crossover, until a group of crossover genes without duplicates is selected or the maximum selection operation is reached.

(6) Mutation. Mutation operation is to select a gene site on the chromosome of an individual in the population to carry out mutation with a certain probability of mutation. The mutation probability in the traditional genetic algorithm is a fixed value, in this way, if the early mutation probability is relatively small, the relatively large richness of the population can't be guaranteed, and the algorithm is easy to fall into the local optimum and reduce the global search ability of the algorithm, and if the mutation probability of the later algorithm is too large, it will lead to the destruction of the excellent individuals, and at the same time, it will cause the disadvantage of low convergence of the algorithm. Aiming at this situation of traditional algorithms, this paper proposes a method to adjust the mutation probability adaptively according to the population fitness. Its basic idea is as follows:

$$P_m = \begin{cases} 0.1 & if\ f_{\max} - f' \geq 0.05 \\ 2 * (f_{\max} - f') & other \end{cases} \quad (1)$$

In (1), $f_{\max}$ represents the maximum fitness of the population, and f' represents the fitness of the individual to be mutated at present. The reason why the dividing line value of 0.05 is chosen is that the accuracy of neural network prediction is mainly concentrated in the relatively small range of 0.65–0.75 through analysis of the simulation experimental data. In the second part of the formula, the reason why the multiplication coefficient 2 is chosen is that when the crossover probability is just transferred to the second stage, the difference between the fitness of the current variation individual and the maximum fitness of the population is about 0.05, which can expand to 0.1 after 2 times, so that's close to the crossover probability of the previous stage and the adaptive adjustment of mutation probability is smooth and won't produce large mutation. When the fitness difference between the individual to be mutated and the optimal individual in the population is relatively large, a fixed mutation probability is selected to increase the possibility that the inferior individual can improve the current fitness through mutation, at the same time, the larger mutation probability is also beneficial to improve the search range of the population, and avoid the precocity of the population and fall into local optimum. At the middle and late stage, the population gradually converges, and the degree of difference between individuals of the population gradually decreases, as the number of iterations is updated, the value of $f_{\max} - f'$ in each generation of the population will get smaller and smaller until it finally approaches 0, in this way, the mutation probability will be reduced adaptively, and the smaller the difference between the fitness value of the individual and the optimal individual is, the smaller the mutation probability will be, this would prevent the genes of good individuals from being destroyed and the convergence ability of the algorithm can be improved.

4 Simulation Results

In order to verify the effectiveness and superiority of the improved genetic algorithm proposed in this paper in finding the significant influencing factors of teacher professional development, this paper makes a comparison with the traditional genetic algorithm, the simulation experiment is realized by Matlab programming. The mutation probability in the traditional genetic algorithm is set to a fixed value of 0.1, due to the relatively small population size setting, and considering that the population becomes more and more convergent in the later iteration, fewer and fewer individuals can be crossed in order to avoid the repetition of individual genes, to ensure the effectiveness of the crossover operation, the crossover probability is no longer set, and the default value is 1. The iterations of the improved genetic algorithm and the original genetic algorithm are set to 100 times, the optimization results of the two algorithms were shown in the table below:

Table 1. Comparison of optimization results of two algorithms

Algorithm	Optimization results	Accuracy
Original genetic algorithm	1 10 11 13 22 23 24 33 40 43	73.26%
Improved genetic algorithm	3 7 18 19 21 24 31 39 41 43	75.43%

The simulation results in Table 1 show that the prediction accuracy of the improved genetic algorithm is significantly higher than that of the traditional genetic algorithm.

Table 1 shows the optimization results of the original genetic algorithm and the improved genetic algorithm after 100 iterations respectively, the data in the table reflects the superiority of the improved genetic algorithm in finding the global optimal solution. As mentioned above, the improved genetic algorithm is obviously better than the original genetic algorithm in terms of algorithm convergence, in order to reflect the superiority of improved genetic algorithm in convergence and we draw the curve of population richness with the number of iterations of the two algorithms. Population richness is the number of distinct genes contained in 10 individuals of a generation population. The stronger the convergence of the algorithm is, the more the population richness approaches 10 with the increase of the number of iterations. In the subject studied in this paper, the maximum population richness is 43. The curve of the population richness of the two genetic algorithms with the number of iterations is recorded in Fig. 1.

Through comparing two curves of Fig. 1, it can be seen that the improved genetic algorithm not only converges faster in the early stage, but also in the later period of species the richness of the improved genetic algorithm is closer to 10, and it tends to be stable. The original genetic algorithm in the later richness is much higher, and its richness is not very stable. This directly reflects the superiority of the improved algorithm in terms of convergence.

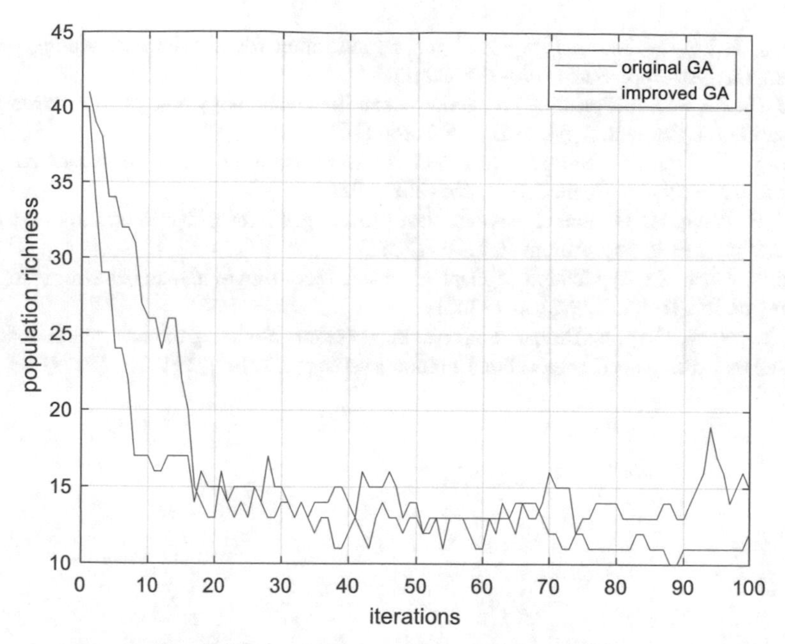

Fig. 1. The variation curve of population richness

5 Conclusion

In this paper, an improved genetic algorithm is proposed based on the traditional genetic algorithm by introducing the elite solution retention strategy and adaptive mutation probability, and we compare the improved algorithm with the traditional algorithm, the experimental results show that the improved genetic algorithm is more effective than the traditional genetic algorithm in finding the significant influencing factors.

Acknowledgements. This research was supported in part by the National Natural Science Foundation of China (Project No. 61901301), in part by the key project of the National Social Science Foundation of China (Project No. AFA70008), and in part by the Tianjin Higher Education Creative Team Funds Program.

References

1. Kang, Y., Li, J.: Research on the Factors of core qualities and ability development of teachers: an empirical investigation based on questionaires nationwide. Contemp. Teac. Educ. **12**(04), 17–24 (2019)
2. Shi, S., Li, Q., Wang, X.: Design optimization of brushless direct current motor based on adaptive genetic algorithm. J. Xi'an Jiaotong Univ. **12**, 1215–1218 (2002)
3. Zhou, Z., Chen, P.: 3D path planning of UAV based on improved adaptive genetic algorithm. J. Project. Rockets Missiles Guidance (2021, to appear)
4. Sang, H., Song, S., Xing, X., Meng, Y., Zhang, Z., Tang, M.: Application of adaptive genetic algorithm in robot path planning. J. Xi'an Polytechnic Univ. **35**(01), 44–49 (2021)

5. Chen, J., Ma, L., Ma, L.: Improved genetic algorithm for job shop scheduling problem. Comput. Syst. Appl. **30**(5), 190–195 (2021)
6. Ma, X., He, F.: Label printing production scheduling technology based on improved genetic algorithm. J. Comput. Appl. **41**(03), 860–866 (2021)
7. Wang, C., Pang, X.: Software unit security test simulation based on improved genetic algorithm. Comput. Simul. **38**(01), 265–268 (2021)
8. Bai, P., Wang, H.: BP neural network blast furnace gas forecast based on improved genetic algorithm. Mech. Eng. Autom. **2**, 77–79 (2021)
9. Bai, Y., Chen, Z.: Application of adaptive genetic algorithm in human behavior recognition. Comput. Inf. Technol. **29**(2), 4–7 (2021)
10. Li, Y., Ye, H., Yang, Y.: Research on customer service scheduling of construction machinery based on hybrid genetic algorithm. Comput. Syst. Appl. **28**(07), 191–198 (2019)

A New Augmented Method for Processing Video Datasets Based on Deep Neural Network

Wei Wang[1,2], Haiyan Wang[1,2(✉)], and Fuchuan Ni[1,2]

[1] College of Informatics, Huazhong Agricultural University, Wuhan 430070, China
wanghaiyan@mail.hzau.edu.cn

[2] Hubei Engineering Technology Research Center of Agricultural Big Data, Wuhan 430070, Hubei, China

Abstract. Large datasets are required for deep learning to achieve good performance. However, there is a lack of sufficient training datasets in many research fields, which may become a shortcoming of computer vision applications. This article provided a new data augmentation method for making training small datasets, which could be divided into two steps: 1. Unbalanced sampling based on information density. 2. Splicing images to form a dataset. Different information density dataset combinations had been used for testing the model generalization. The enhanced loss function which consisted of label smoothing loss and cross-entropy loss had been used to minimize the model preference during training models. Finally, with the same amount of data, the Mean Absolute Error (MAE) of the model with our sampling method could get 55% increase compared with the traditional sampling method. The best MAE could reach 0.98 if the splicing method had been adopted. The results showed that this augmented method was suitable for scenarios with small sample size, especially video datasets. To get the best performance, the splicing method was a nice choice to optimal model generalization performance.

Keywords: Sampling methods · Video datasets · Deep learning · Small samples · Data augmentation

1 Introduction

In recent years, deep learning methods have been widely deployed in agricultural fields, such as yield estimation, plant disease detection, fruit counting [1]. According to a recent survey [1], there is no fast, accurate and lightweight methods to be mentioned for counting livestock. However, rapid and efficient counting methods have always been an urgent need in production, which can help farmers cost less for real-time monitoring pig numbers in each pen. Accurate counting is a very complicated problem because of pigs overlapping, occlusion and deformable shapes. Manual counting easily misses or repeats several pigs. So, counting pig numbers is time-consuming and expensive endeavor [2].

Q. Liang et al. (Eds.): Artificial Intelligence in China, LNEE 854, pp. 125–134, 2022.
https://doi.org/10.1007/978-981-16-9423-3_16

Some researchers had applied deep learning methods to count animal populations based on aerial photography [3–5]. Tian et al. [6] had achieved 42 ms per picture that adopted a modified CNN and combined ResNeXt framework to deal with counting problem. Baweja et al. [7] had achieved R-squared correlation of 0.88 for stalk count. Chen et al. [8] developed a pipeline for pig counting but it still needs to label each pig in the pen. [6] had amplified 373 images into millions of patches to avoid overfitting and achieved 1.67 MAE per image. However, these methods focused more on the number of images rather than the images themselves. This neglect would cause the relationship between the objects of interest to be lost. To avoid overfitting, different methods would be adopted in deep learning, such as flipping, cropping rotation and etc.[9]. Deep learning augmented methods such as feature space augmentation[10] and adversarial training[11] are also very popular in data augmentation.

Here, we proposed an augmented method for deep learning. All original datasets had been divided into frames which were labeled by pig numbers. These frames had been just labeled by their numbers but not including positions. This labelling method could greatly reduce the amount of work required for marking. Because of the small sample datasets, we proposed a new sampling method to balance the original dataset and formed a dataset as original dataset for following augment. Then, two enhanced ways had been adopted to make the trained model more generalized: pixel combination and splicing method. The two augmented methods had been applied into the dataset to get four training datasets. The random erasure is to make the model not depend on the entire target object, and the original shape can be inferred from part of the body, thereby improving the generalization ability of the model. Basic amplifying methods such as random flipping and random erasure had been used for training. The enhanced loss function including cross entropy loss and label smoothing loss had been used as the loss function to get better generalization. At last, a model with strong generalization ability and good performance on small sample data sets is obtained.

2 Experiments Details

2.1 Datasets Augment

The open source dataset originating from Psota et al. [12], which contained 15 different daily videos of pigs, had been used for analyzing. We had reduced the requirements for labeling data, and replaced the most time-consuming and labor-intensive labeling with a labeling method that only provided quantity but not a specific location. For every frame, there was only one quantity as a unique identifier. Therefore, the neural network was expected to give the same result as the data annotation. These 15 datasets with a very strong data preference, which contained only 7 quantity categories, would make themselves to be unbalanced.

High-activity videos were considered to have richer information content than that of the low-activity videos. We divided all videos into three levels according to information density: high, medium and low and they corresponded to dense information density, medium information density and sparse information density, respectively. The rule to determine the information density of a video was the number of active pigs. The ratio of three levels was 1:2:2. In other words, fifteen original videos had been divided to three high information density videos, six medium information density videos and six low information density videos.

To alleviate the imbalance of the data, we sample the data in different ways according to the dataset information density. Therefore, in the sampling process, we set the sampling ratio of high-activity, medium-activity, and low-activity as 6:3:1.

2.1.1 Sampling from Videos

All videos have been separated into frames. Three different ways are used to generate datasets.

1. All frames are chosen one for every ten frames (111 dataset).
2. The frames are selected according to high information density: medium information density: low information density referring to 6:3:1 (631 dataset).
3. All frames are chosen one for every ten frames but just select the same number with the 631 dataset (111 short dataset).

2.1.2 Methods to Augment the Original Dataset

We also had tried two ways to amplify the 631 dataset.

1. One image was chosen from the training dataset as the basic image, and then another image was selected as a candidate image to combine. We selected each pixel at the possibility of 0.5 for basic image and candidate image to synthesize one new image. The label of the basic image was set as label of the new synthesized image [13]. Finally, the number of the original dataset was selected to randomly pick synthesized images (Combination dataset).
2. Four pictures were randomly picked from the training dataset. Then, the four pictures were randomly spliced into a new picture. The label of new picture was the sum of the original four image labels. Finally, all new synthesized images were resized to the original image size and the same amount as the original dataset were randomly chosen to form a new dataset (Mixture dataset).

2.1.3 Obtaining Final Training Datasets

We combined two newly datasets with the 631 dataset thus to get four augmented datasets:

1. Raw training dataset (Raw).
2. Raw training dataset + the combination dataset (Com).
3. Raw training dataset + the mixture dataset (Mix).
4. Raw training dataset + the combination dataset + the mixture dataset (Rcm).

These mixed datasets had been sampled equally to ensure the same size as Raw dataset.

2.2 Training Methods

We used mobilenetv3 [14] as the neural network skeleton. Label smoothing was a common method to training a deep learning model when faced with a multi-classification task [15]. In order to alleviate the problem of category imbalance, we used the label smoothing loss and the cross-entropy loss as the loss function, with the toolbelt [16] to combine, and evaluated the double loss function in a 1:1 manner.

When we training the dataset, we used randomly horizontal flip to augment dataset and randomly erasing [17] to improve the accuracy of occlusion recognition to prevent model overfitting. All images had been resized to 198*112 from 2688*1520 throughout the training and testing process. Two datasets had been selected for test and the rest chosen for training.

3 Results

We compared the difference between the sampling method and the sequential sampling method according to the information density under the same data size. respectively. The result has been shown in Fig. 1. The result showed MAE of the 111 short dataset was 7.32 but the MAE of 631 dataset was 3.29.

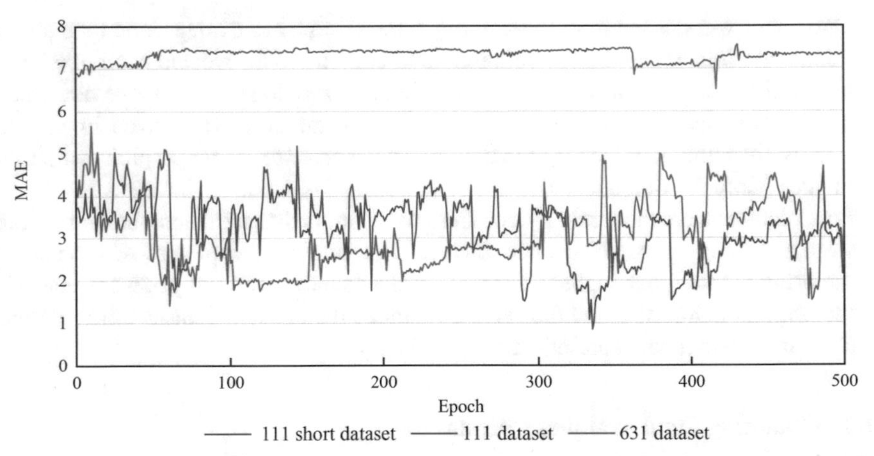

Fig. 1. The MAE of three datasets. The MAE of 111 short dataset is significantly higher than the rest two datasets. The MAE of 111 dataset is similar with the 631 dataset but the 111 dataset is composed of 8797 images where the 631 dataset is 3232.

The MAE of the average sampling method under the same data volume is significantly higher than that of the sampling method based on the information density. In order to achieve a MAE level similar to the original data, the amount of data is nearly as three times as required, that means longer training time and more resources.

In Fig. 2, we plot the test loss between the models with label smoothing and without it. Obviously, compared with the model that does not use label smoothing and only uses

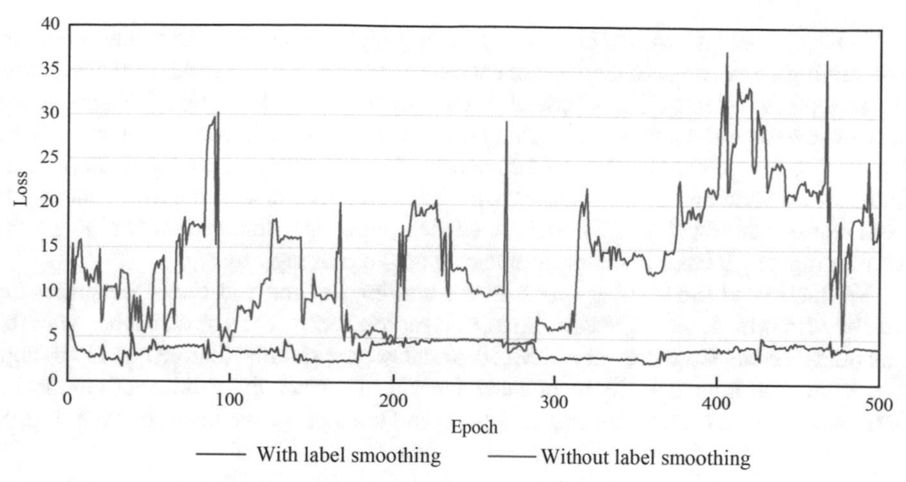

Fig. 2. The loss of the model with label smoothing and not. The model with label smoothing achieves lower loss when compared with the model just using cross-entropy loss as loss function.

cross-entropy loss, as the number of model training increases, the overall loss of the model with label smooth is better than the model that only uses cross-entropy loss.

To ensure the training dataset perform best, two training methods were adopted here:

- Training way a: Training the model with the same target datasets throughout all epochs.
- Training way b: Training the model with the original dataset for 50 epochs then followed by 350 epochs using the target dataset and using the final 100 epochs of the original dataset for fine-tuned training.

The MAE results for three mixed datasets have been shown in Fig. 3.

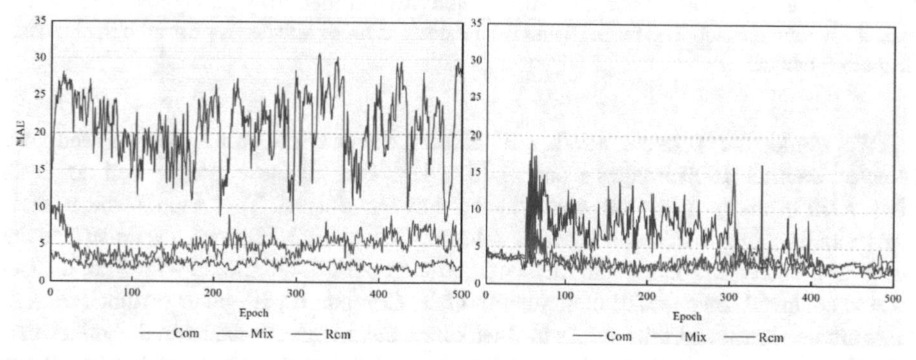

Fig. 3. Comparation with training way a and training way b. The left picture is the MAE of training way a and the right is training way b.

In Fig. 3, the MAE of Mix dataset is obviously higher than other augmented datasets. For training way a, none of augmented datasets converges and the MAE of these dataset can be separated clearly. The Mix dataset is the highest MAE throughout 500 epochs and the Com dataset is the lowest. For training way b, three datasets have similar performance for the first 50 epochs and the last 100 epochs. But the Mix dataset experienced large fluctuations when augmented dataset was added. In contrast with the Mix dataset, the Com dataset stabilized quickly after experiencing the early fluctuation. On the whole, the training way b has a better performance than the training way a.

We have used the training way b as the training method and Fig. 4 illustrates the change of training losses. The Rcm dataset has more effect on the loss than other datasets. The models using Com dataset and Mix dataset become stabilized more quickly though they have been influenced by augmented datasets. To check the validity of the model, four datasets have been tested and the MAEs and the losses have been shown in Fig. 5.

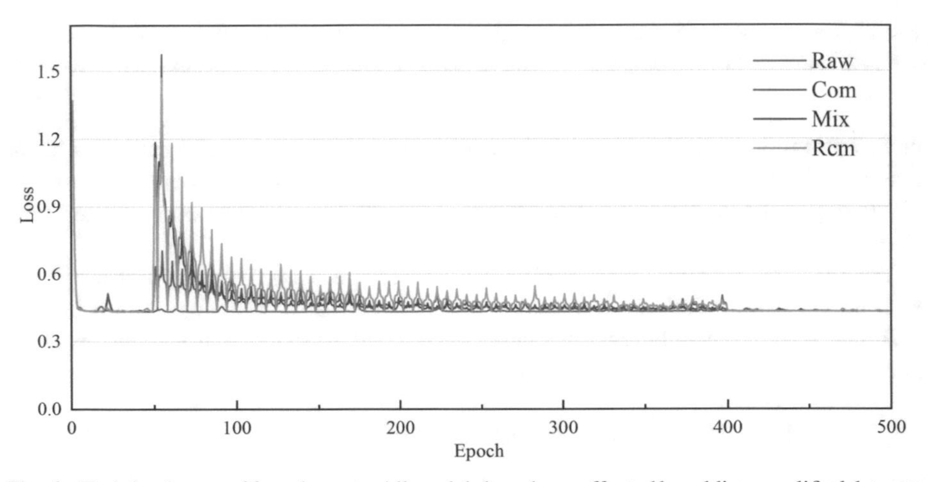

Fig. 4. Training losses of four datasets. All models have been affected by adding amplified datasets except for Raw dataset. The model using Rcm dataset is more affected by the amplified dataset than other models.

We would like to know whether all training datasets are equal. So, we randomly selected another 1000 images from all 15 videos that all these images had not been chosen for training or test and made them a new test dataset. Then we let four models that trained by four different datasets test the new test dataset. Figure 6 has illustrated the result of the new test dataset. In the last fine-tuned period, the model with Rcm dataset trained achieved lower MAE than other models. Compared with other results in Fig. 5, there are no significant differences in whether the test images come from the videos that some images have been trained or not. This result implies that the models have strong generalization ability.

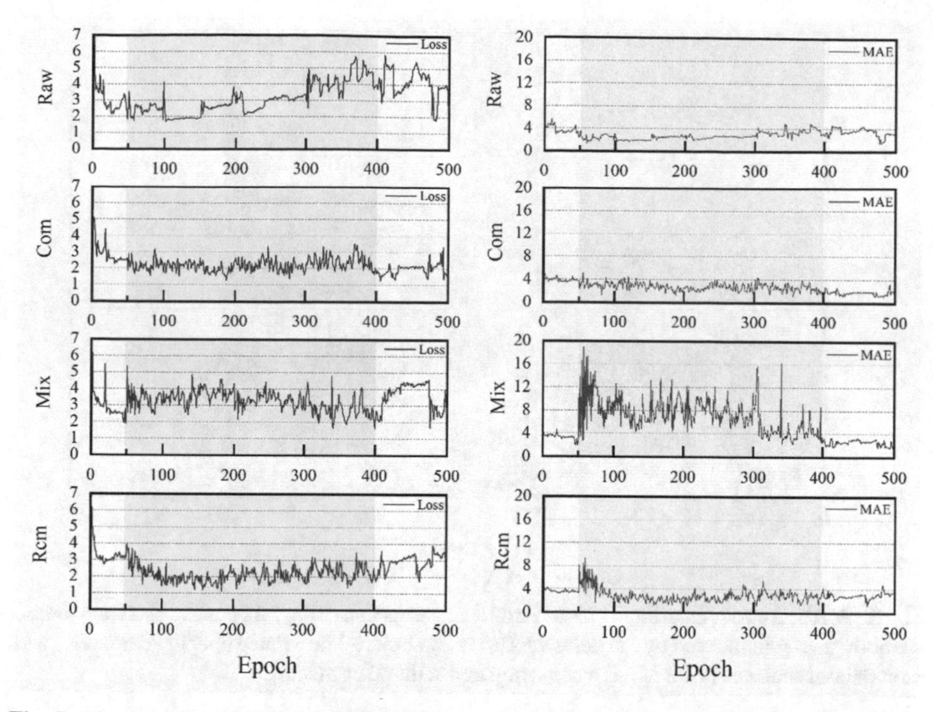

Fig. 5. The losses and MAEs of the four datasets. The left four images show the losses of Raw, Com, Mix, Rcm datasets while the right show the MAEs. The MAE of the model that using Mix dataset has been greatly affected by the augmented dataset and the MAE of the model that using Rcm dataset is slightly influenced by the dataset while the datasets nearly have not effect on the models that using Raw dataset and Com dataset

In order to check whether the model has a preference for the dataset, four kinds of datasets were selected to test the model compatibility according to the information density, which were: high and high, low and low, low and high, medium and medium information density datasets. In other words, four different information density dataset combinations were adopted as the test data set. RMSE and MAE were chosen to measure the model compatibility and they were calculated by the average of last five epochs.

As is shown in Table 1, the RMSE and MAE of the dataset composed of two low information density videos all significantly above other test datasets. The model with Mix dataset obtains the best RMSE and MAE. If the information density of the test data is balanced, the data processing using the mix method will achieve better results (MAE of 0.98), and it has better performance in a variety of test environments (top 1 for tow low information density datasets and top 2 for the rest). The dataset obtained by the pixel combination method is not the best in the 4 test methods, but it is not the worst. To our surprise, the mixed dataset Rcm formed by both amplification methods has the worst performance in all dimensions, which shows that the two amplification methods do not have the function of coordinating and increasing the stability of the model.

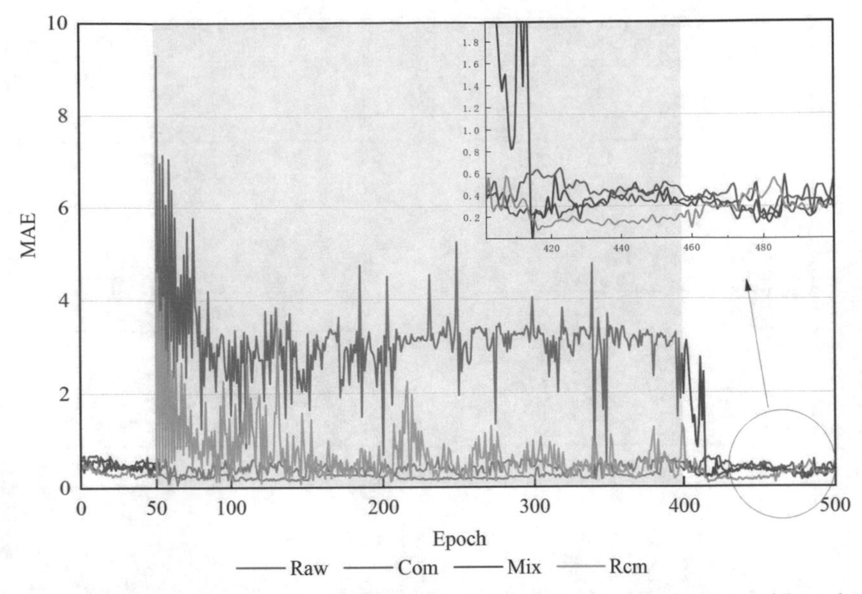

Fig. 6. MAEs for four datasets. The MAEs of the models that using Mix dataset and Rcm dataset are both greatly influenced by the datasets. But in the last 100 fine-tuned epochs, the model with Rcm dataset achieves lower MAE when compared with other models.

Table 1. RMSEs and MAEs of four datasets

Dataset	Type	RMSE	MAE
High + High	Com	2.881125	1.554313
	Mix	2.81124	1.463338
	Raw	3.506515	1.98976
	Rcm	4.523978	2.578607
Low + Low	Com	4.892533	4.180384
	Mix	4.884242	4.026753
	Raw	4.396287	3.337739
	Rcm	5.890768	5.313681
Low + High	Com	3.875839	1.993519
	Mix	3.718505	2.223834
	Raw	4.674497	3.130053
	Rcm	5.011536	3.49456
Medium + Medium	Com	3.027788	1.865885
	Mix	1.865799	0.980143
	Raw	2.968811	1.767904
	Rcm	3.089548	1.963867

4 Conclusion

In this article, we provided a new augmented method to visualize the prior knowledge and applied it to the video datasets. These information density sampling method was compared with the traditional data sampling methods. Sampling according to the information density could greatly improve the accuracy of the model. Our sampling method achieved 55% lower MAE than traditional sampling method. Based on traditional data amplification, two new the data processing methods, pixel mixing amplification and image splicing amplification, had been compared respectively. When the original data category was severely unbalanced, label smoothing had significant influence on the network. The spliced datasets from original dataset behaved itself and achieved good performance than other datasets. The generalization performance of the model which was trained by the Mix dataset is better than other models.

Obtaining the original dataset can be easier for just requiring the number of the pigs rather than labeling where is each pig which greatly reduces the marking workload of counting tasks. The cost for training tasks will decrease to benefit the management of current farms and agricultural production. The method that we proposed improved the performance of the model and gave a reference for actual problems in production. However, for unrecognized images, the accuracy of model recognition still drops to a certain extent. The accuracy and generalization of the model can still be improved by two ways: multiple data augmentation methods to meet the needs of multi-scale identification of objects, and achieve a better adaptive robustness for new environment.

Acknowledgements. This work is supported by China Scholarship Council (No. 201906765023) and Hubei Chenguang Talented Youth Development Foundation.

References

1. Kamilaris, A., Prenafeta-Boldú, F.X.: Deep learning in agriculture: a survey. Comput. Electron. Agric. **147**, 70–90 (2018)
2. Tianhao, Z., Yansen, L., Zhiyi, H.: Applying image recognition and counting to reserved live pigs statistics. Comput. Appl. Software. **12** (2016)
3. Bruijning, M., Visser, M.D., Hallmann, C.A., Jongejans, E.: Trackdem: automated particle tracking to obtain population counts and size distributions from videos in r. Methods Ecol. Evol. **9**(4), 965–73 (2018)
4. Chabot, D., Dillon, C., Francis, C.: An approach for using off-the-shelf object-based image analysis software to detect and count birds in large volumes of aerial imagery. Avian Conserv. Ecol. **13**(1) (2018)
5. Norouzzadeh, M.S., et al.: Automatically identifying, counting, and describing wild animals in camera-trap images with deep learning. Proc. Natl. Acad. Sci. **115**(25), E5716–E25 (2018)
6. Tian, M., Guo, H., Chen, H., Wang, Q., Long, C., Ma, Y.: Automated pig counting using deep learning. Comput. Electron. Agric. **163**, 104840 (2019)
7. Baweja, H.S., Parhar, T., Mirbod, O., Nuske, S.: Stalknet: a deep learning pipeline for high-throughput measurement of plant stalk count and stalk width. In: Hutter, M., Siegwart, R. (eds.) Field and Service Robotics, pp. 271–284. Springer International Publishing, Cham (2018). https://doi.org/10.1007/978-3-319-67361-5_18

8. Chen, G., Shen, S., Wen, L., Luo, S., Bo, L. (eds.): Efficient pig counting in crowds with keypoints tracking and spatial-aware temporal response filtering. In: 2020 IEEE International Conference on Robotics and Automation (ICRA). IEEE (2020)
9. Shorten, C., Khoshgoftaar, T.M.: A survey on image data augmentation for deep learning. J. Big Data **6**(1), 1–48 (2019)
10. DeVries, T., Taylor, G.W.: Dataset augmentation in feature space. arXiv preprint arXiv:170205538. (2017)
11. Moosavi-Dezfooli, S.-M., Fawzi, A., Frossard, P. (eds.): Deepfool: a simple and accurate method to fool deep neural networks. In: Proceedings of the IEEE Conference on Computer Vision and Pattern Recognition (2016)
12. Psota, E.T., Mittek, M., Pérez, L.C., Schmidt, T., Mote, B.: Multi-pig part detection and association with a fully-convolutional network. Sensors. **19**(4), 852 (2019)
13. Inoue, H.: Data augmentation by pairing samples for images classification. arXiv preprint arXiv:180102929. (2018)
14. Howard, A.G., et al.: Mobilenets: Efficient convolutional neural networks for mobile vision applications. arXiv preprint arXiv:170404861. (2017)
15. Lukasik, M., Bhojanapalli, S., Menon, A., Kumar, S. (eds.): Does label smoothing mitigate label noise?. In: International Conference on Machine Learning. PMLR (2020)
16. BloodAxe. pytorch-toolbelt. (2021)
17. Zhong, Z., Zheng, L., Kang, G., Li, S., Yang, Y.: Random erasing data augmentation. Proc. AAAI Conf. Artif. Intell. **34**(07), 13001–13008 (2020). https://doi.org/10.1609/aaai.v34i07.7000

Small-Object Detection with Super Resolution Embedding

Wenyi Tang(✉), Qiucheng Xu, Hui Ding, Lianghao Wu, and Yanyang Shi

State Key Laboratory of Air Traffic Management System and Technology, Nanjing Research Institute of Electronic Engineering, Nanjing 210007, China
tangwenyi@cetc.com.cn

Abstract. The detection performance of ground vehicles on the airport surface in video-sensor based air traffic control surveillance has not been satisfactory compared to larger planes, especially in panoramic images of remote tower applications. Inspired by the success of GAN-based super resolution and oversampling augmentation methods, we apply a new super-resolution network to resize cropped small samples and augment each of those images by resize-and-pasting small objects many times. It allows us to trade off the quantity of the detector on large objects with that on small objects. We propose an architecture with two components: ESRGAN and Detection network. We use residual dense blocks for ESRGAN, and for the detector network, we use one state-of-the-art detector (YOLOv3). Extensive experiments on a public (car overhead with context) dataset and another self-assembled airport surface dataset show superior performance of our method compared to the standalone state-of-the-art object detectors.

Keywords: Object detection · Super resolution · Data augmentation

1 Introduction

Recently, high-density air traffic is becoming a challenge to airport operation. Traditionally, airport surface surveillance methods are based on radar, whose performance suffers from ground clutter, occlusion and multipath effect. In addition, MLAT and ADS-B could only recognize cooperative target. As a result, there is increasingly demand to video-based surveillance at airport surface movement control, which explores in particular the potential of artificial intelligence (AI) and deep learning methods to detect and recognise non-cooperative targets operating on the airfield.

There are several major factors which should be taken into consideration during modeling the detection algorithm. Most of the airfield scenes comprises of a varieties of vehicles with extremely small in size and several relatively larger planes. Moreover, they have variable shapes, multiple scales, orientations, and with complex background which lead to interclass similarities between target and nontarget objects.

Q. Liang et al. (Eds.): Artificial Intelligence in China, LNEE 854, pp. 135–143, 2022.
https://doi.org/10.1007/978-981-16-9423-3_17

To improve the detection of small objects, an approach that draws attention of researchers is to perform super-resolution (SR) to increase the spatial resolution of the images (and thus the size and details of the objects) before performing the detection task. To deal with the lack of details in the low-resolution images, the latest neural network super-resolution (SR) techniques such as CNN-based SR (SR-CNN) [1], Enhanced Deep Residual Super-resolution (EDSR) [2] etc. aim at significantly increasing the resolution of an image much better than the classical and simple bicubic interpolation.

Those aforementioned detectors, which use highly-deep convolution layers, simply ignore the lack of training samples in small object detector training phase. It should be noted that there are relatively fewer images that contain small objects in the dataset, which potentially biases any detection model to focus more on the types of medium and large objects. Secondly, the area covered by small objects is much smaller, implying the lack of diversity of small objects. We conjecture this makes it difficult for the object detection model to generalize to small objects in the test time when their appearance changes.

It is quite straightforward for us to tackle the first issues by copy-pasting those small objects as [3]. The second issue is addressed by high quality resize-and-paste small objects multiple times in each image containing small objects. When pasting each object, we ensure that pasted objects do not overlap with any existing object and increase the resolution of small objects with super resolution methods. This increases the diversity in the locations and scales of small objects while ensuring that those objects appear in correct context.

The contribution of our method can be summarised as:

(1) We propose a super resolution based resize-and-paste algorithm to augment the small objects in the detector training phase.
(2) We propose a novel detector training framework with improved data augmentation, which has better performance in small object detection.

The rest of the paper is organized as follows. In Sect. 2, we first review related work. The details of the proposed method are illustrated in Sect. 3. In Sect. 4, we would presents and discuss the experimental results on two detection benchmark, a public (car overhead with context) dataset and a self-assembled airport surface dataset. Section 5 provides conclusions.

2 Related Work

In this section, we briefly present different deep learning-based methods for super-resolution and detection before focusing on the proposed architecture.

Super Resolution. As mentioned in our Introduction, the two articles [4] give a relatively complete overview of super-resolution techniques based on deep learning. A neural network specialized in super-resolution receives as input a low-resolution image LR and its high-resolution counterpart HR as reference. The network outputs an detail-enhanced higher resolution image $SR = f(LR)$ by minimizing the gap between $f(LR)$ and HR. The simplest architectures are

CNNs consisting of a stack of convolutional layers followed by one or more pixel shift layers. A rearrangement layer allows a change in dimension of a set of layers from the dimension (B, Cr^2, H, W) to (B, C, Hr, Wr), where B, C, H, W represent the number of batches, number of channels, the height and the width of the feature maps respectively and r is the up-sampling factor. Many studies such as [1] already showed a clear improvement of the output image, compared to a classical and simple solution using the bicubic interpolation. An improvement of these networks is to replace the convolutional layers with residual blocks. Part of the input information of a layer is added to the output feature map of that layer. We focus here on the EDSR approach [2] which proposes to exploit a set of residual blocks to replace simple convolutional blocks.

Small Object Detection. Some efforts have been made to tackle the small object detection task by adapting the existing detectors. In [5], Deconvolutional R-CNN was proposed by setting a deconvolutional layer after the last convolutional layer in order to recover more details and better localize the position of small targets. This simple but efficient technique helped to increase the performance of ship and plane detection compared to the original Faster R-CNN. In [6], UAV-YOLO was proposed to adapt the YOLOv3 to detect small objects from unmanned aerial vehicle (UAV) data. Slight modification from YOLOv3 was done by concatenating two residual blocks of the network backbone having the same size.

3 Super Resolution Embedding Augmentation Training

We propose an detection training framework based on super resolution resize-and-paste data augmentation. The proposed framework mainly follow Kisatntal's work [3], which oversample and paste small objects in the training dataset. We would provide an high-level summary of our approach as shown in Fig. 1, leaving the additional details about the more important steps later. Before the training phase, we build the super resolution training pairs from small objects which are cropped in the training frames. Given the proper fine-tuned super resolution model, we resize the small object in different scale and paste the synthetic objects in the original frame.

3.1 Super Resolution Embedding Network

We propose a improved EDSR network, namely ESRGAN. ResNet [7] is proposed by He et al. which suppresses gradient vanishing of deep CNN.

Based on the EDSR network, We replace activation function ReLU [8] with Parametric Rectified Linear Unit [9], which adaptively learns the parameters of the rectifiers and improves accuracy at negligible extra computational cost. Further, we remove residual scaling module [10] from EDSR residual blocks, which plays an important role in making the training procedure numerically stable, as the number of feature maps in EDSR [2] comes up to 256. ESRGAN does not need such trick as it just has 64 filters in each convolution layer (Fig. 2).

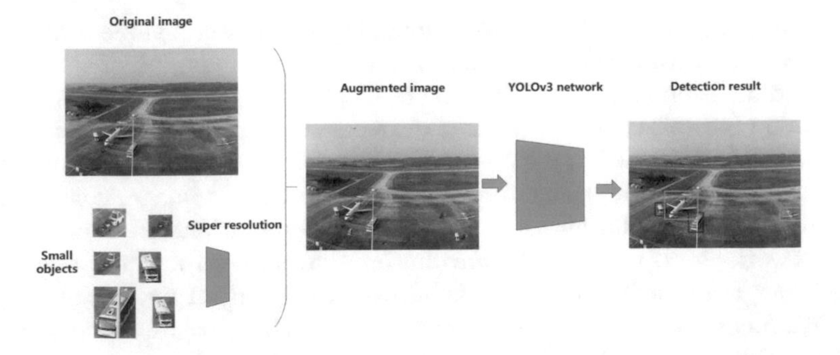

Fig. 1. The flowchart of small object detection based on super resolution embedding data augmentation in airport surface images

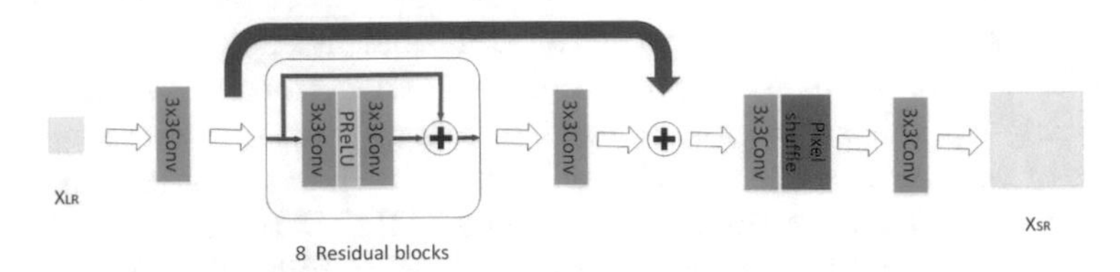

Fig. 2. The network architecture of our prediction network (ESRGAN). It has one 64-filter convolution layer before eight identical residual blocks. Each residual block has two 64-filter convolution layers and a PReLU between them at its residual path. The bending arrow inside residual block is the identity mapping path. One 64-filter convolution layer is appended after residual blocks. An identity mapping the bending arrow) connects the layer before residual blocks and the second layer after eight residual blocks. All the convolution layers above is in 3×3 kernel size. The layer before last convolution layers is combined by one 256 3×3 filters and one pixel shuffle layer. The last layer's filter number agrees to frame channel, which is three for RGB color video.

To make the scaled object consistent with its ground truth under pixel and perceptual criteria, intensity, perceptual constraints are used.

The mean square error (MSE) intensity loss L_{pixel} guarantees the pixel-wise similarity in the pixel RGB space. Specifically, we minimize the $l2$ pixel distance between the resized frame X_{SR} and its ground truth X_{HR} as follows:

$$L_{pixel}(X_{HR}, X_{SR}) = ||X_{HR} - X_{SR}||_2^2 \tag{1}$$

According to Yang [11], the poor perceptual quality of the images obtained by optimizing MSE directly demonstrates a fact: difference in pixel intensity space is not Gaussian additive noise. To tackle this problem, solution is proposed that transferring the image to some feature space where the difference is closer to Gaussian noise. The perceptual loss is designed to optimize network by minimizing the MSE in the feature space produced by VGG-16 [12] as follows:

$$L_{pl} = ||\Psi(X_{SR};\theta) - \Psi(X_{HR})||^2 \tag{2}$$

where Ψ refers to feature maps at layer conv2_2 in our paper, which is set according to experimental results. Generally speaking, successful supervised networks used for high-level tasks can produce very compact and stable features. In these feature spaces, small pixel-level variation and many other trivial information can be omitted, making these feature maps mainly focusing on human-interested pixels. At the same time, with the deep architectures, the most specific and discriminative information of input are retained in feature space because of the great performance of the network in various tasks. From this perspective, using MSE in perceptual feature space will focus more on the parts which are attractive to human observers with little loss of original contents, so perceptually-pleasing predicted frame can be obtained.

3.2 Adversarial Training

Generative adversarial network was introduced by Goodfellow [13], where images patches are generated from random noise using two networks trained simultaneously. In that work, the authors used a discriminative network D to estimate the probability that a sample comes from the dataset instead of being produced by a generative model G. The two models are iteratively trained so that G learns to generate frames that are difficult to be classified by D, while D learns to tell weather the frames are generated by G. In this work, we follow Ledig's work [14] using a deep CNN binary classifier, which is illustrated in Fig. 3.

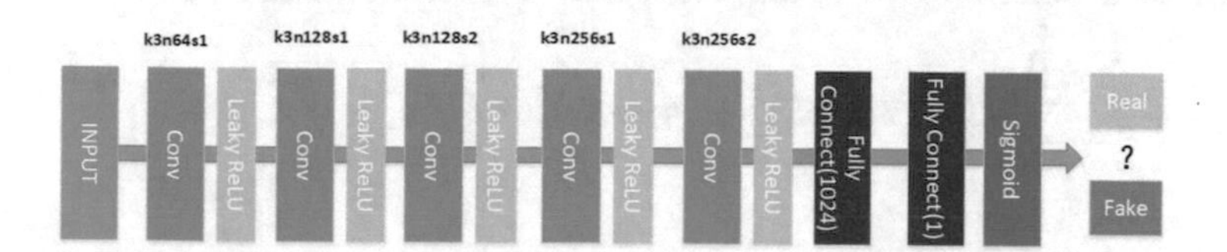

Fig. 3. Architecture of Discriminator Network with corresponding kernel size (k), number of filters (n) and stride (s) indicated for each convolutional layer. The last two Fully connected layers have 1024 nodes and 1 node respectively.

The training of the pair (G, D) consists of two alternated steps, described below.

Training D: We do one SGD step to D, while keeping the weights of G fixed. Therefore, the loss function we use to train D is

$$L^D_{adv}(X_{HR}, X_{SR}) = L_{bce}(D(X_{HR}), 1) + L_{bce}(D(X_{SR}), 0) \tag{3}$$

where L_{bce} is the binary cross-entropy loss, defined as

$$L_{bce}(x, y) = -\sum_i ylog(x) + (1-y)log(1-x) \tag{4}$$

Training G: Keeping the weights of D fixed, we perform one SGD step on G to minimize the adversarial loss:

$$L^G_{adv}(X_{SR}) = L_{bce}(D(X_{SR}), 1) \tag{5}$$

3.3 Loss Function

We combine all above constraints regarding appearance, perceptual and adversarial training, into our final loss function of generator:

$$L_G = \lambda_{pixel} L_{pixel}(X_{HR}, X_{SR}) + \lambda_{pl} L_{pl}(X_{HR}, X_{SR}) + \lambda_{adv} L^G_{adv}(X_{HR}, X_{SR}) \tag{6}$$

When training D, we use the following loss function:

$$L_D = L^D_{adv}(X_{HR}, X_{SR}) \tag{7}$$

3.4 YOLO Detector

After the appropriate ESRGAN is trained according to related objects, we exploit the YOLOv3 object detector as our detection network. It should be noted that the choice of YOLOv3 is optional and could be replaced with other state-of-the-art object detection models. During the standard learning process, the images synthesized by the ESRGAN generator are thus passed to the input of YOLOv3 which calculates the predictions and the loss function (L_{YOLO}) from predicted bounding boxes, whose gradient is back-propagated to update the weights of YOLOv3 network, which seeks to minimize the difference between the detected bounding boxes and the ground truth bounding boxes:

$$L_{YOLO} = \sum_{i=0}^{S^2} \sum_{j=0}^{B} \mathbf{1}^{obj}_{ij} ((x_i - \hat{x}_i)^2 + (y_i - \hat{y}_i)^2 + (\sqrt{w_i} - \sqrt{\hat{w}_i})^2 + (\sqrt{h_i} - \sqrt{\hat{h}_i})^2) \tag{8}$$

4 Experiments

4.1 Experiment Setup

The proposed ESRGAN is implemented using Tensorflow. All of networks are trained on NVIDIA Titan V GPU with Adam solver. We set minibatch size to 6 and do random horizontal flip and 0,90,180,270 rotation for each training pair. Each pair consists of one clip and its 0.5 resized counterpart. To obtain every loss of a scale that is comparable to others, we set λ_{pixel} to 0.001, λ_{pl} to 0.006, λ_{adv} to 0.05. We firstly do trivial training by setting λ_{adv} to 0 and start adversarial training after 500 iterations warm up. Discriminator net is trained at one tenth of generator's learning rate.

The detector training part is conducted as the standard YOLOv3 routine except the data augmentation which is implemented by resizing 6 random small objects from each training frame and pasting them to different locations without overlapped with existing objects.

4.2 Results on Benchmark Datasets

We evaluate the behavior and the performance of the approaches discussed here using the ISPRS 2D Semantic Labeling Contest dataset [15] and a self-assembled airport surface dataset. The former dataset could be exploited for vehicle detection tasks by retaining only the related components of pixels belonging to the vehicle class. The generated dataset thus contains nearly 10,000 vehicles. The latter dataset is constructed from Jiuhuashan Airport surface surveillance video, which totally is recorded from 2 h of busy flight surface operations. Such video sequences have about 1000 well-labelled objects in three types– planes, trucks and buses (Fig. 4) (Table 1).

Fig. 4. Some frame samples and their ground-truth annotations of the dataset.

Table 1. Quantitative evaluation of ISPRS and airport detection performance using YOLOv3 (confidence threshold of 0.25 and an IoU threshold of 0.25): ESRGAN-2 super-resolution compared to the simple bicubic interpolation and the original size dataset.

Method	ISPRS mAP	Airport mAP
Original	81.82	70.01
Bicubic	71.82	65.01
Proposed	**91.69**	**79.36**

We evaluated this model on an augmented validation set, instead of the original one or the simple bicubic interpolation augmentation one. We saw an increase in the overall object detection performance, suggesting that the trained model effectively "overfit" to small objects. The proposed model yielded better results than the bicubic augmentation, confirming the effectiveness of the proposed strategy of ESRGAN-based resizing small objects.

5 Conclusion

In this paper, we have proposed a robust detection training framework based on the super-resolution resize-and-paste method. To make the detector sensitive to the representation of small objects in a training data, a super resolution resize network is learned from the training dataset and is used to improve the small object diversity in scale and location. We combine the above augmentation method with a state-of-the-art detector YOLOv3 and test our framework on two different challenging benchmarks and experiment results show the relative superiority of proposed algorithm compared to the baseline detector.

Acknowledgments. This work is financially supported by the National Key R&D Program of China, Project Number 2018YFE0208700.

References

1. Dong, C., Loy, C.C., He, K., Tang, X.: Image super-resolution using deep convolutional networks. IEEE Trans. Pattern Anal. Mach. Intell. **38**(2), 295–307 (2015)
2. Lim, B., Son, S., Kim, H., Nah, S., Mu Lee, K.: Enhanced deep residual networks for single image super-resolution. In: Proceedings of the IEEE Conference on Computer Vision and Pattern Recognition Workshops, pp. 136–144 (2017)
3. Kisantal, M., Wojna, Z., Murawski, J., Naruniec, J., Cho, K.: Augmentation for small object detection. arXiv preprint arXiv:190207296 (2019)
4. Yang, W., Zhang, X., Tian, Y., Wang, W., Xue, J.H., Liao, Q.: Deep learning for single image super-resolution: a brief review. IEEE Trans. Multimed. **21**(12), 3106–3121 (2019)
5. Zhang, W., Wang, S., Thachan, S., Chen, J., Qian, Y.: Deconv R-CNN for small object detection on remote sensing images. In: IGARSS 2018-2018 IEEE International Geoscience and Remote Sensing Symposium, pp. 2483–2486. IEEE (2018)
6. Liu, M., Wang, X., Zhou, A., Fu, X., Ma, Y., Piao, C.: UAV-YOLO: small object detection on unmanned aerial vehicle perspective. Sensors **20**(8), 2238 (2020)
7. He, K., Zhang, X., Ren, S., Sun, J.: Identity mappings in deep residual networks. In: Leibe, B., Matas, J., Sebe, N., Welling, M. (eds.) ECCV 2016. LNCS, vol. 9908, pp. 630–645. Springer, Cham (2016). https://doi.org/10.1007/978-3-319-46493-0_38
8. Nair, V., Hinton, G.E.: Rectified linear units improve restricted Boltzmann machines. In: Proceedings of the 27th International Conference on Machine Learning (ICML 2010), pp. 807–814 (2010)
9. He, K., Zhang, X., Ren, S., Sun, J.: Delving deep into rectifiers: surpassing human-level performance on ImageNet classification. In: Proceedings of the IEEE International Conference on Computer Vision, pp. 1026–1034 (2015)

10. Szegedy, C., Ioffe, S., Vanhoucke, V., Alemi, A.A.: Inception-v4, inception-ResNet and the impact of residual connections on learning. In: Thirty-First AAAI Conference on Artificial Intelligence (2017)
11. Yang, W., Zhang, X., Tian, Y., Wang, W., Xue, J.H.: Deep learning for single image super-resolution: a brief review. arXiv preprint arXiv:180803344 (2018)
12. Simonyan, K., Zisserman, A.: Very deep convolutional networks for large-scale image recognition. arXiv preprint arXiv:14091556 (2014)
13. Goodfellow, I., et al.: Generative adversarial nets. Adv. Neural Inf. Process. Syst. **27**, 2672–2680 (2014)
14. Ledig, C., et al.: Photo-realistic single image super-resolution using a generative adversarial network. In: Proceedings of the IEEE Conference on Computer Vision and Pattern Recognition, pp. 4681–4690 (2017)
15. Rottensteiner, F., et al.: The ISPRS benchmark on urban object classification and 3D building reconstruction. ISPRS Ann. Photogram. Remote Sens. Spat. Inf. Sci. I-3 **1**(1), 293–298 (2012)

Graph-Based Anomaly Detection of Wireless Sensor Network

Qianwen Zhu, Jinyu Zhou, Shiyu Zhao, and Wei Wang(✉)

Tianjin Key Laboratory of Wireless Mobile Communications and Power Transmission, Tianjin 300387, China
weiwang@tjnu.edu.cn

Abstract. Anomaly detection in wireless sensor networks plays a vital role to ensure the accuracy and reliability of network data. In view of the complexity of sensor networks, the process of detecting anomalies using raw data is complicated, which also requires abundant storage spaces. This paper proposes a WSN anomaly detection algorithm based on graph signals. Use the graph model of the WSN to construct the covariance matrix, so as to realize the anomaly detection of the sensor network. Also, this paper selects a subset of edges with high correlation to construct a new graph to optimize the topology of the sensor network graph. The performance of the algorithm is tested and analyzed on the Intel_Lab_Data.

Keywords: Wireless sensor network (WSN) · Graph construction · Graph spectrum theory · Anomaly detection

1 Introduction

With the continuous development and growth of computer science and technology and the Internet, WSN systems have gradually developed towards intelligence. The working principle of WSN is to sense the environment and collect data from it [1]. However, the sensor node may be affected by its own life, battery power shortage and surrounding environment, etc., which will make the collected data abnormal. If the abnormal information is not detected and analyzed, it will have serious consequences for decision-making. Therefore, to ensure the accuracy and reliability of WSN data, anomaly detection is an indispensable step.

In WSN, outliers can be defined as a set of values that deviate from the data sample standard. There are many types of methods for abnormal data detection. The method based on statistics needs to know the prior knowledge of the sensor data flow distribution to stablish a corresponding mathematical statistical model, as stated in Ref. [2]. However, in most application scenarios, the prior parameters are difficult to get. The classification-based method builds a classification model through a set of data. The test set data will be considered as an outlier when it does not belong to the model, computational complexity is its main disadvantage, as stated in Ref. [3]. The nearest neighbor-based method using the concept of distance as a measure of similarity between each node is described in Ref. [4], which does not need to know the prior of the data flow knowledge,

Q. Liang et al. (Eds.): Artificial Intelligence in China, LNEE 854, pp. 144–150, 2022.
https://doi.org/10.1007/978-981-16-9423-3_18

but its computational complexity increases exponentially in high-dimensional data sets. In addition, a detection algorithm based on clustering is described in Ref. [5]. Although these methods have a high detection rate in some respects, most of them only consider the temporal and spatial correlation of the data.

In recent years, the graph processing theory has provided a new model for network anomaly detection based on topological structure, which is to construct a weighted graph through the topological structure of the network [6]. The method using both the vertex domain smoothness of sensor nodes and the low-frequency characteristics of the graph frequency domain to calculate the smoothness of the graph signal before and after low-pass filtering, and established smoothness judgment criteria to realize the abnormal nodes was proposed in Ref. [7], which can detect the data with small deviation of nodes, but the calculation is relatively complicated. Considering the above problems, this paper proposes a graph-based anomaly detection algorithm. By establishing a graph model for WSN, construct the statistical test quantity to realize the network anomaly detection.

The rest of the paper is as follows. Section 2 details the graph spectrum theory and the anomaly detection theory. Section 3 introduces a graph-based WSN anomaly detection algorithm and an optimization algorithm of graph topology. Section 4 analyzes the experimental results. Section 5 is the conclusion.

2 Theoretical Basis

2.1 Graph Theory

For a WSN with N nodes, use G = (V, E) to represent its graph model, where V denotes the node set of G, and E denotes the edge set. For a general graph, the relationship between the nodes can be described by the adjacency matrix A, which only includes 0 and 1, where 1 means a connection between nodes u and v, and 0 means no connection between u and v, the definition of A is

$$a_{uv} = \begin{cases} 1, & \textit{if u and v are adjacent} \\ 0, & \textit{otherwise} \end{cases} \tag{1}$$

For a weighted undirected graph, use w_{uv} (or w_{vu}) to denote the weight of the edge between nodes u and v, which can form a weighted matrix W. At this time, the adjacency matrix A can be defined by the weight matrix as

$$a_{uv} = \begin{cases} w_{uv}, & \textit{if u and v are adjacent} \\ 0, & \textit{otherwise} \end{cases} \tag{2}$$

In this way, for any weighted graph G, its Laplacian matrix L can be defined as

$$L = D - A \tag{3}$$

Where D is the degree matrix of the graph, denotes the number of edges that connect a node with other nodes, and it is a diagonal matrix, is defined as

$$d_v = \sum_u w_{uv} \tag{4}$$

Orthogonal decomposition of the Laplacian matrix has

$$L = U\lambda U^T \tag{5}$$

Where U is the eigenvector matrix of L, and λ is the diagonal matrix formed by the eigenvalues of L.

2.2 Anomaly Detection Theory

For a WSN with N nodes, N groups of sample signals with length L($y_i \in R^L$, i = 1,……, N) are collected in the sample matrix $Y = \left[y_1, y_2 \ldots\ldots y_N\right]^T$, whose covariance matrix $\Sigma_y = (1/L)YY^T$. Take the first K ($K \ll L$) main eigenvalues $\lambda_1, \lambda_2, \ldots\ldots, \lambda_K$ corresponding main eigenvectors of Σ_y to form the feature matrix $U = [U_1, U_2 \ldots\ldots U_K]$, so the projection matrix $P = (I - UU^T)$, where I is the identity matrix [8, 9]. Therefore, for any tested matrix x, the squared prediction error (SPE) can be obtained

$$SPE = ||Px||^2 \tag{6}$$

The judgment of whether the data is abnormal can be based on whether the value of SPE has changed significantly.

3 Graph-Based Anomaly Detection of Wireless Sensor Network

3.1 Graph Model and Network Anomaly Detection

For a WSN with N nodes, take a data value of length L for each node to form a sample matrix $Y = \left[y_1, y_2, \ldots\ldots, y_N\right]^T$, $Y \in R^{N \times L}$, using the correlation between the data of each node to construct the graph model. Here we first construct the sample covariance matrix R_y as

$$R_{ij} = \mathrm{cov}\left(Y_i, Y_j\right) \quad i, j = 1, 2, \ldots\ldots n \tag{7}$$

Combining the definition of weighted undirected graph, construct the adjacency matrix A as

$$a_{ij} = R_{ii} + R_{jj} - 2R_{ij} \tag{8}$$

According to the formulas in Sect. 2.1, the Laplacian matrix L is obtained. Assuming $U = [U_1, U_2, \ldots\ldots, U_k]$ is the eigenvectors corresponding to the first K maximum eigenvalues $\lambda_1, \lambda_2, \ldots\ldots, \lambda_k$ of L. The magnitude of K can be obtained from the fact that the energy of $\lambda_1, \lambda_2, \ldots\ldots, \lambda_k$ account for about 90% of the energy of the total eigenvalues. In view of the principle in Sect. 2.2, SPE can be obtained. When the value of SPE changes significantly, it is considered that the data is abnormal.

3.2 Optimized Graph Construction Algorithm

For a complete graph G (V, E) with N nodes, in order to remove the redundant information in the data, the edges of the graph G are de-redundant. Specifically, for the sample matrix $Y = [y_1, y_2, \ldots\ldots, y_N]$, $Y \in R^{N \times L}$, our goal is to select K edges among those nodes with the highest correlation, that is, if y_i is strongly correlated with y_j. An edge will be added between nodes i and j. Combined with Eq. (8), it can be seen that we need minimum the value of Eq. (8)

$$\arg\min a_{ij} = \arg\min\left(R_{ii} + R_{jj} - 2R_{ij}\right) \tag{9}$$

Among the Ry is the covariance matrix of the samples [10]. The edge subset obtained by the method contains most of the information in the original network, and we further bring it into Sect. 3.1 to achieve the purpose of anomaly detection.

4 Experimental Results and Analysis

4.1 Intel_Lab_Data

Intel_Lab_Data [11] is the actual data set collected indoors by the Intel Berkeley Research Lab sensor network laboratory. The 54 mica2 sensor nodes collected temperature, humidity, light intensity, and node voltage readings every 30 s from February 28 to April 5, 2004. In this paper, we selected temperature data from February 28 to March 3 for a total of 5 days for the experiment. Due to the particularity of the experimental data, the frequency of abnormalities in the sensor 15 is extremely high. To facilitate the experiment and display clearly, we only take the first 3000 data points, and reorganize the temperature data of the sensor 15, so that we can get the premise that the data length of each node is 10, The data set has 26 anomalies; if the value length is 20, there are 17 anomalies; and if the value length is 50, there are 9 anomalies. The sensor node layout is shown in the Fig. 1.

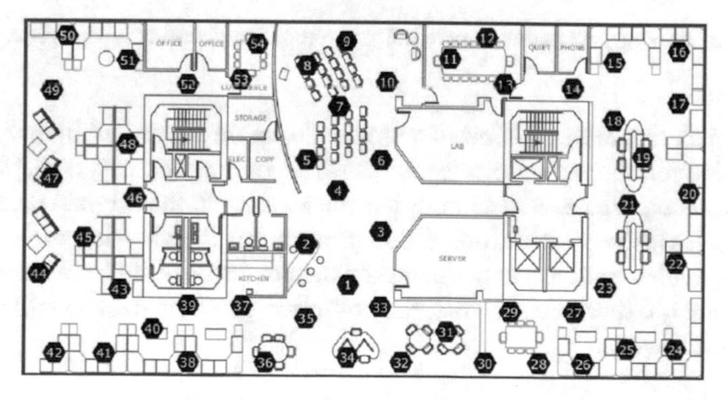

Fig. 1. Intel Lab Data node layout

4.2 Experimental Results and Analysis

In the experiment, first select the normal temperature data value of length 10 for each node of the sensor network to calculate its SPE. Here, X is a test matrix with a size of N*L (L is 10). Figure 2(a) and (c) show the anomalies in the original data model and its anomalies after algorithm analysis. The points marked with circles are abnormal data. It can be seen that, as observed by Lakhina et al. [8], abnormal data is usually not easy to be found in the original data pattern, as shown in Fig. 2(a). Similarly, after analyzing by the proposed algorithm, the abnormality of the data matrix can easily to be caught, as shown in Fig. 2(c). Besides, this paper also tests the effectiveness of the algorithm with matrices of different sizes (L is different). Figure 3(a) and (b) respectively show the abnormality in the tested matrix size of 54*20 and 54*50, and corresponding results with analysis of the algorithm are shown in Fig. 3(c) and (d). The results show that the algorithm proposed in this paper detects abnormalities still valid for data matrices with different sizes.

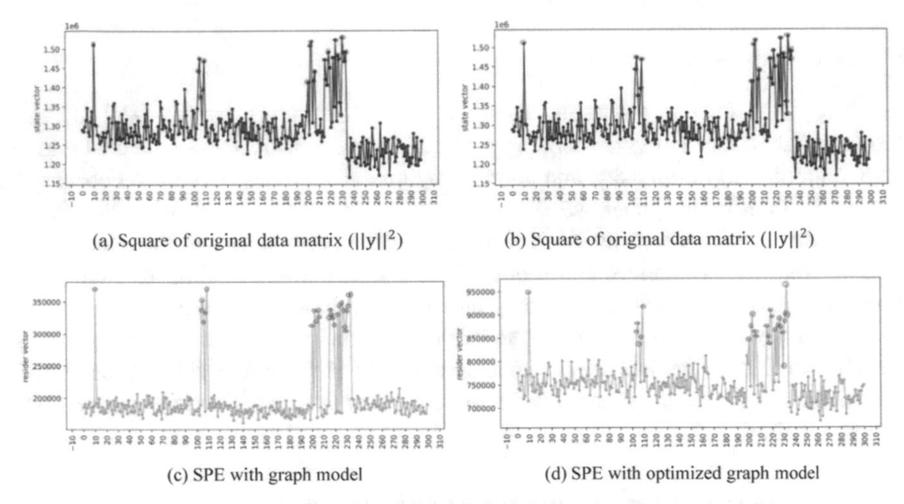

(a) Square of original data matrix ($||y||^2$)

(b) Square of original data matrix ($||y||^2$)

(c) SPE with graph model

(d) SPE with optimized graph model

Fig. 2 Anomaly detection in original graph model and optimized graph model

We also experiment the optimized anomaly detection algorithm. In this paper, 2600 edges with high correlation are selected for construction, and anomaly detection is carried out for experiment. Figure 2(b) denotes the abnormality in the original data mode, and Fig. 2(d) denotes the result of the optimized graph model anomaly detection. At the same time, this paper also conducts experiments on data matrices of different sizes, as shown in Fig. 3(e) and (f). It can be seen that the optimized graph model can still identify and detect anomalies well.

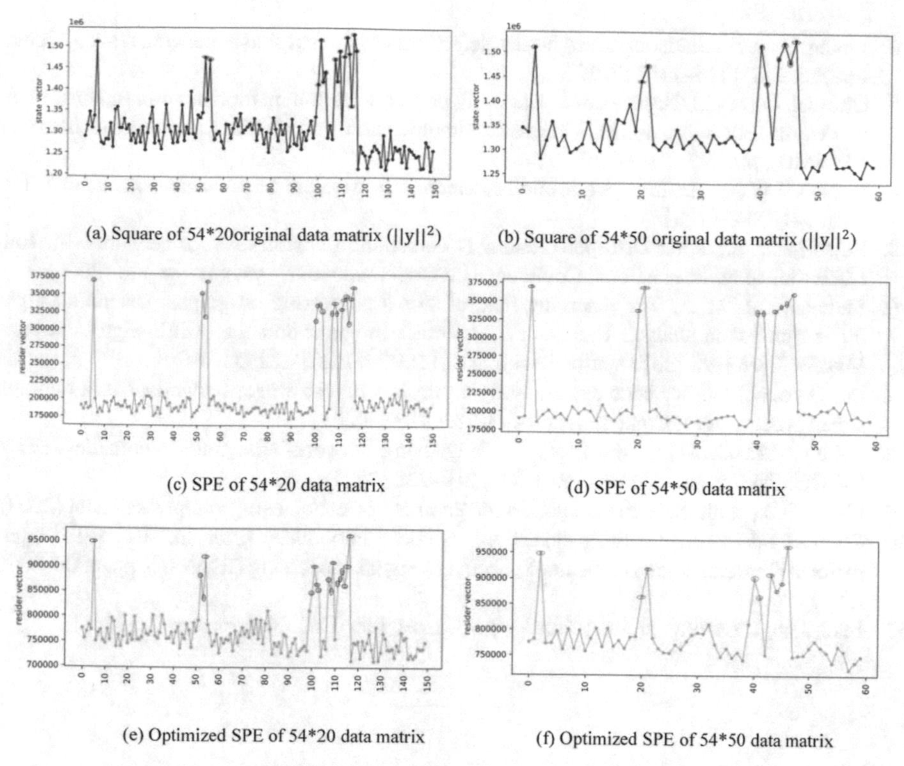

(a) Square of 54*20original data matrix ($||y||^2$)

(b) Square of 54*50 original data matrix ($||y||^2$)

(c) SPE of 54*20 data matrix

(d) SPE of 54*50 data matrix

(e) Optimized SPE of 54*20 data matrix

(f) Optimized SPE of 54*50 data matrix

Fig. 3 Anomaly detection under data matrices of different sizes

5 Conclusion

This paper proposes a graph-based anomaly detection algorithm for WSN, which uses graph theory to measure the topology of WSN to realize anomaly detection. Also, we optimize the selection of the edges of the graph to remove network redundancy. In order to verify the effectiveness of the proposed algorithm, a series of comparative experiments were carried out. The experimental results show that the algorithm can effectively detect abnormal data. However, the proposed algorithm is only tested in small sensor networks. For larger WSNs with more complex network nodes, the algorithm needs to be further expanded.

Acknowledgements. The work was supported by the Natural Science Foundation of China (61731006, 61971310)

References

1. Shukla, D.S., Pandey, A.C., Kulhari, A.: Outlier detection: a survey on techniques of WSNs involving event and error based outliers. In: 2014 Innovative Applications of Computational Intelligence on Power, Energy and Controls with their impact on Humanity (CIPECH), pp. 113–116 (2014)

2. Zhang, Y., et al.: Statistics-based outlier detection for wireless sensor networks. Int. J. Geogr. Inf. Sci. **268**, 1373–1392 (2012)
3. Ghorbel, O., et al.: Classification data using outlier detection method in wireless sensor networks. In: 13th International Wireless Communications and Mobile Computing Conference (IWCMC), pp. 699–704 (2017)
4. Branch, J.W., et al.: In-network outlier detection in wireless sensor networks. Knowl. Inf. Syst. **341**, 23–54 (2013)
5. Rajasegarar, S., et al.: Distributed anomaly detection in wireless sensor networks. In: 10th IEEE Singapore International Conference on Communication Systems, pp. 1–5 (2006)
6. Shuman, D.I., et al.: The emerging field of signal processing on graphs: Extending high-dimensional data analysis to networks and other irregular domains. IEEE Signal Process. Mag. **30**(3), 83–98 (2013). https://doi.org/10.1109/MSP.2012.2235192
7. Lu, G., et al.: Outlier node detection algorithm in wireless sensor networks based ongraph signal processing. J. Comput. Appl. **403**, 783–787 (2020)
8. Lakhina, A., Crovella, M., Diot, C.: Diagnosing network-wide traffic anomalies. ACM SIGCOMM Compt. Commun. Rev. **344**, 219–230 (2004)
9. Pham, D.S., et al.: Scalable network-wide anomaly detection using compressed data (2009)
10. Chepuri, S.P., et al.: Learning sparse graphs under smoothness prior. In: 2017 IEEE International Conference on Acoustics, Speech and Signal Processing (ICASSP), pp. 6508–6512 (2017)
11. Intel_Lab_Data. https://github.com/jzxywpf/Intel_Lab_Data/blob/master/data.zip

Identifying Important Attributes for Secondary School Student Performance Prediction

Ke Chen, Xin Zhang(✉), and Xiu Zhang

Tianjin Key Laboratory of Wireless Mobile Communications and Power Transmission, Tianjin Normal University, Tianjin 300387, China

ecemark@tjnu.edu.cn

Abstract. In traditional teaching, teachers cannot effectively provide targeted guidance to students to improve thier performance. Moreover, the current student performance prediction model is complicated in design, and there are a lot of irrelevant data input. In this paper, taking secondary school students as an example, we have selected a number of important attributes that can influence the prediction of student performance from the attribute set. The attribute set includes past school grades, and several attributes related to individuals, society and school. The two core courses (i.e. mathematics and portuguese) were modeled under Binary/Five-level classification and regression tasks. Three feature selection methods are tested under three input configurations, the accuracy of the model is verified by decision tree algorithm. This makes it more possible for teachers to provide professional guidance to students, and can effectively reduce the dimension of the dataset of the student performence prediction model to improve the prediction efficiency.

Keywords: Feature selection · Importance measure · Student performance prediction

1 Introduction

Today's education has reached to a new height, Throughout the education of China, although there has been rapid development in recent years, the common problem is that teachers can not do targeted prediction and guidance for students.

The vigorous development of information technology has provided infinite possibilities for the intelligent education field, such as online education, etc. Among them, Data Mining (DM) technology plays an important role.

After obtaining data from various channels, we can use DM to analyze and deal with the dataset to obtain more advanced information, so as to solve many problems, such as: which course has the greatest impact on the performance of student? How to predict the test results of a certain classmate? What types of courses are more attractive to students? What are the most important attributes that affect student performance? In this paper, we focus on the last problem.

In fact, there have been many studies on student performance prediction. They mainly perform modeling under classification and regression tasks, and adopt Decision Trees

Q. Liang et al. (Eds.): Artificial Intelligence in China, LNEE 854, pp. 151–158, 2022.
https://doi.org/10.1007/978-981-16-9423-3_19

(DT), Random Forest (RF), Neural Network, Association Rules and other algorithms for model training [1–5]. Taking seventh grade mathematics as an example, Wang et al. used DT C5.0 algorithm to construct a prediction model of academic record [4]. In the study, six sets of online summer mathematics homework (30 questions in each set) were used as data source. The final dataset includes the answer data of 752 students. Wu et al. proposed a student performance analysis and prediction model based on DT algorithm [5]. The model convert the problem of college student performance prediction into the problem of student learning status classification, and the accuracy of student performance prediction reaches 94%. To sum up, the above studies can predict student performance and other behaviors better, but the accuracy of prediction and the key attributes affecting student performance need further research to improve and determine. Therefore, the aim of this paper is to explore the core attributes that affect student performance, that is, the importance of data sample attributes.

Importance measure analysis can find out the important feature variables, so as to reduce the dimension of the input space, reduce the difficulty of the learning ta-sk and improve the efficiency of the model. In the existing research, the main algorithms used include: RF, DT, Random Generation Sequence Selection algorithm, Simulated Annealing Algorithm, Genetic Algorithm, etc. [6–11]. Song et al. improved the importance measure based on RF, and verified the effectiveness of the proposed importance measure index system through calculation examples [10]. Song et al. took the quadratic polynomial output model without cross terms and with cross terms as examples to verify and elaborate the advantages and disadvantages of different importance measures [11].

From the research results of Paulo et al., it can be understood that most of the 33 attributes contained in the student attribute dataset have little or even negligible impact on student performance, and only a few of them have a great impact [12]. For performance prediction tasks, a large part of the dataset is redundant, which greatly increases computational complexity and reduces task efficiency.

In this paper, on the basis of the operation dataset and research results of Paulo, we use feature selection method to further process its dataset, we have selected a number of important attributes that can influence the prediction of student performance from the attribute set. The attribute set includes past school grades, several attributes related to individuals, society and school. For the selection of important attributes, three sequential selection algorithms (i.e. Sequential forward, Sequential backward and Stepwise regression) were tested. By comparing the feature selection results of the three feature selection methods under different classification/regression tasks and input configuations, several important attributes that have the most influence on student performance are finally obtained.

2 Dataset and Feature Selection Methods

2.1 Student Data and Processing Method

The dataset used in this study was collected by Paulo et al. from two public schools in Alentejo region of Portugal. Under the background of that time, the school information system was poor, so the dataset was mainly established by two methods: school reports, including three period grades (i.e. G1, G2, G3) and the number of school absences; and questionnaires, used to obtain statistical data of several attributes related to individuals (e.g. gender), demography (e.g. mother's education level), society (e.g. alcohol consumption), school (e.g. school address), the final real dataset contains a total of 33 attributes. Finally, 677 students' real and effective questionnaire answering data were obtained. The data was intergrated into two datasets related to mathematics (with 395 examples) and the portuguese language (649 records) classes to get the final dataset.

Mathematics and portuguese are modeled under classification and regression tasks, using three supervisory methods:

a. Binary classification–if G3 $\geq$ 10, pass; else failed;
b. Five-level classification–divide the full score of 20 into five levels (Table 1);
c. Regression–the G3 value (numeric output between 0 and 20).

Table 1. The Five-level classification system

Country	I (excellent/very good)	II (Good)	III (satisfactory)	IV (sufficient)	V (fail)
Portugal/ France Ireland	16–20 A	14–15 B	12–13 C	10–11 D	0–9 F

Using three sequential selection algorithms for model training, which aims to select the important attributes that affect the prediction of student performance. Because G1 and G2 are suspected to be the most core attributes of student performance prediction, three input configurations were tested for each feature selection model.

Setup 1. contains all variables except G3 (the output);
Setup 2. similar to 1 but without (the second period grade);
Setup 3. similar to 2 but without G1 (the first period grade);

2.2 Feature Selection Method

All experiments involved in our research are completed by using Matlab, sequential feature selection method is used to complete the selection of important attributes. Sequential feature selection is a commonly used method, which has two components:

i. An objective function, called the criterion, which the method seeks to minimize over all feasible feature subsets. Common criteria are mean squared error (for regression models) and misclassification rate (for classification models).
ii. A sequential search algorithm, which adds or removes features from a candidate subset while evaluating the criterion. There are three feature selection algorithms in this part:

A. Sequential forward selection (SFS), the feature subset X starts from the empty set, and each time a feature x is selected to add to the feature subset X, so that the feature function J(X) is optimal.
B. Sequential backward selection (SBS), starting from the full feature set O, a feature x is removed from the feature set O each time, so that the evaluation function value reachs the optimal after the feature x is removed.
C. Stepwise regression, which is a sequential feature selection technique designed specifically for least-squares fitting. The functions stepwise and stepwisefit make use of optimizations that are only possible with least-squares criteria. Unlike generalized sequential feature selection, stepwise regression may remove features that have been added or add features that have been removed.

Both SFS and SBS are simple greedy algorithms.

3 Simulation Results

This paper aims to pick out the important attributes that affect student performance. So the 33 attributes in the dataset are preprocessed and coded as follows (Table 2):

Table 2. Student related attributes coding

Attribute	Code	Attribute	Code	Attribute	Code
sex	1	famrel	12	higher	23
age	2	reason	13	romantic	24
school	3	traveltime	14	freetime	25
address	4	studytime	15	goout	26
Pstatus	5	failures	16	Walc	27
Medu	6	schoolsup	17	Dalc	28
Mjob	7	famsup	18	health	29
Fedu	8	activities	19	absence	30
Fjob	9	paidclass	20	G1	31
gardian	10	internet	21	G2	32
famsize	11	nursery	22	G3	33

The two core courses (i.e. mathematics and portuguese) are modeled under Binary/Five-level classification and regression tasks. Three feature selection methods (i.e. important attributes selection) were tested under three different input configurations of setup 1, 2, and 3; Finally, the accuracy of the three methods was verified through the decision tree algorithm. Because the portuguese dataset is about 1.6 times that of mathematics, which indicates that the training based on portuguese is more convincing, here we mainly show the feature selection results under four different situations (mostly about portuguese), as shown in Tables 3, 4, 5 and 6, and the results are output in the coded form corresponding to student attributes.

Table 3. Binary classification results based on portuguese

Method	Setup	Selected features
A	1	9,10,13,27,30,31,32
	2	4,6,8,14,15,21,24,26,30,31
	3	3,6,9,11,15,16,17,22,25,29,30
B	1	1,3,10,11,12,14,23,24,25,26,27,28,29,30,31,32
	2	6,7,8,9,10,11,12,14,15,16,19,23,24,25,26,27,28,29,30,31
	3	1,5,6,9,11,13,14,15,16,18,20,21,22,23,24,25,26,27,28,30
C	1	1,10,13,31,32
	2	1,13,15,31
	3	1,15,21,28

Table 4. Five-level classification results based on portuguese

Method	Setup	Selected features
A	1	2,8,9,13,22,24,25,29,30,31,32
	2	1,2,3,5,7,8,9,11,12,15,27,29,30,31
	3	1,3,4,5,6,8,9,11,12,13,15,16,23,24,25,26,27,29,30
B	1	1,3,4,5,6,7,8,9,10,12,13,14,15,18,22,23,24,25,26,27,28,29,30,31,32
	2	1,2,3,4,5,6,7,9,10,11,12,13,14,15,16,18,20,21,22,25,26,27,28,29,30,31
	3	1,3,5,6,7,9,10,11,12,13,14,15,17,18,19,21,22,24,25,26,27,28,29,30
C	1	3,7,9,11,14,16,27,29,30,31,32
	2	1,2,3,7,15,16,21,28,29,30,31
	3	1,2,3,7,12,14,15,16,19,21,27,29,30

No matter which feature selection method is used in classification or regression tasks, attribute 32 (i.e. G2) will always be selected as the most important feature for the setup 1. The result shows that the latest school grade plays the most important role in performance prediction. Based on this, we found that under the setup 2, can always pick out the attribute 31, while no attribute could be stably selected under the setup 3. The above all results greatly prove the importance of the past school grades for the performance prediction. By comparison, it is not difficult to find that in the absence of past school grades, attribute 30 is the most likely to be selected as an important attribute, second only to G1 and G2, that is, the importance of number of school absence is second only to past school grades. In addition, we find that weekly study time, school, number of

Table 5. Regression results based on portuguese

Method	Setup	Selected features
A	1	3,4,5,6,8,9,12,14,15,16,18,20,21,22,24,26,27,28,29,30,31,32
	2	1,2,3,4,5,7,8,11,12,13,15,16,18,19,23,24,25,28,29,30,31
	3	3,4,5,7,8,10,13,14,15,16,18,20,21,23,25,26,27,30
B	1	1,2,4,5,6,7,9,11,12,13,14,15,16,17,18,20,22,23,24,25,26,27,28,29,30,31,32
	2	1,2,4,7,8,9,10,11,12,13,14,15,16,17,18,19,20,21,22,23,24,27,28,29,30,31
	3	1,2,3,4,5,6,7,8,9,10,11,13,14,15,16,19,20,21,22,23,24,25,26,27,28,29,30
C	1	1,7,14,15,16,27,29,31,32
	2	1,3,7,14,15,16,21,27,29,31
	3	1,2,8,14,15,16,21,27,29

Table 6. Regression results based on mathematics

Method	Setup	Selected features
A	1	1,2,3,7,10,11,12,19,24,25,29,30,32
	2	3,4,6,9,11,15,16,17,18,22,24,26,27,28,30,31
	3	2,3,5,6,9,10,13,14,15,16,18,19,20,23,24,26,28,29,30
B	1	3,4,5,6,7,8,9,10,13,14,15,16,17,18,20,21,22,24,26,27,28,29,30,31,32
	2	1,2,3,4,5,6,7,8,9,10,11,12,15,16,17,18,20,22,23,24,25,26,28,30,31
	3	1,2,3,5,6,7,8,9,10,11,12,13,14,15,17,18,24,25,26,28,30
C	1	15,23,30,31,32
	2	15,23,31
	3	2,7,11,15,23,26

failures, going out with friends, mother's job, father's job and health are more important than other attributes.

We use the decision tree to obtain the accuracy of the model. In classification, the evaluation standard of model quality is measured by the percentage of correct classification (PCC). The closer the PCC is to 100%, the better the model is proved. In regression, the root mean square (RMSE) is used to measure, that is, the closer the RMSE is to 0, the better the model is provedr. The specific calculation method is as formulas 1 to 3:

$$\phi(i) = \begin{cases} 1, \ if \ y_i = \hat{y}_i \\ 0, \ otherwise \end{cases} \tag{1}$$

$$PCC = \sum\nolimits_{i=1}^{N} \varphi(i)/N * 100(\%) \tag{2}$$

$$RMSE = \sqrt{\sum_{i=1}^{N} (y_i - \hat{y}_i)^2 / N} \quad (3)$$

Figures 1, 2, 3 and 4 show the accuracy and comparison results of each model under the classification and regression tasks.

We found that method B combined with input configuration 1 can always get the best results, whether classified or regressed.

Binary/Five-level classcification results (in %):

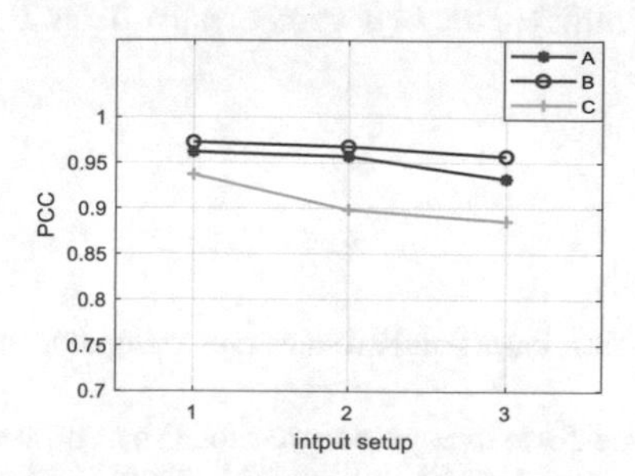

Fig. 1. Bin-portuguese PCC values

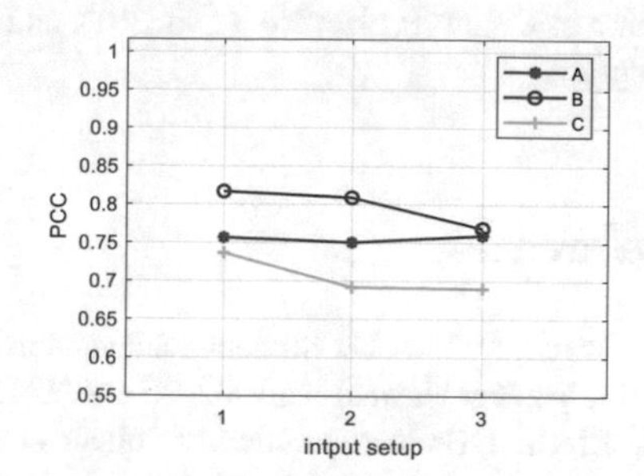

Fig. 2. 5L-portuguese PCC values

Regression results:

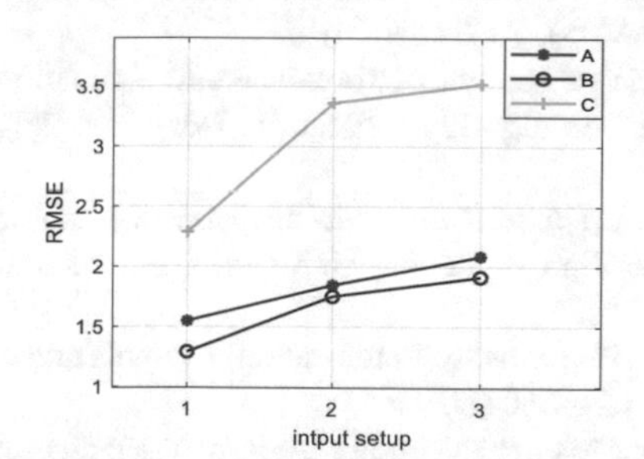

Fig. 3. Bin-portuguese RMSE values

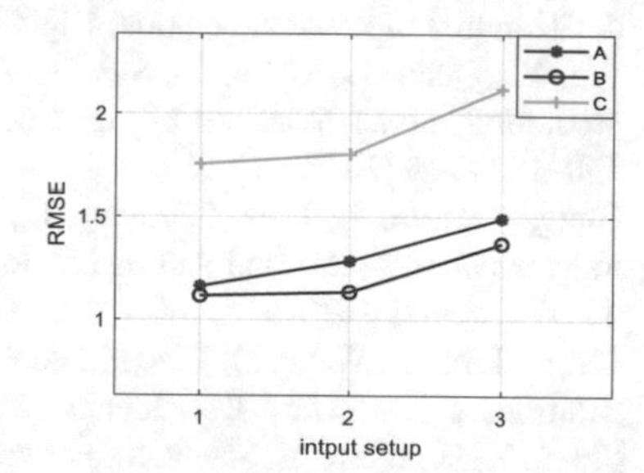

Fig. 4. 5L-mathematics RMSE values

4 Conclusion

In this paper, The basic dataset includes 33 attributes, three feature selection methods are used to select important features under three different input configurations and classification/regression tasks, have picked out the important attributes that influence student performance prediction in the end. Then the validity of the model is verified by DT. The results showed that past school grades are essential to the prediction of student performance. If without past school grades, number of school absence, weekly study time, school, number of failures, going out with friends, mother's job, father's job and health are the most critical attributes that affected the prediction of student performance.

This research can greatly reduce the dimension of the dataset required in the student performance prediction task and improve the computational efficiency. For teachers and parents, self-improvement can be carried out according to important attributes and targeted guidance can be given to students to improve their performance. In addition, there are some shortcomings in this study: the dataset is small and the data source is relatively limited, so the universality of the experimental results needs to be verified; More advanced algorithms need to be explored. Next, further expand the dataset to verify the experimental results; seeking more advanced algorithms to determine important attributes.

Acknowledgements. This research was supported in part by the National Natural Science Foundation of China (Project No. 61901301), and the Tianjin Higher Education Creative Team Funds Program.

References

1. Wang, J., Yan, D.: Student achievement evaluation based on K-means clustering algorithm. J. Mudan. Normal Univ. **02**, 20–22 (2021)
2. Liu, A.: The construction of college students performance predictive model based on data mining technology. Changchun Inst. Tech. (Nat. Sci. Edi) **21**(02), 98–101 (2020)
3. Hu, C., Ma, Y., Jiang, J.: An empirical study on the influence of internal factors on students academic achievement. J. Mudan. Normal Univ. **02**, 71–76 (2021)
4. Wang, X., Wu, X.: Research on the prediction of students' academic record based on the decision tree of knowledge point. Digit. Educ. **6**(05), 70–74 (2020)
5. Wu, X., Quan, L., Chen, C., Shi, L.: Analysis of student performance and simulation of prediction model based on big data decision tree algorithm. Electron. Des. Eng. **28**(24), 138–141+146 (2020)
6. Gong, X., Lyu, Z., Sun, T., Zhang, L., Feng, L.: A new moment-independent importance measure analysis method and its efficient algorithm. J. Beijing Univ. Aeronaut. Astronaut. **45**(02), 283–290 (2019)
7. Xu, L., Lyu, Z., Wang, F., Xiao, S.: Global sensitivity analysis for multiple outputs and their solutions. J. Natl. Univ. Def. Technol. **39**(04), 154–160 (2017)
8. Shuai, Y., Honghui, Y., Sheng, S.: Forward order feature selection algorithm based on mutual information. Technical Acoustics **33**(04), 359–362 (2014)
9. Zuo, J., Lyu, Z., Liu, H., Gong, X.: New importance measure for multivariate output and its effective solution. J. Nanjing Univ. Aeronaut. Astronaut. **49**(03), 441–446 (2017)
10. Song, S., He, R.: Importance measure index system based on random forest. J. Natl. Univ. Def. Technol. **43**(02), 25–32 (2021)
11. Song, J., Lyu, Z.: Investigation of the relation of importance analysis indices for model with correlated inputs. Chin. J. Theor. Appl. Mech. **46**(04), 601–6 (2014)
12. Cortez, P., Silva, A.: Using data mining to predict secondary school student performance. In: European Concurrent Engineering Conference and Future Business Technology Conference, Porto, Portugal, April 9–11, pp. 5–12 (2008)

A Cross-Modal Attention and Multi-task Learning Based Approach for Multi-modal Sentiment Analysis

Yunfeng Song, Xiaochao Fan(✉), Yong Yang, Ge Ren, and Weiming Pan

Xinjiang Normal University, Ürümqi, China
fxc1982@mail.dlut.edu.cn

Abstract. With the rapid development of multimodal research, deep multimodal learning model can effectively improve the accuracy of sentiment classification, and provide important decision support for many applications, such as product reviews, emotion analysis, rumor detection et. Aiming at the problem of feature representation and feature fusion in multimodal sentiment analysis, a model based on multi-task learning and attention mechanism is proposed. Firstly, bi-directional LSTM unit is used to extract the intra-modality representation of single modality. Secondly, the attention mechanism is used to model inter-modality dynamics. Finally, we introduce multi-task learning by predicting sentiment and emotion simultaneously. Experimental results on CMU-MOSEI dataset shows that proposed model outperforms baselines in terms of accuracy and F1-score.

Keywords: Multimodal · Sentiment analysis · Attention mechanism · Multi-task learning

1 Introduction

Sentiment analysis is the process of analyzing, processing, summarizing and reasoning about subjective data with emotional color. The main modality studied in traditional sentiment analysis research are text. With the rapid development of social media in recent years, a large amount of multi-modal data has emerged on the Internet. The way humans understand the world is multi-modal. Compared with text, multi-modal data contains a lot of sentimental information. Sentiment analysis research based on multi-modal data has important application value in the fields of public opinion monitoring, product analysis, product recommendation and human-computer interaction.

Multi-modal sentiment analysis [1] refers to the process of using multi-modal data containing sentimental information to achieve sentiment classification. The rapid development of social media has provided us with massive amounts of multi-modal emotional data. Take video as an example. The video contains three information carriers: text, audio and image. Text can carry semantic information,

Q. Liang et al. (Eds.): Artificial Intelligence in China, LNEE 854, pp. 159–166, 2022.
https://doi.org/10.1007/978-981-16-9423-3_20

and audio can carry the tone of the speaker. Pitch, accent and other information, the facial expression, gestures and other information of the speaker can be extracted from the image. Each information carrier can be called a modality, and each modal carries corresponding sentimental information. Different modalities can explain each other. Using multi-modal data to solve sentiment analysis tasks has advantages that traditional methods cannot match.

This paper proposes a multi-modal sentiment analysis model based on multi-task learning and cross-modal attention mechanism. A number of studies [3,4] show that sentiments and emotions are highly correlated, so learning these two tasks at the same time can improve the generalization performance of the model. Secondly, using the cross-modal attention mechanism to explore the dynamic interaction between different modalities is conducive to model inter-modality features. The contributions of this paper are as follows:

1. Multi-task learning framework: Incorporate the prediction target into the multi-task learning framework, share parameters at the bottom layer, and perform dual recognition of emotions and emotions, which effectively avoids over-fitting and can learn more generalized feature representation.
2. Cross-modal attention method: through the cross-modal attention mechanism to obtain the interactive information between the various modalities.
3. Experiments results show that the proposed model outperformed baselines on sentiment classification and surpass baseline on emotion recognition on anger, disgust, sad, and surprise.

2 Related Work

2.1 Multi-task Learning

Multi-task learning [5] is inspired by the way humans learn things. Humans can learn new things through the knowledge they have learned. Therefore, multi-task learning refers to improving the overall generalization performance of the model by learning useful information contained in multiple associated tasks [6]. Deep learning models usually require a large number of training samples to achieve high classification accuracy, but collecting a large number of training samples is usually time-consuming and labor-intensive. Therefore, in the case of a limited number of samples, multi-task learning is to learn multiple related tasks. Multi-task learning can effectively improve the performance of a single task by sharing layer parameters and prevent model overfitting. In deep learning, there are two main ways of parameter sharing. Hard parameter sharing shares most of the hidden layer parameters in the model, and only the hidden layer of the corresponding task layer is retained. The parameters are independent. Hard parameter sharing can well prevent the problem of overfitting, but it may also cause the model to perform poorly on single task [7]. Soft parameter sharing means that different tasks have their own hidden layer parameters, but regularization constraints are added to the parameters [8]. Multi-task learning acts as data enhancement, because the data generally contains noise. When the model learns multiple tasks at the same time, it can learn more generalized representations.

2.2 Multi-modal Sentiment Analysis

According to the fusion method used, we can divide multi-modal models into two categories: The first category is models used to model single-modal sequence data, such as hidden Markov models, RNN, and LSTM. Early fusion is the simple concatenation of the obtained different modality features and regards them as single modalities. This type of model cannot learn the relationship between modalities well. In contrast to the early fusion, the late fusion votes after the output of each sub-model. Since the fusion process of this method has nothing to do with features, the error of the sub-model is usually also irrelevant [9,10]. Hybrid fusion combines the early and late fusion methods. Although it combines the advantages of the two methods, it increases the structural complexity and training difficulty of the model [11,12]. Poria et al. [13] proposed the BC-LSTM (Bi-directional Contextual Long Short-Term Memory) model, which uses a bidirectional LSTM to capture global context information. Chen et al. [14] proposed GME-LSTM, which combines LSTM with gating mechanism and attention mechanism to perform modal fusion at the word level. However, the above research methods ignore the fusion between single-modal internal information and multi-modal information.

The other is a model specifically designed to handle multi-modal information. Zadeh et al. [2] proposed Tensor Fusion Network for multi-modal sentiment analysis, using a multi-view gated memory module to synchronize multi-modal sequences. The multi-view gated memory module records changes over time. The internal view of the modal and the view interaction between the modals. Zadeh et al. [15] proposed MFN (Memory Fusion Network). In multi-view sequence modeling, there are interactions within views and interactions between views. Memory fusion networks continue to model these two interactions on a time scale. Graph-MFN (Graph Memory Fusion Network) [16] is a dynamic fusion graph built on MFN. It is a fusion method used to solve cross-modal interaction in multi-modal languages to model each mode and change its structure according to the importance of each mode to select a suitable fusion map.

3 Proposed Method

Figure 1 shows the multi-task learning model architecture we propose in this paper. There are intra-modality dynamics and inter-modality dynamics in multi-modal sequences modeling problem. In this framework, it aims to use contextual information and cross-modal attention to simultaneously predict the sentiment and emotion of a utterance in a multi-task learning framework. There are four main components.

1) In the feature extraction layer, the dialogue is composed of a series of utterances, and there are interdependent semantic relations between them. BiLSTM is used to capture the contextual information of sequence features.
2) Cross-modal feature fusion layer, through the cross-modal attention mechanism, the three modalities of text, sound and image are modally interacted in pairs to obtain six groups of modal interaction matrices.

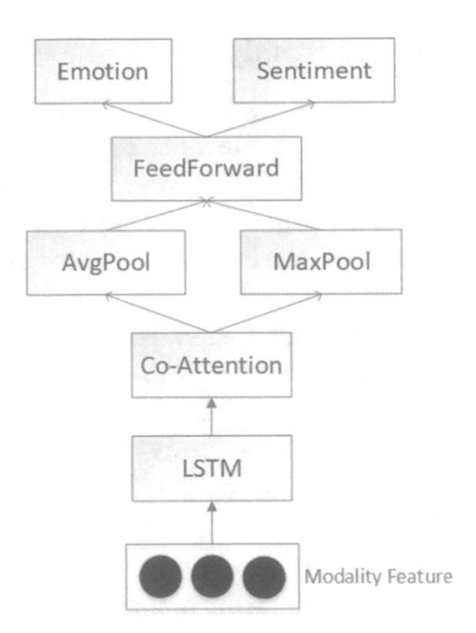

Fig. 1. Proposed model architecture

3) To get global feature and salient feature from the output of cross-modal attention layers. We apply average pooling and maximum pooling to the concatenated output of attention layers.
4) To increase the generalization and avoid overfitting, we resort to the idea of multi-task learning by let the model predict sentiment label and emotion label.

3.1 Modality Representation Extraction

Multimodal sequences usually involve three main forms: language (L), video (V) and audio (A). CNN can be used to extract the local information of the sequence, with similar functionality to N-Gram. Different modality sequences are collected at different sampling rates.

In order to make the input sequence have a common dimension and contain contextual information. The input sequence is passed to the two way LSTM layer: The LSTM model consists of reset gate and update gate, which is denoted as follows:

$$r_t = \delta\left(U_r x_{it} + W_r h_{t-1} + b_r\right) \tag{1}$$

$$z_t = \delta\left(U_z x_{it} + W_z h_{t-1} + b_z\right) \tag{2}$$

$$h_t = \tanh\left(U_h x_{it} + W_h\left(r_t * h_{t-1}\right) + b_h\right) \tag{3}$$

$$h_t = z_t * h_{t-1} + (1 - z_t) * h_t \tag{4}$$

where h_t is the state of the hidden layer at time t of the utterance; U,W, and b are the weight and the paranoia coefficient, respectively. x_{it} is the input feature value of the t-th element in sequence i. BiLSTM can remember not only the forward sequence dependency, but also the backward context information. Therefore, using a two-way LSTM has more advantages in capturing context than a one-way LSTM. In addition, using BiLSTM to project features of different modalities to the same dimension d can ensure that the cross-modal attention mechanism can be used later.

4 Co-attention

How to model the interaction between different modalities has always been a core issue in the field of multimodal research. Vanswani et al. [17] proposed Transformer, a brand-new language model that relies purely on the attention mechanism. The attention used in Transformer is called scaled dot product attention which is also called the self-attention mechanism. Self-attention has three input matrices, query matrix (Q), key matrix (K), and value matrix (V).

For self-attention, Q, K, and V are project from the same input matrix. If the input size is N × K, the operation $QK^\top$ will get an attention weight matrix. The expression $\sqrt{k}$ is a scale factor used for scaling. The concept of Multi-Head Attention is to stack multiple self-attentions from different locations that represent information in different sub-spaces:

The difference between cross-modal attention and self-attention lies in replacing K and V with the key matrix and value matrix of the modality that you want to attend to. For instance, given modality X_α and X_β, the cross-modal attention is formulated as

$$Y_\alpha = CM_{\beta\to\alpha}(X_\alpha, X_\beta) \tag{5}$$

$$= \text{Softmax}\left(\frac{Q_\alpha K_\beta^T}{\sqrt{d_k}}\right) V_\beta \tag{6}$$

$$= \text{Softmax}\left(\frac{X_\alpha W_{Q_\alpha} W_{K_\beta}^T X_\beta^T}{\sqrt{d_k}}\right) X_\beta W_{V_\beta} \tag{7}$$

5 Experiment

In this section, we describe the dataset used in the experiment and experiment details, and provide the experimental results and analysis.

5.1 Dataset

CMU-MOSEI [16] was proposed by Zadeh et al. The dataset was extracted from 23000 conversations from more than 1000 videos on the YouTube website. The training set, validation set and test set contain 16216, 1835, and 4625 utterances respectively. At coarse-grained, the sentimental label of each utterance is divided into a value from −3 to +3, when the value is less than zero, it is regarded as negative, and when the value is greater than or equal to zero, it is regarded as positive. At the same time, each utterance is divided into 6 emotional levels at a fine-grained level, namely anger, disgust, fear, happiness, sadness, and surprise. A conversation may contain the emotional changes of the speaker, for example, from happy to worry. Therefore, when multiple emotions are included in the utterance, it is necessary to distinguish these emotions at the same time and transform the problem into a multi-label classification problem. If there is no emotion label in a certain conversation, add the no emotion label is added into the six emotions to form seven emotion labels, and use the BinaryCrossEntropy optimization function to optimize the emotion labels on all time series at the same time. We use CMU-Multi-modal Data SDK to download and extract features. The text, audio, and video are extracted by GloVe, CovaRep, and Facet, respectively, and the obtained features are averaged according to the dimension of the word, and finally a sentence-level feature representation is obtained.

5.2 Baselines

The following classic multi-modal analysis models are mainly used as benchmark models:

1) **EF-LSTM** (Early Fusion LSTM) concatenated different modality features and simply treat them as single modality features. LSTM is used to capture contextual information.
2) **LF-DNN** (Late Fusion DNN) simply concatenated outputs from sub-networks and vote for classification.
3) **MFN** (Memory Fusion Network) [15]: In the multi-view sequence modeling, there are interactions within the views and interactions between the views, and the memory fusion network continuously models these two interactions on the time scale.
4) **Graph-MFN** (Graph Memory Fusion Network) [16] replaces the attention network in the MFN model with a dynamic fusion graph to model the interaction between modalities.

5.3 Results and Discussion

Table 1. The experimental results on the dataset.

	Emotion														Sentiment	
	Anger		Disgust		Fear		Happy		Sad		Surprise		Average		F1	Acc
	F1	W-Acc	F1	W-Acc	F1	W-Acc	F1	W-Acc	F1	W-Acc	F1	W-Acc	F1	W-Acc		
EF-LSTM	11.3	50.3	57.5	57.9	87.8	57.4	38.9	50.5	11.2	50.1	85.8	56.5	39.4	53	62.9	72
LF-DNN	29	51.7	22.3	50.7	34.9	56.8	57.9	59.1	26	51	30.4	50.8	26.3	51.1	59.7	58.1
Graph-MFN	**72.8**	62.60	76.6	69.1	**89.9**	**62**	**66.3**	**66.3**	66.9	**60.4**	85.5	53.7	**76.3**	**62.3**	77.0	76.9
Proposed	72.53	**63.72**	**77.46**	**69.74**	87.82	57.89	55.98	50.47	**68.50**	59.14	**85.89**	**56.45**	74.70	59.57	**77.28**	**77.77**

Table 1 shows the comparison of F1-value, accuracy and weighted accuracy of the model proposed in this paper on the CMU-MOSEI dataset. The best results are in bold. In terms of sentiment classification, the proposed model outperforms all baselines with 77.77% accuracy and 77.28% F1-score. In terms of emotion recognition, the W-Acc of anger improved 1.12% from 62.6% to 63.72%. The F1-score and W-Acc of disgust are both improved by 0.86% and 0.64% respectively. The F1-score of sad improved by 1.6% from 66.9% to 68.5%. In term of surprise, both evaluation metrics are improved by 0.39% and 2.75%. From the above experimental results, we can see that our model outperforms baselines on sentiment, and outperforms strong baselines in terms of emotion such as disgust and surprise. These experimental results demonstrated the good performance of our proposed model.

6 Conclusion

Multimodal sentiment analysis is an emerging hot spot in the computer field in recent years, which brings new opportunities and challenges to traditional text-based sentiment analysis research. It not only requires the ability to understand the interaction between different modalities, but also the ability to correctly find the modal feature representation that contributes the most to the task. This paper proposes a multi-task learning model based on attention mechanism, which BiGRU to obtain contextual representation of uni-modal sequence data, and then uses a cross-modal attention mechanism to effectively fuse the modalities. The experimental results on the public data set CMU-MOSEI show that the classification performance of the model has been greatly improved whether it is in the coarse-grained sentiment classification problem or in the fine-grained emotion recognition. However, it can also be seen from the experimental results that certain emotions, such as fear and happiness, do not perform well. Future work will continue to study and improve the accuracy of emotion recognition.

References

1. Huddar, M.G., Sannakki, S.S., Rajpurohit, V.S.: A survey of computational approaches and challenges in multimodal sentiment analysis. Int. J. Comput. Sci. Eng. **7**(1), 876–883 (2019)
2. Zadeh, A., Chen, M., Poria, S., et al.: Tensor Fusion Network for Multimodal Sentiment Analysis. arXiv:1707.07250 (2017)
3. Poria, S., Peng, H., Hussain, A., et al.: Ensemble application of convolutional neural networks and multiple kernel learning for multimodal sentiment analysis. Neurocomputing **261**, 217–230 (2017)
4. Soleymani, M., Garcia, D., Jou, B., et al.: A survey of multimodal sentiment analysis. Image Vis. Comput. **65**, 3–14 (2017)
5. Caruana, R.: Multitask learning. Mach. Learn. **28**(1), 41–75 (1997)
6. Zhang, Y., Yang, Q.: A Survey on Multi-Task Learning. arXiv:1707.08114 (2018)
7. Ruder, S.: An Overview of Multi-Task Learning in Deep Neural Networks. arXiv:1706.05098 (2017)

8. Duong, L., Cohn, T., Bird, S., et al.: Low resource dependency parsing: cross-lingual parameter sharing in a neural network parser. In: Proceedings of the 53rd Annual Meeting of the Association for Computational Linguistics and the 7th International Joint Conference on Natural Language Processing (Volume 2: Short Papers), pp. 845–850. Association for Computational Linguistics, Beijing (2015)
9. Snoek, C.G.M., Worring, M., Smeulders, A.W.M.: Early versus late fusion in semantic video analysis. In: Proceedings of the 13th Annual ACM International Conference on Multimedia - MULTIMEDIA 2005, Hilton, Singapore, p. 399. ACM Press (2005)
10. Vielzeuf, V., Pateux, S., Jurie, F.: Temporal Multimodal Fusion for Video Emotion Classification in the Wild. arXiv:1709.07200 (2017)
11. Wu, H., Mao, J., Zhang, Y., et al.: Unified visual-semantic embeddings: bridging vision and language with structured meaning representations. In: 2019 IEEE/CVF Conference on Computer Vision and Pattern Recognition (CVPR), pp. 6602–6611 (2019)
12. Andreas, J., Rohrbach, M., Darrell, T., et al.: Learning to compose neural networks for question answering. In: Proceedings of the 2016 Conference of the North American Chapter of the Association for Computational Linguistics: Human Language Technologies, pp. 1545–1554. Association for Computational Linguistics, San Diego (2016)
13. Poria, S., Cambria, E., Hazarika, D., et al.: Context-dependent sentiment analysis in user-generated videos. In: Proceedings of the 55th Annual Meeting of the Association for Computational Linguistics (Volume 1: Long Papers), pp. 873–883. Association for Computational Linguistics, Vancouver (2017)
14. Chen, M., Wang, S., Liang, P.P., et al.: Multimodal sentiment analysis with word-level fusion and reinforcement learning. In: Proceedings of the 19th ACM International Conference on Multimodal Interaction, pp. 163–171. Association for Computing Machinery, New York (2017)
15. Zadeh, A., Liang, P.P., Mazumder, N., et al.: Memory fusion network for multi-view sequential learning. In: Proceedings of the AAAI Conference on Artificial Intelligence, vol. 32, no. 1 (2018)
16. Bagher Zadeh, A., Liang, P.P., Poria, S., et al.: Multimodal language analysis in the wild: CMU-MOSEI dataset and interpretable dynamic fusion graph. In: Proceedings of the 56th Annual Meeting of the Association for Computational Linguistics (Volume 1: Long Papers), pp. 2236–2246. Association for Computational Linguistics, Melbourne (2018)
17. Vaswani, A., Shazeer, N., Parmar, N., et al.: Attention Is All You Need. arXiv:1706.03762 (2017)

Research on Problem Matching Classification Based on Artificial Intelligence Technology

Linpeng Ban[1], Bo Zhang[1(✉)], Jing Zhang[2(✉)], Junda Lian[1], Xiaonan Zhao[1], and Cheng Wang[1]

[1] Tianjin Key Laboratory of Wireless Mobile Communications and Power Transmission, College of Electronic and Communication Engineering, Tianjin Normal University, Tianjin 300387, China
b.zhangintj@outlook.com

[2] Basic Course Department, Tianjin Vocational Institute, Tianjin 300350, China
rebecca_0228@126.com

Abstract. In recent years, due to the continuous development of the Internet, electronic text information based on natural language expression has gradually increased. Among so many electronic text information, one of the biggest goals based on natural language processing is to effectively obtain and manage this information. For the questions raised by users, natural language processing technology needs to be used for in-depth research and sorting. The classification of questions will be extra important here, and it is the basis of information retrieval. Research on the theory and engineering application of problem classification tasks will also be of great significance. After preprocessing and vectorization of the problem, this paper uses the long and short-term memory model to complete the classification of the problem.

Keywords: Artificial intelligence · Machine learning · Natural language processing · Problem matching classification · Text processing

1 Introduction

Questions expressed by users can generally be divided into different categories according to their different characteristics. The methods of problem classification are mainly based on pattern matching methods and machine learning methods, respectively. The method based on pattern matching is the earliest adopted method. Since questions can be divided into different types, corresponding pattern matching templates need to be established, and the question categories are divided according to the degree of matching between the question sentence and

The work was partially supported by the Natural Science Foundation of Tianjin (18JCYBJC86400), the National Natural Science Foundation of China (62101383) and Doctor Fund of Tianjin Normal University (No. 52XB1604).

Q. Liang et al. (Eds.): Artificial Intelligence in China, LNEE 854, pp. 167–172, 2022.
https://doi.org/10.1007/978-981-16-9423-3_21

the template. The method based on machine learning is to classify the problem through a data-driven method, by extracting the relevant feature set corresponding to the problem, and by establishing a large number of training sets for testing to train the classifier. Meanwhile, the quality of features is also related to the classification effect. In terms of natural language processing, it uses the original text to automatically learn low-latitude and dense feature vectors containing its syntactic and grammatical features through a multi-layer network structure, which solves the shortcomings of sparse and high-dimensionality in the short text feature representation in traditional methods.

The remaining part of this paper is structured as follows. Problems encountered by the problem matching classification is given in Sect. 2. In Sect. 3, research of question matching classification is given. Experimental results are shown in Sect. 4 and conclusions are drawn in Sect. 5.

2 Problems Encountered by the Problem Matching Classification

Question matching is widely used in natural language processing tasks such as information retrieval, automatic question answering, machine translation, dialogue systems, and retelling questions. These natural language processing tasks can be understood as question matching problems to a certain extent. For example, information retrieval can be used as query and document matching, automatic answers can be used as matching questions and answers, and machine translation can be used as matching between two languages. The problems encountered by the problem matching classification technology are mainly manifested in the following aspects:

(1) The diversity of word matching
Different words may have the same meaning, such as synonyms; and the same words may have different meanings in different situations.
(2) The structure of phrase matching
Different words can be combined into phrases according to a specific structure. When matching two phrases, the structure information of the two phrases should also be taken into consideration. For example, "I like you" and "You like me" only match words, but the order is reversed, so the meaning is different.
(3) The level of text matching
The text is organized in a hierarchical manner, using multiple words to form sentences, and then using multiple sentences to form paragraphs. Therefore, when we do problem matching classification, we need to consider different levels of matching information, and organize our text matching information in a hierarchical manner.

3 Research of Question Matching Classification

3.1 Text Preprocessing

3.1.1 Word Segmentation Technology

Word segmentation is to separate the text into individual words. Word segmentation technology is conducive to feature extraction. When understanding the content of text, we must perform word segmentation. This paper uses a statistical-based word segmentation method for word segmentation which has a long-term development based on the traditional statistical model learning method, and it must be manually fixed and manual feature extraction [1]. The main idea is to treat each word as composed of characters. If connected characters appear multiple times in different texts, it proves that these connected characters are likely to be one word.

3.1.2 Delete Stop Words

It will be automatically filtered out that have little effect on the text, such as: conjunctions, modal particles, punctuation marks, etc. The common method is to use a list of stop words and delete the stop words through string matching after completing the first word segmentation. The removal of stop words also reduces the redundancy of the data and improves the accuracy of classification [2].

3.2 Vectorization

Since the text is unstructured data, the computer cannot process it. Therefore, it is necessary to vectorize the text into a data form that the computer can process, and then input the classification model for training.

3.3 Classification Algorithm

Text classification occupies an important position in the field of Natural Language Processing (NLP). At present, the commonly used methods of text classification can be divided into three categories: (1) classification methods based on statistics, such as Bayesian classifier; (2) classification methods based on association network learning, such as neural networks; (3) rule-based The resulting method, such as decision tree classification [3]. This paper uses the second type of Long Short-Term Memory (Long Short-Term Memory, LSTM) to classify text.

As a deformed model of RNN, LSTM model is one of the most widely used solutions [4].

As shown in Fig. 1, the LSTM model is mainly composed of four parts: input gate, output gate, forget gate and memory unit. The function of the input gate is to selectively record new information in the cell state, the output gate is to control which information of the memory cell is output, and the forget gate is to selectively forget some information. x_{t-1}, x_t, x_{t+1} are the three inputs at the previous moment, at the current moment, and at the next moment, respectively.

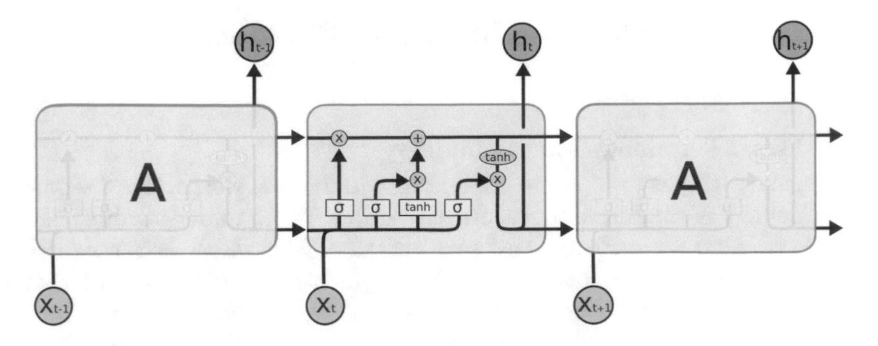

Fig. 1. LSTM model

test_id	question1	question2
0	How does the Surface Pro himself 4 compare with iPad Pro?	Why did Microsoft choose core m3 and not core i3 home Surface Pro 4?
1	Should I have a hair transplant at age 24? How much would it cost?	How much cost does hair transplant require?
2	What but is the best way to send money from China to the US?	What you send money to China?
3	Which food not emulsifiers?	What foods fibre?
4	How "aberystwyth" start reading?	How their can I start reading?
5	How are the two wheeler insurance from Bharti Axa insurance?	I admire I am considering of buying insurance from them
6	How can I reduce my belly fat through a diet?	How can I reduce my lower belly fat in one month?
7	By scrapping the 500 and 1000 rupee notes, how is RBI planning to fi	How will the recent move to declare 500 and 1000 denomination lewin illegal will curb black money?
8	What are the how best books of all time?	What are some of the military history books of all time?
9	After 12th years old boy and I had sex with a 12 years old girl, wit	Can a 14 old guy date a 12 year old girl?
10	What is the best slideshow app for Android?	What are the best app for android?
11	What services are from Google: Facebook, YouTube betray Twitter?	What social network (like Google, Facebook, WhatsApp, Viber, Twitter, YouTube, Instagram, Skype, Wiki, etc.) made huge i
12	What if a cricket hits a batsman' s helmet and then goes to the boun	Should carbonated red balls and 8 yellow balls. If 5 balls are drawn what is the probability of getting 2 red balls and
13	Just how do you learn fruity loops?	How do Fruity Wrappers work?
14	Why does Batman get kill in Batman v Superman?	In Batman v Superman, why reduce Lex Luthor pit Superman against Batman?
15	When can I buy a SpaceX stock?	Should I sell or buy LNKD stock?
16	Is it gouging and price fixing?	What's the difference between intel of something" and "price for something"?

Fig. 2. Data set

h_{t-1}, h_t, h_{t+1} are the three outputs of the previous moment, the current moment, and the next moment, respectively.

The input gate, output gate and forget gate of LSTM are a kind of non-linear summation unit. In the summation unit, a series of multiplication operations are performed on the data from the inside and outside of the unit to generate the gate. The activation conditions, the input gate and output gate are obtained by multiplying the corresponding input data and output data, and the forgetting gate is obtained by multiplying the previous state.

4 Experimental Results

As shown in Fig. 2, the data used in this article is obtained through web crawler technology, which is used to collect specific questions on FaceBook, and aggregate and display the collected data in the form of Excel. The basic facts are a set of labels provided by experts, but the basic fact labels are a subjective idea, because the true meaning of the sentence cannot be accurately known. Therefore, the true label of the data set is not necessarily accurate, and some of them may not be accurate. Therefore, the goal of this research is to predict whether the collected questions have the same meaning for the two questions, so as to increase the accuracy of matching classification.

By preprocessing the collected data set, we use the info() function to conduct an overview of the number of rows, columns, column types and other information of the sample data, and use the Tokenizer() function to analyze the data

	id	qid1	qid2	question1	question2	is_duplicate	seq1	seq2
0	0	1	2	What is the step by step guide to invest in sh...	What is the step by step guide to invest in sh...	0	[2, 3, 1, 1222, 57, 1222, 2581, 7, 576, 8, 763...	[2, 3, 1, 1222, 57, 1222, 2581, 7, 576, 8, 763...
1	1	3	4	What is the story of Kohinoor (Koh-i-Noor) Dia...	What would happen if the Indian government sto...	0	[2, 3, 1, 559, 10, 14300, 13598, 5, 4565, 0, 0...	[2, 43, 182, 25, 1, 82, 237, 11296, 1, 14300, ...
2	2	5	6	How can I increase the speed of my internet co...	How can Internet speed be increased by hacking...	0	[4, 13, 5, 217, 1, 440, 10, 17, 361, 1827, 200...	[4, 13, 361, 440, 24, 3338, 57, 1344, 219, 109...
3	3	7	8	Why am I mentally very lonely? How can I solve...	Find the remainder when [math]23^{24}[/math] i...	0	[16, 72, 5, 2774, 312, 2757, 4, 13, 5, 649, 19...	[87, 1, 4170, 37, 230, 2234, 1343, 230, 3, 245...
4	4	9	10	Which one dissolve in water quikly sugar, salt...	Which fish would survive in salt water?	0	[23, 49, 7131, 8, 231, 1891, 2047, 10570, 12, ...	[23, 1945, 43, 1242, 8, 2047, 231, 0, 0, 0, 0,...

Fig. 3. Data preprocessing

```
Model: "functional_1"
____________________________________________________________________________________________
Layer (type)                    Output Shape         Param #     Connected to
============================================================================================
input_1 (InputLayer)            [(None, None)]       0
____________________________________________________________________________________________
input_2 (InputLayer)            [(None, None)]       0
____________________________________________________________________________________________
embedding (Embedding)           (None, None, 64)     1280000     input_1[0][0]
____________________________________________________________________________________________
embedding_1 (Embedding)         (None, None, 64)     1280000     input_2[0][0]
____________________________________________________________________________________________
lstm (LSTM)                     (None, 32)           12416       embedding[0][0]
____________________________________________________________________________________________
lstm_1 (LSTM)                   (None, 32)           12416       embedding_1[0][0]
____________________________________________________________________________________________
concatenate (Concatenate)       (None, 64)           0           lstm[0][0]
                                                                 lstm_1[0][0]
____________________________________________________________________________________________
dense (Dense)                   (None, 64)           4160        concatenate[0][0]
____________________________________________________________________________________________
dense_1 (Dense)                 (None, 1)            65          dense[0][0]
============================================================================================
Total params: 2,589,057
Trainable params: 2,589,057
Non-trainable params: 0
```

Fig. 4. Parameters of the output model

The set performs text preprocessing, segmentation of the text, and deletes some meaningless symbols or words, and then vectorizes the data set through the text-to-sequences method to convert each word of each character into Numbers, and finally through the np.asarray() function to convert the numbers into the corresponding array sequence, so that it can be input into the model for experimentation. As shown in Fig. 3, in the first row, id is the data field, qid1, qid2 are question pairs of the data set, question1, question2 are the full text of each question, is_duplicate is the objective function. If question1 and question2 have basically the same meaning, we set it to 1, otherwise it is 0. As shown in Fig. 4, this figure shows the parameters of the output model. This article uses a two-layer LSTM model, in which the input function is used to instantiate the

```
Epoch 1/7
3159/3159 [==============================] - 386s 122ms/step - loss: nan - accuracy: 0.6307
Epoch 2/7
3159/3159 [==============================] - 389s 123ms/step - loss: nan - accuracy: 0.6308
Epoch 3/7
3159/3159 [==============================] - 393s 124ms/step - loss: nan - accuracy: 0.6308
Epoch 4/7
3159/3159 [==============================] - 386s 122ms/step - loss: nan - accuracy: 0.6308
Epoch 5/7
3159/3159 [==============================] - 388s 123ms/step - loss: nan - accuracy: 0.6308
Epoch 6/7
3159/3159 [==============================] - 407s 129ms/step - loss: nan - accuracy: 0.6308
Epoch 7/7
3159/3159 [==============================] - 409s 130ms/step - loss: nan - accuracy: 0.6308
```

Fig. 5. Experimental results

Keras tensor, and embedding is the embedding layer. The input dimension of this layer is 64, which is the number of words. The data of LSTM is encoded, and the resulting array is spliced with the concatenate function. The dense is a fully connected layer, and the output of LSTM can be transformed into the desired output through dimensional transformation.

As shown in Fig. 5, a total of seven test experiments were performed on the collected experimental data this time. And the final accuracy rate obtained by training the model is 63%.

5 Conclusions

This paper analyzes and researches the problem matching classification technology based on artificial intelligence technology, proposes the corresponding research method, and verifies the good effect of the method through related experiments. Three important parts of the problem classification research task are introduced, including text preprocessing, text vectorization and text classification. Through these three technologies, the obtained data set was processed, and the final experimental results were obtained.

References

1. Cheng, Y.: Jiyu shendu xuexi de ziran yuyan chuli zhong wenti fenxi de yanjiu [Research on problem analysis in natural language processing based on deep learning]. Xi'an University of Posts and Telecommunications, Master's thesis (2018)
2. Guo, B.: Zidong wenda xitong zhong wenti wenben fenlei [Research on question text classification and answer extraction technology in automatic question answering system]. Kunming University of Science and Technology, Master's thesis (2018)
3. Guo, J.: Jiyu shuangxiang lstm de wenben fenlei fangfa yanjiu [Research on text classification method based on bidirectional lSTM]. Hubei University of Technology, Master's thesis (2010)
4. Zhou, X.: Jiyu shendu xuexi de wenti fenlei de yanjiu [Research on problem classification based on deep learning]. Harbin Institute of Technology, Master's thesis (2016)

Research on the Application of Natural Language Processing in the Virtual Comment's Classification

Junda Lian[1], Bo Zhang[1(✉)], Jing Zhang[2(✉)], Linpeng Ban[1], Xiaonan Zhao[1], and Cheng Wang[1]

[1] Tianjin Key Laboratory of Wireless Mobile Communications and Power Transmission, College of Electronic and Communication Engineering, Tianjin Normal University, Tianjin 300387, China
b.zhangintj@outlook.com

[2] Basic Course Department, Tianjin Vocational Institute, Tianjin 300350, China
rebecca_0228@126.com

Abstract. Natural language processing (NPL) is a booming technical field in artificial intelligence, and it is closely related to our lives. NPL studies how to enable computers to understand and use human language to achieve information interaction between humans and computers, and then extract effective text information from a large amount of text. This includes not only allowing computers to understand natural language texts, but also extending the ability to enable computers to express deep intentions based on natural language texts. This research briefly outlines some processes of natural language processing in practical applications. Network virtual comment classification problem is considered with a two-layer LSTM model to train the original data set, and the results show the effectiveness of the proposed design.

Keywords: Artificial intelligence · Natural language processing · Comment classification · Deep learning · Neural network

1 Introduction

Natural language processing emerged in the United States, and it studies various theories that can realize effective communication between humans and computers in natural language and method [1]. After the Second World War, in the 1950s, when electronic computers were in their infancy, the idea of using computers to process human language has already emerged. After the 1990s, due to the substantial increase in computer speed and storage capacity, NPL research has entered a high-speed period; at the beginning of the 21st century, as various

The work was partially supported by the Natural Science Foundation of Tianjin (18JCYBJC86400), the National Natural Science Foundation of China (62101383) and Doctor Fund of Tianjin Normal University (No. 52XB1604).

Q. Liang et al. (Eds.): Artificial Intelligence in China, LNEE 854, pp. 173–179, 2022.
https://doi.org/10.1007/978-981-16-9423-3_22

technologies began to merge with each other, the research of natural language processing renewed its brilliance. With the development of sentiment analysis, studies have been gradually classified based on different researched candidates. Among them, aspect-based sentiment analysis plays an important role in subtle opinion mining for online reviews [2].

The remaining part of this paper is structured as follows. Significance of Natural Language Processing Research is given in Sect. 2. In Sect. 3, the methods which use natural language processing technology to solve problem classification are given, and conclusions are drawn in Sect. 4.

2 Significance of Natural Language Processing Research

At the theoretical and technical level, natural language processing technology provides theoretical support for applications based on the joint innovation of the Internet and the education system; at the technical application level, the development of natural language processing guides the development of future network technologies such as information and communications technology.

3 Application of Natural Language Processing in the Virtual Comment's Classification

3.1 Data Collection and Classification

The tool used to obtain the data in this study is a web crawler. The crawler tool is used to collect specific target comments on FaceBook and Twitter, and the collected data is summarized in the form of Excel, as shown in Fig. 1.

Based on the thinking of "authenticity", this research uses these foreign social platforms to collect commentary text data, and uses clustering analysis as a data mining method to identify and characterize the learning experience of online learners. Since the format and content of the English texts collected are uneven, the text should be cleaned up first. Some garbled and meaningless data cleared directly, and some highly overlapping data selectively deleted. In order to better classify and select the appropriate tags, we use the NLTK tool to segment the data first. The advantage is that the independence of the sentence can be preserved, that is, the generated result is a two-dimensional list. We count the word frequency of the processed text and store it in the dictionary. The statistical result is stored as a list type, sorted by word frequency from high to low, and some meaningless articles (a, the, etc.), prepositions (at, of, etc.), and conjunctions are removed. Here, the top 6 most frequently occurring words are given in Table 1.

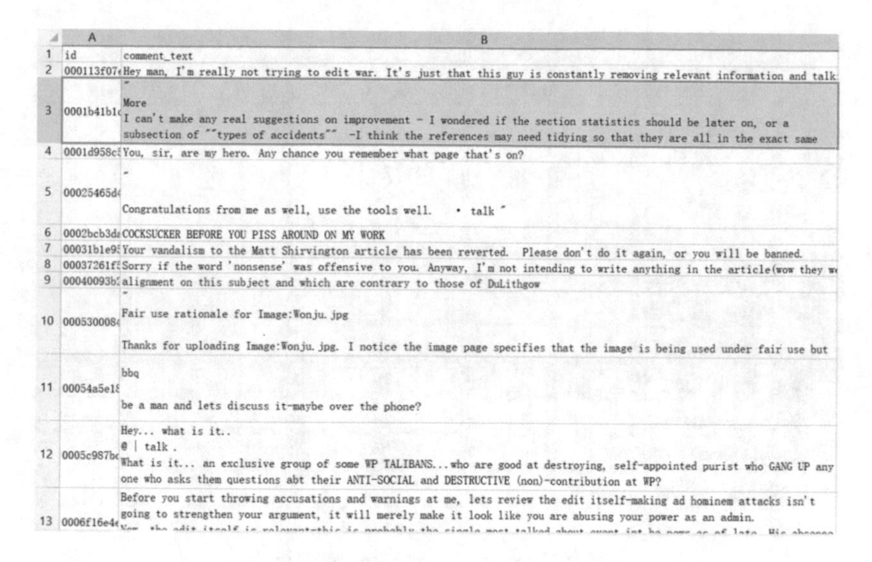

	A	B
1	id	comment_text
2	000113f07	Hey man, I'm really not trying to edit war. It's just that this guy is constantly removing relevant information and talk
3	0001b41b1	" More I can't make any real suggestions on improvement - I wondered if the section statistics should be later on, or a subsection of ""types of accidents"" -I think the references may need tidying so that they are all in the exact same
4	0001d958c	You, sir, are my hero. Any chance you remember what page that's on?
5	00025465d	" Congratulations from me as well, use the tools well. • talk "
6	0002bcb3d	COCKSUCKER BEFORE YOU PISS AROUND ON MY WORK
7	00031b1e9	Your vandalism to the Matt Shirvington article has been reverted. Please don't do it again, or you will be banned.
8	00037261f	Sorry if the word 'nonsense' was offensive to you. Anyway, I'm not intending to write anything in the article(wow they w
9	00040093b	alignment on this subject and which are contrary to those of DuLithgow
10	000530008	" Fair use rationale for Image:Wonju.jpg Thanks for uploading Image:Wonju.jpg. I notice the image page specifies that the image is being used under fair use but
11	00054a5e1	bbq be a man and lets discuss it-maybe over the phone?
12	0005c987b	Hey... what is it.. @ | talk . What is it... an exclusive group of some WP TALIBANS...who are good at destroying, self-appointed purist who GANG UP any one who asks them questions abt their ANTI-SOCIAL and DESTRUCTIVE (non)-contribution at WP?
13	0006f16e4	Before you start throwing accusations and warnings at me, lets review the edit itself-making ad hominem attacks isn't going to strengthen your argument, it will merely make it look like you are abusing your power as an admin.

Fig. 1. Comment data collection example

Table 1. Top 6 word frequency table

Order	Words	Word frequency
1	shit	662
2	fk	588
3	serious	343
4	race	311
5	shame	288
6	menace	274

Since some words contain insulting meanings, this research uses labels with similar meanings to replace them with synonyms. According to the results of word frequency analysis, text comments can be roughly divided into the following six categories: toxic, server toxic, obscene, threat, insult, identity hate. Then, we import the text data into Python to generate the word frequency matrix, some of the word frequency matrix is shown in Fig. 2. Here, we can see number 0 and 1 in the word frequency matrix. The number 1 in the word frequency matrix represents the label to which the keyword in this comment belongs, and the number 0 means that this comment does not belong to this category. If these 6 tags do not appear in a comment, then for this comment, the corresponding matrix is all 0. With this method, we have completed the preparation of the original data set, and the next step is to further preprocess the data set.

	A	B	C	D	E	F	G	H	I
1	id	comment_te	toxic	severe_to	obscene	threat	insult	identity_hate	
2	0005c987bc	Hey... what is it.. @ | talk	1	0	0	0	0	0	
3	00070ef96	Oh, and th	0	0	0	0	0	0	
4	001dc38a8	GET FUCKEI	1	0	1	0	0	0	
5	001e89eb3	Are you th	0	0	0	0	0	0	
6	001ee16c4	Thanks! Ur	0	0	0	0	0	0	
7	001ffdcc3	Awesome! 1	0	0	0	0	0	0	
8	0020e7119	Stupid pea	1	1	1	0	1	0	
9	0020fd96e	#NAME?	1	0	1	0	1	0	

Fig. 2. Word frequency matrix

3.2 Text Preprocessing

Here we need to convert text into vectors that can be directly recognized by computers.

3.2.1 Delete Stop Words and Case Conversion

We use the stopwords statement in the nltk.corpus library to delete stop words. The specific method is to compare each word in the text with the dictionary to find the stop words and delete them. For the case conversion of English text, we use lower() and replace() functions to unify all words into lowercase. Similarly, we use re.sub() to replace the matching items in the string.

3.2.2 Participle

Tokenizer is a class for converting text into a sequence. The job of the tokenizer is to decompose the text stream into tokens. In this text, each token is a subsequence of these characters.

This study sets num words to 40000 (words = 40000), and returns 4000 words with the highest frequency after processing all words by default; we uses the "fit on texts" method in the tokenizer to learn the text dictionary to train the text list; Use the "texts to sequences" method in the token generator to convert every word of each character into a number label.

3.2.3 Text Vectorization

Nowadays, the commonly used methods are word2vec/doc2vec. We first select k words, where the probability of these words co-occurring is higher than a few random words. Then, we use one word to predict the surrounding words (skip-gram), maximize the probability of neighboring words being selected in the vocabulary, and use gradient descent to update the weights.

3.3 Build a Model

3.3.1 Embedding

The purpose of the embedding layer is to project a high-dimensional, relatively sparse data in each dimension to a relatively low-dimensional data, and each dimension can be operated on data from a real number set. Essentially, it replaces discrete space with continuous space to increase space utilization and reduce unnecessary parameters.

The embedding layer is used as the first layer of the model. The input data is one hot encoding, and the input dimension is the number of words, which means that the space utilization is extremely low. In order to make full use of the model space, we connect the LSTM after embedding, so that the input dimension is high and the output dimension is low.

This study uses model.add (Embedding(40000, 128)), sets the embedding layer as the first layer, the maximum input value is 4000, and the output dimension is set to 128.

3.3.2 LSTM

In this paper, we present bidirectional Long Short Term Memory (LSTM) networks, and a modified, full gradient version of the LSTM learning algorithm [3]. The key to LSTM is the state of the cell. The state of the cell is similar to a conveyor belt. The state of the cell runs on the entire chain. Only some small linear operations are applied to it. Information can easily flow through the entire chain without changing. There are many versions of LSTM, one of which is an important version is GRU (Gated Recurrent Unit). According to Google test, the most important one in LSTM is Forget gate, followed by Input gate, and last is Output gate [4].

Based on these advantages, we adopts the construction of two LSTM layers after the Embedding layer as the transition layer. A sequential model was established using LSTM. The output dimension of the LSTM layer was set to 64 during training: units = 64; in order to solve the problem of overfitting, 20% neurons were automatically deleted (dropout = 0.2), and the Boolean value of the first layer LSTM was set to TRUE.

3.3.3 Dense

After the Boolean value of the second layer of LSTM is set to False, the system will only return the last output of the output sequence, and this output is a unit-dimensional vector, and a fully connected layer must be connected to convert the output of the LSTM into the desired output. Here we can simply understand the last Dense layer as a dimensional transformation.

```
Model: "sequential"
_________________________________________________________________
Layer (type)                 Output Shape              Param #
=================================================================
embedding (Embedding)        (None, None, 128)         5120000
_________________________________________________________________
lstm (LSTM)                  (None, None, 64)          49408
_________________________________________________________________
lstm_1 (LSTM)                (None, 64)                33024
_________________________________________________________________
dense (Dense)                (None, 6)                 390
=================================================================
Total params: 5,202,822
Trainable params: 5,202,822
Non-trainable params: 0
_________________________________________________________________
```

Fig. 3. Processed data set

Since the data output from the LSTM layer is non-linearly distributed, and the general neural network can only calculate linear data, we need to use the activation function to introduce non-linearity in the neural network to strengthen the learning ability of the network. The Sigmoid activation function is used to processing probability values. After the model is built, we can see it completely in the output column, as shown in Fig. 3.

3.4 Training Model

Train test-split() is a function provided in sklearn, which can divide the original data set into training set and test set according to a certain proportion. X is one subsets of the original data set, x-train is the original data set divided into training models, and x-val is the test set during the training process, so that we can see the results while training. The experimental code is shown in Fig. 4.

```
model.summary()
model.compile(loss = "binary_crossentropy", optimizer = "adam", metrics = ["AUC"])
x_train, x_val, y_train, y_val = train_test_split(train_padded, train_label, shuffle = True, random_state = 123)
model.fit(x_train, y_train, batch_size = 32, epochs = 1, validation_data = (x_val, y_val))

predict = model.predict(test_padded)
print("Predicted values are",predict)
```

Fig. 4. Code diagram

```
3740/3740 [==============================] - 750s 200ms/step - loss: 0.1030 - auc: 0.8892 -
val_loss: 0.0483 - val_auc: 0.9788

<tensorflow.python.keras.callbacks.History at 0x7fea00263dd0>
```

Fig. 5. Training result

When shuffle = True, random state = integer, the divided subset is out of order, so that randomness can be guaranteed, the accuracy of the test can be improved, and the subset obtained will not change when the statement is run multiple times. This research set the gradient descent to 32, the result is shown in Fig. 5. Since the collimation is 97.88%, it seems that the accuracy is pretty decent for a basic attempt.

4 Conclusions

This research shows that in the field of natural language processing, using the LSTM model to deal with the problem of problem matching works well. The core of dealing with this type of problem lies in defining classification labels for the original review text and choosing a suitable model to train the original data set. In the further experiments, this research is going to test some adjustable parameters to see if the accuracy can be improved to a new level.

References

1. Liu, X.: Zi ran yu yan chu li ji shu zai yuan dai ma yu chu li ying yong de yi yi yu zhan wang [significance and prospects of natural language processing technology applied in source code preprocessing]. Digit. World **2018**(10), 11 (2018)
2. Tian, Y., Yang, L., Sun, Y., Liu, D.: Cross-domain end-to-end aspect-based sentiment analysis with domain-dependent embeddings. Complexity **2021**(10), 1–11 (2021)
3. Graves, A., Schmidhuber, J.: Framewise phoneme classification with bidirectional LSTM and other neural network architectures. Neural Networks **18**(5–6), 602–610 (2005)
4. Sundermeyer, M., Schlüter, R., Ney, H.: LSTM neural networks for language modeling. In: Proceedings of the 13th Conference in the Annual Series of Interspeech events, Portland, September 2021

Design and Application of Endangered Animal Monitoring System Based on Mobile APP

Tian Qiu[1(✉)], Huanshan Wang[2], Lin lin[3], and Cong Chen[1]

[1] Wuhan Library, Chinese Academy of Science, Wuhan 430074, China
34337176@qq.com
[2] Institute of Hydrobiology, Chinese Academy of Science, Wuhan 430074, China
[3] Wuhan Wuchang Hospital, Wuhan 430080, China

Abstract. In view of the wide distribution area of Yangtze finless porpoise, the small number of existing population and the insufficient number of professional monitors, a mobile app-based finless porpoise information collection system is proposed and designed. The app obtains the number, status and location information of finless porpoise by collecting photos or short videos uploaded by social workers in each river section, and utilizes them through management processing. The paper describes the design scheme of the system and introduces the module functions of the program such as photo uploading, background information confirmation, and porpoise news pushing. The adoption of mobile app makes the system "easy to install and refundable", which facilitates more people to participate in finless porpoise conservation.

Keywords: Mobile app · Data acquisition · Animal protection

1 Introduction

Today in China, mobile Internet development has entered the era of all people, the number of mobile applications is growing rapidly, almost all walks of life are trying mobile clients, and in recent years there is a trend of massive development, rapid release and universal use of mobile applications. According to the survey statistics of the Internet Society of China, by the end of 2019, mobile Internet users reached 1.319 billion, accounting for 32.17% of the total number of global Internet users, and mobile Internet access traffic consumption was as high as 122 billion GB, in addition to the number of APPs reaching a theoretical 3.67 million [1]. Compared with the most commonly used native App at present, mobile App is the new trend that needs to be grasped more in the future development.

The Yangtze finless porpoise is one of the 13 global flagship species identified by the World Wide Fund for Nature. It is the only mammal in the Yangtze River. Since the 1980s, the population of Yangtze finless porpoise has been declining rapidly. Surveys show that the population size of the Yangtze finless porpoise is about 1012, which is only half of that of the giant panda, according to the results of the ecological and scientific study of the Yangtze finless porpoise announced in December 2017 [2]. However, with

Q. Liang et al. (Eds.): Artificial Intelligence in China, LNEE 854, pp. 180–190, 2022.
https://doi.org/10.1007/978-981-16-9423-3_23

the implementation of the "10-year fishing ban on the Yangtze River", the ecology of the Yangtze River basin has continued to improve, and the population of finless porpoise has a tendency to rebound. Monitoring the population and distribution of finless porpoise can help protect them. In fact, finless porpoises are widely distributed in the middle and lower reaches of the Yangtze River, so it is difficult to rely on researchers for comprehensive monitoring, and it is imperative to use social forces to assist in the monitoring of finless porpoises.

The application for monitoring porpoise through mobile app design not only does not need extra installation program, but also can collect photo-video data anytime and anywhere, and the user group is huge and the amount of uploaded data is huge. This can greatly improve the efficiency of finless porpoise monitoring and make due contribution to the protection of finless porpoise.

2 Requirements Analysis of Mobile App

2.1 Stronger Self-adaptation

With the advent of more and more smart devices, there is a uniform visual standard for use on different devices. This allows users to reduce adaptation time and consciously have the ability to adapt themselves when applying. In addition to this, users expect more interoperability and wider adaptability across different systems or devices [3].

2.2 Visualization of Data

We call it information visualization when we categorize and summarize a lot of information and display it to users in the form of pie, bar, curve, pattern, etc. in a visual and vivid way. Visualization design has been shown visually in many mobile applications. In recent years, the presentation of data information has also shown an increasingly rich trend [4]. This visualization has led to enhanced recognition and interest, making data more quickly and easily accessible to users.

2.3 Maximize Rate

In today's society, people spend more time on receiving information in fragments, so the efficiency of APP usage is particularly important. In response to this objective fact and usage demand, the design of hierarchy setting and structure in APP development will be more simplified and flatter in the visual interface [5]. Both horizontally and vertically, from depth to breadth, there are fewer entrances in the design to help users locate to their needs faster and complete information inclusion in an efficient way. With the reduction of entry points, another new requirement is to be better guided; a user who is new to the application needs a better guide to help get familiar with the application.

2.4 High Demand for Pictures

As we all know, this is an era of picture reading, and pictures are far more attractive to people than textual expressions. Compared with textual expressions, pictures have stronger visual expression [6]. Nowadays, the application of images in APPs tends to be large background high quality images, and this design has obvious effect in expressing emotion or product positioning.

2.5 Mobile Geolocation Services

Mobile geolocation services integrate mobile devices, GPS, GIS, and wireless communication, and can use navigation satellites or base stations to obtain their location, and then send the location information to a server that outputs it to an electronic map, which can be embedded into multiple application software platforms [7].

3 Porpoise Monitoring Mobile App Requirements Analysis

3.1 Determination of the Time and Location of the Appearance of Porpoises

Reporting the time and location of porpoise emergence is the core requirement for monitoring porpoise, which can most visually reflect the migration path and activity status of the population. At present, the time and location information of porpoise emergence is tracked by scientific researchers through research vessels. Due to the limited research staff, it is difficult to capture the time and location of porpoise emergence in time.

3.2 Status Determination of Porpoise

The number and health status of porpoises are also important data to monitor. Finless porpoises are active in the water, and it is difficult to observe their numbers and health status. The number of porpoises needs to be observed at close range for a long time and confirmed by the observer. The health condition of finless porpoises is more complicated, and various diseases or disabilities are difficult to confirm through short observation. It is difficult for researchers to determine the status of finless porpoises in the field.

3.3 Aggregation and Storage of Collected Information

At present, amateurs mainly rely on traditional methods such as telephone and mail to inform wildlife protection departments about the field observation data of finless porpoise, which is not regular and comprehensive, and it is difficult to collect data and storage efficiency is not high. It is difficult for scientific researchers to analyze valuable porpoise data through reported information.

3.4 Release of Information

At present, there is a lack of a public-oriented porpoise information release platform. The release platform can not only announce the latest research progress and ecological conservation achievements of the finless porpoise to the society, but also stimulate the public's enthusiasm for environmental protection and actively participate in wildlife conservation actions.

4 Mobile App Design Ideas

4.1 System Architecture

The system is mainly divided into two major parts from the overall architecture point of view: mobile app front-end and Web server back-end, the front-end of the system uses mobile app development to realize, the core function is to provide users with photo upload and browse information, the back-end of the Web server is built by Apache HTTP Server, using PHP language and MySQL database to develop text and photo video management system The backend of the web server is built by Apache HTTP Server, and the text and photo video management system is developed by using PHP language and MySQL database, which is convenient for the program manager to manage the operation of related data, and develop API and interface files to enable data interaction and communication

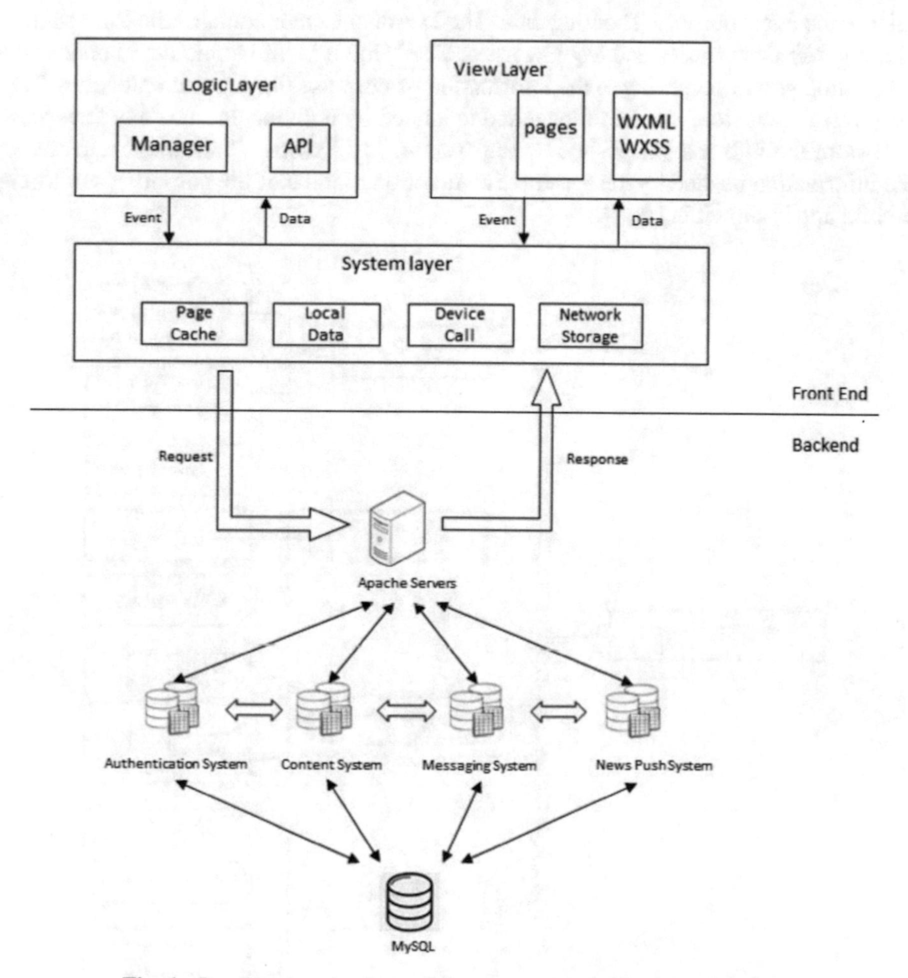

Fig. 1. Porpoise monitoring mobile app system architecture diagram

between the frontend of the mobile app and the backend of the server [8]. The system architecture diagram is shown in Fig. 1.

4.2 Function Modules

The Porpoise Monitor mobile app has four major modules, which are Home, Discovery, News and Personal Center. In the home page module, the extremely simplified design is used to highlight the most core function of the app: shooting and uploading. Users can quickly and intuitively enter the shooting interface to collect porpoise information, which not only avoids missing the shooting time, but also reduces the learning cost for users. The discovery module calls on the phone's photo function, short press the button to take a photo and long press the button to take a 15-s short video. The news module will call news related to porpoise for pushing and update news regularly. It will also release the latest porpoise shooting data. The Personal Center contains modules such as Honors, My Discoveries and My Favorites. The "Honor" will increase the honor value by gaining points according to the information of porpoise filming and uploading, "My Discovery" will store the data filmed and uploaded by individuals, and "My Favorites" will store the information pushed by the system. My Favorites" allows you to collect the information pushed by the system. The function module of the porpoise monitoring mobile app is shown in Fig. 2.

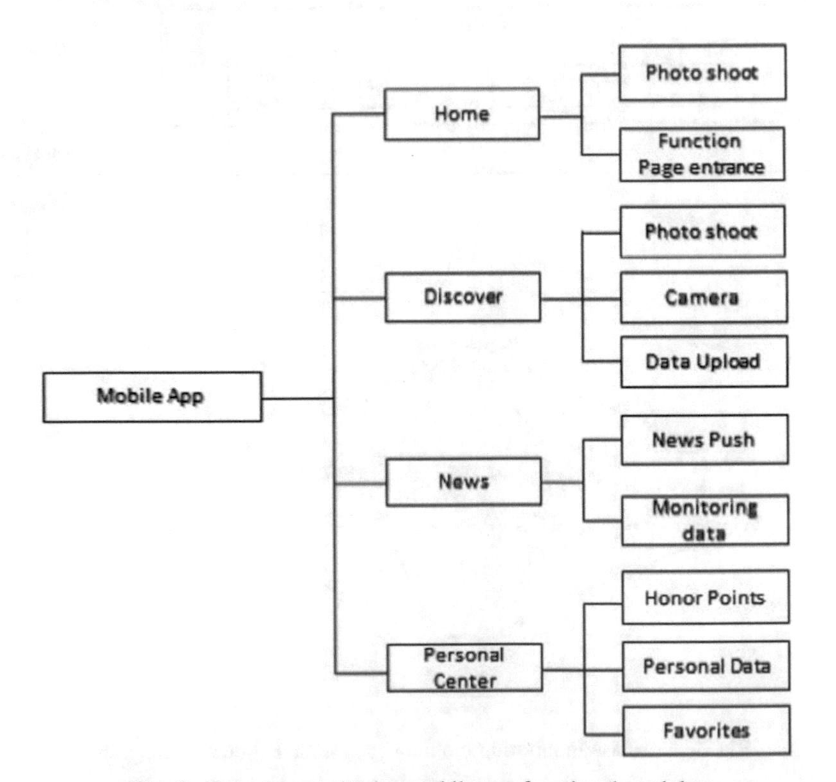

Fig. 2. Porpoise monitoring mobile app functional modules

4.3 Front-End Design

The front-end development of this system uses the mobile appMINA framework. The framework encapsulates the basic functions provided by the WeChat client, including file system, network communication, task management, etc. Developers can use the API to quickly complete application development. JSON, WXML, WXSS, and JAVASCRIPT are the main technologies used in the front-end of the system; JSON is mainly used to store the global configuration information of the applet and the configuration information of each page of the system; WXML, the WeChat Markup Language, is used to represent the structure and content of the applet pages, similar to HTML but very different from HTML because WXML has its own components It is similar to HTML but very different from HTML because WXML has its own components and syntax [9]; WXSS, the Wechat Style Sheet, is used to represent the applet page style, which is almost identical to CSS in WEB front-end development and can be used universally; JAVASRIPT is used to represent the logical structure of the applet, including page operation processing and API calls of the applet.

The js, json, wxml and wxss files in the index directory are used to implement the home page module; the app.json file is used to save the global configuration information of the applet, where the page path list in the pages property is used to specify the loading order of the applet pages, the Windows property is used to set the global default window of the applet, including the parameter settings of the navigation bar and the window, and the property tabBar is used to realize the top or bottom page switching of the applet window. The Windows property is used to set the global default window of the applet, including the parameters of the navigation bar and window settings, and the tabBar property is used to realize the top or bottom page switching in the applet window [10]. The pages is shown in Fig. 3.

Discovery module calls the phone photo function, short press the button to take a photo, long press the button to take a short video of 15 s. The specific code is as follows:

```
WeixinJSBridge.invoke('chooseVideo', {
sourceType : sourceType,
maxDuration : '8',//限制录制时间
camera : camera,
isShowProgressTips : 0
}, function(res) {
alert(JSON.stringify(res));
if (res.err_msg === "chooseVideo:ok") {
window.localId = res.localId;
callback();
}
});
```

Fig. 3. Pages display

4.4 Back-End Systems

The functions of the back-end of the system are mainly to facilitate data management and system maintenance, including data statistics, system settings, content audit, user management and other modules. The operation page, as shown in Fig. 4.

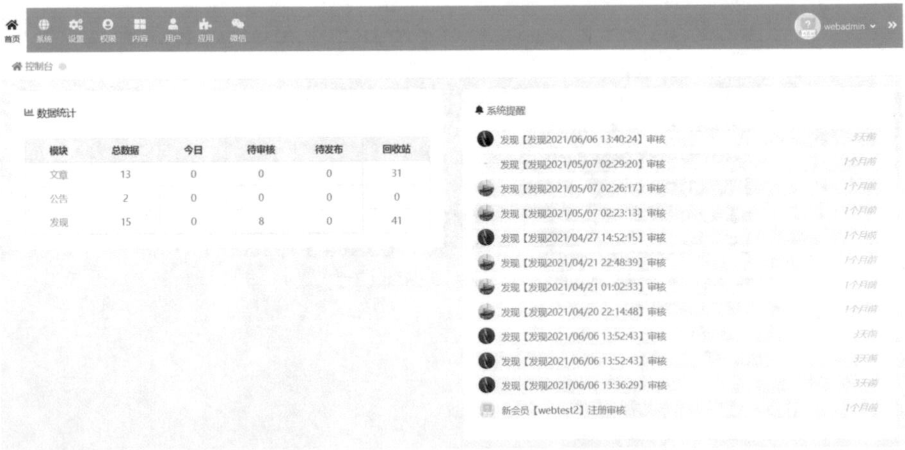

首页 系统 设置 权限 内容 用户 应用 微信 webadmin

控制台

数据统计

模块	总数据	今日	待审核	待发布	回收站
文章	13	0	0	0	31
公告	2	0	0	0	0
发现	15	0	8	0	41

系统提醒

发现【发现2021/06/06 13:40:24】审核 3天前
发现【发现2021/05/07 02:29:20】审核 1个月前
发现【发现2021/05/07 02:26:17】审核 1个月前
发现【发现2021/05/07 02:23:13】审核 1个月前
发现【发现2021/04/27 14:52:15】审核 1个月前
发现【发现2021/04/21 22:48:39】审核 1个月前
发现【发现2021/04/21 01:02:33】审核 1个月前
发现【发现2021/04/20 22:14:48】审核 1个月前
发现【发现2021/06/06 13:52:43】审核 3天前
发现【发现2021/06/06 13:52:43】审核 3天前
发现【发现2021/06/06 13:36:29】审核 3天前
新会员【webtest2】注册审核 1个月前

Fig. 4. Porpoise monitoring mobile app back-end system operation page

System setting is to standardize the page display, data format and other contents; user management module refers to the addition and deletion of managerial accounts, managerial role assignment, etc.; content audit refers to the setting of adding, modifying, deleting and finding the uploaded porpoise information to confirm the authenticity and reliability of the information, and its operation interface is shown in Fig. 5.

Fig. 5. Porpoise monitoring mobile app back-end content review page

The data statistics is the quantity, time and status of all the back-end data. The core technologies used for back-end development are mainly Apache server, database MySQL, and development language PHP. The indicators of porpoise information database are shown in Table 1.

4.5 Front-End and Back-End Communication

Data communication between the front-end of the system program and the back-end of the server is realized through the program's network API, specifically using the wx.request() method, which works by initiating HTTPS network requests from the small program end to the server, passing data parameters, and returning the server interface processing results. Take the discovery of the photo module and the back-end audit module as an example to introduce the front-end and back-end data communication process, the flow chart is shown in Fig. 6.

Table 1. Porpoise information data sheet

Field Name	Field Remarks	Field Type	Field width	Default
Id	Number	Int	11	Not Null
Image	Pictures	Blob	10000000	Null
Jkzt	Health Status	Decimal	10	Null
Sl	Quantity	Varchar	255	Null
Wzsm	Description	Varchar	10000	Null

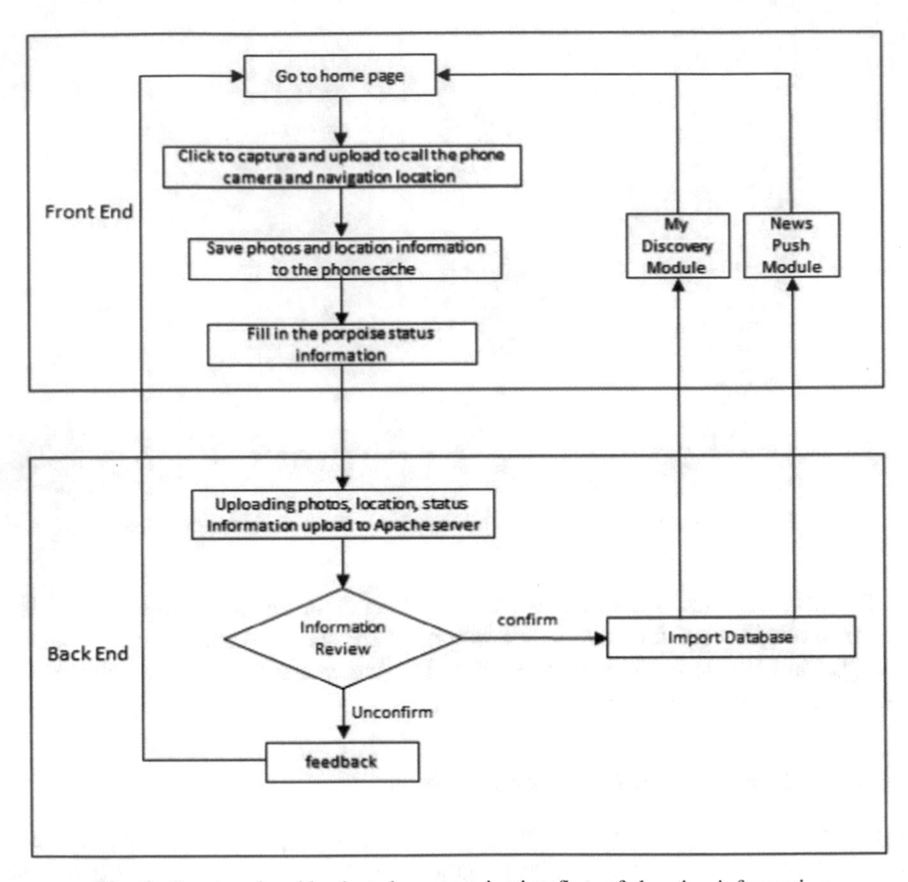

Fig. 6. Front-end and back-end communication flow of shooting information

5 Conclusions

In the context of ecological conservation in China and the growing concern of society for the Yangtze finless porpoise, an endemic aquatic animal in China, research and conservation of the finless porpoise has received a lot of social support. How to use social power more effectively to provide help for finless porpoise monitoring has also become a concern for researchers and the public. In this study, a mobile app-based porpoise information collection system is designed to facilitate the monitoring of porpoises on a large scale. The current features of the system are summarized as follows.

(1) The system adopts various development technologies such as WXML, WXSS, JAVASCRIPT, PHP and MySQL, etc. It can efficiently complete the management of personnel and content in the back-end of the web server, and realize the functions of home page, discovery, news and personal center in the front-end of the mobile app client, which can easily upload the information such as porpoise image and quantity status to the server quickly and publish It is easy to upload images, quantity status and other information to the server, and publish the information and dynamic news of porpoise after audit and confirmation.
(2) After the system was launched, it was proved through practice that the porpoise information collection system has more advantages than traditional wildlife information collection methods, and can meet the participants' needs of "at their fingertips" and "use and go", and the mobile app is a lightweight program system with much lower development and maintenance costs, which will bring a lot of convenience to the system manager.
(3) There are also some problems to be improved in the process of use. At present, the system does not have a large enough amount of data and has not been subjected to stress tests, and only trial accounts are provided for testing. The future will open a small-scale stress test, and eventually achieve a smooth nationwide use.
(4) The R&D team will continue to collect feedback from users, fix the bugs of the system and improve the lack of functions, so that the system will eventually become a universal monitoring tool, not only for finless porpoises, but also to support the information collection of other wild animals.

References

1. Zhang, L.: The development status of mobile Internet technology and future development trend. Electronic Component and Information Technology. **4**(9) (2020)
2. Mei, Z.G., Chen, M., Han, Y., Hao, Y., Zheng, J., Wang, K., Wang, D.: Thresholds of population persistence for the Yangtze finless porpoise: implications for conservation managements. Integr. Zool. **16**(4), 538–547 (2021). https://doi.org/10.1111/1749-4877.12523
3. Tandel, S.: Impact of progressive web apps on web app development. IJIRSET **7**(9) (2018)
4. Friendly, M.: A brief history of data visualization. In: Chen, C., Härdle, W., Unwin, A. (eds.) Handbook of Data Visualization, pp. 15–56. Springer Berlin Heidelberg, Berlin, Heidelberg (2008). https://doi.org/10.1007/978-3-540-33037-0_2
5. Kloumann, I.: The lifecycles of apps in a social ecosystem. In: WW'15: Proceedings of the 24th International Conference on World Wide Web, pp. 581–591 (2015)
6. Nathan, R.S.: A new open source platform for lowering the barrier for environmental web app development. Environ. Model. Softw. **85**, 11–26 (2016)

7. Adeniran, T.C., Anyaegbu, A.O., Olawoyin, L.A., Ajagbe, A.O.: Design of a GPS-based academic personnel clocking system. Niger. J. Technol. **38**(2), 406 (2019). https://doi.org/10.4314/njt.v38i2.17
8. Gong, W., et al.: Keywords-driven web APIs group recommendation for automatic app service creation process. Software Pract. Exper. **51**(11), 2337–2354 (2021). https://doi.org/10.1002/spe.2902
9. Deng, W.X.: Design of a WeChat learning platform for syndrome differentiation. Digit. Chin. Med. **1**(2), 143–154 (2018)
10. Tang, S.: Fortifying web-based applications automatically. In: Proceedings of the 18th ACM Conference on Computer and Communications Security October 2011, pp. 615–626 (2011)

A Rapid Image Semantic Segment Method Based on Deeplab V3+

Hailong Zhang[1(✉)], Hanwu Luo[2], Wenzhen Li[2], and Ruofeng Qin[2]

[1] Wuhan NARI Limited Liability Company, State Grid Electric Power Research Institute, Wuhan, People's Republic of China
8278799@qq.com

[2] East Inner Mongolia Electric Power Co., Ltd., Erdos, People's Republic of China

Abstract. As one of research directions in the fields of image processing and computer vision, image semantic segmentation classifies images by pixel for semantic comprehension. Due to the lack of a general theory of segmentation, conventional image semantic segmentation algorithms are mostly on the strength of specific image models, so a variety of methods develope with their respective range for using and relative merits. What's more, it's a pity that most of them can only use the underlying features to generate simple segmentation results of foreground and background. Since the breakthrough in deep learning at the ImageNet competition in 2012, convolutional neural networks have taken every field of computer vision by storm. Image semantic segmentation has broken through the limits of traditional methods rapidly after the integration of deep learning. After making use of convolutional neural networks, its accuracy for segmentation has been improved effectively so that image semantic segmentation at present can further meet the requirements of segmentation technology in different application scenarios. In recent years, it has developed rapidly and gradually become a research hotspot. This paper mainly studies the image semantic segmentation methods about Deeplab series based on deep learning and raises a lightweight model for semantic segmentation based on Deeplab v3+. Regarding MobileNet as backbone to extract features, it makes use of atrous spatial pyramid pooing to gain global multi-scale features. Then the decoder structure concatenates multi-scale features and low-level features to enhance the expression for spatial features and improve its performance. After sufficient experiments on PASCAL VOC using the fast semantic segmentation model, the result reveals that it obtains a good balance between accuracy and speed.

Keywords: Deep convolutional neural networks · Semantic segmentation · Deeplab

1 Introduction

The task of semantic segment is a pixel-level recognition task. The accuracy and speed of the task is an open challenge in many areas like medical image analysis, automatic driving and in-door reconstruction.

Q. Liang et al. (Eds.): Artificial Intelligence in China, LNEE 854, pp. 191–197, 2022.
https://doi.org/10.1007/978-981-16-9423-3_24

Nowadays, the deep learning model of semantic segment includes a various of methods that differ in their accuracy and speed like R-CNN, Deeplab, Self-Attention etc. And I'll introduce it briefly in the next episode.

In 2018, Maoke Yang in DeepMotion Inc(Beijing) came up with a new network composed of dilated space pyramid pooling ASPP and dense connected DenseASPP, which use a dense connection to combine the expanded convolution thus to have bigger reception field and denser features. The ExFuse framework supposed by Face++ from MEGVII Inc utilized a global convolution network as a basic partition, combined with convolution rearrange to attach more information in low-level semantic features. Also the DFN from Face++, which is composed of Smooth Network and Border Network, can efficiently handle the problem of the inconsistency inner class and the consistency inter class. Yang Zhixiao etc. from Hunan University of Industry put forward ZF-FCN architecture using full-convolution network to construct region network to solve the problem of a sparse result when using typical full-convolution network. Zhang Ligang etc. from Northeast Normal University attach the dilated convolution and dilated pooling to the traditional conv-network which makes the feature map be the same size as the original image. Gu Yifan etc. from Harbin Institute of Technology, Shen Zhen raise a new idea to replace the last few layers in res-net by grouped dilated convolution and take a cascaded up-sampling to get more effective features. Zhang Huabo etc. from University of Electronic Science and Technology combined the dilated-conv with U-net, which use an Encoder-Decoder, to expand the application of U-net.

Besides, Teams from Intelligent Information Research of CAS and Microsoft Research Asia came up with an object context expression OCR, which use a classified information to give weights for each pixel and connect these pixels with the original pixel features to get more robust feature expressions. Tan Lei etc. from Nanjing University of Science and Technology apply a selective conv core to form a res-net module SKAS to extract multi-scale information which makes the model to be lightweight and enhances the relation among the groups. Xi Yifan etc. from Chang 'an University made a change from deeplab v3+ using bottleneck structure and non-bottleneck structure in the backbone and decompose the 3×3 conv to 3×1 conv and 1×3 conv to reduce the calculation amount. Zhou Dengwen etc. from Zhejiang University fine tuned the MoblieNetv2 with a deep dilated pyramid pooling DASPP and dilated residual enhancement module AR in order to handle the features in deep and shallow layer. And Zhao Bing from Harbin Normal University utilized a channel & position attention mechanism to introduce a fine tune module in the resnet and catch global & local features through connected pyramid pooling module.

In world wile 2018, Liang-Chieh Chen etc. from Google raised Deeplab v3+ based on the previous Deeplab v3. This network apply the Xception as its backbone. The backbone is combined with ASPP module and the Encoder-Decoder architecture to both obtain high-level semantic information and low-level space information.

In 2020, Qibin Hou from National University of Singapore raised a new idea to take the long conv into account in order to capture the local context and simulate the long distance relation. In addition, they import the MPM formed by multiple pyramid pooling modules to construct the SPNet.

The main contribution of this paper can be summarized as follows:

1. We give a detailed analysis to the classical semantic segmentation methods based on deep learning.
2. We slightly modified the Deeplab v3+ to reach a balance between accuracy and speed.

2 A Rapid Image Semantic Segment Method Based on Deeplab v3+

The previous methods based on deep-conv network usually have a high demand for computation and storage resources, the complexity of the network itself also leads to a low inference speed. Based on the requirement of limited resources and interactive time limitation, we put forward a new lightweight network based on Deeplab v3+. Our network use the MobileNet v2 to extract features. The MobileNet v2 makes use of deep separable convolution to cut down the parameter amount of the network. Meanwhile, the MobileNet v2 modulate the ResNet to import res-connected bottleneck module. It adds a 1×1 conv before and after the deep conv to first expand the dimensions and then compress the feature map to shrink the dimensions which makes the model to reduce the demand for resources while maintaining the accuracy.

The main modules of Deeplab v3+ above can be listed as follows:

1. *Deep separable MobileNetv2*: This network contains 53 layers formed by deep separable bottleneck res blocks. The deep separable convolution can construct the new features under limited resources while maintain the performance of convolution. Because the information that deep separable convolution can gather largely depends on the input, unlike the res blocks in ResNet, MobileNet v2 import the res blocks in a form of "expand", "convolution", and "compression". It expands the dimension by 1×1 conv and reduce the information loss due to ReLU function, which has a global consideration about the handling of features.
2. *ASPP Module*: It takes the dilated convolution based on the space pyramid pooling of SPP to obtain multi-scale features. The dilated convolution can be charged as an expansion of the standard convolution. It turns to adjust the reception field of conv core to gather multi-scale context. The rate r deter-mines the step of sampling the input image, which means it'll insert r-1 zeros among the sample points. Thus, it can extend the reception field from k*k to $k + (k-1)(r-1)$. The structure of ASPP is shown as Fig. 2.2.

   As you can see in Fig. 2.2, the reception field extends as the dilated convolution inflation r increases while the parameter count of the conv core keep invariant so it can make a tradeoff between the accuracy and the speed. The ASPP module obtains the context of image in multi-scale so it can extract the deeper semantic information. However, due to the large amount channels and the abstraction of information after

combined, we need an encoder to decrease the channel of features and output high-level semantic information.
3. *Decoder*: Because the object border of feature map gained from ASPP is very sharp, the decoder use a 4 time bilinear interpolation to upsample the image and concatenate the output with the low-level features from the feature extraction network which have the same size as the decoder's output.

Compared to the ResNet, the MobileNet can simplify the network by using the deep separable convolution. Supposed the input size of features was D_F*D_F*M, the output size of features was D_F*D_F*N and the size of conv core was K*K, the parameter count and the computation amount of standard convolution can be computed as Formula 1 & Formula 2:

$$K \times K \times M \times N \tag{1}$$

$$K \times K \times M \times N \times D_F \times D_F \tag{2}$$

The same amount of deep separable convolution can be computed as Formula 3 & Formula 4:

$$K \times K \times M + M \times N \tag{3}$$

$$(K \times K \times M + M \times N) \times D_F \times D_F \tag{4}$$

So, the ratio of the same amount between deep separable conv and standard conv can be summarized as Formula 5:

$$\frac{1}{N} + \frac{1}{K^2} \tag{5}$$

Assuming that the output size of feature map is relatively large and the size of conv core is 3×3, then the count of parameters and the amount of computation of MobileNet v2 will be approximately one-nineth of ResNet-101. And that makes the model to be light. Here, we show the comparison in Deeplab v3+ between the MobileNet v2 backbone and the ResNet-101 backbone in Table 1:

Table 1. The parameters comparison

Network	Parameters
ResNet-101	59.34 M
MobileNet v2	5.82 M

As for the loss function, we use the cross entropy loss, which is a widely used loss function in classification tasks.

3 Experiment

3.1 Evaluation Index

In our experiment, we use MIoU to evaluate our model. Suppose there're k + 1 classes including k target classes and 1 background class, p_{ij} means the total count of pixel that is predicted as class i but actually class j. The computation of MIoU is shown in Formula 6:

$$MIoU = \frac{1}{k+1}\sum\nolimits_{i=0}^{k}\frac{p_{ii}}{\sum_{j=0}^{k}p_{ij}+\sum_{j=0}^{k}p_{ji}-p_{ii}} \tag{6}$$

3.2 Environment and Dataset

1. Environment: We use Linux Ubuntu 16.04.6 with the newest miniconda, Python 3.8.5.
2. Dataset: PASCAL VOC 2012 & Semantic Boundaries Dataset(SDB).

The PASCAL VOC dataset includes 20 classes and background marks. The dataset is divides as a training set of 1464 images and a evaluation set of 1449 images.

The SDB dataset is originated from the samples without marks in PASCAL VOC. It is divided as a training set of 8498 images and an evaluating set of 2857 images.

3.3 Training Configuration

The whole experiment use Deeplab v3+ as a basic architecture, we use ResNet-101 or MobileNet v2 as the feature extraction network and the MIoU as the evaluation index.

The training configuration is listed in Table 2:

Table 2. Training configuration

Epoch	GPU count	Batch size	Thread count	Learning rate	LR policy	LR in poly	Weight Decrease	Loss function	Optimizer	Momentum
50	1	8	4	0.007	poly	3	0.0005	CE	SGD	0.9

4 Result

After adding the SBD to PASCAL VOC, we trained our network and get the following result shown in Fig. 1.

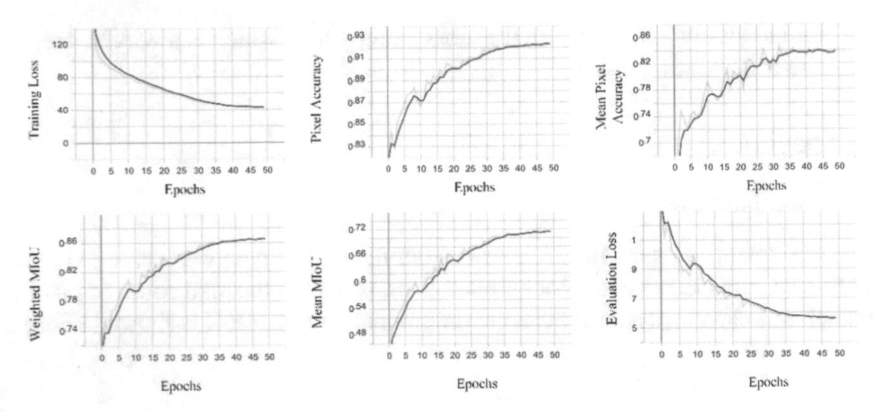

Fig. 1. Result in mobilenet backbone

And the training cost of ResNet & MobileNet is listed in Table 3:

Table 3. Training cost

Feature extract network	Parameters	Training time	Computation amount
ResNet	59.34 M	9.93	185.61B
MobileNet	5.82 M	6.34	54.13B

The comparison of accuracy between our methods and other classical methods is listed in Table 4.

Table 4. Comparison of MIoU

Model	Backbone	MIoU
RRM[45]	ResNet	66.30%
Deeplab v1	VGG	67.64%
Deeplab v3+	MobileNet	71.61%
Deeplab v3+	ResNet	77.53%
DFN[4]	ResNet	79.67%

The semantic segment results of our method reaches a good balance between accuracy and speed. But there're still some work to be done:

1. The batch size and scaling ratio can be further adjusted to make a better tradeoff.
2. The backbone can be optimized to get a more accurate result.
3. Considering a migration leaning in real scenes for a better matching.

Acknowledgements. This work was Funded by the State Grid Science and Technology Project (Research on Key Technologies of Intelligent Image Preprocessing and Visual Perception of Transmission and Transformation Equipment).

References

1. Minaee, S., Boykov, Y.Y., Porikli, F., et al.: Image segmentation using deep learning: a survey. IEEE Trans. Pattern Anal. Mach. Intell. (2021)
2. Yang, M., et al.: Denseaspp for semantic segmentation in street scenes. In: Proceedings of the IEEE Conference on Computer Vision and Pattern Recognition, pp. 3684–3692 (2018)
3. Zhang, Z., Zhang, X., Peng, C., Xue, Sun, J.: Exfuse: Enhancing feature fusion for semantic segmentation. In: Ferrari, V., Hebert, M., Sminchisescu, C., Weiss, Y. (eds.) Computer Vision – ECCV 2018: 15th European Conference, Munich, Germany, September 8-14, 2018, Proceedings, Part X, pp. 273–288. Springer International Publishing, Cham (2018). https://doi.org/10.1007/978-3-030-01249-6_17
4. Yu, C., et al.: Learning a discriminative feature network for semantic segmentation. In: Proceedings of the IEEE Conference on Computer Vision and Pattern Recognition, pp. 1857–1866 (2018)
5. Chen, L.-C., Zhu, Y., Papandreou, G., Schroff, F., Adam, H.: Encoder-decoder with atrous separable convolution for semantic image segmentation. In: Ferrari, V., Hebert, M., Sminchisescu, C., Weiss, Y. (eds.) Computer Vision – ECCV 2018: 15th European Conference, Munich, Germany, September 8–14, 2018, Proceedings, Part VII, pp. 833–851. Springer International Publishing, Cham (2018). https://doi.org/10.1007/978-3-030-01234-2_49
6. Zhang, H., et al.: Context encoding for semantic segmentation. In: Proceedings of the IEEE Conference on Computer Vision and Pattern Recognition, pp. 7151–7160 (2018)
7. Huang, Z., et al.: Ccnet: criss-cross attention for semantic segmentation. In: Proceedings of the IEEE International Conference on Computer Vision, pp. 603–612 (2019)
8. Takikawa, T., et al.: Gated-scnn: gated shape cnns for semantic segmentation. In: Proceedings of the IEEE International Conference on Computer Vision, pp. 5229–5238 (2019)
9. Hou, Q., et al. Strip pooling: rethinking spatial pooling for scene parsing. In: Proceedings of the IEEE Conference on Computer Vision and Pattern Recognition, pp. 4003–4012 (2020)
10. Kamann, C., Rother, C.: Benchmarking the robustness of semantic segmentation models. In: Proceedings of the IEEE Conference on Computer Vision and Pattern Recognition, pp. 8828–8838 (2020)
11. He, K., et al.: Deep residual learning for image recognition. In: Proceedings of the IEEE Conference on Computer Vision and Pattern Recognition, pp. 770–778 (2016)

Analysis and Research on China's New Energy Vehicles Industry Policy Based on Policy Subjects, Tools and Objectives

Rui Yang[1,2], Min Zhang[1,2](✉), Yanli Zhou[1,2], Jun Chen[1,2], and Jingling Xu[1,2]

[1] Wuhan Library of Chinese Academy of Sciences, Wuhan 430071, China
zhangmin@mail.whlib.ac.cn

[2] Hubei Key Laboratory of Big Data in Science and Technology (Wuhan Library of Chinese Academy of Sciences), Wuhan 430071, China

Abstract. New energy vehicles are the future trend of the automotive industry, and the rapid development of China's new energy vehicles industry is inseparable from the strong support of government policies. Sorting out and analyzing relevant industrial policies is an extremely important task, which is conducive to the overall grasp and correct judgment of the policies. This paper collects and organizes China's new energy vehicles industry policies from 2009 to 2020, and analyzes them from the three dimensions of policy subjects, policy tools, and policy objectives with the help of text mining technologies such as text preprocessing, named entity recognition, keyword extraction, and automatic classification. A total of 155 policy sample data were analyzed, it was found that 41 policy subjects were involved, and most policy subjects participated in joint publication. There were more environment-based policy tools and fewer demand-based and supply-based policy tools. The policy objectives were mainly related to industrial development, promotion and application, and energy saving and emission reduction.

Keywords: New energy vehicles · Policy subjects · Policy tools · Policy objectives · Text mining

1 Introduction

With the rapid development of China's economy, the problems of energy shortage and environmental pollution have become increasingly prominent. The automobile industry is an important pillar industry of the national economy and plays an important role in the development of the national economy and society. Under the pressure of energy and environmental protection, new energy vehicles will undoubtedly become the development trend of the future automobile industry. The development of new energy vehicles can not only alleviate the problem of environmental pollution, but also reduce energy consumption and protect the earth's ecological environment. The rapid development of China's new energy vehicles industry is inseparable from the strong support of government policies. This paper sorts out China's new energy vehicles industry policies from 2009 to 2020, constructs the new energy vehicles industry policy analysis framework,

Q. Liang et al. (Eds.): Artificial Intelligence in China, LNEE 854, pp. 198–208, 2022.
https://doi.org/10.1007/978-981-16-9423-3_25

and analyzes these policies from the three dimensions of policy subjects, policy tools and policy objectives with the help of text mining technology.

2 Research Design

2.1 Policy Analysis Framework Design

The new energy vehicles industry policies are a series of policy measures adopted by the government to promote the development of the new energy vehicles industry. Policy subjects, policy tools and policy objectives are important factors that the government must consider when designing, selecting, applying and evaluating policies [1]. This paper constructs the new energy vehicles industry policy analysis framework from the three dimensions of policy subjects, policy tools, and policy objectives, as shown in Fig. 1.

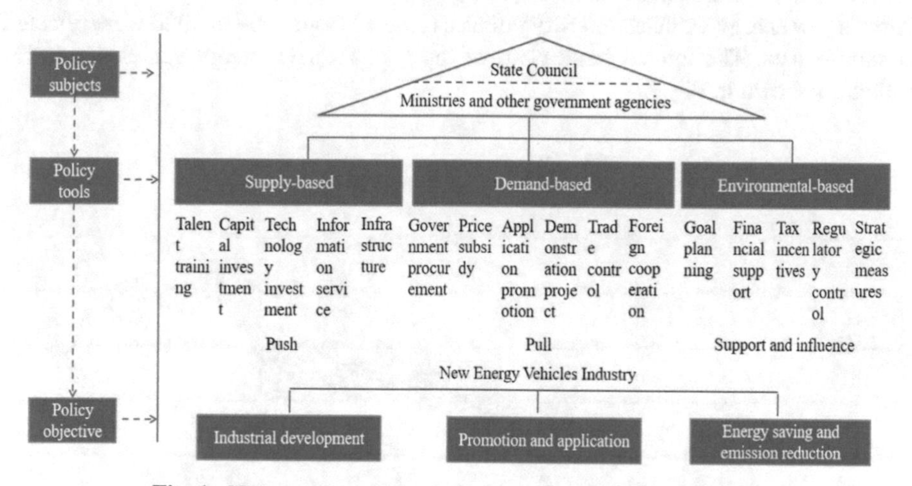

Fig. 1. New energy vehicles industry policy analysis framework

Policy subjects are groups that actively formulate, publish, and implement public policies, they act as decision-makers and managers in the policy operation system, determining and controlling policy directions, policy issues, policy content, policy objects, etc. Policy tools are means to achieve policy objectives, they are the management behaviors of policy subjects to achieve policy objectives. Policy objectives are the goals and effects that the policy subjects expect to achieve when formulating and implementing policies.

2.2 Data Sources

This paper collects and sorts out the relevant policies of the new energy vehicles industry through multiple channels such as government websites, Wanfang Data Knowledge Service Platform, etc., based on the principles of openness, authority, and relevance. Government websites mainly include the Chinese government website, the Ministry of

Finance of the People's Republic of China, the Ministry of Science and Technology of the People's Republic of China, the Ministry of Industry and Information Technology of the People's Republic of China, the National Development and Reform Commission of the People's Republic of China, the National Energy Administration, the Ministry of Transport of the People's Republic of China, etc. It is mainly searched by keywords in the information disclosure, policy release, notice announcement, policy and other columns of the government websites. The legal resources of Wanfang Data Knowledge Service Platform cover national laws, administrative regulations, departmental rules, judicial interpretations and other normative documents, and the sources of information are authoritative and professional [2]. Part of the sample data is obtained from the regulations database of Wanfang Data Knowledge Service Platform by searching keywords.

The above policy data obtained from different sources were de-duplicated by using fields such as file number, release time, and policy title. After screening and sorting, 155 China's new energy vehicles industry policies released from 2009 to 2020 were selected as sample data. The annual trend chart of the China's new energy vehicles industry policies is shown in Fig. 2.

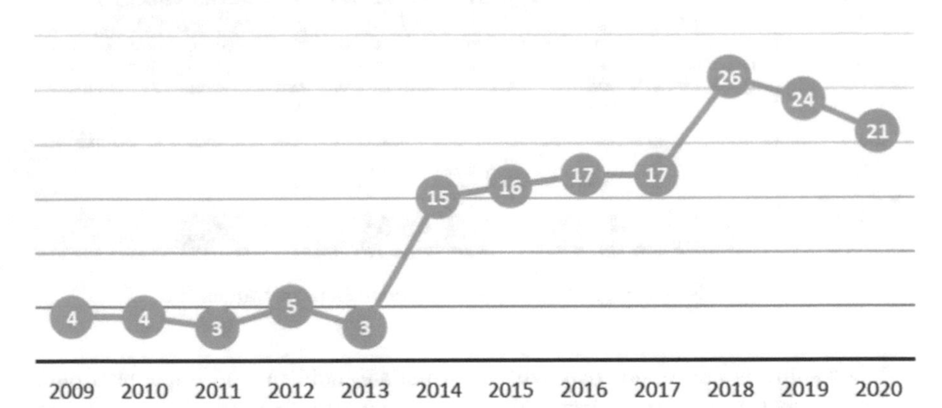

Fig. 2. Annual trend chart of the China's new energy vehicles industry policies

3 Three Dimensional Analysis Based on Policy Subjects, Policy Tools and Policy Objectives

3.1 Policy Subjects Analysis

Policy subjects are the policy issuing agencies. With the help of named entity recognition technology, extract the issuing agencies of the new energy vehicles industry policies, clean the names of the policy issuing agencies, and merge the different writing methods of the same issuing agency in the policy documents. According to statistics, there are

41 policy subjects, and 35 policy subjects participated in the joint publication. Make statistics on the number of policies issued by policy subjects, the top 10 policy subjects in terms of publication volume is shown in Table 1.

Table 1. The top 10 new energy vehicles industry policy subjects in terms of publication volume

Policy subject	Number of issued	Number of joint issued
Ministry of industry and information technology	111	95
State administration of taxation	58	57
Ministry of finance	46	42
National development and reform commission	31	31
Ministry of science and technology	30	27
State administration for market regulation	14	14
Ministry of commerce	14	14
Ministry of transport	12	11
Ministry of ecology and environment	10	9
National energy administration	8	7

Gephi [3] is an open source and free cross-platform JVM-based complex network analysis software, suitable for various graphics and networks. It is mainly used for various networks and complex systems, and is a supplementary tool for traditional statistics. Using the Gephi network analysis software to visually analyze the joint publication of policy subjects. The size of the node represents the number of policies issued by the agency, and the thickness of the edge represents the number of policies issued by cooperation between agencies. The network relationship of joint publication between policy subjects is shown in Fig. 3.

3.2 Policy Tools Analysis

3.2.1 Constructing Policy Tools Analysis Framework

In view of the complexity of policy tools themselves, scholars have divided the types of policy tools in various forms according to different standards. The rational use of policy tools is also an important foundation for policy subjects to successfully achieve policy objectives. Rothwell & Zegveld divided policy tools into three types: supply-based, demand-based, and environmental-based according to the degree of policy impact they produced [4]. Howlett & Ramesh divided policy tools into three categories: compulsory, mixed, and voluntary according to the degree of policy intervention by policy subjects [5]. Mcdonnell & Elmore divided the policy tools into four types: commands, incentives, capacity building, and institutional changes according to the different policy objectives achieved [6]. This paper uses Rothwell & Zegveld's supply-based, demand-based, and environmental-based classification criteria to analyze the policy tools used in the policy

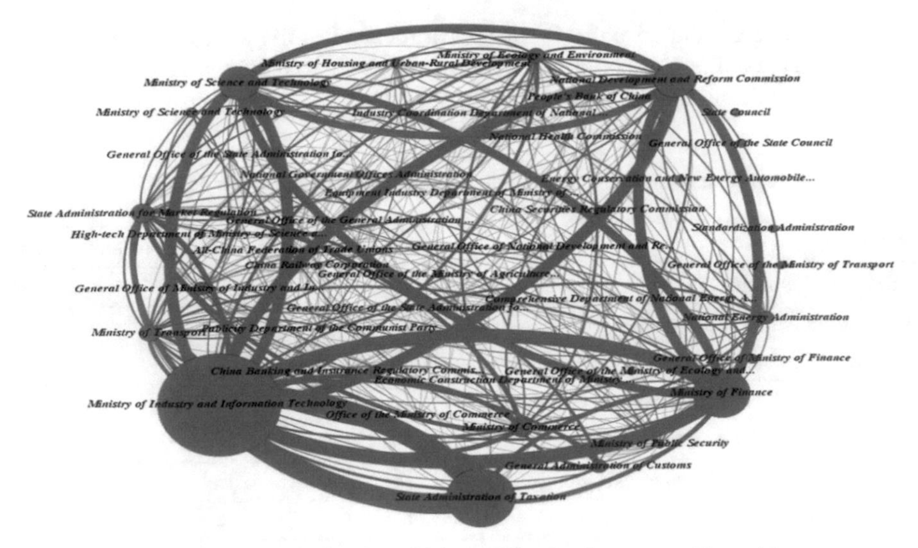

Fig. 3. Network relationship of joint publication between policy subjects

texts. Combining the characteristics of the new energy vehicles industry policy, construct a policy tools analysis framework from the perspective of policy tools [7], as shown in Fig. 4.

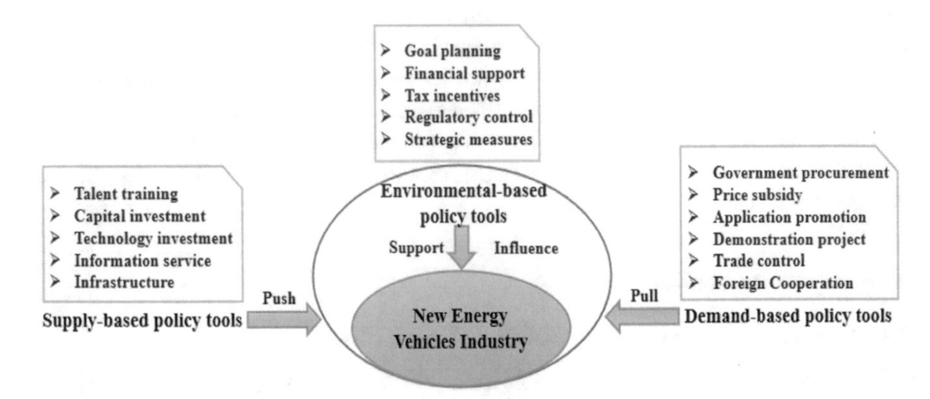

Fig. 4. Policy tools analysis framework

Supply-based policy tools refer to the government's direct element support for the new energy vehicles industry to promote the development of the industry. Supply-based policy tools can be subdivided into talent training, capital investment, technology investment, information service and infrastructure [8].

Demand-based policy tools refer to the government's adoption of various measures to expand the market for the new energy vehicles industry, thereby stimulating the development of the industry. Demand-based policy tools can be subdivided into government procurement, price subsidy, application promotion, demonstration project, trade control, and foreign cooperation.

Environmental-based policy tools have an indirect influence on the industry, which mainly refers to the government influences the development environment of the industry through a series of policies, and indirectly affects and promotes the development of the industry. Environmental-based policy tools can be subdivided into goal planning, financial support, tax incentives, regulatory control, and strategic measures.

3.2.2 Determine Analysis Unit and Analysis Category

The analysis unit is the most important and smallest element in the content analysis method. It refers to the object of calculation, which can be a word, a sentence, a paragraph, a topic, a comment, etc. [9]. The selection of the analysis unit is mainly based on the information needed to achieve the research goals. The specific clauses in the policy text are generally divided into paragraphs. This article uses the paragraphs in the policy text as the analysis unit according to the research needs.

According to the policy tools analysis framework to construct analysis categories, the so-called construction of analysis categories is to determine the category attribution of the analysis unit. This paper constructs the analysis categories and summarizes the subject vocabularies of each category [10]. The classification, name and subject vocabularies of the policy tools dimensions are shown in Table 2.

Table 2. Classification, name and subject vocabularies of policy tools dimensions

Policy tools type	Policy tools name	Subject vocabularies
Supply-based policy tools	Talent training	Talent training, talent development and training, talent team, outstanding leading talent, technical team, expert team, …
	Capital investment	Capital investment, special funds, reward funds, capital management, financial support, investment projects, …
	Technology investment	Technology investment, technology R&D, technological innovation, core technology, major technology, …
	Information service	Information service, information support, information platform, information management system, …
	Infrastructure	Infrastructure, supporting facilities, supporting infrastructure, infrastructure construction, expansion, …

(*continued*)

Table 2. (*continued*)

Policy tools type	Policy tools name	Subject vocabularies
Demand-based policy tools	Government procurement	Government procurement, procurement system, direct procurement, procurement demand, procurement share, …
	Price subsidy	Price subsidy, purchase subsidy, subsidy fund, subsidy standard, subsidy amount, consumption subsidy, …
	Application promotion	Application promotion, promote application, promote use, promotion information, promotion status, …
	Demonstration project	Demonstration project, pilot demonstration, pilot work, demonstration work, demonstration run, …
	Trade control	Trade control, license authorization, foreign trade, customs law, foreign exchange management law, …
	Foreign Cooperation	Foreign cooperation, foreign exchange, cooperation and exchange, international cooperation, …
Environmental-based policy tools	Goal planning	Goal planning, strategic planning, development planning, development goals, goals, planning, …
	Financial support	Financial support, financial institution, financial service, credit loans, credit support, bond, credit, equity, …
	Tax incentives	Tax incentives, tax exempt, exempt, tax cuts, additional deduction, accelerated depreciation, deducted income, …
	Regulatory control	Law, regulation, rule, standard, system, norm, standard system, institutional system, guarantee system, …
	Strategic measures	Policy, strategy, measures, supporting policy, policy measures, policy environment, industrial policy, …

3.2.3 Policy Tool Coding

According to the policy tool analysis category, the policy tool classification corpus is collected according to the subject vocabularies, and the corpus training is performed to generate a training model. The policy text to be analyzed is divided into content analysis units according to paragraphs, and the content analysis units are automatically classified and automatically coded and manually verified and modified. Some examples of content analysis units coding [11] are shown in Table 3.

Table 3. Policy text content analysis units coding

No	Policy file name	Content analysis units	Policy tool coding	Policy tool type	Policy tool name
1	Notice on Continuing the Promotion and Application Work of New Energy Vehicles	1. Relying on cities to promote the application of new energy vehicles	[1-3]	Demand-based	Application promotion
		……	……	……	……
2	Notice on Several Measures to Stabilize and Expand Automobile Consumption	1. Adjust the relevant requirements for the implementation of the National Sixth Emission Standard…	[2-3]	Environmental-based	Regulatory control
		……	……	……	……
……	……	……	……	……	……
155	Notice on Issuing the Development Plan for the New Energy Automobile Industry (2021–2035)	By 2025, the competitiveness of our country's new energy vehicle market will be significantly enhanced…	[155-24]	Environmental-based	Goal planning
		……		……	……

3.2.4 Policy Tool Statistics

Based on the coding of the content analysis unit of policy tools, a total of 1232 items of 155 policies (3500 items in total, of which 2268 items are classified as other) content analysis unit coding are counted. The statistics of the count and proportions of the three types of policy tools are shown in Fig. 5.

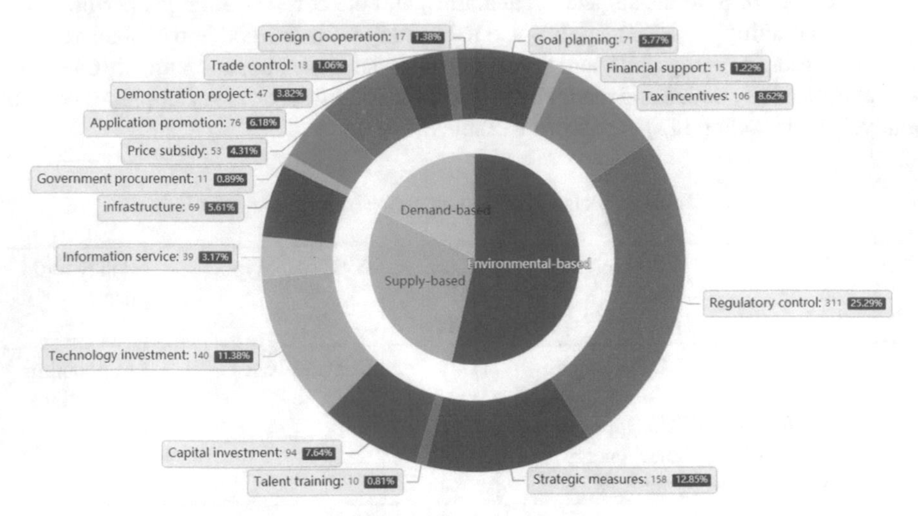

Fig. 5. Statistics on the proportions of policy tools

3.3 Policy Objectives Analysis

The correct industrial policy depends on the correct choice of objectives. In order to better adapt to the stage of industrial development, formulate correct industrial policy goals and adjust industrial structure. Whether the policy objectives of the new energy vehicles industry policy is clear or not affects whether the policy can achieve satisfactory results in its implementation.

In some policy documents, the policy objectives appear at the beginning of the policy documents, in some policy documents the policy objectives appear in the guiding ideology and work objectives, and some policy documents set more detailed objectives. Through sorting out the content of the policy text, it is found that the policy objectives often contain some commonly used vocabularies, such as “to implement”, “to promote”, “by xx year, to achieve”, “further strengthen” and so on. With the help of computer technology, the policy text is first divided into paragraphs, and then the above-mentioned vocabularies are used to initially extract and determine the location of the policy objective, and then make manual judgment and adjustment, and finally the keywords of the policy objective items are extracted and the policy objective are classified. After sorting out the policy objectives, it is found that the policy objectives of the new energy vehicles industry involve industrial development, automobile power, transformation and upgrading, strategic transformation, sustainable development, technological innovation, improve technical level, technological research and development, promotion and application, tax incentives, and energy saving, energy saving and emission reduction, energy saving and environmental protection, energy saving, recycling, etc. The policy objectives are classified, and the three main new energy vehicles policy objectives of industrial development, promotion and application, and energy saving and emission reduction are extracted [12].

Sorting out and categorizing the policy objectives in the new energy vehicles industry policy text, extracting a total of 202 policy objectives, a policy objective can involve multiple categories. It was found that there were 103 times involving industrial development, 102 times involving promotion and application, and 43 times involving energy saving and emission reduction. The proportions of the three types of policy objectives are shown in Fig. 6.

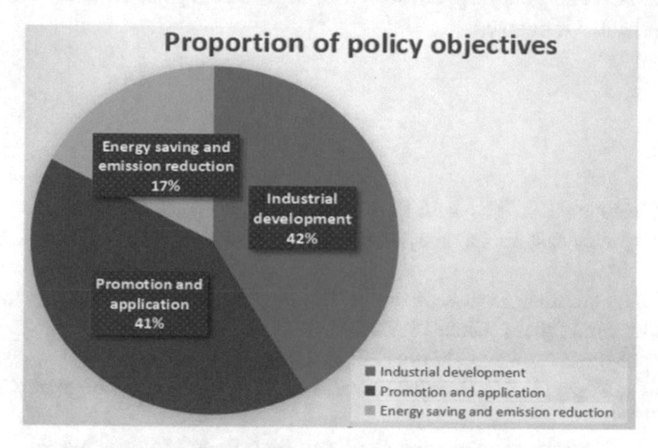

Fig. 6. Statistics on the proportion of policy objectives

4 Conclusions and Suggestions

This paper sorts out China's new energy vehicles industry policies from 2009 to 2020, follows the three-dimensional analysis framework of policy subjects, policy tools, and policy objectives, and makes a comprehensive analysis of China's new energy vehicles industry policies with the help of text mining technology, and draws the following conclusions.

(1) There are many policy subjects participating in joint publication. The 155 policy documents involved 41 policy subjects, and 35 policy subjects participated in the joint publication.
(2) The use of policy tools is uneven. More environmental policy tools are used, among them, there are too many regulatory policy tools.
(3) Policy objectives are scattered. The development of China's new energy vehicles has too many missions, leading to too many policy objectives, which are easy to restrict each other and increase the difficulty of policy coordination.

According to the research results, this paper puts forward the following suggestions for the policymakers of the new energy vehicles industry.

(1) Policy subjects need overall coordination. The high-quality development of the new energy vehicles industry requires the joint efforts of all departments in the industry. The relevant departments must strengthen overall coordination, establish a ministerial coordination mechanism, and form a joint force for policy formulation.
(2) The structure of policy tools needs to be optimized. The use of demand-type and supply-type policy tools should be increased, and the proportion of the use of regulatory policy tools should be reduced.
(3) Policy objectives need to be further focused. Formulate corresponding policies around the overall goal of promoting the high-quality development of China's new energy vehicles industry.

References

1. Zeng, J.P., Zhang, S.Z., Zhang, L.P.: Comparative study of China and the U.S. artificial intelligence policy system: an analysis framework based on policy subjects. E-Government **06**, 13–22 (2019)
2. Wanfang Data: Wanfang Data Knowledge Service PlatformV2.0. https://www.wanfangdata.com.cn/index.html (2017). Cited 18 Mar 2020
3. Gephi: Gephi-0.9.2. https://gephi.org (2017). Cited 14 Apr 2020
4. Rothwell, R., Zegveld, W.: Reindusdalization and Technology. Longman Group Limited, London (1985)
5. Howlett, M., Ramesh, M.: Studying Public Policy: Policy Cycles and Policy Subsystems. Oxford University Press, Oxford (1995)
6. McDonnell, L., Elmore, R.: Getting the job done: alternative policy instruments. Educ. Eval. Policy Anal. **9**(2), 133–152 (1987)
7. Zhang, Y.A., Zhou, Y.Y.: Policy instrument mining and quantitative evaluation of new energy vehicle subsidies. China Popul. Resour. Environ. **27**(10), 188–197 (2017)
8. Zhang, L., Liu, Y.Q.: Analysis of new energy vehicles industry policy in China's cities from the perspective of policy instruments. Energy Procedia **104**, 437–442 (2016)
9. Wang, P., Gao, L.Z., Fei, Z.L.: Study on China's policy tool for advanced material industry based on content analysis. Mod. Chem. Ind. **38**(10), 6–11 (2018)
10. Sheng, Y., Dai, J.X.: A comparative study of industrial policies under the "internet+ manufacturing" mode: take China, Germany and the united states as examples. Sci. Technol. Prog. Policy **36**(17), 114–121 (2019)
11. Wang, G.H., Li, W.J.: Analysis of China's network media policy from the perspective of policy tools: based on the national policy texts from 2000 to 2018. J. Intell. **38**(9), 90–98 (2019)
12. Zhang, Z., Zhao, F.: The comparative study on the development strategies of the new energy automobile industry between China and America: based on the differences of the goal orientation. Studies Sci. Sci. **32**(4), 531–535 (2014)

Research on Semantic Retrieval System of Scientific Literature Based on Deep Learning

Min Zhang[1,2,3] and Rui Yang[1,2](✉)

[1] Wuhan Library, Chinese Academy of Sciences, Wuhan 430071, China
yangr@mail.whlib.ac.cn
[2] Hubei Key Laboratory of Big Data in Science and Technology, Wuhan 430071, China
[3] Department of Library, Information and Archives Management, School of Economics and Management, University of Chinese Academy of Sciences, Beijing 100190, China

Abstract. Based on the background of the application of scientific literature retrieval, this paper puts forward that the current retrieval systems can not meet the semantic retrieval needs of the majority of scientific researchers. In order to solve this problem, the semantic retrieval effect is improved by deep learning. First, the deep learning technology is used to annotate the deep information of semantic knowledge and rich knowledge relationship in scientific literature. Then, the rich semantic information is displayed to users through the friendly retrieval interface. In the process of practical application, the system brings users more fine-grained knowledge level of association and navigation, and improves the semantic retrieval experience of the retrieval system.

Keywords: Semantic retrieval of scientific literature · Deep learning · Semantic annotation · Semantic index · Fine-grained

1 Introduction

Scientific literature retrieval is an indispensable task in the research process of the majority of scientific researchers. Traditional scientific literature retrieval systems mainly use keyword-based grammatical matching, and only return several document links related to the user's query. Users are more eager to discover scientific literature from the perspective of semantic understanding, and to effectively obtain some valuable deep knowledge. With the application of semantic technology in scientific literature retrieval, how to discover and reveal the fine-grained knowledge and knowledge relationship in scientific literature, make full use of knowledge, is the focus of scientific literature retrieval today.

The design goal of the semantic retrieval system of scientific literature based on deep learning proposed in this paper is to improve the existing single-keyword-oriented literature retrieval system, and use deep learning technology for semantic annotation of deep knowledge content, and fully reveal it during the retrieval.

Q. Liang et al. (Eds.): Artificial Intelligence in China, LNEE 854, pp. 209–217, 2022.
https://doi.org/10.1007/978-981-16-9423-3_26

2 Related Research at Home and Abroad

T. Berners-Lee [1] proposed that semantic retrieval is to make retrieval intelligent and humanized, so that the retrieval engine can understand the real needs of users, so as to feed back more accurate and comprehensive retrieval results to users.

At present, semantic retrieval mainly has two directions: retrieval of semantic web resources and semantic expansion of traditional retrieval systems. The research on semantic retrieval of scientific literature is mainly biased towards the latter, using semantic technology to improve the traditional literature retrieval system, fully mining the fine-grained semantic knowledge contained in the literature, which makes the retrieval system more intelligent [2]. According to the semantic technology applied in the retrieval, the semantic retrieval system of scientific literature can be divided into the following three categories:

2.1 Semantic Retrieval System for Scientific Literature Based on Ontology

Ontology's ability to describe concepts and their relationships makes it the key to achieving semantic retrieval. GoPubMed uses Gene ontology and MeSH to build index for PubMed literature, and then submits them to PubMed by entering search terms, and finally allows users to see the main biomedical concepts related to the query [3]. Kleio uses the ontology to semantically mark the semantic concepts of the text, and uses the marked named entity type to facet the retrieval results, which enables the retrieval system to query more accurate results [4].

2.2 Semantic Retrieval System of Scientific Literature Based on NLP

Natural language processing (NLP) provides efficient and intelligent semantic processing, which can solve the problem of semantic labeling in the process of literature retrieval. LitVar [5] is a semantic search engine based on genomic mutation data in 27 million PMC abstracts and 1.8 million PMC full texts. It uses natural semantic processing of Bioc XML format to process all abstracts and full texts, and then uses entity tags to extract all mutations and related entities and other information. Quertle [6] is a relation-driven biomedical literature retrieval tool. It uses NLP to extract subject-verbal-object relationships from biomedical literature collections and discover the relationships between biomedical entities.

2.3 Semantic Retrieval System of Scientific Literature Based on Knowledge Graph

This type of system uses the knowledge graph to give rich semantic information to the query. Springer Nature launched scientific research graph service of SciGraph [7], which organically integrates various information such as scientific research funding agencies, scientific research projects, conferences, scientific research units, publications, etc., so that users can quickly retrieve the most relevant and important information. The knowledge graph introduced by Elsevier [8] involves three fields of scientific research, life sciences, and health care. The knowledge graph connects concepts, data, literature, and software tools.

In summary, it can be found that the third retrieval system only enriches the user's input semantics, reveals the existing knowledge in the knowledge graph, but cannot discover the potential knowledge of the scientific literature itself; The first one needs to use the ontology technology to realize the knowledge content, which has its domain limitations; The second one can not only discover the potential knowledge of scientific literature itself, but also has a certain degree of versatility.

In addition, the vigorous development of natural language processing has benefited from the emergence of deep learning methods. As a learning method based on unsupervised feature learning and feature hierarchical structure, deep learning [9] analyzes and learns by simulating the neural network of the human brain, which solves many complex pattern recognition problems. Besides, it is very valuable in semantic understanding. Faced with the urgent needs of users for semantic retrieval, it is necessary to explore the application of deep learning technology in the semantic retrieval system of scientific literature.

3 The Overall Architecture of the System

The system architecture designed in this paper is divided into three parts: semantic annotation based on deep learning, multi-dimensional semantic indexing and semantic retrieval interface. Among them, the semantic annotation based on deep learning is oriented to knowledge mining, the multi-dimensional semantic index is oriented to the organization of knowledge, and the semantic retrieval interface is oriented to the disclosure of knowledge. As shown in Fig. 1.

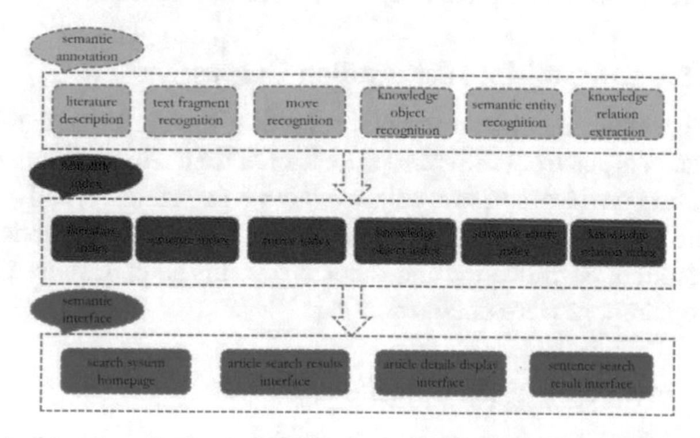

Fig. 1. Architecture diagram of semantic retrieval system for scientific literature

3.1 Semantic Annotation Based on Deep Learning

Semantic annotation based on deep learning mainly determines different semantic annotation content from the four levels of paragraphs, texts, sentences, and concepts of scientific literature. It includes the following parts: literature description, text segment recognition, move recognition [10], knowledge object recognition, semantic entity recognition and knowledge relation extraction.

3.2 Multidimensional Semantic Index

Semantic indexing work is to construct a multi-dimensional semantic indexing system, which is convenient for the disclosure and application of the semantic retrieval system of scientific literature. It is based on the literature metadata, sentence, move, knowledge object, semantic entitity, and knowledge relation.

3.3 Semantic Search Interface

The retrieval system includes four interactive interfaces: search system homepage, article retrieval result interface, sentence retrieval result interface, and article details display interface. The search system homepage provides entry for article retrieval and sentence retrieval, which is concise and quick. The article retrieval interface displays the article retrieval results, and the sentence retrieval results interface displays the sentence retrieval results. The article details display interface is used to reveal fine-grained deep knowledge content.

4 Key Technology Research

4.1 Move Recognition

Three steps are required to complete this work, as described below:

4.1.1 Build a Large-Scale Move Recognition Corpus

First of all, extensively investigate various scientific literature databases, and collect structured abstract corpus from different professional fields. In addition, carry out sufficient pre-processing on these corpora, merge related types of move, and build a database. Among them, the merged types of move mainly include Purpose, Methods, Results, Conclusions, Background, Significance and other types. In the end, more than 40 million move recognition corpora were obtained.

4.1.2 Build a Move Recognition Model Based on Deep Learning

Use the trained natural language processing model of BERT [11] to uniformly select structured abstract corpora of the same scale under different types of move for different professional fields, and perform the move recognition experiment through the Fine-Tuning method under the same test set. The following tests are performed to compare the effect of move recognition under different conditions. If the BERT is directly used for move recognition, the F1 score is 0.8398. If the collected corpus is used to pre-train the BERT, the F1 score is 0.8713.

4.1.3 Improve the Effect of Move Recognition Through Model Optimization

By transforming the input layer of the BERT model, the Mask Sentence Model is proposed. Firstly, learn the content information of the sentence itself. Then, mask a sentence in the abstract to learn its context information. Finally, the two model inputs are merged, and content features and context features are learned at the same time. The model has raised the F1 score from 0.8713 to 0.9208, which achieves an increase of 4.9%.

4.2 Semantic Entity Recognition

This paper takes the recognition of semantic entities in the field of physics as an example, and extracts semantic entities based on the ScienceWISE [12] ontology, which includes physics concepts, categories and semantic relations.

To begin with, this paper uses the BIO labeling strategy to label the concepts in the physics science literature and their categories to form a data set for fine-tuning the model, including 100,000 training sets and 3000 test sets.

What's more, the pre-trained language model of BERT is migrated to semantic entity recognition tasks in the field of physics, and the BERT is fine-tuned using the data set. A feedforward neural network is used on top of BERT to calculate category scores. The semantic entity recognition model is a two-layer model. It uses the second-level category model to make predictions. If the second-level category label is not "None", then use the results obtained by the second-level category model to search for its first-level category in the database; If the label is "None", then uses the first-level category model to make predictions, and set the second-level category label to "None".

4.3 Semantic Index Construction

The goal of semantic index design is to change the current single-dimensional indexing method, use multiple index trees to integrate and work together, and present semantic content in multiple dimensions. It is mainly divided into the following six dimensions:

1. literature index, which is for the metadata of the literature.
2. knowledge object index, which is for the terms extracted from the text.
3. semantic entity index, which is for the entities extracted from the text.
4. move index, which is for the contents of move recognition.
5. knowledge relation index, which is for the knowledge relations including the explicit relations of the ontology and extracted relations from the text.
6. sentence index, which is for sentences fragments.

The above indexs are not isolated, but form a framework which is centered on knowledge objects. Firstly, the input keywords are identified and semantically disambiguated through the knowledge object index; Secondly, the knowledge network is traversed to filter the related knowledge through the knowledge relation index; Thirdly, the move index determines where it appears in the abstract; Fourthly, the category to which it belongs through the semantic entity index; Fifthly, the sentence segment where it is determined through the sentence index; Finally, the literature information where it is located is queried through the literature index. The overall framework is shown in Fig. 2.

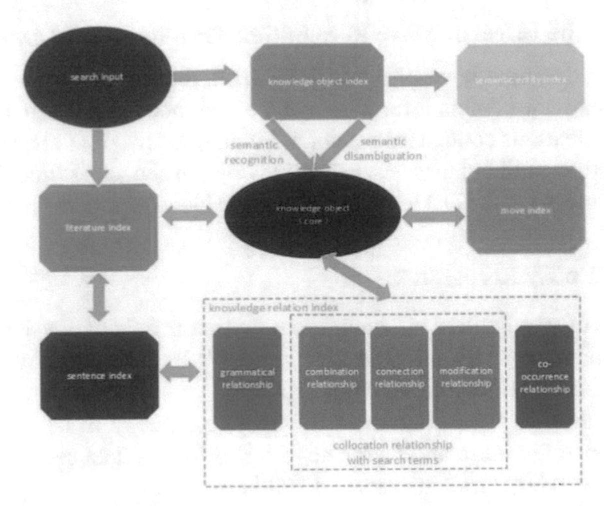

Fig. 2. Multi-dimensional semantic index system

5 System Realization

In order to be able to reveal the service functions and effects of the retrieval system, this paper constructs an automatic semantic annotation retrieval system for scientific research papers in the field of physics. It is centered on users, which reveals around user's knowledge application needs. Compared with other retrieval systems, this system has four characteristics: rich retrieval points, multi-module faceted navigation, humanized results display, and fine-grained article details. The system selects 100,000 scientific research papers in the field of physics from the arXiv database as the initial data sets. For the search term "dark matter", a total of 4643 papers are found. The search result interface is shown in Fig. 3.

5.1 Rich Access Points

As illustrated in Fig. 3, the system adds five move tags to the basic access points of the original retrieval system, and expands the basic access points to: full text, title, abstract, background, purpose, methods, results, and conclusions. Recognizing retrieval points by type labels of move can limit retrieval results so that users can customize them according to their needs. Compared with full-text retrieval and abstract retrieval, the retrieval results are further refined. In addition, this system also provides sentence retrieval function.

5.2 Multi-module Faceted Navigation

As shown in Fig. 3, the left column is the faceted navigation function column, which provides further refinement and screening of the search results. It mainly contains four modules: knowledge relationship revealing module, move feature revealing module, entity characteristic revealing module, and potential knowledge discovery module.

Fig. 3. Search result interface

The knowledge relation revealing module uses the knowledge relation index to count the collocation words combined with the search terms in the results, including four types of relations: combination relations, modification relations, connection relations, and grammatical relations. It is convenient for users to further view the collocation relationship related to the search term and understand the general expression of the search term in the field of physics.

The move feature revealing module uses the move index in the result to count the move position of the search term in the abstract in the search result. Users can further click on a certain move position to further filter the results, and view the search results of the search word appearing under the position of the move, so that users can further understand the expressive intention of the paper.

The entity feature disclosure module provides a reference for users to understand the physics subject category to which the search term belongs and more general application categories.

In the potential knowledge discovery module, the literature index and sentence index are used to count the co-occurring physical terms of the article and the sentence. Users can click on a term to realize the faceted browsing of related physics terms, which is convenient to find valuable content from potential related physics terms, judge the strength of association between co-occurrence terms, and provide navigation function to filter out these papers.

5.3 Humanized Results Display

Figure 3 has shown that the right column displays the search results. For the article search results interface, each result includes the title, publication time, author, subject, and abstract. Compared with the traditional literature retrieval system, the system not only displays the original abstract, but also provides the display of structured abstracts, which enhance the readability of documents so that users can quickly grab the information they need. In order to save page space, and take into account user's browsing habits, it is more convenient to switch between the two types of absrtact by the control of tab.

5.4 Fine-Grained Article Details

The specific article details display interface not only displays metadata information such as title and abstract, but also displays the deep and fine-grained semantic knowledge in the paper, including the semantic entities and its secondary and primary categories, the semantic relations and co-occurrence terms, which allow users to have a more comprehensive understanding of the paper.

6 Conclusions

Under the background that the traditional literature retrieval system cannot meet the semantic retrieval needs of scientific researchers, the scientific literature retrieval system based on deep learning has important practical value to meet the urgent needs of semantic retrieval. This paper focuses on this issue, gives the overall framework and key technologies of the system, realizes a scientific literature retrieval system based on the field of physics, and demonstrates its good effect in the practical application of semantic retrieval. In future work, the system will be further optimized and extended to other different fields.

Acknowledgements. This work is supported by the project "Research and application of key technologies for semantic intelligent search" (No. XQYF0302) from National Science and Technology Library.

References

1. Berners-Lee, T., Handler, J., Lassila, O.: The semantic web. Sci. Am. **284**(5), 34–43 (2003)
2. Wang, Y., Wu, Z.: Summarization of research on semantic retrieval system for scientific and technological documents. Mod. Libr. Inf. Technol. **5**, 1–7 (2015)
3. Sheng, D., Sun, J.: Research on subject knowledge service based on semantic search engine-taking GoPubMed as an example. Books Inf. Knowl. **4**, 113–120 (2015)
4. Nobata, C., Cotter, P., Okazaki, N.: Kleio: a knowledge-enriched information retrieval system for biology. In: the 31st Annual International ACM SIGIR Conference on Research and Development in Information Retrieva (SIGIR '08), New York, pp. 787–788 (2008)
5. Allot, A., Peng, Y., Wei, C.: LitVar: A semantic search engine for linking genomic variant data in PubMed and PMC. Nucleic Acids Res. **46**(Suppl. 8), 530–536 (2018)

6. Coppernoll, B.: Quertle: the conceptual relationships alternative search engine for PubMed. J. Med. Libr. Assoc. **99**(2), 176–177 (2011)
7. SciGraph, [online] Available: https://www.springernature.com/cn/researchers/scigraph
8. Elsevier knowledge graph, [online] Available: http://www.slideshare.net/anitawaard/
9. Hinton, G., Osindero, S., Teh, Y.: A fast learning algorithm for deep belief nets. Neural Comput. **18**(7), 1527–1554 (2006)
10. Zhang, Z., Liu, H., Ding, L.: Comparative study on the effect of move recognition in abstracts of scientific and technological papers based on different deep learning models. Data Anal. Knowl. Discov. **3**(12), 1–9 (2019)
11. Devlin, J., Chang, M., Lee, K.: BERT: Pre-training of deep bidirectional transformers for language understanding. arXiv:1810.04805 [cs.CL] (2018)
12. ScienceWISE. Available: http://sciencewise.info/

Research on the Evaluation System Construction of County Innovation Driven Development-Based on Evaluation and Measurement Model

Li Lanchun[1], Chen Wei[1](✉), Guo Kaimo[1], Yue Fang[1], Tang Yun[1], and Zhao Ke[2]

[1] Wuhan Library, Chinese Academy of Sciences, Wuhan 430071, China
chenw@mail.whlib.ac.cn

[2] Hubei Academy of Scientific and Technical Information, Wuhan 430071, China

Abstract. County territory is the cornerstone of China's economic and social development, and county innovation is an important part of the national innovation system. The implementation of innovation driven development strategy is based on the county, vitality in the county, and difficulties are also in the county. In order to further grasp the current situation of county innovation driven development and clarify the obstacles of county innovation development, based on the county innovation activity data of Hubei Province, this paper constructs the county (city) innovation ability evaluation index from the aspects of innovation input, innovation environment, enterprise innovation, and innovation performance, aiming at comprehensively and accurately reflecting the basic law and evolution characteristics of county innovation ability. This paper puts forward the countermeasures and suggestions to improve the innovation ability of counties (cities), and provides reference for regional innovation policy formulation and innovation work.

Keyword: County economy · Regional evaluation · Innovation system · Calculation model

1 Introduction

County level is the cornerstone of China's economic and social development, and county innovation is an important part of the national innovation system. In 2019, the national level and Hubei Provincial Government successively issued guiding documents on the Implementation Opinions on County Innovation-driven Development, further clarifying the supporting role of county innovation in encouraging regional coordinated and high-quality development. The implementation of the innovation-driven development strategy is based on the county level, vitality and difficulties in the county level [1]. Compared with other types of regions, it is difficult to form a complete scientific and engineering innovation system for counties, most counties lack universities, research institutes, and large backbone enterprises, and seldom conduct scientific and research activities [2]. Secondly, the county is an open system, scientific and technological innovation needs

Q. Liang et al. (Eds.): Artificial Intelligence in China, LNEE 854, pp. 218–225, 2022.
https://doi.org/10.1007/978-981-16-9423-3_27

to utilize a large number of external talents, capital, and technology. Thirdly, the county is subordinate to the state, the province and the city, so there is relatively little room for innovation in institutional mechanisms and other aspects.

Therefore, considering of the characteristics of the county, carrying out the evaluation and research of the county innovation ability can grasp the current situation of the county innovation, clarify the obstacle factors of the county innovation and development, and as a result it can provide strong support and guidance to break down the problems and difficulties prevailing in the county innovation and open through the endings of the county innovation and management.

2 Research Background

Innovation ability evaluation is a comprehensive analysis, comparison and judgment of innovation ability such as national, regional, enterprises, research institutions, colleges and universities, and innovation-intensive areas through the construction of an index system. At present, the domestic research on county innovation evaluation mainly includes: Lei [3] established the evaluation index system of county science and technology innovation ability based on science and technology input, science and technology output, science and technology environment. Sun [4] evaluates the index system of county innovation ability from four dimensions of innovation input ability, innovation economic performance, innovation output ability and innovation environment support. In addition, some scholars took Schumpeter's innovation theory [5] as a starting point to elaborate on the composition characteristics of regional innovation ecosystems [6].

To sum up, in the process of determining evaluation indicators indexesthe academia is usually based on a specific research background or specific field, thus showing the overall difference among the constructed evaluation index systems, which may easily lead to the evaluation results of the same object varying based on the final choice of the index system. It is mainly evident in the following aspects: firstly, the index connotation is similar and the evaluation differentiation is not obvious; secondly, the index definition is vague and the evaluation positioning is inaccurate; thirdly, the index level is misplaced and the evaluation scope is not visible; fourthly, the index structure is poor and the evaluation correlation is not strong. Considering of this, in order to further grasp the county innovation-driven development, this paper constructed a set of county (city) innovation ability evaluation system composed of innovation input, innovation environment, social output, and 20 indexes, by consolidating the statistical data of relevant indexes that are in line with the county's economy, society and technological innovation and development, comprehensively, objective and accurately reflecting the county innovation ability.

3 Evaluation Model Construction

3.1 Principles of Index Design

In the design of county index system, the system needs to comply with the following principles, comprehensive and objective, accuracy and reliability, overall consideration, conciseness and easy operation.

Regional innovation system is an ecosystem of multi-dimensional interaction and integration of innovation elements, including direct elements (innovation activity talents, infrastructure, capital investment, etc.), indirect elements (scientific and technological innovation policies and regulations, environmental construction, etc.), and outcome elements (economic and social output and scientific and technological innovation efficiency, etc.). Therefore, the evaluation index system should cover all the elements that impact the regional innovation level, and emphasize the correlation between the elements, so as to comprehensively and objectively represent the effectiveness of the regional scientific and technological innovation activities. In view of this, on the basis of comprehensive analysis and absorption of the research experience from home and abroad, plus the characteristics of county-level innovation activities, the county (city) innovation ability evaluation indexconsists of innovation input, innovation environment, innovation output, social contribution and 20 secondary indexes.

3.2 Determine the Index Weight

On the basis of following the principle of objectivity, operability and effectiveness, and combining the statistical characteristics of county innovative science and technology, the international universal benchmark analysis method (Benchmarking), namely Lausanne International competitiveness evaluation method, is adopted to ensure the scientific nature and objectivity of the results.

According to the index acquisition data, the min–max standardization method is adopted to conduct the index linear dimensionless normalization processing. The conversion function is as follows:

$$y_{ij} = \frac{x_{ij} - x_{i\min}}{x_{i\max} - x_{i\min}} \tag{1.1}$$

x_{ij} means the index i value of the j county (city), the y_{ij} means the standardization value of index i on j county (city); x_{imax} means the maximum value of sample data and x_{imin} means the minimum value of sample data.

According to the calculation formula of the information entropy, the information entropy E_i of each index:

$$E_i = -\ln(n)^{-1} \sum_{j=1}^{n} p_{ij} \ln p_{ij} \tag{1.2}$$

The index weight is calculated by the information entropy:

$$w_i^E = \frac{1 - E_i}{k - \sum E_i} (i = 1, 2, \cdots, k) \tag{1.3}$$

Calculate the combined weights:

$$w_i = \alpha w_i^E + (1 - \alpha) w_i^D \tag{1.4}$$

w_i^E for the objective weight, w_i^D for the subjective weight.

This study takes the average mean value of the weights obtained by the master and objective empowerment methods, that is, the determined index weights are as shown in Table 1.

Table 1. Evaluation index system of county innovation ability

Primary indexes	Secondary indexes	Weight
Innovation input	Financial expenditure on science and technology at the corresponding level X_1	5.11
	Proportion of expenditure on science and technology in the general public budget X_2	4.16
	R&D expenditure of industrial enterprises above scale in main business income X_3	5.46
	R&D personnel of industrial enterprises above scale: employees of industrial enterprises above scale X_4	4.61
Innovation environment	Innovation intensive Zone X_5	8.58
	The number of popular science bases owned by ten thousand people X_6	2.56
	Unit GDP above the provincial level scientific research platform X_7	3.14
	Technology and finance support scale X_8	6.52
Innovative output	Portion of new product sales income of enterprises above scale in the income of the main business X_9	5.95
	Amount of invention patent application X_{10}	4.04
	Number of invention patents owned by ten thousand people X_{11}	3.46
	Number of high-tech enterprises X_{12}	4.20
	Value-added of high and new technology industries X_{13}	8.35
	Proportion of added value of high-tech industries in GDP X_{14}	7.32
	Number of technology SMEs X_{15}	4.30
	Number of listed technology enterprises on the New Third Board X_{16}	7.98
	Number of scientific and technological achievements introduced and transformed in the current year X_{17}	3.72
Social contribution	Energy consumption reduction rate of unit GDP X_{18}	3.51
	GDP per capita X_{19}	3.46
	Residents' per capita disposable income X_{20}	3.57

3.3 Evaluation and Measurement Model

Based on the above evaluation method and index weight results, the comprehensive evaluation of scientific and technological innovation ability in Hubei province is adopted. The efficacy coefficient scoring method is based on the efficacy coefficient, adopting a percentile system and subject to the average value of Hubei province county (city). If the average of a certain index in a certain area reaches its average value of 60%, then the other 40% is determined according to the specific index value of a certain region.

Index effect coefficient score. Select the value of x_{ij}^h and x_{ij}^s according to the positive and negative efficiency of the index i, and then the efficacy coefficient score of the j county (city) under the i index is calculated according to the following formula:

$$d_{ij} = \frac{x_{ij} - x_{ij}^s}{x_{ij}^h - x_{ij}^s} \times 40 + 60 \tag{1.5}$$

By this analogy, the efficacy coefficient score of the j county (city) under all indexes is figured out, namely $d_{1j}, d_{2j}, \ldots\ldots, d_{20j}$.

Weighted average comprehensive score: due to the different weights of each index, the comprehensive score of each county (city) is calculated using the weighted average method based on the each index weight w_i obtained by the entropy weight method. The comprehensive score of the j county (city) is:

$$E_j = \frac{\sum_{i=1}^{20} w_i d_{ij}}{\sum_{i=1}^{20} w_i} = \frac{w_1 d_{1j} + w_2 d_{2j} + \ldots\ldots + w_{20} d_{20j}}{w_1 + w_2 + \ldots\ldots + w_{20}} \tag{1.6}$$

Similarly, the comprehensive score of each county (city) is obtained and ranked from high to low according to the score results.

4 Empirical Study on Innovation Capacity Evaluation

4.1 Study Objects

At present, the integration of "the Belt and Road Initiative", the Yangtze River Economic Belt, Guangdong-Hong Kong-Macao Greater Bay Area, Yangtze River Delta and other national strategies have boosted the high quality development of middle areas into the fast lane with anobvious later-mover advantage, turning into the key area of our national new round of industrialization, urbanization, informatization and coordinated development of agricultural modernization, which is also an important area to support China to maintain rapid economic growth.

As a large province of scientific and educational resources, Hubei's scientific and technological innovation has continued to maintain the development momentum of the top central region and the first square in China, and the counties innovation and development has also made obvious progress. Yidu, Daye and Xiantao are the first batch of national innovative counties (cities), and many counties (cities) have been shortlisted for the top 100 industrial economy.

4.2 Analysis of the Evaluation Results

From the overall results, Xiantao, Daye, Qianjiang, Yicheng, Anlu, Zhongxiang, Gucheng, Wuxue, Yicheng, Laohekou, Shayang and Zaoyang ranked in the top 12 respectively and in the top 20% of counties (cities) in Hubei Province, with strong comprehensive scientific and technological innovation ability. From the perspective of total score, only 6 counties (cities) such as Xiantao, Daye and Qianjiang were above 80 points, 25 counties (cities) below 70 points, most counties (cities) scientific and technological innovation ability is at the medium level, and the overall development is unbalanced. As shown in Table 2.

Table 2. Evaluation results of innovation-driven development in 63 counties (cities)

Category	Area	County (City)
First square	Innovation-Leading area	Xiantao, Daye, Qianjiang, Yicheng, Anlu, Zhongxiang, Gucheng, Wuxue, Yicheng, Laohekou, Shayang, Zaoyang, Tianmen, Jingshan, Yuan'an, Yingcheng, Hanchuan, Zhijiang, Gong'an, Chibi
Second square	Innovation demonstration area	Dangyang, Macheng, Xishui
Third square	Innovation agglomeration area	Enshi, Luotian, Xingshan, Honghu, Tongcheng, Yunmeng
Fourth square	Innovative growth area	Xiaochang, Danjiangkou, Songzi, Guangshui, Baokang, Yangxin, Jiayu, Shishou, Qichun, Nanzhang, Changyang, Tuanfeng, Zigui, Huangmei, Jianli, Hong'an, Fangxian, Zhushan, Yingshan, Zhuxi, Dawu, Tongshan, Wufeng, Chongyang, SuiXian, Xuanen, Lichuan, Jiangling, Yunxi, Badong, Laifeng, Jianshi, Xianfeng, Hefeng

4.3 Regional Distribution Characteristics

In order to more intuitively reflect the innovation ability of counties (cities) in Hubei Province, taking the sum of the county (city) innovation input as the horizontal axis, the sum of the innovation output and social contribution of each county (city) as the vertical axis, the distribution space is divided into four quadrants, and 63 counties (cities) in Hubei Province are distributed in the four quadrants. As shown in Fig. 1.

From the perspective of innovation investment, innovation environment and innovation output and regional distribution of social contribution, it shows as a overall ellipticalness, with comparatively dense double low area distribution. It can be seen that in most counties whether in innovation capital, talents, platform basic investment and innovation ecology, or intellectual property, industrial cultivation and other output and support economic and social development, there are certain development bottlenecks and difficulties. The first quadrant is a high-high area, namely invention investment, innovation environment and innovation output, social contribution are all in high rank, belonging to the hot spot area. It has good regional innovation ability, distributed by Xiantao, Daye, Qianjiang, Yicheng, Anlu, and other 18 counties (cities), which mainly distributed in

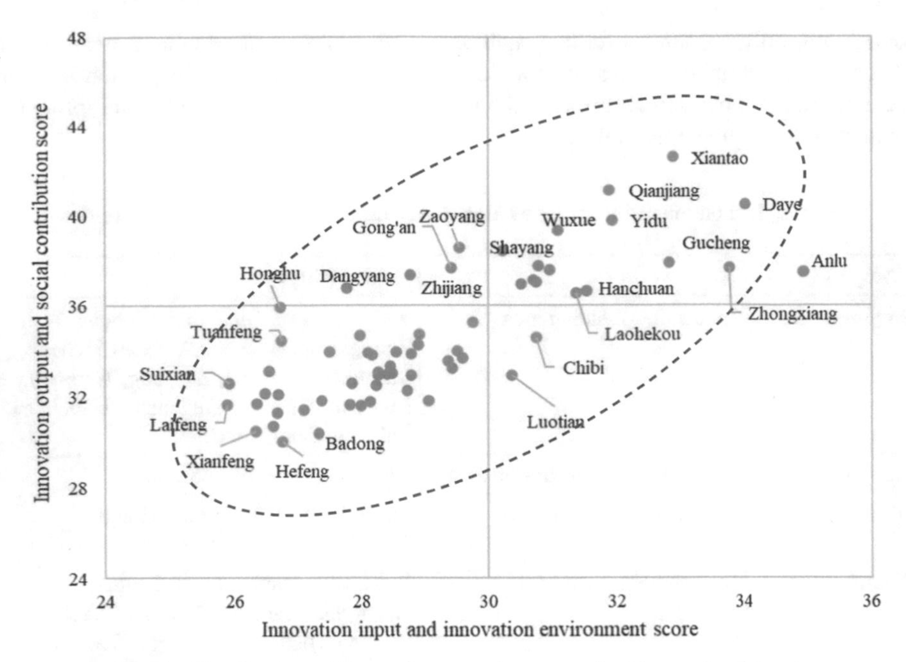

Fig. 1. Distribution of innovation capacity scatter in 63 counties (cities)

Wuhan, xiangyang, yichang comprehensive innovation center radiation area, indicating the significant spillover effect on the scientific and educational resources and innovation resources, namely *"Innovation-Leading Area"*. The second quadrant is low–high area, namely low innovation investmentand innovation atmosphere, high innovation output and social contribution, high regional innovation effectiveness, good innovation ability, with Zaoyang City, Dangyang, Honghu, Zhijiang, public security county distributed in the quadrant, five counties (cities) in multiple regional innovation center radiation cross position. The traditional industry has a good development foundation, and has a good comparative advantage, while in the rapid transformation and upgrading stage, namely *"Innovation Demonstration Area"*. The fourth quadrant is the high-low region, that is, high innovation investmentand innovation environment, low innovation output and social contribution. The regional innovation efficiency is slightly weak and the overall innovation ability level is good. Chibi City, Luotian City are in the period of innovation elements agglomeration and development, namely *"Innovation Agglomeration Area"*. The third quadrant is low-low area, namely innovation investment, innovation environment and innovation output, social contribution are low, belonging to the blind spot, with weak regional innovation ability. 38 counties (cities) are distributed in the quadrant, mainly concentrated in the Dabie mountain area, Mufu mountain, qinba mountain, wuling mountain and other four contiguous poor areas, where scientific and technological innovation level is relatively weak, and still in the innovation-driven development period, namely *"Innovative Growth Area"*.

5 Conclusion

Starting from the connotation of regional innovation ecosystem and the actual characteristics of county innovation, this paper comprehensively considers the integration and development perspectives of innovation chain, industrial chain, value chain and service chain from the perspective of high-quality development path of county innovation. Based on the analysis of existing scientific and technological innovation evaluation literature, this paper analyzes innovation input, innovation environment and innovation output, a set of evaluation system composed of 20 indexes is constructed from four dimensions of social contribution. On this basis, we will carry out the evaluation and research of science and technology innovation ability in Hubei Province, objectively and accurately describe the current situation and crux of science and technology innovation in Hubei County, and explore the correct mode and development of county innovation and path.

The index system should reflect the innovation ability of the county from "many perspectives" as possible. Integrating the innovation ecology of innovation carriers, platforms and policies in Hubei County, the innovation investment of Hubei County itself and its main innovation subject enterprises in capital, personnel, infrastructure, as well as the output and performance of innovation support industrial structure enhancement, innovation subject cultivation, and science and technology benefit the people. In addition, the index design layout has both total indicators, and proportional indicators, and can fully consider the objective conditions of cities and development foundations and other factors.

Empirical model adopts the international universal competitiveness evaluation method, uses max–min extreme difference method to outline linearity the collected statistical index data, uses objective index assignment method to determine the weight coefficient of each index, the evaluation results are more accurate and reliable, in line with the actual characteristics of county innovation in Hubei Province.

References

1. Liu, X., Ge, S.: Explore the internal mechanism of innovation driven by china's economic growth over 20 years -based on the perspective of new Schumpeter growth theory. Sci. Sci. Manag. S&T **39**, 3–18 (2018)
2. Guo, S., Guo, J., Zhao, G.: Analysis on the spatiotemporal transition path and convergence evolution law of county innovation level. Sci. Technol. Prog. Policy **36**(4), 50–57 (2019)
3. Lei, Y.: County Science and Technology Innovation Capacity Assessment Research-Take Hunan as an example. Hunan Normal University (2009)
4. Sun, C.: Study on the Evaluation Index System and Model of the County Innovation Ability. Jilin University (2010)
5. Shi, L., Chen, H.: Research on evaluation index system of innovative counties (city and district) in Shaanxi Province. Statis. Manag. **3**, 125–128 (2019)
6. Wang, Z., Liang, C.: Efficiency evaluation and incentive policy of county innovation-driven development. Sci. Manag. Res. **36**, 48–51 (2018)

Research on the Method of Policy Text Knowledge Mining Based on Neural Network

Yanli Zhou[1,2], Rui Yang[1,2](✉), Min Zhang[1,2](✉), Jingling Xu[1,2], and Jun Chen[1,2]

[1] Wuhan Library of Chinese Academy of Sciences, Wuhan 430071, China
{yangr,zhangmin}@mail.whlib.ac.cn

[2] Hubei Key Laboratory of Big Data in Science and Technology (Wuhan Library of Chinese Academy of Sciences), Wuhan 430071, China

Abstract. In view of the large amount of information in domestic policy texts and the rich semantic information, it is difficult for researchers to quickly and effectively sort out and compare and analyze the content when reading. This paper takes the policy text data resources in the energy field as an example, and sorts them out according to the types of resources. On the basis of sorting out the policy resources, it combines the characteristics of policy text information with deep learning technology, explores new ways and methods to interpret policy text content in a deep learning environment, and further evaluates and analyzes the effect of the neural network models for entity recognition and relationship extraction of domestic policy texts. This paper uses Bi-LSTM to extract text features in both directions to solve problems such as long-distance dependence, introduces the attention mechanism to obtain important features, and effectively reveals various entities and concepts in the policy text and their relationships, thereby assisting researchers in better policy implementation in-depth interpretation and analysis of text content.

Keywords: Deep learning · Policy text · Neural network · Entity recognition · Relationship classification

1 Introduction

Energy policy is a political behaviour or a prescribed code of conduct adopted by state agencies and other authoritative organizations to achieve specific goals in a certain period of time. It includes laws, plans, measures, methods, regulations, notices, opinions, etc., and is used in various energy industries. It is the development program of energy industry. The policy text is rich in semantics, covers a wide range of content, and has a certain timeliness, which poses certain problems for researchers to quickly interpret and analyze. In response to this problem, it is of great significance to quickly extract policy subjects and subject relationships from energy policy texts to help researchers interpret and analyze of policy texts deeply. We use the Bi-LSTM neural network model based on the attention mechanism, which extracts the features of the sentence through the bidirectional LSTM neural network, combines the attention mechanism adds the

Q. Liang et al. (Eds.): Artificial Intelligence in China, LNEE 854, pp. 226–234, 2022.
https://doi.org/10.1007/978-981-16-9423-3_28

weight to the output feature vector, generates a biased vector, and then enters the fully connected neural network layer. Eventually we realize the classification of relations, and effectively reveal the various entities and concepts and their association relations in the policy text.

2 Related Research

Information extraction has been a hot topic in NLP field, and it is a technology to extract specific information from text data. Named entity recognition and relation extraction are two tasks of information extraction. Text data is composed of specific units, such as sentences, paragraphs, and chapters. Text information is composed of small specific units, such as words, phrases, sentences, paragraphs or combinations of these specific units. The task of named entity identification refers to the entity with specific meaning in the text, including the name of person, place name, organization name, and proper noun. In short, it is to identify the boundary and category of entity reference in natural text. Once all named entities in the text are extracted, they can be linked to the collection corresponding to the actual entity. Relation extraction is mainly to discover and classify semantic relations between text entities. These relationships are usually binary relationships, such as correlation, part-whole relationship etc. At present, the most popular deep learning models in the field of named entity recognition are convolutional neural network (CNN) and recurrent neural network (RNN) and BERT.

CNN, a kind of feedforward neural network with convolution computation and depth structure, is one of the representative algorithms of deep learning. CNN can simplify complex problems, reduce a large number of parameters into a small number of parameters, and ensure that the effective information is not lost. CNN is mainly composed of three parts: convolution layer, pooling layer and full connection layer. The convolution layer is responsible for extracting local features in the images. The pooling layer is used to reduce the order of parameters (dimension reduction). The whole connection layer is similar to the part of traditional neural network, which is used to output the desired results.

Zeng et al. used convolutional neural networks to extract word and sentence-level features, and connected the two-level features to form the final feature vector, and finally input it into the SoftMax classifier to predict the relationship between two labeled nouns [1]. Linlin Wang et al. designed a novel CNN architecture for sentence level relation extraction, using two-layer attention mechanism to extract entity and relation respectively, and the effect is remarkable [2].

RNN is a kind of recurrent neural network, which takes sequence data as input, recurses in the evolution direction of sequence, and all nodes (cycle units) are connected by chain. The biggest difference between RNN and traditional neural network is that it will bring the previous output to the next hidden layer and train together.

Makoto Miwa et al. used the neural network model to capture the substructure information of word sequence and dependency tree by superimposing bidirectional tree structure LSTM RNNs on bidirectional sequence LSTM RNNs [3]. Peng Zhou et al. used bidirectional LSTM and attention mechanism to deal with the related problems of text classification, in order to solve the problem that CNN model is not suitable for learning long-distance semantic information [4]. Suncong Zheng et al. used the underlying representation of shared neural network for joint learning. Based on LSTM's end to end model, the two tasks would update the shared parameters through back propagation algorithm during training to achieve the dependency between the two subtasks [5].

Later, in response to the shortcomings of word2vec and other models, google proposed BERT [6], which combines the pre-training model with the downstream task, and gradually moves the downstream specific NLP tasks to the pre-training generated word vectors. Shanchan Wu et al. took the lead in applying BERT to relation extraction, and proposed that the location of the entity can be indicated by adding an identifier before and after the entity, instead of the traditional location vector [7].

3 Methodology

The overall research idea of this paper is shown in Fig. 1. In the selection of data sources, we select the policy notification section of the National Energy Administration which is the authoritative policy issuing agency in the field of energy, to obtain energy related notices, announcements and interpretation data. After data processing, removing irrelevant and redundant information, we enter the corpus preparation stage. The data are divided into training set, verification level and test set. The training set is used for model training, the verification set is used to adjust the model, and the test set is used to evaluate the model.

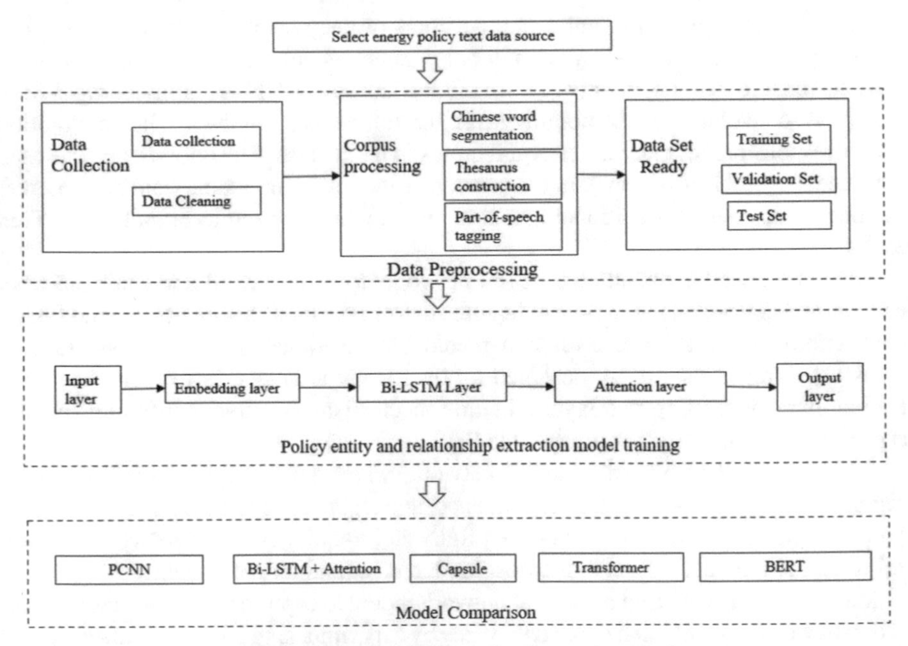

Fig. 1. Overall thinking of policy text analysis

3.1 Theoretical Model

This paper uses the Bi-LSTM model based on the attention mechanism to analyze the energy policy text. The model is shown in Fig. 2. LSTM (Long-Short Term Memory) is a special type of RNN. It solves the problem of gradient explosion or gradient disappearance during RNN training. LSTM cleverly uses the concept of gating to achieve long-term memory, and it can also capture sequence information. Bi-LSTM is the abbreviation of Bi-directional Long Short-Term Memory, which is a combination of the forward LSTM and the backward LSTM. The realization of the attention mechanism is to retain the intermediate output results of the LSTM encoder on the input sequence, and then train a model to selectively learn these inputs and associate the output sequence with it when the model outputs.

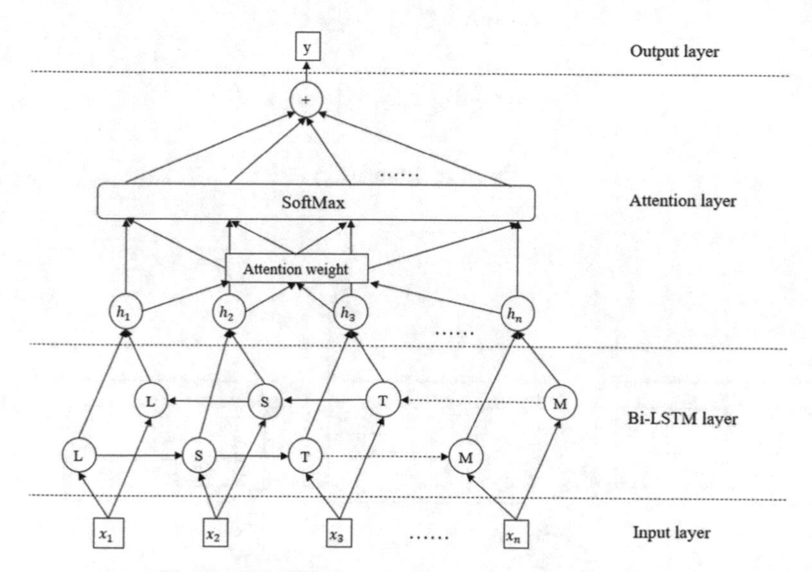

Fig. 2. Bi-LSTM based on attention network model

3.1.1 Bi-LSTM Layer

Bi-LSTM is mainly composed of a combination of the forward LSTM and the backward LSTM [8]. The LSTM model (as shown in Fig. 3) is composed of the input word x_t at time t, the cell state C_t, the temporary cell state $\tilde{C}_t$, the hidden layer state h_t, the forgetting gate f_t, the memory gate, and the output gate o_t. The calculation process of LSTM can be summarized as: by forgetting and remembering new information in the cell state, the useful information for subsequent calculations can be transmitted, and useless information is discarded, and the hidden layer state h_t is output at each time step. Forgetting, memory and output are controlled by calculating the forgetting gate f_t, the memory gate i_t, and the output gate o_t through the hidden layer state h_{t-1} and the current input x_t at the last moment [3]. A single LSTM cannot take into account the long-distance context information, so the reverse LSTM is used to read the same sequence, and the forward and backward LSTMs are combined to form a Bi-LSTM model, which can save the previous context and take into account the back context, and this brings benefits to the processing of long-distance information.

The calculation process is as follows:

$$f_t = \sigma\left(W_f \cdot \left[h_{t-1}, x_t\right] + b_f\right) \tag{1}$$

$$i_t = \sigma\left(W_i \cdot \left[h_{t-1}, x_t\right] + b_i\right) \tag{2}$$

$$\tilde{C}_t = \tanh\left(W_C \cdot \left[h_{t-1}, x_t\right] + b_C\right) \tag{3}$$

$$C_t = f_t * C_{t-1} + i_t * \tilde{C}_t \tag{4}$$

$$o_t = \sigma\left(W_o\left[h_{t-1}, x_t\right] + b_o\right) \tag{5}$$

$$h_t = o_t * \tanh(C_t) \tag{6}$$

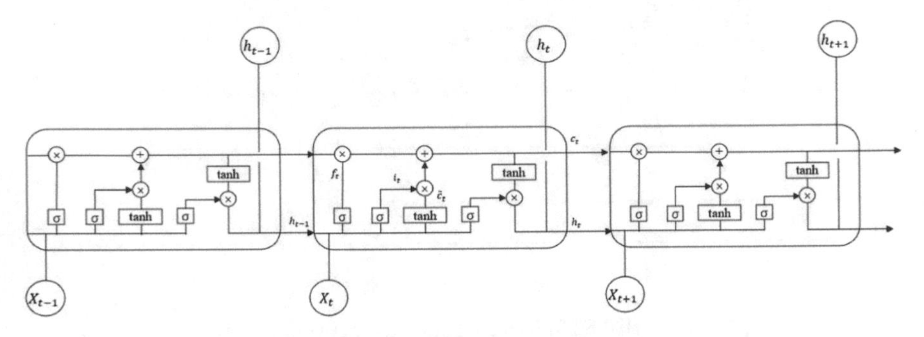

Fig. 3. LSTM framework

3.1.2 Attention Layer

The core logic of attention mechanism is from focusing on the whole to focusing on the key. If the model needs to remember all the information, it will be extremely complicated, resulting in a huge amount of calculation, and ultimately limited by computing power. From focusing on everything to focusing on the key points, it can alleviate the complexity of the model and improve the ability of the neural network to process information, just like the human visual system, focusing limited attention on key information, thereby saving resources and quickly obtaining the most effective information. The basic principle of attention is to consider different weight parameters for each element of the input, thereby paying more attention to the parts similar to the input element, while suppressing other useless information. Its biggest advantage is that it can consider global and local connections in one step, and can parallelize calculations, which solve the calculation problems of large data volume.

The attention layer is added to the Bi-LSTM model. The input of the Bi-LSTM layer is calculated by LSTM in the forward and backward directions, and then the two hidden states are connected together to label the context information to form a global feature output. Then the output input to the attention layer, where the target word is re-encoded through the correlation between the target word and all vocabulary in the sentence. The main process is to calculate the relevance between words, normalize the relevance, and obtain the code of the target word through the weighted summation of the relevance and the coding of all words.

Attention architecture [9] and calculation process are as follows (Fig. 4).

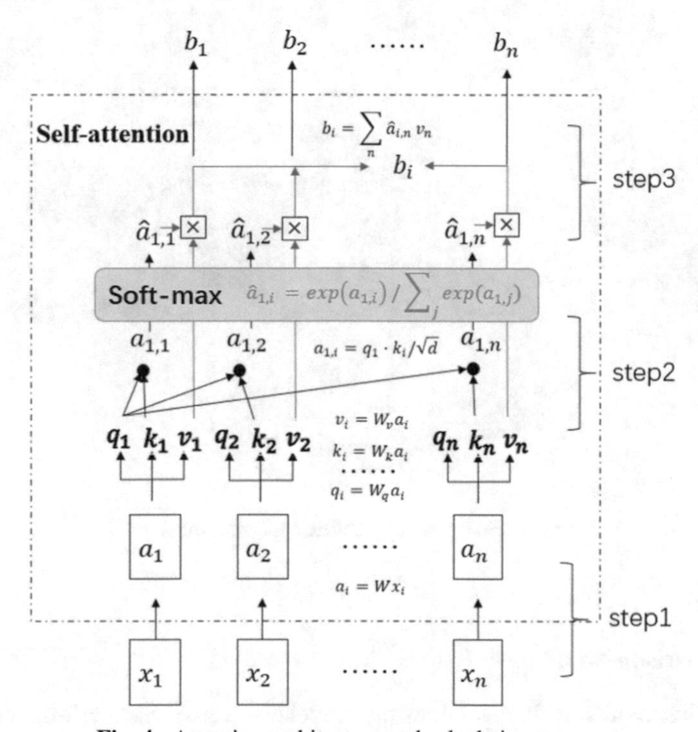

Fig. 4. Attention architecture and calculation process

The overall process can be summarized as:

Step1: calculate the similarity between q and each k to get the weight, the commonly used similarity functions are dot product, splicing, perceptron, etc.

Step2: use the soft-max function to normalize these weights to obtain directly usable weights.

Step3: perform a weighted summation of the weight and the corresponding key value v to obtain the final attention.

4 Experimental Data and Result Analysis

4.1 Data Set

The experimental data in this paper obtained 469 texts related to energy policy notifications, announcements and interpretations from the National Energy Administration website. According to the content of the policy texts, 10 types of energy policy entities and 36 types of entity relationships (Fig. 5) were manually sorted out. We have processed 1498 pieces of corpus data. Then the corpus data has been divided into 839 pieces of training set, 329 pieces of verification set, and 330 pieces of test set.

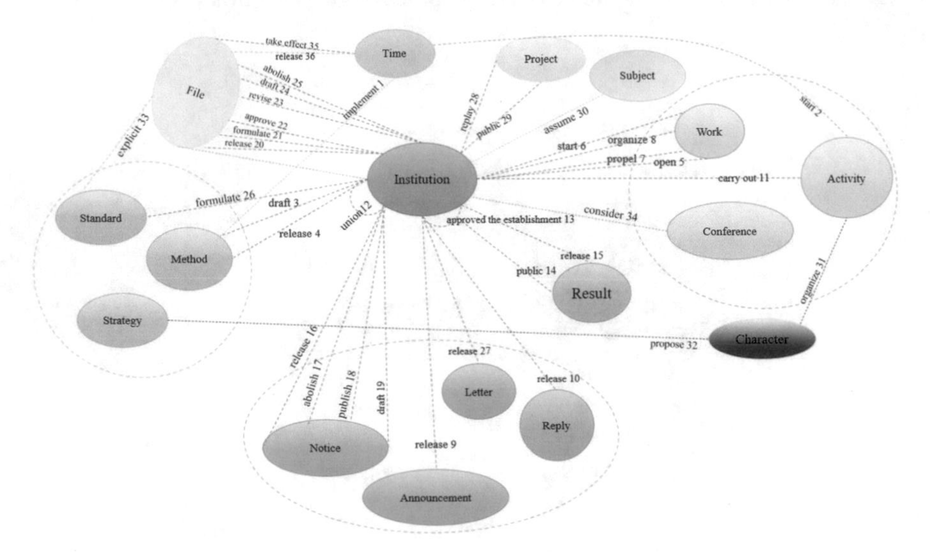

Fig. 5. Energy policy entities and relationships

4.2 Comparison Model

Compare this model with the following models to verify the performance of this model. The following model uses the open-source tool DeepKE of Knowledge Engine Laboratory of Zhejiang University for calculation.

1. PCNN model [10]: the segmented CNN extracts the most important features in each sentence, obtains the feature vector representation of the sentence and uses it for the final classification.
2. Capsule model: the model uses the Bi-LSTM network to obtain a coarse-grained sentence feature representation, then inputs the results into the capsule network to construct a primary capsule, which is dynamically routed. The method obtains the output capsule that matches the classification result [11];

3. Transform model: the model uses the encoder part of transformer to encode sentence information. The multi-head attention module is used to continuously extract important features in the sentence, and the superposition method of the residual network is used to splice the output obtained by the attention layer with the input and regularize it. Finally, it connects the result of transformer to the fully connected layer to get the final classification effect [9].
4. BERT pre-training model: the model uses the Bert-Base-Chinese language pre-training model, then inputs sentences into BERT. Finally, the results are input to the fully connected layer for relation extraction tasks. Using the bidirectional transformer, the context information of the current word can be used at the same time.

4.3 Result

The comparison results of the experimental models in this paper are shown in Table 1.

Table 1. Model comparison experiment results

Model	Precision (P/%)	Recall (R/%)	F1/%
PCNN	97.34	98.22	97.17
Bi-LSTM attention	99.97	98.25	98.93
Capsule	89.70	88.31	88.83
Transformer	90.42	92.90	90.30
BERT	96.05	98.22	96.43

Through the analysis of experimental data, it is found that the methods with better effects include PCNN, Bi-LSTM based on the attention mechanism, and Bert. But in contrast, using the method Bi-LSTM based on the attention mechanism in this paper to analyze the policy text, the experimental effect is better than other methods. It is a classic algorithm for extracting relationships between entities. It has achieved better results by weighting the results of Bi-LSTM. At the same time, from experiments and data, it is found that if the distance between the two entities in the policy text is closer, the possibility of identifying the relationship through the PCNN is greater. Conversely, if the distance between two entities is farther, the possibility of identifying the relationship through the PCNN is less. The PCNN model pays much attention to the distance information between entities, position information, and information between entities. In addition, it was discovered during the experiment that when the amount of data is small, if the number of BERT layers is set to 12, the effect is not ideal. In the experiment, we tried to reduce the number of layers. When the number of layers is changed to 3, the effect was improved significantly.

5 Conclusion

In summary, this paper uses the Bi-LSTM model based on the attention mechanism to analyze the policy text, and compares the current four popular entity naming recognition algorithms. The experiment proves that the Bi-LSTM model based on the attention mechanism in the policy analysis gets good results. Bi-LSTM can learn what information is remembered and what information is forgotten through the training process, can better capture longer-distance dependencies, and can capture bidirectional semantic dependencies better. Introducing the attention mechanism, the system can obtain local features of the text and suppress useless information, which effectively solves the problem of incomplete feature extraction due to the focus on extracting global features in the traditional method in the feature extraction process. Experiments have verified that using this model can effectively analyze policy texts, extract entities and relationships in policy texts, and help interpret policy texts.

References

1. Zeng, D., et al.: Relation classification via convolutional deep neural network. In: Proceedings of the 25th International Conference on Computational Linguistics, pp. 2335–2344 (2014)
2. Wang, L., Cao, Z., de Melo, G., Liu, Z.: Relation classification via multi-level attention CNNs. In: Proc. of the Meeting of the Association for Computational Linguistics, pp. 1298−1307 (2016)
3. Miwa, M., Bansal, M.: End-to-end relation extraction using LSTMs on sequences and tree structures. arXiv preprint, arXiv: 1601.00770 (2016)
4. Huang, Z., Wei, X., Kai, Y.: Bidirectional LSTM-CRF models for sequence tagging. Comput. Sci. (2015)
5. Zheng, S., et al.: Joint entity and relation extraction based on a hybrid neural network. Neurocomputing 257 (2016)
6. Devlin, J., Chang, M.-W., Lee, K., Toutanova, K.: BERT: Pre-training of deep bidirectional transformers for language understanding. axXiv preprint, arXiv:1810.04805 (2018)
7. Wu, S., He, Y.: Enriching pre-trained language model with entity information for relation classification. axXiv preprint, arXiv:1905.08284 (2019)
8. Schuster, M., Paliwal, K.K.: Bidirectional recurrent neural networks. Signal Process. IEEE Trans. **45**(11), 2673−2681 (1997)
9. Vaswani, A., et al.: Attention is all you need. arXiv preprint, arXiv:1706.03762 (2017)
10. Phi, V.T., et al.: Distant supervision for relation extraction via piecewise attention and bag-level contextual inference. IEEE Access **PP**(99), 1 (2019)
11. Zhang, N., et al.: Attention-based capsule networks with dynamic routing for relation extraction. In: EMNLP 2018 (2018)

Concept Drift Detection Based on Restricted Boltzmann Machine in Multi-class Classification System

Jinyu Zhou, Qianwen Zhu, Ruirui Shi, and Wei Wang(✉)

College of Artificial Intelligence, Tianjin Normal University, Tianjin 300387, China
weiwang@tjnu.edu.cn

Abstract. With the development of information technology, more and more data are generated from social life. Concept drift detection in multi-class classification system has gradually become a research hotspot in the field of data mining. To solve this problem, a multi-class concept drift detection algorithm based on Restricted Boltzmann Machine is proposed in this paper. Based on the probability distribution of RBM, the KL divergence and concept drift detection coefficients are constructed to detect concept drift and judge its type. The performance of the algorithm is tested and analyzed on simulation and real data sets.

Keywords: Restricted Boltzmann machine · Concept drift · KL divergence

1 Introduction

With the development of information technology, the rate of data generation in anomaly detection, sensor network and many other fields are getting faster and faster. The information contained in the data stream [1] changes over time and contains multiple attributes. In order to obtain useful information, it is necessary to carry out processing in the limited storage space for these data streams. At present, the problem of the detection of multi-class data streams with concept drift has been paid much attention by the academic circles.

In the late 1980s, Schlimmer and Granger proposed the term "concept drift" for the first time and proposed STAGGER algorithm to solve the problem of concept drift [2]. Since then, many academic researchers have become more and more interested in the study of concept drift. Zhang et al. detected the concept drift by estimating the confidence interval of the true error rate of new concepts [3]. However, this algorithm has the disadvantage of lag. Later, Katakis introduced the "concept vector" to detect concept drift by its distance from each cluster center [4], but this method has limitations on data sample classes. Li adopted IKNN Model algorithm for classification detection of concept drift of mixed model data streams [5], but this algorithm model was complex and had a large amount of computation. Sun et al. proposed to use the data stream mining algorithm of MJD4 to realize the detection of concept drift [6], but this method couldn't detect the sudden drift in time. Liu used the method of dynamically adjusting

Q. Liang et al. (Eds.): Artificial Intelligence in China, LNEE 854, pp. 235–241, 2022.
https://doi.org/10.1007/978-981-16-9423-3_29

the data window by overlapping data window to detect concept drift [7], but the size of the data window in this method still needs to be set by experience. Zhang et al. proposed an effective method of difference measurement to judge the concept drift [8], but this method was not accurate in the discrimination between concept drift and noise.

To solve the above problems, this paper proposes a Concept Drift Detection Based on RBM in Multi-Class Classification System. By constructing concept drift detection coefficient, drift detection is realized.

The rest of this article is organized as follows. Section 2 details the theory and implementation of concept drift detection. Section 3 analyzes the experimental results. Section 4 is the conclusion.

2 Measurement Mechanism of Concept Drift Based on Restricted Boltzmann machine

2.1 Restricted Boltzmann Machine Theory

Restricted Boltzmann Machine (RBM) is a model based on energy function. It describes the random network by the input of the visible layer and the output of the hidden layer, and represents the connection relationship between different nodes through the weight. Parameters include connection weight *W* of neurons in the visible layer and neurons in the hidden layer, neuron offset *a* of the visible layer, neuron offset *b* of the hidden layer. Each node of the visible layer and the hidden layer has a deviation, as shown in the figure below (Fig. 1).

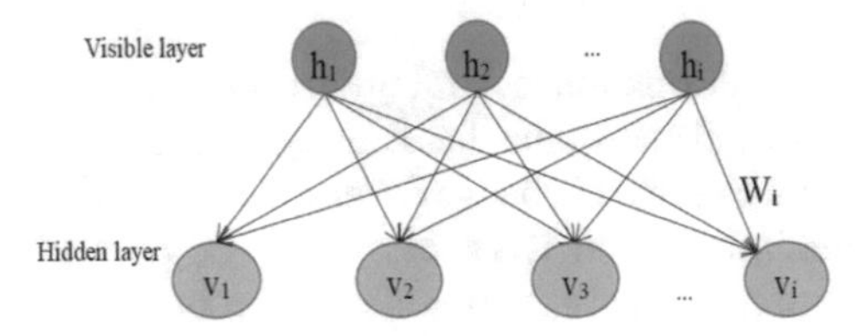

Fig. 1. Restricted Boltzmann machine

2.2 Probability Distribution Measurement Process Based on RBM

By training the RBM model, the energy function and probability distribution variation trend of each class can be obtained. The specific process is as follows:

The total energy in RBM is the sum of the energies between interconnected neurons, which can be expressed as:

$$E(v,h) = -\sum_{i}\sum_{j} E(v_i,h_j) = -(a^T v + b^T h + v^T Wh) \tag{1}$$

According to the energy function, the probability distribution of the input data can be obtained:

$$P(v) = \frac{exp(-E(v))}{Z(\theta)} = \frac{\sum_k exp(-E(v))}{Z(\theta)} \tag{2}$$

Where, Z can make the sum of the probabilities equal to 1. But it cannot be directly calculated. For a well-trained RBM, the value of Z remains unchanged. Therefore, the probability distribution trend of the input data can be obtained.

2.3 Concept Drift Detection Process Based on KL Divergence

We divide the data stream into multiple Windows, so that the data blocks of each window are respectively B_1, B_2, ..., B_i, ..., each data block has the same length, and its class is set as y_{new}, a new base classifier BC_{new} is constructed. Let BC_y and BC_{new} recognize the output value of Bi as $p_y^i(0 \le p_y^i \le 1; 1 \le y \le Y)$ and $p_{new}^i(0 \le p_{new}^i \le 1)$, calculated from the KL divergence, which is used to measure the difference between two probability distributions.

$$\mathrm{p}_{\mathrm{new}}^i = \sum_j \frac{P_{new}(j)}{Q_B(j)} \tag{3}$$

$$\mathrm{p}_{\mathrm{y}}^i = \sum_j \frac{P_{\mathrm{y}}(j)}{Q_B(j)} \tag{4}$$

We define the following concept drift detection coefficient:

$$\omega_{new}^i = \frac{1}{p_{new}^i(1 - p_{new}^i)^2 + \xi} \tag{5}$$

$$\begin{aligned} \omega_y^i &= \frac{1}{p_y^i(1 - p_y^i)^2 + \xi} \\ \omega_y^i &= \beta\omega_y^{i-1} \end{aligned} \tag{6}$$

Where, ξ is a minimal positive number to prevent the coefficient from being meaningless, and $\beta < 1$ represents the time attenuation coefficient.

Equations (3) and (5) represent KL divergence and concept drift detection coefficients of BC_{new} on B_i; When B_i is considered to belong to class y, the KL divergence and detection coefficient on B_i are expressed by Eq. (4) and (6).

For the sudden drift, the detection coefficient of the class corresponding to the data in the previous window will greatly decrease, while the corresponding coefficients of other classes will not increase, only w_{new}^i will increase abruptly. For gradual drift, the detection coefficient of the class corresponding to the previous window will gradually decrease and w_{new}^i will gradually increase. If the corresponding coefficient of a base classifier gradually decreases to 0, it means that the class has not appeared in the data stream for a long time, and the concept can be determined to disappear. In addition, we can also find out whether there is abnormal state by observing the change of the detection coefficient.

3 Experimental Results and Analysis

3.1 Gaussian Datasets

Python built-in functions were used to randomly generate Gaussian distribution data with various variances. Three types of data are taken as three known classes, which are recorded as A, B and C in turn, and the sample size of each class was set as 10. The RBM was used to train the energy function, and the probability distributions of the three classes were calculated.

The data to be tested is composed of five data blocks of length 10. After the probability distribution of three classes and each data block are obtained, calculate the KL divergence between them respectively, which is defined as p_y^i. Substitute the results and calculate the concept drift detection coefficient. The variation trend of KL divergence value and coefficient was observed in the plot, and the results were shown in the figure below.

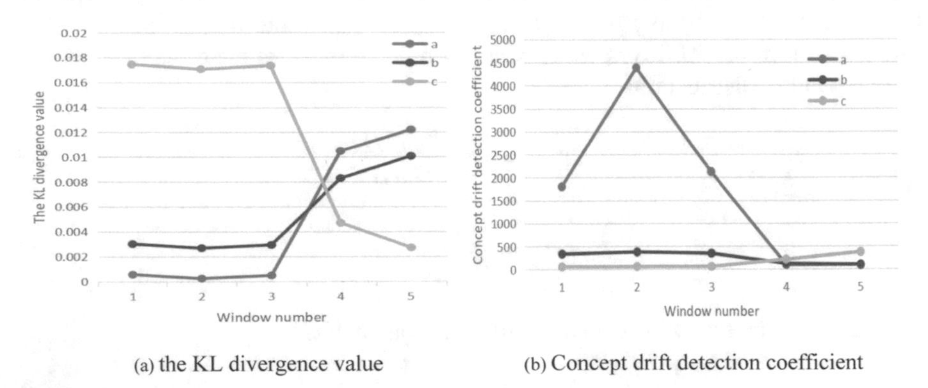

(a) the KL divergence value (b) Concept drift detection coefficient

Fig. 2. Concept drift detection for three classes of Gaussian datasets

According to the curve variation trend in Fig. 2(a), it can be seen that p_c^i is large, but suddenly decreases in window 4. On the contrary, p_a^i and p_b^i are small, and then suddenly increases. Moreover, in the first three Windows, p_a^i approaches 0, indicating that the data to be tested belongs to class A. In the last two Windows, p_c^i is the smallest, indicating that it is gradually approaching to class C.

We need to observe the previous window corresponding class of detection coefficient. known from the analysis of the above, the first three window data corresponding to class A, so we observe the change trend of the corresponding blue curve in Fig. 2(b), clear testing coefficient values at a greatly reduced, and other classes corresponding coefficient does not increase, so the judge sudden drift. In addition, w_b^i gradually decreases to 0, indicating that this class has not appeared in the data to be tested for a long time, and it is determined that the concept of class B has disappeared.

3.2 MNIST Datasets

MNIST datasets was organized by the National Institute of Standards and Technology (NIST). The handwritten digital images were taken from 250 different people, 50 percent were high school students, and 50 percent were Census Bureau workers. In 1998,

Yanlecun et al. proposed the LeNet-5 network for the first time and realized handwriting font recognition by using the above datasets.

The images in datasets are all grayscale images with the size of 28 × 28, and its samples are handwritten digital pictures with different handwriting between 0 and 9. We take the numbers 1, 3 and 7 of MNIST for the experiment, and set them as three classes A, B and C respectively, and set the number of pictures in each class as 200. Images under test are divided into five windows, and the number of images in each window is also set to 200. The result was plotted as follows.

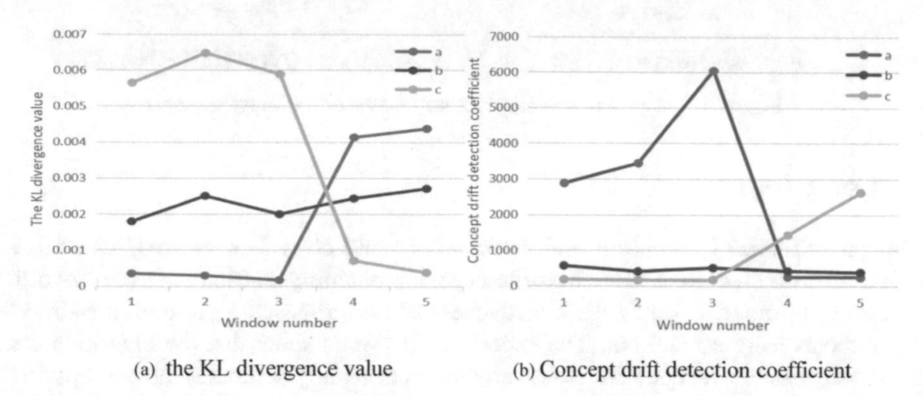

(a) the KL divergence value (b) Concept drift detection coefficient

Fig. 3. Concept drift detection for three classes of MNIST datasets

In Fig. 3(a), we find that under test images by near class A gradually changes into near class C. According to Fig. 3(b), the data of the blue curve decreases significantly, while the coefficients of other classes do not increase. So the judgment is sudden drift. In addition, the corresponding coefficients gradually decrease to 0, so the concept of class A is determined to disappear.

3.3 Satimage Datasets

Satimage datasets consist of the multi-spectral values of pixels in 3 × 3 neighborhood in satellite images. Each example contains the pixel values in the four spectral bands and a number indicating the classification label of the central pixel. The number is a code for the class: 1: Red soil; 2: Cotton crops; 3: grey soil; 4: moist gray soil; 5: soil and vegetation stubble; 6: Very moist grey soil.

We set the above classes as A, B, C, D, E and F respectively, and set the sample number of each class as 50. The result was plotted as follows.

In Fig. 4(a), the data to be tested gradually approaches from class A to class B.

Therefore, the data in the previous window corresponds to class A. By observing Fig. 4(b), it can be seen that the detection coefficient value of class A gradually decreases, so gradual drift is judged.

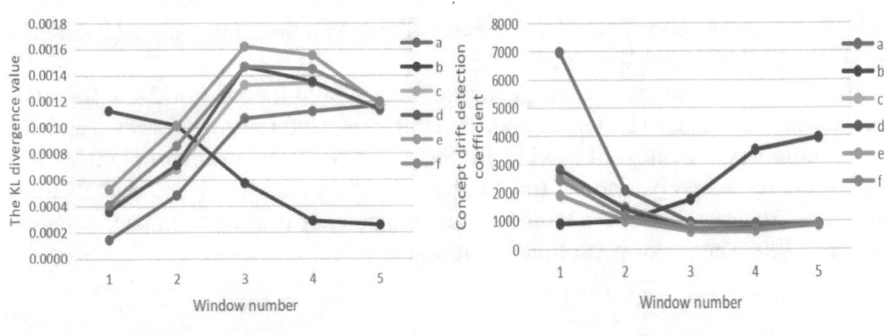

(a) the KL divergence value (b) Concept drift detection coefficient

Fig. 4. Concept drift detection for six classes of Satimage datasets

4 Conclusion

This paper studies the concept drift detection of multi-class data streams, which uses detection coefficient to measure the difference of probability distribution to realize drift detection. In order to verify the effectiveness of the proposed algorithm, a series of experiments were carried out. The experimental results show that the algorithm can effectively detect concept drift. However, this paper mainly addresses the concept drift detection of data and images. For data streams containing audio and video, the algorithm needs to be further expanded.

Acknowledgements. The work was supported by the Natural Science Foundation of China (61731006, 61971310)

References

1. B Zhang G Chen X Wang 2010 Fuzzy clustering based on data flow model Comput. Eng. Appl. 46 124 126
2. J Schlimmer RH Granger Jr 1986 Incremental learning from noisy data Mach. Learn. 1 317 354
3. C Zhou 2011 Detection and classification of concept drift on data stream Minicomputer Sys. 32 421 425
4. G Tsoumakas I Katakis 2009 Multi-label classification: an overview Int. J. Data Warehous. Min. 3 1 13
5. N Li 2013 Concept Drift Data Stream Classification Algorithm and its Application Fujian Normal University Fuzhou
6. Y Sun 2008 Mining concept drift in data stream based on multiple classifiers J. Automation 34 93 97
7. M Liu 2014 Online Concept Drift Detection Based on Data Window Xiangtan University Xiangtan
8. Y Zhang Y Chai L Wang 2013 Martingale based concept drift detection method for data stream Minicomputer Syst. 34 1787 1792
9. P Li X Wu X Hu Q Liang Y Gao 2010 A random decision tree ensemble for mining concept drifts from noisy data streams Appl. Artif. Intell. 24 680 710

10. Y Sun K Tang LL Minku S Wang X Yao 2016 Online ensemble learning of data streams with gradually evolved classes IEEE Trans. Knowl. Data Eng. 28 1532 1545
11. J Lu A Liu F Dong F Gu J Gama G Zhang 2018 Learning under concept drift: a review IEEE Trans. Knowl. Data Eng. 31 2346 2363

Classroom Speech Emotion Recognition Based on Multi-channel Convolution and SEnet Network

Ke-Jin Liang, Hai-Jun Zhang(✉), Ya-Qing Liu, Yu Zhang, and Yue-Yang Wang

School of Computer Science and Technology, Xinjiang Normal University, Urumqi 830054, China
zhjlp@163.com

Abstract. Speech emotion recognition is one of the research hotspots in the current artificial intelligence background. It has applications in many scenarios and fields. In order to solve the problems of low accuracy and complex training parameters of traditional speech emotion recognition models. This paper intends to introduce speech emotion into classroom teaching, and proposes a multi-channel convolution combined with SEnet network as an emotion recognition model by extracting the emotional characteristics of speech. In the self-made emotional data set, the accuracy rate, accuracy rate, F1 value, and recall rate all have good performance. The effectiveness of the model is verified through comparative experiments.

Keywords: Deep learning · Speech emotion recognition · Neural network

1 Introduction

In classroom teaching, the combination of emotion in teaching and teaching has always been what we are pursuing [1], but emotion is often the easiest to be ignored in teaching. With the continuous deepening of the new curriculum reform, the combination of student emotion and knowledge has become a difficult point in teaching. Introducing voice emotion into teaching as a classroom evaluation standard can help teachers grasp the rhythm of the classroom, mobilize students' enthusiasm for learning, and improve learning efficiency [2]. However, there are relatively few studies on classroom speech emotion, and related public data sets are rare. Therefore, this article intends to use web crawler technology to capture excellent web teaching videos as a data set, and use multi-channel convolution combined with SEnet network as a model to obtain better recognition results.

2 Related Works

In the era of artificial intelligence, emotions are an indispensable part. How to make computers possess human emotions is still one of the main research hotspots at present.

Q. Liang et al. (Eds.): Artificial Intelligence in China, LNEE 854, pp. 242–248, 2022.
https://doi.org/10.1007/978-981-16-9423-3_30

The current emotion research mainly focuses on expressions, text body movements and speech [3]. As one of the most effective and convenient ways for humans to express their ideas and thoughts, voice is also a process of emotional communication while transmitting information. In the early SER, most of the machine learning methods were used for classification. When extracting parameters of speech emotion features, Mingzhu Liu et al. [4] joined the Support Vector Machine classifier to select the appropriate SVM kernel parameters, and finally passed Established SVM for classification modeling; Sheng Wang [5] extracted the speech emotion feature parameters, combined with Hidden Markov Model for modeling; Jixiang Ye et al. [6] constructed a three-layer The random forest network of the new model solves the problem of uneven distribution of emotional data sets, and also improves the accuracy of recognition; Hao Ren et al. [7] used Principle Component Analysis for feature dimensionality reduction, and then combined multiple Level SVM algorithm for modeling, and finally determine the emotion type of the voice. With the continuous development of deep learning, it has applications in face recognition, image recognition, etc. Huilian Lu et al. [8] directly use deep neural networks to process and classify the original speech emotion signals, and train them in an end-to-end manner., Achieved good results; Pengxu Jiang et al. [9] proposed a CNN-based speech emotion recognition model, based on the Lenet-5 model, by adding a network layer to change the two-dimensional convolution kernel to a one-dimensional convolution kernel. Performing feature transformation and representation realizes the classification of speech emotion; Desheng Hu et al. [10] used the main and auxiliary network for deep feature fusion speech emotion recognition algorithm, and achieved good results by using Mel spectrogram as the input of the model. However, these studies all choose to increase the effectiveness of the model by increasing the number of layers of the network and extracting features from depth. As everyone knows thc higher the network layer, the larger the receptive field, the stronger the ability to represent semantic information, but it weakens the representation of geometric information; vice versa, the enhancement of geometric representation also weakens the representation of voice information. At the same time, due to the diversity of the speech emotion sequence, the ability to connect the speech emotion context feature information is insufficient, and the features cannot be fully utilized for classification, resulting in low recognition accuracy. Therefore, multi-channel convolution is combined with the SEnet network model, and 3 branches are used to deepen the depth and width of the network. At the same time, the SEnet network is added to enhance the nonlinear ability of the model, better fit the input voice data, and improve the generalization ability of the model.

3 Multi-channel Convolution Combined with SEnet Model

This paper uses multi-channel convolution combined with SEnet as an emotion recognition model to solve the problem of insufficient feature extraction in traditional deep neural networks. The 3 deep convolutional neural network branches are spliced together, and a Gaussian noise layer is added between each convolutional layer and the hidden layer to alleviate the impact of data imbalance. Add the attention mechanism (SEnet) to enhance the nonlinear ability of the model and improve the fit of the model. The specific model is shown in Fig. 1 below. In the figure below, Conv1, Conv2, and Conv3 represent different convolution kernel sizes (3×3, 5×5, 7×7). The number of convolution kernels starts from 32, and each layer is increased by 2 times, and so on, a total of 4 convolution layers are stacked, and finally sent to the fully connected layer for sentiment classification. In order to prevent over-fitting, add Dropout after the fully connected layer and set it to 0.3. Regularization is added to each layer of convolution, which improves the generalization ability of the model; the optimizer uses Adam to not only learn the parameters adaptively, but also accelerate the convergence speed of the model; after each layer of convolution, batch normalization is added to learn independently The weight of each layer of convolution; the learning rate is set to 1e−4.

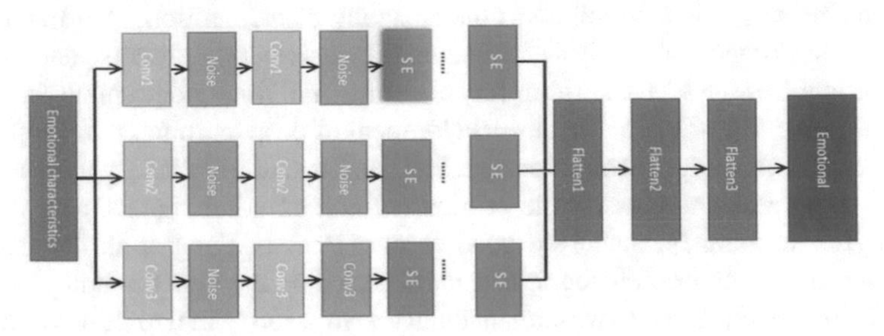

Fig. 1. Multi-channel convolution combined with SENET neural network model.

4 Data Set Source and Processing

This data set uses web crawlers to crawl the teaching videos of the third grade Chinese open class of Jiaoshi.com. Jiaoshi.com contains national high-quality classroom teaching videos and high-quality lectures from various provinces and cities, as well as classroom records of famous teachers, and it is collected from the Internet. The high-quality lesson videos provide a high-quality classroom platform for teachers and students. Since the article is mainly for the recognition of emotion classification of speech, the Audiofileclip package that comes with Python is used to extract the speech in the teaching video, and then split_on_silence is used to segment the speech. The speech with silence exceeding 1000 ms is divided, and the speech with silence is less than −70 dB. It is considered as silence, abandoning the segments less than 500 ms in the speech, adding a 20 ms interval after each segmentation and splicing to segment the sentence, and finally segmenting the speech into 10 s speech segments, generally ensuring that each segment of speech is a complete sentence, and finally Perform voice screening and classification. After the above operations, 1041 pieces of voice data are obtained, and the voice is planned to be divided into several emotions: calm, high, nervous, satisfied, and doubtful. The distribution of speech emotion data is shown in Table 1:

It can be seen from Table 1 that the distribution of various voice emotion types is uneven. Calm and high emotions account for the majority, while nervous emotions are obviously not enough. The weight distribution method is used to set the proportion of each emotion in order to achieve the purpose of balancing the data. After the weight is allocated, the proportion of each voice is shown in Table 2:

Table 1. The distribution of the number of emotions in the speech emotion data set

Emotion	Calm	High	Tension	Satisfaction	Doubt
Quantity	438	394	91	114	265

Table 2. The proportion of each emotion in the speech emotion data set

Emotion	Calm	High	Tension	Satisfaction	Doubt
Weights	0.65	0.65	1.0	1.4	0.80

4.1 Extraction of Voice Features

By extracting commonly used voice emotion features, such as pitch, zero crossing rate, spectral centroid, Mel spectrum coefficient, Mel frequency cepstral coefficient, root mean square energy, chromatogram, spectrum plane and other low-level description features, calculate the high-level statistics of these features Function to get a total of 312 dimensions of emotional features. The extracted and calculated features are shown in Table 3:

Table 3. Low-level description and high-level statistical function characteristics

Features	Low-level description	High-level statistical functions
Prosodic features	Pitch	Mean, Std, Max
	Zero Crossing Rate	Mean
Spectral characteristics	MFCC, Amplitude, Centroid	Mean, Std, Max
	Flatness, Chroma, Mel, Contrast	Mean

5 Experiments and Analysis

This section mainly compares the experiments from three aspects: 1. The influence of the unbalanced data set on the experimental results; 2. The influence of the Gaussian noise layer between the convolutional layer and the hidden layer on the results; 3. The influence of the depth of the convolutional layer on the experimental results. Compare and verify the effectiveness of the model from many aspects.

5.1 The Influence of Weights on Experimental Results

In order to verify whether setting weights on the data set has an impact on the results of the experiment, a confusion matrix is used as the evaluation standard of the experiment.

The experimental results are shown in the following figure: 0 means High; 1 means Nervous; 2 means Satisfaction; 3 means Calm; 4 means Doubt.

It can be seen from Fig. 2 and Fig. 3 that the effect of increasing the weight is more obvious. The recognition accuracy of the three emotions of high, nervous, and calm is higher than the accuracy of the unweighted, but it is slightly inferior in the recognition of satisfaction and doubt. One point, but generally speaking, the effect of increasing the weight on the experimental results is positive and effective.

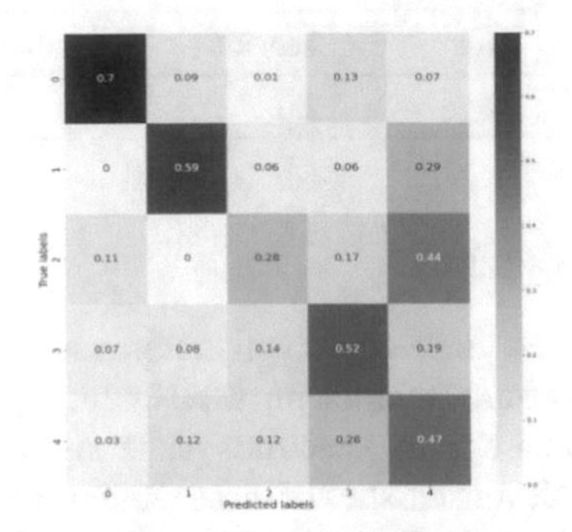

Fig. 2. Unweighted experimental results

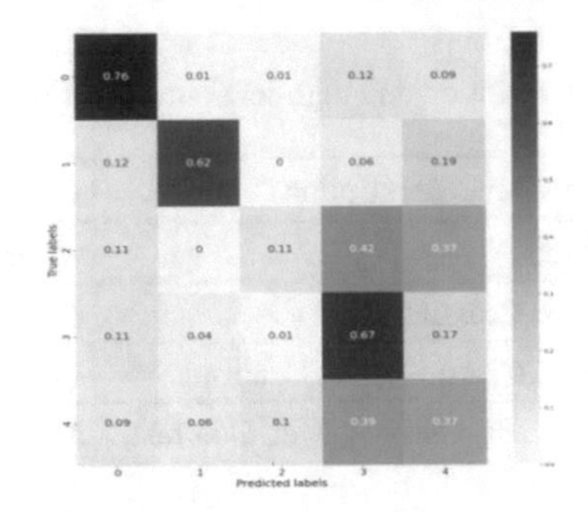

Fig. 3. Weighted experimental results

5.2 The Influence of Gaussian Noise Layer on Experimental Results

In order to verify the influence of the Gaussian noise layer on the experimental results, the noise-free layer and Gaussian noise parameters (0.1, 0.2, 0.3) are set, and the recall rate, F1 value and accuracy rate are used as evaluation indicators. Table 4 below shows the influence of the Gaussian noise layer on the experimental results.

It can be seen from the table that the evaluation index with noise layer added is higher than that without noise layer. The recall rate, F1 value, and accuracy rate are up to 11.97%, 8.61%, and 6.34% higher than those without noise layer. The impact is also positive and effective. It is verified that adding a noise layer can improve the accuracy of the model, and can effectively solve the problem of unstable model recognition caused by data imbalance.

Table 4. The influence of noise floor on experimental results

Influencing factors	Recall rate (%)	F1 value (%)	Accuracy rate (%)
No noise floor	50.96	53.84	57.08
Noise floor (0.1)	52.82	57.14	62.16
Noise floor (0.2)	54.26	57.50	63.26
Noise floor (0.3)	62.93	62.45	63.42

5.3 The Influence of Gaussian Noise Layer on Experimental Results

In order to further verify the effectiveness of the model, RestNet (34), Multi-channel convolution + LSTM, DCNN + LSTM and other models are compared with the model in this paper on the self-made data set, and the AUC value is used as the evaluation standard of the model. The experimental results are as follows shown Table 5:

Table 5. Comparison results of various methods in the self-made data set

Approach	Recall rate	F1 value	Accuracy rate	AUC
RestNet(34)	56.32%	59.60%	61.46%	0.78
Multi-channel convolution + LSTM	54.33%	56.01%	55.65%	0.75
DCNN + LSTM	55.60%	60.34%	59.37%	0.74
DCNN + SEnet	57.30%	58.03%	60.45%	0.83
Multi-channel convolution + SEnet	62.93%	62.45%	63.42%	0.87

It can be seen from the above table that increasing the number of network layers can improve the accuracy of recognition, but it is not the main influencing factor. Most experimental models maintain accuracy at about 50% on the self-made data set, but the model in this article can reach 60%; by introducing AUC as the evaluation standard of the model, the larger the AUC value, the better the performance of the model, which verifies the effectiveness of the model in this paper.

6 Conclusion and Future Works

This article focuses on emotion recognition in class lectures. Using multi-channel convolution combined with attention mechanism as a model, experiments on self-made data sets have achieved good results. Due to different voice sources, the sampling frequency is not uniform, and the imbalance of the data set will also have a certain impact on the performance of the model. Although the loss weight processing improves the accuracy of the model on the self-made data set, it does not fundamentally solve the impact of the imbalance of the data set on the experimental results and the model. In future research, the scale of the data set will be expanded, the sampling frequency will be unified, and the problem of data set imbalance will be further solved.

References

1. Jiamei, L.: Classification of emotional goals in classroom teaching. Psychol. Sci. **06**, 1291–1295 (2006)
2. Chong, N.: Classroom Speech Emotion Recognition and Analysis Based on Deep Learning. Central China Normal University (2020)
3. Gao, F., et al.: Multi-feature speech emotion recognition based on DBM-LSTM. Comput. Eng. Design **41**(2), 465–470 (2020)
4. Liu, M.Z., Li, X., Chen, H.: Research on speech emotion recognition algorithm based on support vector machine. J. Harbin Univ. Sci. Technol. **24**(04), 118–126 (2019)
5. Sheng, W.: Speech emotion recognition based on hidden Markov model. Heihe Sci. Technol. Inform. **28**, 2 (2010)
6. Jixiang, Y., Qingyu, T.: Multi level random forest network speech emotion recognition based on importance score 6(3), 77–83 (2019)
7. Hao, R., Yue, L., Xuejun, S.: Speech emotion recognition algorithm based on multi-level SVM classification. Comput. Appl. Res. **34**(6), 1682–1684 (2017)
8. Lu, H., Hu. W.: Research on speech emotion recognition based on end-to-end deep neural network. J. Guangxi Normal Univ. (NATURAL SCIENCE EDITION) 1–7 (2021)
9. Pengxu, J., Hongliang, F., Huawei, T., Peizhi, L.: A method of speech emotion recognition based on feature representation of convolutional neural network. Electronic Devices **42**(4), 998–1001 (2019)
10. Desheng, H., Xueying. Z., Baoyun, L.: Speech emotion recognition based on main and auxiliary network feature fusion. J. Taiyuan Univ. Technol. 1–9 (2021)

Fake News Detection Based on Attention Mechanism and Convolutional Neural Network

Fei Wang, Haijun Zhang(✉), Rui Cheng, and Pengxu Wang Rongrong Xu

School of Computer Science and Technology, Xinjiang Normal University, Urumqi 830054, China
zhjlp@163.com

Abstract. The rapid development of social media has made Internet news flood all aspects of our lives, and also facilitated the spread of fake news, so it is important to detect fake news to reduce social impact. In this paper, we propose a fake news detection method based on the combination of attention mechanism and convolutional neural network, which obtains the internal relationship of sentences by self-attention mechanism, and then uses convolutional neural network to process the extracted features for classification. In order to verify the effectiveness of the model, this model is experimented on three different public datasets, and good detection results are achieved under ten-fold cross-validation.

Keywords: Attention mechanism · Convolutional neural network · Ten-fold cross-validation · Fake news detection

1 Introduction

With the progress of society, people's access to news has changed from the original way of paper newspapers to social media and other news websites. The convenient access has also contributed to the spread of fake news, which seriously misleads the public to get the real information and causes some negative impacts on society. For example, in the 2020 global epidemic, there are a lot of fake news, including but not limited to "microwave ovens can disinfect masks", "use fans to blow your nose and mouth to kill the virus", "drink Banlangen to prevent the new crown virus", and other such fake news. Such fake news has caused a lot of negative impact on society. The traditional fake news relies on manual review, which obviously cannot cope with the current huge amount of information and has a great lag, so it is of great significance to quickly and automatically identify fake news and kill it in the cradle before it spreads.

The detection of fake news somewhat relies on manual extraction of features, which is labor-intensive and labor-intensive, poorly generalized, and has major limitations. In this paper, we propose a model based on attention mechanism and convolutional neural network, fusing the advantages of both to extract sentence representations and efficiently detect fake news based on the content of the news only, with good results on three different public datasets. The rest of the paper is organized as follows: in the second part, we explore the state of the art in fake news research and propose our solution;

Q. Liang et al. (Eds.): Artificial Intelligence in China, LNEE 854, pp. 249–256, 2022.
https://doi.org/10.1007/978-981-16-9423-3_31

in the third part, the structure of the proposed model is detailed; in the fourth part, the dataset, experimental parameters are explained and experiments are compared with some baseline models; in the fifth part, the conclusions are summarized and the next part of the work is proposed.

2 Related Works

2.1 Fake News Detection

Fake news has been a topic of great concern for a long time with its characteristics of fast, wide and deep dissemination. Over the years, a lot of research has been conducted on fake news, which is broadly divided into two categories,on the one hand fake news detection based on news content features; on the other hand,detection based on the network features between users and content.

Detection based on content features is a more traditional approach to detection. Shin et al. [1] analyzed four aspects from false knowledge, writing style, dissemination pattern, credibility of news creator and disseminator by investigating interdisciplinary grounded theory, which is effective but requires more features to be extracted and relatively more workload. Karimi H et al. [2] performed deception detection by extracting headlines or sentences of different textual features and then using traditional machine learning methods for fake news detection. Feng et al. [3] investigated deception detection by context-independent grammar parse trees, but later according to Shlok Gilda report [4],claimed that this approach was not effective in improving model performance.

The network characteristics between users and content are mainly expressed in group interaction, information sharing, sentiment evaluation, and information deliberation, that is, users' behaviors such as discussion, comments, retweets, and likes on the corresponding news. Ahmed et al. [5] detected fake content through user comments on articles and study six different machine learning models to detect it. Wu et al. [6] used the dissemination path of news and social media user profiles to track fake news. This approach based on the detection of network features between users and content improves the detection effect to some extent, but it is more dependent on content-assisted information and cannot quickly detect the truth or falsity of news.

In this paper, we address the shortcomings in previous studies and use the attention mechanism to quickly capture the global connection of news content, and then use convolution to extract keyword features, and the whole process can quickly detect the truth or falsity of news based on the content of news only.

3 Fake News Detection Model Based on Attention Mechanism and Convolutional Neural Network

For the knowledge background of our research, we propose a model framework based on attention mechanism plus convolutional neural network to detect fake news. The model puts the news content coding and fusion into the attention layer through the coding layer, extracts the global features of the sentence through the attention layer, and finally extracts the local features of the sentence and classifies them through the convolutional neural network, so that the model can capture the global features and local features, which can fully extract the news features and improve the effectiveness of the model, and the model framework diagram is shown in Fig. 1.

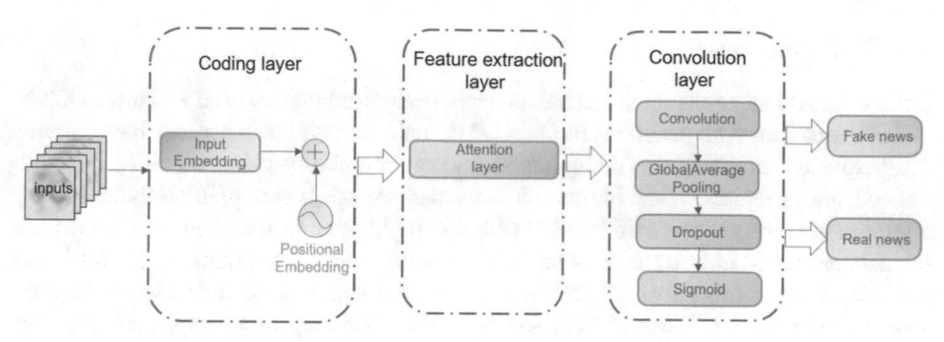

Fig. 1. Model framework diagram

3.1 Input Layer

The input layer is mainly to clean and serialize the news data. The dataset in this paper comes from three different public datasets, so the dataset is first formatted to remove the empty data to ensure that the content of the news corresponds to the tags one by one, followed by cleaning the news data, data cleaning is to remove the useless data from the news, such as punctuation, URL links, deactivation words, etc. Next is to serialize the data. Data serialization is to first convert each data into a token dictionary, then convert each data into a vector form with word subscripts, and finally fill the length less than 300 with zeros to get the processed data set as the input of the model.

3.2 Coding Layer

The coding layer consists of a word embedding layer and a location coding layer, which are used to do vectorization operations on news content, and the word embedding used in this paper is the Glove algorithm [7], which is different from the LSA and Word2Vec algorithms, combining the characteristics of LSA using global features and Word2vec using local contextual features, and is a word characterization algorithm based on global word frequency statistics, which can make full use of the statistical information in the corpus. This model incorporates positional encoding [8], mainly because the model lacks a method of word order in the input sequence that captures the contextual connection of the sentence, so it needs to be used to record the current word position information in a way that uses sine and cosine functions of different frequencies:

$$\begin{aligned} PE_{(pos,2i)} &= \sin\left(\mathrm{pos}/10000^{2i/d_{\mathrm{model}}}\right) \\ PE_{(pos,2i+1)} &= \cos\left(pos/10000^{2i/d_{\mathrm{model}}}\right) \end{aligned} \quad (1)$$

where pos is the position of each word and i is the dimension, meaning that each dimension of the position encoding corresponds to a corresponding sine wave or a cosine wave, so that the model can learn the relative position of each word. Finally, the model stitches the results of position encoding and word embedding and feeds them to the next layer of the network for further processing.

3.3 Attention Layer

Before the advent of the attention mechanism, many researchers used variants of CNN and RNN to deal with problems in the NLP domain, but all these variants have natural limitations, either insufficient computational power or inability to deal with long-distance dependence problems. While the attention mechanism was first used in computer vision, attention gradually blossomed in NLP with the application of the attention mechanism to machine translation by Bahdanau et al. [9]. Inspired by Vaswania et al. [10], the attention layer in this paper adopts a multi-headed attention mechanism to capture the global connections of news content and learn sentence long-range dependencies. The multi-headed attention mechanism used in this paper is as follows Fig. 2:

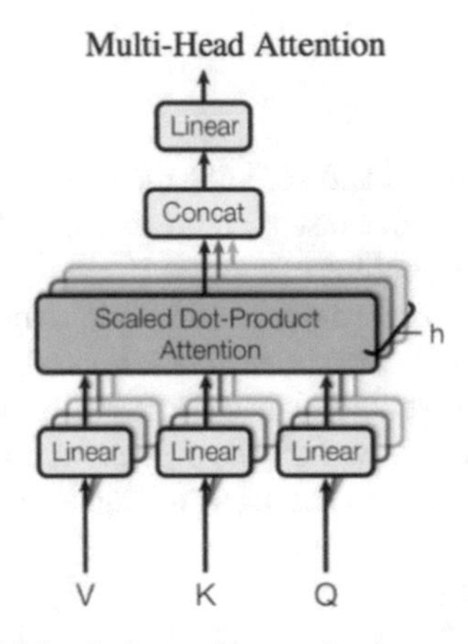

Fig. 2. Multi-headed attention mechanism model diagram

The multi-headed attention mechanism can let the model form multiple subspaces and take the information of different regions of attention respectively, this part is mainly realized by making the process of Scaled Dot-Product Attention as h times, and finally the extracted features will be united as the overall features.

3.4 Convolutional Layer

The features fused in the attention layer are put into the convolution layer, and different sensory fields are obtained after convolution, keyword features are extracted, and then the results of convolution are transformed into uniform length by global pooling, and finally put into the sigmoid layer for binary classification, and the news is output true or false.

4 Experimental Analysis

4.1 Dataset

In this paper, three publicly available datasets are used for experiments, the first one is the publicly available Liar dataset by Wang et al. [11], which contains 12.8k human-labeled short statements with six different classes of labels, namely, fiery, false, barely true, semi-true, mostly true and true, which are mainly in some political domains, and since this study is mainly about distinguishing true from false news, it is combined with the needs of this paper, this dataset is divided into two categories of true and false; the second one is the COVID-19 dataset made public by Parth Patwa et al. [12] in 2020, which is a manually labeled dataset consisting of 10,700 social media posts and true and false news articles. The third one is the fake news dataset from the Kaggle competition, which was collected mainly from some unreliable websites and labeled by Politifact (a fact-checking organization in the USA) and Wikipedia, Table 1 shows the distribution of the three datasets after processing:

Table 1. Dataset distribution

	Fake news size	Real news size
Liar	5657	7134
COVID-19	4080	4480
Fake and real news Datasets	23481	21417

4.2 Experimental Parameter Setting

This model uses the Glove algorithm for word vector partitioning, the word vector dimension is 100, the maximum length of sequence processing maxlen is 300, where the head size in the multi-headed attention model is 8, the head_size is 16, the number of filters in the convolutional neural network is 128, the convolutional kernel size is 4, the learning rate of the dropout layer is 0.5, and in addition this model also uses ten-fold cross-validation, i.e., the parameter n_splits is 10, and the parameters are shown in Table 2:

4.3 Comparison Experiments

To prove that the models are comparable, this paper implements SVM by Ahmed et al. [5], CNN by Wang et al. [11], C_LSTM model by Adam [13], and the base model Naive Bayes classifier and LSTM model, and applies these models on the three different datasets used in this paper, and compares the effect of Precision(P), Accuracy(A), Recall(R), the following attention mechanism plus convolutional neural network model (A-CNN) with other models respectively:

Table 2. Model parameters

Parameters	Description
WordVector dimension	100
Maxlen	300
head	8
Head_size	16
filters	128
Kernel_size	4
rate	0.5
n_splits	10

Table 3. Comparison of A-CNN with each baseline model

Model	Datasets								
	Liar			COVID-19			Fake and real news		
	P	A	R	P	A	R	P	A	R
Naive Bayes	0.58	0.58	0.58	0.84	0.84	0.84	0.79	0.79	0.79
SVM	0.57	0.56	0.56	0.85	0.85	0.85	0.84	0.84	0.84
Glove+CNN	0.58	0.58	0.58	0.88	0.88	0.88	0.99	0.995	0.99
Glove+LSTM	0.54	0.54	0.54	0.52	0.52	0.52	0.99	0.991	0.99
Glove +C_LSTM	0.54	0.54	0.54	0.52	0.52	0.52	0.99	0.995	0.99
Glove +A-CNN	**0.58**	**0.59**	**0.58**	**0.89**	**0.89**	**0.89**	**0.99**	**0.996**	**0.99**

Through the experiments, we can see that the proposed A-CNN model is better than other baseline models on all data sets except for Precision(P), Accuracy(A), and Recall(R) on the Liar data set, which are similar to other models.

As can be seen from Table 3, among the baseline models, the traditional machine learning models, Naive Bayes is better than SVM on the Liar dataset, while SVM is better than Naive Bayes on the other two datasets. among the deep learning models, the detection effects of CNN, LSTM, and C_LSTM perform the same on the large datasets, and on the small C_LSTM and LSTM are inferior to CNN single models on small datasets, mainly due to the fact that they are prone to overfitting or just cannot pay attention to the local connections of the sentences. From the whole experiment, the traditional machine learning models are better than some deep learning models on small data sets, while the opposite is true on large data sets. As can be seen from Fig. 3, the accuracy of our proposed model is quite high on both large and small data sets, which proves the effectiveness of our proposed model.

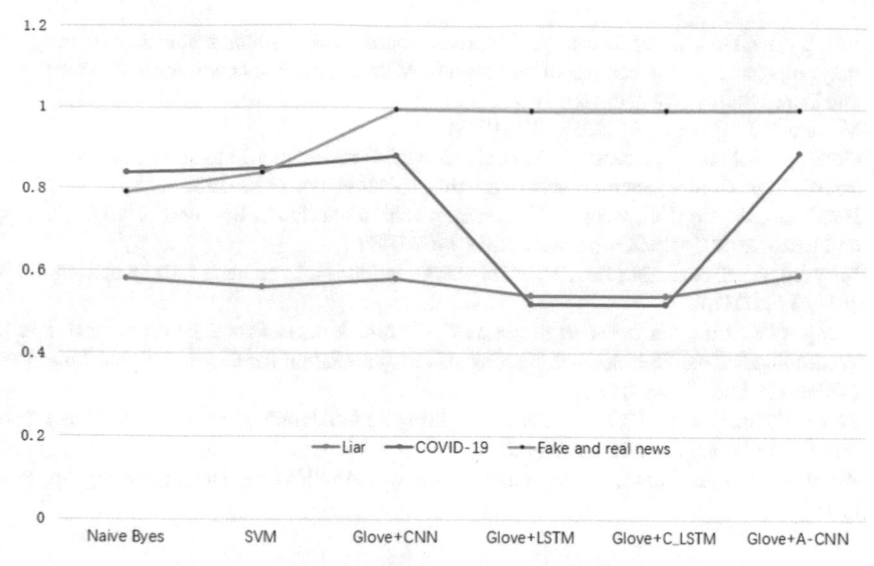

Fig. 3. Accuracy of each model

5 Conclusion and Future Works

This paper constructs an A-CNN model based on the attention mechanism and convolutional neural network, captures the global connection inside the sentence using the attention mechanism, extracts the keyword features with convolutional neural network, and detects them on three different datasets, and achieves good performance. The experimental results show that the method proposed in this paper is effective and feasible, but the method is only suitable for text news. In the future work, we will design a A new model to adapt to multimodal type news.

References

1. Shin, J., Jian, L., Driscoll, K., Bar, F.: The diffusion of misinformation on social media: temporal pattern, message, and source. Comput. Hum. Behav. **8**, 278–287 (2018)
2. Karimi, H., Roy, P.C., Saba-Sadiya, S., et al.: Multi-source multi-class fake news detection (2018)
3. Feng, S., Banerjee, R., Choi, Y.: Syntactic stylometry for deception detection. In: Proceedings of the 50th Annual Meeting of the Association for Computational Linguistics (Volume 2: Short Papers), pp. 171–175 (2012)
4. Gilda, S.: Notice of violation of IEEE publication principles: evaluating machine learning algorithms for fake news detection. In: 2017 IEEE 15th Student Conference on Research and Development (SCOReD). IEEE (2017)
5. Ahmed, H., Traore, I., Saad, S.: Detection of online fake news using N-gram analysis and machine learning techniques. In: Traore, I., Woungang, I., Awad, A. (eds.) ISDDC 2017. LNCS, vol. 10618, pp. 127–138. Springer, Cham (2017). https://doi.org/10.1007/978-3-319-69155-8_9

6. Wu, L., Liu, H.: Tracing fake-news footprints: characterizing social media messages by how they propagate. In: Proceedings of the eleventh ACM international conference on Web Search and Data Mining, 637–645 (2018)
7. Vincent, S.J.: Glove: US1456580[P] (1923)
8. Gehring, J., Auli, M., Grangier, D., et al.: Convolutional sequence to sequence learning. In: International Conference on Machine Learning. PMLR, pp. 1243–1252 (2017)
9. Bahdanau, D., Cho, K., Bengio, Y.: Neural machine translation by jointly learning to align and translate (2014). arXiv preprint arXiv:1409.0473
10. Vaswani, A., Shazeer, N., Parmar, N., et al.: Attention is all you need (2017). arXiv preprint arXiv:1706.03762
11. Wang, W.Y.: Liar, liar pants on fire: a new benchmark dataset for fake news detection. In: Proceedings of the 55th Annual Meeting of the Association for Computational Linguistics (Volume 2: Short Papers) (2017)
12. Patwa, P., Sharma, S., PYKL, S., et al.: Fighting an infodemic: Covid-19 fake news dataset (2020). arXiv preprint arXiv:2011.03327
13. Adam: Classifying News, Satire, and "Fake News": An SVM and Deep Learning Approach (2018)

Research on Attention Analysis Based on Vision

Ge Changyun, Han Ti, Zhang Yuanyuan, Yao Yifei, Li Ning, Zhou Yuhao, Zhang Xiaomeng, and Shen Jian(✉)

Department of Electronic Engineering, Dalian Neusoft University of Information, Liaoning 116023, China
gechangyun@neusoft.edu.cn

Abstract. Attention analysis technology refers to the analysis of people's attention during work or study through monitoring various data. This paper proposed a visual-based attention analysis method, in which a dataset for attention analysis was built and the neural network was adopted to analyze attention. Good results had been achieved on the self-built dataset with a 93% recognition rate.

Keywords: Attention analysis · Neural network · Vision

1 Introduction

In the wake of the rapid development of artificial intelligence, the application of artificial intelligence has gradually become increasingly extensive and neural networks have already been widely used in our daily lives. Through the simulation of the human brain with the neural network structure, the computer was enabled to recognize and judge the behavior of people or other creatures. The recognition of facial expressions could recognize human emotions to a certain extent, thereby improving human-computer interaction, which has high research value. This paper proposed a method for attention analysis [1, 2] based on neural network using facial information.

2 Dataset Establishment and Network Structure Design

2.1 Dataset

The needs to be carried out before the data processing. In the acquisition of the dataset, the facial expression videos of people at work and study needs to be recorded. After the video had been obtained, the video frame acquisition software was used to export the frames in the video as pictures. After that, the resulting pictures were marked.

The label dataset adopted a lightweight open source software labelImg. In this paper, the facial state of the collected object under the working, learning, and entertaining states

This work is supported by Research and Development of Attention Device Based on Brain Science and Machine Vision, Dalian Youth Science and Technology Star 2019(2019RQ125); Key R & D projects of Liaoning Province Education Department (Research on Sensors for Pelvic Floor Rehabilitation of Puerpera Based on Artificial Intelligence).

Q. Liang et al. (Eds.): Artificial Intelligence in China, LNEE 854, pp. 257–260, 2022.
https://doi.org/10.1007/978-981-16-9423-3_32

was obtained through the video stream, and then the video stream was converted into frame photos. In the end, each frame of image was tagged manually to generate an.xml file. Every picture will be marked into 6 levels from A to G according to the attention state of the person been monitored.

2.2 Model

The AlexNet model [3] has made breakthroughs in deep learning. Through its excellent performance, it can be found that the recognition ability of the network could be improved by increasing the depth of the network. Based on this idea, VGG, a research group at the University of Oxford in the UK, proposed a VGGNet network model. There were two main types of the model, one of which was the VGG-16 model studied in this paper ranking highly in the 2014 ImageNet competition [4] positioning task. A deeper network level might improve the performance of the network, but it would increase the amount of huge parameters and the computing power requirements of the computer exponentially [5]. VGG-16 [6] adopted multiple 3×3 small convolution kernels to replace large convolution kernels similar to 5×5, 7×7 etc. Under the condition of determining the same receptive field, the number of parameters was greatly reduced. The total training parameters were 138,357,544, which greatly reduced the difficulty of training.

VGG-16 mainly included 13 convolutional layers, 3 fully connected layers and 5 pooling layers, as shown in Fig. 1. Among them, only the convolutional layer and the fully connected layer needed to update the weight parameters, a total of 16 layers, so it was named as VGG-16.

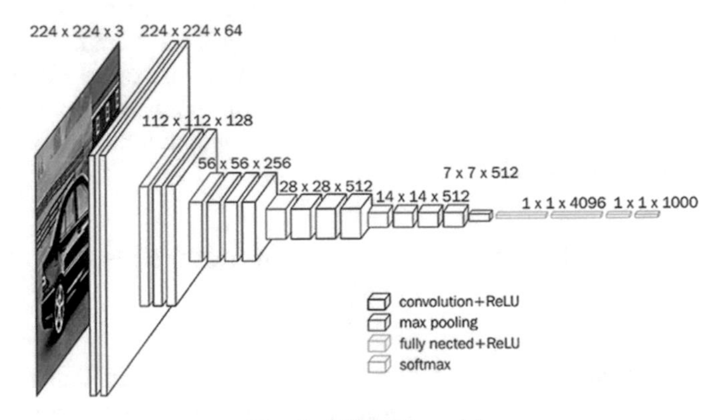

Fig. 1. VGG-16 model

During the training process of the VGG-16 network, the mini-batch gradient descent SGD was adopted for optimization. The batch size was set to 256 and the momentum 0.9. The training was normalized by L2 weight decay (the decay size was set to 5e − 4), and the dropout regularization of the first two fully connected layers (the drop ratio is set to 0.5) were normalized. The learning rate was initially set to 1e − 2. When the accuracy of the validation set stopped improving, the learning rate was reduced by 10 times.

Because it was a classification problem, the last layer was set as the Softmax layer, and the classification results were divided into 6 categories. Therefore, the loss function adopted cross entropy [7], as shown in the following formula.

$$\text{Loss} = -\sum_{i=1}^{N} y^i * log\hat{y}^i \tag{1}$$

In the formula, y^i represented the actual output, and $\hat{y}^i$ represented the expected output.

3 Experiment and Result Analysis

The training process lasted 100 epochs. Shown below were the recognition results and output of the facial expression recognition module. As shown in Fig. 2, in different expression states, different attention recognition results were displayed in the window.

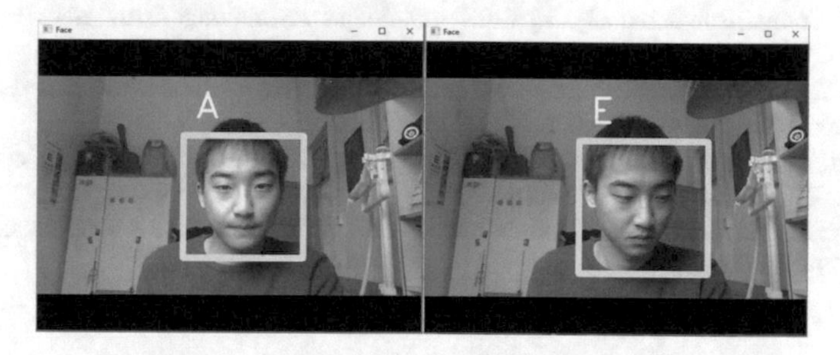

Fig. 2. Recognition result of face attention

The trained model can correctly recognize the pictures in the test set with a certain accuracy rate. In order to facilitate viewing, this article added a face detection algorithm to the mark, and drew a frame. In addition, the attention level was marked in the upper left corner of the frame.

Table 3–1 shows the accuracy results during the model training process.

Table 1. Model training accuracy

Number of iterations	10	30	50	70	90	100
Accuracy	0.49	0.70	0.82	0.89	0.92	0.94

4 Conclusion

The face recognition technology is still under development, and the analysis of expression attention differs from the common basic expressions of human faces. For the judgment of the degree of attention, the expression level alone cannot get a satisfactory result. The experiment in this article achieved a good recognition result for the self-built dataset. However, from the practical perspective, a combination of all aspects was needed in the achievements of a more accurate and reliable attention recognition result. For the optimization of attention recognition, the possible improvements currently foreseeable are as follows: recognition and analysis of the eyes; optimization of the model; expansion of the dataset.

References

1. Min, X., Zhai, G., Gu, K.: Visual attention analysis and prediction on human faces. Inf. Sci. **420**, 417–430 (2017)
2. Chen, Y., Song, M.: An audio-visual human attention analysis approach to abrupt change detection in videos. Sig. Process. **143**(12), 110 (2015)
3. Moon, H., Browatzki, B., Blais, C., Wallraven, C.: Deep neural networks process similar facial features compared to humans in facial expression recognition. IBRO Rep. **6**, S193–S194 (2019)
4. Zhao, J.: Comparison and analysis of different deep convolutional neural networks based on tensorflow. Electron. World **06**, 25–26 (2018)
5. Wencui, Y., Shi, K.: Convolutional neural network research based on tensorflow deep learning framework. Microcomput. Appl. **2**(34), 29–32 (2018)
6. Hammad, I., El-Sankary, K.: Impact of Approximate Multipliers on VGG Deep Learning Network. IEEE Access. **6**, 60438–60444 (2018)
7. Brink, A.D., Pendock, N.E.: Minimum cross-entropy threshold selection. Pattern Recogn. **29**(1), 179–188 (1996)

Research on the Labor Education Practice Project of Normal Students Under the Background of Artificial Intelligence

Yi Wang(✉)

Tianjin Normal University, Tianjin, China
16781026@qq.com

Abstract. Labor is the most primitive essential feature of mankind. In the traditional sense, labor represents the physical labor that human beings must perform in order to survive. From the perspective of social development, any work we do today embodies the meaning of labor. The content and form of labor education are also changing with the development of society. Today, artificial intelligence is sweeping the world. On the one hand, some traditional types of work are gradually replaced by machines. On the other hand, the era of intelligence has stimulated new fields and job opportunities, and the skills of workers also present the characteristics of digitalization, diversification, and creativity. The epoch-making changes in labor education have led to changes in the core literacy of workers, the content of labor education, and the scene of labor education.

Keywords: Artificial intelligence · Labor education · Normal student

1 Comparison of Labor Education at Home and Abroad

Foreign labor education is mainly based on "livelihood education" as the implementation goal. As early as the end of the 19th century to the beginning of the 20th century, American educators such as Dewey, Parker, and Kerberch put forward that "school is society, life and education" and "learning by doing". A series of thoughts such as this laid the foundation for the tradition of combining education and labor in the United States. Open production and labor courses in primary and secondary schools, learn one or even several vocational skills according to each child's hobbies, qualifications, and specialties, and then learn about the skills and characteristics of different industries in society through a series of training. At the same time, it also makes necessary preparations for some children in the future study process if they quit school and choose careers. The specific setting is as follows: in grade1 to grade 6, the main task of labor education is "professional knowledge". The purpose is to cultivate children's professional awareness and self-awareness through unit teaching, cultivate hands-on ability, recognize the value of labor, and improve children Understanding of occupations; in grades 7 to 10, the main task of labor education is to explore the classification of occupations and "occupation groups" through occupational exploration, and begin to make tentative choices among

Q. Liang et al. (Eds.): Artificial Intelligence in China, LNEE 854, pp. 261–267, 2022.
https://doi.org/10.1007/978-981-16-9423-3_33

occupation groups; in grades 10 to 12, the main task of labor education is to help students to make more in-depth explorations in certain career fields of their choice and determine their future development direction. After graduating from high school, some students will enter the university to continue their academic education, while the other students will enter the community college to study and engage in technical work after graduation. The students who enter the community college begin to receive systematic vocational education; the students who enter the university begin to learn academic knowledge, and some engage in further high-end vocational and technical training.

Labor education in China is basically "technical education" as the main content. At the basic education stage, it is mainly based on voluntary physical labor and part of "technical" learning. At the same time, it cultivates good labor concepts and labor habits to make necessary preparations for future social life. In middle schools and universities, it pays more attention to vocational-related technical learning, so that students not only need to master basic labor knowledge and skills in labor education but also courses offered by some schools include students' understanding of raw material production and products. The manufacturing process and marketing process, etc. To enable students to understand the specific production process of the modern industrial information society. In the final analysis, it is through labor education to enable students to deeply understand the relationship between man and nature, man and society, labor and society. To lay a solid foundation in society in the future.

However, some colleges and universities lack scientific management and institutional guarantee in the process of education. There are some problems, such as the weak construction of labor education places and practice bases, the shortage of labor education teachers, and the inadequate institutional guarantee. However, for normal universities, especially for the professional quality training of future "teachers", there is a relative lack of labor practice projects commensurate with teachers' posts, and the evaluation standards for the labor quality of normal students still need to be improved. Many normal universities simply equate labor education with voluntary physical labor on campus. As a result, a lot of formalist labor education activities appeared.

2 The Significance of Labor Education in the Context of Artificial Intelligence

2.1 Meet the Requirements of My Country's National Development Strategy

The Central Committee of the Communist Party of China and the State Council recently issued the "Opinions on Comprehensively Strengthening Labor Education in Universities, Middle Schools and Primary Schools in the New Era", emphasizing that labor education should reflect the characteristics of the times, " adapt to the development of science and technology and industrial change, focus on the new forms of labor, and pay attention to the support of emerging technologies and the new changes of social services". Adhering to the spirit of the important exposition of general secretary Xi Jinping on labor education, we should adhere to the basic task of Li De Shu, and comprehensively cultivate students' mental outlook, labor value orientation, and labor skill level, stimulate the spirit of innovation, cultivate the practical ability, achieve unity of knowledge and practice, and comprehensively promote the expansion from technology application to ability and quality. To lay a foundation for the cultivation of innovative talents with international competitiveness.

2.2 In The Era of Artificial Intelligence, a New Stage of Talent Training

In the intelligent age, educational scientific research requires us to further strengthen cross-integration, combine the research methods of natural sciences and social sciences, combine educational scientific research with cutting-edge technology research, and combine theoretical and technical research with the actual needs of education and teaching, and integrate into our country's cultural characteristics, gather intelligence, innovate, build and share, and promote the realization of the comprehensive, free, and individualized development of people. In the future teaching scene, technology will be able to take on more knowledge transfer tasks, and teachers' focus will shift more to the cultivation of students' abilities, literacy cultivation, psychological intervention, and personality shaping. As a teacher, the standard of his professional work-ability will be redefined, that is, the teacher's professional quality will be fully updated, and the comprehensive development of morality, intelligence, physical education, and labor is one of the goals of teacher training. Labor education fully integrates the educational elements of moral education, intellectual education, physical education, and aesthetic education, which not only embodies the intersection of disciplines but also integrates a variety of technologies and educational methods.

2.3 The Needs for Professional Ability Training for the Normal Student

Under the background of artificial intelligence, human-computer collaboration has become a new form of labor. This form of labor not only requires the thinking, creativity, and communication skills of normal students but also includes human-computer collaboration and human-computer dialogue capabilities. Therefore, the labor education significance of normal students embodies the literacy of science and technology, the literacy of artificial intelligence, and the literacy of new teaching skills. On the one hand, from the perspective of the goal of talent training, to cultivate students' need for innovation awareness and innovation ability as the goal, it is necessary to integrate new technology into labor education to bring reforms to labor. The teaching concept is combined with STEM education through labor education practice courses., The latest development achievements of maker education and artificial intelligence education, design the practical project as a real experimental project with multi-disciplinary integration, contextualization, task system, and difficulty gradient, while taking into account the differentiation of different individuals, implementing personalized teaching, At the same time, around the reserve of future vocational skills requirements, strengthen artificial intelligence to empower labor education-related disciplines, enhance the scientific and technological vision and innovative thinking of normal students, and integrate the comprehensive teaching ability of new technologies. On the other hand, from the perspective of cultivating labor awareness and establishing correct labor values, labor education and moral education need to be penetrated all aspects of student training, adding labor education content to daily teaching, and designing teaching with labor practice elements.

activity. Guide the teacher students' sense of responsibility for their work in the future teaching practice process, and urge them to take the initiative to improve their workability through this sense of responsibility, and at the same time allow teachers' labor value to be fully reflected.

2.4 Serving Basic Education, a New Fulcrum for the Integration of Production and Education

First of all, normal colleges, as institutions of higher learning that serve basic education, have a large number of scientific research resources and scientific and technological education resources. A comprehensive artificial intelligence + labor education curriculum system should be constructed to build a professional teaching team and cultivate human resources. Intelligent and labor literate students help the development of regional basic education, form an effective connection with higher education stage education, and build a multi-level and stepped talent system. The purpose of the proposed project is to promote the integration and sharing of the whole social resources, effectively connect universities and primary and secondary schools, complement each other's advantages, promote the popularization of science and technology and the cultivation of labor quality, and gradually form an integrated social service ecosystem from university training to primary and secondary school practice.

3 The Construction of Labor Education Practice Projects in Normal Colleges

3.1 Design Goals

Since the artificial intelligence major of domestic normal colleges has not been certified as a teacher major, the labor education major is also in the plan to open, and the practical project is based on the new requirements of the times for teacher positions to carry out more extensive and more comprehensive training for normal students. Level and multi-field labor skills training, in order to deliver high-quality, high-conscious, and high-skilled talents for future teaching positions.

The practical project plan is to formulate an interdisciplinary training model, break the boundaries between theoretical and practical courses, increase student learning methods through project training, establish the scientific and technological vision, cultivate labor literacy, and rely on off-campus internship practice cooperation units to provide teacher students Social practice base, selected from normal students to participate in the teaching of labor education practice courses in primary and secondary schools. Use theory to guide practice and develop theory in practice. Combining the characteristics of normal universities, reduce the cost of talent training, cultivate excellent graduates with good scientific and technological literacy, high teaching level, strong innovation ability, and meet the needs of basic education teachers in the country, supplement the higher level of artificial intelligence labor education teachers, and enhance the intelligence The overall level of the training of innovative technology teaching talents will eventually form a sustainable talent advantage.

3.2 Design Ideas

Combining with the characteristics of normal universities and using artificial intelligence as the entry point, labor education practice projects should include all-around knowledge such as artificial intelligence theory, underlying technology, design thinking, engineering assembly, curriculum design skills, classroom teaching management, educational technology, and other comprehensive knowledge. Robots are the main learning carrier, emphasizing project-based experiments throughout the entire knowledge system, composed of three types of modules: core, support, and training. It emphasizes both theoretical study and practical application, teaches all aspects of artificial intelligence + education from the three levels of perception, decision-making, and execution, and cultivates students' practical ability through stepped project practice.

The creation of practical projects should be combined with various scenarios in all walks of life. Include typical artificial intelligence application cases into the curriculum to help students understand the background and development of various industries, and at the same time learn the design logic behind each artificial intelligence application system and the social and economic value brought by it, enhance students' awareness of artificial intelligence, and cultivate students' computational thinking, problem-solving ability, hands-on practice ability, innovative application ability, and other qualities.

Improve the "Course-Activity-Evaluation" system and guide normal students to improve their scientific cognition and professional labor skills through participation, experience, practice, and hands-on production. Based on the project-based learning method, knowledge construction, skill training, and thinking development are integrated into the process of solving problems and completing tasks in the creative labor education of normal students.

With the help of the cooperative relationship between normal colleges and basic education, and relying on the artificial intelligence science activities of normal colleges, establish and improve the effective mechanism of artificial intelligence labor education and school science education. To form a closed loop of the ecological environment for youth innovative labor practice education. Send the trained college students to the science and technology labor education activities of elementary and middle schools in our city, develop rich and diverse curriculum resources, and finally build a high-quality and shareable curriculum system and training program for normal students' artificial intelligence labor practice. In order to facilitate the cultivation of outstanding graduates with good scientific and technological literacy, high teaching level, and strong innovation ability, it will provide teachers for the popularization of science and technology and artificial intelligence education in primary and secondary schools in our city. At the same time, the completion of the popularization of artificial intelligence courses will also play a pioneering role in the national basic education comprehensive practice curriculum reform, and enhance the social benefits and educational impact of this project.

3.3 Project Focus

It will focus on the cultivation of the innovative ability of normal students from the perspective of artificial intelligence education, combined with the construction of labor education experimental courses, carry out the practice of integrating artificial intelligence and education and teaching, and strive to take the lead in the city's schools based on artificial intelligence education pilot schools. Artificial intelligence popularizes labor education practice, promotes the integration of artificial intelligence and labor education practice courses, enhances the scientific and technological vision and innovative thinking of normal students, and integrates the comprehensive teaching ability of new technologies.

Practical project design adopts the STEAM concept, and an interdisciplinary and integrated curriculum system must be constructed based on "one body", that is, the construction of a normal student training curriculum system that integrates knowledge, ability, and quality structure. The main difficulty in the construction of the curriculum system lies in the following three aspects: First, pay attention to the logical interdisciplinary reorganization of disciplines; Second, pay attention to ensuring the integrity of basic knowledge during reorganization; Third, pay attention to maintaining theoretical and practical knowledge. Pay attention to the creation of real problem situations.

The structure of the course module is shown in Table 1:

Table 1. .

Course module structure	Module name
Core curriculum	Artificial Intelligence Fundamentals
Supporting curriculum	Design of science and technology activities in primary and secondary schools
	Case Analysis of Intelligent Design Courses in Primary and Secondary Schools
	Primary and secondary school science and technology curriculum resource development
	Frontier Technology Progress in Developed Countries
	Overview of STEM Education in the United States
Practical training	Primary and secondary school curriculum micro-video competition
	World Robot Contest
	National College Student Engineering Training Comprehensive Ability Competition

4 Concluding Remarks

Establish and improve an effective mechanism for the connection between artificial intelligence labor education and school science education, and form a closed loop of the ecological environment of innovative labor practice education for young people with universities——elementary and middle schools——universities. After a period of time, a high-quality, shareable curriculum system and training program for normal students' artificial intelligence labor practice will be finally built. Develop new-type labor practice exploration activities, and gradually form a new era labor education brand curriculum for teaching students with "bold exploration, disciplinary integration, applied science, and courage to innovate".

References

1. Yuedan, L.: Research on the status quo and causes of labor education in colleges and universities. Ind. Sci. Educ. Forum **21**, 105–106 (2020)
2. Suqian, C.: Research and analysis on the innovation of university labor education under the background of artificial intelligence. J. Qiqihar Univ. **11**, 61–62 (2020)
3. Xianyou, Z.: Analysis of the path to practice labor education in the training of normal students. Sci. Educ. Forum **1**, 72–73 (2021)
4. Xianyou, Z.: Construction of labor education evaluation system based on the training of normal students. Sci. Educ. Forum **2**, 68–69 (2021)

A Method of Remote Sensing Image Object Detection Based on Attention

Menglin Qi[1,2] and Bingcai Chen[1,2](✉)

[1] School of Computer Science and Technology, Xinjiang Normal University, Urumqi 830054, China
china@dlut.edu.cn

[2] School of Computer Science and Technology, Dalian University of Technology, Dalian 116024, Liaoning, China

Abstract. Remote sensing image object detection is an important and challenging task in the field of computer vision. Remote sensing image object detection has important applications in agriculture, meteorology, military and other fields. The traditional method of remote sensing image object detection and recognition is difficult to deal with a large amount of data, and its feature expression relies on artificial design. This paper uses a remote sensing image object detection method based on attention mechanism, uses SSD object detection model with resnet50 as feature extraction network, and further improves the accuracy by adding attention module in the feature extraction network, The results compared with the SSD algorithm without attention module, the proposed model has higher recognition accuracy for remote sensing image objects, and the average accuracy is 92.1% on the RSOD dataset.

Keywords: Object detection · Remote sensing image · Resnet50 · SSD · Attention mechanism

1 Introduction

With the continuous development of China's aerospace industry, more and more remote sensing images are used in military, agriculture, environmental monitoring, smart cities and many other fields. Remote sensing image object detection has a wide range of application requirements in many fields. In recent years, object detection algorithm has made remarkable achievements, and some excellent object detection algorithms have continuously emerged. [1–3] Currently, the commonly used object detection algorithms, such as Ross girshick et al. [4, 5], proposed R-CNN and Fast R-CNN algorithms in 2013 and 2015 respectively. The algorithm consists of two stages, the core of which is the generation of candidate regions. In order to eliminate the dependence of candidate frames on artificial design, he Kaiming et al. [6] proposed Faster R-CNN in 2016. The main improvement of Faster R-CNN is to use RPN (region proposal network) network instead of Fast R-CNN search algorithm to automatically extract candidate frames, so that the quality of candidate frames obtained by the network is higher, and the end-to-end feature

Q. Liang et al. (Eds.): Artificial Intelligence in China, LNEE 854, pp. 268–274, 2022.
https://doi.org/10.1007/978-981-16-9423-3_34

extraction is completely realized. In 2015, long et al. [6] proposed FCN (full convolution network). FCN replaces the full connection layer of CNN network with the convolution layer. Because there is no full connection layer, FCN network can receive input pictures of any size and greatly reduce the number of parameters. He et al. [7] proposed Mask R-CNN in 2017, Liu et al. [8] proposed SSD (single shot multibox detector) object detection algorithm in 2016, all of which are pre trained on Imagenet and other public data sets. Due to the particularity of remote sensing image, the detection effect of the pre trained classifier on the conventional data set is not good. Liu Feng et al. [9] believe that the remote sensing image has the characteristics of scale diversity, and this kind of image is usually shot from high altitude, resulting in the characteristics of similar object size diversity. Therefore, the object detection algorithm of remote sensing image still needs to be improved. In this paper, SSD model based on resnet50 is used to extract network features, and attention mechanism is added to improve the accuracy while ensuring the speed. By adding attention mechanism, the average accuracy is improved by 1.7%.

2 SSD Object Detection Model and Attention Mechanism

2.1 SSD Object Detection Model

As shown in Fig. 1, this paper proposes a remote sensing image object detection algorithm (MSSD) integrating attention mechanism to improve SSD algorithm, replacing feature extraction network with modified resnet50 and adding attention mechanism. Like SSD algorithm, MSSD algorithm also adopts the strategy of extracting small objects from large feature map and extracting large objects from small feature map, and the size of input image is 300 × 300 A total of six layers of feature maps with sizes of 38 × 38, 19 × 19, 10 × 10, 5 × 5, 3 × 3, 1 × 1. Generate the detection frame on the feature map of different scales to achieve the purpose of multi-scale object detection.

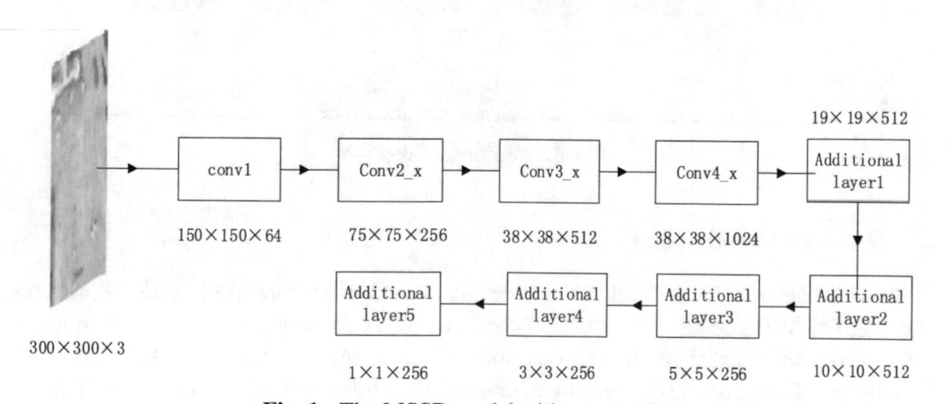

Fig. 1. The MSSD model without attention

Among them, conv4_x The stride of the identity map and the second convolution layer in block1 are 1, and the other blocks remain unchanged. The structures of additional layer 1, additional layer 2 and additional layer 3 are shown in Fig. 2, and the structures of additional layer 4 and additional layer 5 are shown in Fig. 3.

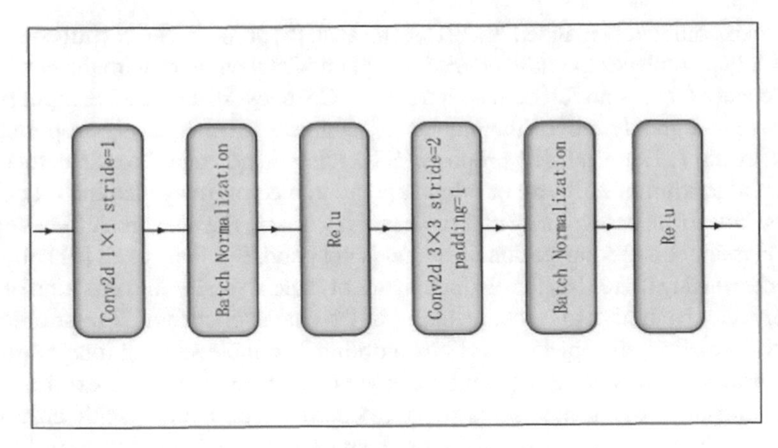

Fig. 2. Additional layer1–3

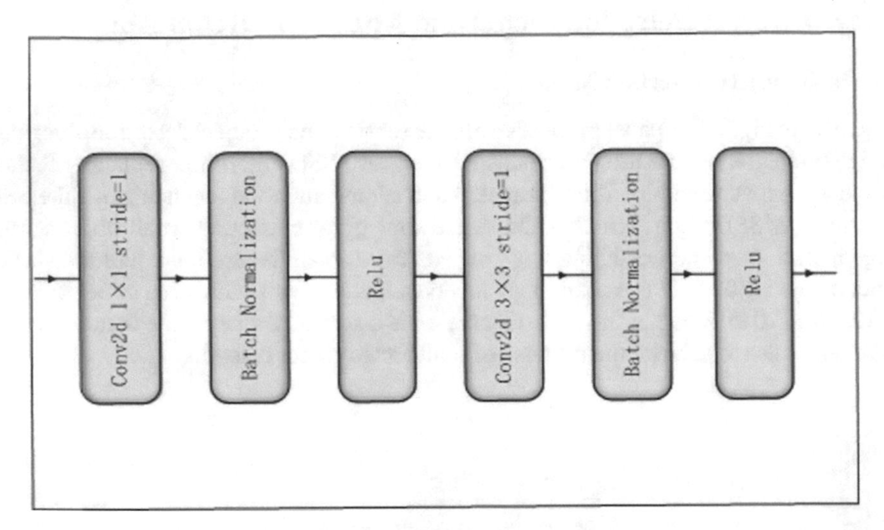

Fig. 3. Additional layer4–5

2.2 Attention Mechanism

In order to get more attention area in the image, attention module is added based on the original SSD model. The attention module makes the key parts of the image more attention by adding weight. The convolutional block attention module (CBAM) proposed by woo et al. [10] is a kind of attention mechanism module combining spatial and channel. Compared with SEnet [11], only focusing on the attention mechanism of channel can achieve better results. The location of attention mechanism is shown in Fig. 4.

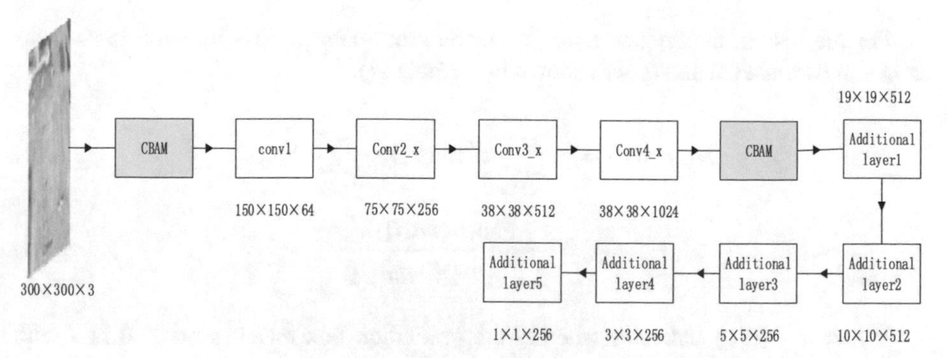

Fig. 4. MSSD model

3 Experiment and Result Analysis

3.1 Experimental Environment

In the experimental platform environment, the CPU is AMD ryzen r5 1400, the GPU is NVIDIA Geforce gtx1060, the size of video memory is 6 GB, the operating system is Ubuntu 16.04, and the deep learning framework is Python 1.6.0.

3.2 Experimental Data Set

The experimental dataset uses the RSOD dataset [12], which is used for object detection in remote sensing images. It contains four types of objects: aircraft, playground, overpass and oil. The number of objects is 4993 aircraft, 191 playground, 180 overpass, 1586 oiltank. The data set was released by Wuhan University in 2015.

3.3 Loss Function Design

The loss function of SSD target detection algorithm consists of two parts: location loss and classification loss. As shown in formula (1)

$$L(x, c, l, g) = \frac{1}{N}\left(L_{conf}(x, c) + \alpha L_{loc}(x, l, g)\right) \tag{1}$$

Among them, $L_{conf}(x, c)$ is the classification loss, $L_{loc}(x, l, g)$ is the location loss, x is the prediction category, c is the confidence level of the prediction box category, g is the location of the real tag, N is the number of prior boxes, and α is the coefficient. In the loss function $L_{loc}(x, l, g)$, (l) represents the prediction box and (g) represents the real box, using smoothl1loss, as shown in formula (2)

$$loss(x, y) = \frac{1}{n}\sum_{i=1}^{n}\begin{cases} 0.5 * (y_i - f(x_i))^2, & if\, |y_i - f(x_i)| < 1 \\ |y_i - f(x_i)| - 0.5, & otherwise \end{cases} \tag{2}$$

In formula (2), y_i and $f(x_i)$ are vectors, where y_i is the real coordinate and $f(x_i)$ is the predicted coordinate.

For the loss of confidence $L_{conf}(x, c)$, the cross entropy loss function is used, as shown in formula (3) and q_i was shown in formula (4).

$$L_{conf}(x, c) = -\sum_{i \in Pos}^{N} x_i^p \log(q_i) - \sum_{i \in Neg} \log(C) \tag{3}$$

$$q_i = \frac{\exp(input[i])}{\sum_{j=0}^{N} \exp(input[j])} \tag{4}$$

Where $x_i^p \in \{0,1\}$ indicates whether the prediction box matches the real box with respect to category p, 0 indicates no match, 1 indicates match, q_i which is the probability value calculated by softmax function and C is the probability of negative samples.

3.4 Evaluating Indicator

In this paper, Mean Average Precision (mAP) was used as the evaluation index. In the process of evaluation, each category can draw a P-R curve according to the accuracy and recall. The average accuracy (AP) is the area under the curve. The formula is shown in Eq. (5). mAP is the average value of multiple categories of AP.

$$AP = \int_0^1 p(r)dr \tag{5}$$

Figure 5 shows the detection results of MSSD algorithm on various types of objects. It can be seen that MSSD algorithm can detect small objects in remote sensing images very well.

3.5 Contrast experiment

In order to prove that the detection performance of our algorithm is better than that of Faster R-CNN and SSD algorithm, we compare it with our algorithm, and the results are shown in Table 1.

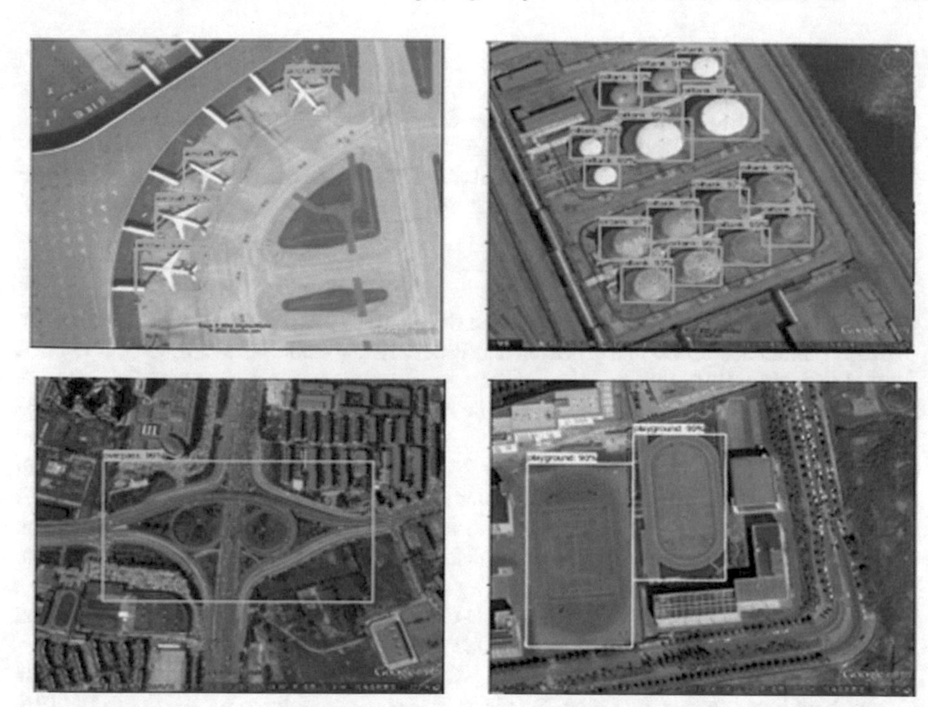

Fig. 5. Test results by category

Table 1. Comparison of test results

Method	Backbone	mAP/(%)
Faster R-CNN	ResNet50	88.7
SSD	ResNet50	90.4
MSSD	ResNet50	92.1

4 Conclusion

Aiming at the problem of different scales of the same object in remote sensing image, the traditional SSD algorithm is improved, and a remote sensing image object detection algorithm based on attention mechanism is proposed. On the basis of SSD algorithm, the feature extraction network is modified and attention mechanism is added to enhance the feature extraction ability of the network, Finally, an average accuracy of 92.1% is achieved on the public remote sensing image data set RSOD. Compared with the Faster R-CNN algorithm and the conventional SSD algorithm, the average detection accuracy of the proposed algorithm is improved. In the next step, we will continue to optimize the network structure and consider joining the FPN network to further improve the detection accuracy.

References

1. Dai, J., Li, Y., He, K., et al.: R-FCN: object detection via region-based fully convolutional networks[EB/OL]. arXiv:1605.06409 (2016)
2. Redmon, J., Divvala, S., Girshick, R., et al.: You only look once: unified, real-time object detection. In: IEEE Conference on Computer Vision and Pattern Recognition. Las Vegas, NV, USA, 779–788. IEEE (2016)
3. Law, H., Deng, J.: CornerNet: Detecting objects as paired keypoints. Int. J. Comput. Vis. **128**(3), 642–656 (2020)
4. Cai, Z., Vasconcelos, N.: Cascade R-CNN: delving into high quality object detection. In: IEEE Conference on Computer Vision and Pattern Recognition. Salt Lake City, UT, USA, pp. 6154–6162. IEEE (2018)
5. Girshick, R.: Fast r-CNN. In: IEEE International Conference on Computer Vision, pp. 1440–1448 (2015)
6. Ren, S., He, K., Girshick, R., et al.: Faster R-CNN: towards real-time object detection with region proposal networks. IEEE Trans. Pattern Anal. Mach. Intell. **39**(6), 1137–1149 (2017)
7. Long, J., Shelhamer, E., Darrell, T.: Fully convolutional networks for semantic segmentation. In: 2015 IEEE Conference on Computer Vision and Pattern Recognition (CVPR). Boston, MA, USA, pp. 3431–3440 (2015)
8. He, K., Gkioxari, G., Dollár, P., et al.: Mask R-CNN. In: Proceedings of the IEEE International Conference on Computer Vision. Venice, Italy, pp. 2980–2988. IEEE (2017)
9. Liu, W., Anguelov, D., Erhan, D., Szegedy, C., Reed, S., Fu, C.-Y., Berg, A.: SSD: Single shot MultiBox detector. In: Leibe, B., Matas, J., Sebe, N., Welling, M. (eds.) ECCV 2016. LNCS, vol. 9905, pp. 21–37. Springer, Cham (2016). https://doi.org/10.1007/978-3-319-46448-0_2
10. Liu, F., Shen, T., Ma, X., et al.: Ship recognition based on multi-band deep neural network. Opt. Precis. Eng **25**, 166–173 (2017)
11. Woo, S., Park, J., Lee, J.-Y., Kweon, I.S.: CBAM: Convolutional Block Attention Module. In: Ferrari, V., Hebert, M., Sminchisescu, C., Weiss, Y. (eds.) ECCV 2018. LNCS, vol. 11211, pp. 3–19. Springer, Cham (2018). https://doi.org/10.1007/978-3-030-01234-2_1
12. Hu, J., Shen, L., Sun, G.: Squeeze-and-excitation networks. In: Proceedings of the IEEE Conference on Computer Vision and Pattern Recognition, pp. 7132–7141 (2018)
13. Yang, L.O.N.G., Yiping, G.O.N.G., Zhifeng, X.I.A.O., et al.: Accurate object localization in remote sensing images based on convolutional neural networks. IEEE Trans. Geosci. Remote Sens. **55**(5), 2486–2498 (2017)

Attention Word Embedding Network-Based Lightweight Automatic Essay Scoring Model

Xianbing Zhou(✉), Xiaochao Fan, Yong Yang(✉), and Ge Ren

School of Computer Science and Technology, Xinjiang Normal University, Ürümqi, China
1783696285@qq.com, 68523593@qq.com

Abstract. In recent years, automatic essay scoring technology has received more and more attention in the fields of education and research, and its purpose is to use natural language processing technology to automatically analyze and score the essay. In this paper, we proposes a lightweight automatic essay scoring model based on attention word embedding network. Using randomly initialized and self-trained word embedding and attention/max/mean pooling operation, we are able to score essays automatically. Our models match or outperform optimal QWK (Quadratic Weighted Kappa) performance on all eight datasets using significantly fewer parameters compared to baseline methods.

Keywords: Automatic Essay Scoring · Attention mechanism · Pooling technology · Natural language processing

1 Introduction

Automatic Essay Scoring (AES) is a technology that uses linguistics, statistics and natural language processing techniques to automatically score essays, and is often used in large-scale examinations [1]. Since AES was put forward in 1966, it has been successfully applied to various major exams, such as IELTS and TOEFL exams abroad and the College English Test (CET) in China. In addition, AES is an important part of the Automatic Essay Evaluation (AEE) system. Accurately scoring the essay can make the AEE system evaluate the essay more objectively and with reference significance, which can help teachers to better evaluate the students' writing and improve the quality of teaching. AES can effectively avoid the influence of the teacher's subjective factors on the essay scoring, thereby greatly improving the fairness and accuracy of the essay scoring. At the same time, the automatic essay scoring technology can greatly reduce the workload of the marking teacher and save material and manpower costs.

Early automatic essay scoring methods constructed shallow features that could reflect the characteristics of the essay [3,18], such as vocabulary, grammar, syntax, and text structure features, and then used machine learning methods for feature mining to indirectly evaluate essay scores. This type of traditional

Q. Liang et al. (Eds.): Artificial Intelligence in China, LNEE 854, pp. 275–282, 2022.
https://doi.org/10.1007/978-981-16-9423-3_35

machine learning methods based on feature engineering ignores the latent semantic features of the text and does not truly understand the essay from the semantic level. Therefore, this method cannot achieve satisfactory results. Recently, neural network methods based on deep learning have achieved better performance in essay scoring tasks [2,14]. The current deep learning methods are based on the word embedding representation obtained from large-scale corpus training, and use neural networks for feature extraction, crossover and fusion, which can extract the high-dimensional latent semantic features of the essay from a deeper level. However, most of the commonly used pre-training word vectors use corpus from social media on the Internet, and they are quite different from the essay corpus in terms of semantic expression, logical structure, and language style. Using these word vectors may not improve the performance of essay scoring. Instead, it will cause a certain degree of semantic deviation probably. On the other hand, most of the existing automatic essay scoring methods based on deep learning tend to use models with many parameters and complex structures, but this may not effectively improve essay scoring performance and make the system operating efficiency too low. Affect the widespread use of the AES system.

In response to the above problems, we propose a lightweight automatic essay scoring model based on the attention word embedding network, which is a modified Simple Word Embedding-based Model [13], to automatically score the essay. Our model features:

- A simple deep learning approach with few parameters enabling quick and robust training;
- Using random initialization and self-training word vectors instead of the commonly used pre-training word vectors for embedding layer;
- Significantly better performance than all baseline methods on publicly available datasets.

In the following sections, we discuss related work on automatic essay scoring, followed by a description of the methods, experimental data and results of our study.

2 Related Work

Automatic essay scoring is an important auxiliary tool in the field of education and research, and there have been many research results at home and abroad. For the research of AES, according to the different methods of use, this section will sort out the previous work from two aspects: the traditional machine learning method based on feature engineering and the method based on deep learning.

The traditional machine learning method based on feature engineering constructs artificial features according to the grammar and syntax rules of the essay score, and uses traditional machine learning methods such as Logistic Regression (LR) and Support Vector Machine (SVM) to score the essay. Abroad, PEG (Project Essay Grade) [10] is one of the earliest automatic scoring systems, which automatically score essays by constructing the structure of writing and other shallow semantic features. Pramukantoro et al. [11] proposed an unsupervised automatic essay scoring method based on cosine similarity. Domestically, Liang Maocheng et al. [7] first proposed the AES scoring method, using grammar, syntax and language expression and other essay features, and scoring the essay by using the linear regression method. Zhou Ming et al. [8] extracted essay features from three levels of words, sentences, and paragraphs, used a variety of machine learning algorithms to classify the essay of the text, and used a linear regression model to score the structure of the text.

In recent years, deep learning methods based on neural networks have achieved many research results in the field of AES. Taghipour et al. [14] uses a convolutional neural network and a Long Short-Term Memory network in series to automatically extract essay features. Dong et al. [5] used a hierarchical convolutional neural network model based on the attention mechanism to automatically learn features from two levels of sentence structure and text structure and score the essay. The SkipFlow neural network model proposed by Tay et al. [15] can better model the semantic connection of long texts. Rodriguez et al. [12] applied the BERT (Bidirectional Encoder Representation from Transformers) and XLNet model to the field of automatic essay scoring and achieved good performance. Uto et al. [16] integrated item response theory (IRT) into the deep network model to eliminate the prejudice of the rater when scoring the essay, thereby improving the scoring performance of the model. Li et al. [6] proposed a deep neural network model for cross-topic knowledge transfer, and achieved the best performance in scoring cross-topic essays. Ormerod [9] aims at the problem that large-scale language models are difficult to train, and proposes an efficient language model based on the transformer structure, which can accurately and efficiently predict essay scores. From the perspective of hierarchical semantics, Zhou Xianbing et al. [17] constructed a essay scoring model based on multi-level semantic features.

3 Attention Word Embedding Network

This section will introduce the framework of Attention Word Embedding Neteork Model (ATWEM). The method is an improvement of SWEM-concat [13]. First, we use word embedding technology to map each word of the essay to a low-dimensional dense space; then use three pooling operations to fuse each word vector to form a low-dimensional essay semantic vector, and concatenate the three essay semantic vectors; After the fully connected network, features are blended; finally, the Sigmoid function is used to output the essay score.

3.1 Input Layer

This layer maps each word of the input essay to a low-dimensional feature space to obtain a low-dimensional dense vector of the word while maintaining the semantic information of the word. We use a randomly initialized 300-dimensional vector as the word vector of each word, and set the embedding representation of the essay $W = \{x_1, x_2, ..., x_L\}$, $x_i \in \mathbb{R}^d$ is Word embedding representation, d is the dimension of word vector.

3.2 Pooling Layer

After the embedding layer, we take advantage of three pooling methods to obtain the semantic representation of the essay. we use maximum pooling to capture the most significant features of each word embedding dimension, which makes some specific words in the essay have higher value in the word embedding space, and use the average pooling operation to average the embedding space of each word to obtain the overall essay semantics. At the same time, we also make use of the attention mechanism to dynamically focus on each word in the essay, and give higher weight to the words that can make the essay get higher scores, thereby improving the performance of automatic scoring of the essay.

The three pooling operations can capture different semantic information and complement each other. Therefore, we use fusion technology to fuse three kinds of semantic information to further improve the performance of essay scoring.

3.3 Fully Connection Layer

This article uses the feature vector after the three kinds of pooling vectors fusing as the semantic representation of the essay, and then it is passed into a two-layer fully connected network. At the same time, We use ReLU as the activation function of the network to enhance nonlinear representation learning.

3.4 Output Layer

As mentioned above, we use three pooling operations to extract the semantic information of the essay, and use the fusion technology to merge the three feature vectors, and finally use the fully connected network for nonlinear representation to obtain the final essay feature vector Si. And then, the Sigmoid function is used to predict the essay score.

4 Experiment and Result Analysis

This chapter first introduces the datasets and evaluation method, then explains the experimental setup, compares the performance of the proposed model and the baseline model in detail. At the same time, we also analyze the pre-trained word vector and model efficiency.

4.1 Data and Evaluation Indexes

We use the publicly available datasets of the Kaggle ASAP (Automated Student Assessment Prize) competition, which are widely used in the field of automatic essay scoring. The essays included in ASAP are all written by students in grades 7–10. According to the topic content of the essay, the datasets are divided into 8 subsets. Each subset contains a essay prompt document and multiple related topic essays. The details of the data set are shown in Table 1.

Table 1. Statistics of the ASAP dataset

Prompt	Essays	Avg length	Scores
D1	1783	350	2–12
D2	1800	350	1–6
D3	1726	150	0–3
D4	1772	150	0–3
D5	1805	150	0–4
D6	1800	150	0–4
D7	1569	250	0–30
D8	723	650	0–60

4.2 Experimental Setup

Since the test set of the Kaggle ASAP competition has not been made public, we only use its training datasets as the experimental data for this article. Consistent with the work of [5,14,16], we use a 5-fold cross-validation method in the experiment to evaluate the model we proposed. Each fold has a training data ratio of 60% and a validation set of 20%, The test set accounts for 20%.

In the training process, we don't use pre-trained word vectors, all of which are initialized randomly, with a dimension of 300. The output dimension of the attention mechanism is 300, and the number of fully connected network nodes in both layers is 300 dimensions. Our optimization function is RMSProp, the decay rate is set to 0.9, and the learning rate is set to 0.001. In order to prevent overfitting, an early stopping mechanism is used during training, and We add dropout with a drop rate of 0.1 in the final layer.

Table 2. Performance comparison between ATWEM and other baseline methods.

Model	D1	D2	D3	D4	D5	D6	D7	D8	Avg QWK (%)
CNN [14]	79.70	63.40	64.60	76.70	74.60	75.70	74.60	68.70	72.25
LSTM [14]	77.50	68.70	68.30	79.50	81.80	81.30	80.50	59.40	74.63
SkipFlow LSTM [15]	83.20	68.40	69.50	78.80	81.50	81.00	80.00	69.70	76.51
CNN + LSTM [14]	82.10	68.80	69.40	80.50	80.70	81.90	80.80	64.40	76.08
CNN+LSTM+ATT [5]	82.20	68.20	67.20	81.40	80.30	81.10	80.10	70.50	76.38
Topic+BiLSTM+ATT [4]	82.70	69.60	69.10	81.60	81.10	82.30	80.90	70.70	77.30
BERT + XLNet [12]	80.78	69.67	70.31	81.9	80.82	81.45	80.67	60.46	75.78
Electra+Mobile-BERT [9]	83.10	67.90	69.00	**82.50**	81.70	82.20	**84.10**	74.80	78.20
ATWEM (Ours)	**85.24**	**73.02**	**72.62**	80.78	**82.62**	**82.32**	80.92	**74.88**	**79.05**

(* indicates a direct reference to the original results.)

4.3 Result Analysis

In order to verify the effectiveness of the lightweight automatic essay scoring method based on the attention word embedding network, we compared with the baseline methods. Table 2 lists the comparison between the method we proposed and previous work. The experimental results show that:

(1) The mixed model of CNN and LSTM can effectively improve the overall performance of essay scoring. Compared with a single model, the six subsets have improved. The performance of the model is further improved after using the attention mechanism, which shows that the attention mechanism can effectively obtain the semantics of the essay. The fusion of topic features in the BiLSTM model based on the attention mechanism can effectively improve the performance of essay scoring, indicating that the topic relevance of the essay plays an important role in essay scoring.
(2) Using a large pre-trained language model for automatic essay scoring can achieve certain results. The overall performance of BERT+XLNet is slightly higher than LSTM by 1.15%, indicating that large-scale pre-trained language models are not effective in essays on specific topics. Electra+Mobile-BERT is better than the BERT+XLNet model on 7 subsets, and the overall performance is 2.42% higher, indicating that the lightweight language model is better than the large pre-training language model when performing essay scoring tasks.
(3) The ATWEM model we proposed achieves the best performance in six subsets, and the average performance of ATWEM in all datasets is higher than that of the latest Electra+Mobile-Bert. This is due to insufficient training samples, which leads to insufficient fine-tuning of the large pre-training model, thus making the model performance poor. Compared with CNN, each subset has been greatly improved and the average performance has increased by 6.8%. Therefore, the method proposed in this article has better generalization and has the best overall performance compared to the all baseline methods.

4.4 Analysis for Word Embedding

The pre-training word vector is trained from a large-scale corpus and can contain rich semantic information. For most NLP tasks, using word vectors based on large-scale corpus pre-trained can effectively improve model performance. To verify the effect of pre-trained word vector on the performance of ATWEM, we conducted a brief experiment on the ATWEM model from four aspects: the use, type, dimension and scale of the pre-training word vector. The pre-trained word vectors we used include Word2Vec (w2v-50d[1]) and Glove (glv-6B-50d; -glv-6B-300d; -glv-42B-300d)[2], where B stands for scale, d stands for dimension, and -w/o word-vec means that no pre-trained word vector is used. We conducted related experiments on the data set D1, and the experimental results are shown in Fig. 1.

It can be seen from the figure that when no pre-trained word vector is used, the performance of the model is much higher than using any other pre-trained

[1] https://ai.stanford.edu/~wzou/mt/biling/_mt/_release.tar.gz.

[2] https://nlp.stanford.edu/projects/glove/.

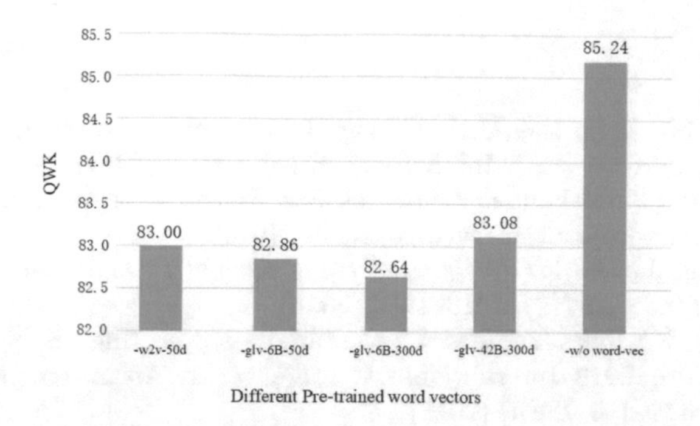

Fig. 1. Model performance using different pre-trained word vectors

word vectors. Compared with using the largest-scale pre-trained word vector -glv-42B-300d, the -w/o word-vec scoring performance is still 2.16% higher. This may be because the training corpus of the pre-trained word vector and the essay corpus are quite different in terms of semantic expression, logical structure, and language style. Therefore, using the word vector that pre-trained on this corpus might result in a certain degree of semantic deviation, which reduces the scoring performance of the model. The experimental results show that due to the lack of large-scale essay corpus, using randomly initialized and self-training word vectors can effectively improve the scoring performance of ATWEM for essay.

5 Conclusions

We propose a lightweight model based on the attention word embedding network to automatically score the essay. We use random initialization and self-training word vectors instead of the commonly used pre-training word vectors for embedding layer. After embedding the layer, abandoning the commonly used CNN and RNN structures, we directly use the three pooling operations. The experimental results of 8 datasets in the Kaggle ASAP competition show that the method in this paper can accurately and efficiently score the essay, and compared with the baseline method, it achieves the best performance on this datasets.

Although the model proposed in this paper can accurately and efficiently score the essays that distinguish the subject, there are differences in many aspects between the essays of different themes. How to automatically score cross-topic essays will be one of the important directions of future research. We will continue to try different methods for further research on cross-topic essay.

Acknowledgments. This work was supported by grant from Xinjiang Uygur Autonomous Region Natural Science Foundation Project No. 2021D01B72. This work was also supported by the Natural Science Foundation of China No. 62066044.

References

1. Chen, Y.Y., Liu, C.L., Lee, C.H., Chang, T.H., et al.: An unsupervised automated essay-scoring system. IEEE Intell. Syst. **25**(5), 61–67 (2010)
2. Alikaniotis, D., Yannakoudakis, H., Rei, M.: Automatic text scoring using neural networks. arXiv preprint arXiv:1606.04289 (2016)
3. Chen, H., Xu, J., He, B.: Automated essay scoring by capturing relative writing quality. Comput. J. **57**(9), 1318–1330 (2014)
4. Chen, M., Li, X.: Relevance-based automated essay scoring via hierarchical recurrent model. In: 2018 International Conference on Asian Language Processing (IALP), pp. 378–383. IEEE (2018)
5. Dong, F., Zhang, Y., Yang, J.: Attention-based recurrent convolutional neural network for automatic essay scoring. In: Proceedings of the 21st Conference on Computational Natural Language Learning (CoNLL 2017), pp. 153–162 (2017)
6. Li, X., Chen, M., Nie, J.Y.: SEDNN: Shared and enhanced deep neural network model for cross-prompt automated essay scoring. Knowl.-Based Syst. **210**, 106491 (2020)
7. Liang, M., Wen, Q.: A critical review and implications of some automated essay scoring systems. Technol. Enhanced Foreign Lang. Educ. **5**, 18–24 (2007)
8. Ming, Z., Yan-ming, J., Cai-lan, Z., Ning, X.: English automated essay scoring methods based on discourse structure. Comput. Sci. **46**(03), 240–247 (2019)
9. Ormerod, C.M., Malhotra, A., Jafari, A.: Automated essay scoring using efficient transformer-based language models. arXiv preprint arXiv:2102.13136 (2021)
10. Page, E.B.: Grading essays by computer: progress report. In: Proceedings of the Invitational Conference on Testing Problems (1967)
11. Pramukantoro, E.S., Fauzi, M.A.: Comparative analysis of string similarity and corpus-based similarity for automatic essay scoring system on E-learning gamification. In: 2016 International Conference on Advanced Computer Science and Information Systems (ICACSIS), pp. 149–155. IEEE (2016)
12. Rodriguez, P.U., Jafari, A., Ormerod, C.M.: Language models and automated essay scoring. arXiv preprint arXiv:1909.09482 (2019)
13. Shen, D., et al.: Baseline needs more love: on simple word-embedding-based models and associated pooling mechanisms. arXiv preprint arXiv:1805.09843 (2018)
14. Taghipour, K., Ng, H.T.: A neural approach to automated essay scoring. In: Proceedings of the 2016 conference on empirical methods in natural language processing, pp. 1882–1891 (2016)
15. Tay, Y., Phan, M., Tuan, L.A., Hui, S.C.: Skipflow: Incorporating neural coherence features for end-to-end automatic text scoring. In: Proceedings of the AAAI Conference on Artificial Intelligence, vol. 32 (2018)
16. Uto, Masaki, Okano, Masashi: Robust neural automated essay scoring using item response theory. In: Bittencourt, Ig Ibert, Cukurova, Mutlu, Muldner, Kasia, Luckin, Rose, Millán, Eva (eds.) AIED 2020. LNCS (LNAI), vol. 12163, pp. 549–561. Springer, Cham (2020). https://doi.org/10.1007/978-3-030-52237-7_44
17. Zhou, X., Fan, X., Re, G., Yang, Y.: English automated essay scoring methods based on multilevel semantic features. Comput. Appl. 1–8 (2021)
18. Yannakoudakis, H., Briscoe, T., Medlock, B.: A new dataset and method for automatically grading ESOL texts. In: Proceedings of the 49th Annual Meeting of the Association for Computational Linguistics: Human Language Technologies, pp. 180–189 (2011)

Development and Application of Water Purification Process in Changchun Hydro-Ecological Garden Based on BIM+VR Technologys

Xiangshu Peng[1], Zhiming Ma[1](✉), Ping Wang[2], Xuanli Mo[2], and Menglin Qi[1]

[1] School of Computer Science and Technology, Xinjiang Normal University, Urumchi 830054, China
406287175@qq.com

[2] School of Electricity and Computer, Jilin JianZhu University, Changchun 130000, China

Abstract. Using BIM frontier technology and Unity real-time 3D development platform, based on the analysis of relevant technical characteristics, BIM intelligent platform and VR technology are applied to the water purification process display, and the VR experience Hall of Water Culture Museum of Changchun Water Culture Ecological Park is developed. After the project is put into application, it gives the experiencer a new personalized immersive interactive experience, and is widely praised by the experiencer and concerned and reported by the news media.

Keywords: VR technology · BIM · Unity · Immersed interaction · Changchun Water Culture Ecological Park

1 Introduction

The content of the traditional exhibition hall mainly uses traditional expressions and techniques such as text, pictures, videos, and sculptures. It has the characteristics of universality singleness and obsolescence. It is a formality, lacks innovation, freshness and attractiveness. Obviously, it violates the original intention and original intention of the exhibition experience hall to popularize knowledge.

With the development of science and technology and the continuous advancement of virtual reality technology (VR), the virtual reality exhibition hall has gradually come into people's sight. The virtual reality exhibition hall uses computer graphics technology to construct a digital exhibition hall. It is an immersive virtual three-dimensional interactive experience. Based on the core requirements such as outstanding experience effects and concise creation process, it breaks through the traditional exhibition demonstration method and innovate to create "immersive" Immersive interactive experience.

Q. Liang et al. (Eds.): Artificial Intelligence in China, LNEE 854, pp. 283–289, 2022.
https://doi.org/10.1007/978-981-16-9423-3_36

2 Project Development Background

Changchun Water Culture Ecological Park was originally the site of the 80-year-old Changchun No. 1 Water Purification Plant (hereinafter referred to as Yijing). In order to display and inherit the water purification process, people can better understand, approach and cherish water resources. The water culture factor is perfectly integrated with the public art design. In 2016, the Changchun Municipal Government and Changchun Urban Construction Investment (Group) Co Ltd. carried out the renovation and reconstruction of the old water plant site, and entrusted Jilin Jianda Information Technology Co, Ltd. and Jilin Jianzhu University to develop Changchun water culture Ecological Garden Water Culture Museum Water Purification Process VR Experience Hall Project, in order to create an ecological park with the theme of collecting water cultural elements, and make it a new landmark of Changchun city culture as "Northern Spring City".

3 Development of Water Purification Process Based on BIM+VR Technology

3.1 Demand Analysis

In order to popularize the knowledge of water conservation and water purification processes, and establish the awareness of cherishing water resources; based on the concept of "Water and City" Theme Exhibition Outline in the Museum District of Changchun Water Culture Ecological Park and "Water and City" Changchun Water Culture Ecological Park "Museum District Exhibition Design Plan". Discussed and planned the virtual reality process and interactive content, and initially formulated three thematic plans "I am a water purification staff", "I am a museum tour guide", and "Review of history through time and space". While maintaining progress and meeting customer requirements. We share ideas in real time and collect and implement feedback. After continuous improvement and comprehensive consideration and evaluation. The evaluation form is as follows, and the final selection plan 1: "I am a water purification staff" Use virtual reality technology Restore the change process of Yijing, simulate the three-generation water purification process of the water purification plant, and create an immersive interactive experience (Table 1).

3.2 BIM Smart Platform

BIM (Building Information Modeling) is a landmark smart innovation platform in the construction industry, The core of BIM is to establish a virtual three-dimensional model of construction engineering through Revit, and use digital technology to provide a complete and consistent model for the actual situation, The construction engineering information database improves efficiency, reduces costs, and creates value for the entire life cycle of a building and it has eight characteristics of information integrity, relevance, consistency, visualization, coordination, simulation, optimization, and drawing.

Table 1. VR theme program evaluation

<table>
<tr><th colspan="3">Evaluation factors</th><th>Option I</th><th>Option II</th><th>Option III</th></tr>
<tr><td>Prime factor</td><td colspan="2">Estimated completion time</td><td>2–4 weeks</td><td>5–7 weeks</td><td>4–5 weeks</td></tr>
<tr><td rowspan="8">Comprehensive factors</td><td colspan="2">Basic information</td><td>Less</td><td>Adequate</td><td>Rarely</td></tr>
<tr><td rowspan="2">Implement step</td><td>3D modeling quantity</td><td>Less</td><td>Many</td><td>Medium</td></tr>
<tr><td>VR programming difficulty</td><td>Medium</td><td>Medium</td><td>Easy</td></tr>
<tr><td colspan="2">Commissioning and trial operation</td><td>Medium</td><td>Complex</td><td>Easy</td></tr>
<tr><td colspan="2">Visitors difficulty handling</td><td>Easy</td><td>Easy</td><td>Easy</td></tr>
<tr><td colspan="2">Interactive effect</td><td>Better</td><td>Good</td><td>General</td></tr>
<tr><td colspan="2">post-maintenance</td><td>Easy</td><td>Easy</td><td>Easy</td></tr>
<tr><td colspan="2">Investment</td><td>Medium</td><td>High</td><td>Medium</td></tr>
</table>

3.3 Unity Real-Time 3D Development Platform

Unity is a real-time 3D interactive content creation and operation platform, which can used to create visualization products and build immersive interactive virtual experiences. Real-time rendering can achieve "seeing is what you get", allowing developers to maximize their creativity. The VR technology is used to simulate the process flow, so that the experiencer can experience the real environment without touching the physical object, with-out being restricted by physical boundaries. Once this was just a dream in the future, it has now become a reality. Unity Reflect is an official innovation tool between Unity and Autodesk. It supports one-click lossless transmission of BIM models and data and real-time synchronization to Reflect Viewer, realizing visualized real-time design changes and seamlessly optimized BIM data integration, and has the ability to transmit model data flexibly and the powerful functions connected to other programs have great potential in reducing the time required for the design review and verification process, allowing all parties to communicate efficiently and conveniently, speed up iteration, and promote precise decision-making. The AEC field h-as a disruptive significance.

3.4 Revit 3D Modeling

On the basis of field investigation and measurement, CAD software is used to draw model drawings, and modeling is described in accordance with CAD base drawings to improve work efficiency and accuracy; through the establishment of BIM 5D relational data base a net model volume can be calculated quickly and accurately, which is convenient Data analysis management. In the modeling stage, Revit software is used for modeling work and 3ds Max three-dimensional modeling software is used for partial and detailed modification and improvement in the later period. The Japanese puppet period, post-liberation and current Yijing pavilion models are created respectively. The real scene of the period reappears, allowing people to feel the past and present life. Figure 1 shows the Yijing real scene, and Fig. 2 shows the 3D model.

Fig. 1. Real view of the first water purification plant

Fig. 2. The first water purification plant 3D model

3.5 Interactive Content Development Based on Unity

We import the original BIM model and data into the Unity 3D engine through Unity Reflect to realize real-time interconnection and collaborative development; use a powerful C# script system and sound API interface functions to meet the needs of development work. It mainly includes scene construction; head control, split screen display (three-dimensional effect configuration, simulated human eye pupil distance); interactive design (head control visual selection, touch feedback, Microsoft X-BOX gamepad) and rendering work.

External scene construction: rendering of the building model of the first water purification workshop, couplet mapping, Green plant model, four sided curved screen synchronously playing the opening video, building a 360-degree surrounding shadow effect, and entering the interactive scene through the Cube trigger component. Interactive scene construction of the internal water purification workshop: add components to the architectural model, such as light sources, railings, water pumps, water particles, tables, on-off valves, display cards, explanatory text, various objects placed on the table, and the rise of partition doors Falling animation, game link prompt sound, background music various special effects, etc. As shown in Fig. 3.

Unity has universal rendering pipeline (URP), high-definition rendering pipeline (HDRP), Shader Graph and other tools to meet system graphics rendering needs. Native VR rendering processes the image twice for the left and right eyes, renders the shadow twice, and initiates twice the draw call, which is called multi-channel rendering. This method takes up twice the CPU power consumption and consumes a lot of system

Fig. 3. Water purification scene

hardware resources. To this end, we use the programmable high definition rendering pipeline HDRP, support a single-channel forward rendering loop, and use tools for HTC VIVE head-mounted displays (HMD) and optimize them.

In order to reduce the dizziness of VR helmets, create a good user experience, reduce latency, and increase frame rate, build a Windows platform oriented VR SDK, A well-performing SDK achieves low latency and high frame rate; to reduce image aliasing, multi-sampling is adopted Anti-aliasing, support forward rendering, change the number of samples (8x), balance quality and performance; in order to improve the refresh rate and resolution of the binocular display, disable unnecessary functions in the HD-RP resource settings to reduce the energy consumption of VR rendering, We get input from the VR controller and refer to the input information t-able of the VR Input document to input information such as triggers and gamepad buttons; in order to achieve excellent visual effects, the particle system can be used to achieve better texture combination and batch processing functions. Including bright lines and trajectories, plus a high fidelity physical collision module, to achieve more realistic effects and animations, and create high-level graphics fidelity and realistic immersive VR projects.

4 Project Application

4.1 Project Experience Process

The experience process is as follows, the first scene: 4 pieces of curved screens play the opening video. In the second scene, I came to the flocculation tank and opened the valve to see the sewage in the tank. This link requires the experiencer to add flocculants to flocculate the impurities in the sewage; then enter the advection sedimentation tank, settle some impurities, and then enter the filter tank. Put the filtered water into the disinfection tank, add disinfectant to disinfect it; finally, enter the clean pool, the water at this time is our daily drinking tap water, and the entire water purification process is completed.

During the whole process, the experiencer wears the HTC VIVE professional virtual reality smart headset, turns on the Steam VR connection device, uses two locators to collect the experiencer's behavior, and the Microsoft X-BOX gamepad performs interactive operations, insisting on taking the experiencer as the center. The protagonist character completes the water purification operation and interaction, with a sense of accomplishment in his heart, and truly achieves an immersive interactive experience.

4.2 Project Application Effect

The project was successfully completed and put into use on schedule, carrying forward the spirit of Lingnan 1932 with water as a cultural element, and has great social significance in promoting the inheritance of traditional water purification technology; it has been widely praised by the municipal government and relevant departments of the urban investment group, Changchun citizens are passionate It is full of hearts and an endless stream of people come to experience the project, which has won unanimous praise from the experiencers, which has improved the happiness of the citizens.

Guangming Daily reported: "The operation is complete, please proceed to the next link. The fourth link, disinfection pool, please add disinfectant…". Visit the exhibition, and then walk into the interactive experience area, the citizen Bie Yujie in the staff Under the guidance, wear 3D glasses to complete the simulation experience of the water purification process. Take off the glasses and smile, "This park is very cultural! I plan to let relatives and friends come around". The "cultural taste" felt by uncle Bie is one of the construction and development concepts of the ecological park. Fu Qiang, general manager of Changchun Urban Construction Group, responsible for the management and operation of the ecological park, said: "Changchun Water Culture Ecological Park is built in the urban area. Prosperous area, convenient transportation, quiet in the middle of noisy. It provides a leisure choice for citizens during the National Day holiday." – Excerpt from literature [7] "Guangming Daily".

5 Conclusion

The new media era has come as promised, and diversified thematic exhibitions have enriched our lives. It is an inevitable trend to apply virtual reality technology to the construction of digital exhibition halls. This article integrates BIM + VR technology into new media and innovates exhibition expression forms and techniques; From concept to data integration, visualization and prototyping, BIM models and data are imported into Unity through Unity Reflect, real-time and seamless collaboration, easy and immediate modification, rapid iteration, and immersive interactive experience. The development and application of this project will become a model for the combination of urban construction and development and technological innovation, and provide a reference for restoring more old-generation factories and technological processes, and restoring more historical footprints.

References

1. Wu, B., Liu, L., Huang, T., Bai, Y.: BIM+VR technology in engineering research on innovative applications in the project. Urban Resid. House **27**(05), 242–243 (2020)
2. Song, C., Yu, Q., Gu, C.: Research on the key technology of 3D scene construction of waterway based on Unity3D+BIM. China Water Transp. Waterway Div. Technol. (02), 46–49 (2020)
3. Wang, C.: Architecture design of BIM cloud platform based on unity 3D and function research and development. Qingdao Technological University (2019)
4. Liang, Y.: Analysis of VR technology in the thematic exhibition in the new media era application in China. China Media Technol. (12), 72–74 (2020)

5. Wang, L., Li, Z.: Urban public art based on interactive experience design innovation research. Design **32**(17), 146–148 (2019)
6. Yang, Y.: Research on the application of immersive experience landscape—moment factory's interactive creative space as an example. Green Sect. Technol. (11), 53–55 (2020)
7. Ren, S., Jilin, C.: Enjoy leisure in the ecological garden. Guangming Day Newspaper, 04 October 2018
8. Li, J.: BIM integrated with VR technology applied in construction engineering. J. Beihua Aerosp. Ind. Inst. **30**(05), 18–20 (2020)
9. Hu, X., Nong, R., Li, J., Lv, L.: Innovative application of Bim and VR technology in construction engineering. Urban Hous. **28**(04), 189–190 + 192 (2021)
10. Li, J.: Research on the combination method of BIM technology and VR technology. Shanxi Archit. **46**(17), 191–193 (2020)
11. Ye, X.: Assembly (superposition) and connection design of water treatment structures. China Water Supply Drain. **30**(18), 36–39 (2014)

Research on Object Detection Algorithm Based on Improved Yolov5

Shuping Chen[1] and Bingcai Chen[1,2](✉)

[1] School of Computer Science and Technology, Xinjiang Normal University, No. 102 Xinyi Road, Urumqi 830054, Xinjiang, China
[2] School of Computer Science and Technology, Dalian University of Technology, No. 2 Linggong Road, Dalian 116024, Liaoning, China
china@dlut.edu.cn

Abstract. Object detection is an important fundamental problem in computer vision research, and is also the basis for other high-level visual tasks such as object tracking, behavioral analysis, and image description. Since target recognition is the simultaneous recognition and assessment of multiple targets, the accuracy of target recognition is generally low. In order to improve the accuracy of target detection, this paper introduces the Efficient Channel Attention (ECA) module to optimize the original yolov5 model, which has fewer parameters and significantly improved performance. After the ECA module performs channel-by-channel global average pooling without reducing the dimensionality, it considers each channel and its k neighbors to capture local cross-channel interactions, which enhances the discrimination of features. Experiments on the coco data set show that the improved yolov5 model based on the ECA module proposed in this paper can be used 68.84% accuracy to detect these object. Compared with the comparison algorithm, the training accuracy of the yolov5 model integrated with the ECA module is improved 3.02%, which proves that the ECA module can further improve the accuracy of yolov5, and can meet the requirements of reliability and stability of target detection.

Keywords: Object detection · yolov5 · Efficient channel attention module · Computer vision

1 Introduction

In recent years, the development of deep learning has not only improved the ordinary two-dimensional image classification problem to a new level, but also provided a variety of ideas and methods for the field of object detection. The main task of object detection is to find out all interested targets in a given image accurately and efficiently, and to locate the position and size of the detected object by using rectangular bounding box. Object detection plays an important role in scientific development and actual industry, such as UAV navigation, unmanned driving [1] and Internet of things. In the process of object detection, due to the appearance, posture, shape and quantity of all kinds of target in the image are different, and also affected by various factors such as illumination

Q. Liang et al. (Eds.): Artificial Intelligence in China, LNEE 854, pp. 290–297, 2022.
https://doi.org/10.1007/978-981-16-9423-3_37

and occlusion, the target is distorted, which makes the object detection more difficult. Therefore, many scholars are still studying and exploring the object detection algorithm.

Using deep convolution network features and SVM regression model can get more accurate boundary box [2]. On the basis of R-CNN, researchers propose a fast R-CNN, which does not convolute candidate regions, and the classification and frame fitting of candidate regions are completed synchronously. The training and testing time of this method is faster than that of R-CNN [3]. Both R-CNN and Fast R-CNN use selective search to obtain candidate regions. The time of processing an image is about 2S, which is time-consuming. Therefore, researchers propose a Fast R-CNN which can generate candidate regions directly through convolution neural network, which is composed of two convolution neural networks. That is, the candidate Region Proposal Network (RPN) and Faster R-CNN, which can classify and return the border, shorten the training and test time by nearly 10 times compared with Fast R-CNN [4]. Although R-CNN series models have improved the accuracy and speed of object detection to a new level, these methods only generate candidate regions, which makes the process of object detection more complicated and time-consuming. To solve this problem, researchers propose YOLO [5], which abandons the intermediate steps of generating candidate regions, divides the input image into S rows and S columns of grid, predicts the boundary box in each grid cell, and realizes the boundary box regression and category prediction through convolution neural network, which truly realizes the one-step decision recognition. However, because YOLO uses high-level features after multi-layer convolution to complete target recognition, the detection accuracy, especially for small targets, is not ideal.

2 Related Work

Yolov5 is characterized by flexibility, which introduces depth_Multiple and width_Multiple coefficient is used to get different size models. Depth_Multiple is the scaling factor of channel, which controls the series number of C3 modules in the model. Similarly, width_multiple controls the coefficient of channel size in the model. Each structure size model is only different in depth_multiple and width_multiple, these two parameters control the size or complexity of the model. The overall architecture of the improved yolov5 model is shown in Fig. 1. In order to reduce the parameter growth, this paper embeds ECA module into yolov5 model including focus module, SPP module and C3 module.

2.1 SPP Module [7]

The spatial pyramid pool layer is the core of SPP module, as shown in the Fig. 2(a), its main purpose is to produce fixed size output for any size input. The idea is to divide a feature map of any size into 16, 4, and 1 blocks, and then maximize the pooling on each block. The pooled features are spliced to get a fixed dimension output. SPP module makes CNN feature no longer a single scale through spatial pyramid pooling, which means that the network can introduce any size of input, that is to say, it can easily train in multiple sizes. Secondly, spatial pyramid pooling can produce fixed output for any size of input, which makes it possible to extract primary features from multiple regions of a

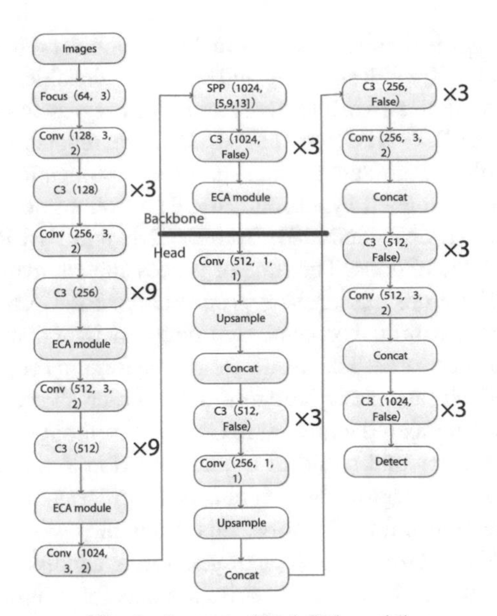

Fig. 1. Improved YoloV5 module

picture. As shown in the Fig. 2(b), the SPP module consists of two standard convolution modules and X maximum pooling layers. First, a standard convolution module is used to reduce the number of channels. Then, after passing through X different scale maximum pooling layers, the others are concatenated with the data that has not been pooled. Then, a standard convolution module is used to restore the number of channels to the output.

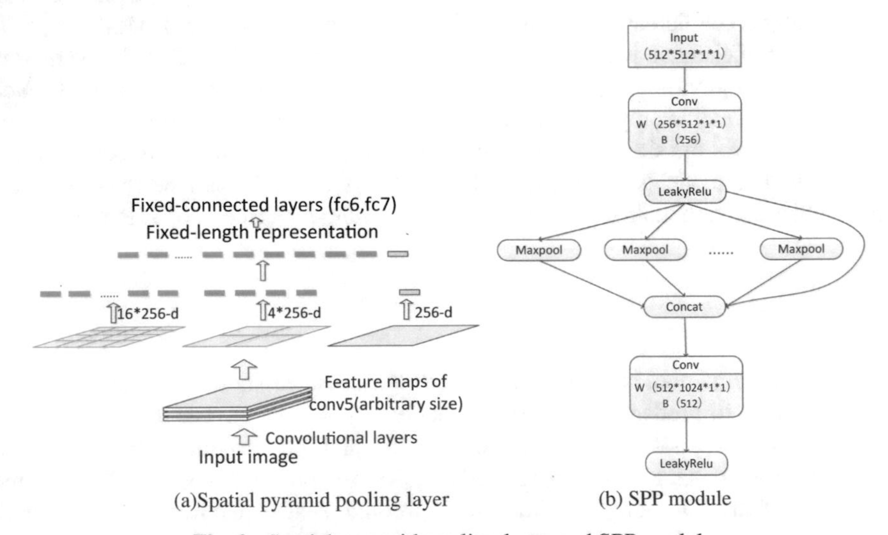

(a)Spatial pyramid pooling layer (b) SPP module

Fig. 2. Spatial pyramid pooling layer and SPP module

2.2 C3 Module

The C3 module is similar to the Bottleneck CSP, but is missing a convolution layer. C3 module, including three standard convolution layers and X Bottleneck modules. As shown in the Fig. 3, after the input feature enters, it is divided into two lines. In route 1, a revolutionary layer is used to reduce the number of channels, and then several bottlenecks are used. The number of bottlenecks is determined by the second parameter of the layer and depth_multiple in the configuration file that defines the model. Route 2 only passes through one convolution layer to reduce the number of channels.

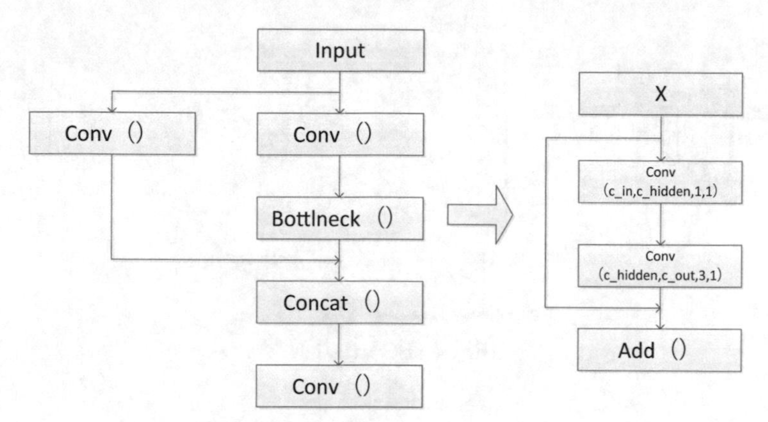

Fig. 3. C3 module

2.3 ECA Module

Since the emergence of Squeeze and Exception Networks [8], attention mechanism is more and more widely used in neural networks. Earlier SE and CBAM [9] modules usually adopt the method of reducing dimensions, and capture all channel interactions. The attention mechanism can only choose between efficiency and accuracy. On the one hand, ECA module avoids dimension reduction, on the other hand, it effectively captures cross channel interaction. As shown in Fig. 4, ECA captures local cross channel interactions by considering each channel and its K nearest neighbors after global average pooling on a channel by channel basis without reducing dimensions, so as to ensure computational performance and model complexity.

The high-dimensional (low dimensional) channel is proportional to the long-distance (short-distance) convolution when the number of grooves is fixed. Similarly, the coverage of cross channel information interaction, that is, the kernel size k of one-dimensional convolution, should also be proportional to the channel dimension C. In other words, there may be a mapping φ between K and C. The simplest mapping method is linear mapping, but due to the limitation of linear functions on certain correlation characteristics, and the channel dimension is usually an exponential multiple of 2, it is based on 2 The exponential function is used to express the nonlinear mapping relationship:

$$C = \phi(\mathrm{k}) = 2^{(\gamma * \mathrm{k} - \mathrm{b})} \tag{1}$$

Therefore, given the channel dimension C, the convolution kernel size k can be calculated according to the following Formula 2:

$$k = \varphi(C) = \left| \frac{\log_2(C)}{\gamma} + \frac{b}{\gamma} \right|_{odd} \tag{2}$$

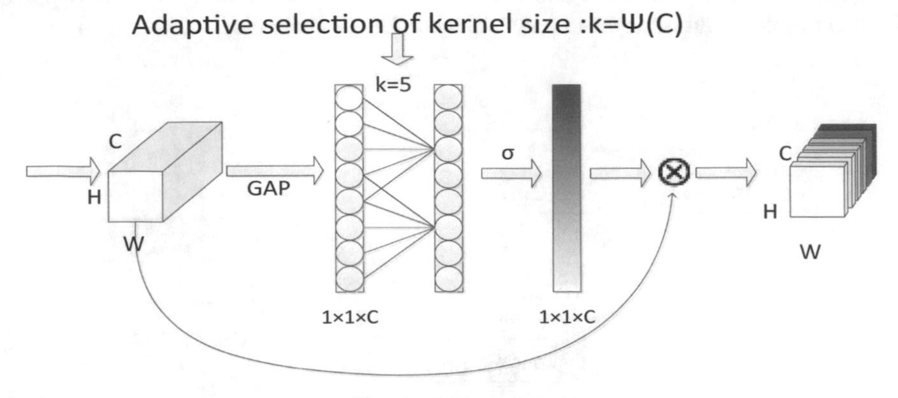

Fig. 4. ECA module

3 Experimental Evaluation

3.1 COCO Data Set

Because the coco object detection data set used in this paper covers 80 categories, including 118287 training sets, 5000 verification sets and 40670 test sets, and the total number of detected targets is much higher than other models, reaching 886266 targets. On this basis, because the distribution of large, medium and small targets in the coco data set is very balanced, all above 30%, the coco datasets has the most difficult object scale to frame. All these factors lead to the low recognition rate of the current coco data set.

3.2 Data Preprocessing

The mosaic data enhancement used in this paper refers to the Cut Mix data enhancement method proposed at the end of 2019. Cut Mix data enhancement uses two images for stitching, but mosaic uses four images. As shown in the Fig. 5, this kind of data enhancement method is simply to splice four images by random scaling, random cutting and random arrangement. Its advantages are that it enriches the background and small targets of the detected objects, and enhances the robustness of recognition. In addition, when calculating the batch normalization, the data of four images are calculated at a time, so that the mini batch size does not need to be very large, Using a single GPU can also achieve better results.

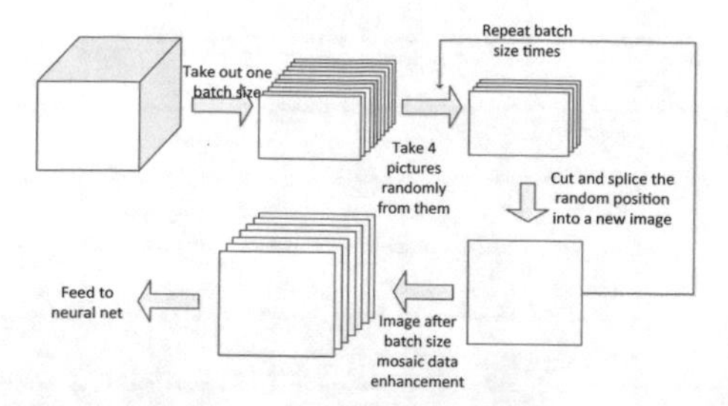

Fig. 5. Data multiprocessing flow chart

3.3 Comparison of Model Accuracy Before and After Improvement

Under the above conditions, the accuracy curve and loss function curve of the training set is shown in the Fig. 6 and Fig. 7. It is obvious from the figure that the accuracy of the original yolov5 model is 65.86%, and the accuracy of the improved yolov5 model is 68.84%, which is an increase of 3.02%. The original yolov5 model loss function is 0.04944, and the improved yolov5 model loss function is reduced to 0.04902, which can prove that the improved model converges faster than the original model.

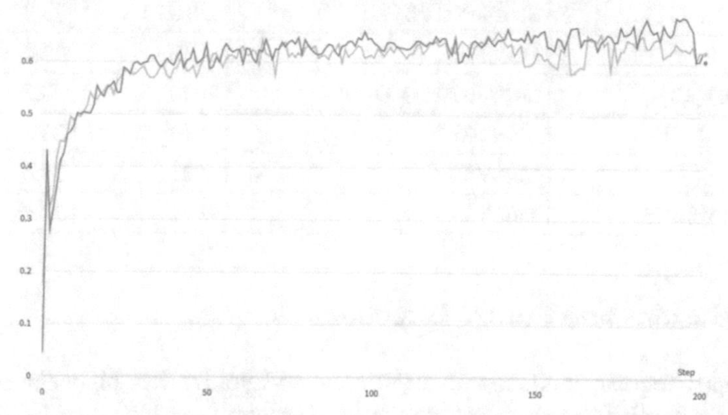

Fig. 6. YoloV5 and improved YoloV5 accuracy on coco data set

3.4 Comparison with Other Models

mAP is a measure of recognition accuracy in target detection. In multi category target detection, each category can draw a curve according to recall and accuracy. AP is the area under the curve. MAP means to average the AP of each class (Table 1).

The table shows the performance of different methods in the coco data set, including MobleNetV3 and other popular and novel target detection networks in recent years. It can be seen from the table that the result of 34.54 is higher than that of MobleNetV3. Horizontally, it is proved that the performance of this model is better than the current mainstream target recognition network.

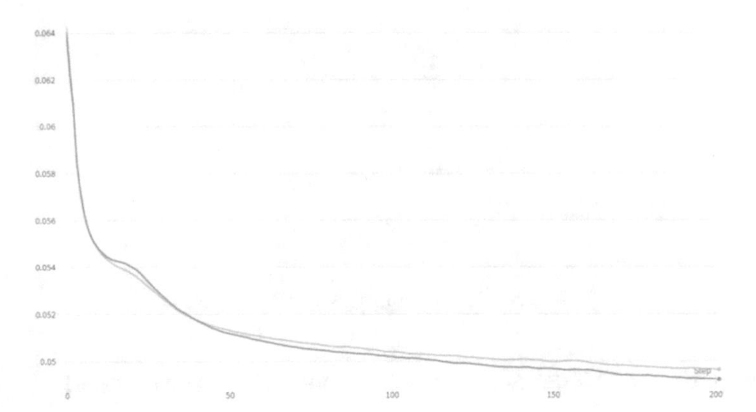

Fig. 7. YoloV5 and improved YoloV5 loss on coco data set

Table 1. Comparison of the mAP between Improved YoloV5 and other models

Method	Backbone network	Year	mAP/%
MobileNetV3 [10]	MobileNetV3	2019	22.0
ShuffleNetV2 [11]	ShuffleNetV2	2018	23.2
Pelee [12]	PeleeNet	2018	22.4
Tiny-DSOD [13]	DDB + Net + D-FPN	2018	23.2
ThunderNet [14]	SNet146	2019	23.7
YoloV5	YoloV5	2021	34.3
Improved YoloV5	**YoloV5**	**2021**	**34.5**

4 Conclusions and Future Directions

In this paper, experiments show that the improved yolov5 has higher accuracy and can be better applied to target detection tasks. Through the addition of ECA attention mechanism, the feature extraction ability of the network is enhanced, and the small target object in the target detection can be better recognized. In the coco data set, which is the most complex and difficult data set at present, it has achieved the ideal effect, reaching the accuracy rate of 68.84%. The experimental results of other object detection networks on the coco data set are compared horizontally, which proves that the changes made in this paper have a positive effect on the original model. In the future work, we will be more committed to small target detection and recognition, to enhance the application scenarios of the model.

Acknowledgments. This work was supported in part by the Tianshan Young Talent Program, Xinjiang Uygur Autonomous Region under Grant 2018Q024, in part by the Natural Science Foundation of China under Grant 61771089 and Grant 61961040, and in part by the Regional Cooperative Innovation Program of Autonomous Region (Aid Program of Science and Technology to Xinjiang) under Grant 2020E0247 and Grant 2019E0214.

References

1. Lan, D.U., Di, W.E.I., Lu, L.I., et al.: SAR target detection network via semi-supervised learning. J. Electron. Inf. Technol. **42**(1), 154–163 (2020)
2. Girshick, R., Donahue, J., Darrell, T., Malik, J.: Rich feature hierarchies for accurate object detection and semantic segmentation. In: Proceedings of the IEEE Conference on Computer Vision and Pattern Recognition, pp. 580–587 (2014)
3. Girshick, R.: Fast R-CNN. In: Proceedings of the IEEE International Conference on Computer Vision, pp. 1440–1448 (2015)
4. Ren, S., He, K., Girshick, R., Sun, J.: Faster R-CNN: towards real-time object detection with region proposal networks. arXiv preprint arXiv:1506.01497 (2015)
5. Redmon, J., Divvala, S., Girshick, R., et al.: You only look once: unified, real-time object detection. In: Proceedings of the IEEE Conference on Computer Vision and Pattern Recognition, pp. 779–788 (2016)
6. Shi, W., et al.: Real-time single image and video super-resolution using an efficient sub-pixel convolutional neural network. In: Proceedings of the IEEE Conference on Computer Vision and Pattern Recognition, pp. 1874–1883 (2016)
7. He, K., Zhang, X., Ren, S., Sun, J.: Spatial pyramid pooling in deep convolutional networks for visual recognition. IEEE Trans. Pattern Anal. Mach. Intell. **37**(9), 1904–1916 (2015)
8. Hu, J., Shen, L., Sun, G.: Squeeze-and-excitation networks. In: Proceedings of the IEEE Conference on Computer Vision and Pattern Recognition, pp. 7132–7141 (2018)
9. Woo, S., Park, J., Lee, J.-Y., Kweon, I.S.: CBAM: convolutional block attention module. In: Ferrari, V., Hebert, M., Sminchisescu, C., Weiss, Y. (eds.) ECCV 2018. LNCS, vol. 11211, pp. 3–19. Springer, Cham (2018). https://doi.org/10.1007/978-3-030-01234-2_1
10. Howard, A., Sandler, M., Chu, G., et al.: Searching for MobileNetV3. In: Proceedings of the IEEE/CVF International Conference on Computer Vision, pp. 1314–1324 (2019)
11. Ma, N., Zhang, X., Zheng, H.T., et al.: ShuffleNet V2: practical guidelines for efficient CNN architecture design. In: Proceedings of the European Conference on Computer Vision (ECCV), pp. 116–131 (2018)
12. Wang, R.J., Li, X., Ling, C.X.: Pelee: a real-time object detection system on mobile devices. In: Proceedings of the 32nd International Conference on Neural Information Processing Systems, pp. 1967–1976 (2018)
13. Li, Y., Li, J., Lin, W., et al.: Tiny-DSOD: lightweight object detection for resource-restricted usages. In: Proceedings of BMVC, p. 59 (2018)
14. Qin, Z., Li, Z., Zhang, Z., et al.: ThunderNet: towards real-time generic object detection on mobile devices. In: Proceedings of the IEEE/CVF International Conference on Computer Vision, pp. 6718–6727 (2019)

New Trend in Front-End Techniques of Visual SLAM: From Hand-Engineered Features to Deep-Learned Features

Yue Wang[1,2(✉)], Yu Fu[2], Ruixue Zheng[2], Le Wang[3], and Jianzhong Qi[3]

[1] Key Laboratory of Electronics and Information Technology for Space Systems, Chinese Academy of Sciences, Beijing 100190, China
wangyue@nssc.ac.cn

[2] Tianjin Key Laboratory of Wireless Mobile Communications and Power Transmission, Tianjin Normal University, Tianjin 300387, China

[3] Department of Electronic and Information Engineering, North China University of Technology, Beijing 100144, China

Abstract. Visual simultaneous localization and mapping (V-SLAM) technique plays a key role in perception of autonomous mobile robots, augmented/mixed/virtual reality, as well as spatial AI applications. This paper gives a very concise survey about the front-end module of a V-SLAM system, which is charge of feature extraction, short/long-term data association with outlier rejection, as well as variable initialization. Visual features are mainly salient and repeatable points, called keypoints, their traditional extractors and matchers are hand-engineered and not robust to viewpoint/illumination/seasonal change, which is crucial for long-term autonomy of mobile robots. Therefore, new trend about deep-learning-based keypoint extractors and matchers are introduced to enhance the robustness of V-SLAM systems even under challenging conditions.

Keywords: 3D vision · Data association · Deep learning · Feature extraction · Keypoints · Simultaneous localization and mapping (SLAM) · Visual odometry (VO)

1 Introduction

The visual simultaneous localization and mapping (V-SLAM) system adopts cameras (usually with inertial measurement units (IMUs)) to simultaneously inference its own state (e.g. pose) and build a consistent surrounding map [1]. V-SLAM systems are widely used by autonomous mobile robots, especially the

Y. Wang—This work was supported by the Pre-Research Project of Space Science (No. XDA15014700), the National Natural Science Foundation of China (No. 61601328), the Scientific Research Plan Project of the Committee of Education in Tianjin (No. JW1708), and the Doctor Foundation of Tianjin Normal University (No. 52XB1417), and the Beijing Natural Science Foundation (No. 4204095).

Q. Liang et al. (Eds.): Artificial Intelligence in China, LNEE 854, pp. 298–307, 2022.
https://doi.org/10.1007/978-981-16-9423-3_38

resource-limited agile micro drones [2], augmented/mixed/virtual reality devices [3], as well as spatial AI applications [4,5]. Recently, with the huge tide of deep learning (DL) techniques [6,7], some end-to-end DL-based V-SLAM or visual odometry (VO) systems are proposed [8–11]. In this paper, we focus on the traditional de-facto framework of V-SLAM systems with standard cameras, as shown in Fig. 1, which contains two modules: a front-end and a back-end. The front-end is responsible for processing visual and IMU data, including feature extraction, short/long-term data association with outlier rejection, as well as initial estimation of the current pose and newly detected landmark positions. The back-end takes the initialization information from the front-end, and makes a maximum-a-posteriori (MAP) estimation based on a factor graph to refine both poses and landmark positions [12]. This paper focuses on the new trend in front-end techniques of V-SLAM systems.

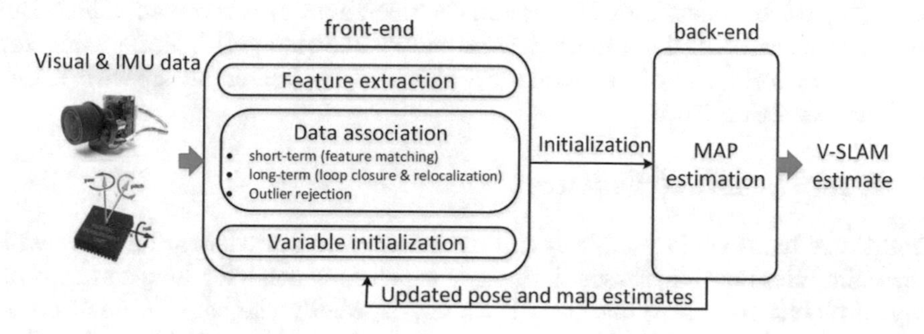

Fig. 1. The de-facto framework of a V-SLAM system.

2 Hand-Engineered Features

Traditionally, visual features are mainly restricted to salient and repeatable points (called keypoints), due to the unreliable extraction of high-level geometric features (e.g. lines or edges) using unlearning methods. Besides keypoints, dense methods using all pixel information (e.g. optical flow [13] or correspondence-free method [14]) have also been adopted to inference ego-motion with small motion assumption. In the early research of stereo VO for Mars rovers [15–18], keypoints were tracked among successive images from nearby viewpoints. Nister proposed that keypoints could be independently extracted in all images and then matched in his landmark paper [19]. This method then became the dominant approach because it can successfully work with a large motion/viewpoint change [20].

Keypoint-based feature extraction includes two stages: keypoint detection and keypoint description. The keypoint detectors can be divided into two categories: corner detectors and blob detectors [21]. Traditional keypoint extractors are hand-engineered: corner detectors include Moravec [22], Forstner [23], Harris [24], Shi-Tomasi [25], and FAST/FASTER [26,27]; blob detectors include SIFT [28], SURF [29], SENSURE [30], RootSIFT [31], and KAZE [32,33]; keypoint descriptors include SSD/NCC [34], census transform [35], SIFT [28], GLOH [36], SURF [29], DAISY[37], BRIEF [38], ORB [39], BRISK [40], and KAZE [32,33].

Data association (also called feature matching) is commonly conducted by comparing similarity measurements between keypoint descriptors along with a mutual consistency check procedure [21]. With a prior knowledge of motion constraints (e.g. from IMU sensors or constant velocity assumptions) or stereo's epipolar line constraints, the time used for data association can be shorten by restricting the searching space [17,41,42]. Due to the visual aliasing, wrong data associations (called outliers) are unavoidable to both short-term feature matching and long-term loop closure. Therefore, consensus set search (e.g. RANSAC) [43–48] and geometric information (i.e. pose and map estimations) from the back-end are usually adopted to remove outliers.

Since traditional back-end of V-SLAM systems relies on local iterative optimization algorithms (e.g. Gauss-Newton) [12], a fairly good initialization of variables (e.g. 6-DoF poses and 3D coordinate of keypoints) is required. The 6-DoF poses can be estimated via multiview geometry knowledge [18,34,49] using keypoints. 3D coordinate of keypoints are obtained by triangulation with careful keyframe selection [50,51].

3 Deep-Learned Features

Traditional hand-engineered keypoint extractors and matchers are not robust to viewpoint/illumination/seasonal change, which is crucial for long-term autonomy of mobile robots. While deep learning, especially the convolutional neural network (CNN) [52], is good at feature extraction and processing. Therefore, new trend about DL-based keypoint extractors and matchers are proposed to enhance the robustness of V-SLAM systems even under challenging conditions.

The DL-based keypoint extractors can be divided into three categories: detect-then-describe, jointly detect-and-describe, describe-to-detect. For the detect-then-describe method, keypoint detectors include TILDE [53], covariant feature detector [54], TCDet [55], MagicPoint [56], Quad-Networks [57], texture feature detector [58], and Key.Net [59]; keypoint descriptors include convex optimized descriptor [60], Deepdesc [61], TFeat [62], UCN [63], L2-Net [64], HardNet [65], averaging precision ranking [66], GeoDesc [67], RDRL [68], LogPolarDesc [69], ContextDesc [70], SOSNet [71], GIFT [72], and CAPS descriptor [73]. Jointly detect-and-describe methods include LIFT [74], DELE [75], LF-Net [76], SuperPoint [77], UnsuperPoint [78], GCNv2 [79], D2-Net [80], RF-Net [81], R2D2 [82], ASLFeat [83], and UR2KiD [84]. Different from the above two categories, Tian proposed a describe-to-detect (D2D) method [85], which selects keypoints based on the dense descriptors information.

After feature extracting, robust keypoint matching methods include content networks (CNe) [86], deep fundamental matrix estimation [87], NG-RANSAC [88], NM-Net [89], Order-Aware Network [90], ACNe [91], and SuperGlue [92].

Besides above extract-then-match methods, some end-to-end keypoint extract-and-match methods have also been proposed, they include NCNet [93], KP2D [94], Sparse-NCNet [95], LoFTR [96], and Patch2Pix [97].

Some datasets and benchmarks are released to train and evaluate different DL networks, such as Brown and Lowe's dataset [98], HPatches [99], matching in the dark (MID) dataset [100], image matching challenge [101], and long-term visual localization benchmark [102,103].

4 Conclusions

In this paper, we give a very concise survey of the front-end techniques of V-SLAM systems, emphasizing the DL-based keypoint extracting and matching techniques. Future research directions include extraction and association of high-level geometric features (e.g. lines [104,105], symmetry [106], and holistic 3D structures [107]) as well as semantic object-level features [108–110].

References

1. Cadena, C., et al.: Past, present, and future of simultaneous localization and mapping: toward the robust-perception age. IEEE Trans. Robot. **32**(6), 1309–1332 (2016)
2. Vidal, A.R., Rebecq, H., Horstschaefer, T., Scaramuzza, D.: Ultimate SLAM? Combining events, images, and IMU for robust visual SLAM in HDR and high-speed scenarios. IEEE Robot. Autom. Lett. **3**(2), 994–1001 (2018)
3. Microsoft HoloLens 2. http://www.microsoft.com/en-us/hololens
4. Davison, A.J.: FutureMapping: the computational structure of spatial AI systems. ArXiv Preprint arXiv:1803.11288 (2018)
5. Davison, A.J., Ortiz, J.: FutureMapping 2: Gaussian belief propagation for spatial AI. ArXiv Preprint arXiv:1910.14139 (2019)
6. LeCun, Y., Bengio, Y., Hinton, G.: Deep learning. Nature **521**, 436–444 (2015)
7. Bengio, Y., LeCun, Y., Hinton, G.: Deep learning for AI. Commun. ACM **64**(7), 58–65 (2021)
8. Kendall, A., Grimes, M., Cipolla, R.: PoseNet: a convolutional network for real-time 6-DOF camera relocalization. In: IEEE 12th International Conference on Computer Vision (ICCV), 7–13 December 2015, Santiago, Chile, pp. 1–9 (2015)
9. Wang, S., Clark, R., Wen, H., Trigoni, N.: DeepVO: towards end-to-end visual odometry with deep recurrent convolutional neural networks. In: IEEE International Conference on Robotics and Automation (ICRA), 29 May–3 June 2017, Singapore, pp. 1–8 (2017)
10. Yang, N., Stumberg, L.V., Wang, R., Cremers, D.: D3VO: deep depth, deep pose and deep uncertainty for monocular visual odometry. In: IEEE/CVF Conference on Computer Vision and Pattern Recognition (CVPR), 13–19 June 2020, Seattle, WA, USA, pp. 1–12 (2020)
11. Chen, C., Wang, B., Lu, C.X., Trigoni, N., Markham, A.: A survey on deep learning for localization and mapping: towards the age of spatial machine intelligence. ArXiv Preprint arXiv:2006.12567v2 (2020)
12. Wang, Y., Peng, X.: New trend in back-end techniques of visual SLAM: from local iterative solvers to robust global optimization. In: International Conference on Artificial Intelligence in China (CHINAAI), 21–22 August 2021, pp. 1–10 (2021)
13. Engel, J., Koltun, V., Cremers, D.: Direct sparse odometry. IEEE Trans. Pattern Anal. Mach. Intell. **40**(3), 611–625 (2018)

14. Makadia, A., Geyer, C., Daniilidis, K.: Correspondence-free structure from motion. Int. J. Comput. Vis. **75**(3), 311–327 (2007)
15. Moravec, H.: Obstacle avoidance and navigation in the real world by a seeing robot rover. Ph.D. dissertation, Standford University, USA (1980)
16. Cheng, Y., Maimone, M.W., Matthies, L.: Visual odometry on the mars exploration rovers. IEEE Robot. Automat. Mag. **13**(2), 54–62 (2006)
17. Maimone, M.W., Cheng, Y., Matthies, L.: Two years of visual odometry on the mars exploration rovers: field reports. J. Field Robot. **24**(3), 169–186 (2007)
18. Scaramuzza, D., Fraundorfer, F.: Visual odometry: part I: the first 30 years and fundamentals. IEEE Robot. Autom. Mag. **18**(4), 80–92 (2011)
19. Nister, D., Naroditsky, O., Bergen, J.: Visual odometry. In: IEEE Conference on Computer Vision and Pattern Recognition (CVPR), 27 June–2 July 2004, Washington, DC, USA, pp. 1–8 (2004)
20. Mur-Artal, R., Tardós, J.D.: ORB-SLAM2: an open-source SLAM system for monocular, stereo, and RGB-D cameras. IEEE Trans. Robot. **33**(5), 1255–1262 (2017)
21. Fraundorfer, F., Scaramuzza, D.: Visual odometry: part II: matching, robustness, optimization, and applications. IEEE Robot. Autom. Mag. **19**(2), 78–90 (2012)
22. Moravec, H.: Towards automatic visual obstacle avoidance. In: 5th International Joint Conference on Artificial Intelligence, p. 584, August 1977
23. Forstner, W.: A feature based correspondence algorithm for image matching. Int. Arch. Photogram. Remote Sens. **26**(3), 150–166 (1986)
24. Harris, C., Pike, J.: 3D positional integration from image sequences. In: 3rd Alvey Vision Conference, Cambridge, pp. 233–236, September 1987
25. Tomasi, C., Shi, J.: Good features to track. In: IEEE Conference on Computer Vision and Pattern Recognition (CVPR), 21–23 June 1994, Seattle, WA, USA, pp. 1–8 (1994)
26. Rosten, E., Drummond, T.: Machine learning for high-speed corner detection. In: Leonardis, A., Bischof, H., Pinz, A. (eds.) ECCV 2006. LNCS, vol. 3951, pp. 430–443. Springer, Heidelberg (2006). https://doi.org/10.1007/11744023_34
27. Rosten, E., Porter, R., Drummond, T.: Faster and better: a machine learning approach to corner detection. IEEE Trans. Pattern Anal. Mach. Intell. **32**(1), 105–119 (2010)
28. Lowe, D.: Distinctive image features from scale-invariant keypoints. Int. J. Comput. Vis. **20**(2), 91–110 (2003)
29. Bay, H., Tuytelaars, T., Van Gool, L.: SURF: speeded up robust features. In: Leonardis, A., Bischof, H., Pinz, A. (eds.) ECCV 2006. LNCS, vol. 3951, pp. 404–417. Springer, Heidelberg (2006). https://doi.org/10.1007/11744023_32
30. Agrawal, M., Konolige, K., Blas, M.R.: CenSurE: center surround extremas for realtime feature detection and matching. In: Forsyth, D., Torr, P., Zisserman, A. (eds.) ECCV 2008. LNCS, vol. 5305, pp. 102–115. Springer, Heidelberg (2008). https://doi.org/10.1007/978-3-540-88693-8_8
31. Arandjelovic, R., Zisserman, A.: Three things everyone should know to improve object retrieval. In: IEEE Conference on Computer Vision and Pattern Recognition (CVPR), 16–21 June 2012, Providence, RI, USA, pp. 1–8 (2012)
32. Alcantarilla, P.F., Bartoli, A., Davison, A.J.: KAZE features. In: Fitzgibbon, A., Lazebnik, S., Perona, P., Sato, Y., Schmid, C. (eds.) ECCV 2012. LNCS, vol. 7577, pp. 214–227. Springer, Heidelberg (2012). https://doi.org/10.1007/978-3-642-33783-3_16

33. Alcantarilla, P.F., Nuevo, J., Bartoli, A.: Fast explicit diffusion for accelerated features in nonlinear scale spaces. In: British Machine Vision Conference (BMVC), 9–13 September 2013, Bristol, UK, pp. 1–11 (2013)
34. Ma, Y., Soatto, S., Kosĕcká, J., Sastry, S.S.: An Invitation to 3-D Vision: From Images to Geometric Models. Springer, New York (2004). https://doi.org/10.1007/978-0-387-21779-6
35. Zabih, R., Woodfill, J.: Non-parametric local transforms for computing visual correspondence. In: Eklundh, J.-O. (ed.) ECCV 1994. LNCS, vol. 801, pp. 151–158. Springer, Heidelberg (1994). https://doi.org/10.1007/BFb0028345
36. Mikolajczyk, K., Schmid, C.: A performance evaluation of local descriptors. IEEE Trans. Pattern Anal. Mach. Intell. **27**(10), 1615–1630 (2005)
37. Tola, E., Lepetit, V., Fua, P.: DAISY: an efficient dense descriptor applied to wide-baseline stereo. IEEE Trans. Pattern Anal. Mach. Intell. **32**(5), 815–830 (2010)
38. Calonder, M., Lepetit, V., Strecha, C., Fua, P.: BRIEF: binary robust independent elementary features. In: Daniilidis, K., Maragos, P., Paragios, N. (eds.) ECCV 2010. LNCS, vol. 6314, pp. 778–792. Springer, Heidelberg (2010). https://doi.org/10.1007/978-3-642-15561-1_56
39. Rublee, E., Rabaud, V., Konolige, K., Bradski, G.: ORB: an efficient alternative to SIFT or SURF. In: IEEE International Conference on Computer Vision (ICCV), 6–13 November 2011, Barcelona, Spain, pp. 2564–2571 (2011)
40. Leutenegger, S., Chli, M., Siegwart, R.: BRISK: binary robust invariant scalable keypoints. In: IEEE International Conference on Computer Vision (ICCV), 6–13 November 2011, Barcelona, Spain, pp. 2548–2555 (2011)
41. Davison, A.: Real-time simultaneous localization and mapping with a single camera. In: IEEE International Conference on Computer Vision (ICCV), 14–17 October 2003, Nice, France, pp. 1403–1410 (2003)
42. Scharstein, D., Szeliski, R.: A taxonomy and evaluation of dense two-frame stereo correspondence algorithms. Int. J. Comput. Vis. **47**(1–3), 7–42 (2002)
43. Fishler, M.A., Bolles, R.C.: Random sample consensus: a paradigm for model fitting with applications to image analysis and automated cartography. Commun. ACM **24**, 381–395 (1981)
44. Latif, Y., Cadena, C., Neira, J.: Robust loop closing over time for pose graph SLAM. Intl. J. Robot. Res. **32**(14), 1611–1626 (2013)
45. Mangelson, J.G., Dominic, D., Eustice, R.M., Vasudevan, R.: Pairwise consistent measurement set maximization for robust multi-robot map merging. In: IEEE International Conference on Robotics and Automation (ICRA), 21–25 May 2018, Brisbane, Australia, pp. 1–8 (2018)
46. Yang, J., Huang, Z., Quan, S., Qi, Z., Zhang, Y.: SAC-COT: sample consensus by sampling compatibility triangles in graphs for 3-D point cloud registration. IEEE Trans. Geosci. Remote Sens. (Early Access)
47. Yang, H., Antonante, P., Tzoumas, V., Carlone, L.: Graduated non-convexity for robust spatial perception: from non-minimal solvers to global outlier rejection. IEEE Robot. Autom. Lett. **5**(2), 1127–1134 (2020). (Best Paper Award Honorable Mention)
48. Shi, J., Yang, H., Carlone, L.: ROBIN: a graph-theoretic approach to reject outliers in robust estimation using invariants. In: IEEE International Conference on Robotics and Automation (ICRA), 30 May–5 June 2021, Xi'an, China, pp. 1–16 (2021)
49. Hartley, R., Zisserman, A.: Multiple View Geometry in Computer Vision, 2nd edn. Cambridge University Press, UK (2003)

50. Mourikis, A.I., Roumeliotis, S.I.: A multi-state constraint Kalman filter for vision-aided inertial navigation. In: IEEE International Conference on Robotics and Automation (ICRA), 10–14 April 2007, Roma, Italy, pp. 1–8 (2007)
51. Leutenegger, S., Lynen, S., Bosse, M., Siegwart, R., Furgale, P.: Keyframe-based visual-inertial odometry using nonlinear optimization. Intl. J. Robot. Res. **34**(3), 314–334 (2015)
52. LeCun, Y., Bottou, L., Bengio, Y., Haffner, P.: Gradient-based learning applied to document recognition. Proc. IEEE **86**(11), 2278–2324 (1998)
53. Verdie, Y., Yi, K.M., Fua, P., Lepetit, V.: TILDE: a temporally invariant learned DEtector. In: IEEE/CVF Conference on Computer Vision and Pattern Recognition (CVPR), 7–12 June 2015, Boston, MA, USA, pp. 1–10 (2015)
54. Lenc, K., Vedaldi, A.: Learning covariant feature detectors. In: Hua, G., Jégou, H. (eds.) ECCV 2016. LNCS, vol. 9915, pp. 100–117. Springer, Cham (2016). https://doi.org/10.1007/978-3-319-49409-8_11
55. Zhang, X., Yu, F.X., Karaman, S., Chang, S.-F.: Learning discriminative and transformation covariant local feature detectors, In: IEEE/CVF Conference on Computer Vision and Pattern Recognition (CVPR), 21–26 July 2017, Honolulu, HI, USA, pp. 1–9 (2017)
56. DeTone, D., Malisiewicz, T., Rabinovich, A.: Toward geometric deep SLAM. ArXiv Preprint arXiv:1707.07410v1 (2017)
57. Savinov, N., Seki, A., Ladický, L., Sattler, T., Pollefeys, M.: Quad-networks: unsupervised learning to rank for interest point detection. In: IEEE/CVF Conference on Computer Vision and Pattern Recognition (CVPR), 21–26 July 2017, Honolulu, HI, USA, pp. 1–9 (2017)
58. Zhang, L., Rusinkiewicz, S.: Learning to detect features in texture images. In: IEEE/CVF Conference on Computer Vision and Pattern Recognition (CVPR), 18–22 June 2018, Salt Lake City, UT, USA, pp. 1–9 (2018)
59. Laguna, A.B., Riba, E., Ponsa, D., Mikolajczyk, K.: Key.Net: keypoint detection by handcrafted and learned CNN filters. In: IEEE/CVF International Conference on Computer Vision (ICCV), 27 October–2 November 2019, Seoul, Korea (South), pp. 1–9 (2019)
60. Simonyan, K., Vedaldi, A., Zisserman, A.: Learning local feature descriptors using convex optimisation. IEEE Trans. Pattern Anal. Mach. Intell. **36**(8), 1573–1585 (2014)
61. Simo-Serra, E., Trulls, E., Ferraz, L. Kokkinos, I., Fua, P., Moreno-Noguer, F.: Discriminative learning of deep convolutional feature point descriptors. In: IEEE International Conference on Computer Vision (ICCV), 7–13 December 2015, Santiago, Chile, pp. 1–9 (2015)
62. Balntas, V., Riba, E., Ponsa, D., Mikolajczyk, K.: Learning local feature descriptors with triplets and shallow convolutional neural networks. In: British Machine Vision Conference (BMVC), 19–22 September 2016, York, UK, pp. 1–11 (2016)
63. Choy, C.B., Gwak, J., Savarese, S., Chandraker, M.: Universal correspondence network. In: Conference on Neural Information Processing Systems (NeurIPS), 5–10 December 2016, Barcelona, Spain, pp. 1–9 (2016)
64. Tian, Y., Fan, B., Wu, F.: L2-net: deep learning of discriminative patch descriptor in euclidean space. In: IEEE/CVF Conference on Computer Vision and Pattern Recognition (CVPR), 21–26 July 2017, Honolulu, HI, USA, pp. 1–9 (2017)
65. Mishchuk, A., Mishkin, D., Radenović, F., Matas, J.: Working hard to know your neighbor's margins: local descriptor learning loss. In: Conference on Neural Information Processing Systems (NeurIPS), 4–9 December 2017, Long Beach, NY, USA, pp. 1–12 (2017)

66. He, K., Lu, Y., Sclaroff, S.: Local descriptors optimized for averaging precision. In: IEEE/CVF Conference on Computer Vision and Pattern Recognition (CVPR), 18–22 June 2018, Salt Lake City, UT, USA, pp. 1–10 (2018)
67. Luo, Z., et al.: GeoDesc: learning local descriptors by integrating geometry constraints. In: European Conference on Computer Vision (ECCV), 8–14 September 2018, Munich, Germany, pp. 170–185 (2018)
68. Yu, X., et al.: Unsupervised extraction of local image descriptors via relative distance ranking loss. In: IEEE/CVF International Conference on Computer Vision Workshops (ICCVW), 27 October–2 November 2019, Seoul, Korea (South), pp. 1–10 (2019)
69. Ebel, P., Trulls, E., Yi, K.M., Fua, P., Mishchuk, A.: Beyond cartesian representations for local descriptors. In: IEEE/CVF International Conference on Computer Vision (ICCV), 27 October–2 November 2019, Seoul, Korea (South), pp. 1–10 (2019)
70. Luo, Z., et al.: ContextDesc: local descriptor augmentation with cross-modality context. In: IEEE/CVF Conference on Computer Vision and Pattern Recognition (CVPR), 16–20 June 2019, Long Beach, CA, USA, pp. 1–10 (2019)
71. Tian, Y., Yu, X., Fan, B., Wu, F., Heijnen, H., Balntas, V.: SOSNet: second order similarity regularization for local descriptor learning. 16–20 June 2019, Long Beach, CA, USA, pp. 1–10 (2019)
72. Liu, Y., Shen, Z., Lin, Z., Peng, S., Bao, H., Zhou, X.: GIFT: learning transformation-invariant dense visual descriptors via group CNNs. In: Conference on Neural Information Processing Systems (NeurIPS), 8–14 December 2019, Vancouver, Canada, pp. 1-12 (2019)
73. Wang, Q., Zhou, X., Hariharan, B., Snavely, N.: Learning feature descriptors using camera pose supervision. In: Vedaldi, A., Bischof, H., Brox, T., Frahm, J.-M. (eds.) ECCV 2020. LNCS, vol. 12346, pp. 757–774. Springer, Cham (2020). https://doi.org/10.1007/978-3-030-58452-8_44
74. Yi, K.M., Trulls, E., Lepetit, V., Fua, P.: LIFT: learned invariant feature transform. In: Leibe, B., Matas, J., Sebe, N., Welling, M. (eds.) ECCV 2016. LNCS, vol. 9910, pp. 467–483. Springer, Cham (2016). https://doi.org/10.1007/978-3-319-46466-4_28
75. Noh, H., Araujo, A., Sim, J., Weyand, T., Han, B.: Large-scale image retrieval with attentive deep local features. In: IEEE International Conference on Computer Vision (ICCV), 22–29 October 2017, Venice, Italy, pp. 1–10 (2017)
76. Ono, Y., Trulls, E., Fua, P., Yi, K.M.: LF-net: learning local features from images. In: Conference on Neural Information Processing Systems (NeurIPS), 2–8 December 2018, Montreal, Canada, pp. 1–11 (2018)
77. DeTone, D., Malisiewicz, T., Rabinovich, A.: SuperPoint: self-supervised interest point detection and description. In: IEEE/CVF Conference on Computer Vision and Pattern Recognition Workshops (CVPRW), 18–22 June 2018, Salt Lake City, UT, USA, pp. 1–13 (2018)
78. Christiansen, P.H., Kragh, M.F., Brodskiy, Y., Karstoft, H.: UnsuperPoint: end-to-end unsupervised interest point detector and descriptor. ArXiv Preprint arXiv:1907.04011 (2019)
79. Tang, J., Ericson, L., Folkesson, J., Jensfelt, P.: GCNv2: efficient correspondence prediction for real-time SLAM. IEEE Robot. Autom. Lett. **4**(4), 3505–3512 (2019)
80. Dusmanu, M., et al.: D2-Net: a trainable cnn for joint description and detection of local features. In: IEEE/CVF Conference on Computer Vision and Pattern Recognition (CVPR), 16–20 June 2019, Long Beach, CA, USA, pp. 1–10 (2019)

81. Shen, X., et al.: RF-net: an end-to-end image matching network based on receptive field. In: IEEE/CVF Conference on Computer Vision and Pattern Recognition (CVPR), 16–20 June 2019, Long Beach, CA, USA, pp. 1–9 (2019)
82. Revaud, J., Weinzaepfel, P., Souza, C.D., Humenberger, M.: R2D2: repeatable and reliable detector and descriptor. In: Conference on Neural Information Processing Systems (NeurIPS), 8–14 December 2019, Vancouver, Canada, pp. 1–11 (2019)
83. Luo, Z., et al.: ASLFeat: learning local features of accurate shape and localization. In: IEEE/CVF Conference on Computer Vision and Pattern Recognition (CVPR), 14–19 June 2020, pp. 1–10 (2020)
84. Yang, T.-Y., Nguyen, D.-K., Heijnen, H., Balntas, V.: UR2KiD: unifying retrieval, keypoint detection, and keypoint description without local correspondence supervision. ArXiv Preprint arXiv:2001.07252 (2020)
85. Tian, Y., Balntas, V., Ng, T., Barroso-Laguna, A., Demiris, Y., Mikolajczyk, K.: D2D: keypoint extraction with describe to detect approach. In: Asian Conference on Computer Vision (ACCV), 30 November–4 December 2020, pp. 1–18 (2020)
86. Yi, K.M., Trulls, E., Ono, Y., Lepetit, V., Salzmann, M., Fua, P.: Learning to find good correspondences. In: IEEE/CVF Conference on Computer Vision and Pattern Recognition (CVPR), 18–22 June 2018, Salt Lake City, UT, USA, pp. 1–9 (2018)
87. Ranftl, R., Koltun, V.: Deep fundamental matrix estimation. In: European Conference on Computer Vision (ECCV), 8–14 September 2018, Munich, Germany, pp. 292–309 (2018)
88. Brachmann, E., Rother, C.: Neural-guided RANSAC: learning where to sample model hypotheses. In: IEEE/CVF International Conference on Computer Vision (ICCV), 27 October–2 November, Seoul, Korea (South), pp. 1–10 (2019)
89. Zhao, C., Cao, Z., Li, C., Li, X., Yang, J.: NM-net: mining reliable neighbors for robust feature correspondences. In: IEEE/CVF Conference on Computer Vision and Pattern Recognition (CVPR), 15–20 June 2019, Long Beach, CA, USA, pp. 1–10 (2019)
90. Zhang, J., et al.: Learning two-view correspondences and geometry using order-aware network. In: IEEE/CVF International Conference on Computer Vision (ICCV), 27 October–2 November 2019, Seoul, Korea (South), pp. 1–10 (2019)
91. Sun, W., Jiang, W., Trulls, E., Tagliasacchi, A., Yi, K.M.: ACNe: attentive context normalization for robust permutation-equivariant learning. In: IEEE/CVF Conference on Computer Vision and Pattern Recognition (CVPR), 14–19 June 2020, pp. 1–10 (2020)
92. Sarlin, P.-E., DeTone, D., Malisiewicz, T., Rabinovich, A.: SuperGlue: learning feature matching with graph neural networks. In: IEEE/CVF Conference on Computer Vision and Pattern Recognition (CVPR), 14–19 June 2020, pp. 1–10 (2020)
93. Rocco, I., Cimpoi, M., Arandjelović, R., Torii, A., Pajdla, T., Sivic, J.: Neighboorhood consensus networks. In: Conference on Neural Information Processing Systems (NeurIPS), 2–8 December 2018, Montreal, Canada, pp. 1–12 (2018)
94. Tang, J., Kim, H., Guizilini, V., Pillai, S., Ambrus, R.: Neural outlier rejection for self-supervised keypoint learning. In: International Conference on Learning Representations (ICLR), 26 April–1 May 2020, pp. 1–14 (2020)
95. Rocco, I., Arandjelović, R., Sivic, J.: Efficient neighbourhood consensus networks via submanifold sparse convolutions. In: Vedaldi, A., Bischof, H., Brox, T., Frahm, J.-M. (eds.) ECCV 2020. LNCS, vol. 12354, pp. 605–621. Springer, Cham (2020). https://doi.org/10.1007/978-3-030-58545-7_35

96. Sun, J., Shen, Z., Wang, Y., Bao, H., Zhou, X.: LoFTR: detector-free local feature matching with transformers. In: IEEE/CVF Conference on Computer Vision and Pattern Recognition (CVPR), 19–25 June 2021, pp. 1–10 (2021)
97. Zhou, Q., Sattle, T., Leal-Taixé, L.: Patch2Pix: epipolar-guided pixel-level correspondences. In: IEEE/CVF Conference on Computer Vision and Pattern Recognition (CVPR), 19–25 June 2021, pp. 1–10 (2021)
98. Brown, M., Lowe, D.: Automatic panoramic image stitching using invariant features. Int. J. Comput. Vis. **74**, 59–73 (2007)
99. Balntas, V., Lenc, K., Vedaldi, A.: A benchmark and evaluation of handcrafted and learned local descriptor. In: IEEE/CVF Conference on Computer Vision and Pattern Recognition (CVPR), 21–26 July 2017, Honolulu, HI, USA, pp. 1–10 (2017)
100. Song, W., Suganuma, M., Liu, X.: Matching in the dark: a dataset for matching image pairs of low-light scenes. ArXiv Preprint arXiv:2109.03585 (2021)
101. Jin, Y., et al.: Image matching across wide baseline: from paper to practice. Int. J. Comput. Vis. **129**, 517–547 (2021)
102. Sattler, T., et al.: Benchmarking 6DOF outdoor visual localization in changing conditions. In: IEEE/CVF Conference on Computer Vision and Pattern Recognition (CVPR), 18–22 June 2018, Salt Lake City, UT, USA, pp. 1–10 (2018)
103. Toft, C., et al.: Long-term visual localization: revisited. IEEE Trans. Pattern Anal. Mach. Intell. (Early Access)
104. Zhou, Y., Qi, H., Ma, Y.: End-to-end wireframe parsing. In: IEEE/CVF International Conference on Computer Vision (ICCV), 27 October–2 November 2019, Seoul, Korea (South), pp. 1–10 (2019)
105. Dai, X., Yuan, X., Gong, H., Ma, Y.: Fully convolutional line parsing. ArXiv Preprint arXiv:2104.11207 (2021)
106. Zhou, Y., Liu, S., Ma, Y.: NeRD: neural 3D reflection symmetry detector. In: IEEE/CVF Conference on Computer Vision and Pattern Recognition (CVPR), 19–25 June 2021, pp. 1–10 (2021)
107. Zhou, Y., et al.: HoliCity: a city-scale data platform for learning holistic 3D structures. ArXiv Preprint arXiv: 2008.03286 (2020)
108. Salas-Moreno, R.F., Newcombe, R.A., Strasdat, H., Kelly, P.H.J, Davison, A.J.: SLAM++: simultaneous localisation and mapping at the level of objects. In: IEEE Conference on Computer Vision and Pattern Recognition (CVPR), 23–28 June 2013, Portland, OR, USA, pp. 1–8 (2013)
109. Mccormac, J., Clark, R., Bloesch, M., Davison, A., Leutenegger, S.: Fusion++: volumetric object-level SLAM. In: International Conference on 3D Vision (3DV), 5–8 September 2018, Verona, Italy, pp. 1–10 (2018)
110. Sucar, E., Wada, K., Davison, A.: NodeSLAM: neural object descriptors for multi-view shape reconstruction. In: International Conference on 3D Vision (3DV), 25–28 November 2020, Fukuoka, Japan, pp. 1–10 (2020)

New Trend in Back-End Techniques of Visual SLAM: From Local Iterative Solvers to Robust Global Optimization

Yue Wang[1,2(✉)] and Xiaodong Peng[1]

[1] Key Laboratory of Electronics and Information Technology for Space Systems, Chinese Academy of Sciences, Beijing 100190, China
wangyue@nssc.ac.cn

[2] Tianjin Key Laboratory of Wireless Mobile Communications and Power Transmission, Tianjin Normal University, Tianjin 300387, China

Abstract. Visual simultaneous localization and mapping (V-SLAM) technique plays a key role in perception of autonomous mobile robots, augmented/mixed/virtual reality, as well as high-level geometric perception and spatial AI applications. This paper gives a very concise survey about the back-end module of a V-SLAM system, which is essentially a nonlinear least square (NLLS) problem. This problem is traditionally solved by a local iterative linearized optimization algorithm with a good initial guess, such as the Gauss-Newton algorithm. Due to the nonconvexity of the NLLS problem, these local iterative solvers cannot provide any guarantee on the global minimum convergence, which is crucial for active SLAM systems and life-critical applications, such as autonomous driving. Therefore, new trend about robust global optimization algorithms for the pose graph are introduced, which adopt duality theory, convex relaxation, and robust cost functions to provide a certified global optimal solution.

Keywords: 3D vision · Convex relaxation · Global optimization · Lagrangian duality · Robust cost function · Simultaneous localization and mapping (SLAM)

1 Introduction

The simultaneous localization and mapping (SLAM) system simultaneously estimates its state using on-board sensors and constructs a consistent environmental map [1]. This perceptual capability is essential for autonomous mobile

This work was supported by the Pre-Research Project of Space Science (No. XDA15014700), the National Natural Science Foundation of China (No. 61601328), the Scientific Research Plan Project of the Committee of Education in Tianjin (No. JW1708), and the Doctor Foundation of Tianjin Normal University (No. 52XB1417).

Q. Liang et al. (Eds.): Artificial Intelligence in China, LNEE 854, pp. 308–317, 2022.
https://doi.org/10.1007/978-981-16-9423-3_39

robots, supporting functions as planning, navigation, and control, as well as augmented/mixed/virtual reality devices. SLAM is also a fundamental problem to achieve high-level spatial perception [2–4] and spatial AI [5–11] applications.

Since SLAM is an umbrella concept which can utilize multimodal sensors and their fusion, such as camera, inertial measurement unit (IMU), lidar, radar, and GPS. In this paper, we focus on the visual SLAM (V-SLAM) system, which adopts cameras (usually with IMUs) to inference its own state (e.g. pose) and build a consistent surrounding map [1,12–15]. V-SLAM systems are widely used by autonomous mobile robots, especially the resource-limited agile micro drones [16], as well as augmented/mixed/virtual reality devices [17]. Three similar concepts related to V-SLAM are visual-inertial navigation systems (VINS) [18], visual/visual-inertial odometry (VO/VIO) [18–20] and structure from motion (SfM) [21,22]. VINS is almost identical to a V-SLAM system with IMU sensors; VO/VIO can be considered a simplified V-SLAM system without loop closure, relocalization and mapping ability; and the V-SLAM system without IMU sensors can be considered as a special SfM system making real-time estimations based on sequential ordered images.

Through over 30 years development, we saw classical age (1986–2004) and algorithmic-analysis age (2004–2015) of V-SLAM techniques, now we arrive at the robust-perception age [1]. There are a lot of successful open-source V-SLAM systems, such as the sparse ORB-SLAM system [23–25], the semi-dense SVO system [26,27], and the dense DSO system [28]. Meanwhile, new types of cameras, such as range/depth cameras [29], light-field/plenoptic cameras [30], thermo cameras [31], and event cameras [32,33], have been adopted for some V-SLAM systems. Recently, with the huge tide of deep learning (DL) techniques [34,35], some end-to-end DL-based V-SLAM/VO systems have been proposed [36–39]. In this paper, we focus on the traditional V-SLAM system with standard cameras. The de-facto framework of a traditional V-SLAM system is given in Fig. 1, which contains two modules: a front-end and a back-end. The front-end is responsible for processing visual and IMU data, including feature extraction, short/long-term data association with outlier rejection, as well as initial estimation of the current pose and newly detected landmark positions [40]. The back-end takes the initialization information from the front-end, and makes a maximum-a-posteriori (MAP) estimation based on a factor graph to refine both poses and landmark positions. This paper focuses on the new trend in back-end techniques of V-SLAM systems.

2 Local Iterative Solvers

Generally, the back-end is responsible for constructing and solving a MAP problem, which can be converted to a nonlinear least squares (NLLS) problem [1,41]. Historically, during the early development period of V-SLAM systems, due to the limited on-board resource, various filtering techniques were adopted to realize real-time V-SLAM systems [42,43], such as extend Kalman filter (EKF) [44–46], iterative EKF (iEKF) [47], and multi-state constraint Kalman filter (MSCKF)

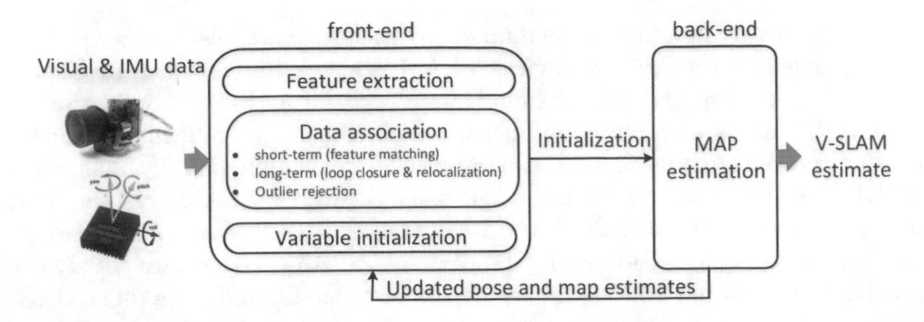

Fig. 1. The de-facto framework of a V-SLAM system.

[48]. Then, keyframe-based V-SLAM systems were introduced using either keypoints or direct methods [49,50] and became the predominate choice since they have more accurate results compared with filter-based methods [51].

Local iterative linearized optimization solvers to the NLLS problem in Euclidean Space were mature, such as Gauss-Newton (G-N) and Levenberg-Marquardt (L-M) algorithms [52], and they were extended to the optimization problem on smooth manifolds using Lie theory [53–55]. However, the high computation cost of large-scale matrix manipulations kept these algorithms away from real-time V-SLAM systems. By introducing factor graphs [56,57] to construct the NLLS problem, Dellaert and Kaess found its sparse structure and adopted sparse matrix factorization to realize optimization-based real-time solvers [58]. They then utilized temporal structure of the factor graph and Bayes trees to realize incremental NLLS inference [59,60]. After that, several open-source graph optimization libraries were provided [61–64]. Reader are recommended to reference [65–68] for detailed information. Based on these theoretical breakthroughs as well as the IMU preintegration method [69], optimization-based inference algorithms (also called smoothers) were widely adopted to tackle the real-time NLLS problem, such as fixed-lag smoother [70] and sliding-window keyframe-based V-SLAM systems [71,72]. Recently, it has been mathematically proved that various filtering techniques are equivalent to the sliding-window optimization techniques with different number of G-N iterations combined with different marginalization strategies [41,73].

Considering the nonconvexity of the NLLS problem, local iterative algorithms (e.g. G-N or L-M) can only converge to a local minima, thus a fairly good initialization of variables is required to attain the global minimum, and different initialization methods had been proposed in [74–76]. While these heuristics are often effective in practice, they cannot provide any guarantee on global minimum convergence, which is crucial for active SLAM systems [77] and life-critical applications, such as autonomous driving.

3 Robust Global Optimization

The nonconvexity of the NLLS problem in V-SLAM systems and the local minima problem associated with traditional local iterative solvers were initially discussed by Huang and his colleagues [78–81], Carlone gave a convergence analysis of the G-N algorithm [82], and a theoretical review was given by Huang and Dissanayake [83]. Meanwhile, Carlone and his colleagues tried to address the nonconvexity problem and empirically demonstrated the strong duality for the 2D/3D pose graph optimization (PGO) problem, which is the problem of estimating a set of poses from pairwise relative measurements, and they developed global optimal verification techniques [84–87]. Then, Rosen and his colleagues mathematically proved the strong duality of the PGO problem under sufficiently small noise conditions, and developed a certifiably global optimal algorithm (called SE-Sync) based on the low-rank Riemannian semidefinite optimization [88–91].

Since outliers (usually caused by wrong data association with visual aliasing) can dramatically degrade the performance of optimization algorithms, consensus set search with minimal solvers were usually adopted to remove outliers of loop closures [92–95]. For non-minimal solvers, differentiable robust cost functions were usually adopted to achieve robust estimations [96–100]. For high outlier-rate (e.g., >60%) scenarios, Yang and his colleagues proposed two algorithms to provide the certifiable global optimal solution. One is called graduated non-convexity (GNC) algorithm [101,102], which can often succeeds in providing global optimal solution for non-minimal solvers. The other algorithm adopts truncated least squares estimation and polynomial optimization to provide the certifiable optimal solution [102–104].

The robust global optimization algorithms [105] introduced above had also been adopted to tackle the rotation averaging problem [106–108] as well as multi-robot distributed V-SLAM problems [109–111].

4 Conclusions

In this paper, we give a very concise survey of the back-end techniques of V-SLAM systems, emphasizing the robust global optimization algorithms. Future research directions include the generalization from PGO to optimization on arbitrary factor graphs, fast and incremental semidefinite programming algorithms, and the fusion between deep learning and Bayesian inference frameworks.

References

1. Cadena, C., et al.: Past, present, and future of simultaneous localization and mapping: toward the robust-perception age. IEEE Trans. Robot. **32**(6), 1309–1332 (2016)
2. Rosinol, A., Abate, M., Chang, Y., Carlone, L.: Kimera: an open-source library for real-time metric-semantic localization and mapping. In: IEEE International Conference on Robotics and Automation (ICRA), Paris, France, 31 May–31 August, pp. 1–8 (2020)

3. Rosinol, A., et al.: Kimera: from SLAM to spatial perception with 3D dynamic scene graphs. ArXiv Preprint arXiv:2101.06894 (2021)
4. Talak, R., Hu, S., Peng, L., Carlone, L.: Neural trees for learning on graphs. ArXiv Preprint arxiv:2105.07264v1 (2021)
5. Davison, A.J.: FutureMapping: the computational structure of spatial AI systems. ArXiv Preprint arXiv:1803.11288 (2018)
6. Davison, A.J., Ortiz, J.: FutureMapping 2: Gaussian belief propagation for spatial AI. ArXiv Preprint arXiv:1910.14139 (2019)
7. Wada, K., Sucar, E., James, S., Lenton, D., Davison, A.J.: MoreFusion: multi-object reasoning for 6D pose estimation from volumetric fusion. In: IEEE/CVF Conference on Computer Vision and Pattern Recognition (CVPR), Seattle, WA, USA, 13–19 June, pp. 1–10 (2020)
8. Sucar, E., Wada, K., Davison, A., NodeSLAM: neural object descriptors for multi-view shape reconstruction. In: International Conference on 3D Vision (3DV), 25–28 Nov, pp. 1–10 (2020)
9. Czarnowski, J., Laidlow, T., Clark, R., Davison, A.J.: DeepFactors: real-time probabilistic dense monocular SLAM. IEEE Robot. Autom. Lett. **5**(2), 721–728 (2020)
10. Ortiz, J., Pupilli, M., Leutenegger, S., Davison, A.J.: Bundle adjustment on a graph processor. In: IEEE/CVF Conference on Computer Vision and Pattern Recognition (CVPR), Seattle, WA, USA, 13–19 June, pp. 1–10 (2020)
11. Ortiz, J., Evans, T., Davison, A.J.: A visual introduction to Gaussian belief propagation. ArXiv Preprint arXiv:2107.02308 (2021)
12. Ma, Y., Soatto, S., Kosěcká, J., Sastry, S.S.: An Invitation to 3-D Vision: From Images to Geometric Models. Springer, New York (2004). https://doi.org/10.1007/978-0-387-21779-6
13. Hartley, R., Zisserman, A.: Multiple View Geometry in Computer Vision, 2nd edn. Cambridge University Press, Cambridge (2003)
14. Wu, Y., Tang, F., Li, H.: Image-based camera localization: an overview. Visual Comput. Ind. Biomed. Art **2018**(8), 1–13 (2018)
15. Long, X.X., et al.: Recent progress in 3D vision. J. Image Graph. **26**(6), 1389–1428 (2021)
16. Vidal, A.R., Rebecq, H., Horstschaefer, T., Scaramuzza, D.: Ultimate SLAM? Combining events, images, and IMU for robust visual SLAM in HDR and high-speed scenarios. IEEE Robot. Autom. Lett. **3**(2), 994–1001 (2018)
17. Microsoft HoloLens 2. http://www.microsoft.com/en-us/hololens
18. Huang, G.: Visual-inertial navigation: a concise review. In: IEEE International Conference on Robotics and Automation (ICRA), Montreal, QC, Canada, 20–24 May, pp. 1–11 (2019)
19. Scaramuzza, D., Fraundorfer, F.: Visual odometry: part I: the first 30 years and fundamentals. IEEE Robot. Autom. Mag. **18**(4), 80–92 (2011)
20. Fraundorfer, F., Scaramuzza, D.: Visual odometry: part II: matching, robustness, optimization, and applications. IEEE Robot. Autom. Mag. **19**(2), 78–90 (2012)
21. Agarval, S., Snavely, N., Simon, I., Seitz, S.M., Szeliski, R.: Building Rome in a Day. In: IEEE International Conference on Computer Vision (ICCV), Kyoto, Japan, 29 Sept.–2 Oct., pp. 1–8 (2009)
22. Ozyesil, O., Voroninski, V., Basri, R., Singer, A.: A survey of structure from motion. Acta Numer. **26**, 305–364 (2017)
23. Mur-Artal, R., Montiel, J.M.M., Tardós, J.D.: ORB-SLAM: a versatile and accurate monocular SLAM system. IEEE Trans. Robot. **31**(5), 1147–1163 (2015)

24. Mur-Artal, R., Tardós, J.D.: ORB-SLAM2: an open-source SLAM system for monocular, stereo, and RGB-D cameras. IEEE Trans. Robot. **33**(5), 1255–1262 (2017)
25. Campos, C., Elvira, R., Rodríguez, J.J.G., Montiel, J.M.M., Tardós, J.D.: ORB-SLAM3: an accurate open-source library for visual, visual-inertial, and multimap SLAM. IEEE Trans. Robot. (Early Access)
26. Forster, C., Pizzoli, M., Scaramuzza, D.: SVO: fast semi-direct monocular visual odometry. In: IEEE International Conference on Robotics and Automation (ICRA), Hong Kong, China, 31 May–7 June, pp. 1–8 (2014)
27. Forster, C., Zhang, Z., Gassner, M., Werlberger, M., Scaramuzza, D.: SVO: semidirect visual odometry for monocular and multicamera systems. IEEE Trans. Robot. **33**(2), 249–265 (2017)
28. Engel, J., Koltun, V., Cremers, D.: Direct sparse odometry. IEEE Trans. Pattern Anal. Mach. Intell. **40**(3), 611–625 (2018)
29. Kerl, C., Sturm, J., Cremers, D.: Dense visual SLAM for RGB-D cameras. In: IEEE/RSJ International Conference on Intelligent Robots and Systems (IROS), Tokyo, Japan, 3–7 November, pp. 1–7 (2013)
30. Zeller, N., Quint, F., Stilla, U.: From the calibration of a light-field camera to direct plenoptic odometry. IEEE J. Sel. Top. Signal Process. **11**(7), 1004–1019 (2017)
31. Saputra, M.R.U., et al.: DeepTIO: a deep thermal-inertial odometry with visual hallucination. IEEE Robot. Autom. Lett. **5**(2), 1672–1679 (2020)
32. Gallego, G., et al.: Event-based vision: a survey. IEEE Trans. Pattern Anal. Mach. Intell. (Early Access)
33. Kim, H., Leutenegger, S., Davison, A.J.: Real-time 3D reconstruction and 6-DoF tracking with an event camera. In: Leibe, B., Matas, J., Sebe, N., Welling, M. (eds.) ECCV 2016. LNCS, vol. 9910, pp. 349–364. Springer, Cham (2016). https://doi.org/10.1007/978-3-319-46466-4_21
34. Wick, C.: Deep learning. Informatik-Spektrum **40**(1), 103–107 (2016). https://doi.org/10.1007/s00287-016-1013-2
35. Bengio, Y., LeCun, Y., Hinton, G.: Deep learning for AI. Commun. ACM **64**(7), 58–65 (2021)
36. Kendall, A., Grimes, M., Cipolla, R.: PoseNet: a convolutional network for real-time 6-DOF camera relocalization. In: IEEE International Conference on Computer Vision (ICCV), Santiago, Chile, 7–13 December, pp. 1–9 (2015)
37. Wang, S., Clark, R., Wen, H., Trigoni, N.: DeepVO: towards end-to-end visual odometry with deep recurrent convolutional neural networks. In: IEEE International Conference on Robotics and Automation (ICRA), Singapore, 29 May–3 June, pp. 1–8 (2017)
38. Yang, N., Stumberg, L.V., Wang, R., Cremers, D.: D3VO: deep depth, deep pose and deep uncertainty for monocular visual odometry. In: IEEE/CVF Conference on Computer Vision and Pattern Recognition (CVPR), Seattle, WA, USA, 13–19 June, pp. 1–12 (2020)
39. Chen, C., Wang, B., Lu, C.X., Trigoni, N., Markham, A.: A survey on deep learning for localization and mapping: towards the age of spatial machine intelligence. ArXiv Preprint arXiv:2006.12567v2 (2020)
40. Wang, Y., Fu, Y., Zheng, R., Wang, L., Qi, J.: New trend in front-end techniques of visual SLAM: from hand-engineered features to deep-learned features. In: International Conference on Artificial Intelligence in China (CHINAAI), 21–22 August, pp. 1–10 (2021)

41. Saxena, A.: Simultaneous localization and mapping through the lens of nonlinear optimization. Master thesis, University of California, Berkeley, USA (2021)
42. Durrant-Whyte, H., Bailey, T.: Simultaneous localization and mapping: part I. IEEE Robot. Autom. Mag. **13**(2), 99–110 (2006)
43. Bailey, T., Durrant-Whyte, H.: Simultaneous localization and mapping: part II. IEEE Robot. Autom. Mag. **13**(3), 108–117 (2006)
44. Anderson, A.J.: Real-time simultaneous localization and mapping with a single camera. In: IEEE International Conference on Computer Vision (ICCV), Nice, France, 13–16 October, pp. 1–8 (2003)
45. Solá, J., Monin, A., Devy, M., Lemaire, T.: Underlayed initialization in bearing only SLAM. In: IEEE/RSJ International Conference on Intelligent Robots and Systems (IROS), Edmonton, AB, Canada, 2–6 August, pp. 1–6 (2005)
46. Solá, J.: Simulataneous localization and mapping with the extended Kalman filter: a very quick guide with Matlab code. https://www.iri.upc.edu/people/jsola/JoanSola/objectes/curs_SLAM/SLAM2D/SLAM
47. Tully, S., Moon, H., Kantor, G., Choset, H.: Iterated filters for bearing-only SLAM. In: IEEE International Conference on Robotics and Automation (ICRA), Pasadena, CA, USA, 18–23 May, pp. 1–7 (2008)
48. Mourikis, A.I., Roumeliotis, S.I.: A multi-state constraint Kalman filter for vision-aided inertial navigation. In: IEEE International Conference on Robotics and Automation (ICRA), Roma, Italy, 10–14 April, pp. 1–8 (2007)
49. Klein, G., Murray, D.: Parallel tracking and mapping for small AR workspaces. In: 6th IEEE and ACM International Symposium on Mixed and Augmented Reality, Nara, Japan, 13–16 November, pp. 1–10 (2007)
50. Newcombe, R.A., Lovegrove, S.J., Davison, A.J.: DTAM: dense tracking and mapping in real-time. In: IEEE International Conference on Computer Vision (ICCV), Barcelona, Spain, 6–13 November, pp. 1–8 (2011)
51. Strasdat, H., Montiel, J.M.M., Davison, A.J.: Real-time monocular SLAM: why filter? In: IEEE International Conference on Robotics and Automation (ICRA), Anchorage, AK, USA, 3–7 May, pp. 1–8 (2010). Best Vision Paper Award
52. Nocedal, J., Wright, S.J.: Numerical Optimization, 2nd edn. Springer, New York (2006). https://doi.org/10.1007/978-0-387-40065-5
53. Absil, P.-A., Mahony, R., Sepulchre, R.: Optimization Algorithms on Matrix Manifolds. Princeton University Press, Princeton (2007)
54. Barfoot, T.D.: State Estimation for Robotics. Cambridge University Press, Cambridge (2017)
55. Solá, J., Deray, J., Atchuthan, D.: A micro lie theory for state estimation in robotics. ArXiv Preprint arXiv:1812.01537 (2018)
56. Kschischang, F.R., Frey, B.J., Loeliger, H.-A.: Factor graphs and the sum-product algorithm. IEEE Trans. Inform. Theory **47**(2), 498–519 (2001)
57. Loeliger, H.-A.: An introduction to factor graphs. IEEE Signal Process. Mag. **21**(1), 28–41 (2004)
58. Dellaert, F., Kaess, M.: Square root SAM: simultaneous localization and mapping via square root information smoothing. Int. J. Robot. Res. **25**(12), 1181–1203 (2006)
59. Kaess, M., Ranganathan, A., Dellaert, F.: iSAM: incremental smoothing and mapping. IEEE Trans. Robot. **24**(6), 1365–1378 (2008)
60. Kaess, M., Johannsson, H., Roberts, R., Ila, V., Leonard, J.J., Dellaert, F.: iSAM2: incremental smoothing and mapping using the Bayes tree. Int. J. Robot. Res. **31**(2), 216–235 (2012)

61. GTSAM (Georgia Tech Smoothing and Mapping). https://gtsam.org
62. Kümmerle, R., Grisetti, G., Strasdat, H., Konolige, K., Burgard, W.: g^{2}o: a general framework for graph optimization. In: IEEE International Conference on Robotics and Automation (ICRA), 9–13 May, Shanghai, China, pp. 1–7 (2011)
63. Seres Solver, Google Inc. https://ceres-solver.org
64. Blanco-Claraco, J.L.: A modular optimization framework for localization and mapping. In: The Robotics: Science and Systems (RSS), Freiburg, Germany, 22–26 June, pp. 1–10 (2019)
65. Grisetti, G., Kümmerle, R., Stachniss, C., Burgard, W.: A tutorial on graph-based SLAM. IEEE Intell. Transp. Syst. Mag. **2**(4), 31–43 (2010)
66. Dellaert, F.: Factor graphs and GTSAM: a hands-on introduction. Technical report number GT-RIM-CP&R-2012-002 (2012)
67. Dellaert, F., Kaess, M.: Factor graphs for robot perception. Found. Trends Robot. **6**(1–2), 1–139 (2017)
68. Dellaert, F.: Factor graphs: exploiting structure in robotics. Annu. Rev. Control Robot. Auton. Syst. **4**, 141–166 (2021)
69. Forster, C., Carlone, L., Dellaert, F., Scaramuzza, D.: On-manifold preintegration for real-time visual-inertial odometry. IEEE Trans. Robot. **33**(1), 1–21 (2017)
70. Sibley, G., Matthies, L., Sukhatme, G.: Sliding window filter with application to planetary landing. J. Field Robot. **27**(5), 587–608 (2010)
71. Leutenegger, S., Lynen, S., Bosse, M., Siegwart, R., Furgale, P.: Keyframe-based visual-inertial odometry using nonlinear optimization. Int. J. Robot. Res. **34**(3), 314–334 (2015)
72. Qin, T., Li, P., Shen, S.: VINS-mono: a robust and versatile monocular visual-inertial state estimator. IEEE Trans. Robot. **34**(4), 1004–1020 (2018)
73. Chiu, C.-Y.: Simultaneous localization and mapping: a rapprochement of filtering and optimization-based approaches. Master thesis, University of California, Berkeley, USA (2021)
74. Carlone, L., Tron, R., Daniilidis, K., Dellaert, F.: Initialization techniques for 3D SLAM: a survey on rotation estimation and its use in pose graph optimization. In: IEEE International Conference on Robotics and Automation (ICRA), Seattle, USA, 26–30 May, pp. 1–8 (2015)
75. Rosen, D.M., DuHadway, C., Leonard, J.J.: A convex relaxation for approximate global optimization in simultaneous localization and mapping. In: IEEE International Conference on Robotics and Automation (ICRA), Seattle, USA, 26–30 May, pp. 1–8 (2015)
76. Arrigoni, F., Rossi, B., Fusiello, A.: Spectral synchronization of multiple views in SE(3). SIAM J. Imaging. Sci. **9**(4), 1963–1990 (2016)
77. Zhang, Z.: Active robot vision: from state estimation to motion planning. Ph.D. thesis, University of Zurich, Switzerland (2020)
78. Huang, S., Lai, Y., Frese, U., Dissanayake, G.: How far is SLAM from a linear least squares problem? In: IEEE/RSJ International Conference on Intelligent Robots and Systems (IROS), Taipei, Taiwan, 18–22 October, pp. 1–6 (2010)
79. Huang, S., Wang, H., Frese, U., Dissanayake, G.: On the number of local minima to the point feature based SLAM problem. In: IEEE International Conference on Robotics and Automation (ICRA), Saint Paul, MN, USA, 14–18 May, pp. 1–6 (2012)
80. Wang, H., Hu, G., Huang, S., Dissanayake, G.: On the structure of nonlinearities in pose graph SLAM. In: The Robotics: Science and Systems (RSS), Sydney, Australia, 9–13 July, pp. 1–8 (2012)

81. Khosoussi, K., Huang, S., Dissanayake, G.: Novel insights into the impact of graph structure on SLAM. In: IEEE/RSJ International Conference on Intelligent Robots and Systems (IROS), Chicago, USA, 14–18 September, pp. 1–8 (2014)
82. Carlone L.: A convergence analysis for pose graph optimization via Gauss-Newton methods. In: IEEE International Conference on Robotics and Automation (ICRA), Karlsruhe, Germany, 6–10 May, pp. 1–8 (2013)
83. Huang, S., Dissanayake, G.: A critique of current developments in simultaneous localization and mapping. Int. J. Adv. Rob. Syst. **13**(5), 1–13 (2016)
84. Carlone, L., Dellaert, F.: Duality-based verification techniques for 2D SLAM. In: IEEE International Conference on Robotics and Automation (ICRA), Seattle, WA, USA, 26–30 May, pp. 1–8 (2015)
85. Carlone, L., Rosen, D.M., Calafiore, G., Leonard, J.J., Dellaert, F.: Lagrangian duality in 3D SLAM: verification techniques and optimal solutions. In: IEEE/RSJ International Conference on Intelligent Robots and Systems (IROS), Hamburg, Germany, 28 Sept.–2 October, pp. 1–8 (2015)
86. Tron, R., Rosen, D.M., Carlone, L.: On the inclusion of determinant constraints in Lagrangian duality for 3D SLAM. In: IEEE/RSJ International Conference on Intelligent Robots and Systems (IROS) Workshop "The Problem of Mobile Sensors: Setting Future Goals and Indicators of Progress for SLAM", Hamburg, Germany, 28 September–2 October, pp. 1–6 (2015)
87. Carlone, L., Calafiore, G.C., Tommolillo, C., Dellaert, F.: Planar pose graph optimization: duality, optimal solutions, and verification. IEEE Trans. Robot. **32**(3), 545–565 (2016)
88. Rosen, D.M., Carlone, L., Bandeira, A.S., Leonard, J.J.: A certifiably correct algorithm for synchronization over the special Euclidean group. In: Algorithmic Foundations of Robotics XII. SPAR, vol. 13, pp. 64–79. Springer, Cham (2020). https://doi.org/10.1007/978-3-030-43089-4_5
89. Rosen, D.M., Carlone, L., Bandeira, A.S., Leonard, J.J.: SE-Sync: a certifiably correct algorithm for synchronization over the special Euclidean group. Computer Science and Artificial Intelligence Laboratory Technical Report MIT-CSAIL-TR-2017-002 (2017)
90. Rosen, D.M., Carlone, L.: Computational enhancements for certifiably correct SLAM. In: IEEE/RSJ International Conference on Intelligent Robots and Systems (IROS) Workshop "Introspective Methods for Reliable Autonomy", Vancouver, Canada, 24–28 September, pp. 1–8 (2017)
91. Rosen, D.M., Carlone, L., Bandeira, A.S., Leonard, J.J.: SE-Sync: a certifiably correct algorithm for synchronization over the special Euclidean group. Intl. J. Robot. Res. **38**(2–3), 95–125 (2019)
92. Latif, Y., Cadena, C., Neira, J.: Robust loop closing over time for pose graph SLAM. Int. J. Robot. Res. **32**(14), 1611–1626 (2013)
93. Latif, Y., Cadena, C., Neira, J.: Robust graph SLAM back-ends: a comparative analysis. In: IEEE/RSJ International Conference on Intelligent Robots and Systems (IROS), Chicago, IL, USA, 14–18 September, pp. 2683–2690 (2014)
94. Mangelson, J.G., Dominic, D., Eustice, R.M., Vasudevan, R.: Pairwise consistent measurement set maximization for robust multi-robot map merging. In: IEEE International Conference on Robotics and Automation (ICRA), Brisbane, Australia, 21–25 May, pp. 1–8 (2018)
95. Shi, J., Yang, H., Carlone, L.: ROBIN: a graph-theoretic approach to reject outliers in robust estimation using invariants. In: IEEE International Conference on Robotics and Automation (ICRA), Xi'an, China, 30 May–5 June, pp. 1–16 (2021)

96. Huber, P.J., Ronchetti, E.M.: Robust Statistics, 2nd edn. Wiley, New York (2009)
97. Sunderhauf, N., Protzel, P.: Switchable constraints for robust pose graph SLAM. In: IEEE/RSJ International Conference on Intelligent Robots and Systems (IROS), 7–12 October, Vilamoura, Portugal, pp. 1–6 (2012)
98. Agarwal, P., Tipaldi, G.D., Spinello, L., Stachniss, C., Burgard, W.: Robust map optimization using dynamic covariance scaling. In: IEEE International Conference on Robotics and Automation (ICRA), Karlsruhe, Germany, 6–10 May, pp. 1–8 (2013)
99. Olson, E., Agarval, P.: Inference on networks of mixtures for robust robot mapping. Int. J. Robot. Res. **32**(7), 826–840 (2013)
100. Wang, Y.: A review of robust cost functions for M-estimation. In: Liang, Q., Wang, W., Liu, X., Na, Z., Li, X., Zhang, B. (eds.) CSPS 2020. LNEE, vol. 654, pp. 743–750. Springer, Singapore (2021). https://doi.org/10.1007/978-981-15-8411-4_99
101. Yang, H., Antonante, P., Tzoumas, V., Carlone, L.: Graduated non-convexity for robust spatial perception: from non-minimal solvers to global outlier rejection. IEEE Robot. Autom. Lett. **5**(2), 1127–1134 (2020). (Best Paper Award in Robotic Vision at ICRA 2020, Best Paper Award Honorable Mention from RAL 2020)
102. Antonante, P., Tzoumas, V., Yang, H., Carlone, L.: Outlier-robust estimation: hardness, minimally-tuned algorithms, and applications. ArXiv Preprint arXiv:2007.15109 (2020)
103. Yang, H., Carlone, L.: One ring to rule them all: certifiably robust geometric perception with outliers. In: Conference on Neural Information Processing Systems (NeurIPS), 6–12 December, pp. 1–14 (2020)
104. Yang, H., Carlone, L.: Certifiable outlier-robust geometric perception: exact semidefinite relaxations and scalable global optimization. ArXiv Preprint arXiv:2109.03349 (2021)
105. Rosen, D.M., Doherty, K.J., Espinoza, A.T., Leonard, J.J.: Advances in inference and representation for simultaneous localization and mapping. Annu. Rev. Control Robot. Auton. Syst. **4**, 215–242 (2021)
106. Eriksson, A., Olsson, C., Kahl, F., Chin, T.-J.: Rotation averaging and strong duality. In: IEEE/CVF Conference on Computer Vision and Pattern Recognition (CVPR), Salt Lake City, UT, USA, 18–23 June, pp. 1–9 (2018)
107. Bustos, A.P., Chin, T.-J., Eriksson, A., Reid, I.: Visual SLAM: why bundle adjust? In: IEEE International Conference on Robotics and Automation (ICRA), Montreal, Canada, 20–24 May, pp. 1–7 (2019)
108. Dellaert, F., Rosen, D.M., Wu, J., Mahony, R., Carlone, L.: Shonan rotation averaging: global optimality by surfing $SO(p)^n$. ArXiv Preprint arXiv:2008.02737 (2020)
109. Tian, Y., Khosoussi, K., Rosen, D.M., How, J.P.: Distributed certifiably correct pose-graph optimization. IEEE Trans. Robot. (Early Access)
110. Tian, Y., Chang, Y., Arias, F.H., Nieto-Granda, C., How, J.P., Carlone, L.: Kimera-multi: robust, distributed, dense metric-semantic SLAM for multi-robot systems. ArXiv Preprint arXiv:2106.14386 (2021)
111. Lajoie, P.-Y., Ramtoula, B., Wu, F., Beltrame, G.: Towards collaborative simultaneous localization and mapping: a survey of current research landscape. ArXiv Preprint arXiv:2108.08325v1 (2021)

Research on Malicious TLS Traffic Detection Based on Spatiotemporal Feature Fusion

Mingyue Qin[1], Mei Nian[1(✉)], Jun Zhang[1,2], and Bingcai Chen[1]

[1] School of Computer Science and Technology, Xinjiang Normal University, No. 102, Xinyi Road, Urumqi 830054, Xinjiang, China
2468830639@qq.com

[2] Xinjiang Institute of Physical and Chemical Technology, Chinese Academy of Sciences, Urumqi 830011, China

Abstract. Aiming at the problem that traditional machine methods rely on expert experience and the effect of malicious traffic identification is not ideal, a deep learning hybrid model is proposed to detect malicious TLS traffic. The model combines one-dimensional convolutional neural network and two-way long-term and short-term memory network to compress and extract network traffic features from two dimensions of space and time series. At the same time, attention score of output information is extracted by attention mechanism, and traffic identification is carried out by using mixed features obtained by fully connected neural network. Based on the open data set, the experimental results show that the accuracy, recall and F1 value of the model on the test set reach 94.67%, 89.66% and 91.08% respectively, which has good recognition effect.

Keywords: Malicious TLS traffic · Convolution neural network · BiLSTM · Attention mechanism

1 Introduction

In recent years, with the wide use of encryption technology, traffic encryption has become a standard practice. In 2015, 21% of the website traffic was encrypted. By 2019, more than 80% of the website traffic was encrypted, with a year-on-year growth of more than 90%. Malicious software uses encrypted channel and traffic encryption technology to achieve deep hiding and frequent variation of malicious behavior, resulting in a large number of malicious traffic characteristics in the Internet have not been found. Cisco released a security report in 2018, which pointed out that more than 70% of malicious software used TLS (Transport Layer Security) encryption technology to avoid exposure of attack behavior [1]. Traffic encryption not only protects users' privacy, but also provides convenience for many malware to hide their attacks. As a basic project of network defense, port obfuscation and port hopping technology used in encrypted traffic lead to a sharp decline in the accuracy of the traditional DPI and DFI methods based on plaintext [2], and the decryption behavior consumes a lot of computing resources and time. Therefore, how to accurately identify and quickly classify malicious TLS traffic without decrypting it has become a challenge.

Q. Liang et al. (Eds.): Artificial Intelligence in China, LNEE 854, pp. 318–325, 2022.
https://doi.org/10.1007/978-981-16-9423-3_40

2 Related Work

In view of the new challenges brought by the abuse of encryption technology to network management and security, academia and industry have turned their research focus to the use of machine learning technology based on load or behavior [3]. The workflow is as follows: first, design features manually (such as traffic features or grouping features), then extract and select appropriate features from the original traffic, and finally use classifiers (decision tree, naive Bayes and random forest, etc. [4–6]) to classify traffic.

Recently, researchers pay more attention to deep learning methods. Wang [7] proposed for the first time to convert the traffic into gray image and use two-dimensional CNN to extract the spatial characteristics of the traffic, but this method only uses the first 784 bytes of the whole flow, does not fully combine with other information of the traffic, and lacks the anonymization of the IP address. Cheng Hua et al. [8] proposed to transform the traffic load into sentence vector using word2vec model, and realized the identification of malicious encrypted C&C traffic through CNN.

The existing research on the identification and classification of encrypted traffic mainly focuses on the temporal or spatial characteristics of traffic. Usually, only one feature dimension of encrypted traffic is studied, which leads to the lack of robustness of the model. In the face of complex network traffic, the recognition effect may decline seriously.

To solve this problem, this paper proposes a malicious TLS traffic detection model based on CNN-BiLSTM-Attention. The model uses one-dimensional CNN and BiLSTM-Attention to compress and extract the spatial and temporal features of the traffic, and stitches the processed temporal and spatial features together to get a mixed feature vector, The full connected neural network is used to complete the recognition task. Experiments show that the effect of the model has been significantly improved compared with the existing research.

3 Model Design

3.1 Model Overview

In this paper, we use deep learning algorithm to automatically extract the temporal and spatial features from malicious encrypted traffic and train the classifier. In the spatial dimension feature learning module, one-dimensional CNN algorithm is used to extract the spatial features. The processed TLS streams are converted into two-dimensional gray image, and then the images are converted into a byte sequence of CSV file. 1D-CNN is used to learn the spatial features; In the time dimension feature learning module, BiLSTM-attention algorithm is used to learn time features in the field of time series classification, and attention mechanism is used to extract the attention score of the output information of BiLSTM. Finally, the features mined by the two deep neural networks are spliced and input into the softmax classifier to complete the identification and classification of malicious TLS traffic. The model architecture is shown in Fig. 1.

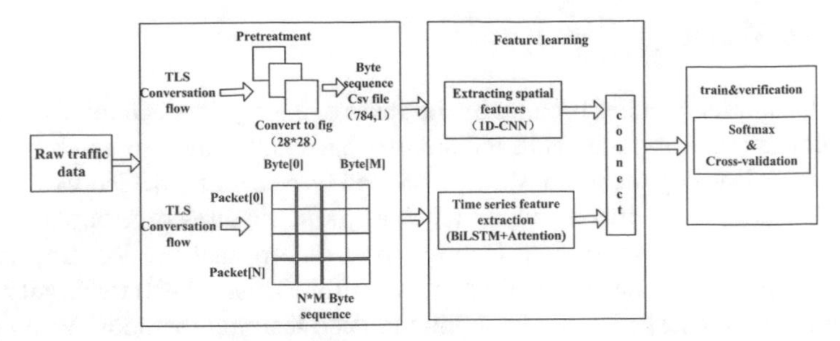

Fig. 1. Model architecture diagram

3.2 Data Preprocessing

The original data set file is in pcap format, which is composed of multiple data packets. The feature learning object of this model is data stream, so we first preprocess the data set and aggregate it into data stream. Firstly the streamdump tool is used to aggregate the packets into a data stream according to the packet quintuple. Secondly, the special MAC address, IP address and other specific information in the packet that interfere with the classification results are deleted.

3.3 Spatial Feature Extraction Model Based on 1D-CNN

Literature [7] shows that 1D-CNN, which is suitable for sequence data classification, can achieve better classification effect for traffic classification. Therefore, 1D-CNN model is used to compress and extract spatial features in traffic dataset.

CNN requires that the input dimension size is the same, and the data stream connects the information and content exchange part of the front part, which can better reflect the main characteristics of the whole data stream. Therefore, the first 784 bytes of data of each data stream are intercepted, and the data stream whose length is less than 784 bytes is supplemented with 0x00, and its category is marked.

The flow chart of one-dimensional CNN model in this paper is shown in Fig. 2, which is divided into two parts: 1) a cyclic structure composed of a convolution layer and a maximum pooling layer, which repeat two rounds; 2) Dropout layer, flatten layer and their connected fully connected neural networks.

3.4 Time Series Feature Extraction Model Based on BiLSTM-Attention

Network traffic has obvious hierarchical characteristics. The chain structure of bytes, packets and data streams is almost the same as the composition of words, sentences and paragraphs in natural language. Therefore, this paper uses the variant model of LSTM, bi-directional LSTM (BiLSTM), which is excellent in the field of natural language processing, to process the data. At the same time, considering the different importance of each packet in the session traffic, in order to highlight this difference and further improve the recognition effect of the model, attention mechanism is used to calculate the weight of the hidden layer output and weighted sum on the basis of BiLSTM model. The structure of the whole BiLSTM-Attention model is shown in Fig. 3.

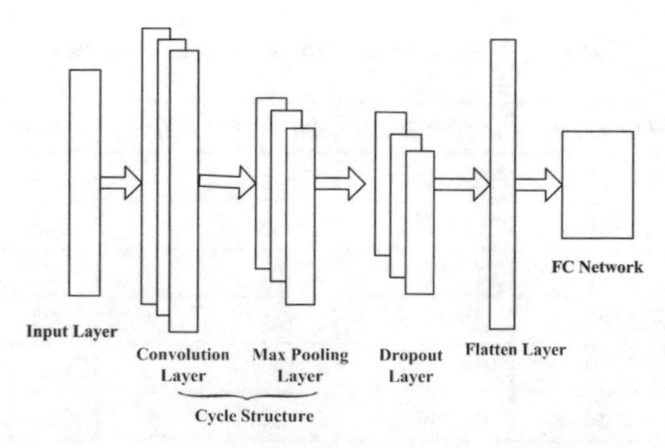

Fig. 2. CNN model

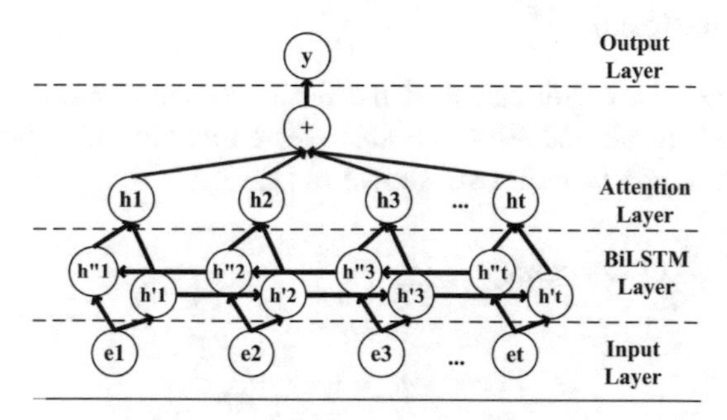

Fig. 3. BiLSTM-attention model

The input data needs to be normalized in format. The first 8 packets of each data stream are intercepted. The first 100 bytes of each packet are taken. The length of less than 100 bytes is supplemented with 0x00 at the end.

4 Experimental

4.1 Experimental Environment

In this paper, the experimental environment is windows 10 system, the CPU is i7-6700, the main frequency is 3.7 GHz, the memory is 8 g, and the environment is Python 3.6. Use keras to build the model.

4.2 Data Sources

In this experiment, the flow in CICIDS2017 [9] data set is used as the normal flow; Malware traffic analysis [10] and stratosphere IPS [11] are merged as malicious traffic. The content distribution of the experimental data set is shown in Table 1.

Table 1. Distribution of malicious data samples

Traffic category	Quantity	Proportion	Traffic category	Quantity	Proportion
Dridex	20429	36.14%	Neris	218	0.39%
Vawtrak	19260	34.07%	Tofsee	232	0.41%
Miuref	6771	11.97%	Shifu	322	0.57%
Razy	1141	2.02%	Htbot	631	1.13%
Emotet	53	0.09%	Zeus	2032	3.59%
Reposfxg	84	0.15%	Normal	5352	9.47%

4.3 Evaluating Indicator

In order to evaluate the performance of the detection model proposed in this paper, accuracy, precision, recall and F1 value are selected as the evaluation indexes of the model. These calculation formulas are shown in (1)–(4):

$$Accuracy = \frac{TP + TN}{TP + TN + FP + FN} \tag{1}$$

$$\Pr ecision = \frac{TP}{TP + FP} \tag{2}$$

$$\text{Re}call = \frac{TP}{TP + FN} \tag{3}$$

$$F_1 = \frac{2TP}{2TP + FN + FP} \tag{4}$$

Among them, TP means to correctly identify the encrypted traffic belonging to a certain classification as the classification, FP means to identify the encrypted traffic not belonging to a certain classification as the classification, TN means to identify the encrypted traffic not belonging to a certain classification as non classification, FN means to identify the encrypted traffic belonging to a certain classification as non classification.

4.4 Experimental Results and Analysis

In order to evaluate the rationality of CNN-BiLSTM-Attention model design, this paper conducts multi classification experiments with CNN, LSTM and other basic models to verify the generalization ability of the model. For the training model, early stopping strategy is used to dynamically control the number of training iterations. The data set is randomly divided into training set and test set, accounting for 80% and 20% respectively. The experiment was conducted by using categorical_ Cross entropy function is used as the loss function and Adam as the optimizer. The evaluation indexes of each model on the open data set are shown in Table 2. Figure 4 shows the 12 dimensional confusion matrix output by the four models after classification on the test set. The abscissa represents the

Table 2. Evaluation indexes of each mode

Model	Accuracy	Precision	Recall	F1
LSTM	97.40	75.92	72.67	73.33
BiLSTM	98.34	78.91	79.33	78.83
CNN	99.41	88.50	85.83	85.91
CNN-BiLSTM-A	99.65	94.67	89.66	91.08

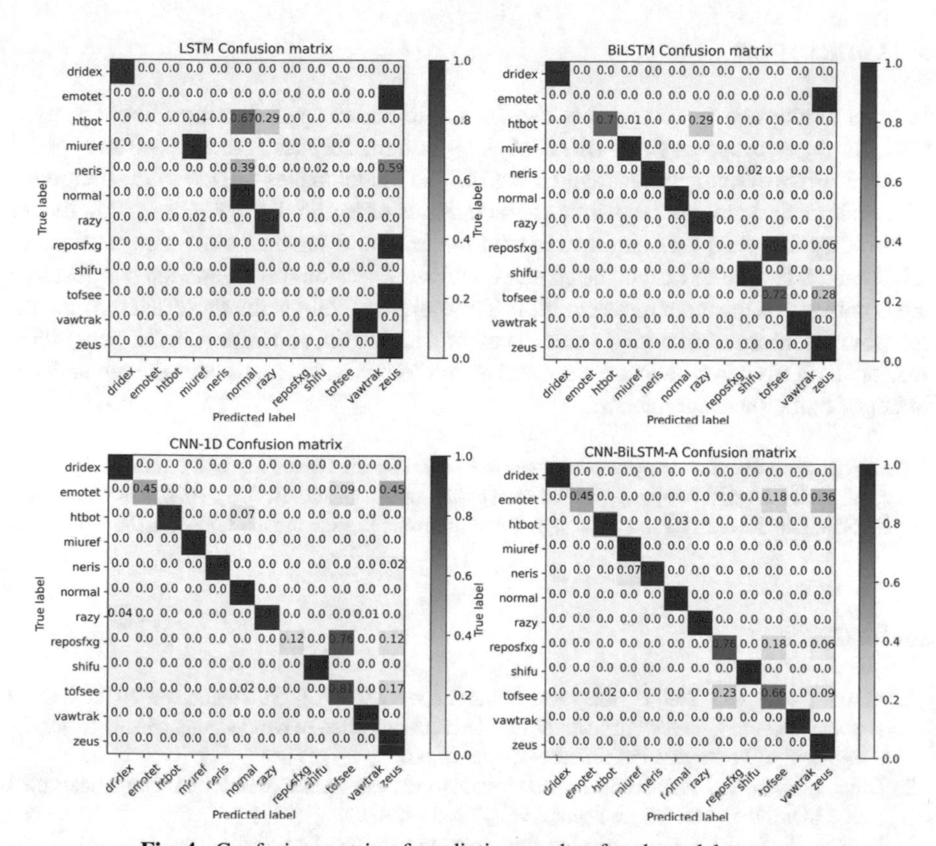

Fig. 4. Confusion matrix of prediction results of each model test set

prediction label, the ordinate represents the real label, and the value on the main diagonal of the confusion matrix is the correct proportion of each category predicted.

It can be seen from Table 2 that the accuracy of the model based on spatial features is higher than that based on temporal features when using single dimension features, which indicates that the spatial features used in this paper can better reflect the characteristics of malicious encrypted traffic compared with temporal features. In terms of temporal characteristics, the recognition effect of BiLSTM is better than that of LSTM.

After several training iterations, the performance of CNN-BiLSTM-Attention is improved compared with that of single feature. In addition, it can be seen from Table 1 that the malicious encrypted traffic data samples obtained in this paper are unbalanced. According to the confusion matrix in Fig. 4, the base model mistakenly discriminates small sample data into large sample data during training, which affects the overall accuracy of the model. In the case of sample imbalance, the F1 value of the proposed model reaches 91.08%, which is 5%–20% higher than other base models. This shows that the proposed model can also get better recognition rate in the case of unbalanced samples.

5 Conclusion

In order to improve the recognition effect of malicious encrypted traffic without decryption, this paper proposes CNN-BiLSTM-Attention model based on temporal and spatial characteristics of malicious encrypted traffic. The model makes full use of the hierarchical structure and temporal dynamic characteristics of traffic. BiLSTM-Attention model is introduced in the packet layer to extract the temporal features of session traffic, and CNN model is used to extract the spatial features of session traffic, and the two features are combined as the input of recognition. The experimental results show that the method has obvious improvement in F1 value, recall rate and so on, and improves the recognition rate of small sample data to a certain extent, and effectively strengthens the recognition effect of malicious encrypted traffic.

Acknowledgments. This project was supported in part by the Open Research Fund of Key Laboratory of Data Security, Xinjiang Normal University, under Grant XJNUSY102017B04 and University Scientific Research Project, Xinjiang Autonomous Region under Grant XJEDU2017S032.

References

1. Cisco. Encrypted Traffic Analytics White Paper [EB/OL], 31 December 2018. https://www.cisco.com/c/dam/en/us/solutions/collateral/enterprise-networks/enterprise-network-security/nb-09-encryted-traf-anlytcs-wp-cte-en.pdf
2. Chen, L., Gao, S., Liu, B., et al.: Research status and development trends on network encrypted traffic identification. Netinfo Secur. **19**(3), 19–25 (2019)
3. Rezaei, S., Liu, X.: Deep learning for encrypted traffic classification: an overview. IEEE Commun. Mag. **57**(5), 76–81 (2019)
4. Meng, P., Zhou, G.P., Meng, J.: Fast identification of encrypted traffic via large-scale sparse screening. In: Proceedings of International Conference on Advanced Cloud & Big Data, pp. 273–278. IEEE Press, Washington D.C. (2017)
5. Okada, Y., Ata, S., Nakamura, N., et al.: Comparisons of machine learning algorithms for application identification of encrypted traffic. In: Proceedings of International Conference on Machine Learning and Applications and Workshops, pp. 358–361. IEEE Press, Washington, D.C. (2011)
6. Callado, A.C., Kamienski, C.A., Szabo, G., et al.: A survey on internet traffic identification. IEEE Commun. Surv. Tutor. **11**(3), 52 (2009)

7. Wang, W., Zeng, X., Ye, X., et al.: Malware traffic classification using convolutional neural networks for representation learning. In: The 31st International Conference on Information Networking (ICOIN), pp. 712–717 (2017)
8. Cheng, H., Chen, L., Xie, L.: CNN-based encrypted C&C communication traffic identification method. Comput. Eng. **45**(8), 31–34, 41 (2019)
9. University of New Brunswick. ICIDS2017 [EB/OL] (2017). http://www.unb.ca/cic/datasets/ids-2017.html
10. BradDuncan. Malware-traffic-analysis [EB/OL] (2019). https://www.malware-traffic-analysis.net
11. Straosphere Lab. Malware Capture Facility Project [EB/OL] (2019). https://www.stratosphereips.org/datasets-malware

Research on the Application of Artificial Intelligence Technology in Cross-Cultural English Teaching

Chi Lina(✉)

Tianjin Vocational Institute, Tianjin, China
lina870624@163.com

Abstract. Language and culture are inextricably linked. Language is the carrier of culture, and culture is the base of language. The purpose of college English teaching is not only to help students grasp the language system but to disseminate human civilization and cultural knowledge so that they can have the awareness of cultural differences and cross-cultural communicative competence. Under the background of "Internet+" development, artificial intelligence brings new direction and momentum to cross-cultural English teaching. This paper tries to combine artificial intelligence technology with cross-cultural English teaching and explores methods of applying artificial intelligence in teaching process, teaching materials and assessment system to effectively improve college students' cross-cultural communication skills.

Keywords: Artificial intelligence · Cross-cultural English teaching · Teaching materials · Assessment system

1 Introduction

In today's world, modern information technology is constantly being updated, and multimedia and Internet technologies are being improved. Everyone has more or less entered a network that connects different nations, races and cultures around the world. How to improve cross-cultural communication skills and how to effectively communicate with people of different cultural backgrounds are the challenges of the times that today's society poses to everyone. College English, as an important and basic course of college students, should plays its role in cultivating students' linguistic and cross-cultural abilities. As a semiotic system, language can not only reflect cultural phenomena, but also restrict the formation of culture. Culture permeates language, and any language communication (oral or written) is the expression, interpretation and new generation of culture. College English teaching should put the cultivation of students' cross-cultural communication awareness and the cultivation of students' language knowledge and language skills in an equally important position. Only by fully understanding the history and culture of the target language and native language, can students truly and effectively master language knowledge, improve language skills, and apply them more effectively in practice.

Q. Liang et al. (Eds.): Artificial Intelligence in China, LNEE 854, pp. 326–335, 2022.
https://doi.org/10.1007/978-981-16-9423-3_41

The development and application of artificial intelligence makes the computer complete intelligence work that can only be completed by human in the past, thereby greatly expanding human intelligence and physical strength. With the development of modern information and internet technology, different kinds of hardware and software techniques are being applied into education. As early as 1980s, CALL (Computer Assisted Language Learning) is the earliest practice of information technology in cultural teaching. With the popularity artificial intelligence, U-Learning (Ubiquitous Learning) became an important language and culture learning method which is to use information technology to provide students with a 3A (Anytime, Anywhere, Anydevice) learning model to carry out learning activities.

The wide use of artificial intelligence has brought opportunities of classroom revolution in foreign language teaching, and has played an essential role for Chinese students to learn foreign cultures and to understand the contrast and integration of Chinese and Western cultures. Artificial intelligence realizes not only novel changes in the surface form of teaching, but more importantly, it brings students a new knowledge through the interactive experience of informatization and intelligence, so that language teaching and cultural teaching are integrated and connected. This is the meaning in the combination of artificial intelligence and cross-cultural teaching.

2 The Significance and Problems of Applying Artificial Intelligence in Cross-Cultural English Teaching

2.1 The Significance of Applying Artificial Intelligence in Cross-Cultural English Teaching

The theoretical basis of the application of artificial intelligence in cultural teaching can be traced back to Vygotsky's sociocultural theory, which emphasizes the decisive influence of the social environment on the development of human cognition and the mediating role of tools and language. Cross-cultural English teaching should provide students with corresponding culture learning materials under a certain social and cultural background, using tools and language as an intermediary, so as to promote the development of students' cultural awareness and communication skills. Artificial intelligence provides vivid and realistic teaching scenarios for the teaching of culture in the form of pictures, texts, sounds and images. Students can strengthen the cognition of the target language and experience the target culture more directly (shown in Fig. 1).

In the latest edition of "Guide for College English Teaching" promulgated by the Ministry of Education, it points out that "The application of modern information technology in college English teaching has not only modernized, diversified and facilitated teaching techniques, but promoted the change and innovation of teaching concepts, teaching content and teaching methods." In the "Work of 2019 Education Informationization and Network Safety" issued by the General Office of the Ministry of Education, it is further clearly indicated that modern information technology such as AI (artificial intelligence), big data, information system and so on has provided a new way for foreign language teaching.

Due to the visibility, interest, figurativeness, and interactivity of AI technology, teachers can use modern information technology to make cultural teaching content in the form of pictures, audio, video, animation and so on to attract students' attention, guiding students to actively participate in cultural learning and communication and strengthen students' emotional experience. Through the combination of information and network technology, mobile intelligent terminals, network platforms and various teaching software can be utilized to improve the teaching efficiency and effect, thus cultivating talents with emotional vision. A variety of network resources also bring new opportunities for cultural teaching and learning. Through the sharing of network resources, people at home and abroad can share cultural resources and communicate with each about the common and different features of the native culture and target culture.

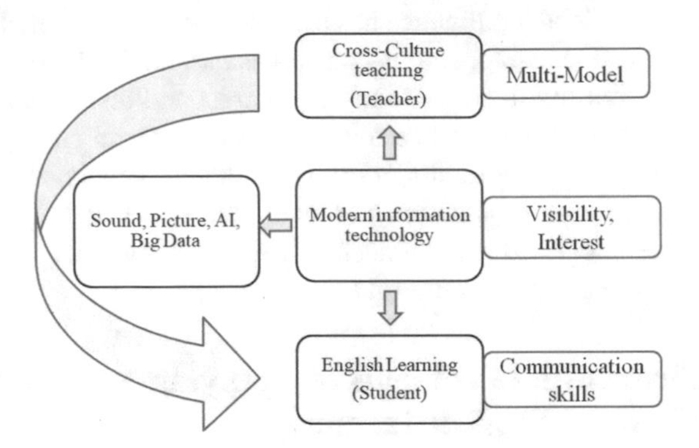

Fig. 1. Significance of AI

2.2 The Problems of Applying Artificial Intelligence in Cross-Cultural English Teaching

The topics about the application of artificial intelligence in cross-cultural English teaching like the application methods, advantages and practical problems have been discussed by many scholars. Cao Mengyue and Wang Jun (2020) investigated the application of information technology in the process of college English culture teaching in the intelligent age and analyzed its existing problems and reasons. Wang Jijun and Meng Zhaokuan (2014) analyzed the theoretical basis and practical use of information technology in culture teaching from the perspective of training in intercultural communicative competence. Deng Yanling (2017) explored the integration methods of information technology and intercultural communication course in vocational college. There is no doubt that modern information technology has played an important role in solving the problems of passive learning of students, weak cross-cultural knowledge of teachers and students, and lack of high-quality textbook resources. However, information technology still has many problems in assisting cross-cultural English teaching, which are mainly manifested in the following aspects.

Firstly, the application range of artificial intelligence in college cross-cultural English teaching is not wide enough.

Most teachers who undertake college English teaching have used modern information technology platforms such as QQ, WeChat, and Xuexitong to assist teaching in cross-cultural teaching. However, this utility only stays in the surface form of sending and receiving homework or quiz, and supplementing resources. Most of the teachers still use the method of lecturing in the process of cultural English teaching, instead of using the interactive non-traditional teaching method for cultural teaching with the assistance of modern hardware or software technology. Although teachers and students have gradually realized the significance of information technology for cultural teaching and have begun to explore and experiment it, it has not been widely used. The specific teaching mode also needs further expansion and standardization, and most students have not really accepted and adapted to the application of modern information technology in the cross-cultural teaching of college English.

Secondly, the application of artificial intelligence in college cross-cultural English teaching has not achieved significant results.

Although many teachers tried to use modern information technology for cross-cultural teaching, the feedback from students was not very satisfactory. Most of the students relied on the teacher-centered teaching model, and they lack of the autonomous exploration and independent learning by using modern information technology. In the context of technological development and epidemic prevention and control, teachers may use some modern information technology to assist teaching in cross-cultural teaching, such as multimedia courseware. Although some pictures or sound special effects are added, they only have decorative effects, and some even have nothing to do with the content of teaching. Although some have some relations with the teaching content, it has little effect on improving the teaching effect. Although modern information technology is regarded as a modern educational tool and has been recognized by teachers and students to a certain extent, its application effect is still not satisfactory.

Finally, the application of artificial intelligence in college cross-cultural English teaching has great potential.

The content of cross-cultural English teaching is rich and diverse, including literary common sense, historical events, customs, etiquette and so on. The traditional teacher-centered explanation cannot arouse students' interest. Since the beginning of the 21st century, the Internet, search engines, big data, cloud computing, communication technology, and the Internet of Things have promoted the in-depth development of AI (artificial intelligence). The fourth industrial revolution with typical characteristics of intelligence, networking, and data is in progress.Teaching software based on artificial intelligence and related hardware of smart classrooms provide a good medium for teachers and students' self-learning, human-computer interaction, and knowledge updating. Based on the characteristics of culture and the needs of teaching, it is a general trend to assist cultural teaching with scientific and technological means, so that the teaching and learning of culture can be more direct and vivid.

3 The Application Model of Modern Technology in Cross-Cultural English Teaching

3.1 The Application of AI in Teaching Process

In the opening of the Education 2030 Work Report published by the UNESCO—"AI in Education: challenges and opportunities for sustainable development", it is pointed out that "Artificial intelligence is a booming technical field, which will change different aspects of social interaction, and artificial intelligence has begun to propose new teaching and learning solutionsin education, which are currently under test in different contexts." As shown in Fig. 2, artificial intelligence is changing the whole teaching process.

From the aspect of subject, learning takes place on the learner, teaching can happen in teachers, media (textbooks or videos) and learners themselves, even on artificial intelligence agents (teachers). Teaching is mainly undertaken by teachers, but teaching materials, machines, media and learners can partially undertake the function of teaching. It can be understood from this sense that the teachers' presence is not necessarily the main basis for teaching. In the process of cross-cultural English teaching, With the assistance of big data and artificial intelligence technology, teachers and students can communicate through new media platforms or chatting software and apps at any time online. What's more, language learners can communicate with people from other nations and cultures at any time, and learn more about the culture as well as the language. Under the promotion of technology, robots or artificial intelligence assisting education will gradually become a normal thing.

From the aspect of space, traditional classroom is definitely no longer a place for lecture and it becomes a learning room and counseling room which is a place for students' exploration of the world and a stage for cooperation and competition. The occurrence of teaching processes not only relies on textbooks, but also video, machine assistance and students' autonomous learning. Teachers can collect, compare and process the data, images and voice by AI technology in the process of students' cross-cultural English learning to finish the task of teaching, test and monitoring.

The use of AI techniques in the teaching process not only improves students' learning initiative and learning effects, but also helps improve teaching efficiency, relieve teachers' teaching pressure, and realize personalized counseling for individual student. At present, some college teachers, especially those who major in computer science, use their professional advantages to cooperate with foreign language teachers to build a foreign language learning network platform. Teachers will transfer electronic resources related to major English-speaking countries such as geography, history, political system, medical care, customs, etc. to the platform. At the same time, there are a large number of exercises to facilitate students' self-study. The cultural content of the self-learning platform has strengthened the cultivation of students' cross-cultural knowledge and communication skills. This space is also convenient for teachers to conduct online and offline mixed teaching, which can not only reduce the burden on teachers, save classroom time, but also cultivate students' independent learning ability. It is not only a simple individual learning, but also a people-oriented learning based on the current situation, which closely integrates the individual and the society.

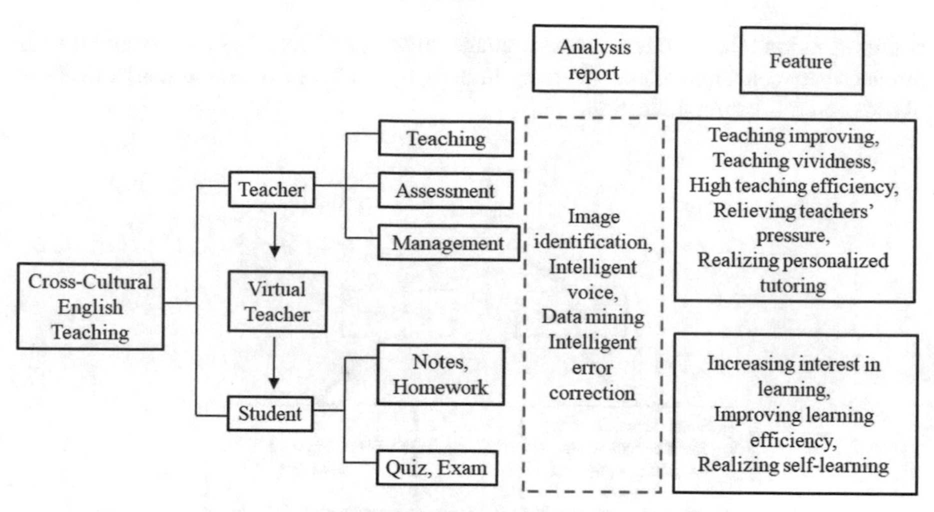

Fig. 2. AI in teaching process

3.2 The Application of AI in Teaching Materials

The application of AI in cross-cultural English teaching is inseparable from hardware equipment and software development. In the "Internet+" era, the universality and mobility of the Internet enable people to use smart phones, tablet computers and other smart terminals to access the Internet anytime and anywhere, which has changed the ways and means for people to obtain knowledge and information. Based on the voice interaction of Baidu, text recognition, face recognition, human body recognition, common corpus, AR and other AI technologies, the software and hardware teaching products are empowered to achieve a better human-computer interaction teaching experience and obtain high-quality educational effect with lower teacher costs. At the same time, a smart campus, campus safety, school attendance, classroom effect monitoring and other key tasks can be carried out with the assistant of AI technology so as to improve campus life experience and safety, and reduce management costs (shown in Fig. 3).

In many cases, culture teaching in English-teaching classes only relies on publications, books and related materials prepared by teachers. Sometimes, audio and video are played for students to introduce cultural knowledge, although these materials can enrich classroom teaching to some extent, it is difficult to meet the students' needs of personal experience for foreign culture and the classroom teaching environment lacks of interest. AI technology provides teachers and students with authentic and rich cultural corpus. Common corpus is mainly in the form of audio, video and web pages. These resources should be divided into different categories and levels for learners to use according to the individual needs of students. For example, by watching some places of interest and e-kooks in the form of VR through the Internet, students can experience the lifestyles, behavioral norms and values of Western countries. At the same time, through human-computer interaction, students learn a lot of knowledge and methods of language communication related to the target language culture, and correctly understand the differences between Chinese and Western cultures, and cultivate the awareness and ability of cross-cultural communication.

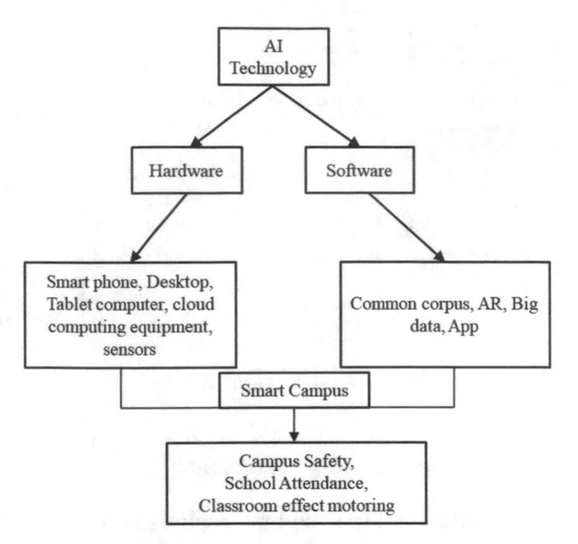

Fig. 3. AI in teaching materials

3.3 The Application of AI in Assessment System

Based on the development of artificial intelligence technology and the actual needs of classroom teaching evaluation reform, the design of cross-cultural English teaching evaluation under artificial intelligence can be divided into five levels: listening, speaking, reading, writing and translation (shown in Fig. 4) Classroom teaching evaluation under artificial intelligence brings feedback to teachers' classroom teaching through the analysis of students' language and behavior. Through the collection of students' voice, posture, facial and physiological signal data, the design of recognition and analysis of algorithms can be conducted, including speech recognition, natural language processing, gesture recognition, facial expression recognition, and EEG-based emotion recognition. The whole process is directed towards classroom language analysis, classroom behavior analysis, classroom sentiment analysis, and classroom teaching evaluation system.

Through the use of voice recognition technology to process the sound information collected by students in the process of cross-cultural language communication, language structure and emotional information are analyzed. Human body gesture recognition refers to the recognition of students' gestures, movements, and postures in the process of listening, reading and writing, drawing out the S-T curve. Through modeling, positioning and segmentation of human body structures, the emotional orientation conveyed in gesture actions can be analyzed to help teachers understand students' learning status. In the process of cross-cultural English learning, students will show changes in facial expressions and physiological signals. In professional smart classrooms, AI technology can be used to provide more accurate results of classroom teaching emotional performance analysis. CAISBNU which is a classroom teaching effect automatic monitoring prototype system developed by Beijing Normal University uses facial expression recognition to identify and analyze the expressions of students when they are in class, so as to evaluate the classroom teaching effect.

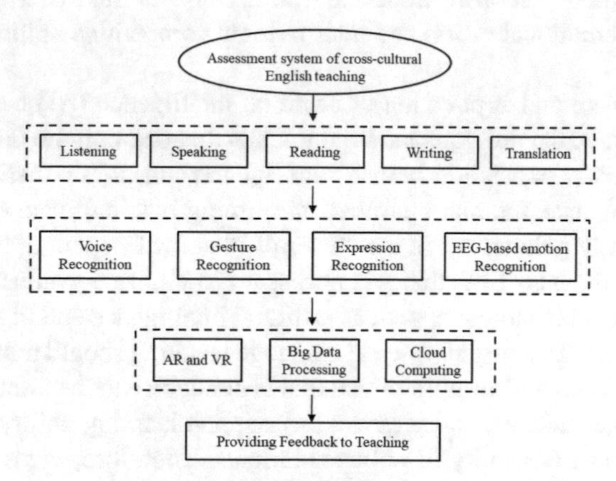

Fig. 4. AI in assessment system

With the deeper integration of artificial intelligence and education, it will be possible to collect and analyze image, video, space, posture and multi-modal data, track the trajectory of the classroom in all directions, extract and analyze the multi-dimensional data in the classroom, and realize the integrated classroom teaching and research of intelligent diagnosis and intelligent feedback. Classroom teaching assessment is based on teachers' teaching and students' learning, focusing on improving teachers' education and teaching ability, improving classroom teaching quality, and evaluating design, process and results of classroom teaching. The development of artificial intelligence technology has brought a revolution to teaching reform. Driven by the development of modern science and technology, artificial intelligence will promote the reform of the subject, content, method and result of classroom teaching assessment, relying on the data possessed by artificial intelligence. Mining, high-speed calculation, automatic analysis and other functions assist in the efficient development of evaluation work, record the dynamic changes of emotional information, achieve accurate collection of real classrooms, and track teachers to generate long trajectories.

4 Conclusion

At present, more and more colleges and universities have begun to introduce artificial intelligence into campuses, trying to carry out classroom teaching under artificial intelligence. The combination of AI technology with traditional language teaching, and the using of innovative models such as mobile smart terminals, teaching software, independent network platforms and other network resources, not only helps stimulate students' interest in learning, but also effectively promotes the learning of cultural knowledge and the improvement of students' cultural capabilities. Meanwhile, this model is helpful for teachers to innovate their teaching concepts, build a student-centered classroom, and establish a dynamic evaluation system. In short, the new teaching model is a useful supplement to traditional culture teaching and multimedia teaching. It can effectively improve the effectiveness of college English and cultural teaching, enhance students' cross-cultural communication awareness and communication skills, which can help them have a certain international vision and international competitive ability in future career development.

The development and application of artificial intelligence (AI) technology enables computers to replace the previous intelligent tasks that can only be done by the human brain, thereby greatly expanding human intelligence and physical strength. In the era of artificial intelligence, the main purpose of learning is to cultivate core qualities and abilities. It is clearly pointed out in the "Overall Plan for Deepening Educational Evaluation Reform in the New Era" that it is necessary to innovate evaluation tools and use modern information technologies such as artificial intelligence and big data to carry out teaching at all levels. The current foreign language teaching should make full use of artificial intelligence to create an environment that is conducive to the intelligent learning of English for college students, and improve the English learning ability and comprehensive English application ability of college students. Therefore, vigorously developing artificial intelligence technology and actively exploring the integration and development of artificial intelligence and education evaluation are inevitable measures for out country to promote smart education.

Acknowledgment. This article is the research achievement of 2019 Tianjin Vocational Institute Scientific Research Fund Project "The Investigation and Countermeasures of 'Chinese Culture Aphasia' in College Students' Cross-cultural Communication from the Perspective of Cultural Ecology" (Project number: 20192111), and the "14th Five-Year Plan" Educational and Teaching Reform Project of Tianjin Vocational Institute- "The Construction of a Scaffolding Teaching Model of English Writing in Higher Vocational Colleges from the Perspective of Sociocultural Theory". (Project number: JY2021017).

References

Mengyue, C., Jun, W.: A survey of the application of modern information technology in college English culture teaching in the age of intelligence. Educ. Teach. Forum **10**(3), 95–96 (2020)

Bing, F.: The origin, impact and countermeasures of artificial intelligence fever in Chinese universities. Mod. Educ. Technol. **29**(4), 33–38 (2019)

Xinrong, H.: Philosophical reflections on the upsurge of artificial intelligence. J. Shanghai Norm. Univ. **4**, 34–42 (2018)

Ashwin T.S., Guddeti, R.M.R.: Automatic detection of students' affective states in classroom environment using hybrid conventional neural networks. Educ. Inf. Technol. **25**(2), 1387–1415 (2020)

Lu, J., Wang, D.: Automatic evaluation of teacher-student interaction based on dialogue text. In: 2017 2nd International Conference on Education, Management and Computing Technology, pp. 1517–1522. Atlantis Press, Paris (2017)

Libao, W., Yanan, C., Yiming, C.: Reform and practical paths of classroom teaching evaluation under artificial intelligence. China Educ. Technol. **5**, 94–100 (2020)
Qingtang, L., Haoyi, H.: Classroom teaching behavior analysis method and its application based on artificial intelligence. China Educ. Technol. **9**, 13–21 (2019)
Xiaoru, W., Zheng, W.: The development trend and practical cases of artificial intelligence education applications. Mod. Educ. Technol. **2**, 5–11 (2018)
Hannan, A., Silver, H.: Innovating in Higher Education Teaching, Learning and Institutional Cultures. Open University Press, Buckingham (2000)
Yanju, T.: Research on smart education-oriented effective teaching model for college English. J. Beijing City Univ. **2**(2), 85–89 (2019)

A Remote Sensing Object Detection Algorithm Based on the Attention Mechanism and Faster R-CNN

Lixia Zhang, Zhiming Ma(✉), and Xiangshu Peng

School of Computer Science and Technology, Xinjiang Normal University, Urumchi 830054, China
13319824228@qq.com, 406287175@qq.com

Abstract. For the multi-scale, diversity and complex background characteristics of remote sensing targets, To improve the detection effect of the Faster R-CNN algorithm, an Faster R-CNN algorithm that introduces the attention mechanism is proposed, By improving traditional deep residual networks, introducing CBAM attention mechanisms in ResNet50 and ResNet101 networks, The improvement brought its mAP value to 89.2% and 94.43%, up by 2.1% and 1.92%, respectively. It is shown that the Faster R-CNN algorithm introducing the attention mechanism not only enhances the feature extraction ability of multi-scale complex targets, but also is suitable for the detection and identification of complex remote sensing targets.

Keywords: Remote sensing targets · Faster R-CNN · Attention mechanism · Detection and identification · Residual networks

1 Introduction

Remote sensing image technology has great space from military to civil fields. It has great application value in urban planning, resource investigation and environmental monitoring. Remote sensing images are characterized by large size, diversity and complex background, so it is very difficult to detect remote sensing targets. In the target detection field, Zhu et al. [1]. Proposed the introduction of attention mechanisms in the SSD algorithm and Malini et al. [2]. Proposed the automatic evaluation of the improved VGG architecture under the Faster RCNN framework, Cui et al. [3]. Proposed intelligent crack detection based on attention mechanisms in convolutional neural networks, which improves the detection accuracy. In this study, the three channels of RGB are differently weighted and experimental and tested by introducing the CBAM attention module to improve the accuracy of remote sensing image detection.

2 Principles and Method

2.1 ResNet Network Framework

ResNet, also known as a deep residual network, is a convolutional neural network proposed by 4 scholars from Microsoft Research [4], Residual networks are easy to optimize

Q. Liang et al. (Eds.): Artificial Intelligence in China, LNEE 854, pp. 336–344, 2022.
https://doi.org/10.1007/978-981-16-9423-3_42

and can improve accuracy by increasing some depth. Its internal residual block uses jump connections, adding a x constant mapping (identity mapping) to ordinary convolution, alleviating the gradient vanishing problem caused by increasing depth in deep neural networks. ResNet network relative to the VGG network, reduces the number of parameters, increases network depth and better image recognition. Figure 1 is the residual structure in ResNet. Assuming that the residual structure the output result is G(x) = F(x) + x, where x is the output of the previous layer, the result F(x) of the residual structure, if the network gradient vanishes, then F(x) = 0, G(x) = x, can avoid the problem of network performance degradation.

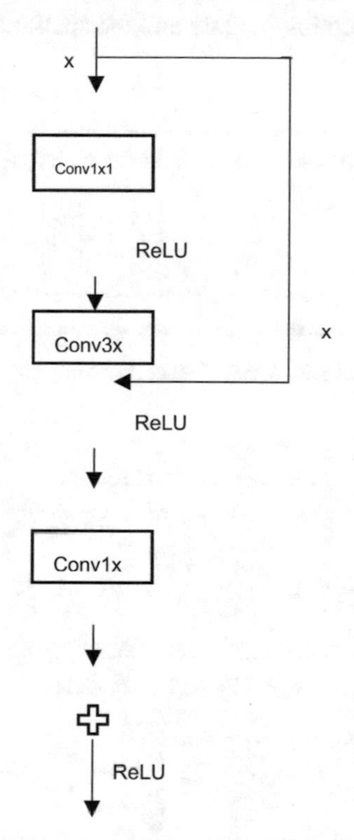

Fig. 1. Residual block structure in ResNet

2.2 Attention Mechanism Model

The basic idea of attention mechanism is to allow the system to learn that attention—— can ignore irrelevant information and focus on key information [5]. Attention mechanism can be divided into Hard-attention and Soft-attention. Hard-attention, The 0 or 1, the area must be focused or not, commonly used for image cropping. Hard attention, also known as strong attention, differs from soft attention in that every point in the image may extend the attention, and hard attention is a stochastic prediction process that emphasizes dynamic change. The key thing is that hard attention is a non-differentiable attention, and the training process is often done through reinforcement learning.

Soft-attention, It is the problem of continuous distribution between [0, 1], and each region has a different degree of attention, expressed in terms of the real number between 0–1. The key point of soft attention is that this attention is more focused on the region or channel, and soft attention is deterministic attention, learning can be generated directly through the network, the most important place is that soft attention is differentiable, which is a very important place. Because differentiable attention can learn the weight of attention through neural network and forward propagation and backward feedback. The CBAM [6] attention model is introduced in this study, whose overall structure is shown in Fig. 2, and the channel attention module and the spatial attention module are shown in Fig. 3.

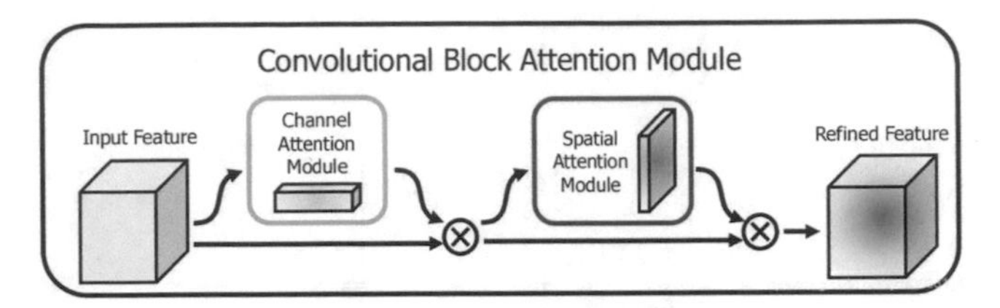

Fig. 2. CBAM structure diagram

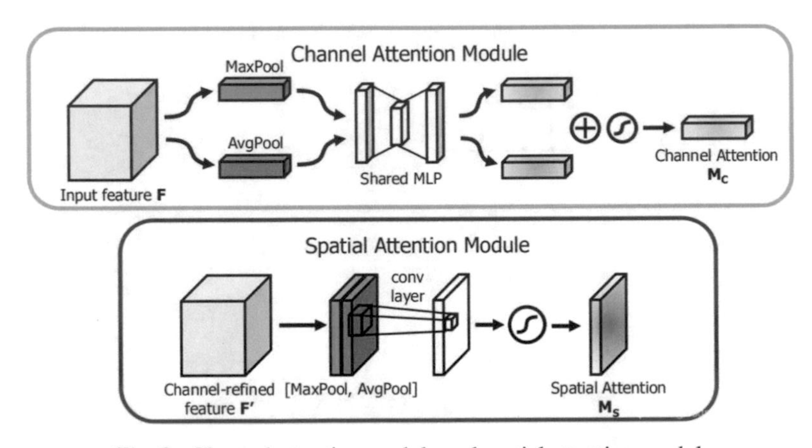

Fig. 3. Channel attention module and spatial attention module

The attention module in Figs. 2 and 3 mainly uses the following formula, where formula (3) F is input feature map as input, ⊗ represents element by element multiplication, formula (4) M_c is attention extraction on the channel dimension, and in formula (5) M_s is attention extraction on the spatial dimension.

$$F' = M_c(F) \otimes F \tag{1}$$

$$F'' = M_c\left(F'\right) \otimes F' \tag{2}$$

$$F \in \mathbb{R}^{C \times H \times W} \tag{3}$$

$$M_c \in \mathbb{R}^{C \times 1 \times 1} \tag{4}$$

$$M_s \in \mathbb{R}^{1 \times H \times W} \tag{5}$$

Channel attention formula:

$$M_C(F) = \sigma(MLP(AvgPool(F)) + MLP(MaxPool(F))) = \sigma\left(W_1\left(W_0\left(F^c_{avg}\right)\right) + W_1\left(W^c_{max}\right)\right) \tag{6}$$

Spatial attention formula: (Space domain attention is obtained by performing AvgPool and MaxPool on the channel axis)

$$M_s(F) = \sigma\left(f^{7\times7}([AvgPool(F); MaxPool(F)])\right) = \sigma\left(f^{7\times7}\left(\left[F^s_{avg}; F^s_{max}\right]\right)\right) \tag{7}$$

Where σ represents the sigmoid function, $f^{7\times7}$ represents the convolution operation, the convolutional kernel size of 7×7, $AvgPool(F)$ represents the mean pooling for F, and $MaxPool(F)$ represents the maximum pooling for F.

2.3 A Remote Sensing Object Detection Algorithm Based on the Attention Mechanism and Faster R-CNN

In order to improve the recognition accuracy, transfer the key image information in the network effectively, and help the network effectively capture the image important features, this paper applies the attention mechanism module to each residual structure in the ResNet network, forms a residual module based on CBAM module, and the residual structure after introducing the attention module is shown in Fig. 4.

The ResNet50 algorithm flow based on the attention mechanism is shown in Fig. 5.

2.4 Transfer Learning

The main idea of transfer learning is to migrate labeled data or knowledge structures from related fields, to complete or improve the learning effects of target areas or tasks. Cross-domain learning problems can be solved by extracting favorable information from relevant domains [7].Transfer learning technology mainly applies image recognition, image classification, object detection and other tasks [8]. If the number of existing data sets is small or the data distribution is equal, you can learn the models trained from the source data. Because transfer learning techniques retain the feature extraction capabilities of pre-trained models, the target recognition accuracy and generalization capabilities can be improved.

3 Experimental Analysis and Result Analysis

This study designs the results of different models, as well as channel attention and spatial attention ablation experiments for attention mechanisms.

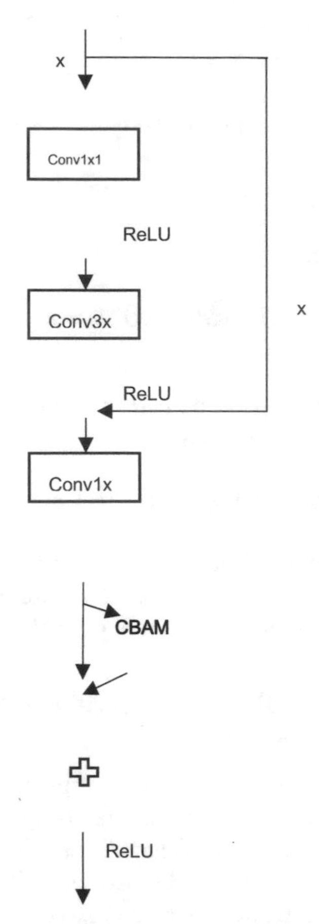

Fig. 4. Residual block structure after introducing attention module

3.1 Experimental Environment

The operating system for the experimental environment in this institute is Windows10, framework is PyTorch, compilation environment is the Python3.7, processor is Intel (R) Xeon (R) CPU E5–2630 v4 @ 2.20 GHz, and the graphics card is NVIDIA Tesla K80, explicitly stored as 12G.

3.2 Model Preparation and Hyperparameter Settings

First, build the ResNet network, load the officially provided pre-training weights of ResNet50 and ResNet101, and introduce the CBAM attention mechanism in each block, allowing the different features to obtain different weights. As a pretraining model for transfer learning, some hyperparameters are set before training, and the initial learning rate is set to 0.005. As a very important hyperparameter, it indicates the rate of the model weight update, and the loss function fluctuates too much and the experimental results are inaccurate; the network model convergence effect is too slow, resulting in

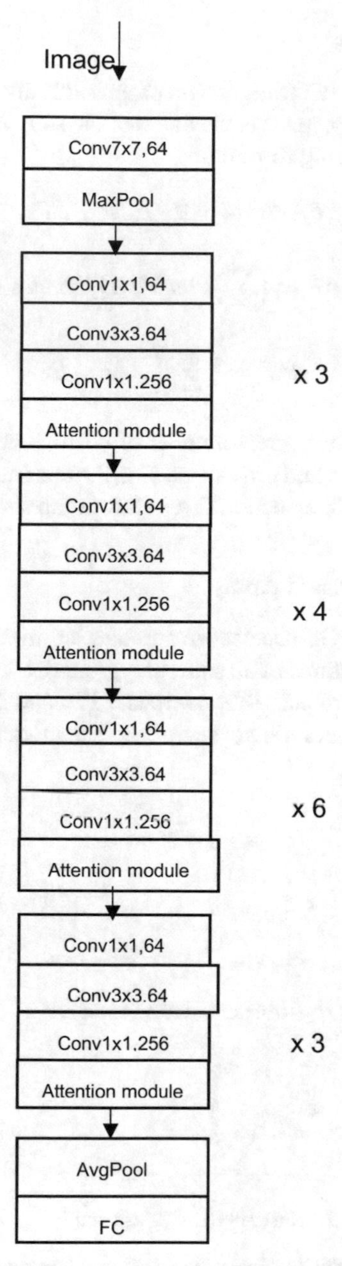

Fig. 5. Algorithm flow chart after introducing attention module.

longer training. Therefore, this experiment uses the SGD optimizer with momentum 0.9, and the learning rate iterated every 15 Epochs, 0.3 times the previous. To facilitate the viewing of the model training process, set the accuracy to output each Epoch. Finally, only the model weight with the highest accuracy is saved.

3.3 Evaluation Indicators

To evaluate the performance of the algorithm proposed in this paper, the evaluation index is the mean average accuracy. mAP is the mean of the sum of the average accuracy. Can be represented by the following formula.

$$Precision = TP/(TP + FP) \tag{8}$$

$$AP = \left(\sum Precision\right)/images \tag{9}$$

$$mAP = \left(\sum\nolimits_{i=0}^{N} AP_i\right)/N \tag{10}$$

Where TP is the predicted correct number of positive samples, FP is the number of positive samples for the predicted errors, and N is the total number of categories. TP and FP are judged by the IOU (Intersection Over Union) threshold.

3.4 Experimental Results and Analysis

Using a remote sensing RSOD-Dataset image dataset, marked by the team of Wuhan University, includes 976 pictures of aircraft, playground, overpass, oil bucket, including aircraft: 446 aircraft, playground: 189, overpass: 176 and 165 oil bucket. The experimental training set and test set ratio are set to 7:3. The model training Loss and lr curves are shown in Fig. 6.

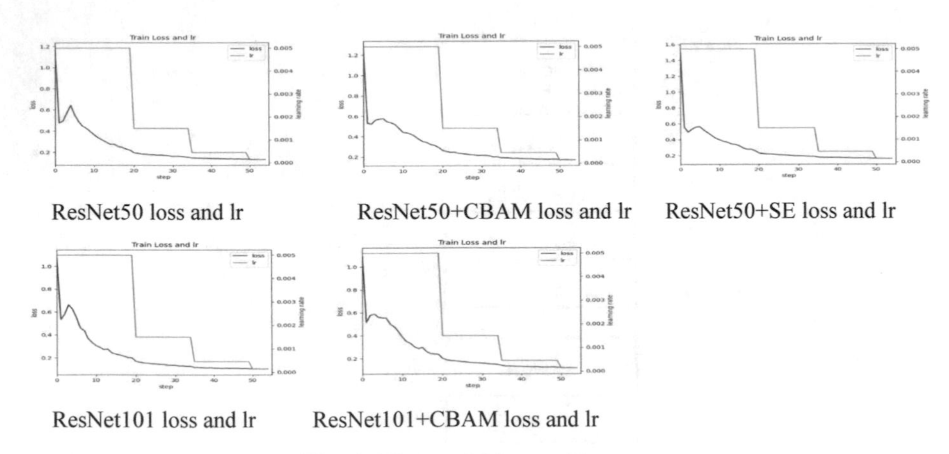

Fig. 6. Five model loss and lr curves

Transfer learning based on attention mechanism is done separately using ResNet of different depths, the remote sensing image model detection effect is shown in Fig. 7 and Table 1, The ResNet model without an attention module has a faster training time for each epoch due to the small number of parameters. The ResNet model with the attention mechanism improves the detection accuracy on the test set, The ResNet50+

CBAM model is 89.2%, and the ResNet101+ CBAM is 94.43%. In training time, since the ResNet101 is deeper than ResNet50, the training time is longer than ResNet50. It is shown that the model after introducing the attention mechanism outperformed others of the control group, considering mAP and model training time.

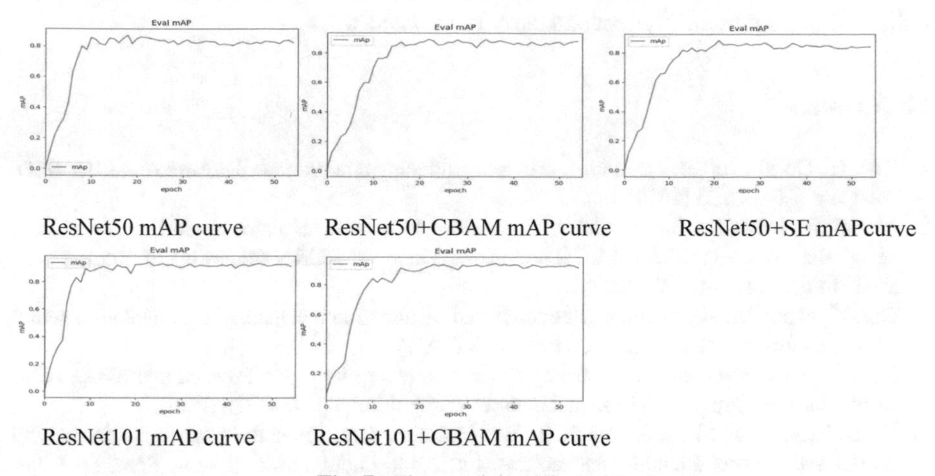

Fig. 7. Five model mAP curves

Table 1. Model detection results

Model	mAP/%	Time of each Epoch of training
ResNet50	87.1	344 s
ResNet50 + ca	88.19	450 s
ResNet50 + sa	88.2	349 s
ResNet50 + SE	88.5	384 s
ResNet50 + CBAM (ours)	89.2	508 s
ResNet101	92.51	540 s
ResNet101 + ca	94.2	735 s
ResNet101 + sa	92.98	565 s
ResNet101 + CBAM (ours)	94.43	756 s

4 Conclusion

In this paper, for remote sensing image detection problems, A detection model of the attention mechanism based on the Faster R-CNN algorithm is proposed, First, the image features are used by transfer learning method to extract, and then introduce the attention mechanism into the ResNet residual structure to give the features different weights.

After experimental comparison, the method proposed in this paper has a good effect in remote sensing image detection, The ResNet50+ CBAM and ResNet101+ CBAM models reached the mAP values of 89.2% and 94.43% on the test set, respectively. Due to the few images of the remote sensing dataset used today, the network training is insufficient. Therefore, the next step will enhance the dataset image, compare the enhanced dataset training effect and carry out related work.

References

1. Zhu, H., Gu, C.: Target detection algorithm introducing attention mechanism: Attention_SSD. Int. Core J. Eng. **6**(7) (2020)
2. Malini, A., Priyadharshini, P., Sabeena, S.: An automatic assessment of road condition from aerial imagery using modified VGG architecture in faster-RCNN framework. J. Intell. Fuzzy Syst. **40**(6), 11411–11422 (2021)
3. Cui, X., et al.: Intelligent crack detection based on attention mechanism in convolution neural network. Adv. Struct. Eng. **24**(9), 1859–1868 (2021)
4. He, K., et al.: Deep residual learning for image recognition. In: Proceedings of the IEEE Conference on Computer Vision and Pattern Recognition, pp. 770–778 (2016)
5. Tovar, P., Adarme, M.O., Feitosa, R.Q.: Deforestation detection in the amazon rainforest with spatial and channel attention mechanisms. Int. Arch. Photogramm. Remote Sens. Spat. Inf. Sci. **XLIII-B3–2021**, 851–858 (2021)
6. Huixuan, F., Song, G., Wang, Y.: Improved YOLOv4 marine target detection combined with CBAM. Symmetry **13**(4), 623 (2021)
7. Kilvisharam Oziuddeen, M.A., Poruran, S., Caffiyar, M.Y.: A novel deep convolutional neural network architecture based on transfer learning for handwritten Urdu character recognition. Tehnički vjesnik, **27**(4), 1160–1165 (2020)
8. Yosinski, J., et al.: How transferable are features in deep neural networks. CoRR, abs/1411.1792 (2014)

Route Planning of Flight Diversion Based on Genetic Algorithm and Gene Recombination

Yinfeng Li[1,2](✉), Ping Chen[1], Yi Mao[2], Tong Wei[2], Chang Ruan[3], and Ranran Shang[2]

[1] State Key Laboratory of Air Traffic Management System and Technology, Nanjing 210007, China
15030576997@126.com

[2] College of Civil and Architectural Engineering, North China University of Science and Technology, Tangshan 063210, China

[3] China Civil Aviation Air Traffic Management Bureau in North China, Beijing 100621, China

Abstract. In this paper, genetic algorithm and gene recombination algorithm are used to search the optimal path or suboptimal path by constructing the navigation environment model in two-dimensional plane. Then, the gene recombination quadratic optimization path is used to optimize the large rotation angle to make it close to the smooth path of plane angle under the premise of not intersecting with any flight restricted area. The experimental results show that the combination of GA and GA is better than GA in blocky and scattered flight restricted area environment, which reduces the complexity of GA, shortens the path length, and obtains relatively optimal collision free path.

Keywords: Genetic algorithm · Flight diversion · Path planning · Smooth path

1 Introduction

The essence of detour diversion due to weather is route planning. Genetic algorithm (GA), ant colony algorithm, annealing algorithm and other heuristic algorithms are used in path planning. Because genetic algorithm has the good ability of global optimization and processing constraints, it solves the problems that traditional optimization methods cannot solve. Therefore, more and more scholars have studied genetic algorithm. Li Yongxi et al. proposed an improved genetic algorithm with a new crossover operator for the whole network flight path planning problem, which makes the algorithm closer to the global optimum in the fully connected network [1]. Based on the theory of genetic algorithm, Davies C and Lingras P proposed a new genetic coding and genetic operator, and applied it to the dynamic network path planning problem to realize the route re-planning to adapt to the change of network information [2]. For dynamic spatial division, Chen Yangzhou et al. fixed mutation operator and crossover operator. Compared with traditional genetic algorithm, this method can produce better individuals [3]. Considering the safety and energy consumption of diversion movement, this paper proposes a genetic algorithm based on gene recombination to optimize the path, reduce the rotation angle of the diversion, increase the smoothness of the path, and then reduce the energy consumption to obtain a relatively short optimal path.

Q. Liang et al. (Eds.): Artificial Intelligence in China, LNEE 854, pp. 345–352, 2022.
https://doi.org/10.1007/978-981-16-9423-3_43

2 Route Planning of Flight Diversion Based on Genetic Algorithm

Through the application of genetic algorithm, chromosome coding, initial population, fitness function, selection, crossover and mutation operators to select the feasible path. According to the principle of survival of the fittest, the expected optimal value is obtained by global search.

2.1 Environment Modeling

The goal of route planning is to find the best route for flight diversion from S to Q. The global path planning is used to obtain constrained and conflict free paths. The movement space of flight diversion is represented by two-dimensional plane graphics, and the restricted area is set as static and known arbitrary irregular polygon. In this paper, the environmental models under different scenarios are designed. It is assumed that the safe distance between diversion and flight restricted area is d, and the process of diversion from starting point to terminal point is regarded as a particle.

2.2 Chromosome Coding

Chromosome coding includes real number coding, binary coding and tree coding, which affects the efficiency of path planning to a certain extent [4]. In this paper, the real number coding method is used to represent the coordinates of points, and the path coordinates are encoded directly. Compared with binary coding, real number coding has several advantages: (1) Using real number as chromosome to participate in genetic operation, which saves time-consuming decoding process; (2) Eliminating Hamming-cliff problem in binary coding; (3) Using real number coded chromosome, there are many kinds of crossover mutation operators to cultivate new individuals, which improves the search efficiency of the algorithm.

2.3 Initial Population

In order to design the optimal route for flight diversion, the population is initialized and a relatively optimal path is found by using genetic operators. Because the initial path is generated randomly, the reachable path and the unreachable path are included in the continuous domain. In the global search, the unreachable path will reduce the search efficiency. When chromosomes cross, the two unreachable paths are likely to produce unreachable paths.

In this paper, the initial population is set in the path generation, the path avoiding the flight restricted area is reserved, and the path encountering the flight restricted area is re selected to accelerate the convergence speed. The initialization method uses a single intersection, setting (x_0, y_0) as the starting point and (x_n, y_n) as the end point, meeting the requirements of $x_n \geq x_0, y_n \geq y_0$. The current point is i, and the $i + 1$ point is generated in the following two cases:

$$\begin{cases} x_{i+1} = (x_n - x_i) * rand + x_i & x_{i+1} \in [x_i, x_n] \\ y_{i+1} = (y_n - y_0) * rand & y_{i+1} \in [y_0, y_n] \end{cases} \tag{1}$$

$$\begin{cases} x_{i+1} = (x_n - x_0) * rand & x_{i+1} \in [x_0, x_n] \\ y_{i+1} = (y_n - y_i) * rand + y_i & y_{i+1} \in [y_i, y_n] \end{cases} \tag{2}$$

The *rand* is a random number between 0 and 1 generated in MATLAB.

2.4 Fitness and Genetic Operators

Fitness is the evaluation of the adaptability of an individual in the environment. The fitness of the path through the restricted area is low. After mutation crossover, the fitness operator can be divided into two situations: if an individual passes through the restricted flight area, it can obtain the fitness through (3); if it doesn't pass through the restricted flight area, it can pass through (4).

$$F_j = 0 \tag{3}$$

$$F_j = \frac{C}{\sum_{i=1}^{n-1} \sqrt{(x_{i+1} - x_i)^2 + (y_{i+1} - y_i)^2}} \tag{4}$$

When C is any constant, the greater the individual's adaptability, the higher the fitness. The fitness value is directly proportional to the probability of the individual being retained to the next generation.

The genetic operators mainly include selection operator, crossover operator and mutation operator. (1) Selection operator: by using fitness function to select better genes to retain and inherit to the next generation; (2) Crossover operator: single point crossover method is used to ensure whether a single individual crosses by generating random numbers; (3) Mutation operator: determines whether an individual mutates by creating a random number.

This paper uses roulette selection operator, single point crossover operator and random mutation operator.

3 Path Optimization Operator Based on Gene Recombination

In this part, the original operator of optimization path is used to determine, select and redistribute individual genes according to the characteristics of individual genes and environment, so as to further improve the individual fitness of traditional genetic algorithm and improve the convergence speed of the algorithm. The optimization operator consists of three parts: original optimization operator, immutable gene judgment operation and variable gene redistribution operation. Through the loop operation of the above three steps, the path optimization is realized, as shown in Fig. 1.

3.1 Original Optimization Operator

When the line of the end point of the multi segment line does not pass through the restricted area, the original optimization operator corrects the multi segment line in the

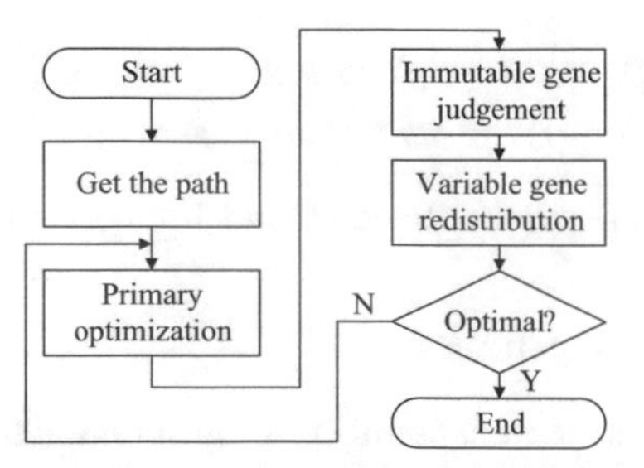

Fig. 1. The optimization flow chart

path to reduce the path length. The process is shown in Fig. 2. Suppose that the current path is a folded segment of S, p_1, p_2, Q, and segment SQ does not pass through the restricted flight area, and then the path is modified to segment SQ, two points are evenly distributed between segments SQ, and two points are set as p'_1, p'_2.

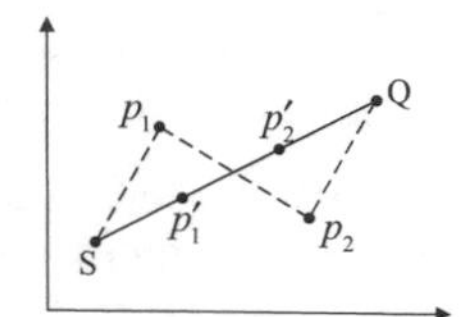

Fig. 2. The path correction diagram

Set the path $P = \{p_1, p_2, p_3 \ldots p_n\}$, where the coordinate point is $(x_i, y_i)i = 1, 2, 3, \ldots n$, and then optimize the path after point i completes the following steps:

Step 1: detect the line segments between p_i and $p_n, p_{n-1}, \ldots, p_{i+2}$ in turn. If $p_k(k = i+2, i+3, \ldots, n)$ doesn't pass through restricted area, continue to the next step;
Step 2: the connecting line between p_k and p_i is regarded as the new path of the individual, and the other points between the two points are evenly distributed on the straight line. Then the coordinates of point j on the line can be expressed as follows:

$$\begin{cases} x_j = \frac{j*(x_k - x_i)}{k-i} + x_i \\ y_j = \frac{j*(y_k - y_i)}{k-i} + y_i \end{cases} \tag{5}$$

Step 3: after traversing all the points on the path in step 2, the path obtained through one optimization is obtained. The vertex of the broken line in the path is an immutable gene, and the point on the line segment is a variable gene.

3.2 Immutable Gene Judgment Operator

After the initial optimization, the distance of the path is obviously shortened, and the smoothness is also improved. However, there is still the problem of unreasonable distribution of gene points, which leads to the large rotation angle of the path, which not only consumes the movement energy of diversion, but also increases the path length. Therefore, the vertex is determined by calculating the angle between the points in the path and the connection between two adjacent points. If the angle is less than 180, the point is regarded as the vertex, that is, the invariant gene; otherwise, it is the variable gene. In order to determine the coordinates and number of the immutable genes in the path, and improve the probability of the variable genes in the optimal position. The angle between the two sides of point $i + 1$ is calculated according to (6):

$$\theta_{i+1} = \arccos\left[\frac{(x_i - x_{i+1})(x_{i+2} - x_{i+1}) + (y_i - y_{i+1})(y_{i+2} - y_{i+1})}{((x_i - x_{i+1})^2 + (y_i - y_{i+1})^2)((x_{i+2} - x_{i+1})^2 + (y_{i+2} - y_{i+1})^2)}\right] \tag{6}$$

3.3 Variable Gene Redistribution

The control of variable gene distribution is a key step in the whole optimization operator. The best effect can be achieved by controlling the distribution position of variable gene. Suppose that the path after the main optimization operator is shown in Fig. 3(a). The black squares are restricted areas. As can be seen from the figure, the starting point of the path is S, and the end point is Q, in the middle of the path, there are seven nodes, namely k_1, k_2, p_1, p_2, k_3, k_4, k_5, p_1 and p_2 are the immutable gene, k_1, k_2, k_3, k_4 and k_5 are the variable gene. It can be seen that due to the unreasonable distribution of variable genes, the operation can not be further optimized. Therefore, this part will redistribute the variable genetic location and control its distribution position. The path after reallocation is shown in Fig. 3(b). The angle of the node p_1 is larger, so the number of variable genes around it is obviously greater than p_2. The path shown in Fig. 3(c) can be obtained by optimizing the path shown in Fig. 3(b) using the main optimization operator. It can be seen that through further optimization, after the redistribution of variable genes, the path becomes shorter and the rotation angle becomes smaller. After several times of optimization according to the above method, the relative optimal path is shown in Fig. 3(d). Compared with Fig. 3(a) and (d), the optimized path is obviously better than the original path.

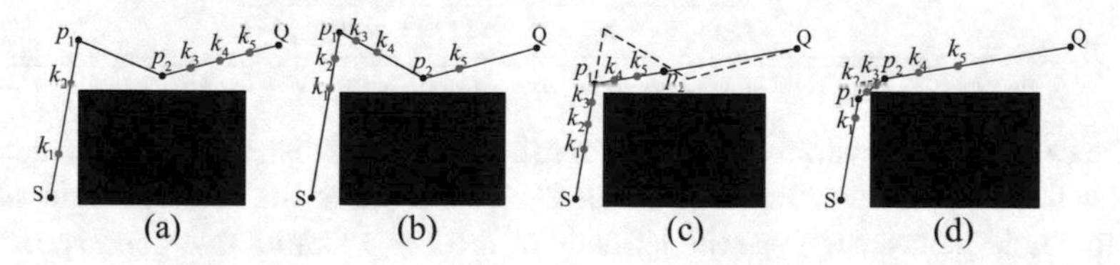

Fig. 3. The optimization process diagram

Therefore, the implementation of the operator is divided into three parts: the distribution of variable genes around the immutable genes, the distribution of variable genes on both sides of the immutable genes, and the calculation of the coordinates of variable genes after redistribution.

(1) The distribution of variable genes around immutable genes: the distribution of variable genes mainly considers the influence of path angle on the distribution of variable genes. The greater the path angle, the greater the exercise energy consumption, and the higher the probability of distribution of variable genes. The implementation steps are as follows:

Step 1: obtain variable gene number. Suppose that the number of genes is n (including the starting and ending points), the number of invariant genes is k, and the number of variable genes is $m = n - k$;
Step 2: the distribution of variable genes is randomly assigned by roulette method according to a certain probability. If the rotation angle of the invariant gene is $\theta = [\theta_1, \theta_2, \theta_3, \ldots, \theta_k]$, the probability of assigning the i-th immutable gene to the variable gene is as follows (7):

$$P_i = \frac{\theta_i}{\sum_{i=1}^{k} \theta_i} \tag{7}$$

Step 3: Roulette method can get the number of variable genes randomly assigned, and the probability can be calculated with (7).

(2) Distribution of variable genes on both sides of immutable genes: in order to ensure the optimal distribution of variable genes in the optimal location of path optimization, after determining the number of variable genes, it is necessary to quantitatively distribute the variable genes on both sides of the invariant genes. The number of variable genes was still allocated by roulette. The probability of selecting the left side of the $i + 1$ immutable gene is P_l, and the probability of selecting the right side is P_r. The probability of variable gene distribution on both sides of invariant gene can be calculated by (8).

$$\begin{aligned} P_l &= \frac{\sqrt{(x_i - x_{i+1})^2 + (y_i - y_{i+1})^2)}}{\sqrt{(x_i - x_{i+1})^2 + (y_i - y_{i+1})^2)} + \sqrt{(x_{i+2} - x_{i+1})^2 + (y_{i+2} - y_{i+1})^2}} \\ P_r &= \frac{\sqrt{(x_{i+2} - x_{i+1})^2 + (y_{i+2} - y_{i+1})^2)}}{\sqrt{(x_i - x_{i+1})^2 + (y_i - y_{i+1})^2)} + \sqrt{(x_{i+2} - x_{i+1})^2 + (y_{i+2} - y_{i+1})^2}} \end{aligned} \tag{8}$$

(3) The calculation of variable genetic coordinates after redistribution: the calculation of variable gene coordinates is divided into two parts: the left and right sides of immutable genes. Suppose the left side of the $i + 1$ immutable gene is p_i, p_i and p_{i+1} the midpoint of the line between and is p_l, then the position of the left variable gene is evenly distributed p_i and p_{i+1}. Similarly, suppose the right side of the $i + 1$ immutable gene is p_{i+2}, the midpoint of the line between p_{i+1} and p_{i+2} is p_r. Assuming that the left variable gene number is n and the right variable gene number is m, the coordinates of the k variable gene on both sides of the immutable gene can be calculated by formula (9) and (10).

$$\begin{cases} x_k = \frac{k^*(x_{i+1} - x_i)}{2(n+1)} + \frac{x_{i+1} + x_i}{2} & k = 1, 2, \ldots, n \\ y_k = \frac{k^*(y_{i+1} - y_i)}{2(n+1)} + \frac{x_{i+1} + x_i}{2} & k = 1, 2, \ldots, n \end{cases} \tag{9}$$

$$\begin{cases} x_k = \frac{k^*(x_{i+2}-x_{i+1})}{2(m+1)} + x_{i+1} & k = 1, 2, \ldots, m \\ y_k = \frac{k^*(y_{i+2}-y_{i+1})}{2(m+1)} + y_{i+1} & k = 1, 2, \ldots, m \end{cases} \tag{10}$$

4 Experimental Verification

In order to verify the effectiveness of the algorithm, experiments are carried out in the blocky and scattered restricted area environment. The experimental results are shown in Fig. 4(a) and (b). The initial environment is composed of coordinate system, and the shadow part is obstacle area. Genetic algorithm parameters are set as follows: population size is 100; crossover probability is 0.8; mutation probability is 0.3; evolution algebra is 30.

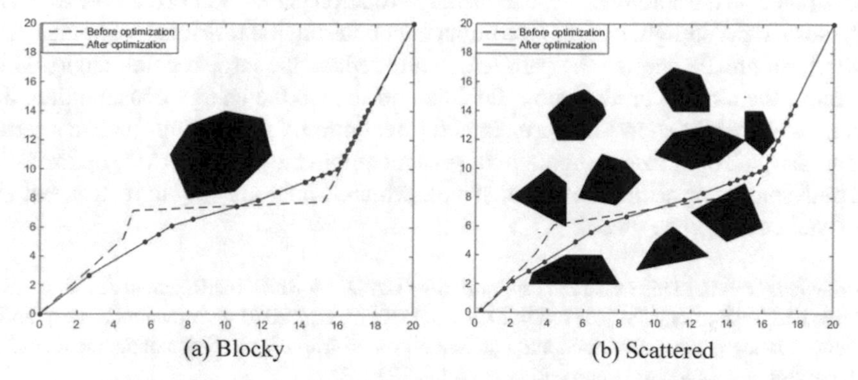

(a) Blocky (b) Scattered

Fig. 4. The smooth path in different flight restricted area environment

The Fig. 4(a) shows the path planning of blocky flight restricted area, in which the blue path is the initial optimized path, and the red path is the twice optimized path after gene rearrangement. The starting point is (0, 0) and the ending point is (20, 20). It can be seen from the figure that after two times of optimization, the path is significantly shorter than the previous path, with higher smoothness, and variable gene points are distributed around the invariant gene points. The Fig. 4(b) shows the path planning in scattered flight restricted area environment. It can be seen that the above algorithm is used to minimize the large rotation angle. It not only improves the safety of diversion, but also reduces the energy consumption. It can be seen from the figure that the larger the corner, the more concentrated the distribution of gene spots.

It can be seen from Table 1 that in the three cases, the path optimization operator can not only shorten the path again, but also greatly reduce the rotation angle and improve the safety of diversion. From the table data, we can see that the algorithm is better than the standard genetic algorithm in path length and corner. Due to the optimization of the path operator, the evolution speed is accelerated and the path length under various environments based on genetic algorithm is reduced.

Table 1. Comparison of data before and after optimization

	The length of the path		Rotational corners sum (degree)	
	Before optimization	After optimization	Before optimization	After optimization
Blocky	29.3169	28.8039	145.0151	58.7275
Scattered	31.0504	29.6238	119.2044	79.1175

5 Conclusion

In this paper, genetic recombination based on genetic algorithm is used to optimize the route of flight diversion. According to the size of the rotation angle, the gene points are redistributed on the chromosome, and the large rotation path is optimized to be a smooth path close to the straight line angle without encountering the restricted flight area. This method can greatly shorten the path length and reduce the large rotation angle, which can make the aircraft turning more flexible and reduce the energy consumption. The experimental results show that the method further optimizes the optimal path to a certain extent. The path optimization operator of gene recombination realizes twice optimization and path smoothing within the safe distance, which is not only easy to realize, but also has faster convergence speed.

Acknowledgement. This research is supported by CAAC North China Regional Administration Science and Technology Project (NO: 201904, 202002), and 2020 civil aviation safety ability building funding project: Research and application on air traffic operation management analysis technologies based on multi-source big data (NO: 202072).

References

1. Li, Y.X., et al.: Network-wide flight trajectories planning in China using an improved genetic algorithm. In: Digital Avionics Systems Conference. IEEE (2016)
2. Davies, C., Lingras, P.: Genetic algorithms for rerouting shortest paths in dynamic and stochastic networks. Eur. J. Oper. Res. **144**(1), 27–38 (2003)
3. Chen, Y., Bi, H., Zhang, D., Song, Z.: Dynamic airspace sectorization via improved genetic algorithm. J. Mod. Transp. **21**(2), 117–124 (2013). https://doi.org/10.1007/s40534-013-0010-2
4. Wang, M.M., et al.: Generate optimal grasping trajectories to the end-effector using an improved genetic algorithm. Adv. Space Res. **66**(7), 1803–1817 (2020)

Power Grid Routing Based on Spatial Constraints of Remote Sensing Images and Power Planning

Hao Chang[1](✉), Liang Guo[1], Minzi Zhang[1], Xian Zhou[1], Xin Li[2], and Xiangwei Zhao[2]

[1] GroupState Grid Taizhou Power Supply Company of Jiangsu Electric Power Co., Ltd., Taizhou 225309, Jiangsu, China
616900289@qq.com

[2] Jiangsu Power Design Institute, Co., Ltd. of China Energy Engineering Group, Nanjing 211102, China

Abstract. A method of power grid line planning using power space planning constraint information is studied. This work uses image recognition technology to identify and classify satellites or aerial images in the planned area, so as to obtain the type of land in the planned area. Combined with the relevant data in the planning database, the classified land range of the planning area is obtained and the planning scene is set, and then the expert evaluation method is used to assign the cost of the classified land features. Subsequently, a step-by-step path finding algorithm based on A* was designed to find a series of economically better routes in the assigned planning area. Finally, the route to be selected was comprehensively evaluated through the analytic hierarchy process, and the optimal route was obtained. Through a series of optimization design, this paper can get the result in a short time and ensure that the algorithm can be applied to the problem of power grid planning.

Keywords: Power planning · Line selection algorithm

1 Introduction

The route selection of transmission line is the premise of the construction of Electrical Grid. The effectiveness of its design affects the investment and operation cost of power grid and running stability. The selection of transmission lines requires line selectors to comprehensively consider the factors such as terrain, land-form, geology, traffic, environment, policies and regulations of various regions and departments. The traditional grid line selection work is to select relatively suitable lines through the electric power staff according to certain principles and experience. This line selection method is not only costly, but also difficult to ensure the rationality of power lines. In order to reduce the difficulty of line selection and reduce the workload of line selection personnel and save the

Q. Liang et al. (Eds.): Artificial Intelligence in China, LNEE 854, pp. 353–359, 2022.
https://doi.org/10.1007/978-981-16-9423-3_44

time and cost of line selection, the use of intelligent algorithm to select transmission lines has become the focus of power designers and related researchers [1]. In recent years, domestic and foreign researchers have made a lot of progress in the research of transmission line routing algorithms. Datta A, et al. [2] is an earlier study on the problem of transmission line selection. This work divides the factors affecting line selection into three categories: natural, social and technical, and evaluates the relative importance of each factor to power line selection. Monteiro et al. [3] used dynamic programming algorithm to generate paths containing cumulative costs, and then selected them according to transmission line construction costs and other economic factors respectively. This method takes into account the influence of terrain on route selection, that is, both the average terrain cost and the local terrain cost considering the direction are considered at the same time. In addition, it also considers the influence of the corner tower position on the cost of the transmission line. Eroglu, et al. [4] used genetic algorithm to optimize the transmission line path and provide alternative schemes for line designers. Shu, et al. [5] proposed a method based on raster GIS environment for transmission line path planning and power grid evaluation in order to solve spatial power grid planning.

2 Problem Description

Transmission line path has certain particularity, and its selection is constrained by many conditions, mainly including:

1) The principle of the least number of path corners;
2) The principle of shortest route;
3) The principle of minimum route crossing;
4) Path spacing (the distance between towers) should be guaranteed to correspond to the voltage grade of the transmission line;
5) Close to vertical when crossing with important linear features such as railways and expressways; When crossing with weak current lines such as communication lines, the crossing Angle of first-level weak current line is greater than or equal to 45°, and the crossing Angle of second-level weak current line is greater than or equal to 30°;
6) Path rotation Angle does not produce obtuse Angle. The value of A cannot be obtuse Angle, that is, it does not cross the thick black line, and the degree of rotation Angle is as small as possible;
7) On the premise of feasible technology and reasonable cost, the route should be selected "along the river, along the road and along the infrastructure" and the boundary areas of provinces, cities and counties, so as to save land resources and reduce the impact on the environment;
8) Try to avoid larger villages, concentrated residential areas and areas with relatively dense houses, and reduce the number of houses across the border;

3 Method Analysis

In order to solve the above problem, this paper first adopts image recognition technology to extract the raster map of land types in the planned area from remote sensing, and then constructs the cost raster map required by the planned route through weighted grouping method. Then, this paper proposes a skipping route selection algorithm based on A* to find the economical optimal route in the region of assignment planning. In this paper, the results can be obtained in a relatively short time through a series of optimization design to ensure that the algorithm can be well applied to the problem. Include the following steps:

3.1 Block Type Raster Map Extraction

Deep learning intelligent classification algorithm based on full convolution method is adopted to recognize remote sensing or aerial images in the planned area as corresponding land parcel classification images. The area to be planned will be classified into a labeled rectangular grid map according to the land type, and vector information such as traffic lines and communication lines will be extracted from it. Where the raster diagram is represented as $M = \{g_k \mid k \in N^{m \times n}\}$.

In the operation, we mainly divide the land types into four categories: one is mostly farmland and other areas, which can stand towers; the other is water system and other areas, which can be crossed, and in general, no towers should be erected as far as possible; Three for residential areas, roads and other areas, can be crossed, not erected towers. Four types of planning prohibited areas, can not be crossed and erect towers. The main purpose of this work is to subdivide the characteristic attributes and control value of different ground objects, and formulate the path search criteria and limiting conditions, so as to meet the standard requirements and criteria of the general route selection.

3.2 Cost Raster Chart Construction

According to different labels of plots and other social, economic and natural factors, different weights are assigned to each grid plot. For four different types of plots, weight assignment was conducted roughly in accordance with the following

Table 1. Land type classification

Land Type	Description	Weight Range
I	Area where tower base can be built	[100,80)
II	Transmission line can be crossed, try not to build tower	[80,60)
III	Transmission line can be crossed, but tower cannot be built	[60,20)
IV	Transmission line cannot be crossed and tower cannot be built	[20,0)

table. At the same time, according to the above constraint conditions, the weight of the plot is adjusted for the neighboring villages, traffic lines and other special sites, and the weighted linear combination and mean filtering are used to smooth the weight of the graph.

3.3 Improved A* Skip Algorithm

The algorithm flow chart is shown below.

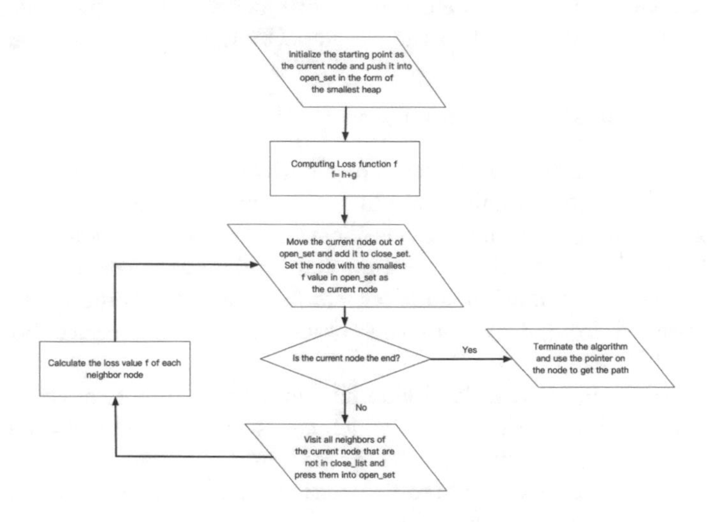

Fig. 1. Flow chart of routing algorithm

In the algorithm, we calculate the cost from the grid to the starting point in the planning path, i.e. $F(\mathrm{g_{origin}}, \mathrm{g}_k)$, and iteratively select the node with the least cost to form the planning path to be selected. Among them, the calculation formula of cost value F is as this: $F(\mathrm{g_{origin}}, \mathrm{g}_k) = \mathrm{G}(\mathrm{g_{origin}}, \mathrm{g}_k) + \mathrm{H}(\mathrm{g}_k, \mathrm{g_{end}})$. Among them, $G(\mathrm{g_{origin}}, \mathrm{g}_k)$ is the tested cost, and g_k represents the actual cost value of the raster distance from the starting point, which will be updated continuously in the iteration process. Its calculation formula is,

$$\mathrm{G}(\mathrm{g_{origin}}, \mathrm{g}_k) = \mathrm{G}(\mathrm{g_{origin}}, \mathrm{g}_{k-1}) + G(\mathrm{g}_{k-1}, \mathrm{g}_k)$$

$$G(g_{k-1}, g_k) = G(g_{k-1}, g_k) + \sum \{w_m \mid g_m \ \mathit{Between}\, \mathrm{g}_{k-1} \ \mathit{and}\ \mathrm{g}_k\ \mathit{straight\,path}\ \}, g_k \in N_f$$

$$N = \{\mathrm{g}_N \mid \mathrm{D}(\mathrm{g}_{k-1}, \mathrm{g}_k) \leq \mathrm{c}\} \xrightarrow{\text{filer}} N_f$$

Where, D represents the Euclidean distance between grids, and $filter$ operation eliminates grids that do not conform to relevant constraints of power lines.

3.4 Planning Path Evaluation and Visualization

Evaluate the multiple paths obtained in Step 3 and output the visual planning path scheme, the intelligent evaluation result of the scheme and the analysis report of the influencing factors of the path. The content includes the path length, the number of turning angles, the information of crossing points and the number of crossing points, and the distance mainly affects the distance of ground objects. Finally, through the analytic hierarchy process, comprehensive evaluation of the selected line, the optimal path.

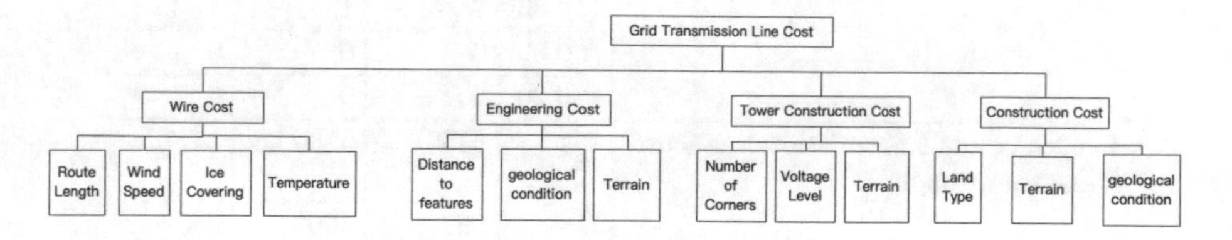

Fig. 2. Influencing factors of line cost

In the use of analytic hierarchy process (AHP) to choose A comprehensive evaluation of transmission line, must be the evaluation factors of shown in Fig. 9, $X = \{x_1, x_2, ...x_n\}$ build paired comparison matrix, with a_{ij} said the factor relative to the first j factor of the comparison results, paired comparison matrix A is shown in the following type:

$$A = (\mathrm{a}_{ij})_{\{\mathrm{n}\times\mathrm{n}\}} = \begin{pmatrix} a_{11} & a_{12} & \cdots & a_{1n} \\ a_{21} & a_{22} & \cdots & a_{2n} \\ \cdots & \cdots & \cdots & \cdots \\ a_{n1} & a_{n2} & \cdots & a_{nn} \end{pmatrix}$$

Calculated by the matrix A, the influence of factors on the total line cost lower value obtained weights $w = (W_1, W_2, ..., W_n)^T$, contrast effect weight w is used to calculate the maximum eigenvalue of matrix λ_{max}, to observe the judgment matrix and calculate the consistency index is reasonable, calculation formula is as follows:

$$AW = \lambda_{\max} w$$

$$CI = \frac{\lambda_{\max} - n}{n - 1}$$

Finally, the weight vector obtained is normalized, and the normalized vector is the weight coefficient of the evaluation factor. Combined with the weight coefficient, the route with the highest score is the selected route.

4 Experiment

The experimental area used in this paper is about 10.5 km long and 8.14 km wide. The experimental area includes residential area, planning area, expressway, complex ground features, transportation convenience, crossing with traffic lines,

Table 2. Experiment result

		Result 1	Result 2	Result 3
Tortuosity coefficient		1.0340	1.0342	1.0371
#Corners		16	16	16
Distribution of degrees of rotations	(0,10]	14	15	15
	(10,30]	2	1	1
	(30,60]	0	0	0
	(60,90]	0	0	0
Length of the transmission line when it passes through	I	5459	5269	5317
	II	987	1002	953
	III	1559	1744	1723
	IV	197	218	226

residential area, general planning area and other influencing factors. Overall Accuracy (OA) was adopted as the evaluation index for the results of image recognition and classification. The accuracy of road and river system recognition can reach more than 97%, and the overall accuracy is 97.15%, which still has a high classification accuracy on large scale data sets. From the perspective of time spent, the running time of our result is 393.06 s.

In the algorithm, the starting point and end point coordinates of the selected line are input, and the weight of each influencing factor is set in parameter setting. Then, in the function of grid generation, the whole line selection region is taken as the grid range, and a cell with a size of 1 m $\times$ 1 m is used to create channel grid.

5 Conclusion

In order to solve the problem of transmission line path planning, an intelligent power line selection system based on GIS is designed and implemented. The experimental results show that given the starting and ending points of the path and the related constraints, the system can quickly select a reasonable transmission line, which can provide decision support for the power designers. This system not only improves the efficiency and quality of line selection, but also greatly reduces the cost of line selection, so it has broad application prospects.

Acknowledgments. This paper is supported by Research on Power Grid Spatial Planning Technology Based on Deep Learning of Geographic Information (J2020003).

References

1. Yizhuo, L.: Design and implementation of three-dimensional power line selection system supported by skyline. Geospatial Inf. **05**, 30–32 (2010)
2. Datta, A.L., Verma, S., Gupta, A.K.: A GIS application for power transmission line sitting. In: Symbiosis Institute of Geoinformatics, pp. 10–13 (1996)
3. Monteiro, C., et al.: Compromise seeking for power line path selection based on economic and environmental corridors. IEEE Trans. Power Syst. **20**(3), 1422–1430 (2005)
4. Eroğlu, H., Aydın, M.: Genetic algorithm in electrical transmission lines path finding problems. In: 2013 8th International Conference on Electrical and Electronics Engineering (ELECO), pp. 112–116. IEEE (2013)
5. Shu, J., Lei, W., Li, Z., Shahidehpour, M., Zhang, L., Han, B.: A new method for spatial power network planning in complicated environments. IEEE Trans. Power Syst. **27**(1), 381–389 (2011)

Extraction Method of Power's Corridor Centre Based on Grid Inspection Image and K-Means Algorithm

Xiangwei Zhao[1(✉)], Xin Li[1], Hao Chang[2], Jun Yao[2], and Beibei Weng[2]

[1] Jiangsu Power Design Institute, Co., Ltd. of China Energy Engineering Group, Nanjing 211102, China
zhaoxiangwei@jspdi.com.cn

[2] GroupState Grid Taizhou Power Supply Company of Jiangsu Electric Power Co., Ltd., Taizhou 225309, Jiangsu, China

Abstract. The acquisition and update of the ground power corridor is of great significance in the power distribution network planning. Two traditional methods are used to achieve this goal, that is, manual mapping and updating of Grid GIS data. The former brings more workload to the planning work and greatly reduces the efficiency, while the later has a greater impact on the planning results due to the current potential of GIS data. In this paper, based on the image data of power grid operation and maintenance inspection, we propose a extraction method of power's corridor centre. Firstly, the location information of inspection image data is extracted. Secondly, the tower image is processed by K-means clustering algorithm, and the clustering centroid of each tower is obtained. Finally, the above ground power corridor is generated. In order to test the validity of the algorithm, the actual inspection data of the high-voltage power grid and distribution network are taken as the experimental data, and the actual coordinates of the poles and towers are taken as the inspection data. The verification results showed that the clustering error of high pressure patrol was basically distributed within ±1.5 m, and the error of some towers was up to 4 m. The clustering errors of distribution network inspection were all distributed within ±2.0 m. There was no clustering error appear in the two experimental results, which could meet the requirements of channel analysis accuracy of distribution network planning, and greatly improve the depth of data mining and data utilization.

Keywords: Distribution network planning · Above ground power corridor · K-means · Network operation and maintenance inspection

1 Introduction

Distribution network planning is a plan for the expansion and transformation of the planning system, which based on the research and analysis of the current

Q. Liang et al. (Eds.): Artificial Intelligence in China, LNEE 854, pp. 360–367, 2022.
https://doi.org/10.1007/978-981-16-9423-3_45

situation of the regional distribution network and the future load growth. It takes the operation economy as the index to select the optimal planning scheme, so that the power company can realize the maximum benefit [1,2].

One of the most important content in distribution network planning is the acquisition of power corridor, which plays an significant role in the reliable and reasonable operation of regional power grid planning [3]. Therefor, Therefore, several scholars have conducted in-depth research on this issue. Pei Huikun et al. [4] used UAV tilt photography to reconstruct the three-dimensional model of power corridor. However, the method makes the workload and cost increase greatly in the distribution network planning. Li baisong et al. [5] used lidar data to extract power corridor of transmission line quickly, which also has the problem of low efficiency and not suitable for large area. On the other hand, the GNSS measurement and positioning equipment is also the main way of manual field work, while this way also increases a lot of field work and increases the cost of data acquisition [6]. In addition, using grid GIS data in the system [7–9] is also a way to quickly obtain power corridor, while the maintenance cycle of Grid GIS is slow and the data update is delayed, which easily leads to the weak current situation of the data and affects the rationality of the distribution network planning scheme.

In recent years, the UAV inspection work of operation and maintenance department has been carried out in depth. Therefore, in view of the limitations of the traditional way, this paper proposes the technology of obtaining the center line of power corridor by using the UAV data information of operation and maintenance inspection combined with K-means clustering algorithm. Specifically, based on the inspection image of power grid, this paper uses K-means clustering algorithm to extract the center of power corridor, and verifies the correctness and applicability of this method with the actual inspection data of a high-voltage power grid and distribution network in Jiangsu Province.

2 The Ground Power Corridor

Obtaining accurate information of the ground power corridor can provide reference and guidance for the development of urban distribution network planning. One of the traditional ways is to export and obtain the corresponding data information through the internal GIS data platform, but the uncertainty of the maintenance and update cycle of GIS information can not guarantee the current situation of the corridor information, which affects the rationality of the planning scheme. Another way is to use manual on-site layout to obtain power corridor information. However, this method has heavy workload and the complex on-site condition makes the corridor information incomplete and affects the overall planning effect.

In recent years, with the development of Unmanned Aerial Vehicle (UAV) inspection in power network, the updating speed of power corridor information is also accelerated. The operation and maintenance unmanned aerial vehicles (uavs) carry different types of sensors to patrol the electric poles and towers in the working area and obtain different types of data of the Poles and towers. The uavs are equipped with GPS positioning modules, the Position Information of the UAV can be recorded at the moment of exposure, and the central position of the operational tower can be obtained by using the data of different positions of the operational track of patrol inspection, so as to provide a ground power corridor with better current situation for the distribution network planning, to assist the distribution network planning work.

(a) Patrol inspection of distribution network

(b) High voltage inspection

Fig. 1. Power grid patrol

3 K-Means Clustering Algorithm

K-means clustering algorithm is an iterative clustering analysis algorithm. Its core idea is to randomly select k objects as the initial clustering center, calculate the Euclidean distance between each object and each seed clustering center, and then assign each object to the nearest clustering center [10,11]. In the concrete realization, the European distance is taken as the similarity evaluation index. The closer the Euclidean distance between two objects, the greater the similarity between them, and the greater the probability of belonging to the same class. The cluster center and the assigned object represent the same cluster. It varies with the number of samples allocated. The cluster center will be recalculated according to the existing objects in the cluster, iterating until a termination condition is satisfied. The termination condition can be that no (or minimum number) objects are reassigned to different clusters, no (or minimum number) cluster centers change again, and the sum of squares of errors is locally minimum. The algorithm considers that the class cluster is composed of close objects, so the final goal is to get a compact and independent class cluster [12,13].

In K-means clustering algorithm, for the sample data defined in Euclidean space, the distance between each point and the centroid is Euclidean distance, which can be expressed as:

$$\boldsymbol{D}_i = \sum_{j=1}^{m} (c_{i,j} - x_j)^2 \tag{1}$$

Where D_i is the Euclidean distance; $c_{i,j}$ is the location of the calculated sample; x_j is the current position of the iteration centroid. The sum of squared error (SSE) is used as the objective and cost function to evaluate the clustering results:

$$\boldsymbol{SSE} = \sum_{i=1}^{k} \sum_{x \in c_i} \boldsymbol{D}_i \left(c_i, x\right)^2 \tag{2}$$

The optimal clustering result should ensure that SSE value reaches the minimum value. For each sample, the centroid closest to all sample data is the centroid with the highest similarity. When all samples can find the centroid with the highest similarity, the above formula (2) differentiates the centroid and obtains the new position of the centroid.

$$c_k = \frac{1}{m_k} \sum_{x \in \boldsymbol{C}_k} x_k \tag{3}$$

Where m_k is the number of centroids, and the new centroid will move to the average coordinate value of m clusters whose centroids are changed before the iteration. When the distance between the new centroid and the original centroid is less than the set value ε, the iteration of K-means clustering algorithm is terminated, and the final centroid position is obtained.

4 Case Study and Analysis

4.1 High Voltage Transmission Inspection

In order to verify the application accuracy of K-means clustering algorithm in distribution network channel statistics, this paper takes the high voltage transmission inspection data as an example. The actual inspection data of 200 tower of 5904 line of a 500 kV transmission line in Jiangsu Province is selected for calculation, and the cluster number k = 1 is selected. The results of cluster calculation are shown as below.

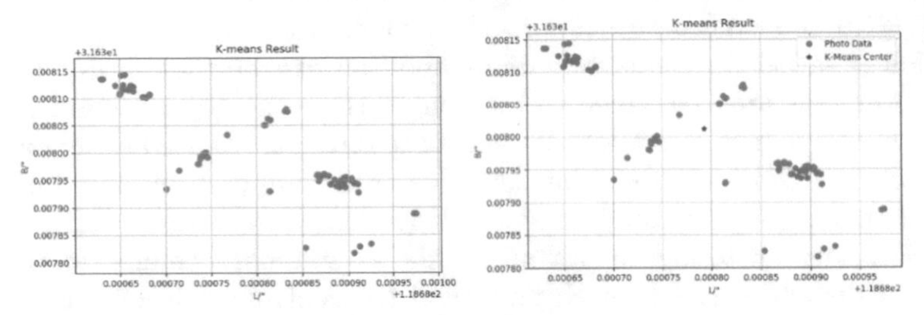

(a) Location distribution of patrol data (b) Clustering results of single file data

Fig. 2. Results of 200 # K-means clustering calculation on line 5904

Figure 2(a) shows the location distribution of patrol inspection. The analysis shows that there are a large number of photos in a single patrol inspection of high-voltage power transmission, and the photos are evenly distributed on both sides of the tower. Through clustering algorithm, the location of clustering centroid is obtained, as shown in the red five pointed star in Fig. 2(b) above. It is obvious that the clustering centroid is located in the distribution center of patrol inspection photos.

On the other hand, due to the large span between high-voltage transmission poles and towers, which is basically two orders of magnitude larger than the

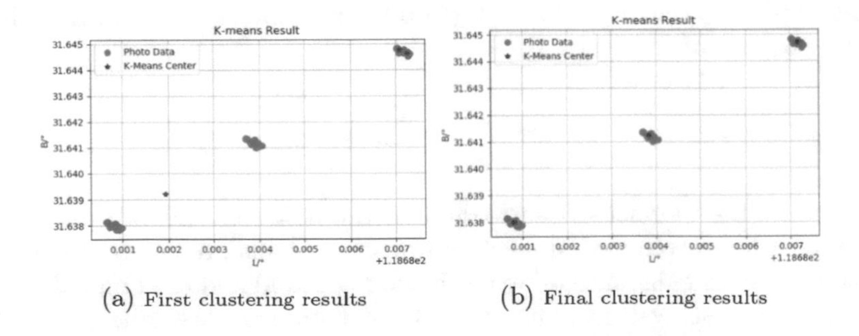

(a) First clustering results (b) Final clustering results

Fig. 3. Overall clustering of 200–202 towers on line 5904

distance between inspection photos of single tower, centralized clustering can be carried out for multi tower inspection. Taking the inspection images of 200 #–202 # - three base towers of 5904 line as the experimental data, the clustering number k = 3, and the clustering results of K-means are shown in Fig. 3.

After several times of clustering calculation, the k-means algorithm can effectively divide the tower position into three categories, and the data in each category is the most similar. Through the clustering centroid, the calculated position of each tower can be obtained, and the transformation from the inspection image data to the tower position data can be realized. Taking the actual position of each tower as the accurate value, the clustering value error of the algorithm can be obtained, The error distribution is shown in Fig. 4 below.

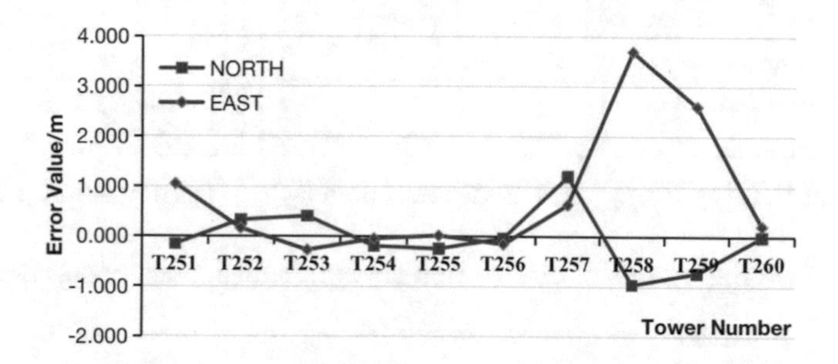

Fig. 4. Error distribution of clustering calculation for high voltage patrol inspection

By analyzing the error sequence, it can be seen that most of the clustering errors of the two coordinate components are within ±1.5 m, and the error of individual tower position is about 4 m, but they can fully meet the accuracy of distribution network power channel analysis.

4.2 Distribution Network Transmission Inspection

Different from the high-voltage power grid inspection, due to the small span between towers, the centralized clustering algorithm is easy to cause the local optimal situation and can not obtain the optimal clustering results. Therefore, the circular single file processing is carried out. Taking the 10 kV distribution network inspection data of a certain area in Jiangsu Province as an example, 1–6 towers are selected for experimental analysis, The clustering results are shown in Fig. 5.

The corresponding position of each tower can be obtained by clustering algorithm, and the error distribution of clustering algorithm results can also be calculated by analyzing the precise position of tower, as shown in Fig. 6 below.

Through the analysis, it can be seen that the error distribution of the coordinate component of the clustering is basically within ±2 m, and there is no gross error jump phenomenon in the clustering results, which can meet the accuracy requirements of channel analysis in distribution network planning.

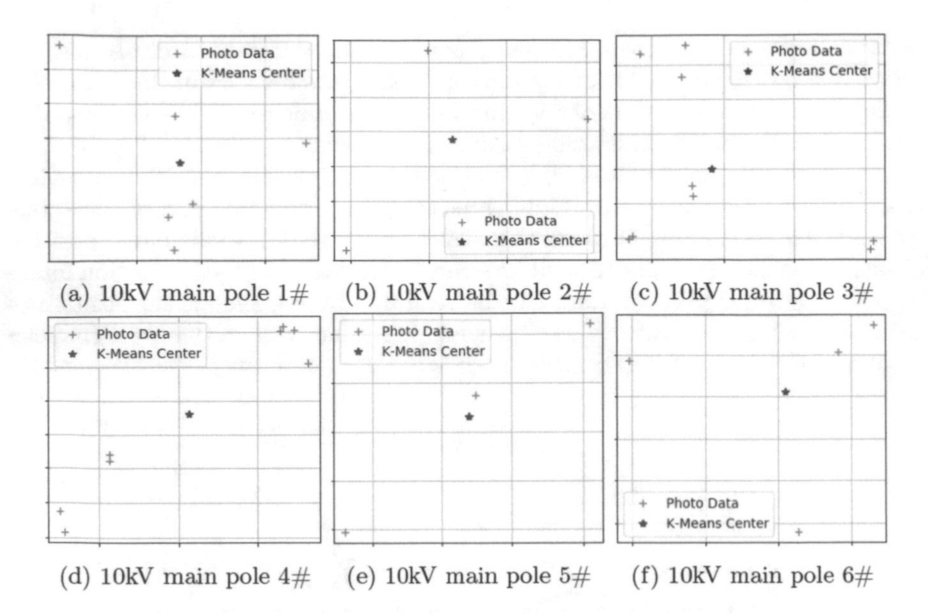

(a) 10kV main pole 1# (b) 10kV main pole 2# (c) 10kV main pole 3#

(d) 10kV main pole 4# (e) 10kV main pole 5# (f) 10kV main pole 6#

Fig. 5. Clustering results of distribution network transmission inspection data

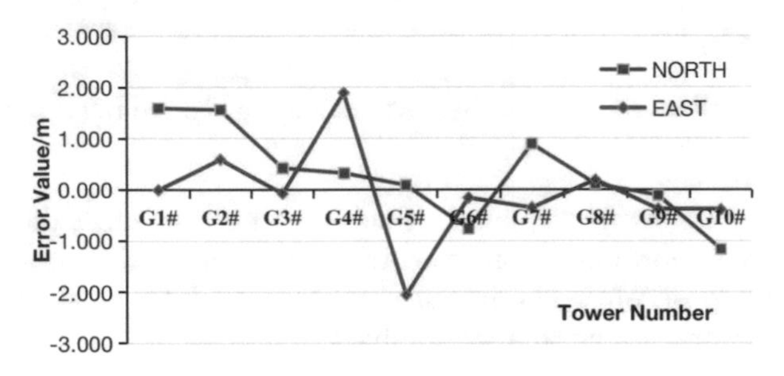

Fig. 6. Distribution network inspection clustering calculation error distribution

5 Conclusion

Determining the current situation of power line corridor is one of the main tasks in distribution network planning. In order to obtain the power line corridor with better current situation, the operation inspection department regularly carries out the power grid UAV inspection work. Taking the power grid operation and maintenance inspection image as the basic data, this paper proposes to use k-means clustering algorithm to realize the accurate clustering calculation of tower coordinates. At the same time, the correctness of the algorithm results is verified by high-voltage inspection data and distribution network inspection data, The results show that the proposed algorithm can provide a reasonable power line corridor for distribution network planning.

Acknowledgments. This paper is supported by Research on Power Grid Spatial Planning Technology Based on Deep Learning of Geographic Information (J2020003).

References

1. Wang, Y., Huiyong, H.: Distribution network planning based on improved particle swarm optimization. China Equip. Eng. (2019)
2. Zhang, H.: Research on urban distribution network planning and existing problems. Eng. Technol. Res. (2019)
3. Li, X.: On the role of medium and low voltage power corridor in regional grid planning. North China Power Technol. (2009)
4. Pei, H., Jiang, S., Lin, G., Huang, H., Jiang, W., Yang, C.: 3D reconstruction of transmission route based on UAV oblique photogrammetry. Sci. Surv. Mapp. (2016)
5. Bai-Song, L.I., Yun-Ni, L.I., Lai, X.D., Yong, D.U., Dong, X.Q., Jiang, S.: On the method of rapid extraction for lidar data of power transmission line corridors. Hubei Electr. Power (2014)
6. Liu, H.J., Liang, H.E., Wang, W.: Application of RTK surveying and mapping technology in power engineering construction (2017)
7. Fang, C.S., Huang, K., Yuan, H.E., Yun-Guang, S.U.: Research and implementation of grid visualization platform based on GIS. Inf. Technol. (2019)
8. Chen, J., Wan, G.C., Sun, G.N., Liu, W.G., Yao, S.: Development and application of GIS platform decision-making support system for power network planning. In: Proceedings of the Chinese Society of Universities for Electric Power System and its Automation (2010)
9. Yushan, Y., Hu, X.: Research on smart grid planning system based on GIS. China Electr. Power Educ. (2009)
10. Lu, L.L., Qin, J.T.: Multi-regional logistics distribution center location method based on improved k-means algorithm. Comput. Syst. Appl. (2019). B. School
11. Zhang, Z., Guo, X., Zhang, K.: Clustering center selection on k-means clustering algorithm. J. Jilin Univ. (Inf. Sci. Edit.) (2019)
12. Tingbo, M.A., Liu, T., Jianguo, X.U.,Liu, X.: Information analysis of auto market competition based on improved k-means cluster algorithm. J. Shandong Univ. Sci. Technol. (Nat. Sci.) (2019)
13. Wang, J., Ma, X., Duan, G.: Improved k-means clustering k-value selection algorithm. Comput. Eng. Appl. (2019)

Improved Stereo Matching Algorithm Based on Adaptive Grid for Fixed-Wing UAVs

Hua Xia, Hongyuan Zheng(✉), and Xiangping Zhai

College of Computer Science and Technology/College of Artificial Intelligence, Nanjing University of Aeronautics and Astronautics, Nanjing 211106, China
11122067@qq.com

Abstract. Modern unmanned aerial vehicles (UAVs) rely on binocular ranging modules to complete tasks such as obstacle avoidance, 3D reconstruction, and terminal strikes. However, the limited computing resources and high flight speed all put forward requirements for the real-time performance of the ranging algorithm. Thus, in this paper, we focus on how to make the algorithm dynamically allocate computing resources according to the changes of the UAV attitude to improve the system efficiency. To this end, we propose an improved semi-global matching method based on adaptive grid for fixed-wing UAVs. Experimental results demonstrate that the proposed method can effectively adapt to the changes in the speed and attitude of the UAV, and improve the real-time performance of the ranging module.

Keywords: UAV · Stereo matching · Adaptive grid

1 Introduction

The booming development of UAV technology makes it rapidly heat up in the civilian and military fields. More advanced flight controllers, vision modules, and energy systems promote the development of UAVs in a multi-purpose and intelligent direction such as transportation, inspection, reconnaissance, and terminal strikes.

Stereo matching technology has become the core of binocular vision algorithms because it can generate dense disparity maps, and has been widely used in 3D reconstruction, robot obstacle avoidance, and automatic driving [1]. Stereo matching algorithms are divided into local [2], global [3], semi-global [4,5] and deep learning-based algorithms [6]. The semi-global stereo matching algorithm is an improvement of the global algorithm. By transforming the two-dimensional image optimization problem into a one-dimensional optimization problem with multiple paths, the computational complexity is reduced while preserving the precision of the algorithm. Li et al. [7] used the census transform [8] to calculate the matching cost, and used the image pyramid to perform 8-path SGM [9] processing on each layer, and established a coarse-fine parallel stereo matching algorithm to reduce the matching time.

Q. Liang et al. (Eds.): Artificial Intelligence in China, LNEE 854, pp. 368–375, 2022.
https://doi.org/10.1007/978-981-16-9423-3_46

This paper proposes a stereo matching algorithm to solve the problem of the poor real-time performance of fixed-wing UAVs to obtain scene depth information. Specifically, we divide the image into several grids, and the stereo matching algorithm will adjust the calculation method according to the attitude of the UAV, and different calculation methods are assigned to different grids.

2 Problem Formulation and Preliminary Definitions

Classical stereo matching algorithms such as PatchMatch and SGM are not specifically optimized for UAV platforms. For example, the real-time changes in the attitude of the UAV are not reflected in the image processing stage. This paper is interested in making the UAV maintain a high-frequency perception of obstacles when its attitude changes. Some assumptions are used to simplify the problem:

- The obstacle is stationary.
- The gimbal is mounted on a fixed-wing UAV.

According to the kinematic constraints on fixed-wing UAVs, we introduce three influencing factors, namely forward velocity $v(t)$, yaw rate $y(t)$, and pitch rate $p(t)$. We are interested in finding a strategy to make the UAV subject to the constraints of these three factors during image processing and produce the corresponding output. At the same time, we divide the image into nine grids, as shown in Fig. 1, each of which represents a receptive field. We define the system as the following states: locked state LS, forward state FS, yaw state YS, pitch state PS. The state of the system is derived from the following equation:

$$state(t) = f(v(t), y(t), p(t)) = \begin{cases} LS, & v(t) = 0 \\ FS, & v(t) > 0 \\ YS, & v(t) > 0, y(t)/Y_{max} > \sigma_1 \\ PS, & v(t) > 0, p(t)/P_{max} > \sigma_2 \end{cases} \tag{1}$$

where Y_{max} represents the maximum yaw rate, P_{max} represents the maximum pitch rate, σ_1 and σ_2 are hyperparameters and need to be adjusted according to the parameters of the UAV flight controller.

When the system is in the FS state, the computing resources are concentrated in the 5th receptive field. In the YS state, the computing resources are concentrated in the $\{1, 4, 7\}$ or $\{3, 6, 9\}$ receptive field, which depends on the yaw direction. In the PS state, the computing resources are concentrated in the $\{1, 2, 3\}$ or $\{7, 8, 9\}$ receptive field. The system can be in multiple states at the same time.

3 Improved Semi-global Matching Based on Adaptive Grid

The stereo matching algorithm is generally divided into 4 steps: cost initialization, cost aggregation, disparity selection, and disparity optimization. The

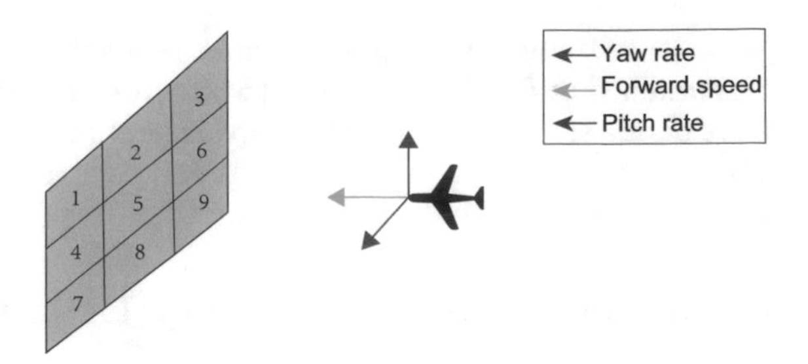

Fig. 1. The image is divided according to the receptive field of the UAV, and the forward speed, yaw rate, and pitch rate determine the system state.

cost initialization of semi-global matching adopts the mutual information algorithm but requires a relatively large amount of calculation. We use a simpler Census transform to replace this method. At the same time, we added a module before cost aggregation to dynamically calculate the grid size. This module reads data from the UAV's sensors and provides grid width information to the cost aggregation stage before each frame. The cost aggregation stage is also carefully designed, multi-path cost aggregation is used to approximate global matching. The algorithm flow chart is shown as in Fig. 2.

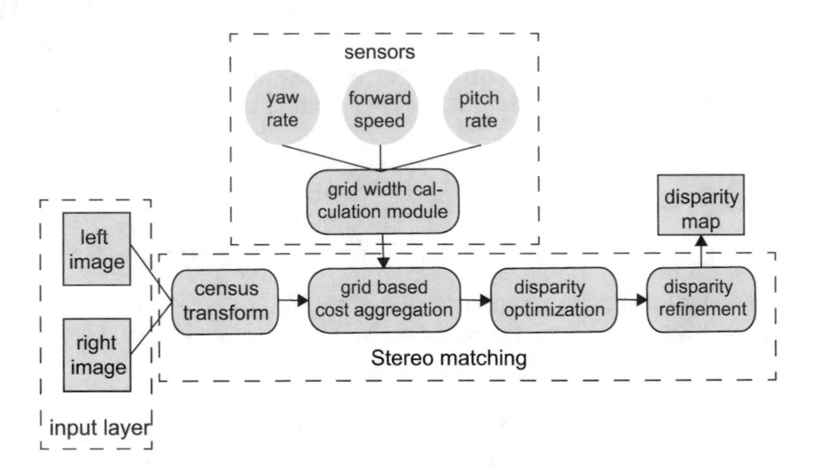

Fig. 2. Summary of processing steps of Sect. 3.1, 3.2, 3.3, and 3.4.

3.1 Cost Initialization

Census transform compares the pixel gray value in the center pixel domain window (the window size is $n*m$, n and m are odd) with the gray value in the center of the window, and maps the boolean value obtained from the comparison to a bit string. Finally, the bit string value is used as the Census transform value C_s of the center pixel, as shown in the following equation:

$$C_s := \bigotimes_{i=-n}^{n} \bigotimes_{j=-m}^{m} \xi(I(u,v), I(u+i, v+j)) \tag{2}$$

where u, v are pixel coordinates, and ξ operation is defined by

$$\xi(x,y) = \begin{cases} 0, x \le y \\ 1, x \ge y \end{cases} \tag{3}$$

The matching cost calculation method based on census transform is to calculate the Hamming distance of the census transform value of the two pixels corresponding to the left and right images, that is, the number of different corresponding bits of the two-bit strings. The calculation method is to carry out NOR operation on two-bit strings, and then count the number of bits that are not equal to 1 in the result of NOR operation as the initial matching cost. The equation is as follows:

$$C(u,v,d) := Hamming(C_{sl}(u,v), C_{sr}(u-d,v)) \tag{4}$$

where $C(u,v,d)$ is the cost value of pixels (u, v) under disparity d.

3.2 Adaptive Grid Size

The fixed grid size cannot adapt to the speed changes of the aircraft. We hope that the UAV's attitude changes can be reflected in the algorithm. We have designed a module specifically for this purpose. This module obtains data from the plane's sensors and aims to provide width and height information of the grid for subsequent calculations. We calculated the values of the three factors introduced in Sect. 2 based on the UAV's accelerometer, magnetometer, gyroscope, and equipped GPS module, and determined the system state, and finally calculated the size of the grid.

When the system is in the FS, YS, or PS state, the width w and height h of the 5th grid are adaptive and calculated by the following equation:

$$\begin{aligned} w &= W - 2v(t)/V_{max} * W \\ h &= H - 2v(t)/V_{max} * H \end{aligned} \tag{5}$$

where W and H represent the width and height of the input image, V_{max} represents the maximum forward speed. Once the system state and grid size are determined, we will use this information in the cost aggregation stage to improve the real-time performance of the algorithm.

3.3 Grid Based Cost Aggregation

The cost of the pixel p is aggregated from the multi-directional path cost, such as 16-path cost aggregation, 8-path cost aggregation, and 4-path cost aggregation. The calculation method of the path cost of pixel p along a certain path r is as follows:

$$L_r(p,d) = C(p,d) + \min \left\{ \begin{array}{l} L_r(p-r,d) \\ L_r(p-r,d-1) + P_1 \\ L_r(p-r,d+1) + p_1 \\ \min\limits_i L_r(p-r,i) + P_2 \end{array} \right\} - \min_i L_r(p-r,i) \tag{6}$$

Among them, the first term is the matching cost value C, the second term is the smoothing term, the third item is to ensure that the new path cost value L_r does not exceed the upper limit of the value. The value accumulated to the path cost takes the least cost value in the three cases of no penalty, P_1 penalty, and P_2 penalty. P_1 penalty is less severe and punishes the situation where the disparity change of adjacent pixels is very small (1 pixel). P_2 penalty is more severe, which punishes the situation where the disparity of adjacent pixels varies greatly (greater than 1 pixel). To protect the discontinuity of disparity in the real scene, P_2 is often dynamically adjusted according to the gray difference of adjacent pixels, as shown below:

$$P_2 = \frac{P_2'}{|I_{bp} - I_{bq}|}, P_2 > P_1 \tag{7}$$

where the P_2' is the initial value of the P_2 and is generally set to a value far greater than that of the P_1, I_{bp} and I_{bq} represents pixel intensity.

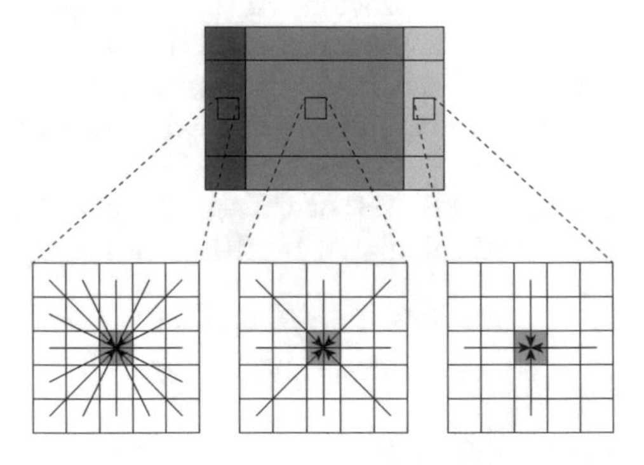

Fig. 3. When the UAV yaws to the left, the left grid adopts 16-path cost aggregation, and the calculations consumed by the middle and right grids only account for 1/2 and 1/4 of the left.

We decide to use several path cost aggregations based on the system state and the location of the pixel. As shown in Fig. 3, when the aircraft is in the

YS state and yaws to the left, the $\{1, 4, 7\}$ receptive field uses 16-path cost aggregation to obtain more accurate depth information, while $\{3, 6, 9\}$ uses 4-path cost aggregation, which makes the calculation amount of the right area only 1/4 of the left. Our method not only ensures the flight safety of the UAV but also improves the real-time performance of the algorithm due to the reduction of the overall calculation amount.

3.4 Disparity Optimization and Disparity Refinement

The disparity optimization step uses the winner-take-all method to calculate the optimal disparity map. WTA selects the disparity corresponding to the smallest cost value as the final disparity according to the DSI (Disparity Space Image) generated after cost aggregation.

We use a variety of methods to refine the disparity. First, the left-right consistency method, the elimination of small connected regions, and the uniqueness detection are used to reduce the mismatch rate, then the quadratic curve fitting is used to calculate the sub-pixel disparity, and finally, the 3×3 median filter algorithm is used to suppress the noise.

4 Experiments

We evaluated our method on Middlebury [10] datasets. We also used the AirSim [11] simulation environment to evaluate the influence of different system states and different flight speeds on the performance of the algorithm.

4.1 Datasets and AirSim Simulation Environment

Since the Middlebury dataset is not a picture taken by a UAV in a real scene, we have pre-defined the parameters of the aircraft. The maximum forward speed of the UAV supports 20 m/s, the maximum yaw rate is 10 rev/s, and the maximum pitch rate is set to 8 rev/s. Select Djembe, Hoops, Piano, and Teddy as the test pictures. AirSim is a cross-platform open-source simulator for drones and other autonomous mobile devices built on Unreal Engine. We use the built-in virtual binocular camera of AirSim to test the performance of the algorithm. The size of the image generated by AirSim is $256 * 144$, and the test scene is CityPark.

4.2 Experimental Results and Analysis

According to Fig. 4, when the system is in the FS state, as the speed of the UAV increases, the size of the 5th grid decreases, the proportion of 16-path cost aggregation decreases, resulting in a decrease in the running time of the algorithm. Our improved method is approximately 25% faster than the SGM algorithm at maximum speed. Table 1 shows the processing time for Teddy pictures when the system is in different states.

AirSim's experimental results show that when the UAV is in the YS state, the left grid adopts 16-path cost aggregation, and the objects in the red box show clearer disparity, as shown in Fig. 5.

Table 1. The running time of our method on Teddy image. We fixed the forward speed to obtain the comparison result.

State	Speed (m/s)	Time (ms)
FS	10	663
YS	10	788
PS	10	764

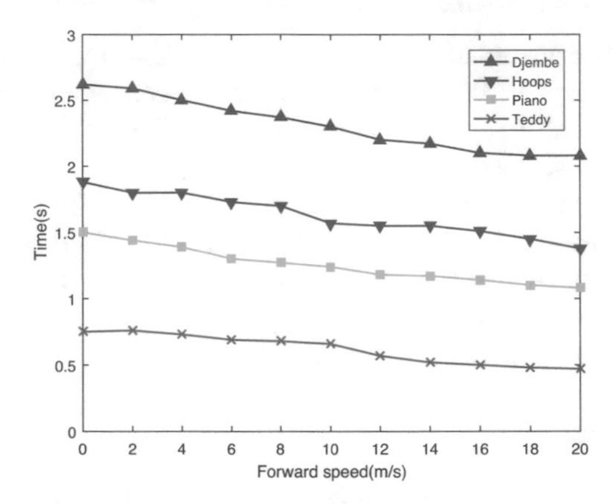

Fig. 4. The test results of the Middlebury dataset show that the running time of the algorithm can be reduced by up to about 25% as the forward speed increases.

(a) (b)

Fig. 5. The disparity map generated in AirSim. (a) Reference image; (b) Disparity map. Our method can well adapt to the attitude change of UAVs. The object in the red box shows high definition when the system is in the YS state.

5 Conclusions

We propose an improved stereo matching algorithm for fixed-wing UAVs. This method is based on the adaptive grid, which can dynamically allocate computing resources to different grids according to the attitude changes of UAV. Our experimental results on the Middlebury dataset show that this method is faster than the SGM algorithm at maximum speed. The experimental results in AirSim show that our method has good robustness in a simulated environment.

Acknowledgements. This work was supported by National Defense Science and Technology Innovation Zone Foundation under Grant No. 19-163-16-ZD-022-001-01.

References

1. Zhao, C., Li, W., Zhang, Q.: Research and development of binocular stereo matching algorithm. J. Front. Comput. Sci. Technol. **14**(7), 1104–1113 (2020)
2. Zeglazi, O., Rziza, M., Amine, A., Demonceaux, C.: A hierarchical stereo matching algorithm based on adaptive support region aggregation method. Pattern Recogn. Lett. **112**, 205–211 (2018)
3. Ma, N., Men, Y., Men, C., Li, X.: Segmentation-based stereo matching using combinatorial similarity measurement and adaptive support region. Optik **137**, 124–134 (2017)
4. Chai, Y., Yang, F.: Semi-global stereo matching algorithm based on minimum spanning tree. In: 2018 2nd IEEE Advanced Information Management, Communicates, Electronic and Automation Control Conference (IMCEC), pp. 2181–2185. IEEE (2018)
5. Hernandez-Juarez, D., Chacón, A., Espinosa, A., Vázquez, D., Moure, J.C., López, A.M.: Embedded real-time stereo estimation via semi-global matching on the GPU. Procedia Comput. Sci. **80**, 143–153 (2016)
6. Zbontar, J., LeCun, Y., et al.: Stereo matching by training a convolutional neural network to compare image patches. J. Mach. Learn. Res. **17**(1), 2287–2318 (2016)
7. Li, Y., Zheng, S., Wang, X., Ma, H.: An efficient photogrammetric stereo matching method for high-resolution images. Comput. Geosci. **97**, 58–66 (2016)
8. Zabih, R., Woodfill, J.: Non-parametric local transforms for computing visual correspondence. In: Eklundh, J.-O. (ed.) ECCV 1994. LNCS, vol. 801, pp. 151–158. Springer, Heidelberg (1994). https://doi.org/10.1007/BFb0028345
9. Hirschmuller, H.: Stereo processing by semiglobal matching and mutual information. IEEE Trans. Pattern Anal. Mach. Intell. **30**(2), 328–341 (2007)
10. Scharstein, D., Szeliski, R.: A taxonomy and evaluation of dense two-frame stereo correspondence algorithms. Int. J. Comput. Vision **47**, 7–42 (2002). https://doi.org/10.1023/A:1014573219977
11. Shah, S., Dey, D., Lovett, C., Kapoor, A.: AirSim: high-fidelity visual and physical simulation for autonomous vehicles. In: Hutter, M., Siegwart, R. (eds.) Field and Service Robotics. SPAR, vol. 5, pp. 621–635. Springer, Cham (2018). https://doi.org/10.1007/978-3-319-67361-5_40

Research on Road Traffic Moving Target Detection Method Based on Sequential Inter Frame Difference and Optical Flow Method

Da Li[1(✉)], Zhi-wei Guan[1,2], Qiang Chen[1], He-ping Shi[1], Tao Wang[1], and Huan-huan Yue[1]

[1] Tianjin University of Technology and Education, Tianjin 300222, China
leedtute@163.com

[2] Tianjin Sino-German University of Applied Sciences, Tianjin 300350, China

Abstract. In order to realize the effective detection of moving objects in intelligent transportation environment, the proposed method of motion detection with higher accuracy and efficiency by combining the sequence frame difference and optical flow algorithm. Firstly, adaptive median filter and histogram equalization are used to preprocess video frames. Then, the preprocessed image is processed by pairwise sequence difference, and the sum of the difference results is calculated. Next, the optical flow method is used to process the differential video sequence to accurately detect and recognize the moving target. Finally, the proposed algorithm is compared with background difference method, optical flow method and inter frame difference method, and the detection effect and accuracy of moving objects in different scenes are analyzed. The experimental results show that the overall performance of the proposed algorithm is stable, the overall contour recognition of moving objects is high, the average true positive rate of moving objects detection is above 0.9.

Keywords: Intelligent transportation · Frame difference · Optical flow · Target detection

1 Introduction

Due to the increasingly serious traffic problems of urban roads and the urgent needs of people to travel, my country's annual investment in intelligent transportation construction has increased, and the pressure on urban intelligent transportation management systems has become increasingly serious. Using artificial intelligence technology to effectively identify vehicle information such as violations and traffic accidents during vehicle driving, and to quickly and accurately detect targets on the road ahead is the key to the development of smart road traffic monitoring systems in the future.

At present, the detection of moving objects in the traffic environment is mainly realized by traditional moving object detection algorithms and moving object detection algorithms based on deep learning. The traditional moving target detection algorithm mainly uses the pixel information of the video image to process the image frame to obtain the

Q. Liang et al. (Eds.): Artificial Intelligence in China, LNEE 854, pp. 376–383, 2022.
https://doi.org/10.1007/978-981-16-9423-3_47

relevant moving target, which is divided into background difference method [1], optical flow method [2], and frame difference method [3] etc. This type of algorithm is relatively simple and easy to implement, and has low requirements on the hardware of the device, but it is greatly interfered by dynamic factors in the background, and the detection effect based on a single motion feature is general. Moving target detection algorithms based on deep learning mainly include deep learning algorithms represented by convolutional neural networks [4] and YOLO [5]. This type of algorithm has a large amount of calculation, high requirements for computer monitoring hardware, and complex model design, which has certain shortcomings in the real-time detection.

In view of the problems that traditional algorithms are susceptible to interference from dynamic environmental factors, deep learning algorithms are too computationally expensive and require high hardware equipment. Based on the improvement of the traditional target detection algorithm, after preprocessing the collected video frames, a new moving target detection algorithm is proposed by combining the improved inter frame difference method and the optical flow method.

2 Video Frame Preprocessing

Adaptive median filter improves the defect that the median filter adopts fixed processing method for different pixels in the process of noise removal, which is easy to damage the information of some pixels. In the case of high noise, the neighborhood size of the median filter is changed according to the preset conditions, which can reduce the noise and ensure the effect of image details.

Histogram equalization determines the transformation function according to the image histogram information, and uses the transformation function to approximately evenly distribute the gray value of the input image, so as to widen the gray value with more pixels in the image and merge the gray value with less pixels, which can make the image have a larger dynamic range of gray and higher contrast, and the details of the image are clearer, Enhance the visual effect of the image.

3 Optical Flow Method

Optical flow represents the instantaneous speed of motion when a moving object in space is shot and projected onto a two-dimensional plane by a camera. The L-K optical flow method [6] is based on the premise that the brightness of the image acquisition is constant, the change of motion within a small time interval will not cause the change of the position of the imaging object, and the change of the optical flow in the neighborhood of the pixel area is constant. The basic optical flow equation in the pixel neighborhood is solved by the least square method. Using the correlation between the current frame and the previous frame of the video image, through the temporal change of the same pixel in two adjacent frames, the movement of the pixel on the image is determined according to the change in the image caused by the movement of the target object in a small time Direction and rate of movement.

4 Sequential Frame Difference Method

The proposed method of motion detection is improved on the basis of the three frame difference method. Through continuous difference of the adjacent three frames of the moving target video, the sequence inter frame difference summation method is proposed to solve the inter frame difference method and the three-frame difference Problems with the difference method. By performing inter frame difference between three consecutive frames of the video image sequence, the difference results are added together, and the small and subtle motion results in the moving image are reflected on the final processing result. The final image processing effect is better than the three frame difference The method has a certain improvement, and a better detection effect is obtained.

Suppose that the gray values of the pixels corresponding to P_{n-1}, P_n and P_{n+1} of frames $n-1$, n and $n+1$ are $f_{n-1}(x, y)$, $f_n(x, y)$ and $f_{n+1}(x, y)$ respectively, and that the difference image obtained by the inter frame difference between the current frame image and the next frame image is Z_n and Z_{n+1}, then:

$$\begin{aligned} Z_n &= |f_n(x, y) - f_{(n-1)}(x, y)| \\ Z_{n+1} &= |f_{n+1}(x, y) - f_n(x, y)| \end{aligned} \tag{1}$$

In this paper, the OTSU is used for image binarization threshold segmentation. Select the appropriate threshold T to binarize the difference image, and the difference image is as follows:

$$\begin{aligned} g_n &= \begin{cases} 1, & Z_n \geq T \\ 0, & Z_n \leq T \end{cases} \\ g_{n+1} &= \begin{cases} 1, & Z_{n+1} \geq T \\ 0, & Z_{n+1} \leq T \end{cases} \end{aligned} \tag{2}$$

When the pixel value of the difference image is greater than or equal to the preset threshold value, the result of binarization is 1, which means the pixel value of the object. Otherwise, it means that the area is the background and there is no object. After the difference between two frames, the sum of the image is calculated:

$$R = g_n + g_{n+1} \tag{3}$$

The advantage of the summation operation is that under the condition that the specific position of the detection target remains unchanged, the frame difference images are added, and the moving objects on the two consecutive frame difference images are extracted respectively, which further expands the contour of the target object and improves the detection accuracy.

5 Sequential Frame Difference Method and Optical Flow Method

Because the image area extracted by the inter frame difference method is incomplete, although the position information of the target contour can be obtained, holes are prone to appear inside, and the extraction results are not accurate. In addition, when the speed of the moving object is too fast or too slow, it is easy to cause the moving object cannot be detected, and the detection result is prone to ghosting, which affects the accuracy of the detection result.

The L-K optical flow method has a large amount of calculation and a slow running speed, resulting in poor real-time detection. In the detection, the hardware requirements of the computer are often high. Moreover, the optical flow method has many constraints, which has a greater impact on the change of light intensity, and its robustness is poor. At the same time, the inter frame difference method has good real-time performance in target detection, good adaptability to environmental changes, small amount of calculation, high detection accuracy of the optical flow method, and wide application range. The sequence inter frame difference method is combined with the optical flow method. The flow method for moving target detection can make up for the shortcomings between the two to a certain extent, and improve the detection efficiency and detection accuracy of traffic moving targets in the smart road environment.

In the algorithm processing, first carry out the sequence frame difference method adopted in this paper to roughly extract the approximate outline of the target object, and obtain the approximate area of the target object. On this basis, perform L-K optical flow calculation processing to reduce the influence of light factors on the detection result. In order to improve the progress of detection, reduce the amount of calculation of optical flow detection. The specific process is shown in Fig. 1.

The specific steps of the algorithm are as follows:

1. The video image is preprocessed, including the adaptive median filter of image sequence frame to reduce the overall noise of the image, and then histogram equalization to make the image have a larger dynamic range of gray and higher contrast;
2. Three consecutive image sequences P_{n-1}, P_n and P_{n+1} are obtained by frame extraction;
3. The obtained images P_n and P_{n-1} are differentiated to obtain a differential image g_n; The P_{n+1} and P_n images are differentiated to get g_{n+1};
4. For the difference images g_n and g_{n+1}, the threshold T corresponding to the maximum inter class variance is obtained by using the OTSU, and the binary threshold segmentation is performed to obtain the moving region I of the target image;
5. Taylor expansion is performed on the pixels in the moving region of the target image, and $\begin{bmatrix} \Delta x \\ \Delta y \end{bmatrix} = \left[\frac{\partial I}{\partial x} \ \frac{\partial I}{\partial y} \right]^{-1} [-\frac{\partial I}{\partial t} \Delta t]$ is performed on each pixel to obtain the optical flow $(\Delta x, \Delta y)$ of each pixel in the moving region, and then the optical flow field of the moving region is obtained;
6. The morphological closed operation is performed on the image that detects the optical flow field of the image, so as to bridge the discontinuity of the image in the narrow area, the whole image is incomplete, and fill the small image hole.
7. According to the binary image obtained after processing, it is judged that the position with large amplitude is the foreground, that is, the moving target, and the area with small amplitude is the background, that is, the static background;

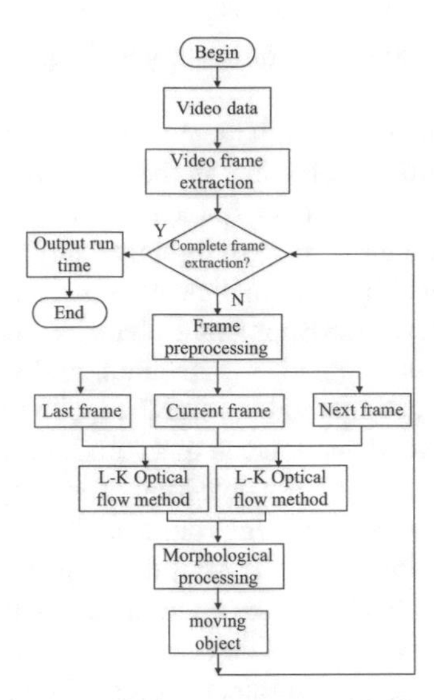

Fig. 1. The algorithm flow chart of this paper

6 Experimental Analysis

The hardware environment of this paper is based on Windows platform, the processor is Intel Core i5-7500, frequency is 3.40 GHz, and the software environment is matlab. Video image is mainly video image in MATLAB software, and the camera installed at high place to take driving image of road vehicles.

Figure 2, 3, 4 shows the typical detection results obtained from three video images using different algorithms. According to the detection results, it can be seen that errors in the comparative analysis of background model and foreground in the process of vehicle detection by background difference method are easy to cause misjudgment of foreground detection results, and the virtual image of vehicle under illumination is judged as a moving target. The detection results of inter frame difference method are incomplete and can not reflect the overall contour of the vehicle. In optical flow detection, the road contour is determined as the foreground, which has a deviation from the actual results. Under the same conditions, the proposed algorithm has good detection effect, the extracted target contour is relatively complete, and the internal cavity is well filled.

In order to further verify the overall performance of the proposed algorithm in target detection, the correct detected foreground pixels (TP), the correct detected background pixels (TN), the wrong detected foreground pixels (FP) and the wrong detected background pixels (FN) are counted respectively [7]. Using the formula:

$$TPR = \frac{TP}{TP + FN} \tag{4}$$

$$FPR = \frac{FP}{FP + TN} \tag{5}$$

The true positive rate (TPR) represents the ratio of all pixels detected as foreground that are correctly judged as foreground. The false positive rate (FPR) represents the ratio

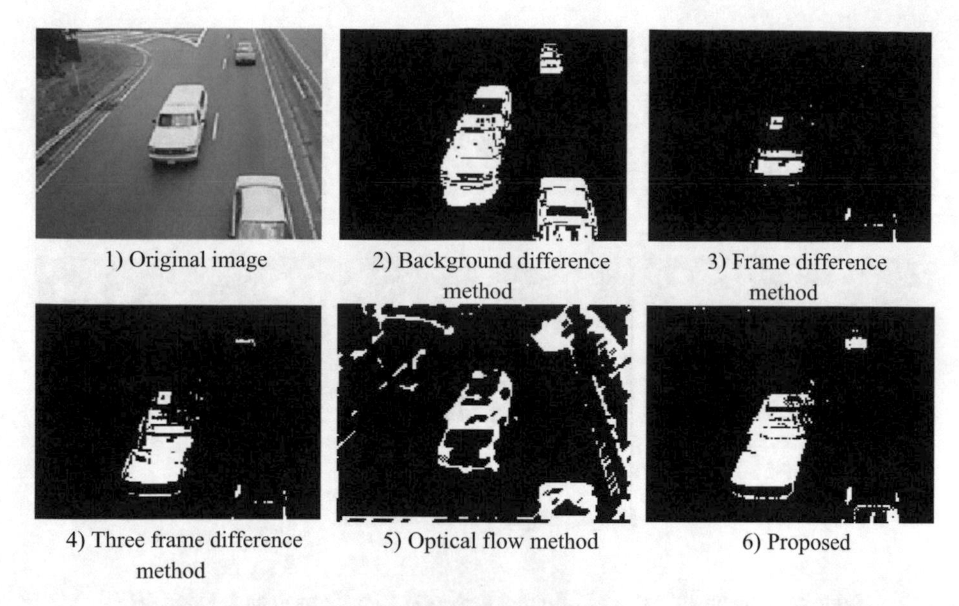

1) Original image
2) Background difference method
3) Frame difference method
4) Three frame difference method
5) Optical flow method
6) Proposed

Fig. 2. Multi-vehicle detection

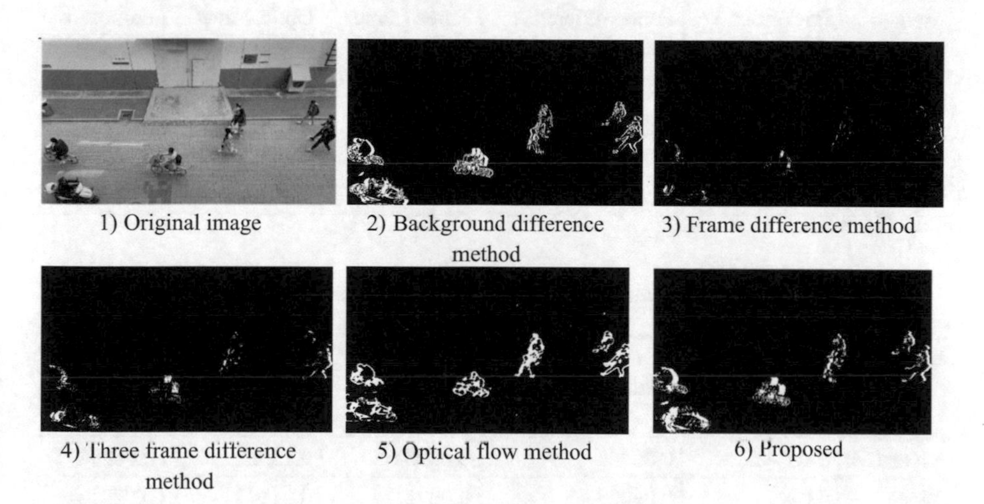

1) Original image
2) Background difference method
3) Frame difference method
4) Three frame difference method
5) Optical flow method
6) Proposed

Fig. 3. Pedestrian target detection

of all detected background pixels that are correctly judged as background. The final statistical results are shown in Tables 1, 2:

In the comparison results of the detection performance of each algorithm in different traffic scenarios, the algorithm proposed in this paper has higher overall recognition than other detection algorithms, good detection stability, and the pixel values of the target foreground pixels that can be correctly detected are all greater than 0.88, The detection

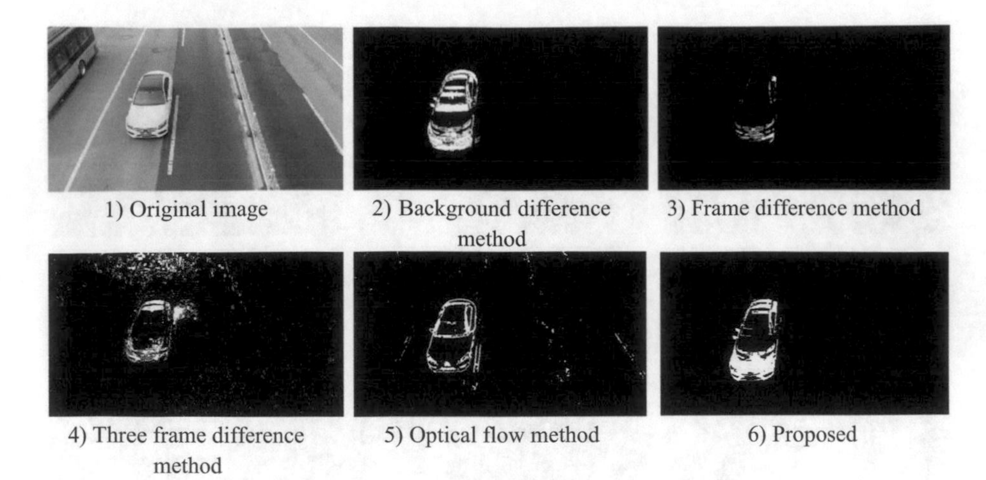

Fig. 4. Single vehicle detection

Table 1. True positive rate of different detection algorithms in each video (TPR)

Method	Background difference method	Frame difference method	Three frame difference method	Optical flow method	Proposed
Video1	0.92603	0.63243	0.84407	0.76074	0.88762
Video2	0.89598	0.64879	0.73412	0.90393	0.92810
Video3	0.86958	0.60368	0.78457	0.89990	0.93962

Table 2. False positive rate of different detection algorithms in each video (FPR)

Method	Background difference method	Frame difference method	Three frame difference method	Optical flow method	Proposed
Video1	0.00547	0.00344	0.00957	0.13522	0.00398
Video2	0.01471	0.06008	0.03721	0.01143	0.02010
Video3	0.02127	0.04326	0.09117	0.03073	0.01361

effect of the target is good. The accurate judgments for background pixels are all below 0.02, which can clearly distinguish the foreground and the background.

In this paper, the moving target detection results based on the combination of sequence frame difference and optical flow have a certain improvement in the overall effect of target detection. In terms of the recognition effect of moving targets, the foreground and background processing is more prominent. Combined with the real-time overall analysis, the algorithm is superior to the comparison algorithm, which has good robustness and further improves the integrity of the target.

7 Conclusion

In this paper, the moving target detection combining the difference between the sequence frames and the optical flow is adopted. After the road traffic video image is preprocessed, the moving target is accurately extracted by combining the difference between the sequence frames and the optical flow method. Before the optical flow detection, the image processing of the frame difference is performed, which reduces the amount of calculation for the subsequent optical flow detection. To a certain extent, the efficiency of video-based target detection is improved. The experimental results show that the detection accuracy of the proposed algorithm is improved to a certain extent compared with the traditional algorithm. The overall detection effect of moving targets is good. The average true positive rate of the detection results is above 0.9. The accuracy of the algorithm has been improved.

Acknowledgements. This work is supported by the Tinajin Research Innovation Project for Postgraduate Students: Research on multi-sensor fusion vehicle detection algorithm in complex weather conditions (2020YJSS086); Tianjin Artificial Intelligence Science and technology major project (17ZXRGGX00070); Key projects of scientific research plan of Tianjin Municipal Education Commission (2019ZD20); Tianjin Science and Technology Plan (20KPH-DRC00030); The University Foundation of the Tianjin University of Technology and Education (KJ1903).

References

1. Zeng, W., Xie, C., Yang, Z., et al.: A universal sample-based background subtraction method for traffic surveillance videos. Multimed. Tools Appl. **79**(31), 22211–22234 (2020)
2. Peng, Y., Chen, Z., Wu, Q., et al.: Traffic flow detection and statistics via improved optical flow and connected region analysis. SIViP **12**(1), 99–105 (2018)
3. Jia, R.D., Chen, Y.G., Zhang, Y.E.: Fusion algorithm of infrared and visible images based on frame difference detection technology and area feature. Int. J. Comput. Appl. **6**, 1–6 (2018)
4. Qin, Z., et al.: How convolutional neural networks see the world–a survey of convolutional neural network visualization methods. Math. Found. Comput. **1**(2) (2018)
5. Du, J.: Understanding of object detection based on CNN Family and YOLO. J. Phys. Conf. Ser. **1004**(1) (2018)
6. Bai, J., et al.: Research on LK optical flow algorithm with Gaussian pyramid model based on OpenCV for single target tracking (2018)
7. Fan, P., Lang, G., Yan, B., et al.: A method of segmenting apples based on gray-centered RGB color space. Remote Sens. **13**(6), 1211 (2021)

Interaction Analysis of "Environment-Behavior" in Community Public Space Under the Background of Micro-renewal

Zhang Ping, Fu Erkang(✉), Zhao Yuqi, and Tang Ruimin

College of Landscape Architecture, Sichuan Agricultural University, Chengdu 611130, China
erkang@sicau.edu.cn

Abstract. As the accessible daily outdoor activity space for residents, community public space has become the concerned object of urban micro-renewal. The interactive analysis of its "environment-behavior" is few. In this paper, field observation and questionnaire survey are used to conduct an empirical study in Jinxiu Community, Chenghua district, Chengdu. The research shows that: (1) residents' activities in community public space are mainly necessity activities and spontaneous activities. (2) The higher the residents' perception and evaluation of community public space landscape, the higher the activity frequency; (3) Gender and residence years are the most important factors affecting the "environment-behavior" of community public space.

1 Introduction

As the most accessible outdoor activity space, community public space is the place where residents use outdoor activities most frequently. It has gradually become an important way to improve the environment of community public spaces. Fu Erkang (2018) based on the view of topological complexity, proposes that the landscape perception of public space in urban open blocks has a positive correlation with addressing performance. MOORE.G.T's research pointed out that "we find out which are the important factors that determine the nature of the objective material environment, and explore the influence of these factors on people's quality of life, using multidisciplinary research methods and means to improve people's living environment (MOORE GT, 1985), (Jan Gehl 2003)." But, there is little research on how the elements and characteristics of environmental space affect people's activity selection preferences and stimulate people to engage in different types of activities. So, this paper taking Jinxiu Community in Chenghua District of Chengdu as the research object, makes an empirical study on its public space by means of behavior trajectory observation and questionnaire survey, trying to analyze the following three questions: ① What is the relationship between the environmental characteristics of community public space and the types and preferences of crowd activities? ② What are the factors that affect the "environment-behavior" of community public space? ③ How does it affect. This research is helpful for a deeper understanding of the interaction between the environmental characteristics of community public space and crowd behavior activities, and provides empirical analysis basis for the micro-renewal of community public space.

Q. Liang et al. (Eds.): Artificial Intelligence in China, LNEE 854, pp. 384–390, 2022.
https://doi.org/10.1007/978-981-16-9423-3_48

2 Theoretical Basis

2.1 Micro-renewal of Community Public Space

It refers to the environmental narrative design intervention and community construction of community public space, to better meet the lifestyle of the subject of community space (Jiang Yubo 2019). It is a tiny and sustainable spatial organic renewal mode. The spatial texture, and community culture are reconstructed in a small organic way (Gong Rui 2019).

2.2 "Environment-Behavior" Interaction Theory

It studies the relationship between people and the surrounding material environment of various scales (MOOREG T 1984). Jan Gehl (2002) pointed out in "Life between Buildings": "Outdoor activities can be divided into necessary activities, spontaneous activities and social activities." The rational design of environment has a very important influence on the behavior in space, and factors such as time, activity group, activity place, frequency, purpose and content also put forward requirements for the external space environment.

3 Methods

In this paper, the public space of Jinxiu Community in Chengdu is taken as the research object, and an empirical study is carried out by means of field observation and questionnaire survey.

3.1 Situation of the Research Area

Jinxiu Community is one of the emerging communities in Chenghua district of Chengdu, with a total area of 0.7 square kilometers. Its public space is mainly composed of three public green spaces and three streets (Fig. 1).

3.2 Data Source

This research USES the data for population activities trajectory data and survey data.

3.2.1 Activities Trajectory Data

Trajectory data for activities, in order to avoid the influence of climate and special date to travel behavior, select the temperature is appropriate, on Sunday and working day in October. Six groups of researchers were assigned to make on-site observations in six major public spaces in the community from October 13 to 15, 2019, and the behavior track analysis method from the population characteristics, the types of activities and activity track three dimensions was used to record the crowd activity status from 07:00 to 21:00 every day in each public space, According to statistics, a total of $6 \times 3 = 18$ people's activity trajectories were obtained.

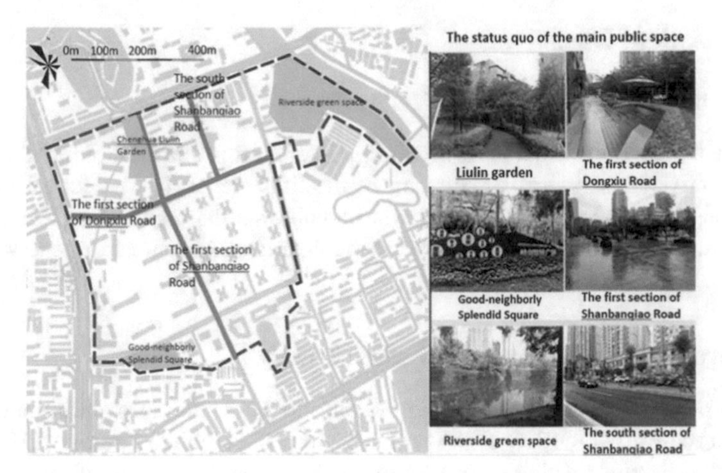

Fig. 1. General situation of public space in the research area (Image source: self-painted/self-photographed)

3.2.2 Questionnaire Design and Pretreatment

Questionnaire in the form of a semi-structured questionnaire, this study includes people attribute characteristics, community space integral feeling, 10 parts etc. preference survey questionnaire mainly focus on the crowd, so there is no reliability and validity test, in order to guarantee the validity of the scale, this study all the multi-item draw lessons from the domestic. In order to further ensure the scientific nature of the scale, through personal interviews, this study of ambiguity and ambition is multi-item after making a formal questionnaire. The researcher distributed 240 questionnaires on community public space activities and preferences to the permanent residents of the community. According to statistics, 233 valid questionnaires were obtained.

4 Discussion

4.1 Correlation Analysis of "Environment-Behavior" in Community Public Space

According to the "crowd activity track map" of six main public spaces in the community, the active groups are classified into retirees, students and office workers, "Environment-Behavior correlation analysis map of crowd activities in community public spaces" (Fig. 2) is obtained. The analysis shows that residents' outdoor activities in community public space are mainly necessary activities and spontaneous activities of retirees and students, while social activities are relatively lacking. It shows that the design of public green space in community lacks effective connection with community slow-moving system to enhance residents' walking experience. From the perspective of behavioral characteristics, social activities such as neighbourhood communication, parties and parent-child games rarely occur in the community, which shows that there is a lack of open space to carry this type of activities in the construction of community public space.

According to the statistical analysis of "Community Public Space Activities and Preferences Questionnaire", it can be seen that people's landscape perception of public space will affect their activities and behaviors in it (Fig. 3). The results show that the higher the evaluation of landscape, the higher the frequency of activities, so a good landscape environment can promote residents' activities in public space.

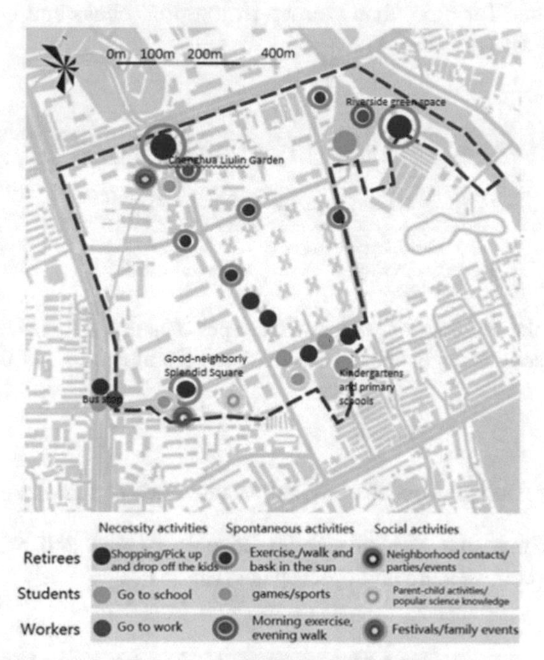

Fig. 2. Correlation analysis chart of "environment-behavior" of crowd activities in public space of Jinxiu community (Image source: self-painted)

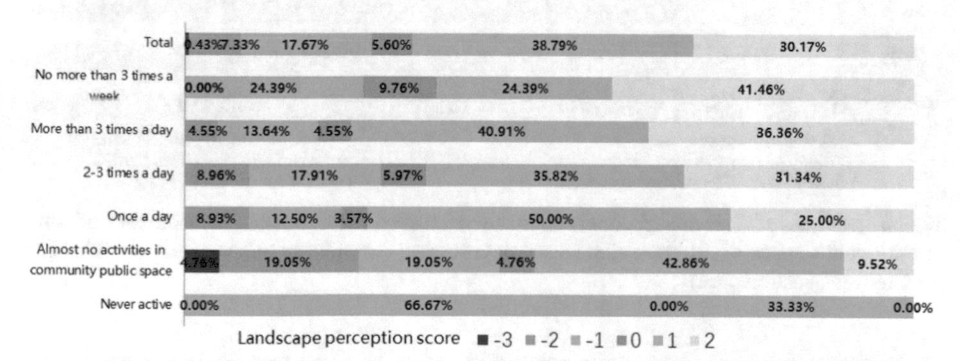

Fig. 3. Analysis of the influence of landscape perception on activity frequency in Jinxiu community (Basic data source: community public space activities and preferences questionnaire)

4.2 Analysis of Influencing Factors of "Environment-Behavior" in Community Public Space

Further analysis shows that "gender" and "residence years" have great influence on the activity types people choose in the corresponding public space. Women accounted for more than men in square dancing, chatting with friends and using fitness equipment, while men accounted for more than women in jogging, chess and card entertainment, rest and daze, and walking (Fig. 4).

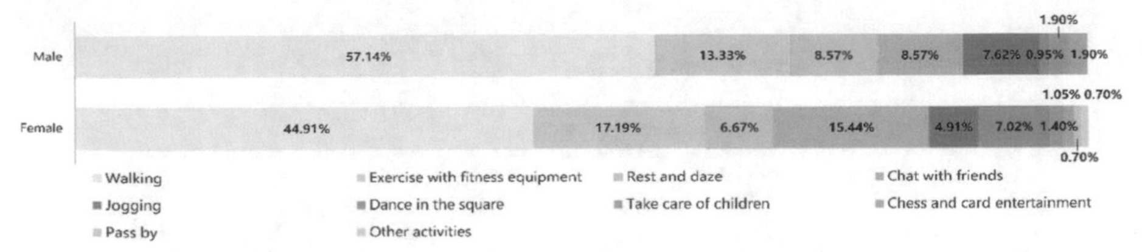

Fig. 4. Analysis of the influence of gender on the types of activities in community public space (Basic data source: community public space activities and preferences questionnaire)

4.3 Analysis of "Environment-Behavior" Preference in Community Public Space

The results also show that residents under 25 prefer quiet and private space, while residents over 25 prefer leisure and open space (Fig. 5).

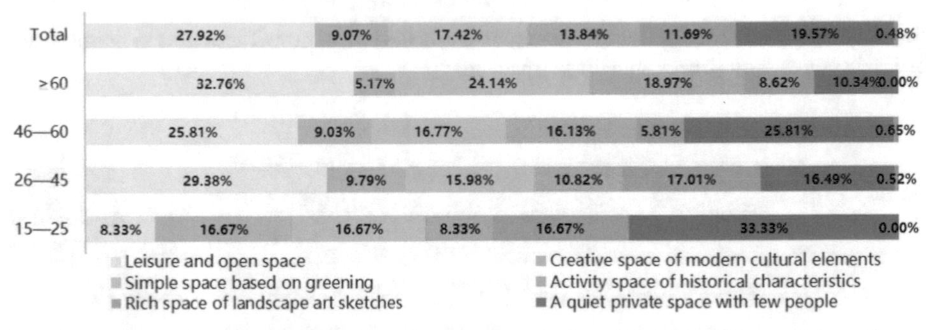

Fig. 5. Analysis of residents' preference for micro-renewal environment characteristics of community public space of different ages (Basic data source: community public space activities and preferences questionnaire)

Residents who have lived for more than 20 years are most looking forward to industrial and cultural activities, residents who have lived for 2–5 years are most looking forward to thematic activities (Fig. 6).

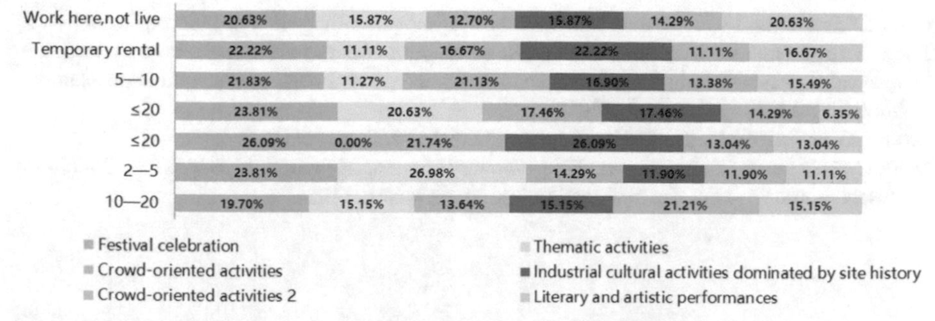

Fig. 6. Analysis of the influence of residence years on the types of activities in community (Basic data source: community public space activities and preferences questionnaire)

5 Conclusion

Under the background of micro-renewal of community public space, this paper starts from three dimensions: "environment-behavior" correlation analysis, "environment-behavior" influence factor analysis and "environment-behavior" preference analysis. The results show that: (1) Residents' activities in community public space are mainly necessary activities and spontaneous activities, necessary activities mostly occur in street space, and spontaneous activities occur in street space and public green space; (2) The higher the residents' perception and evaluation of community public space landscape, the higher the activity frequency; (3) Gender and residence years are the most important factors affecting the "environment-behavior" of community public space. Residents of different ages and years of residence have different preferences for the environmental characteristics and types of carrying activities of community public space micro-renewal.

The exploration of this paper is helpful to deeply understand the interaction between the environmental characteristics of community public space and crowd behavior activities, and provides the theoretical basis for micro-renewal of community public space. However, due to the limitation of research samples, this paper only discusses the community public space in new urban residential areas, involving only street space and public green space, and cannot extend the research conclusions to other types of communities and public spaces. In the future, empirical research on the interaction between "environment-behavior" of community public space can be carried out for more types of objects.

References

Erkang, F.: Research on optimization of public space addressing in Urban open block based on topological complexity. In: Liang, Q., Liu, X., Na, Z., Wang, W., Mu, J., Zhang, B. (eds.) Communications, Signal Processing, and Systems. CSPS 2018. LNEE, vol. 515. Springer, Singapore (2019). https://doi.org/10.1007/978-981-13-6264-4_65

Gong, R.: Research on community public space design from the perspective of micro-renewal. East China Normal University (2019)

Gehl, J.: New city spaces (2001)

Gehl, J.: Translated by He Renke. Life between Buildings, China Building Industry Press (2002)
Jiang, Y.: Research on reconstruction design of old community public space based on organic renewal concept—taking Xinguidong community in Xindu district of Chengdu as an example. Southwest Jiaotong University (2019)
Moore, G.T.: Environmental Design Research Directions. Praeger (1985)
Moore, G.T.: New Directions for Environment-Behavior Research Inachitecture, Architural Research, pp. 95–112. Van Nostrand Reinhold, New York (1984)

Empirical Study on the Impact of Different Types of Urban Built Environment and Physical Activity on Stress Recovery

Ren Yuxin, Fu Erkang(✉), Fan Qiqi, and Wu Yuanhao

College of Landscape Architecture, Sichuan Agricultural University, Chengdu 611130, China
erkang@sicau.edu.cn

Abstract. The development of urbanization has caused serious challenges to people's physical and mental health, and the potential restorative effect of the urban environment has attracted widespread attention. Most researches focus on analyzing the restorability of green spaces, and few people have noticed the potential restorative effects of other types of urban built environments. The influence of age and gender on the restoration of the urban built environment is often overlooked, although it has been confirmed in the experiments of related green space that age and gender can affect the use of the environment and the restoration effect. We used physiological data and POMS scale to measure 60 samples, while observing the restoration effect of the built environment of the four cities. The analysis shows that all urban built environments have healthy restorative effects, and the restorative effects of green space are indeed the best. Women and young people benefited more from the experimental urban built environment. The study confirmed that different types of urban built environments have potential restorative effects, and explored possible factors that cause differences in restorability. This will provide another way of thinking for the planning and design of the urban built environment to improve the health of residents.

Keywords: Urban built environment · Physical activity · Restorative potential

1 Introduction

The rapid development of cities has brought about the explosion of material wealth, but the deterioration of the environment and the change of lifestyle have put people under greater psychological pressure. Pressure is the main factor that causes people to develop mental health diseases (Yang et al. 2013). Increasingly serious health problems cannot be solved by drugs alone, and people are gradually realizing that the urban built environment may provide unexpected opportunities to improve this trend. Erkang et al. 2019 pointed out that a well built environment can stimulate people's physical activities and promote people's physiological conditions and stress relief. Previous studies have made a detailed investigation of the landscape elements that affect the restoration of the urban built environment. Among them, water features, diverse vegetation, open lawns and bright flowers have been proved to have important restoration properties, while

Q. Liang et al. (Eds.): Artificial Intelligence in China, LNEE 854, pp. 391–399, 2022.
https://doi.org/10.1007/978-981-16-9423-3_49

urban roads, building facades and Pavement square is a negative factor (Deng et al. 2020; Ulrich 1991). But some studies had put forward different points of view, Amanda et al., 2015 believes that spending half an hour in a city square can improve people's mental state, and not all the natural environments used have a restorative effect. The reason for this contradiction may be that previous studies chose the environment with recreational purposes when selecting the natural environment, and the selected urban built environment is usually a street-a place for traffic (Staats et al. 2016). In this case, the prejudice to the urban built environment may cause some of the knowledge collected so far to be partial and inaccurate.

Therefore, this study aims to compare the impact of the four types of urban built environment in Chengdu on the restoration effect through wearable physiological measurement devices and subjective psychological scale questionnaires, and to discuss the factors that may cause these effects. Through this research, we can optimize the community built environment in a targeted manner and provide sustainable solutions for the development of the best community built environment.

2 Method

2.1 Experiment Sites

The experiment was conducted in the Jinxiu community in Chenghua District, Chengdu. The community covers an area of 0.7 km^2 and is a modern community that includes schools, parks, residential areas and commercial facilities. There are about 10,313 residents in the community. When evaluating the test site, we included some relevant parameters, such as street view, landscape (such as street trees, colors), public facilities (such as stools and exercise equipment), and the flow of people (Borst H C et al. 2009; Purciel M et al. 2009). In the end, we selected four community built environments (the specific description of these four community built environments can refer to Table 1).

2.2 Measurements

We use the EDA module transmitter and ECG module transmitter in the Biopac MP150 workstation to monitor, measure and record electrical skin activity (EDA) and pulse parameters. The simplified mood scale (Profile of Mood States, POMS) compiled by Mcanir et al. in 1971 was used to assess the mental state of the subjects. POMS consists of 40 adjective phrases, divided into tension-anxiety (T-A), depression (D), Anger-Hostility (A-H), Fatigue (F), Confusion (C) and Vigor (V) With 6 subscales, participants were asked to score according to how they felt at the time, using a 5-level scoring method (0, 1, 2, 3 and 4). For the negative emotion subscale (T-A, D, A-H, F, C), a lower score represents a better emotional state, while for the positive emotion subscale (V), a higher score represents a better emotional state.

Table 1. The descriptions of four experiment sites.

Experiment sites	Jinxiu plaza	Shan banqiao road	Shahe green space	Street sport area
Illustrations				
Image				
Description	Have open line of sight, plant canopy is not high, can see a lot of sky, rest seat is not much, moderate flow of people, the environment is noisy.	Have less open line of sight, plant canopy is medium, can see part of the sky, no rest seat, large flow of people, the environment is very noisy.	Have not open line of sight, plant canopy is very high, it is difficult to see part of the sky, there is no rest seat, the flow of people is small, basically no noise.	Have less open line of sight, plant canopy is medium, part of the sky can be seen, rest seats are many, moderate flow of people, the environment is noisy.

2.3 Procedure

This experiment finally recruited 60 participants (30 males and 30 females; mean age, 30.3 ± 15.46 years), all of whom were residents of the Jinxiu community in good health. They participated in a four-day experiment in October 2020 (temperature were 21.8 ± 0.8 °C; 22.2 ± 1.4 °C; 21 ± 1 °C; 21.6 ± 1.2 °C for the four spaces, respectively). We randomly divided 60 participants into 4 groups (A, B, C and D) with 15 people in each group. Participants experience a space each day, and the individual daily experiment time is consistent to eliminate the influence of light on human circadian rhythms (Table 2).

Participants went to the experiment sites at the agreed time on the experiment day. After arriving at the site, participants wore the ECG and EDA module emitters on the Biopac MP150 workstation. Participants were told not to talk, drink water or play on their mobile phones during the whole experiment. After a 10-min sitting break, participants were directed to an enclosed car for three minutes for noise stimulation (babies crying, dogs barking, metal tearing, etc.). Noise has been shown to cause stress and emotional distress (R.E.M. Lees et al. 1980). Participants were asked to complete the first POMS questionnaire in the noisy environment. Soon afterwards, participants were guided into the experimental space to experience for 3 min according to their daily usage state. During this period, we recorded the behavior activities of participants in the experimental space

Table 2. Experimental arrangement

		Jinxiu square	Shan Banqiao road	Shahe green space	Street sport area
October 11, 2020	9:00AM—12:00AM	A	B		
	2:00PM—5:00PM			C	D
October 12, 2020	9:00AM—12:00AM		A	B	
	2:00PM—5:00PM	D			C
October 13, 2020	9:00AM—12:00AM			A	B
	2:00PM—5:00PM	C	D		
October 14, 2020	9:00AM—12:00AM	B			A
	2:00PM—5:00PM		C	D	

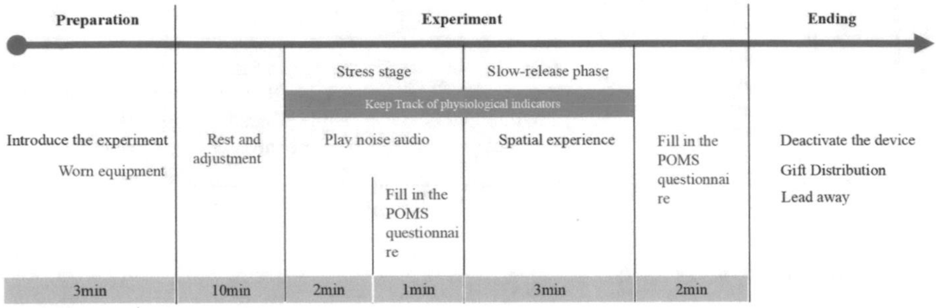

Fig. 1. Experimental procedure.

(without any interference). After the experience, participants were asked to fill in a second POMS questionnaire. AcqKnowledge software was used to continuously record the changes of electrodermal activity (EDA) and heart rate during the whole experiment (Fig. 1).

2.4 Date Analysis

Firstly, the normality test of each data (EDA, heart rate, positive mood, negative mood and TMD) was conducted, then the paired t-test was used to analyze the mean value of physiological data and psychological data, and the variance analysis of multiple factors was used to analyze the differences of relevant indicators. The statistical analysis was performed by SPSS 22.0 and a P-value <0.05 was considered to indicate statistical significance.

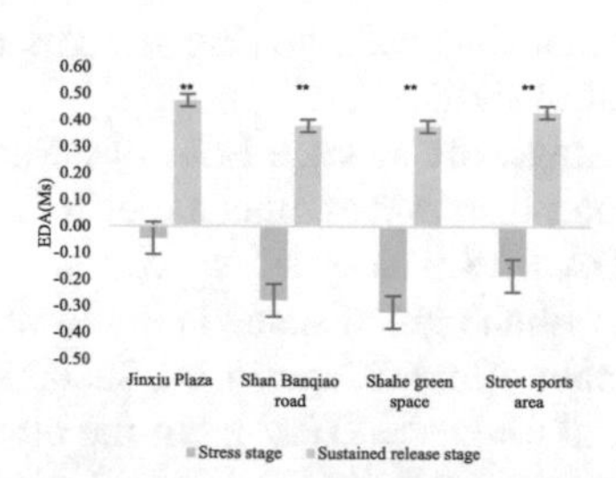

Fig. 2. Comparison of EDA before and after slow release in the four spaces. N = 60; mean ± SD; *P < 0.05; **P < 0.01; verified by paired t-test.

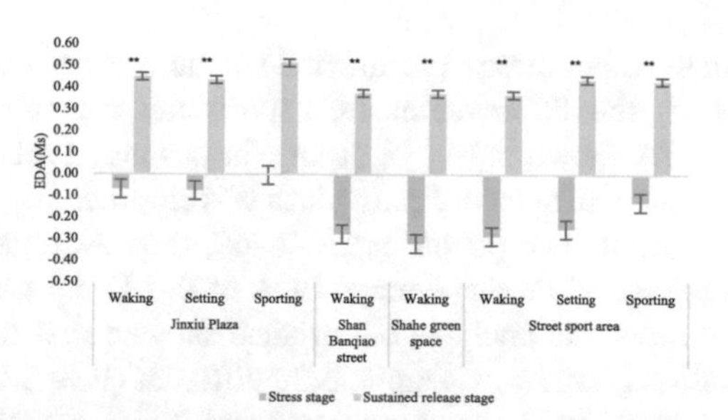

Fig. 3. Comparison of EDA before and after sustained release during different physical activities in four spaces. N = 60; mean ± SD; *P < 0.05; **P < 0.01; verified by paired t-test.

3 Result

3.1 Physiological Results

The paired t-test showed that the EDA in the four different spaces changed significantly before and after the slow release. As shown in Fig. 2, in the Shahe green space, the EDA changed the most (–0.32 ± 0.42 in stress stage, 0.37 ± 0.07, p = 0.000 in sustained release stage). And the EDA changes the most when walking in the green space and when sitting in the street sports area (Fig. 3), Stress stage in green space –0.32 ± 0.42, sustained release stage 0.37 ± 0.07, p = 0.000; stress stage in street sports area –0.25 ± 0.39, sustained release stage 0.44 ± 0.11, p = 0.000. This may indicate that among the four spaces, urban green space has the greatest impact on people, but when different physical activities occur in other urban built environments, it may also have the same restorative effect as green space.

Multivariate analysis of variance showed that the main effect of gender in the sustained release phase of the experiment was not significant (F = 0.087, df = 1, P = 0.769), and the main effect of the experimental site was significantly different (F = 36.355, df = 3, P = 0.000). We found that in addition to the Jinxiu Plaza, the EDA values of women

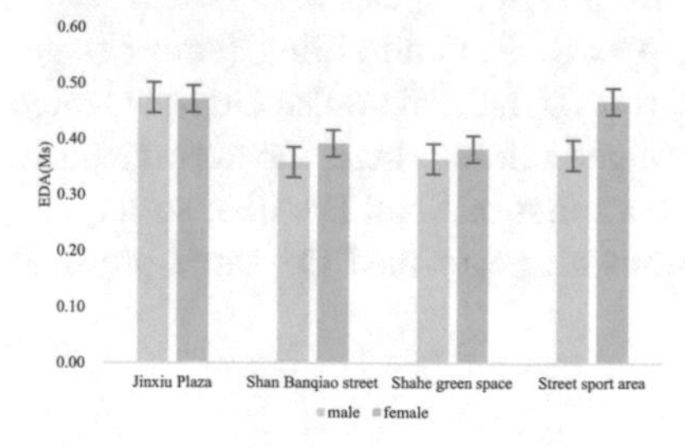

Fig. 4. Comparison of the average value of EDA in the four gender-related spaces in the sustained release stage.

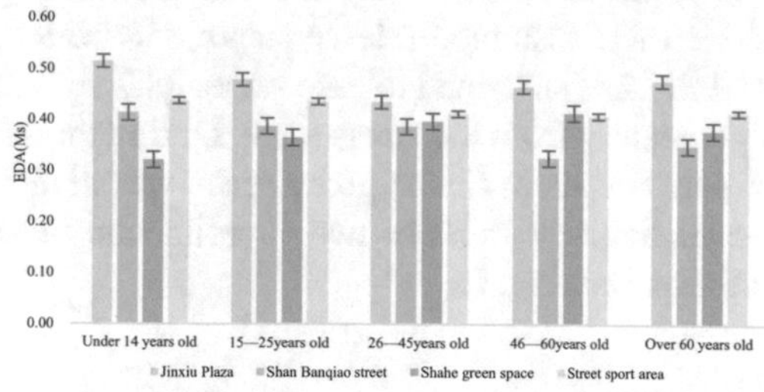

Fig. 5. Comparison of EDA in the sustained release stage in four age-related spaces

in the other three experimental locations were higher than those of men (Fig. 4). Among them, the difference in street sports area was the most obvious.

As shown in Fig. 5, during the sustained release stage, the average EDA of participants of all ages in Jinxiu plaza was significantly greater than that of other experimental locations. For people aged 46–60, they average EDA of Shahe green space was the highest, while the average EDA of 0–14 year olds in Shahe green space is the lowest. Multivariate analysis of variance showed that the main effect of age in the sustained release stage was significantly different ($F = 2.875$, $df = 4$, $P = 0.024$), but the effect between groups was not significant ($F = 0.69$, $df = 11$, $P = 0.747$).

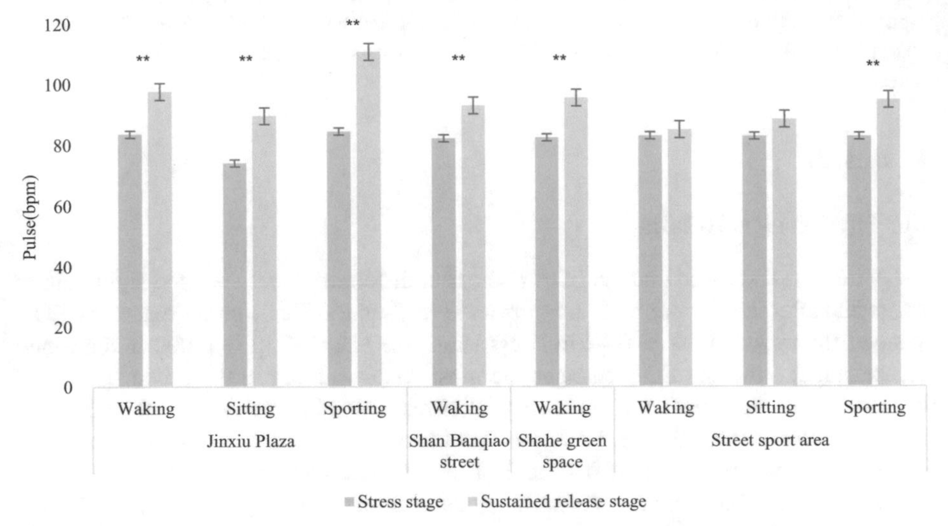

Fig. 6. Comparison of the average pulse rate of people of different genders in the four spaces participating in different physical activities. $^*P < 0.05$; $^{**}P < 0.01$; verified by paired t-test.

Multivariate analysis of variance showed that the main effect of behavior was significant ($F = 10.086$, $d.f = 2$, $P = 0.000$), but the main effects of age and gender were not significant. As shown in Fig. 6, the average pulse rate in the sustained release stage was higher than that in the stress stage, walking in the street sports area (stress stage: 83 ± 14.1, sustained release stage: 85.4 ± 9.6, $p = 0.597$) and sitting (stress stage: 83.1 ± 12.1, sustained release stage: 88.7 ± 12.2, $p = 0.074$). The pulse did not change significantly from the stress stage. During the sustained release stage, the average pulse of participants in Shahe green space was higher than that of Shan Banqiao street, and the participants in these two experimental locations all performed the same physical activity: walking.

3.2 Psychological

As shown in Fig. 7, significant emotional changes were observed under all conditions, with a significant increase in positive emotions and a significant decrease in negative emotions. In Shahe Green space, the score of “Vigor” increased the most ($P = 0.000$),

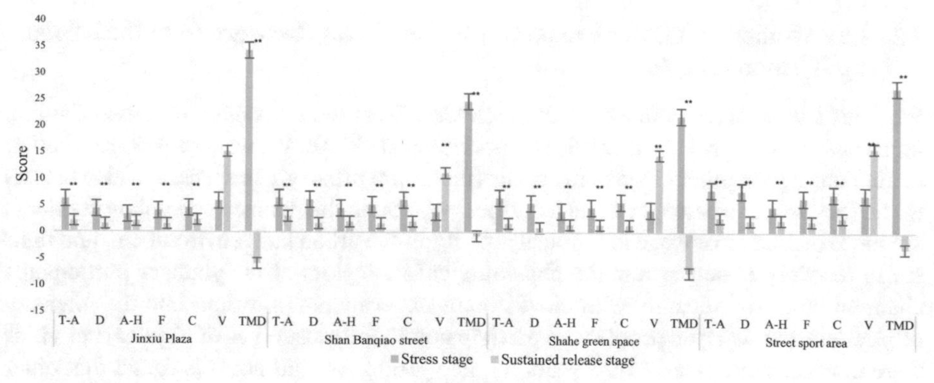

Fig. 7. Comparison of POMS scores before and after sustained release stage. T-A: tension-anxiety; D: depression-dejection; A-H: anger-hostility; F: fatigue; C: confusion; V: vigor. N = 60; mean ± SD; * $P < 0.05$; ** $P < 0.01$; verified by Wilcoxon signed-rank test.

and the score of negative emotion scale decreased significantly: "tension-anxiety" ($P < 0.01$), "depression" ($P < 0.01$), "anger-hostility" ($P < 0.01$), "fatigue" ($P < 0.01$), "confusion" ($P < 0.01$) and total mood disturbance ($P < 0.01$).

4 Discussion

4.1 The Influence of Environmental Characteristics and Physical Activity on Restorative Effects

Pleasant and unpleasant emotions can cause significant changes in the electrical skin of a person (Nasoz et al. 2016). The results of this experiment showed that the average value of the skin electricity in the four experimental environments had changed significantly before and after the sustained release stage. Among them, the average value of the experimental skin electricity in Shahe green space showed a more significant change. However, from the perspective of physical activity, the Shahe Greenbelt in this experiment did not stimulate people to engage in more active physical activity. This is consistent with the previous study by Wang et al. 2016: In a built urban environment with complete facilities, complete functions, and good sanitation, people tend to have multiple activities, and the monotonous environment is indeed unattractive. The mechanism of urban built environment influencing physical activity is very complicated. Among them, urban space design is the main factor (ZIMRING C et al. 2005). For example, more urban furniture, sports equipments and greening can attract people to physical activity. In addition, previous studies have shown that EDA is closely related to the secretion of sweat glands (Darrow C W et al. 1927). Therefore, the reason why the EDA value in the Shahe green space was unlikely to change is because the residents did not engage in high-intensity activities, which affected the secretion of sweat glands.

4.2 Can Women and Older People Get Better Stress Recovery from the Urban Built Environments?

Previous studies have pointed out that gender affects the perception and use of urban green space (Kaczynski et al. 2009; Schipperijn et al. 2010). We want to explore whether gender and age in other urban built spaces (including urban green space) are also factors that affect people's perception and activities. We found that women and young residents are more inclined to engage in physical activity in the urban built environment, and their stress recovery is better. And the physiological indicators of the younger participants changed more significantly with the different experimental locations and the intensity of physical activity. The people who participated in the most types of physical activities were concentrated in the 15–25 years old age group. We did not It is found that older people can get more strong evidence of health restoration from the urban built environment. These were as predicted by some early studies, the services provided based on the type of space differ in the gender and age of the users (Schipperijn et al. 2010; Tyrväinen et al. 2007).

The spatial characteristics of the experimental site may not stimulate the best pressure recovery for all people, but it just reminds us whether we can take into account users of different age groups and different genders when planning and designing the urban built environment. Perhaps we should consider how to improve the safety of the urban built environment and the quality of the landscape, and create as many green spaces as possible to provide the population with opportunities to enhance physical health, mental health, social connections and improve the quality of life (Aspinall et al. 2010).

5 Conclusion

At the beginning of the twentieth century, only 10% of the world's population lived in urban areas. According to recent estimates, by 2050, this number will rise to 66% of the population (United Nations, 2014). Therefore, the role of the urban built environment has become increasingly important. This study showed that not only the urban green space in the urban built environment has a restorative effect, but other types of built space also have huge pressure recovery potential. We also found that demographic factors have a certain impact on the restoration effect of the urban built environment, which provides reference information for the planning and design of the urban built environment. We propose to improve the functions of the urban built environment and improve the quality of the landscape so that the design of the urban built environment can greatly help improve the well-being and quality of life of citizens, reduce their pressure, and restore their health.

References

Erkang, F., et al.: Interactive control coupling analysis of environment-behaviour health utility. In: RACE 2019 Association for Computing Machinery, pp. 78–82 (2019)

Borst, H.C., Vries, S., Graham, J., Dongen, J., Bakker, I., Miedema, H.: Influence of environmental street characteristics on walking route choice of elderly people. J. Environ. Psychol. **29**(4), 477–484 (2009)

Deng, L., et al.: Empirical study of landscape types, landscape elements and landscape components of the urban park promoting physiological and psychological restoration. Urban For. Urban Green. **48**(C) (2020)

Nasoz, F, Lisetti, C.L., Alvarez, K., Finkelstein, N.: Emotion recognition from physiological signals for user modeling of affect. Int. J. Cognit. Technol. Work-Spec. Issue Presence (2016). 2003

Wang, Y., Chau, C.K., Ng, W.Y., Leung, T.M.: A review on the effects of physical built environment attributes on enhancing walking and cycling activity levels within residential neighborhoods. Cities **50**, 1–15 (2016)

Darrow, C.W.: Sensory, secretory, and electrical changes in the skin following bodily excitation. J. Exp. Psychol. **10**(3), 197–226 (1927)

Kaczynski, A.T., Potwarka, L.R., Smale, B.J.A., Havitz, M.E.: Association of parkland proximity with neighborhood and park-based physical activity: variations by gender and age. Leisure Sci. **31**(2), 174–191 (2009)

Schipperijn, J., et al.: Factors influencing the use of green space: results from a Danish national representative survey. Landsc. Urban Plan. **95**(3), 130–137 (2010)

Tyrväinen, L., Mäkinen, K., Schipperijn, J.: Tools for mapping social values of urban woodlands and other green areas. Landsc. Urban Plan. **79**(1), 5–19 (2007)

Aspinall, P.A., Ward Thompson, C.W., Alves, S., Sugiyama, T., Brice, R., Vickers, A.: Preference and relative importance for environmental attributes of neighborhood open space in older people. Environ. **37**, 1022–1039 (2010)

United Nations (2014): World Urbanization Prospects: The 2014 Revision, Highlights (ST/ESA/SER.A/352). New York, NY

Purciel, M., et al.: Creating and validating GIS measures of urban design for health research. J. Environ. Psychol. **29**(4), 457–466 (2009)

Lees, R.E.M., Romeril, C.S., Wetherall, L.D.: A study of stress indicators in workers exposed to industrial noise. Can. J. Public Health Revue Canadienne De Sante'e Publique **71**(4), 261–265 (1980)

Staats, H., Jahncke, H., Herzog, T.R., Hartig, T. : Urban options for psychological restoration: common strategies in everyday situations. PLoS ONE **11**(1), e0146213 (2016)

Ulrich, R.S., et al.: Stress recovery during exposure to natural and urban environments. J. Environ. Psychol. (1991)

Yang, G., Wang, Y., Zeng, Y., Gao, G.F., Liang, X., Zhou, M.: Rapid health transition in China, 1990–2010: findings from the global burden of disease study 2010. Lancet **381**(9882), 1987–2015 (2013)

Research on Restorative Environmental Evaluation of Community Parks Based on Environmental Characteristics

Wang Yihao, Li Xinyun(✉), Hou Tianyu, and Fu Erkang

College of Landscape Architecture, Sichuan Agricultural University, Chengdu 611130, China
lixinyun9301@sicau.edu.cn

Abstract. This study takes eight community parks in Chengdu as the research object, collects and organizes images of their environment, and selects the most representative environmental sample data for quantitative analysis to explore the impact of different physical characteristics of community parks on the evaluation of psychological characteristics of restorative environments, and establish a subjective evaluation model. The result shows: (1) Each physical feature element could explain the psychological feature score, and the richness of landscape plants, landscape naturalness, green vision rate, and activity space abundance all have a positive impact on the evaluation of psychological characteristics of the restorative environment; (2) Each physical feature has different interpretations of the psychological feature score. The naturalness of the landscape explains the psychological feature score the most, followed by the green vision rate and the richness of landscape plants. On the contrary, the richness of the activity space has a negative impact. The research results could provide design method suggestions for restorative environmental construction of community parks, and also provide basic support for future restorative landscape research.

Keywords: Environmental characteristics · Community park · Restorative environment · Evaluation research

1 Introduction

The rapid progress of urbanization and industrialization has gradually separated the close relationship between people and nature in the city, and the physical and mental health of the population has continued to decline. Parks are an important resource to promote the mental health of the population [1]. As the material space carrier most frequently used by people and the most accessible for outdoor activities, community parks have a positive impact on the people's physical and mental health [2, 3]. It is particularly important to deepen the research on its restorative benefits. Studies have shown that watching photos and videos of the park's natural environment could reduce skin conductance, heart rate, and other physiological indicators of stress [4]. Walking in woods and other natural landscapes could reduce cortisol levels [5, 6], and walking 50 in the natural environment of a community park for 50 min could increase the impact of

Q. Liang et al. (Eds.): Artificial Intelligence in China, LNEE 854, pp. 400–408, 2022.
https://doi.org/10.1007/978-981-16-9423-3_50

positive emotions [7]. Approaching to parks and green spaces could promote low levels of "mental stress" and greater mental health [8]. However, the existing research focuses more on the impact of the overall environment of the park on the recovery effect of the population, and seldom explores the difference in the recovery benefits of different environmental factors within the park, especially the lack of quantitative analysis of the recovery effect of the environmental factors on the population. Therefore, this study takes eight community parks in Chengdu as the research object to explore the relationship between the different physical elements of community parks and the evaluation of the psychological characteristics of the restorative environment, constructs a community park restorative environment evaluation model based on environmental characteristics, and put forward the planning and design principles and strategies of community park restorative environment.

2 Research Area and Sample Acquisition

The main urban community park in Chengdu is selected as the case area of this study. The selection principles include covering all districts in the downtown area of Chengdu, high utilization rate, relatively complete landscape types, and area control within 1–5 hectares. Based on the above principles, 8 samples of Chengdu community parks were finally selected based on actual research (Living Water Park, Supo Park, Jiulidi Park, Jinniu Park, Chadianzi Park, Nanzhan Park, Hemei park, Sanliguqiao Park).

Randomly divide 60 photos into 3 groups, numbered A, B, and C, with 20 photos in each group. After that, 60 college students were invited to participate in the restoration benefit experiment, and the restoration overall score was made on 60 sample photos of community park environment. Each group selects 10 sample photos with high scores, of which 3 sample photos are too repetitive. In order to ensure the scientificity of subsequent experiments, only one is kept, and 28 sample photos are finally obtained. After renumbering, as shown in Fig. 1.

Fig. 1. Community park restorative environment sample photo

3 Research Method

3.1 Psychological Characteristics Data Acquisition

The psychological characteristic data of the restorative environment of the community park is obtained from the psychological characteristic evaluation scale, which is composed of four parts, namely, distance, malleability, fascinating, and compatibility. Each item of the scale is performed in the range of 0 to 10 points, where 0 = does not meet the description of the sentence at all, 5 = some agree with the description of the sentence, 10 = completely agree with the description of the sentence.

3.2 Physical Characteristics Data Acquisition

Use the photo grid measurement method to identify and quantify the physical elements of the 28 photos obtained, The steps are shown in Fig. 2. The processed physical elements are presented in grids of different colors (Fig. 3), and the ratio of the number of grids occupied by different elements to the total number of grids can be used to obtain the proportion of the element (Fig. 4).

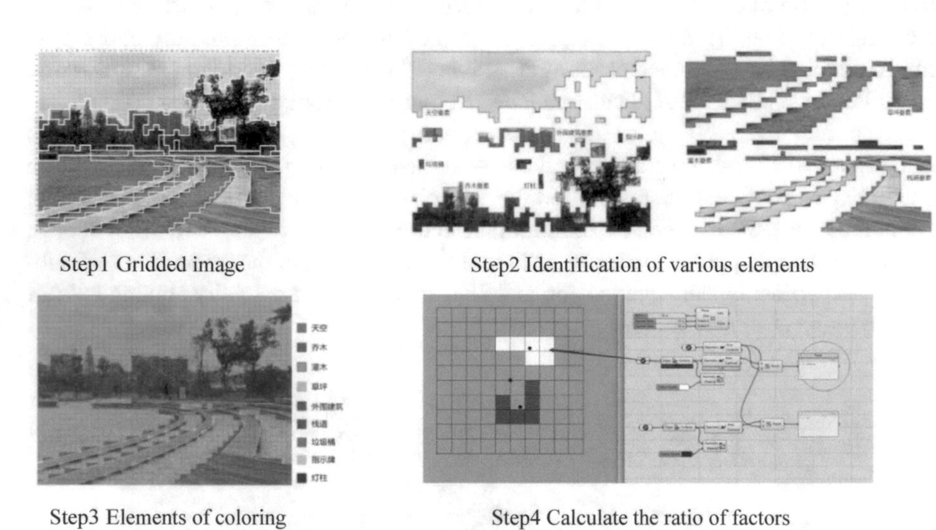

Fig. 2. Steps of photo grid measurement

3.3 Summary of Physical Characteristics

We summarized the physical characteristics of 28 sample photos through field investigations in community parks and combined with the physical elements of the sample photos (Table 1).

Fig. 3. Identify the processed sample image

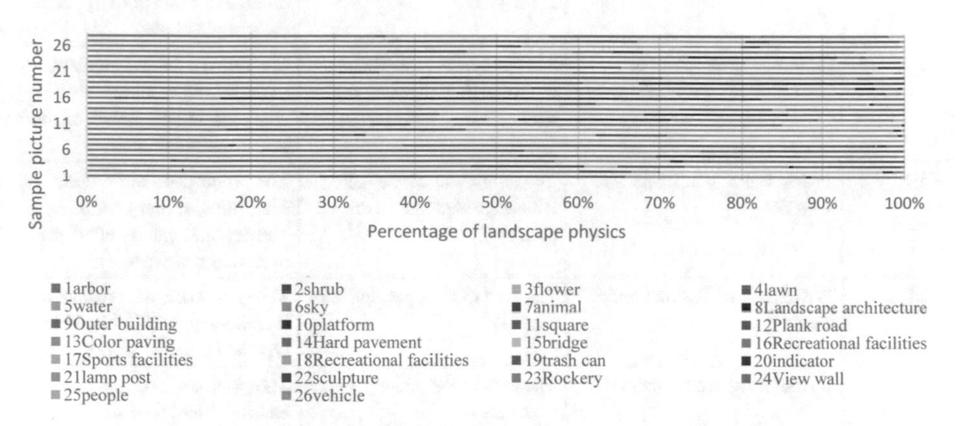

Fig. 4. Distribution of landscape elements in sample photographs

Table 1. Analysis on physical characteristics of restorative environment of Community Park

Number	Physical characteristics	Significance	Measurement methods
1	A Coverage of trees and shrubs	The area of trees and shrubs in the park	The proportion of trees and shrubs in the sample photos
2	B Coverage of flower lawn	The area of lawns and flowers in the park	The proportion of lawn and flower elements in the sample photos
3	C Waterscape coverage	The area of the water body in the park	The proportion of grass and water elements in the sample photos

(*continued*)

Table 1. (*continued*)

Number	Physical characteristics	Significance	Measurement methods
4	D Plant species richness	Number of plant species in the park	Statistics of the number of plant species in the sample photos
5	E The degree of undulation of the terrain	Undulating state of terrain	Divided into 5 levels based on sample photos from low-lying to mountainous areas
6	F Green vision rate	Percentage of green plants within sight	The proportion of trees, shrubs, flowers, and lawn elements in the sample photos
7	G Sky rate	The proportion of the sky in the line of sight	The proportion of sky elements in the sample photos
8	H Color richness	The obvious types of colors in the park	The statistical value of the number of main color types in the sample photos
9	I Landscape naturalness	The ratio of natural elements to artificial elements in the park	The ratio of natural elements to the total area of artificial elements and natural elements in the sample photos
10	J Richness of landscape sketches	The number and types of landscape sketches in the park	The proportions of bridges, sculptures, scenery walls, rockery, and gallery elements in the sample photos
11	K Richness of Recreational Space	The area of the open space in the park	The proportion of elements of recreational facilities and the type and quantity of rest space
12	L Activity space richness	The area of the activity space in the park	The proportion of sports, entertainment facilities, squares, platforms, hard paving, color paving, and human elements in the sample photos
13	M Internal environmental interference	The number and types of other physical elements in the park	The proportion of street lights, signs, trash cans, and vehicles in the sample photos
14	N External environmental interference	The area of the external environment of the park	Percentage of peripheral building elements in sample photos
15	O Total number of elements	Number of all physical elements in the park	Statistics of the number of physical elements in the sample photos

4 Result Analysis

4.1 Analysis of Psychological Characteristics

Calculate the psychological feature score of each sample picture according to the evaluation scale of the psychological characteristics of the restorative environment of the community park (Fig. 5), There are 13 pictures with a psychological feature score above 74.13.

4.2 Analysis of Physical Characteristics

From Fig. 6 we can see that there are 16 photos with physical feature scores above 37.14, among which there are 10 sample photos of community parks with a high score (above 40 points), of which the highest score is 43 points, which is 21 sample photos; there are 12 samples below the average score. The minimum score is 2753 points, which is 16 sample photos.

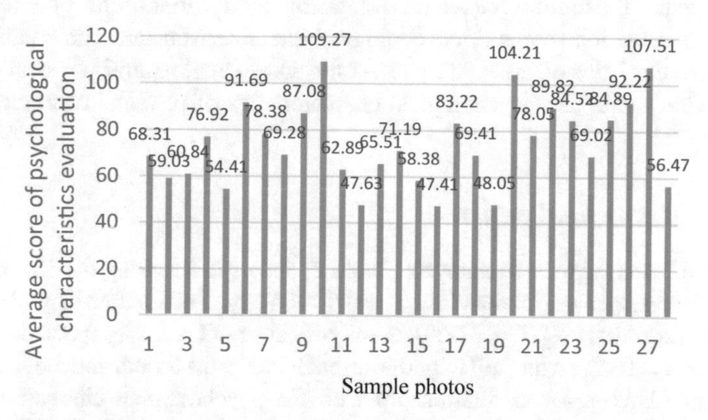

Fig. 5. The average score of psychological characteristics evaluation

5 Establishment of Restorative Environmental Assessment Model for Community Parks

5.1 Correlation Analysis

We use SPSS 22.0 software to analyze the correlation between the psychological feature evaluation score (dependent variable) and the physical feature evaluation score (independent variable): Psychological characteristics evaluation scores are positively correlated with A tree and shrub coverage rate and K resting space richness at the 0.05 level (bilateral), and D plant species richness and G sky rate are significantly positively correlated at 0.01 level (bilateral), and The abundance of J landscape sketches is significantly negatively correlated at the 0.01 level (bilateral), and is not significantly correlated with

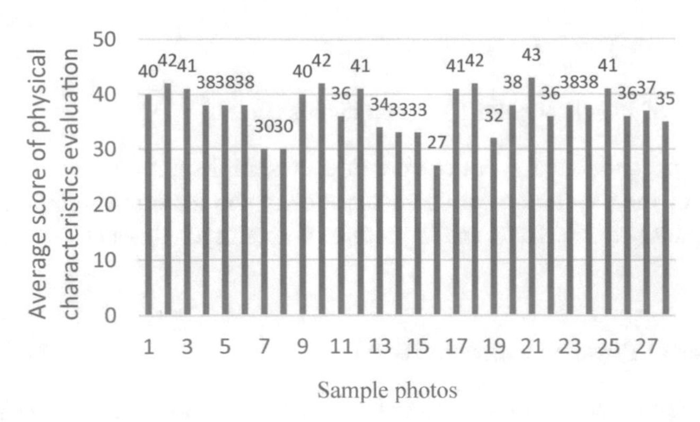

Fig. 6. The average score for physical characteristics evaluation

other landscape elements (P > 0.05). It shows that the evaluation scores of psychological characteristics are closely related to the physical characteristics. The naturalness of the landscape has the highest impact on restoration, but it cannot reflect the relationship between the evaluation scores of psychological characteristics and the combination of physical characteristics of the landscape. At the same time, natural physical characteristics are often more The psychological characteristics of physical characteristics are better evaluated.

5.2 Linear Regression Analysis

The multiple linear regression analysis (Table 2) shows that among the 15 independent variables, D plant species richness (Sig. $P = 0.004$), F green vision rate (Sig. $P = 0.001$), I landscape naturalness (Sig. $P = 0.000$), the significance of L activity space richness (Sig. $P = 0.004) < \alpha = 0.05$, so the null hypothesis that it is 0 is discarded, and the four physical characteristics have a linear relationship with the psychological characteristics. The established subjective evaluation model for the restorative environment of community parks is:

$$Y = 0.145X_D + 0.155X_F + 0.236X_I - 0.07X_L + 3.378$$

Where Y refers to the evaluation of psychological characteristics, X_D refers to the richness score of plant species, X_F refers to the green vision score, X_I refers to the naturalness score of the landscape, and X_L refers to the activity space richness score.

Table 2. Multiple linear regression analysis

	Standardization factor		Standardization factor	T Significance	
	B	Standard error	Beta		
Constant	3.378	.737		10.513	.001
A Coverage of trees and shrubs	−.032	.093	−.066	−.340	.740
B Coverage of flower lawn	−.001	.070	−.002	−.011	.991
C Waterscape coverage	−.043	.071	−.065	−.599	.561
D Plant species richness	.145	.042	.327	3.466	.004
E The degree of undulation of the terrain	−.022	.056	−.032	− .392	.702
F Green vision rate	.155	.076	.470	9.032	.001
G Sky rate	.001	.054	.002	.024	.981
H Color richness	.121	.043	.205	2.775	.017
I Landscape naturalness	.236	.136	.551	11.738	.000
J Richness of landscape sketches	−.009	.055	−.018	−.157	.878
K Richness of Recreational Space	.017	.077	.014	.224	.827
L Activity space richness	−.070	.045	−.324	−3.557	.004
M Internal environmental interference	.033	.081	.072	.411	.688
N External environmental interference	.016	.031	.031	.513	.617
O Total number of elements	.084	.058	.121	1.450	.173

Dependent variable: psychological characteristics.
Independent variable: physical characteristics.

6 Conclusion and Outlook

In this paper, taking community parks in the main urban area of Chengdu as an example, through the subjective evaluation method of restorative environment, the sample photos of restorative environment in selected community parks are collected and mathematically analyzed to establish a subjective evaluation model. The results show that each physical feature element can explain the psychological feature score, and the richness of landscape plants, landscape naturalness, green vision, and activity space abundance have a positive impact on the evaluation of psychological characteristics of the restorative environment. Each physical feature has different interpretations of the psychological feature score. The naturalness of the landscape explains the psychological feature score the most, followed by the green vision rate and the richness of landscape plants. On the contrary, the richness of the activity space has a negative impact. The research results can provide data support for improving the public space quality of community parks and promoting the construction of healthy cities.

This study has certain limitations in the process of restorative environmental evaluation: the physical elements of the evaluation are relatively independent, and there are deficiencies in the combination of physical elements; in addition, the subjective evaluation of the restorative environment cannot reflect the psychological changes of the subjects in the evaluation process. The experimental design should be optimized in the follow-up, and the quantitative research on the restorative environment of community parks should be further carried out.

References

1. Grahn, P., Stigsdotter, U.K.: The relation between perceived sensory dimensions of urban green space and stress restoration. Landsc. Urban Plan. **94**(3), 264–275 (2010)
2. Erkang, F., et al.: Interactive control coupling analysis of environment-behavior health utility. In: RACE 2019 Association for Computing Machinery, pp. 78–82 (2019)
3. Erkang, F., et al.: Study on health utility optimization control of traditional urban block open space based on availability theory. In: Proceedings of the 2019 The 2nd International Conference on Robotics, Control and Automation Engineering, pp. 93–97 (2019)
4. Gladwell, V.F., et al.: The effects of views of nature on autonomic control. Eur. J. Appl. Physiol. **112**(9), 3379–3386 (2012)
5. Park, Y.S.: Family affection as a protective factor against the negative effects of perceived Asian values gap on the parent–child relationship for Asian American male and female college students. Cult. Divers Ethnic Minor Psychol. **15**(1), 18–26 (2009)
6. Tyrvaeinen, L., et al.: The influence of urban green environments on stress relief measures: a field experiment. J. Environ. Psychol. **38**(jun.), 1–9 (2014)
7. Berman, M.G., Kross, E., Krpan, K.M., et al.: Interacting with nature improves cognition and affect for individuals with depression. J. Affect. Disord. **140**(3), 300–305 (2012)
8. Deng, L., et al.: Empirical study of landscape types, landscape elements and landscape components of the urban park promoting physiological and psychological restoration. Urban For. Urban Green. **48**(C) (2020)

Research on Recreational Behavior Characteristics of People in Landscape Space of Rural New Community in the Western Sichuan Plain—Take Nongke Village in Chengdu as an Example

Fu Erkang, Yuqi Zhao, Jiawen Zhou, Yuxin Ren, and Li Xi(✉)

College of Landscape Architecture, Sichuan Agricultural University, Chengdu 611130, China
lixi@sicau.edu.cn

Abstract. This study takes Nongke Village, Pidu District, Chengdu as the research object, on the basis of field investigation and quantitative collection of its landscape spatial information by digital technology, the characteristics of crowd recreation are quantitatively measured by questionnaire survey and participatory evaluation, and the data superposition analysis is carried out based on ArcGIS platform to explore the characteristics of crowd recreation behavior in the landscape space of rural new communities in the Western Sichuan Plain. The results show that: (1) Residents' recreational periods are relatively scattered, while foreign visitors' recreational periods are relatively concentrated, and there are significant differences among different age groups; (2) There are significant differences in the recreational stops between residents and foreign visitors, and there are significant differences between residents of different ages. Farmyards and squares are landscape spaces where most people stop more; (3) The recreational types of foreign visitors are more abundant than those of permanent residents, and walking and rest are the recreational types that most people have. The research results can provide theoretical basis and data support for improving the landscape space quality of rural new communities.

1 Introduction

The rural new community is a relatively concentrated and shared public supporting and municipal infrastructure in a certain area determined by the town (township) planning, which is composed of several groups of different sizes and independent courtyards. It is predicted that in the mature stage of urbanization (70% urbanization rate), there will be 200–400 million people living in rural new communities with complete supporting facilities in China. Therefore, it is extremely urgent to accelerate the construction of beautiful rural living environment (Fu et al. 2019). At the same time, the countryside is rich in landscape resources, which is an important destination for urban residents' leisure and recreation (Luo et al. 2021), so it is very important to optimize its landscape space recreation function. At present, some scholars have constructed the evaluation

Q. Liang et al. (Eds.): Artificial Intelligence in China, LNEE 854, pp. 409–415, 2022.
https://doi.org/10.1007/978-981-16-9423-3_51

system of rural landscape recreation value and recreation service from the perspective of recreation places (Zhang et al. 2020; Wang et al. 2020). Other scholars, starting from the tourists, analyzed the recreational behavior needs of rural tourists and explored the influencing factors of their recreational behavior intention (Markus et al. 2019; Zhang et al. 2020; Sun et al. 2018). However, few scholars focus on the rural new community, linking its landscape space with the characteristics of crowd recreation behavior to carry out research.

The Western Sichuan Plain (also known as Chengdu Plain) is located in the west of Sichuan Basin, China, with superior natural conditions, developed agriculture for generations, and enjoys the reputation of "Land of Abundance". Chengdu, the hinterland of the region, is a pilot area for the national coordinated urban-rural comprehensive supporting reform and a park city construction demonstration area. Several rural new community demonstration sites have been built in Chengdu, forming a spatial form of "small-scale, group-type, micro-pastoral and ecological", which plays a benchmark demonstration role in China. However, due to the blind pursuit of construction progress, many rural new communities did not form a systematic understanding of landscape resources during the construction process, which led to problems such as similar recreational facilities, separation of production and living space, the loss of regional characteristics and other problems, especially did not match the recreational needs of people, and limited utilization of landscape space resources. Therefore, this paper takes Nongke Village, Pidu District, Chengdu as an example, collects data by means of field survey, digital technology, questionnaire survey and participatory evaluation, and inputs it into ArcGIS platform to analyze the superposition of landscape spatial environment elements and people's recreational behavior characteristics. Based on this, puts forward the landscape promotion strategy of rural new communities, which is of great significance for promoting the construction of human living environment in beautiful countryside, helping the transformation and upgrading of rural new communities, and promoting rural revitalization and urban-rural integration development.

2 Method

2.1 Research Area

This study takes Nongke Village in Pidu District of Chengdu as an example for the following reasons: (1) Pidu District of Chengdu is located in the central part of the Western Sichuan Plain, which is a national demonstration area for rural revitalization, and the "rural new community Pidu model" has been extended to the whole Western Sichuan Plain. As one of the national agricultural tourism demonstration sites and the birthplace of "Agritainment", Nongke Village has great research value. (2) Nongke Village has a good natural base, complete infrastructure, rich landscape elements and diverse recreation activities, which provide favorable conditions for the measurement and extraction of landscape recreation in this paper.

Table 1. Landscape spatial attribute table of Nongke Village

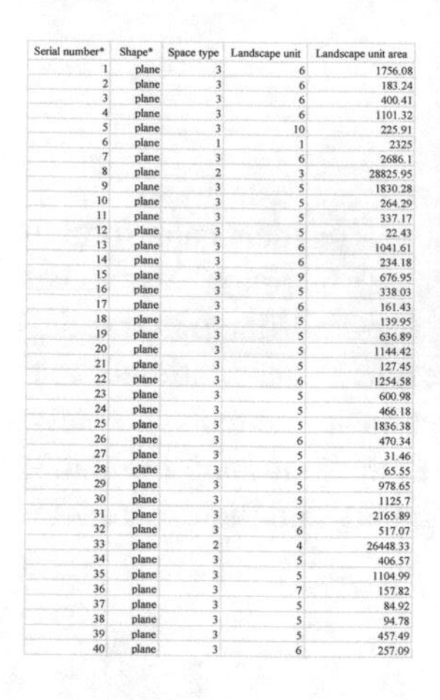

Serial number*	Shape*	Space type	Landscape unit	Landscape unit area
1	plane	3	6	1756.08
2	plane	3	6	183.24
3	plane	3	6	400.41
4	plane	3	6	1101.32
5	plane	3	10	225.91
6	plane	1	1	2325
7	plane	3	6	2686.1
8	plane	2	3	28825.95
9	plane	3	5	1830.28
10	plane	3	5	264.29
11	plane	3	5	337.17
12	plane	3	5	22.43
13	plane	3	6	1041.61
14	plane	3	6	234.18
15	plane	3	9	676.95
16	plane	3	5	338.03
17	plane	3	6	161.43
18	plane	3	5	139.95
19	plane	3	5	636.89
20	plane	3	5	1144.42
21	plane	3	5	127.45
22	plane	3	6	1254.58
23	plane	3	5	600.98
24	plane	3	5	466.18
25	plane	3	5	1836.38
26	plane	3	6	470.34
27	plane	3	5	31.46
28	plane	3	5	65.55
29	plane	3	5	978.65
30	plane	3	5	1125.7
31	plane	3	5	2165.89
32	plane	3	6	517.07
33	plane	2	4	26448.33
34	plane	3	5	406.57
35	plane	3	5	1104.99
36	plane	3	7	157.82
37	plane	3	5	84.92
38	plane	3	5	94.78
39	plane	3	5	457.49
40	plane	3	6	257.09

Fig. 1. Landscape spatial plan of Nongke Village

2.2 Data Collection

2.2.1 Based on Field Survey and Digital Technology to Quantify the Collection of Landscape Spatial Information in Nongke Village

Combined with literature and field investigation, low-altitude unmanned aerial vehicle remote sensing was used to collect the global landscape digital information of Nongke Village, and then the 3D spatial data were generated by importing Agisoft Photoscan software to obtain quantitative data such as spatial attributes, geographical location and morphological scale; Six-legged software and infrared rangefinder were used to collect high-precision and non-destructive information of typical landscape space in the case area, and obtained quantitative data such as landscape elements and landscape factors. The research data were input into ArcGIS platform to generate the landscape layout map of Nongke Village, and inputted the field attributes of the elements in the map according to the hierarchical and progressive way of "landscape type-landscape unit-landscape elements" to form the landscape spatial attribute table of Nongke Village (Fig. 1 and Table 1).

2.2.2 Based on Questionnaire Surveys and Participatory Evaluation to Measure Recreational Characteristics of Crowd

In Nongke Village, 120 permanent residents and 120 foreign visitors were recruited as respondents (divided into four age groups: <18 years old, 18–45 years old, 46–59 years old and >60 years old, and the number of each age group was determined by quota sampling) to participate in the measurement of landscape recreation behavior characteristics. Through the questionnaire survey, the demographic characteristics, living background, health beliefs, recreation preferences and other information of the respondents were obtained; Through participatory evaluation, the characteristics of the respondents' recreation behavior were obtained, including recreation path, recreation type, recreation frequency and recreation duration.

3 Result and Discussion

3.1 Analysis of Recreation Frequency of Crowd

Based on ArcGIS platform, the recreation spatio-temporal data of residents and foreign visitors with spatio-temporal attribute fields are generated (Fig. 2). Through the nuclear density analysis of the frequency of landscape space used in Nongke Village in one day (Fig. 3), the results show that the recreational periods of residents are scattered, and the recreational stops are mainly concentrated in living spaces such as streets, markets, street green spaces and squares, and production spaces such as nurseries and farmland; While the recreational periods of foreign visitors are concentrated, and their recreation stops are mainly concentrated in squares and farmyards.

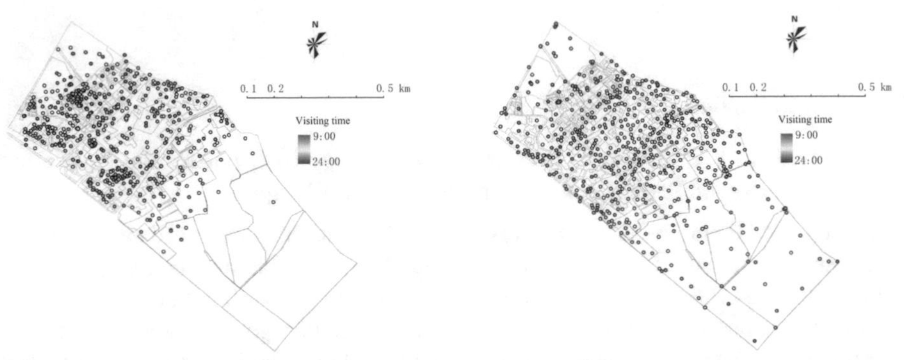

(a) Recreation spatio-temporal point data of foreign visitors (b)Recreation spatio-temporal point data of residents

Fig. 2. Recreation spatio-temporal point data of crowd in the landscape space of Nongke Village

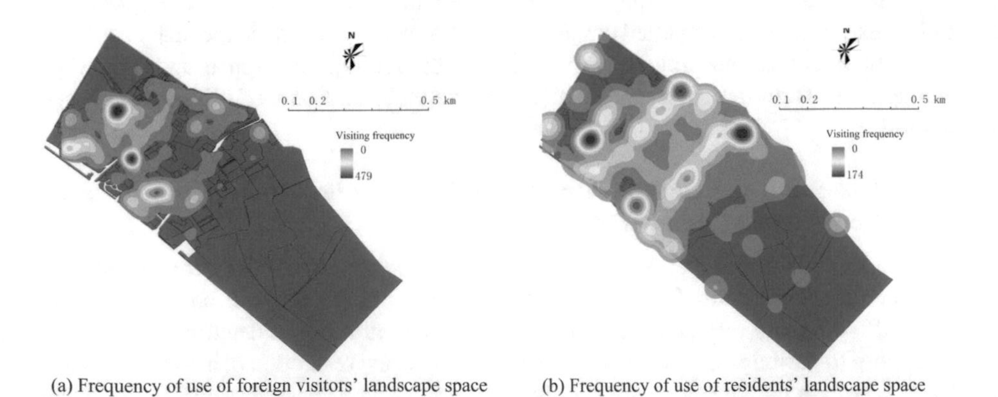

(a) Frequency of use of foreign visitors' landscape space (b) Frequency of use of residents' landscape space

Fig. 3. Analysis of recreation frequency in the landscape space of Nongke Village

3.2 Analysis of Recreation Spatio-Temporal Behavior Path of Crowd

Based on the ArcScene environment and ArcObjects object model of ArcGIS platform, the STpath analysis plug-in was used to process the data of temporal and spatial points to generate the crowd recreation spatio-temporal path map (Table 2) with the plane as the spatial layout of Nongke Village and the Z coordinate as the time axis (unit: h), and classified and extracted the recreation spatio-temporal behavior path of different groups to analyze their recreational spatio-temporal behavior characteristics (Tables 3

Table 2. Analysis of typical individual recreation spatio-temporal behavior

	Recreation spatio-temporal path map	Recreation spatio-temporal behavior characteristics			
Typical individual schematic	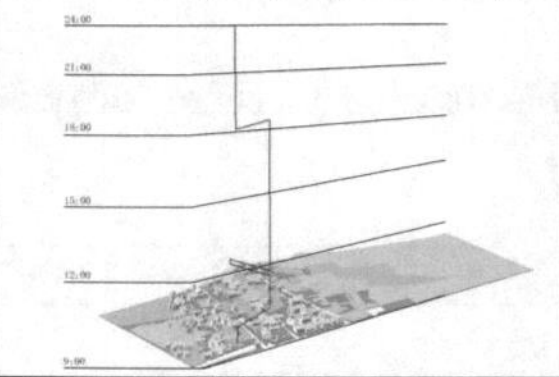	Recreation stops	Recreation time	Recreation frequency	Recreation types
		Farmyards	8h	once	Resting, chess, fishing
		Street green space	0.5h	twice	Walking, chatting

Table 3. Analysis of recreational spatio-temporal behavior characteristics of residents in different age groups

< 18 years old permanent residents		Permanent residents aged 19-45	
Recreation spatio-temporal path	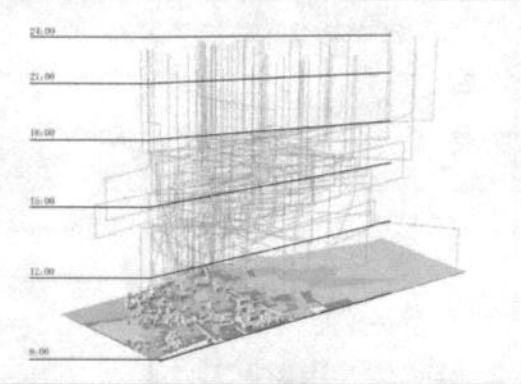	Recreation spatio-temporal path	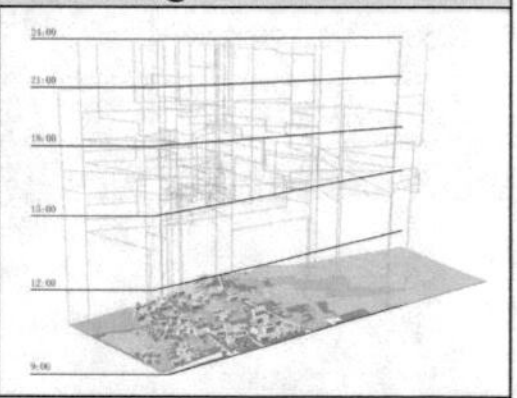
Recreation periods	12 noon-6 p.m.	Recreation periods	3 p.m.-9 p.m.
Recreation stops	Farmyards, children's playgrounds...	Recreation stops	Nurseries, farmyards...
Recreation types	Gathering, playing ball, walking...	Recreation types	Walking,farming, resting...
Permanent residents aged 46-59		**> 60 years old permanent residents**	
Recreation spatio-temporal path	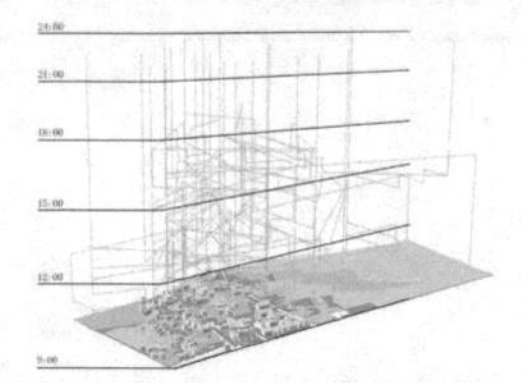	Recreation spatio-temporal path	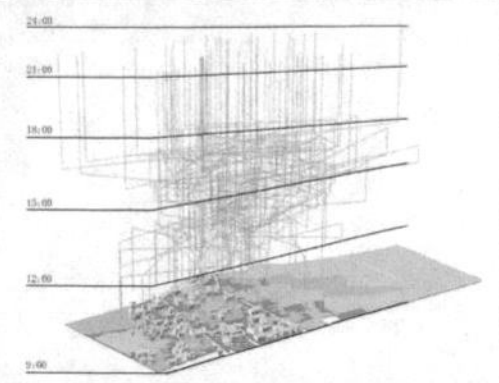
Recreation periods	9 a.m.-6 p.m.	Recreation periods	12 noon-6 p.m.
Recreation stops	Farmland, squares...	Recreation stops	Squares, farmyards...
Recreation types	Walking, chatting, chess...	Recreation types	Walking, chatting, square dancing...

and 4). The results show that: (1) For the leisure time, the leisure time of residents younger than 18 years old and foreign visitors is basically the same, while the leisure time of residents older than 18 years old and foreign visitors is quite different; There are some differences in recreation time among residents of different ages, and the main recreation time of most people occurs in afternoon and evening. (2) There are significant differences between residents and foreign visitors, and there are significant differences between residents of different age groups, farmyards and squares are landscape spaces where most people stop. (3) For recreation types, walking and rest are the recreation types that most people have, and the recreation types of foreign visitors are more abundant than those of permanent residents, such as riding and picking. Therefore, when constructing the rural new community landscape, it is necessary to set up recreation facilities that can meet the needs of the corresponding people in the parking places.

Table 4. Analysis of recreational spatio-temporal behavior characteristics of foreign visitors in different age groups

< 18 years old foreign visitors		foreign visitors aged 19-45	
Recreation spatio-temporal path	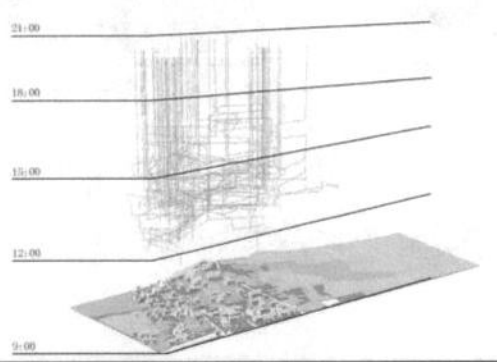	Recreation spatio-temporal path	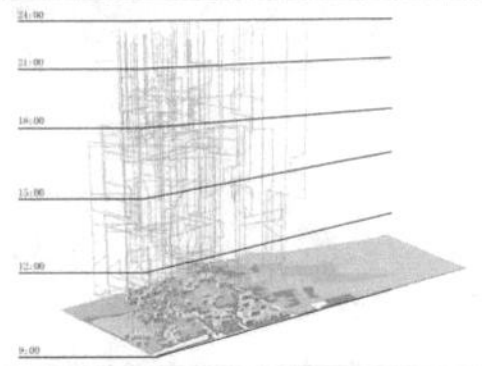
Recreation periods	12 noon-6 p.m.	Recreation periods	9 a.m.-9 p.m.
Recreation stops	Children's playgrounds, farmyards…	Recreation stops	Street green space, farmyards…
Recreation types	Cycling, resting …	Recreation types	Cycling, walking, picking...
foreign visitors aged 46-59		>60 years old foreign visitors	
Recreation spatio-temporal path	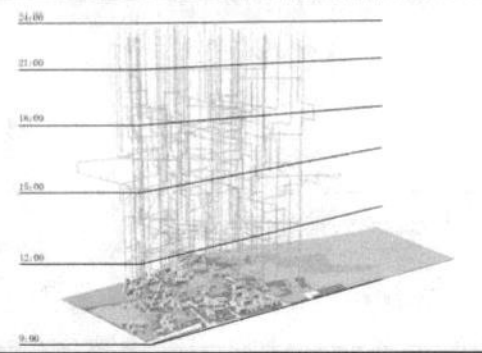	Recreation spatio-temporal path	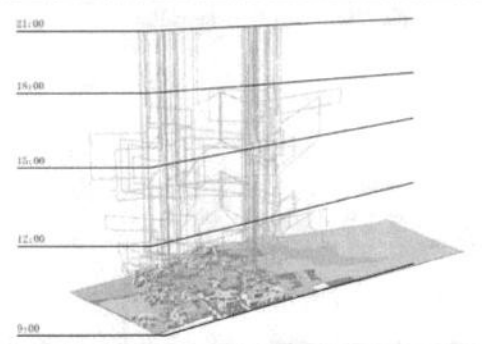
Recreation periods	12 noon-9 p.m.	Recreation periods	9a.m.-6 p.m.
Recreation stops	Farmyards ,squares, …	Recreation stops	Farmland, farmyards…
Recreation types	Walking, picking, resting …	Recreation types	Resting, walking, sightseeing …

4 Conclusion

In this paper, taking Nongke Village, Pidu District, Chengdu as an example, based on ArcGIS platform, the spatial information of rural landscape, the recreation frequency of crowd and recreation spatio-temporal behavior path were superimposed and analyzed, so as to clarify recreational behavior characteristics of different groups of people. The results suggest that the recreation time of residents is relatively scattered, the recreation time of foreign visitors is relatively concentrated, and the leisure time of different age groups is significantly different; There are significant differences between residents and foreign visitors, and there are significant differences between residents of different age groups, farmyards and squares are landscape spaces where most people stop; Walking and rest are the recreation types of most people have, and the recreation types of foreign visitors are more abundant than those of permanent residents. The research results can provide accurate data support for improving the landscape space quality of rural new communities.

The data collection period of this study is the holiday with the most abundant recreation of the crowd. It does not have universal significance for the regularity of crowd recreational behavior characteristics on working days, and has certain research limitations. In the future, it is necessary to further analyze the recreational behavior characteristics of different groups in different time periods.

References

Opinions of the central committee of the communist party of china and the state council on establishing and improving the system, mechanism and policy system for the integrated development of urban and rural areas, p. 4 (2019)

Fu, E., Wang, Y., Ren, Y., Fan, Q., et al.: Study on health utility optimization control of traditional urban block open space based on availability theory. In: Proceedings of the 2019 The 2nd International Conference on Robotics, Control and Automation Engineering, pp. 93–97 (2019)

Luo, H., Deng, L., Jiang, S., Song, C., Fu, E., et al.: Assessing the influence of individual factors on visual and auditory preference for rural landscape: the case of Chengdu, China. J. Environ. Plann. Manage. (2021)

Zhang, L., Ma, C.: Study on the value of rural landscape recreation based on the ternary theory of human settlement environment. Chin. Gard. For. **35**(09), 25–29 (2019)

Wang, Z., Wang, S., Xu, M.: Evaluation of landscape recreation service in county area—taking Wusheng County of Sichuan Province as an example. South. Constr. (03), 27–33 (2020)

Meyer, M.A., Rathmann, J., Schulz, C.: Spatially-explicit mapping of forest benefits and analysis of motivations for everyday-life's visitors on forest pathways in urban and rural contexts. Landscape Urban Plann. **185** (2019)

Zhang, Y., Yu, X.: Influencing factors and paths of tourists' rural recreation in the reconstruction of living space: a qualitative comparative analysis of a fuzzy set. J. Nat. Resour. **35**(07), 1633–1646 (2020)

Sun, B., Fu, Z.: Discussion on the development of rural tourism in Northwestern China based on landscape improvement: a case study of Taochi Village, Yichuan County, Shaanxi Province. Chin. Landscape Archit. **34**(S1), 23–25 (2018)

Ships Target Detection on Water Based on Improved Faster RCNN

Changjun Wang and Cheng Peng(✉)

School of Computer Science and Technology, Xinjiang Normal University, No. 102 Xinyi Road, Urumqi 830054, Xinjiang, China
pcxjnu@163.com

Abstract. The intelligent research of watercraft detection is of great significance to the construction of marine transportation industry. Based on the Faster RCNN algorithm, this paper proposes a watercraft target detection algorithm based on the improved Faster RCNN. Due to the serious loss of low-level feature information during feature extraction. The fusion path between high-level and low-level features is longer, which increases the difficulty of information positioning. Specially increase the side fusion path network to strengthen the fusion of low-level features and high-level information from bottom to top. In view of the situation where the gradient of the Intersection over Union (IoU) bounding box loss is 0 and the intersection method not judged, the Distance-IoU loss function is introduced to improve the process of NMS filtering out repeated target boxes, improve the accuracy of position regression and improve the missed and false detection situations. Experiments show that the improved Faster RCNN algorithm in different weather environments has mean Average Precision (mAP) of 86.12% on ship images, which is an increase of 13.24% compared with the Faster RCNN algorithm, and the average IoU has increased by 6.94%. The overall detection accuracy is rate is improved.

Keywords: Ships target detection · Faster RCNN · Side fusion path network · Distance-IoU

1 Introduction

With the increase of data and the improvement of computing speed, the rapid development of deep learning in target detection, and a series of Faster RCNN [1], YOLO (You Only Look Once) [2–4], SSD (Single Shot MultiBox Detector) [5] is a representative target detection algorithm, which is greatly improved compared with traditional algorithms. Zhang et al. [6] uses a ship detection algorithm for front and back background segmentation. Lu et al. [7]. proposed compression sensing and saliency detection inland moving ship detection algorithms to improve motion trajectory damage and differential image problems. Wang et al. [8] used the Faster R-CNN network to automatically detect ports and inland ships, which greatly improved the detection accuracy compared to traditional methods. Wang et al. [9] used SSD networks to overcome false detections caused

Q. Liang et al. (Eds.): Artificial Intelligence in China, LNEE 854, pp. 416–423, 2022.
https://doi.org/10.1007/978-981-16-9423-3_52

by waves, and used transfer learning technology to improve the generalization ability of the network. The experimental data sets are all integrated by network download, and the image size is uneven, the pixels are different, and the value of practical application is lacking.

Faster RCNN, as a two-stage target detection algorithm, is slightly slower in speed, but the target accuracy is high, the adaptability is stronger, and the network adjustment is flexible. Based on Faster RCNN, this paper adds a side fusion path network in the extraction stage to strengthen the low-level and high-level information fusion. DIoU [10] is introduced to improve the position regression process where the gradient is 0 and the way of judging the intersection.the combination of Soft-NMS and DIoU is used to improve the positioning accuracy and stable real-time detection of ship targets can be achieved.

2 Network Improvement

2.1 Feature Extraction Layer Improvement

The Faster RCNN model extracts low-level feature information and loses severely. Besides, the path between the high-level and low-level features longer.

The feature pyramid network is divided into two parts: bottom-up stage, output $\{C_1, C_2, C_3, C_4, C_5\}$, top-down stage, using subsampled to enlarge the small feature map at the top level with the same size, the two-stage result feature fusion is added and operated, and the output $\{P_1, P_2, P_3, P_4, P_5\}$, finally add the sum, output $\{N_1, N_2, N_3, N_4, N_5\}$, as shown in Fig. 1.

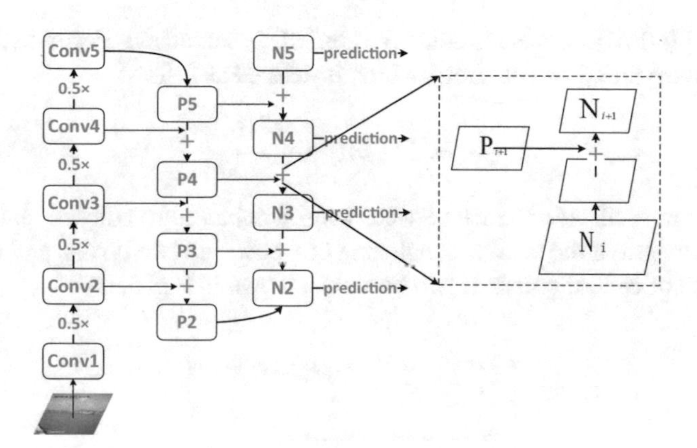

Fig. 1. Side converged path network

3 Loss Function Improvement

IoU (Intersection over Union) is a standard for the accuracy of detecting objects. Formula 1 calculates the loss for the intersection and union of them, but no return gradient. Scale is invariant, not affected by the size of the object. Figure 2, the gradient will be 0, resulting that not optimized. between the two is close or far, as shown in Fig. 2(a) and (b), overlap area is no judged, Fig. 2 in (c) and (d).

$$\mathrm{IOU} = \frac{\mathrm{A} \cap \mathrm{B}}{\mathrm{A} \cup \mathrm{B}} \tag{1}$$

In formula A is the prediction frame, and B as target frame. the intersection area of the two to the combined area of the two is used to evaluate the quality of the prediction frame.

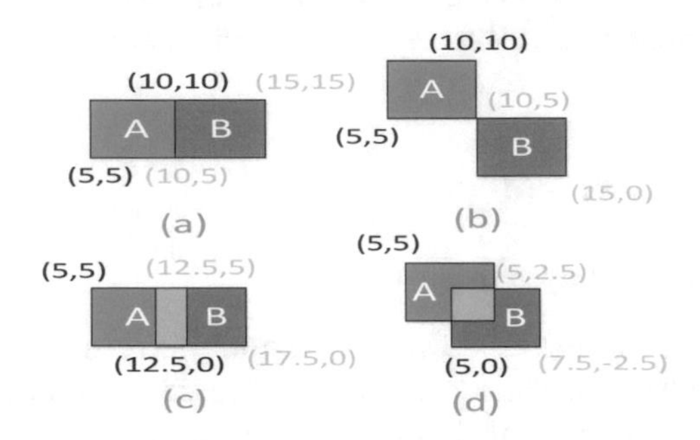

Fig. 2. Intersection of candidate frame and target frame

3.1 Distance-IoU

Distance-IoU (DIoU) Loss is introduced when IoU is calculated. Formula 2, DIoU Loss is a penalty term based on IoU Loss, which is defined as follows:

$$\mathcal{L}_{DIoU} = 1 - IoU + \frac{\rho^2(\mathbf{b}, \mathbf{b}^{gt})}{c^2} \tag{2}$$

where the normalized distance between central points can be directly minimized. c is the diagonal length of the smallest enclosing box covering two boxes, and $d = \rho(b; b^{gt})$ is the distance of central points of two boxes As shown in Fig. 3.

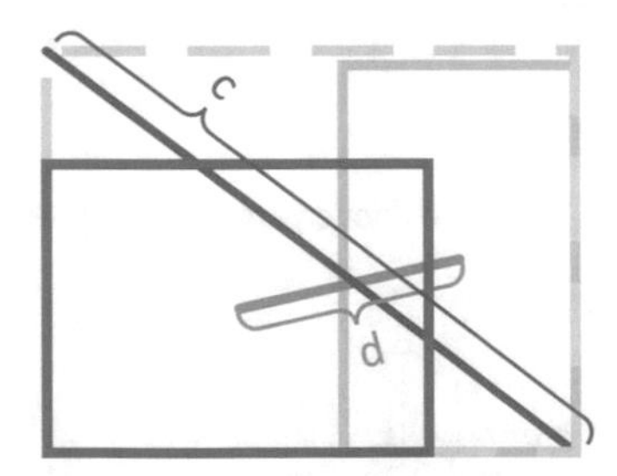

Fig. 3. DIoU loss calculation diagram

4 Simulation

4.1 Lab Environment

Some of the experimental parameters used in the network training phase in this article are shown in Table 1.

Table 1. Experimental data parameter setting

Parameter	Value
Batch size of input data	6
Adjust input image size	418 × 418
Number of iterations	50
Initial learning rate	0.0001
End learning rate	0.000001

Table 1 shows the uniform parameters batch size is 6.

4.2 Data Set

SeaShips [11] is an official ship data set. the scene in SeaShips is daytime, lacking different weather scenes, and the number is augmented with data. The addition of fog and low-illuminance weather improves the robustness and generalization ability of the network. The effect of sample enhancement is shown in Fig. 4.

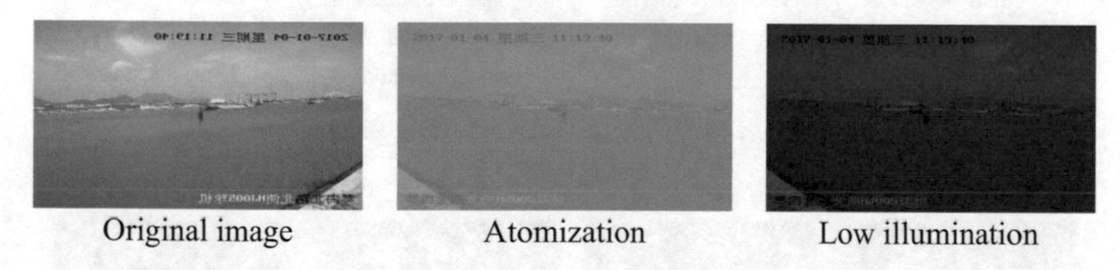

Fig. 4. Simulation of different weather conditions data enhancement renderings

5 Experiment Analysis

5.1 Improved Method Comparison Experiment

In order to obtain a higher-performance training model, different methods will be combined in the network training stage to obtain better results. The influence of different method combinations on the network effect is shown in Table 2.

Table 2 "√" Using the current method, after synthesizing severe weather, the accuracy rate increased by 2.46% points. The addition of the side-converged path network increased the accuracy rate by 7.54%. Using DIoU, the average intersection ratio of the network increased by 6.94%.

Table 2. Different improvement methods improve algorithm performance

Method	Side converged path network	DIoU	Synthetic severe weather	mAP/%	Average IoU/%
Faster RCNN				72.88	53.56
Improve algorithm1			√	75.34	54.52
Improve algorithm2	√		√	82.88	56.78
Improve faster RCNN	√	√	√	86.12	60.5

5.2 Experimental Effect

5.2.1 Forecast Results in Different Scenarios

In the test results of Faster RCNN, as shown in Fig. 5, between the hulls, misdetection and missed detection problems were caused. Figure 6 is improved by adding variance voting, the visual effect of the target frame marking is better.

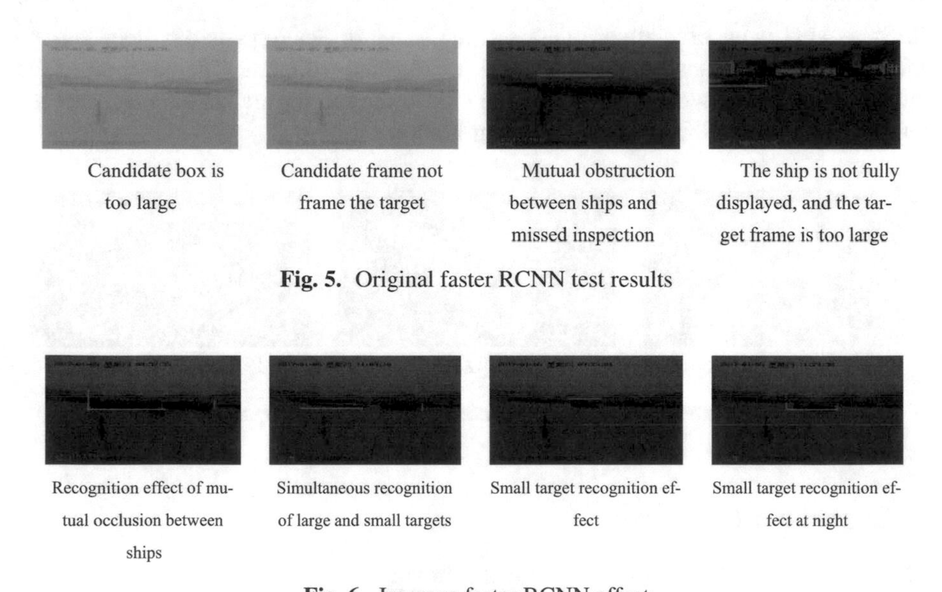

Candidate box is too large | Candidate frame not frame the target | Mutual obstruction between ships and missed inspection | The ship is not fully displayed, and the target frame is too large

Fig. 5. Original faster RCNN test results

Recognition effect of mutual occlusion between ships | Simultaneous recognition of large and small targets | Small target recognition effect | Small target recognition effect at night

Fig. 6. Improve faster RCNN effect

5.2.2 Ships Shows Incomplete Scene

The problem of target occlusion that the effective recognition area becomes smaller, As shown in Fig. 7, the network is able to verify the effect of the side fusion path network in different environments. The experiment is to modify the loss to DIoU on Faster RCNN. Figure 8, this paper satisfies the prediction of overlapping targets and the accuracy is higher.

One-half of the boat and Foreign body obscures the hull

Fig. 7. Target showing incomplete scene

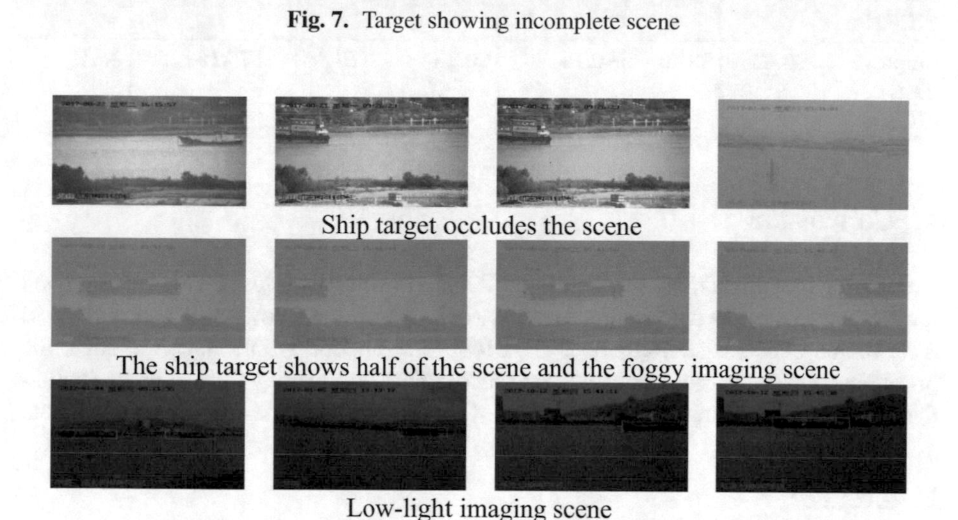

Fig. 8. Improved faster RCNN algorithm for ship target detection results in different scenarios

5.3 Analysis of Experimental Results of Different Algorithms

This model is compared with common deep learning algorithms, including single target detection algorithms YOLOv3 and SSD. The performance of the algorithm is evaluated by the official evaluation algorithm. Table 3 shows the effects of three representative detection algorithms.

Table 3, the improved Faster RCNN has improved accuracy.

Table 3. Comparison of experimental results of different algorithms

Detection algorithm	AP/%						mAP/%
	Ore carrier	Bulk cargo carrier	Container ship	General cargo ship	Fishing boat	Passenger ship	
SSD512	—	—	—	—	—	—	66.71
YOLOv3	68.88	69.24	83.34	70.00	69.53	52.16	68.86
Faster RCNN	81.83	71.24	83.75	71.19	72.61	56.66	72.88
Improve faster RCNN	90.02	90.00	90.10	90.24	81.16	75.18	86.12

6 Conclusion

Combining the official data set, on Faster RCNN, the low-level feature loss is improved during extraction, the side fusion path network is added, the low-level and high-level information fusion is strengthened, the DIoU loss function is introduced to solve the problems of false alarms, occlusion, and incomplete hull display for problems such as the gradient not returned in the regression of the bounding box to the ground truth box.

References

1. Ren, S., He, K., Girshick, R., et al.: Faster R-CNN: towards real-time object detection with region proposal networks. IEEE Trans. Pattern Anal. Mach. Intell. **39**(6), 1137–1149 (2017)
2. Redmon, J., Divvala, S., Girshick, R., et al.: You only look once: unified, real-time object detection. IEEE (2016)
3. Redmon, J., Farhadi, A.: YOLO9000: better, faster, stronger. In: IEEE Conference on Computer Vision & Pattern Recognition, pp. 6517–6525 (2017)
4. Redmon, J., Farhadi, A.: YOLOv3: an incremental improvement. arXiv e-prints (2018)
5. Liu, W., Anguelov, D., Erhan, D., et al.: SSD: single shot multibox detector. In: Leibe, B., Matas, J., Sebe, N., Welling, M. (eds.) European Conference on Computer Vision, pp. 21–37. Springer, Cham (2016). https://doi.org/10.1007/978-3-319-46448-0_2
6. Zhang, Y., Li, Q.Z., Zang, F.N.: Ship detection for visual maritime surveillance from non-stationary platforms. Ocean Eng. **141**(Sep.1), 53–63 (2017)
7. Lu, P., Liu, Q., Fei, T., et al.: Inland moving ships detection via compressive sensing and saliency detection. In: Jia, Y., Du, J., Zhang, W., Li, H. (eds.) Proceedings of 2016 Chinese Intelligent Systems Conference. CISC 2016. Lecture Notes in Electrical Engineering, vol. 404, pp. 55–63. Springer, Singapore (2016). https://doi.org/10.1007/978-981-10-2338-5_6
8. Wang, X., Jiang, F., Ning, F., et al.: Ship detection with improved convolutional neural network. Navig. China (2018)
9. Wang, Y., Yang, Y., Yao, Y.: Single shot multibox detector for ships detection in inland waterway. J. Harbin Eng. Univ. (2019)

10. Zheng, Z., Wang, P., Liu, W., et al.: Distance-IoU loss: faster and better learning for bounding box regression. In: AAAI Conference on Artificial Intelligence (2020)
11. Shao, Z., Wu, W., et al.: SeaShips: a large-scale precisely annotated dataset for ship detection. IEEE Trans. Multimed. **20**, 2593–2604 (2018)

New Energy Vehicles Sales Prediction Model Combining the Online Reviews Sentiment Analysis: A Case Study of Chinese New Energy Vehicles Market

Yu Du, Kaiyue Wei(✉), Yongchong Wang, and Jingjing Jia

Business School, Beijing Language and Culture University, No. 15, Xueyuan Road, Haidian District, Beijing 100083, People's Republic of China
karywwy@126.com

Abstract. Researches on the sales of new energy vehicles (NEVs) can provide theoretical support and practical instructions for the government and the automobile industry. Traditional sales prediction methods often solely includes historical sales data or excludes the impact of public sentiment. In this paper Random Forest Regression (RFR) model is trained on the historical influencing factors and monthly sales using supervised learning method. Then, SARIMA model is used to predict the influencing factors. Lastly, these SARIMA predicted influencing factors are passed into the trained RFR model to acquire the predicted monthly sales and compare with true monthly sales. For further improvement, text sentiment variables are introduced into the regression model. The results indicate that the RFR model shows better performance in predicting monthly sales when taking online comments of the month before last month as text sentiment variables.

Keywords: New energy vehicles (NEVs) · Sales prediction · Random Forest Regression (RFR) · SARIMA · Text sentiment variables

1 Introduction

In order to deal with environmental pollution and energy shortage, many countries have vigorously developed new energy vehicles[1] (NEVs). Data from the China Association of Automobile Manufacturers shows that China annual sales of NEVs are on the rise from 2015 to 2020, with sales hitting a record high in 2020. Researches on NEVs sales can provide theoretical support and practical instructions for the government to improve policies and for the automobile industry to update marketing strategies.

Knowing what factors influence the sales can help us to predict it [1]. Zhang [2] studied the factors affecting NEVs sales from the supply side and found the producer input can affect the sales. Tu and Yang [3] found the number of charging piles and

[1] New energy vehicles refer to vehicles with new structures by using unconventional vehicle fuels as power sources or using conventional vehicle fuels and integrating advanced technologies in power control and drive of vehicles.

Q. Liang et al. (Eds.): Artificial Intelligence in China, LNEE 854, pp. 424–431, 2022.
https://doi.org/10.1007/978-981-16-9423-3_53

the price and life of batteries will affect consumers purchase intention. Wang et al. [1] considered season, oil price change, technologies progress and government policy into sales prediction model. Zheng and Huang [4] introduced the Internet search index into sales forecast model and the result is better. Thus, the main influencing factors include material cost, infrastructure, seasonal factors, policy and consumers' attitude.

As for the model construction, Wang et al. [1] established the Wavelet Neutral Network and BP neural network prediction model considering influencing factors, but didn't take public opinion into account. Zhang [5] classified user reviews by naive Bayes classification method, and predicted the NEVs monthly sales by SARIMA model, but didn't take the comments as variables, too. Zheng and Huang [4] introduced the Internet search index into sales forecasting model but merely considering the influence of historical sales data.

Therefore, a combined Random Forest and SARIMA model is proposed to predict NEVs monthly sales, which includes not only the basic influencing factors, but also the text sentiment variables reflecting the vehicle owners' attitudes and emotions.

2 Methodology and Related Theories

2.1 Principles of SARIMA

The SARIMA (p, d, q) (P, D, Q)s model is used to describe time series with obvious periodic changes in which p and P signify the order of autoregressive and seasonal autoregressive respectively, d and D the order by difference and seasonal difference, and q and Q the order of moving averages and seasonal moving averages [6]. The period of seasonal series is expressed as s. The general expression is as follows.

$$\Phi_p(L)A_p(L^s)(\Delta^d \Delta_s^D y_t) = \Theta_q(L)B_Q(L^s)u_t \tag{1}$$

there into

$$\Phi_p(L) = (1 - \phi_1 L - \phi_2 L^2 - \cdots - \phi_1 L^p) \tag{2}$$

$$A_p(L^s) = (1 - \alpha_1 L^s - \alpha_2 L^{2s} - \cdots - \alpha_p L^{P_s}) \tag{3}$$

$$\Theta_q(L) = (1 - \theta_1 L - \theta_2 L^2 + \cdots + \theta_q L^q) \tag{4}$$

$$B_Q(L^s) = (1 + \beta_1 L^s + \beta_2 L^{2s} + \cdots + \beta_Q L^{Q_s}) \tag{5}$$

Δ and Δ_s represent non seasonal and s-phase seasonal differences respectively, which are used to ensure that y is converted into a stable time series. $u_t \sim IID(0, \sigma^2)$ is white noise. $\Phi_p(L)$ and $A_p(L^s)$ represent non seasonal and seasonal autoregressive feature polynomials, respectively. $\Theta_q(L)$ and $B_Q(L^s)$ represent non seasonal and seasonal moving average feature polynomials, respectively.

2.2 Principles of Random Forest Regression

The Random Forest (RF) algorithm [7] has been widely used for classification and regression. The Random Forest Regression (RFR) in this paper is composed of multiple binary decision trees. The exhaustive method is used for the selection of segmentation variables and points. The quality of segmentation is generally measured by the impurity of nodes after segmentation, that is, the weighted impurity of each sub node, $G(x_i, v_{ii})$.

$$G(x_i, v_{ij}) = \frac{n_{left}}{N_s} H(X_{left}) + \frac{n_{right}}{N_s} H(X_{right}) \tag{6}$$

x_i is a segmentation variable, v_{ij} is a segmentation value of the segmentation variable, n_{left}, n_{right} and N_s are the number of training samples of the left sub node, the right sub node and all the training samples of the current node after segmentation respectively. X_{left} and X_{right} are the training sample set of the left and right sub nodes, $H(X)$ is an impurity function.

This paper we adopt the RandomForestRegressor function in Sklearn module in Python, in which the MSE function ($H(X_m)$) is taken as the impurity function.

$$H(X_m) = \frac{1}{N_m} \sum\nolimits_{i \in N_m} (y - \overline{y_m})^2 \tag{7}$$

$\overline{y_m}$ is the average value of the target variable of the current node sample. The training process of a node is mathematically equivalent to finding the segmentation variable and the segmentation point to minimize G. The prediction result of RFR is obtained by averaging the prediction results of all internal binary decision trees.

2.3 Methodology

In this paper RFR model is trained on the historical influencing factors (x_i) and monthly sales (y) using supervised learning method. Then, SARIMA model is used to predict the influencing factors. Lastly, these SARIMA predicted influencing factors are passed into the trained RFR model to acquire the predicted monthly sales ($\hat{y}$) and compare with true monthly sales (y). This methodology of implementing influencing factors with regression method to predict monthly sales has proven to be better than directly using the SARIMA model to predict future sales based on the historical sales. The methodology's framework is shown as Fig. 1 and the detailed procedures of all experiments follow after Fig. 1.

Firstly, the traditional time series model SARIMA is adopted to predict target monthly sales based on the historical data for further comparison with other models.

Secondly, the RFR model is employed to illustrate the relationship between the NEVs monthly sales volume and its influencing factors which include the number of charging piles, manufacturing capacity of ternary material, manufacturing capacity of LiFePO4 (LFP) material, cost of ternary material, cost of LFP material, gasoline price, vehicle consumption demand index, number of patents, and seasonal indices. Then, the SARIMA model is utilized to predict each influencing factor of the target months; the predicted values of these influencing factors from SARIMA model are returned back into the RFR model to predict the NEVs monthly sales of the same target months.

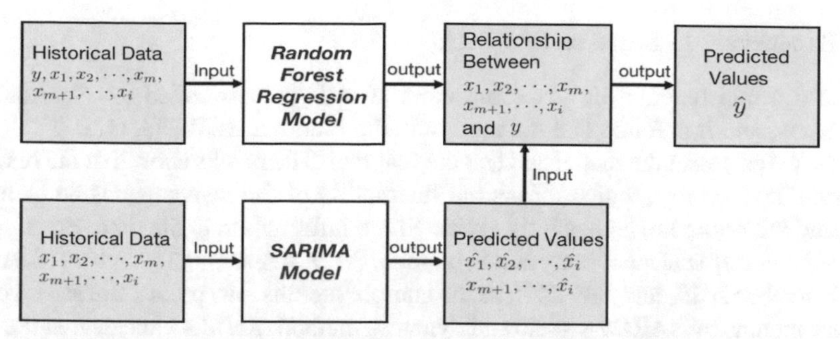

Fig. 1. Methodology framework

Thirdly, consider that public opinion may influence the adoption of new technologies, variables reflecting public sentiment are introduced into the RFR model while holding the other variables constant. As the time lag effect may exist in the influence of online public opinions on the sales, comments of the current month, last month and the month before the last month are each added into the RFR model obtained in the second phase to investigate the relationship between monthly sales and all these influencing factors. The model which includes the current month's comments is employed to predict the target monthly sales based on the predicted values of all variables, while the other two models considering comments of last month or the month before the last month are to predict monthly sales based on the true comments variables and the predicted values of other influencing factors.

For the above models, we adopt Root Mean Squared Error (RMSE) (in the unit of thousands) and R-squared (R^2) to measure and compare their performance. Where m is the total number of samples, $\widehat{y}_J$ represents predicted value of samples and $\overline{y}_J$ represents the mean of true samples.

$$RMSE = \sqrt{\frac{1}{m}\sum\nolimits_{j=1}^{m} (y_j - \widehat{y_J})^2} \tag{8}$$

$$R^2 = 1 - \frac{\sum_{j=1}^{m} (y_j - \overline{y_J})^2}{\frac{1}{m}\sum_{j=1}^{m} (y_j - \widehat{y_J})^2} \tag{9}$$

3 Experiment and Analysis

3.1 Data Preparation

The dataset which collected from Wind Database, PcAuto Website and IPTop Website consists of 65 sample data of China NEVs monthly sales, the number of charging piles, manufacturing capacity of ternary material, manufacturing capacity of LFP material, cost of ternary material, cost of LFP material, gasoline price, vehicle consumption demand index, number of patents and seasonal index, and 15713 sample data of car owners grading and comment texts from January 2016 to May 2021.

3.2 Experiment I - Based on SARIMA

The auto_arima function in pmdarima module in Python is called to select the best parameters, and it is found that the most suitable model is SARIMA (1, 0, 1) (1, 1, 0) [12]. So we establish the model and test the residual. The results show that the residuals are white noise series, which means the fluctuation of the residuals has no statistical rules and the fitting model has fully extracted the information in the time series.

We randomly select February and September 2017, January and March 2018, August and December 2019, and July 2020 as the sample months. We predict the sales volume of these months by SARIMA static and dynamic method. Results indicate that the static RMSE of testing set is 42.421, and static R^2 score of testing set is 0.0354. The dynamic RMSE of testing set is 49.994, and dynamic R^2 score of testing set is −0.3398.

3.3 Experiment II - Based on RFR and SARIMA

Feature Selection. Based on the review of previous studies, we take the number of charging piles (X_1), manufacturing capacity of ternary material (X_2), manufacturing capacity of LFP material (X_3), cost of ternary material (X_4), cost of LFP material (X_5), gasoline price (X_6), vehicle consumption demand index (X_7), number of patents (X_8), and seasonal index (X_9) as the NEVs monthly sales influencing factors.

Model Construction. The dataset are randomly split into training and testing set with ratio 9:1, then the RandomForestRegressor function of the sklearn module is applied to train the training set and test the testing set. Results show that RMSE of testing set is 10.971, R^2 score of training set is 0.9616, and R^2 score of testing set is 0.9570.

Acquisition of Variables for Prediction. SARIMA model is used to predict the influencing factors of February and September 2017, January and March 2018, August and December 2019, and July 2020. The results are shown as the Table 1.

Table 1. The predicted values of influencing factors

	X_1	X_2	X_3	X_4	X_5	X_6	X_7	X_8	X_9
2017-02	160720.19	196685.72	267303.08	99250.97	66552.77	3818.51	123.58	6285.23	2.00
2017-09	196236.30	203146.36	215626.58	190270.85	92314.01	5860.41	169.94	9409.35	9.00
2018-01	225331.68	267846.42	261907.54	234338.03	87131.17	8603.20	9.76	12207.61	1.00
2018-03	249145.40	246964.47	231613.17	201159.83	81739.32	7539.03	77.89	12022.57	3.00
2019-08	456972.34	335621.96	237177.89	115521.76	57144.95	7079.41	119.70	27269.84	8.00
2019-12	517679.35	343289.87	241884.17	126720.20	52344.29	6883.75	74.89	30316.34	12.00
2020-07	583055.42	378737.92	279426.82	111390.07	40733.16	6437.77	61.76	36460.89	7.00

Prediction Results. The above predicted influencing factors are returned back into the RFR model built before, and the predicted values of the target monthly sales are obtained.

The predicted values are 25.499, 85.729, 67.974, 61.495, 100.300, 172.370 and 96.699 with the unit of thousands, respectively. And the true values are 17.596, 78.000, 38.470, 68.000, 85.255, 163.448 and 97.854 with the unit of thousands, respectively. We calculate the RMSE and R^2 based on these predicted and true values of the target months. And it is found that RMSE is 13.848, and R^2 score is 0.8972.

3.4 Experiment III – with Comments Text Sentiment

Feature Selection. The monthly average review score by car owners (X_{10}) and the monthly average comments sentiment (X_{11}) are introduced into the RFR model, with X_1 to X_9 remain constant. 15713 review scores and comments text about NEVs from January 2016 to May 2021 are obtained from the PcAuto Website and are grouped by months. The monthly average car owners' review score ranging from zero to five equals to the average of all review scores by car owners during that month. And the monthly average comments sentiment ranging from zero to one equals to the average of all comments sentiments during that month. The comments sentiment which displays attitudes and emotions of car owners towards NEVs is returned from the sentiment classifier SVM-kernel [8].

Model Construction. The first model (Model_t) takes X_1~X_{11} of the current month as variables. The second model (Model_t–1) takes the X_{10} and X_{11} of last month and X_1~X_9 of the current month as variables. The third model (Model_t–2) takes the X_{10} and X_{11} of the month before last month and the X_1~X_9 of the current month as variables.The dataset are also split into training and testing set with ratio 9:1, then the RFR function is applied to train and test. Performances are shown as Table 2.

Table 2. Performances of models

	Model_t	Model_t–1	Model_t–2
RMSE	30.760	22.348	22.336
Training R^2	0.9479	0.9459	0.9453
Testing R^2	0.8345	0.9149	0.9057

Acquisition of Variables for Prediction. SARIMA model is used to predict X_1~X_9 of February and September 2017, January and March 2018, August and December 2019, and July 2020. X_{10}~X_{11} are predicted based on SARIMA for Model_t, the true values of X_{10} and X_{11} of the last month are taken as variables into Model_t–1, the true values of X_{10} and X_{11} of the month before last month are taken as variables into Model_t–2.

Prediction Results. The prediction results of target monthly sales of three models are shown in Table 3 with the unit of thousands and the corresponding RMSE and R^2 in Table 4. According to the comparison results, Model_t–2 with lowest RMSE and highest testing R^2 shows the best performance among the three models.

Table 3. Prediction results of target monthly sales

	True value	Model_t	Model_t–1	Model_t–2
2017-02	17.596	39.991	42.013	32.507
2017-09	78.000	77.929	73.426	79.231
2018-01	38.470	72.082	61.952	56.375
2018-03	68.000	65.105	67.539	71.586
2019-08	85.255	95.202	92.520	94.966
2019-12	163.448	171.600	168.311	166.033
2020-07	97.854	100.264	96.373	98.748

Table 4. Performances of prediction models

	Model_t	Model_t–1	Model_t–2
RMSE	22.348	13.349	9.703
Training R^2	0.9459	0.9459	0.9453
Testing R^2	0.9149	0.9044	0.9495

Table 5. Performance comparison of different prediction models

	Static SARIMA	Dynamic SARIMA	Basic RFR	RFR with comments
RMSE	42.421	49.994	13.131	9.703
Testing R^2	0.0354	−0.3398	0.9076	0.9495

4 Conclusion and Future Works

In this paper, the RFR model is adopted to predict monthly sales based on the influencing factors predicted by the SARIMA model. Results indicate that this prediction method achieves better performance with lower RMSE and higher R^2 than the traditional SARIMA prediction model which merely depends on the historical sales data. And the RFR prediction model which introduces comments sentiment and review scores by car owners with two-month lagging works better than the model without considering such influencing factors. Results of these models are shown in Table 5.

Overall, this paper provides a better method to predict NEVs monthly sales, but there is still a big margin for improvement in the future work. More monthly data and more influencing factors will be introduced to improve the performance of the prediction model. Furthermore, the comments sentiment analysis will cover online sample data from more platforms and websites, and the sentiment classifier with higher precision rate and recall rate will be utilized.

Acknowledgments. This research project is supported by Science Foundation of Beijing Language and Culture University (supported by "the Fundamental Research Funds for the Central Universities") (Approval number: 20YBB04 and 18PT02).

References

1. Wang, Z., Guo, D., Wang, H.: Sales forecast of chinese new energy vehicles based on wavelet and BP neural network. In: 18th International Symposium on Distributed Computing and Applications for Business Engineering and Science (DCABES), pp. 141–144. IEEE (2019)
2. Zhang, S.: Studying the factors affecting sales of new energy vehicles from supply side. In: 2nd International Conference on Humanities and Social Science Research (ICHSSR 2016), pp. 336–340. Atlantis Press (2016)
3. Tu, J.C., Yang, C.: Key factors influencing consumers' purchase of electric vehicles. Sustainability **11**(14), 3863 (2019)
4. Zheng, S., Huang, J.H.: New energy vehicles sales prediction method and empirical research under the environment of big data. In: Li, X., Xu, X. (eds.) Proceedings of the Fifth International Forum on Decision Sciences, pp. 295–306. Springer, Singapore (2018). https://doi.org/10.1007/978-981-10-7817-0_26
5. Zhang, J.: Research on China's new energy vehicle market analysis and sales forecast based on data mining. Beijing University of Technology (2020)
6. Zhang, X.X., Zhou, B.F., Hui, J.: A comparison study of outpatient visits forecasting effect between ARIMA with seasonal index and SARIMA. In: 2017 International Conference on Progress in Informatics and Computing (PIC) (2017)
7. Breiman, L.: Random forests, machine learning 45. J. Clin. Microbiol. **2**, 199–228 (2001)
8. Lcc, H., Lee, N., Seo, H., Song, M.: Developing a supervised learning-based social media business sentiment index. J. Supercomput. **76**(5), 3882–3897 (2019). https://doi.org/10.1007/s11227-018-02737-x

Multi-label Algorithm Based on the Second and Higher Order Strategies

Chong Sun, Kaijie He(✉), Weiyu Zhou, Zhongshan Song, and Jun Tie

Hubei Provincial Engineering Research Center for Intelligent Management of Manufacturing Enterprises, Wuhan, China
1007692607@qq.com

Abstract. At present, the time complexity of multi-label algorithm based on second-order and higher-order strategies is usually increased due to the increase of the size of the label set. To solve this problem, this chapter proposes a Multi-label algorithm by mining local label correlations (MLMC) with efficiency as the optimization objective. Firstly, from the broad dimension of the correlation between labels, the correlation between labels is used to optimize the classification model. Finally, the nearest neighbor rough set theory is used to consider the positive and negative dependencies between the labels. Experiments show that the MLMC has better accuracy and efficiency than the classical methods.

Keywords: Multi-label learning · Correlation between labels · Label correlation strategies

1 Introduction

Under the rapid development of artificial intelligence, it is making an intensive study of the multi-label learning problem, in the real world, the application of multi-label learning has become more and more extensive and diversified. So far, the main problem in the research of multi-label learning is to consider: improving the accuracy of the classification model and reducing the complexity of the algorithm, on the basis that the algorithm classifies multiple labels correctly [1].

The essence of solving multi-label learning problem is to find a mapping from feature space to label space, the time complexity of traditional algorithm will increase with the increase of the number of the label in the label space, and the global correlation between labels is used by most of the multi-label algorithm, which may affect the prediction of the classifier.

Therefore, a Multi-label algorithm by mining local label correlations (MLMC) is proposed in this paper, which is suitable for the above problems. Firstly, MLMC considers the local correlation between labels through the topic partition. Instances in the same topic share the same correlation between labels. Finally, the nearest neighbor rough set theory is used to obtain the distribution

S. Chong—Major Projects of Technological Innovation in Hubei Province (No. 2019ABA101).

Q. Liang et al. (Eds.): Artificial Intelligence in China, LNEE 854, pp. 432–439, 2022.
https://doi.org/10.1007/978-981-16-9423-3_54

of each instance and label in the multi-label problem, and the positive and negative dependencies between the labels will be obtained, which help the classifier to predict, optimizing the performance of the classifier effectively.

The other parts of this chapter are as follows, the Sect. 2 is related work, the Sect. 3 is the description of the specific implementation of the MLMC algorithm, the Sect. 4 is the experimental verification of the MLMC algorithm, and the Sect. 5 is the summary of this chapter.

2 Related Work

Exploiting the correlation between labels, the multi-label learning algorithm divides the strategies into three categories: first-order strategies, second-order strategies, and higher-order strategies [2].

Boutell et al. [3] proposed a classic first-order algorithm BR, which regards each label in the label space as an individual. The prediction label set is output by the model as a set of correct labels. Zhang et al. [4] proposed a first-order algorithm ML-kNN, which improved based on the K-nearest neighbor algorithm. The above first-order multi-label algorithms are not complicated in the implementation method, but ignore the possible correlation between labels, which affects the accuracy of the prediction results of the model output. When training a multi-label classification model, the second-order multi-label algorithm involves the correlation between broad and narrow labels. Elisseeff et al. [5] proposed RankSVM, a second-order multi-label algorithm based on the narrow correlation between labels. Ghamrawi et al. [6] proposed CRF, a second-order multi-label algorithm based on the broad correlation between labels.

Although the complexity of the second-order multi-label algorithm is lower than that of the high-order multi-label algorithm, it cannot be well applied to real-world problems. The high-order multi-label algorithm can better reflect the label correlation of real-world problems. Read et al. [7] proposed a classifier chain CC algorithm based on an improved BR algorithm, but the classifier chain seriously affects the performance of CC. To solve this problem, Read et al. [8] proposed ECC, a high-order multi-label algorithm improved based on the CC algorithm. In addition, great progress has been made in multi-label algorithms based on the combination of second-order and high-order. Yu et al. [9,10] proposed the multi-label algorithm MLRS based on rough set and the multi-label algorithm MLRS-LC based on the correlation between local labels, according to the rough set theory. But the high-order multi-label algorithm also has the problem of too complicated a model.

Most of the above algorithms make use of the correlation between labels in a global way, which may affect the prediction of the classifier. Therefore, this chapter uses the broad multi-label method based on the nearest neighbor rough set theory to solve this problem.

3 The Proposed Methods

This chapter presents a multi-label algorithm by mining the local label correlations (MLMC) algorithm, which is divided into four steps: (1) problem formulation. (2) range of correlation between local labels, the MLMC divides the sample into several topics, and the examples in each topic share the same correlation of label to exploit local correlation of label. (3) local label correlation matrix, define two matrices: positive correlation matrix and negative correlation matrix. (4) label dependency graph and model prediction, the Bayesian networks based on the matrix is established and the conditional probability of related labels is calculated.

3.1 MLMC Problem Formulation

Set d-dimension label space. $X = \mathbb{R}^d$, label set: $y = \{y_1, y_2, .., y_q\}$, sample set: $D = \{(x_1, Y_1), (x_2, Y_2), ..., (x_m, Y_m)\}$, $|D| = m$, $x_i \in X$, the i-th sample of x_i is represented by d-dimension feature vector. $(x_{i1}, x_{i2}, ..., x_{id})$, $Y_i \subset y$, Y_i is the set of labels corresponding to the $i-th$ instance, $Y_i = [Y_{i1}, Y_{i2}, ..., Y_{iq}]$, $Y_{ij} \in \{0, 1\}$, if $Y_{ij} = 1$, x_i has Y_{ij}, otherwise, x_i does not have Y_{ij} . The goal of MLMC is to learn a multi-label classifier $f : X \rightarrow 2^y$ from sample set D.

3.2 The Range of Correlation Between Local Labels

Definition 1 Topic. For each instance x_i in the training set, similar labels can be divided into the same topic by measuring Y_i whether its label set is similar to other sets. The dataset can be divided into t topics: $\Psi = \{\Psi_1, \Psi_2, ..., \Psi_t\}$.

Definition 2 Edit Distance of Label Sets. The Edit distance [11] of label sequences refers to the minimum number of conversions required between label sets Y_i and Y_j from the label sequence of Y_j to the label sequence of Y_i.

If $min(|Y_i|, |Y_j|) = 0$, $Edits_{Y_i,Y_j} = max(|Y_i|, |Y_j|)$, otherwise, $Edits_{Y_i,Y_j}$ is the minimum of $Edits_{Y_i,Y_j}(|Y_i| - 1, |Y_j|) + 1$, $Edits_{Y_i,Y_j}(|Y_i|, |Y_j| - 1) + 1$ and $Edits_{Y_i,Y_j}(|Y_i| - 1, |Y_j| - 1) + \mathbf{I}_{(|Y_i| \neq |Y_j|)}$. The function $\mathbf{I}_{(|Y_i| \neq |Y_j|)}$ can express the relationship between $|Y_i|$ and $|Y_j|$, if $|Y_i| = |Y_j|$, then $\mathbf{I}_{(|Y_i| \neq |Y_j|)} = 0$, otherwise, if $|Y_i| \neq |Y_j|$, then $\mathbf{I}_{(|Y_i| \neq |Y_j|)} = 1$.

The smaller the edit distance $Edits_{Y_i,Y_j}$ between the label sets, the greater the similarity between sample x_i and x_j.Set XC that the training set data can be divided into C_n^2 label pairwise, MLMC divides the sample into t topics by comparing the size between the threshold μ and $Edits_{Y_i,Y_j}$. For example, the topic Ψ_s in Ψ can be divided as shown in Formula 1:

$$\Psi_s = \{xc_{<i,j>} | Edits_{Y_i,Y_j} \leqslant \mu, xc_{<i,j>} \in XC\} \quad (1)$$

3.3 Local Label Correlation Matrix

Definition 3 Label Positive/Negative Correlation. For instance x_i, when $y_i \in Y_i$,y_p/y_e has a great probability belongs/not belong to Y_i, then the relationship between y_i and y_j, then the relationship between y_i and y_e is called label positive/negative correlation respectively.

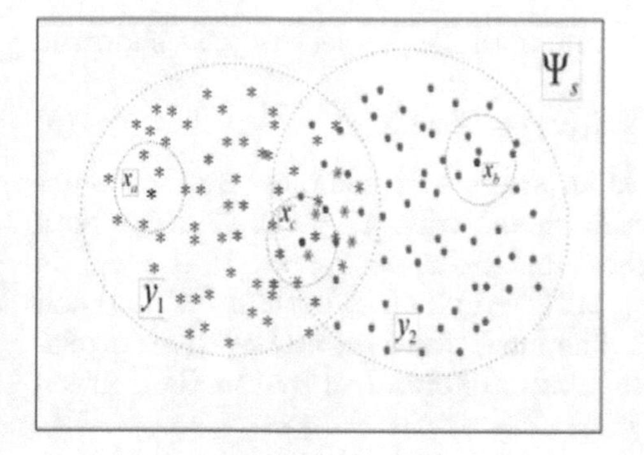

Fig. 1. Classification system based on neighborhood rough set

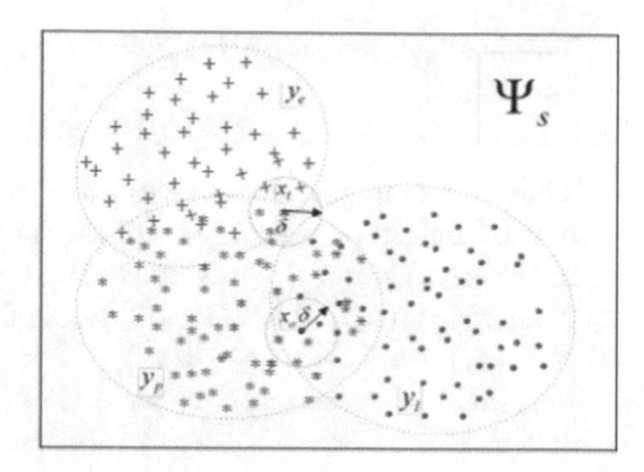

Fig. 2. Label distribution and neighborhood distribution

MLMC calculates the positive and negative correlation between labels based on the nearest neighbor rough set theory. In the single-label classification system, as shown in Fig. 1, it indicates the label distribution and neighborhood distribution in topic Ψ_s, label category y_1 and y_2 are represented by the symbols '*' and '.'. There are three sample points in Fig. 1, the red dotted circle area in Fig. 1 is the nearest sample point of three samples. For the sample point x_a, neighbor sample points are in the area of label y_1, $y_1 \in Y_a$, the same for x_b. For the sample point x_c, its nearest neighbor sample points fall both in the area of label y_1 and in the area of label y_2. Therefore, y_1 and y_2 may belong to Y_c. For instance x_i, $\delta(x_i) = \{x_r | \Delta(x_i, x_r) \leq \delta, x_r \in X\}$, the neighborhood of an attribute. $\delta_j(x_i) = \{x_e | Y_{ej} = 1, x_e \in \delta(x_i)\}$, $|\delta_j(x_i)|$ indicates instances which included in $\delta(x_i)$, y_i represents the number of examples of corresponding label set. In $\delta(x_i)$, $SL(x_i)$ indicate all set of $\delta_j(x_i)$, as shown in Formula 2.

$$SL(x_i) = \bigcup_{q}^{j=1} \delta_j(x_i) \tag{2}$$

Then, the inclusion degrees of sets $SL(x_i)$ and $SY(j)$ can be obtained, as shown in Formula 3. $SY(j) = x_e | Y_{ej} = 1, x_e \in X$,$SY(j)$ is the set composed of all instances with corresponding label y_i value of 1 in the dataset. The function of $Card$ is the sum of the number of element in set.

$$I(SL(x_i), SY(j)) = Card(SL(x_i) \cap SY(j))/Card(SL(x_i)) \tag{3}$$

For instance x_i, the upper and lower approximations of label y_i can be defined, as shown in Formula 4.

$$\begin{aligned} \overline{R}(j) &= \{x_e | I(SL(x_i), SY(j)) \geq 1 - \beta, x_e \in X\}, \\ \underline{R}(j) &= \{x_e | I(SL(x_i), SY(j)) \geq \beta, x_e \in X\} \end{aligned} \tag{4}$$

The positive and negative fields of x_i can be obtained respectively, as shown in Formula 5.

$$PO_j(x_i) = \underline{R}(j), NE_j(x_i) = X - \overline{R}(j) \tag{5}$$

Then we can get the distribution of label space and samples. Set the number of neighbor examples in the neighborhood with x_i as k, if $\delta(x_i)$ both in $PO_j(x_i)$,$|\delta_j(x_i)| \geqslant k \times \beta$, then the label of x_i is $Y_{ij} = 1$; if $\delta(x_i)$ in $NE_j(x_i)$,$|\delta_j(x_i)| < k \times (1-\beta)$, label of x_i is $Y_{ij} = 0$; if $\delta(x_i)$ both in $PO_j(x_i)$ and $NE_j(x_i)$, $k \times (1-\beta) \leq |\delta_j(x_i)| \leq k \times \beta$, The most positive and negative correlation label of x_i concerning for one of its labels are obtained by the distribution of $\delta(x_i)$.

Definition 4 Positive/Negative Correlation Matrix. The two-dimensional matrix $PO \in \mathbb{R}^{m \times q}$ and $NE \in \mathbb{R}^{m \times q}$, used to store label index values. The element PO_{ij} and NE_{ij} of matrix represents the label index value with the maximum positive/negative correlation with a certain label Y_{ij} of example x_i, if $Y_{ij} = 1$, and y_s is most positive/positive correlation label, then $PO_{ij} = p$ and $NE_{ij} = e$, if it does not have most positive/negative correlation label, then $PO_{ij} = 0$ and $NE_{ij} = 0$.

MLMC uses the second-order strategy to explore the label correlation. The theme Ψ_s based on the rough set is shown in Fig. 2. Each instance point x_i takes the circular area drawn with its center and radius δ as its neighborhood. In Fig. 2, for the instance $x_a(x_a \in \Psi_s)$, if $y_l \in Y_a$, then the posterior probability of an event label $y_s \in Y_a(y_s \in y)$ under conditions $x_a \in \Psi_s$ and $y_l \in Y_a$ is $P_{po}(y_s)$, as shown in Formula 6.

$$P_{po}(y_s) = P(y_s \in Y | y_l \in Y_a, x_a \in \Psi_s) \tag{6}$$

The value of $P_{po}(y_s)$ is represented by neighbor instance points and relative label distribution in $\delta(x_i)$, $P_{po}(y_s) = \delta_l(x_i) \cap \delta_s(x_i)/|\delta_l(x_i)|$, $\delta_l(x_i) \cap \delta_s(x_i)$ represent the number of instances of y_l and y_s in the neighborhood of x_i. If there is a maximum positive correlation between label y_i and y_s, then $P_{po}(y_s)$ should be the maximum value, and suppose that the label with the maximum positive correlation with y_l is y_p,$y_p = \max\limits_{y_s, y_s \in Y_i} P_{po}(y_s)$, For $PO_{ij} = p$, the element of positive correlation matrix, if $y_p \notin PO_{ij}$,$PO_{ij} = 0$.

Similar to the above, we can also get y_e,$y_e = \max\limits_{y_s, y_s \in Y_i} P_{ne}(y_s)$. If there is no most negative correlation label y_e exist in $NE_{ij} = e$, the element of negative correlation matrix NE, then $NE_{ij} = 0$. Then we can get all element of the matrix PO and NE.

3.4 Label Dependency Graph and Model Prediction

After obtaining the matrixsPO and NE, the local correlations between labels for each instance is obtained. Depending on the learned local correlation between labels, a label dependency graph can be constructed based on the structure of the Bayesian network [12], where a node represents a label and a directed edge represents the dependency of one label on another. MLMC considers the two kinds of correlation between local labels by using a Bayesian network structure based on a high-order strategy to construct a label dependency graph.

After getting the Bayesian network between labels, the current labels of an instance can be predicted according to the dependency between labels. Set the test instance $x_i = \Psi_s$, and predicts whether a label y_t belongs to Y_i. MLMC constructs the posterior probability formula of y_t by the dependence of labels in Bayesian Networks and also solves whether y_t belongs to Y_i. MLMC adopts linear weighting method to calculate the influence of positive and negative correlation label of y_t on its predicted value, and introduces parameter $\eta(\eta \in [0,1])$ to adjust the positive correlation and negative correlation. Then, the predictive function $f_t(x_i)$ of the label y_t of x_i is shown in Formula 7.

$$f_t(x_i) = \max_{a, a \in \{0,1\}} P_{combine}(Y_{it} = a|x_i) \tag{7}$$

Assuming that the theme of sample x_i is Ψ_s, the row vectors corresponding to other instances in matrix PO and NE in Ψ_s are po_R^t and ne_R^t respectively. Let $po_t(x_i)$ and $ne_t(x_i)$ be the set of all elements of the $t-th$ column vector in po_R^t and ne_R^t, respectively, representing the set of the most positive and negative correlation labels with label t in Ψ. The posterior probability $P_{combine}(Y_{it} = a|x_i)$ traverses each of the most relevant labels in $po_t(x_i)$ and $ne_t(x_i)$ as y_p and y_e respectively when calculating the positive correlation and negative correlation between labels. Finally, Formula 7 can be optimized to Formula 8.

$$\begin{aligned} f_t(x_i) = \max_{a, a\{0,1\}} \frac{1}{N}(\eta \sum_{y_p \in po_t(x_i)} P(Y_{it} = a, y_p \in Y_i|x_i \in \Psi_s)P(y_p \notin Y_i|x_i \in \Psi_s)), \\ +(1-\eta) \sum_{y_p \in ne_t(x_i)} P(Y_{it} = a, y_e \in Y_i|x_i \in \Psi_s)P(y_e \notin Y_i|x_i \in \Psi_s) \end{aligned} \tag{8}$$

MLMC builds a classifier for each label in instance x_i, each of which is $f(x_i)$, as shown in Formula 9

$$f(x_i) = \{f_1(x_i), f_2(x_i), ..., f_q(x_i)\} \tag{9}$$

4 Experimental Results and Analysis

In this chapter, three evaluation criteria are selected to compare five algorithms, including MLMC, and run on five datasets respectively.

Firstly, both MLkNN and MLRS-LC outperform most of the other algorithms in Hamming Loss index of the models trained on the yeast dataset and scene dataset. MLkNN can achieve the best of five algorithms on the yeast dataset because the nearest neighbor algorithm uses a voting prediction mechanism (Table 1).

Table 1. Comparison of experimental results under Hamming loss

Dataset	MLMC	BR	MLkNN	MLRS-LC	ECC
Yeast	0.2084	0.2369	**0.1933**	0.2139	0.2099
Emotions	**0.1907**	0.2471	0.1951	0.2054	0.2030
Scene	0.0962	0.2454	**0.0862**	0.0990	0.0926
Enron	0.0701	0.0508	0.0523	0.0640	**0.0481**
Cal 500	**0.1311**	0.1615	0.1388	0.1642	0.1450

Secondly, MLRS-LC has the best performance index on yeast, scene, and cal 500 datasets, which is also because MLRS-LC is an improved algorithm based on the nearest neighbor algorithm. Finally, ECC is superior to other algorithms on the enron dataset, because the feature vector space dimension of enron dataset is high, then the algorithm based on the nearest neighbor and similarity method will be affected by sparse data in the calculating distance.

It can be seen from the experimental data that the MLMC algorithm proposed in this paper is superior to other algorithms or even the best under the conditions of different data sets and evaluation indexes, indicating that MLMC can optimize the performance of multi-label classification model and improve the accuracy of classifier prediction. The model performance of MLMC is affected by the similarity between the label sequences corresponding to each instance in the dataset because MLMC calculates the edit distance according to the similarity of the label sequences when solving the correlation between local labels (Tables 2 and 3).

Table 2. Comparison of experimental results under Average Precision

Dataset	MLMC	BR	MLkNN	MLRS-LC	ECC
Yeast	0.7594	0.6216	**0.7620**	0.7331	0.7476
Emotions	**0.8207**	0.7014	0.7965	0.8187	0.8004
Scene	0.8041	0.6216	**0.8478**	0.8492	0.8399
Enron	0.5809	0.5929	0.6322	0.5069	**0.6875**
Cal 500	**0.5011**	0.3597	0.4812	0.4534	0.4625

Table 3. Comparison of experimental results under F1-measure

Dataset	MLMC	BR	MLkNN	MLRS-LC	ECC
Yeast	0.6402	0.5635	**0.6204**	0.6426	0.5973
Emotions	**0.7945**	0.5566	0.6138	0.7805	0.5855
Scene	**0.7314**	0.5732	0.6811	0.7209	0.6533
Enron	0.4829	0.5257	0.4288	0.4510	**0.5652**
Cal 500	0.4218	0.3305	0.3228	**0.4524**	0.3385

5 Summary

In the face of the problem that the time complexity of the traditional algorithm increases due to the increase in the scale of the label set, MLMC is proposed by using a broad dimension of the correlation between labels. Finally, the nearest neighbor rough set theory is used to consider the positive and negative dependencies between the labels. Experimental results show that the algorithm in this chapter performs better than four other algorithms on five datasets based on three evaluation criteria, which verified the optimization of MLMC for a multi-label classification model.

Acknowledgments. This work is supported and assisted by the Research Team of Key Technologies of Smart Agriculture and Intelligent Information Processing and Optimization (No. 2019ABA101).

References

1. Yu, Y.: A review of multi label learning. J. Comput. Eng. Appl. **51**(17), 20–27 (2015). (in Chinese)
2. Zhang, M., Zhou, Z.: A review on multi-label learning algorithms. J. IEEE Trans. Knowl. Data Eng. **26**(8), 1819–1837 (2014)
3. Boutell, M.R., Luo, J., Shen, X., et al.: Learning multi-label scene classification. J. Pattern Recogn. **37**(9), 1757–1771 (2004)
4. Zhang, M.L., Zhou, Z.H.: KNN: a lazy learning approach to multi-label learning. J. Pattern Recogn. **40**(7), 2038–2048 (2007)
5. Elisseeff, A., Weston, J.: A kernel method for multi-labelled classification. Advances in Neural Information Processing Systems MIT, pp. 681–687 (2001)
6. Ghamrawi, N., McCallum, A.: Collective multi-label classification. In: Proceedings of the 14th ACM International Conference on Information and Knowledge Management, pp. 195–200. ACM, Breman (2005)
7. Read, J., Pfahringer, B., Holmes, G., Frank, E.: Classifier chains for multi-label classification. In: Buntine, W., Grobelnik, M., Mladenić, D., Shawe-Taylor, J. (eds.) ECML PKDD 2009. LNCS (LNAI), vol. 5782, pp. 254–269. Springer, Heidelberg (2009). https://doi.org/10.1007/978-3-642-04174-7_17
8. Read, J., Pfahringer, B., Holmes, G., et al.: Classifier chains for multi-label classification. J. Mach. Learn. **85**(3), 333–359 (2011)
9. Yu, Y., Pedrycz, W., Miao, D.: Multi-label classification by exploiting label correlations. J. Expert Syst. **41**(6), 2989–3004 (2014)
10. Yu, Y., Miao, D.Q., Zhao, C.R., et al.: Knowledge acquisition method of multi marker decision system based on rough set. J. Comput. Sci. Explor. **9**(1), 94–104 (2015). (in Chinese)
11. Sun, Q.H., Yang, H.B.: Optimization of sequence clustering algorithm based on editing distance. J. Comput. Technol. Dev. **28**(3), 109–113 (2018)
12. Lin, S.M., Tian, F.Z., Lu, Y.C.: Studies on Bayesian classifier in data mining. J. Comput. Sci. **85**(3), 333–359 (2000). (in Chinese). Read, J., Pfahringer, B., Holmes, G., et al.: Classifier chains for multi-label classification. Mach. Learn. **85**(3), 333–359

Named Entity Disambiguation Based on Bidirectional Semantic Path

Zimao Li[1,2], Yue Zhang[1,2](✉), Fan Yin[1,2], and Mengyan Nie[1,2]

[1] College of Computer Science, South-Central University for Nationalities, Wuhan 430074, China
1114279203@qq.com

[2] Hubei Provincial Engineering Research Center for Intelligent Management of Manufacturing Enterprises, Wuhan 430074, China

Abstract. In the existing entity disambiguation algorithms, the full connection method is adopted to measure the association between candidate entities, ignoring the ambiguity of entity mention, and the shortest path feature used for feature extraction does not consider the influence of different path starting and ending points. Therefore, this paper proposes a progressive named entity disambiguation (NED) algorithm PNED, which uses the bidirectional semantic association feature and topic relevance feature extracted from the disambiguated entity to calculate the association relation between entities. The algorithm fully considers the difficulty of different entity mention disambiguation, and gradually completes the NED task from easy to difficult, which improves the implementation efficiency of NED. Experimental results show the effectiveness of the algorithm.

Keywords: Entity disambiguation · Bidirectional semantic path · Progressive disambiguation

1 Introduction

Named entity disambiguation has been extensively studied and has achieved many outstanding results, but the related research still needs to be further improved. Although many entity disambiguation methods have been reported successively, the asymmetry of the shortest path between entities has not been taken into account when using the shortest path feature to measure the degree of association between candidate entities. In fact, different starting point may lead to two shortest paths with different lengths [1]. As shown in Fig. 1, the shortest path length from "Buffalo Bills" to "UCF Knights football" is 2, while the length from "UCF Knights football" to "Buffalo Bills" is 1.

To mine the path association between two entities more accurately, we propose a new method to redefine the shortest path length between two entities. Besides, existing entity disambiguation algorithms use a fully connected

Z. Li—Major Projects of Technological Innovation in Hubei Province (No. 2019ABA101).

Q. Liang et al. (Eds.): Artificial Intelligence in China, LNEE 854, pp. 440–447, 2022.
https://doi.org/10.1007/978-981-16-9423-3_55

method to measure the association between candidate entities, ignoring the degree of discrimination of entities, which increases the complexity of algorithm execution. Therefore, we use the bidirectional semantic association feature and topic relevance feature to calculate the association relation between the disambiguated entity and the remaining entity vertex, which disambiguate the entities gradually.

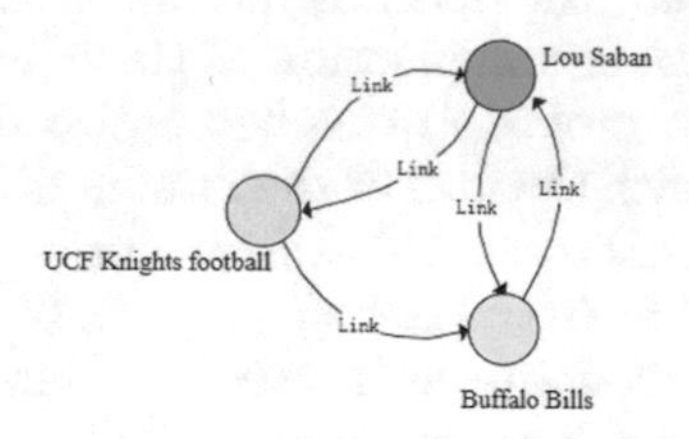

Fig. 1. The shortest path on knowledge graph

2 Related Work

The main idea of NED algorithm is using the extracted features to measure the similarity between the entity mention and the candidate entity. In 2011, Hoffert et al. [2] took entity mention and candidate entity as the vertex of graph structure, and used the greedy algorithm to find the smallest strong connected graph to link entities. In the same year, Han et al. [3] proposed an integrated reasoning algorithm in the graph structure to capture the correlation between vertex. In 2014, Alhelbowy et al. [4] proposed two integrated entity disambiguation algorithms using graph structure. The first algorithm is to sort all vertex by PageRank algorithm [4–7], and select candidate vertex according to PR value and local confidence. The second algorithm is to use the clustering algorithm to find the most important sub-cluster, and expand it until all entity mentions are disambiguated. In 2016, Stefan et al. [6] introduced a new "pseudo" theme vertice in the entity correlation graph to enhance the consistency between entities and the theme vertice. In 2017, Pappu et al. [8] used forward-backward algorithm to only consider the adjacent relation between entities, and achieved the fastest disambiguation result. In 2019, Liu et al. [9] introduced an extension of BERT (KBERT) in his work, in which KG triples are injected into the sentences as domain knowledge. In 2020, Mulang et al. [10] use the context extracted from the knowledge graph to provide enough information for the model, and improves the performance of NED.

In summary, there is a wide variety of approaches in the literature for NED, which provides more possibilities to break the shackles of traditional search engines.

3 Progressive Named Entity Disambiguation (PNED)

3.1 Problem Description and Formalization

In practical application, the ambiguity degree of each entity mention in the text is not always the same. When disambiguating the entities, entity with least ambiguity is firstly disambiguated, and then disambiguating the remaining entities gradually. In particular, the disambiguated entities can provide additional information to assist in the disambiguation of the remaining entities.

This paper proposes a progressive named entity disambiguation algorithm PNED, which combined with the idea of the minimum spanning tree. The entity mentions contained in text D is marked as $M = \{m_1, \ldots, m_n\}$, and its candidate set in knowledge base W is $C = \{\{e_1^1, e_1^2, \ldots\}, ..., \{e_n^1, e_n^2, \ldots\}\}, C_i \in C$. The task of PNED is to find an entity e_i in each candidate subset C_i, so that the sum of edge weights between tree vertex is minimized, which is marked as L.

3.2 Candidate Entity Representation Model

The representation model is presented as graph structure [11], which is recorded as $G(V, E)$. The candidate entities of mentions in text D are formalized as V of the graph, and the relation between entities is expressed as E. We use Stanford NER tool to get entity mention set $M = \{m_1, \ldots, m_n\}$, and obtain the candidate entity set through the Search API provided by Google Knowledge Graph.

Definition 1 Initial Confidence of Candidate Entities. The initial confidence of a candidate entity is the possibility of being linked without considering the context information. In our paper, the Search API of Google Knowledge Graph is used to return the entity's result score as the initial confidence as shown in Formula 1. We use the minimum edit distance to filter the candidate entities, and the top 5 candidate entities are selected to generate the final candidate entity set.

$$CM(v_a) = \frac{ResultScore(v_a)}{\sum_{v_a \in V} ResultScore(v_a)} \tag{1}$$

Definition 2 Bidirectional Path Associations Between Candidate Entities. It is the shortest path length of the hyperlink between entries of candidate entities in Wikipedia:

$$RelPath(v_a, v_b) = \frac{1}{1 + ShortPath(v_a, v_b)} \tag{2}$$

$ShortPath(v_a, v_b)$ is the shortest path length between the candidate entity vertice v_a and v_b, which can be obtained by Formula 3:

$$ShortPath(v_a, v_b) = \frac{FShortPath(v_a, v_b) + BShortPath(v_a, v_b)}{2} \tag{3}$$

$FShortPath(v_a, v_b)$ represents the shortest path length from the vertice v_a to v_b in Wikipedia; while $BShortPath(v_a, v_b)$ represents the length from the vertice v_b to v_a.

Definition 3 Bidirectional Semantic Associations Between Candidate Entities. Using semantic information to weight the shortest path can better express the correlation between two vertex. The calculation formula is shown in Formula 4.

$$RelBsa(v_a, v_b) = (1 - \alpha)SimText(v_a, v_b) + \alpha RelPath(v_a, v_b) \tag{4}$$

$SimText(v_a, v_b)$ is the semantic similarity between the text description of candidate entities, which is calculated by cosine similarity [12]. For the elements in the candidate entity set corresponding to the same entity mention, the $RelBsa$ value is 0.

Definition 4 Topic Relevance. Entity mentions in a text usually have a common topic. Combining the topic relevance features of candidate entities into the disambiguation process can further enrich the features and help to improve the accuracy of algorithm. $Topic(e_i^k)$ represents the topic vector of the kth candidate entity e_k of entity mention m_i. The topic relevance is calculated in Formula 5:

$$RelTopic(e_i^k, e_j) = \frac{Topic(e_i^k) \cdot Topic(e_j)}{\|Topic(e_i^k)\| \, \|Topic(e_j)\|}, e_i^k \in C_i, e_j \in E_d \tag{5}$$

Definition 5 Entity Relevance. The relevance of entities is related to the disambiguation of the remaining entities, which is determined by the bidirectional semantic association of entities and the topic relevance between candidate entities. The entity association degree of disambiguated entities and remaining entities is shown in Formula 6:

$$Rel(e_i^k, e_j) = \beta RelBsa(e_i^k, e_j) + (1 - \beta)RelTopic(e_i^k, e_j), e_i^k \in C_i, e_j \in E_d \tag{6}$$

3.3 Determination of Initial Vertice

To adapt to the task of NED, in this paper, the initial vertice of the minimum spanning tree is not randomly selected as the traditional Prim algorithm, but the entity with the least ambiguity.

Definition 7 Easy Entity Mention (EEM). The entity mention with the smallest amount of information has more certainty, the least ambiguity, and the lowest disambiguation difficulty, which is determined as Easy Entity Mention (EEM). Its real semantics and attributes can help to resolve the ambiguity of other entity mentions. The EEM should satisfy the following two conditions: (1) If the number of candidate entities corresponding to an entity mention is 1, the entity mention is EEM. (2) If the number of candidate entities corresponding to all entity mentions in the entity mention set is more than 1, the entity mention with the smallest amount of information is EEM.

To determine the EEM, the topic vector of the candidate entity is used to represent the information quantity of entity mention. In this paper, we train LDA topic model and construct the entity mention topic matrix through gensim tool. Then, Singular Value Decomposition (SVD) is used to obtain the singular values of the entity mention topic matrix.

Suppose an entity mention m_i has l candidate entities, and the topic matrix $MTM_{i,l \times c}$ of m_i is constructed from the topic vectors of l candidate entities, where c represents the number of topics in LDA topic model. The singular value reflects the information quantity of matrix $MTM_{i,l \times c}$. Formula 7 is used to approximately calculate the singular degree (SD). σ is the diagonal element of the diagonal matrix after the matrix $MTM_{i,l \times c}$ singular value decomposition.

$$SD = \frac{\sigma_1}{\sum_{h=1}^{c} \sigma_h} \tag{7}$$

The amount of information $H(m_i)$ of entity mention decreases with the increase of singular value concentration degree SD of entity mention topic matrix, as shown in Formula 8:

$$H(m_i) \propto \frac{1}{SD} \tag{8}$$

Definition 8 Initial Vertice of Minimum Spanning Tree. The candidate entity that the EEM refers to is regarded as the initial vertice of the minimum spanning tree, which is also added to the disambiguated entity set E_d.

If the number of candidate entities corresponding to the EEM is 1, then the candidate entity is the real linked entity of the EEM. If the number is more than 1, the candidate entities are filtered according to the initial confidence and the text relevance between. The real direction of the EEM is determined as shown in Formula 9:

$$E_d = \left\{ e_i^k | argmax(SimCon(e_i^k) + CM(e_i^k)), e_i^k \in EEM \right\} \tag{9}$$

3.4 Progressive Named Entity Disambiguation (PNED)

Combined with the idea of getting the minimum spanning tree by Prim algorithm, this paper proposes a progressive NED algorithm. The execution process is as follows: firstly, the EEM is selected. After that, the EEM is disambiguated according to the initial confidence of candidate entities and the text relevance. Then, by calculating the bidirectional semantic association and topic relevance, the remaining entities are gradually disambiguated.

Suppose that a text contains five entity mentions, and each entity refers to 2–3 candidate entities. The schematic diagram of PNED candidate entity representation model construction process is shown in Fig. 2.

4 Experiment and Analysis

4.1 Dataset

Wikipedia corpus covers a wide range of information as a comprehensive knowledge base. The knowledge base of NED algorithm in this paper is constructed by

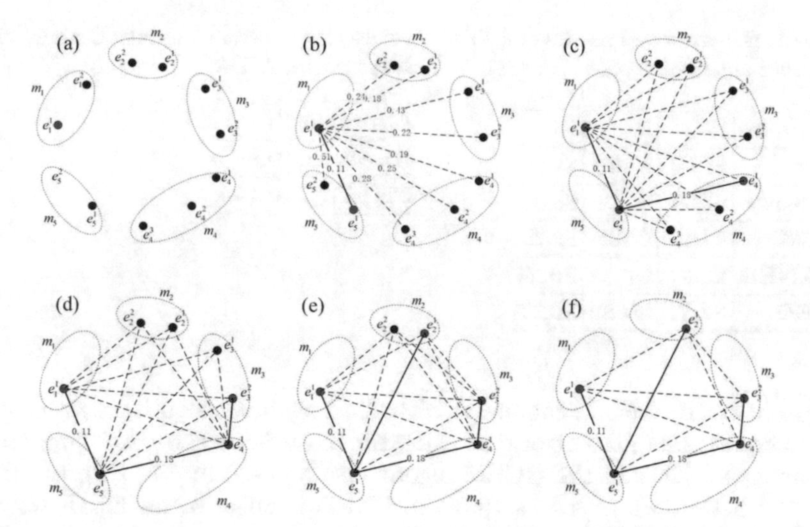

Fig. 2. Schematic diagram of construction process of PNED candidate entity representation model

English Wikipedia, which contains more than 5 million entities and more than 30 million hyperlinks. We use AIDA dataset to test the NED effect of PNED algorithm. There are 1393 documents in AIDA dataset, including 34956 entity mentions, 27820 linkable entity mentions, and 7136 non-linkable (NIL) entity mentions.

4.2 Comparison Algorithm

In order to verify the effectiveness of PNED algorithm in the NED task, it is compared with the current mainstream algorithm on AIDA dataset. The algorithms involved in the comparison include the algorithm proposed by Alhelbawy et al. [4], the algorithm proposed by Hoffert et al. [2], GCEM algorithm proposed by Liu [12] in 2018, and BSANED, our NED method based on bidirectional semantic association but without progressive disambiguation.

4.3 Experimental Parameter Setting

In PNED algorithm, parameter α is used to weigh the weight of edge E in candidate entity representing model. The TestA test set in AIDA dataset is used to determine the value of α, which is set to 0.4, and the maximum length of the shortest path, which is set to 5.

4.4 Analysis of Experimental Results

In order to verify the effectiveness of the BSANED algorithm proposed in this paper, it is compared with four algorithms which also use AIDA dataset.

Table 1. Experimental results of PNED and other algorithms on AIDA (%)

Algorithms	P	R	F1
Hoffart	81.82	81.91	81.86
Alhelbawy	87.59	84.19	85.85
GCEM	84.25	81.36	82.78
BSANED	**90.75**	**87.84**	**89.27**
PNED	**87.76**	**85.82**	**86.78**

Table 2. Average execution time of each text on AIDA

Algorithms	Time
BSANED	**0.51** s
PNED	**0.37** s

As shown in Table 1, compared with the PageRank algorithm proposed by Alhelbowy et al. [4], the algorithm using the most dense sub-graph proposed by Hofhart et al. [2], and the GCEM algorithm proposed by Liu [12], the PNED performs better, which shows that the PNED is effective for NED. However, compared with BSANED, PNED has lower accuracy, recall rate and F1 value. This is mainly because in PNED, only one entity mention is disambiguated in the first disambiguation process. If the disambiguation can be referred to multiple EEMs in the first step, the topic of the text can be captured more accurately. In progressive named entity disambiguation, the PNED makes use of the topic relevance and bidirectional semantic association features of the entities. If the topic vector of the disambiguated entity deviates from the topic vector of the text, it is easy to cause topic drift, which will affect the accuracy of subsequent entity disambiguation.

In Table 2, we compared PNED with BSANED by the average execution time of each text. PNED reduces the time by about 37.8%. This is because BSANED calculates the association among all candidate entities in the same text in batch while PNED reduces the number of vertices in the graph structure by removing invalid candidate entities, which can improve the efficiency of the whole disambiguation process.

5 Summary

In this paper, a progressive named entity disambiguation algorithm is proposed according to the different degrees of ambiguity of the entity mention. The minimum spanning tree is constructed in the graph structure to achieve the purpose of gradually disambiguating the entity mentions from easy to difficult. The task of progressive entity disambiguation is completed by combining the topic relevance feature obtained from EEM and bidirectional semantic relevance feature. The disambiguation of EEM can bring more clear and reliable context information, which can further enrich the feature. Moreover, by removing the remaining invalid candidate entities, the number of vertice in the graph structure and the calculation of the association relation between entities are reduced, which can help to improve the efficiency of the whole disambiguation. Experimental results on AIDA dataset show that the proposed algorithm can improve the implementation efficiency of NED algorithm.

Acknowledgments. This work was supported by the Major Projects of Technological Innovation in Hubei Province (No. 2019ABA101), Industry-University-Research Innovation Fund-"Moss Digital Intelligence Integration" Collaborative Innovation Project (2020QT08), Science and Technology Development Center of Ministry of Education, and the Research Team of Key Technologies of Smart Agriculture and Intelligent Information Processing and Optimization.

References

1. Yang, G.: Research on named entity disambiguation based on graph method. Harbin Industrial University (2015). (in Chinese)
2. Hoffart, J., Yosef, M.A., Bordino, I., et al.: Robust disambiguation of named entities in text. In: Proceedings of the 2011 Conference on Empirical Methods in Natural Language Processing, pp. 782–792. ACL, Edinburgh (2011)
3. Han, X., Zhao, J.: Structural semantic relatedness: a knowledge-based method to named entity disambiguation. In: Proceedings of the 48th Annual Meeting of the Association for Computational Linguistics, pp. 50–59. ACL, Uppsala (2010)
4. Alhelbawy, A., Gaizauskas, R.: Collective named entity disambiguation using graph ranking and clique partitioning approaches. In: Proceedings of COLING 2014, the 25th International Conference on Computational Linguistics: Technical Papers, pp. 1544–1555. ACL, Dublin (2014)
5. Alhelbawy, A., Gaizauskas, R.: Graph ranking for collective named entity disambiguation. In: Proceedings of the 52nd Annual Meeting of the Association for Computational Linguistics, pp. 75–80. ACL, Baltimore (2014)
6. Zwicklbauer, S., Seifert, C., Granitzer, M.: Robust and collective entity disambiguation through semantic embeddings. In: International ACM SIGIR Conference on Research and Development in Information Retrieval, pp. 425–434. ACM, Pisa (2016)
7. Haveliwala, T.H.: Topic sensitive PageRank: a context-sensitive ranking algorithm for web search. IEEE Trans. Knowl. Data Eng. **15**(4), 784–796 (2003)
8. Pappu, A., Blanco, R., Mehdad, Y., Stent, A., Thadani, K.: Lightweight multilingual entity extraction and linking. In: International Conference on Web Search and Data Mining Conference Series: Web Search and Data MiningLink, pp. 365–374. ACM, Cambridge (2017)
9. Liu, W., et al.: K-BERT: enabling language representation with knowledge graph. In: The 10th AAAI Symposium on Educational Advances in Artificial Intelligence, pp. 2901–2908. AAAI, New York (2019)
10. Mulang, I.O., Singh, K., Prabhu, C., Nadgeri, A., Hoffart, J., Lehmann, J.: Evaluating the impact of knowledge graph context on entity disambiguation models. In: Proceedings of the 29th ACM International Conference on Information and Knowledge Management, pp. 2157–2160. ACM, New York (2020)
11. Milne, D., Witten, I.H.: Learning to link with Wikipedia. In: Proceedings of the 17th ACM Conference on Information and Knowledge Management, pp. 509–518. ACM, Napa Valley (2008)
12. Liu, B.: Entity Linking Based on Graph and deep learning. Central China Normal University (2018). (in Chinese)

Improved Yolov5 Algorithm for Surface Defect Detection of Strip Steel

Sun Zeqiang[1] and Chen Bingcai[1,2](✉)

[1] College of Computer Science and Technology, Xinjiang Normal University, Urumqi 830054, China
cbc9@qq.com

[2] School of Computer Science and Technology, Dalian University of Technology, Dalian 116024, Liaoning, China

Abstract. Aiming at the problems of low detection accuracy and slow detection speed in the traditional method of detecting strip steel surface defects, this paper proposes an improved yolov5 algorithm for detecting strip steel surface defects. Firstly, the data set of strip surface defects was constructed, and the K-means algorithm was used to cluster the defect samples, and the prior box parameters of different sizes were obtained. Secondly, the attention-yolov5 algorithm is proposed, which draws on the item-based Attention mechanism, adds channel Attention and spatial Attention mechanism to the feature extraction network, and uses the filtered weighted feature vector to replace the original feature vector for residual fusion. Finally, In order to improve the ability of defect feature extraction, the convolution layer is added after the main feature is extracted from different feature layers of the network output and after the pooling structure of spatial pyramid. The experimental results show that the mAP value of the improved yolov5 algorithm on the test set is as high as 87.3%, which is 5% higher than the original yolov5 algorithm. The average detection time of a single image is 0.0219s, which is basically the same as the original algorithm, and the detection performance is also better than the Faster RCNN and yolov3.

Keywords: Algorithm · yolov5 · Steel strip · Surface defect detection · Deep learning

CLC Number: TP391.4 Document code A

1 Introduction

Strip steel products are widely used in China's national economy and iron and steel industry important products, material performance, geometric size and surface quality is the main judging basis to measure the material efficiency. At present, the material and size of cold-rolled strip products can basically meet the requirements, but the surface quality problem is often the main influencing factor [1]. In recent years, the construction

Q. Liang et al. (Eds.): Artificial Intelligence in China, LNEE 854, pp. 448–456, 2022.
https://doi.org/10.1007/978-981-16-9423-3_56

of infrastructure is higher and higher demand for the accuracy of strip steel, strip steel is affected by the structure and production process led to a variety of surface defects, such as inclusion, patches, pitting, pressure into the scale and scratches, and so on, these defects light to reduce the service life of steel strip and uses, or may affect the construction of the building safety construction accidents. Therefore, this paper hopes to effectively detect the defects on the strip steel surface through the method of computer vision object detection, and then can separate the defective products.

Deep learning as an important technology of image recognition field, has a broad prospect of application and research on image recognition technology to promote the development of computer vision and artificial intelligence has important theory value and practical significance [2], the convolutional neural network and the introduction of deep learning, realized the high precision and high efficiency in the defect detection of advantage, It has gradually become the main research direction of defect detection. Currently, object detection algorithms are mainly divided into two categories. The first category is two-stage object detection algorithms represented by RCNN series, such as RCNN algorithm [3], Fast RCNN algorithm [4], Faster RCNN algorithm[5], Mask RCNN algorithm [6], The other is the one-stage object detection algorithm represented by yolo algorithm, such as OverFeat algorithm [7], yolo algorithm [8], yolov2 algorithm [9], yolov3 algorithm [10], and SSD algorithm [11]. The two-stage object detection algorithm requires the algorithm to generate the target candidate box, that is, the target position, first. Then do the classification and regression to the candidate box. The one-stage object detection algorithm only uses a convolutional neural network CNN to directly predict the categories and locations of different targets. The first kind of method is more accurate, but the speed is slow, but the second kind of algorithm is faster, but the accuracy is lower.

In recent years, many scholars have begun to devote themselves to defect detection. Li Bin et al. [12] used the improved yolo algorithm to detect the surface defects of aerospace engine parts, and the mAP value of the detection result was as high as 82.67%, and the average detection time of a single picture was 0.1240s.Chang Jiang et al. [13] solved the problem of insufficient samples in deep learning by using improved generation adversation network to generate more realistic strip defect images. In order to realize defect detection of bullet appearance, Ma Xiaoyun et al. [14] used k-means++ algorithm to generate sliding window anchor in Faster RCNN. Cheng Song et al. [15] realized image detection and recognition of welds through an improved yoloV4 model, and its mAP value was 88.52%, and the detection speed was also improved to 24.47fps.Weng Yushang et al. [16] improved the data of MaskRCNN in the detection of strip steel defects, which greatly improved the detection accuracy, but the detection speed was only 5.9 fps.

In this study, on the basis of deep learning, attention modules are added to the optimized yolov5 network to add weight to find more important feature channels, and experiments and tests are carried out to improve the accuracy of intelligent identification of strip surface defects under different signal-to-noise ratios.

2 Deep Convolutional Network Model and Attention Mechanism

2.1 Yolo Algorithm

Yolo algorithm regards the object detection framework as a spatial regression problem, and a single neural network can get the prediction of the boundary box and category probability from the complete image after a single operation. Its detection process is shown in Fig. 1. Firstly, the image size is adjusted, and then the image is sent into the convolutional network. Finally, the target detected by the network prediction is processed. The specific operation method is to divide the whole image into S × S grids implicitly, and the prediction can be made according to the grid in which the object center appears.

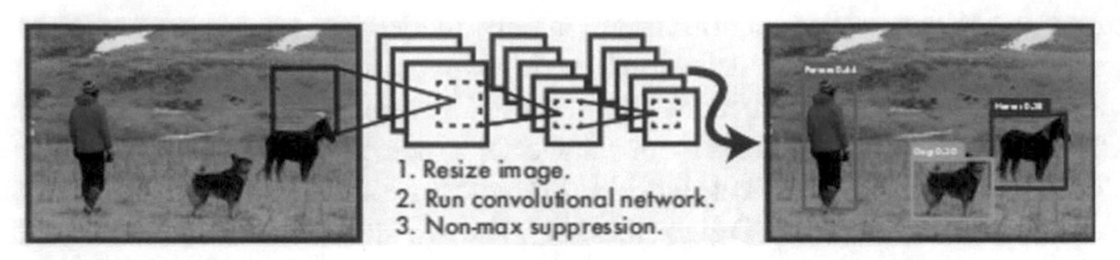

Fig. 1. Detection process of yolo algorithm

2.2 SE Attentional Mechanism

There are two main types of attention mechanisms. One is the channel attention, such as the SE module. The other is the space attention, such as the space converter network. And features can be tuned on a per-channel basis, so that the network can learn to selectively reinforce features that contain useful information and suppress unwanted features through global information. Spatial attention aims to understand more detailed features. Spatial converter network STN explicitly allows spatial manipulation of data and processing of data to enhance geometric invariance of the model. The attention mechanism can be intuitively interpreted using human visual mechanism. For example, our visual system tends to focus on parts of an image that help with judgment and ignore irrelevant information.

The basic structure of the SE Block used in this paper is shown in Fig. 2. The first step is the squeeze operation, which takes the global spatial features of each channel as the representation of the channel to form a channel descriptor. The second step is to learn the dependence on each channel and adjust the feature graph according to the dependence. The adjusted feature graph is the output of the SE Block.

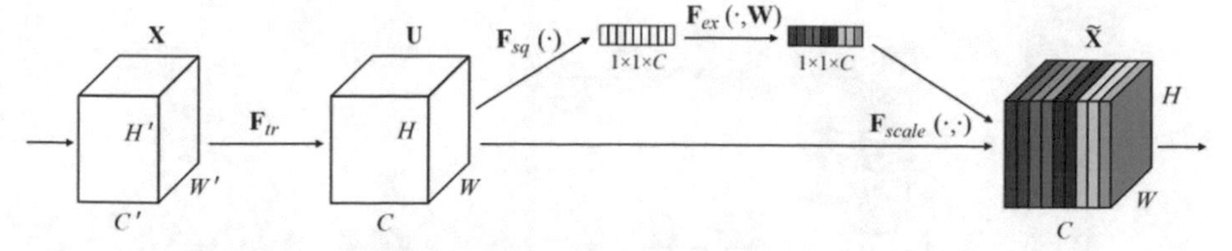

Fig. 2. A squeeze-and-excitation block.

2.3 Improved Network Architecture

The yolov5 algorithm borrowed yolov4's CSPDarkNet53 as the backbone feature extraction network and added some new features, such as FOCUS, CONV, LIBRARY CSP, SPP, etc., in which FOCUS: First, four input copies are copied and divided into four slices by slicing operation. Then, they are spliced by concat layer. Here, splicing refers to the combination of the number of channels to increase the number of features of the image, while the information under each feature remains unchanged. Then through the CBL layer, that is, through the convolution layer (CONV) first, extract the different features of the input, which is helpful to find specific local image features. Secondly, through the Batch Norm layer, the gradient distribution of each batch is controlled near the origin to achieve the normalization of the results, so that the deviation of each batch will not be too large. Finally, the leaky_relu activation function is used to enter the result to the next level of convolution. The Focus module is designed to increase speed by reducing the amount of computation and reducing the number of layers, not by increasing the mAP. SPP: space pyramid pooling consists of three parts: CONV, Maxpooling and CONCAT. First, CONV is used to extract the feature output, and then the maximum pooling layer of three different kernel_sizes is used to conduct the subsampling, and the respective output results are splicing and fused and added to their initial features. Finally, the output is restored to the same as the initial input through conv.

However, in order to solve the problem that the feature distinction between defect types and defect and component structure is not obvious, this paper uses the design idea of convolutional layer in the original yolov5 algorithm to add a layer of SE attention module between CSP2_1 and CBL, and the SE attention mechanism models the dependency relationship between channels. The characteristic response values of each channel can be adjusted adaptively, and the improved network structure is shown in the dotted box in Fig. 3. After the improved network structure processing, on the one hand, SE Block dynamically adjusts the characteristics of each channel according to the input, so as to enhance the expression ability of the network; on the other hand, it only increases a small amount of model complexity and calculation expense, so as to extract the features of defect targets more quickly.

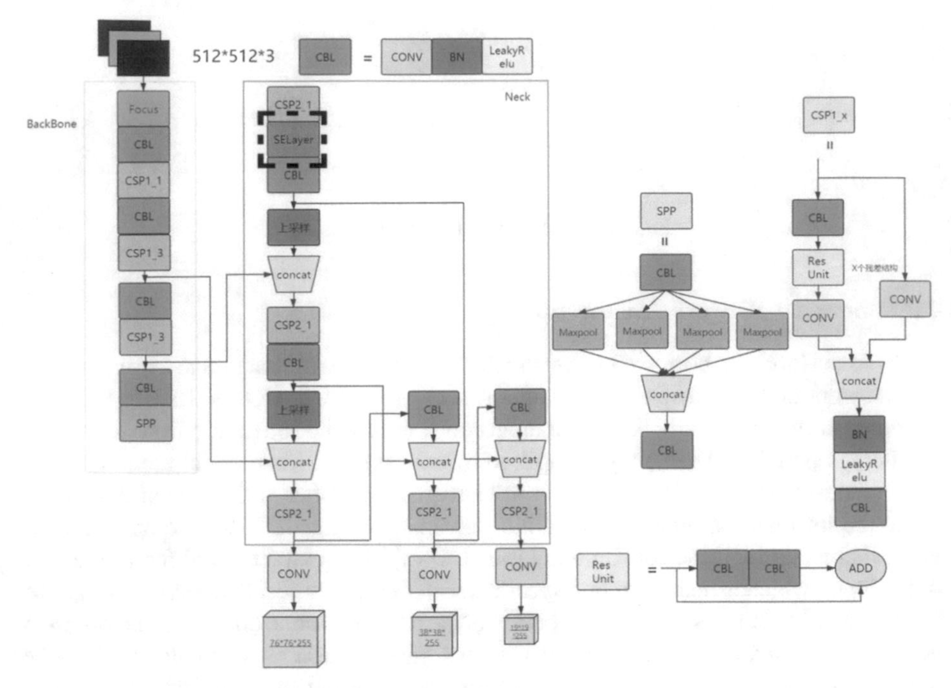

Fig. 3. Improved yolov5 network structure

3 Experimental Results and Analysis

3.1 Experimental Platform

The experimcnt carried out in this paper was completed under the Windows10 operating system. The hardware configuration used in the experiment is as follows: CPU: Intel(R) Core(TM) I3-10100F CPU@3.60GHz;Graphics processor (GPU): NVIDIA GeForce RTX 2080 TI.

3.2 Experimental Data

In this paper, NEU-DET dataset released by Northeastern University was used for experiments. The data set collected 300 pictures of surface defects of 6 types of strip steel. However, the crazing type defect pictures were not obvious, and the defect ambiguity was not good for the experiment, so the crazing type defect was deleted. In the end, the defect pictures used in the experiment in this paper include Inclusion, Patches, Pitted surfaces, Rolled in Scale and Scratches. This data set has a total of 1500 images. The distribution table of strip surface defect data set is shown in Table 1.The defect picture sample of data set is shown in Fig. 4.

Table 1. Distribution table of strip surface defect data set

Types of defects	In	Pa	Ps	Rs	Sc
Number of images	300	300	300	300	300

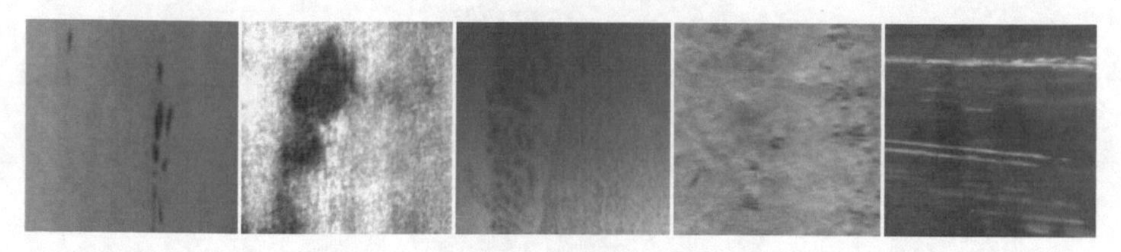

Fig. 4. A graphic example of a data set

3.3 Evaluation Criteria

MAP value was used as the evaluation index in the experiment. mAP is the mean of the average accuracy of all categories, and its calculation formula is

$$mAP = \frac{\sum AP}{N(\text{class})} \quad (1)$$

Where, N is the number of category detection, and AP can comprehensively consider the influence of accuracy and recall rate.PR curve can be obtained by taking accuracy as the vertical axis and recall rate as the horizontal axis. For continuous PR curve, AP in Eq. (1) is

$$AP = \int_0^1 P(R)\,\mathrm{d}R \quad (2)$$

Where P is the accuracy rate and R is the recall rate

$$P = \frac{TP}{TP + FP} \quad (3)$$

$$R = \frac{\mathrm{TP}}{TP + \mathrm{FN}} \quad (4)$$

When training the model, NEU-DET data set is first transformed into COCO data set format. In the training, the transfer learning method is firstly adopted to load the 'yolov5x.pt' pre-weight file to train the NEU-DET data set, so as to generate the weight file of its own model. And set epochs to 100 and batch_size to 8.Set the learning rate as 0.01 and the attenuation coefficient as 0.2.

In this paper, the average detection Time (TIME) of a single image is used as the evaluation index of detection speed. The smaller Time is, the faster the detection speed is, and it is easier to achieve real-time detection. Equation (5) is the calculation formula of Time.

$$Time = \frac{TotalTime}{NumFigure} \tag{5}$$

Where, TotalTime is the TotalTime of detection, and NumFigure is the number of images detected.

3.4 Experiment Settings

In this experiment, SE attention mechanism was added to the original yolov5 model to compare it with the original yolov5 model. The experimental results are shown in Table 2.

Table 2. Comparative results of the improved yolov5 experiment

Model	Single defect recognition accuracy/%					mAP/%	Detection speed/fps
	In	Pa	Ps	Rs	Sc		
yolov5	84.3	90.9	87	48.1	99.5	82.0	**33.299**
yolov5_se	**87.4**	**93.9**	**94.5**	**60.9**	**99.5**	**87.3**	32.91

As can be seen from Table 2 above, after the addition of attention mechanism, the mAP value of yolov5_se generated was increased by 5%, and the detection speed was slightly decreased, but the detection accuracy of each defect was greatly improved.

For better comparison shows that after joining attention mechanism accuracy improvement, this experiment to further introduce other model for comparison are shown in Table 3 below, the one stage detection speed and phase detection speed significantly faster than the two stage detection speed of the algorithm, it also illustrates the stages of the algorithm is better real-time performance, however, It can also be found that the detection performance of yolo algorithm for the pressed oxide skin RS is poor. This is because the pressed oxide skin RS has some defects of small targets, so the accuracy is low. Even the detection accuracy of yolov3_se for RS is only 39.1%, which seriously lowers the average accuracy mAP value. As for the scratch SC defect, it is a big target and easy to detect. Moreover, the contrast between the scratch and the background is obvious, which further reduces the difficulty of detection. Therefore, the detection rate of yolov3_se, yolov5 and yolov5_se on this target reaches 99.5% (Fig. 5).

Table 3. Detection results of different algorithms

Model	Single defect recognition accuracy/%					mAP/%	Detection speed/fps
	In	Pa	Ps	Rs	Sc		
FasterRCNN	70.2	85.6	74.3	63.9	81.7	75.14	2.5
MaskRCNN	72.3	86.5	79.3	**77.8**	89.5	81.02	2.3
ResNetDNN	83.9	89.5	88.3	74.9	89.1	85.14	2.9
SSD	65.1	84.9	69.1	58.8	87.3	73.04	41
yolov3	60.5	83.1	73.2	60	86.1	75.28	50
yolov3_se	**93.1**	85.8	**99.5**	39.1	99.5	83.4	**53.47**
yolov5	84.3	90.9	87	48.1	99.5	82.0	33.299
yolov5_se	87.4	**93.9**	94.5	60.9	**99.5**	**87.3**	32.91

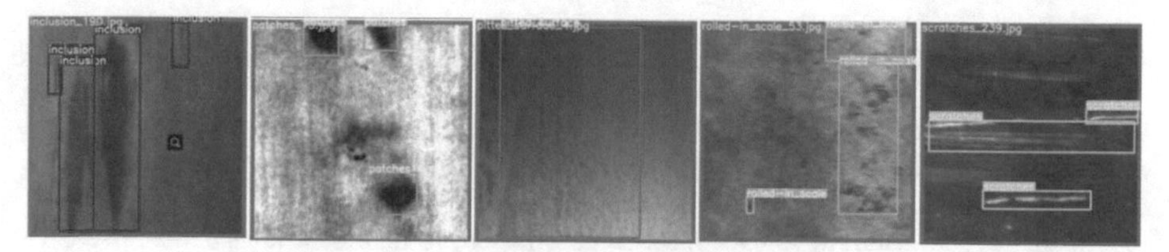

Fig. 5. Improved yolov5 detection results

4 Conclusion

In this paper, a strip surface defect detection method based on improved yolov5 algorithm is proposed. Firstly, k-means algorithm was used to cluster the surface defect data, and nine groups of different prior boxes were obtained to increase the size difference of prior boxes. Then, an SE attention layer is added after the CSP network to realize the channel attention mechanism, and the expression ability of the network is increased by adaptive adjustment of the characteristic response value of each channel. Experimental results show that compared with the original YOLOv5 algorithm, The improved YOLOv5 algorithm can effectively improve the detection accuracy of strip surface defects and realize the intelligent and efficient detection of strip surface defects under the condition that the detection speed is basically flat. Compared with FasterR-CNN and YOLOv3, the detection accuracy of the improved YOLOv5 algorithm is better than both of them, but the detection speed is slower than that of YOLOv3.The next step is to preprocess the original images and expand the data set through data enhancement to further optimize the accuracy of the network. At the same time, the network structure is optimized to improve the speed of defect detection and achieve a better real-time intelligent detection of strip surface defects.

References

1. Li, X.: Study on surface quality defects of cold rolling hood furnace annealing strip, **4**(05), 11–12 (2021)
2. Zheng, Y., Li, G., Li, Y.: A review on the application of deep learning in image recognition. Comput. Eng. Appl. **55**(12), 20–36 (2019)
3. Girshick, R., Donahue, J., Darrell, T., et al.: Rich feature hierarchies for accurate object detection and semantic segmentation. In: Proceedings of the IEEE Conference on Computer Vision and Pattern Recognition, pp. 580–587. IEEE Computer Society Press, LosAlamitos (2014)
4. Girshick, R.: Fast RCNN. In: Proceedings of the IEEE International Conference on Computer Vision, pp. 1440–1448. IEEE Computer Society Press, Los Alamitos (2015)
5. Ren, S.Q., He, K.M., Girshick, R., et al.: Faster RCNN: towards real-time object detection with region proposal networks. IEEE Trans. Pattern Anal. Mach. Intell. **39**(6), 1137–1149 (2017)
6. He, K.M., Gkioxari, G., Dollár, P., et al.: Mask RCNN. In: Proceedings of the IEEE International Conference on Computer Vision, pp. 2961–2969. IEEE Computer Society Press, Los Alamitos (2017)
7. Sermanet, P., Eigen, D., Zhang, X., et al.: OverFeat: integrated recognition, localization and detection using convolutional neworks[OL]. https://arxiv.org/abs/1312.6229. 03 July 2019
8. Redmon, J., Divvala, S., Girshick, R., et al.: You only look once: unified, real-time object detection. In: Proceedings of the IEEE Conference on Computer Vision and Pattern Recognition, pp. 779–788. IEEE Computer Society Press, Los Alamitos (2016)
9. Redmon, J., Farhadi, A.: yolo9000: better, faster, stronger. In: Proceedings of the IEEE Conference on Computer Vision and Pattern Recognition, pp. 7263–7271. IEEE Computer Society Press, Los Alamitos (2017)
10. Redmon, J., Farhadi, A.: yolov3: an incremental improvement[OL]. https://arxiv.org/abs/1804.02767. 03 July 2019
11. Liu, W., Anguelov, D., Erhan, D., Szegedy, C., Reed, S., Fu, C.-Y., Berg, A.C.: SSD: single shot multiBox detector. In: Leibe, B., Matas, J., Sebe, N., Welling, M. (eds.) ECCV 2016. LNCS, vol. 9905, pp. 21–37. Springer, Cham (2016). https://doi.org/10.1007/978-3-319-46448-0_2
12. Li, B., Wang, C., Wu, J., Liu, J., Tong, L., Guo, Z.: An Improved yoloV4 algorithm for surface defect detection of AEROENGINE parts [J/OL]. Laser Optoelectron. Prog. {3}, {4}, {5}, 1–17. http://kns.cnki.net/KCMS/detail/31.1690. 18 July 2021. TN. 20210105.1048.006. HTML
13. Chang, J., Guan, S., Shi, H., Hu, L., Ni, Y.: Defect classification of strip steel based on improved generative adversation network and mobilenetv3. Adv. Laser Optoelectron. 201, **58**(04), 221–226
14. Ma, X., Zhu, D., Jin, C., Tong, X.: Detection of bullet appearance defect based on improved RCNN. Laser Optoelectron. Prog. **56**(15), 117–124 (2019)
15. Cheng, S., Dai, J., Yang, H., Chen, Y.: Weld image detection and recognition based on improved yolov4 [J/OL]. Laser Optoelectron. {3}, {4} {5}, 1–12. http://kns.cnki.net/kcms/detail/31.1690. 18 July 2021 The tn. 20210712.1728.088. HTML
16. Weng, Y., Xiao, J., Xia, Y.: Improved mask RCNN algorithm for surface defect detection of strip steel [J/OL]. Comput. Eng. Use. http://kns.cnki.net/kcms/detail/11.2127.TP.20210420.0937.014.html

Urban Streetscape Tree Density Estimation Algorithm Based on Image Semantic Segmentation

Bin Wang[1], Ping Sun[1], Zhongwang Zhang[2], and Lin Ma[1](✉)

[1] School of Electronics and Information Engineering, Harbin Institute of Technology, Harbin, China
malin@hit.edu.cn

[2] China National Aeronautical Radio Electronics Research Institute, Shanghai, China

Abstract. In the aspect of urban forest density estimation, there is a lack of automatic or efficient estimation methods. For the existing research on urban streetscape trees, mainly uses lidar to process point cloud data or combines deep learning to achieve tree segmentation and detection. However, these methods lead to too much computation, low efficiency, and fail to provide estimation results of urban tree density. By processing the image data, this paper proposes a tree density estimation algorithm based on image semantic segmentation, which deals with the only image data in the whole process, and realizes the estimation of tree density in the city streetscape. This algorithm is more efficient and accurate than the complex point cloud operation or the method combining point cloud with a deep learning algorithm.

Keywords: Semantic segmentation · Urban streetscape · Tree density estimation

1 Introduction

Trees growing in cities or large areas of green forest can be uniformly defined as urban forests. The existing database of urban forests, however, there are insufficient data, updating the problem of not in time, the traditional manual measurement and estimation cannot meet the needs of modern urban forest estimation, people are eager to realize automation and efficient estimation in urban forest density estimation, in order to facilitate real-time monitoring of their time and space distribution.

The diverse composition of cities makes it challenging to use lidar filtering. At the same time, complex 3D point cloud computing makes it more difficult to achieve segmentation and 3D reconstruction of urban trees. In [1], voxel grid filtering is used to reduce the number of point clouds. In this algorithm, grid filtering, spatial clustering, and the k-nearest neighbor method are used in turn to extract the final tree height. The effect is good, but the algorithm design is still complex. In [2], A new method based on point density is proposed to reconstruct the complete three-dimensional structure of a single tree, but this algorithm is very dependent on the composition of the environment,

Q. Liang et al. (Eds.): Artificial Intelligence in China, LNEE 854, pp. 457–464, 2022.
https://doi.org/10.1007/978-981-16-9423-3_57

and if it is too complex, the classification accuracy and reconstruction accuracy will be reduced. With the development of deep learning, tree detection and tree segmentation are realized by combining LIDAR with the target detection algorithm in [3]. The overall segmentation accuracy is very high, but there is still a lack of in-depth research on the statistics of trees. In [4], the LiDAR point cloud was used to detect a single tree and its related measurements, and the measurements of tree height, canopy diameter, and depth were realized. However, there was no statistical study on the distribution density of urban forests. In terms of computational complexity, the use of lidar data combined with deep learning algorithms will be more complex.

Aiming at these problems, this paper proposes an estimation algorithm of urban tree density based on image semantic segmentation. Firstly, we segment the whole image. Secondly, the trees' areas are extracted through a series of image processing techniques, which reduces the number of image pixels that the algorithm needs to traversal in the process of estimating the density of trees. Finally, the size of the tree pixel in the whole image is calculated by the number of pixels to realize the estimation of the density of trees. The algorithm does not use three-dimensional data similar to the lidar point cloud but achieves estimated extraction of tree density in the scene through segmentation and processing of the whole image. Therefore, compared with the processing algorithm of point cloud data, its complexity is reduced and it only needs to realize image processing.

The chapters of this paper are arranged as follows. In Sect. 2, the basic principle of image semantic segmentation is briefly introduced. Section 3 introduces an algorithm of urban tree density estimation based on image semantic segmentation. Section 4 gives the overall process of algorithm implementation and the final detection results. The experimental results are analyzed and the advantages of the proposed algorithm are given. Finally, the thesis is summarized in Sect. 5.

2 System Design

Since the appearance of the FCN network [5] in 2014, image semantic segmentation has realized pixel-level image segmentation. It is an end-to-end neural network based on CNN structure. In the input layer, it is not restricted to the size of the image and the final full connection layer of CNN is removed, making the semantic segmentation using a pre-trained network more efficient. A jump connection is introduced to solve the problem of pixel location with rough up-sampling. But it is insensitive to the details of the image, which restricts the application of data requiring precise statistics such as tree density. After FCN came into being, various forms of neural networks came into being [6–9]. In 2020, the University of Science and Technology of China and Microsoft jointly proposed a High-Resolution Net (HRNET) [10]. The network maintains a high resolution in the whole process. Starting from the high-resolution convolution stream, it gradually increases the high-resolution convolution stream to the low-resolution convolution stream and connects the multi-resolution convolution stream in parallel. The advantage of HRNET is that in terms of feature extraction, image details can always be kept without loss, which is richer in semantic expression and more accurate in space.

Therefore, this paper proposes an urban streetscape tree density estimation algorithm based on HRNET semantic segmentation. Firstly, the open-source cityscapes images are selected as the original input of the whole algorithm, and the semantic region of trees in the image is extracted by using the HRNET network. The extracted results are processed by image grayscale, edge extraction, mathematical morphology, to achieve the extraction of tree contour. Finally, count the number of tree pixels in the region and get the final estimation of urban tree density according to the number of pixels in the whole image. The system block diagram of the whole algorithm is shown in Fig. 1.

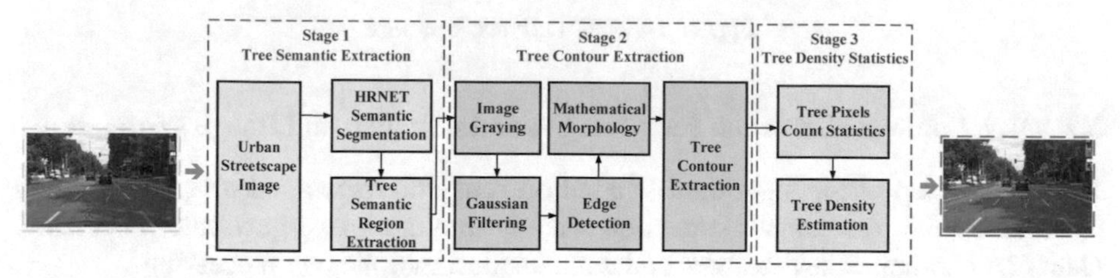

Fig. 1. Block diagram of tree density estimation system based on image semantic segmentation

In Fig. 1, a set of open-source cityscapes image data sets are used as the input of the entire tree semantic extraction and tree density estimation system, and the final output results are used to achieve tree density estimation for each image. Among them, the estimated algorithm of urban tree density based on HRNET semantic segmentation reduces the computational complexity of the original laser radar combined with image processing, improves the computing efficiency of urban tree density estimation, and realizes the automatic tree density estimation. In the algorithm proposed in this paper, through a series of image processing methods, HRNET semantic segmentation results can not only focus on the tree region but also make full use of the pixel number of the region to achieve automatic urban tree density estimation. So that the estimated algorithm used in the actual application has a better real-time interaction.

3 Tree Density Estimation Algorithm

3.1 Semantic Region Extraction Based on HRNET

HRNET network is used for semantic segmentation to achieve segmentation of each type of object in the city streetscape, such as road surface, trees, etc. Through segmentation, each type of object can be analyzed in detail, which lays a foundation for estimating of the density of trees in the city streetscape. Its structure is shown in Fig. 2.

It can be observed that the network is divided into four stages, each of which is composed of channel maps with different resolutions, and each of which has one more branch than the previous one. The new branch is the result of the step convolution fusion of all the feature images of the previous stage. The resolution at the end of the stage will be half of the resolution of the previous branch, and the number of channels will be doubled. On the left is the input image which can be extracted from the semantic region of the city streetscape through the HRNET network.

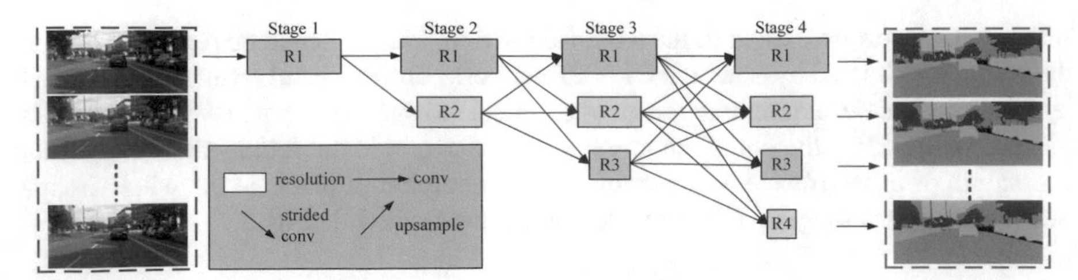

Fig. 2. HRNET network structure

3.2 Tree Contour Extraction Based on Semantic Region and Image Processing

The second step of the algorithm is based on semantic region and image processing to complete the tree contour extraction, mainly through image grayscale processing, Gaussian filtering, edge detection, and mathematical morphology processing.

(1) Image grayscale

The semantic segmentation map of urban trees is grayed to keep only the urban tree area that we are interested in and remove the redundant other areas. The whole image is traversed, and the tree area is assigned a value of 255, that is, white, while the other areas are assigned a value of 0. It can also be processed by the grayscale formula as:

$$\text{Gray} = 0.299\text{R} + 0.587\text{G} + 0.114\text{B} \tag{1}$$

where, RGB represents the image's three channels, red, green, and blue.

(2) Gaussian filtering

The Gaussian filter is a linear filter, which can eliminate the noise generated in the process of digital processing and extract useful features in the image. For the next edge detection, an appropriate Gaussian filter should be selected to suppress a kind of noise whose probability density function follows Gaussian distribution, and the Gaussian kernel size of 7 × 7 is selected. The calculation formula of Gaussian kernel parameters by second-order Gaussian function is shown as follows:

$$g(x, y) = \frac{1}{2\pi\sigma^2} e^{-\left(\frac{x^2+y^2}{2\sigma^2}\right)} \tag{2}$$

where, x^2 and y^2 respectively represent the distance between the pixels in the Gaussian core and the center pixel, and represent the standard deviation. As the standard deviation increases, the effect of Gaussian filtering will be more obvious.

(3) Edge detection

The canny operator is used to realize edge detection. It adopts the method of smoothing first and then derivative. It belongs to the first-order differential filter and is an optimization operator that can achieve a strong filtering effect. By calculating the gradient of the image, the amplitude and angle of the edge of the image are obtained and the change of the gray value is detected. If a drastic change is found somewhere, the area can be determined as the edge of an object. The Sobel operator is selected to obtain the pixel gray value corresponding to the 8-neighborhood region. The template operator and the pixel gray value of 8-neighborhood are shown in (3) and (4) respectively.

$$M_x = \begin{bmatrix} -1\ 0\ 1 \\ -2\ 0\ 2 \\ -1\ 0\ 1 \end{bmatrix}\ M_y = \begin{bmatrix} -1 & -2 & -1 \\ 0 & 0 & 0 \\ 1 & 2 & 1 \end{bmatrix} \tag{3}$$

$$g_{px} = \begin{bmatrix} f(x-1,y+1) & f(x,y+1) & f(x+1,y+1) \\ f(x-1,y) & f(x,y) & f(x+1,y) \\ f(x-1,y-1) & f(x,y-1) & f(x+1,y-1) \end{bmatrix} \tag{4}$$

where, M_x is the x direction template, M_y is the y direction template.

Horizontal and vertical templates are shown in formula (5), and pixel gray value and angle in 8-neighborhood are shown in formula (6) and (7) respectively.

$$\begin{aligned} \mathbf{G}_x =& f(x+1,y+1) - f(x-1,y+1) + 2f(x+1,y) \\ &- 2f(x-1,y) + f(x+1,y-1) - f(x-1,y-1) \\ \mathbf{G}_y =& f(x-1,y+1) - f(x-1,y-1) + 2f(x,y+1) \\ &- 2f(x,y-1) + f(x+1,y+1) - f(x+1,y-1) \end{aligned} \tag{5}$$

$$\mathbf{G} = \sqrt{\mathbf{G}_x^2 + \mathbf{G}_y^2} \tag{6}$$

$$\theta = \arctan\left(\frac{\mathbf{G}_y}{\mathbf{G}_x}\right) \tag{7}$$

(4) Mathematical morphology processing

Mathematical morphology processing is used to further extract the boundary of the tree area, where corrosion and expansion are two basic operations. $B_{(x,y)}$ is the convolution template, B_x is the structural elements, and A is the input image, then the corrosion of B_x to A is shown as follows:

$$A - B_x = \{f(x,y)|B_{(x,y)} \subseteq A\} \tag{8}$$

where, $f(x,y)$ is the pixel value of any pixel point in the image A. Similarly, the expansion of B_x with respect to A is shown as follows:

$$A + B_x = \{f(x,y)|B_{(x,y)} \cap A = \varnothing\} \tag{9}$$

where, $\varnothing$ is the empty set.

3.3 Tree Density Estimation Algorithm Based on Tree Contour Region

The density of an object in an image is defined as the proportion of the number of pixel points of the object to the total number of pixel points in the whole image. Therefore, the density calculation of trees can be expressed as follows:

$$\rho_{\text{tree}} = \frac{n_{\text{tree}}}{M \times N} \tag{10}$$

where, ρ_{tree} is the density of trees; n_{tree} is the number of pixel points in the tree part; and M, N are the height and width of the whole image respectively. To calculate the number of pixels in the tree part, we only need to traverse the whole image and then sum the number of pixels in the tree part.

4 Experiment and Analysis

4.1 Experimental Environment

The implementation of this algorithm is based on Linux-Ubuntu operating system and PyTorch framework under a deep learning environment. The open-source data sets Cityscapes are selected. The data set LeftIng8Bit and Gtfine store image sets and label sets respectively, and each folder contains training sets, verification sets, and test sets respectively. A total of 5,000, 2,975 for training, 500 for verification, and 1,525 for testing. To reduce the time spent on training the network, 1100 pictures of Stuttgart city were selected for this project, and 8 pictures were selected as shown in Fig. 3. The semantic segmentation graphs were selected as in Fig. 4.

Fig. 3. Dataset of Stuttgart

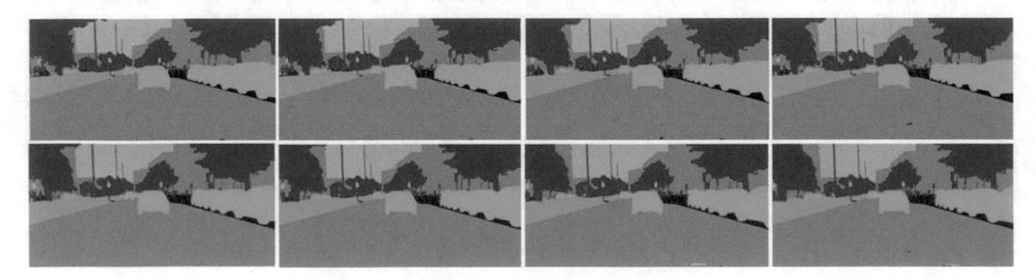

Fig. 4. Semantic segmentation of Stuttgart

4.2 The Experimental Results

According to the results of urban tree distribution in Stuttgart, the criteria for estimating the density of trees in the city can be set, as shown in Table 1. Generally, the density of trees in the city can be divided into three levels: low, medium, and high.

Table 1. Standard for estimating urban tree density

The serial number	Tree density range	The thick degree
1	[0, 0.18)	Low
2	[0.18, 0.28)	Medium
3	[0.28, 1]	High

A total of 6 pictures of 270–280 frames are selected with the span of every two frames to estimate the density of trees. The peripheral contour of trees is marked in yellow and brown. The value of tree density and the degree of trees in each frame of the picture is output in the middle of the picture. The six pictures are shown in Fig. 5.

Fig. 5. Tree extraction and tree density output

Six pictures are selected in the figure above. It can be seen from the analysis that as the vehicle progresses, it generally moves towards the direction of dense distribution of trees, so the value of tree density continues to increase. Meanwhile, the overall value is greater than 0.28, so the thickness of the trees is shown as high (Table 2).

Table 2. Tree density and dense degree table of 270–280 frames

The serial number	Frame	Tree density	The thick degree
1	270	0.289906	High
2	272	0.292167	High
3	274	0.292631	High
4	276	0.294183	High
5	278	0.295065	High
6	280	0.296948	High

5 Conclusion

Aiming at the problems of low efficiency of the algorithm of processing point cloud data by lidar and the lack of automation of urban tree density estimation, this paper proposes an urban streetscape tree density estimation algorithm based on image semantic segmentation. Firstly, the open-source cityscapes data set combined with HRNET semantic segmentation network is used to divide the semantic regions of trees. Secondly, a series of image processing algorithms are used to extract the contours of trees. Finally, the density of trees in the city is estimated based on the statistics of pixels. The whole process is based on image data processing, without the need for other sensor collection point cloud processing or combined with a deep learning algorithm to achieve tree region segmentation, the algorithm enhances the real-time interactivity of tree density estimation, so that the density of trees with faster speed and automation.

Acknowledgment. This paper is supported by National Nature Science Foundation of China (41861134010, 61971162) and National Aeronautical Foundation of China (2020Z066015002).

References

1. Liu, Y., Xing, M., Zhou, X., et al.: Tree height extraction in sparse scenes based on UAV remote sensing. In: IGARSS 2020 - 2020 IEEE International Geoscience and Remote Sensing Symposium, pp. 6499–6502 (2020)
2. Xiangyu, W., Donghui, X., Guangjian, Y., et al.: 3D reconstruction of a single tree from terrestrial LiDAR data. In: 2014 IEEE Geoscience and Remote Sensing Symposium, pp. 796–799 (2014)
3. Alon, A.S., Festijo, E.D., Juanico, D.E.O.: Tree detection using genus-specific retinanet from orthophoto for segmentation access of airborne LiDAR data. In: 2019 IEEE 6th International Conference on Engineering Technologies and Applied Sciences (ICETAS), pp. 1–6 (2019)
4. Tarawally, M., Wenbo, X., Weiming, H., et al.: Effect of deforestation on land surface temperature: a case of Freetown and Bo town in Sierra Leone. In: IGARSS 2018 - 2018 IEEE International Geoscience and Remote Sensing Symposium, pp. 5232–5235 (2018)
5. Long, J., Shelhamer, E., Darrell, T., et al.: Fully convolutional networks for semantic segmentation. In: 2015 IEEE Conference on Computer Vision and Pattern Recognition (CVPR), pp. 3431–3440 (2015)
6. Badrinarayanan, V., Kendall, A., Cipolla, R., et al.: SegNet: a deep convolutional encoder-decoder architecture for image segmentation. IEEE Trans. Pattern Anal. Mach. Intell. **39**(12), 2481–2495 (2017)
7. Chen, L., Papandreou, G., Kokkinos, I., et al.: DeepLab: semantic image segmentation with deep convolutional nets, atrous convolution, and fully connected CRFs. IEEE Trans. Pattern Anal. Mach. Intell. **40**(4), 834–848 (2018)
8. Yu, C., Wang, J., Peng, C., et al.: Learning a discriminative feature network for semantic segmentation. In: 2018 IEEE/CVF Conference on Computer Vision and Pattern Recognition, pp. 1857–1866 (2018)
9. Yu, C., Wang, J., Peng, C., et al.: BiSeNet: bilateral segmentation network for real-time semantic segmentation. In: 2018 IEEE Conference on Computer Vision and Pattern Recognition (CVPR) (2018)
10. Wang, J.D., Sun, K., Cheng, T.H., et al.: Deep high-resolution representation learning for visual recognition. IEEE Trans. Pattern Anal. Mach. Intell. **43**, 3349–3364 (2020)

Forest Fire Detection Algorithm Based on Aerial Image

Menglin Qi[1] and Bingcai Chen[1,2](✉)

[1] School of Computer Science and Technology, Xinjiang Normal University, Urumqi 830054, China
china@dlut.edu.cn

[2] School of Computer Science and Technology, Dalian University of Technology, Dalian 116024, Liaoning, China

Abstract. Forest fire detection is an important application of computer vision in the field of disaster prevention. Forest fire is easy to lose control. With the spread and expansion, it brings harm and loss of life and property. It has the characteristics of strong sudden, destructive, large area of fire, and difficult rescue. Once a large-scale fire occurs, it is easy to cause huge losses. However, in the early stage of forest fire, accurate fire detection and fire identification can give early warning and reduce the loss of life and property. In this paper, a forest fire detection method based on SSD is used. The original vgg16 feature extraction network is replaced by resnet50, and attention mechanism is added to the network structure to further improve the accuracy. The experimental results show that this model has a high average accuracy for forest fire detection, and the mean average precision is 94.3% on the FLAME dataset of North Arizona University, It has the advantages of high accuracy, low false alarm rate and short detection time, so it is feasible to apply it in forest fire detection.

Keywords: Object detection · Forest fire · Resnet50 · SSD · Attention mechanism

1 Introduction

Forest fire is one of the main natural disasters in the world, with the characteristics of fast speed, strong burst and destructive, so the disaster rescue work is more difficult. Once a forest fire occurs, because the forest is very easy to burn, the fire will spread around quickly, which is the biggest threat to forest safety. In recent years, with the improvement of China's ecological environment, the continuous expansion of China's forest area, the difficulty of forest fire prevention management is huge, the cost of forest fire prevention and control is grim, and the detection and early warning of forest fire is becoming more and more important. With the continuous development of deep learning technology and image detection technology, a large number of fire detection technologies using object detection algorithms have emerged. Some scholars use color space rules for flame recognition [2], which has good recognition effect; However, it is one-sided

Q. Liang et al. (Eds.): Artificial Intelligence in China, LNEE 854, pp. 465–472, 2022.
https://doi.org/10.1007/978-981-16-9423-3_58

to extract flame features only by color space rules, and some features are lost, which makes the results limited. Celik et al. [3] proposed a rule-based general color model for flame pixel classification, which uses YCbCr color space to better distinguish the brightness and chroma of the flame, and achieves a higher flame detection rate, but there are still problems such as high false alarm rate, which can not be applied to the actual environment. Wang et al. [4] proposed a fire recognition model based on the blue component, and determined the key threshold by drawing the ROC curve of sample images. The experiment can eliminate part of the noise interference, but the feature extraction method is still relatively simple; At present, the commonly used object detection algorithms are roughly divided into two categories, one is the two-stage object detection algorithm represented by Faster RCNN [5], and the other is the single-stage object detection algorithm represented by Yolo [6] and SSD [7]. Because the two-stage object detection algorithm needs to generate candidate boxes, its precision is usually higher than that of the single-stage object detection algorithm, and the single-stage object detection speed is slower. Forest fires often need to be detected quickly. Therefore, this paper uses the single-stage SSD model to detect forest fires, which can improve the accuracy under the condition of ensuring the speed. By adding attention mechanism, the mAP of 1.3% is improved.

2 Project Development Background

2.1 ResNet Structure

In this paper, we use the classic resnet50 structure. In order to solve the problem of performance degradation caused by the gradient disappearance of deep network, Kaiming he et al. [8] proposed the ResNet model, which effectively solves the problem of gradient disappearance with the deepening of network. The key structure to solve the gradient disappearance problem is residual block, which adds identity mapping to the network, In this paper, we add SE module [9] to the residual block to enhance the correlation between channels. The structure of residual block is shown in Fig. 1.

2.2 SSD Object Detection Model

SSD algorithm is divided into ssd300 and ssd512 according to the size of the input image. This paper proposes a forest fire image object detection algorithm (ALSSD) which integrates attention mechanism. ssd300 algorithm is improved by replacing the feature extraction network with resnet50 and adding attention mechanism. The input image size of ALSSD algorithm is 300×300. A total of six feature maps are generated. Each feature map is independent of each other. Small objects are detected on the large feature map, and large objects are detected on the small feature map to achieve the purpose of multi-scale target detection. The ALSSD model is shown in Fig. 2.

Different from the original resnet50, the stride of identity mapping and the second convolution layer in block 1 of conv4 in ALSSD is 1. The additional layer consists of convolution layerw and batch normalization layers. The structure of additional layer 1–3 is shown in Fig. 3, and the structure of additional layer 4–5 is shown in Fig. 4.

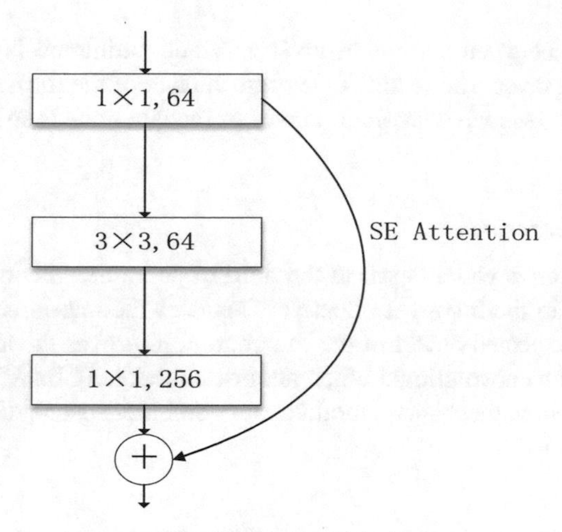

Fig. 1. Residual block

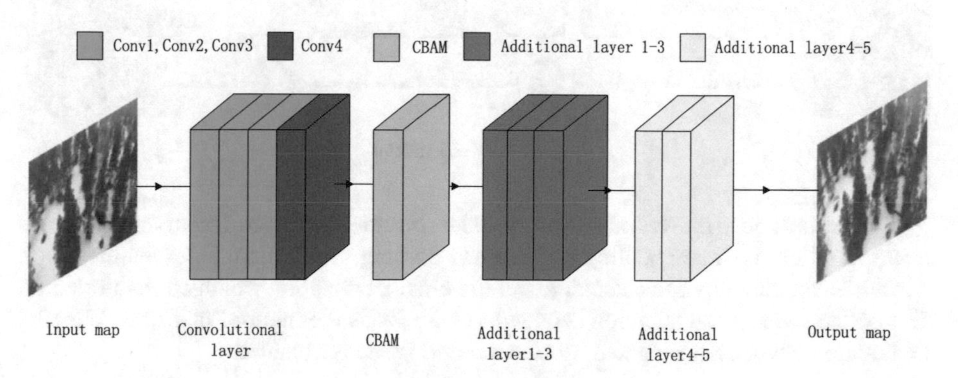

Fig. 2. ALSSD model

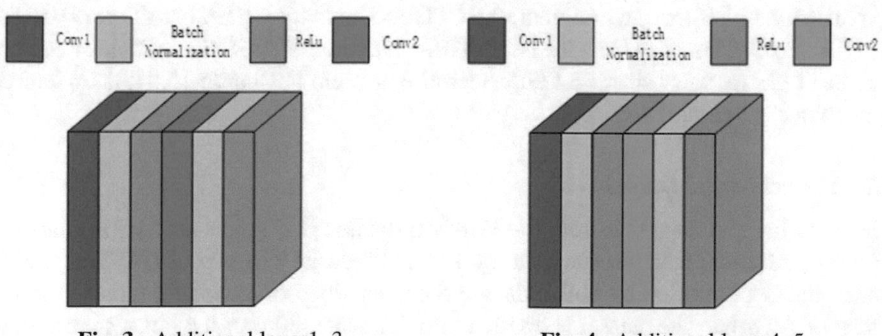

Fig. 3. Additional layer1–3 **Fig. 4.** Additional layer4–5

The difference between additional layer 1–3 and additional layer 4–5 lies in the second convolution layer. The second convolution layer of the former one is stride = 2, padding = 1, and the second convolution layer of the latter one is stride = 1, padding = 0.

2.3 Attention Mechanism

Attention mechanism is widely used in the field of computer vision. By extracting the important features in the image, it effectively reduces the interference factors that are not related to the expected object in the image, and improves the ability to obtain the expected object. The convolutional block attention module (CBAM) proposed by woo et al. [10] is an attention mechanism module that combines spatial attention and channel attention, as shown in Fig. 5.

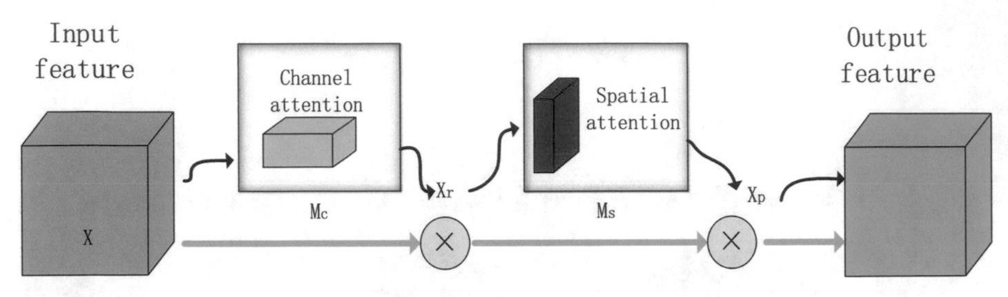

Fig. 5. CBAM model

CBAM is divided into two attention modules, channel and space, to extract the feature network. After maximum pooling and average pooling, the original input feature graph X is added attention by Mc channel to get the channel attention weighted feature graph Xr, and then Ms spatial attention is added to take Xr as the input of this stage. Finally, the feature network Xp weighted by channel and space is obtained.

3 Experiment and Result Analysis

3.1 Experimental Environment

Experimental platform environment, the CPU model is Intel (R) Core (TM) i3-10100F, CPU @ 3.60 GHz, GPU model is NVIDIA GeForce RTX2080ti, graphics memory size is 11 gb, memory size is 32 GB, operating system is Ubuntu 16.04, deep learning framework is Python 1.6.0.

3.2 Experimental Dataset

The experimental dataset is from the FLAME dataset [11] published by the University of North Arizona, which contains a total of 2003 aerial pictures of UAVs. It is a fire image dataset collected by UAVs during the prescribed burning of debris in the pine forest of Arizona. The dataset is divided into 8:2 scale, 80% of the images are training set for training, 20% of the images are test set for testing, and the image size needs to be set to 300 × 300 during training input detection model.

3.3 Loss and Learning Rate

It is not only the network structure that affects the accuracy of the model. As an important super parameter, the learning rate also plays an important role. It represents the rate of weight update. If it is set too much, the cost function will fluctuate too much, and the experimental results are not accurate. If it is set too small, the convergence rate of the model will be slow, and it is easy to fall into the local optimal solution. This model uses SGD optimizer, and the initial learning rate is set to 0.0005, The learning rate decreases by 0.3 times every 15 epochs, and the model converges after training to 50 epochs. The changes of training loss and learning rate are shown in Fig. 6.

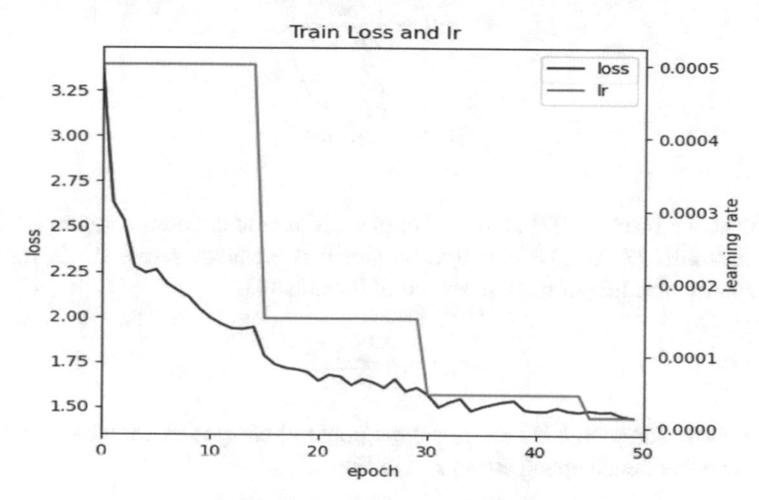

Fig.6. Loss and learning rate

3.4 Evaluating Indicator

Unity is a real-time 3D interactive content creation and operation platform, which can used to create visualization products and build immersive interactive virtual experiences. Real-time rendering can achieve "seeing is what you get", allowing developers to maximize their creativity. The VR technology is used to simulate the process flow, so that the experiencer can experience the real environment without touching the physical object, with-out being restricted by physical boundaries. Once this was just a dream in the future, it has now become a reality. Unity Reflect is an official innovation tool between Unity and Autodesk. It supports one-click lossless transmission of BIM models and data and real-time synchronization to Reflect Viewer, realizing visualized real-time design changes and seamlessly optimized BIM data integration, and has the ability to transmit model data flexibly and the powerful functions connected to other programs have great potential in reducing the time required for the design review and verification process, allowing all parties to communicate efficiently and conveniently, speed up iteration, and promote precise decision-making. The AEC field h-as a disruptive significance.

In this paper, mean average precision (mAP) was used as the evaluation index. In the process of evaluation, a P-R curve is drawn according to the accuracy and recall. The area enclosed by the curve and the coordinate axis is average precision (AP). The accuracy and recall can be calculated according to TP, TN, FP and FN. Precision, recall and AP are shown in formula (1–3)

$$Precision = \frac{TP}{TP + FP} \tag{1}$$

$$Recall = \frac{TP}{TP + FN} \tag{2}$$

$$AP = \int_{0}^{1} p(r)dr \tag{3}$$

In the above formula, *TP* is true example, *TN* is true counterexample, *FP* is false positive example, *FN* is false counterexample, *P* is precision rate and *r* is recall rate. The calculation formula of mAP is shown in formula (4)

$$mAP = \frac{\sum_{i=1}^{k} AP_i}{k} \tag{4}$$

In the above formula, *k* is indicates the number of categories, *i* indicates the class *i*. The map results of each epoch are shown in Fig. 7.

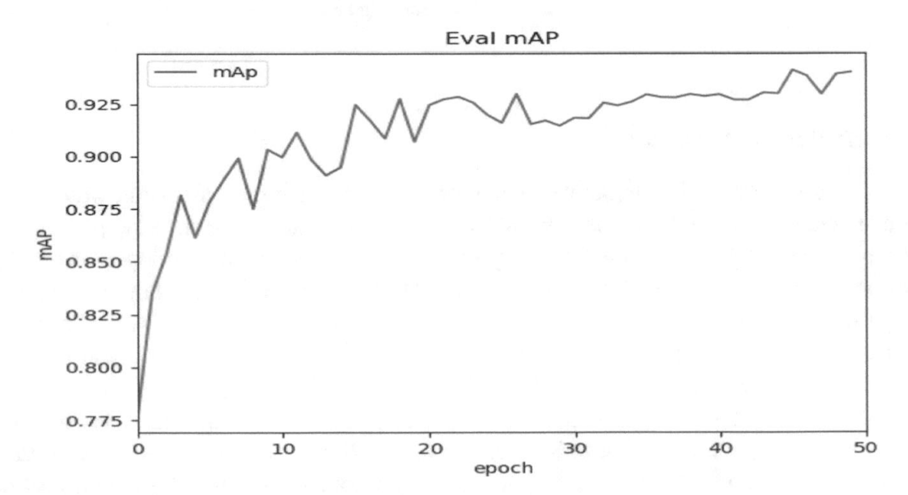

Fig. 7. mAP of each epoch

The forest fire detection results are shown in Fig. 8.

Fig. 8. Detection result

3.5 Comparative Experiment

Compare the mean average precision of ALSSD algorithm with Faster r-cnn and SSD algorithm in the same dataset, and the results are shown in Table 1.

Table 1. mAP comparison of each method

Method	Backbone	mAP/(%)
Faster R-CNN	ResNet50	94.2
SSD	ResNet50	93.0
ALSSD	ResNet50	94.3

4 Conclusion

The experimental results show that this algorithm can successfully warn the fire, and the recognition accuracy is as high as 94.3%. This paper improves the traditional SSD algorithm, and proposes a forest fire image detection algorithm integrating attention mechanism. Based on the SSD algorithm, the feature extraction network is modified and attention mechanism is added to enhance the feature extraction ability of the network, Finally, the average accuracy of 94.3% is achieved on the open forest fire dataset FLAME. Compared with the Faster r-cnn algorithm, the speed of the proposed algorithm is greatly improved and the average detection accuracy is slightly improved. In the next step, we will continue to optimize the network structure, consider using GAN network to enhance the data and further improve the detection accuracy.

References

1. Gao, W.: Problems and suggestions for forest fire prevention. Mod. Agric. Sci. Technol. **21**, 142–143 (2018)
2. Poobalan, K., Liew, S.C.: Fire detection based on color filters and Bag-of-Features classification. In: 2015 IEEE Student Conference on Research and Development (SCOReD), pp. 389–392. IEEE (2015)
3. Celik, T., Ma, K.K.: Computer vision based fire detection in color images. In: IEEE Conference on Soft Computing in Industrial Applications, pp. 258–263. IEEE (2008)
4. Wang, T., Bu, L., Zhou, Q., et al.: A new fire recognition model based on the dispersion of color component. In: 2015 IEEE International Conference on Progress in Informatics and Computing (PIC), pp. 138–141. IEEE (2015)
5. Ren, S., He, K., Girshick, R., et al.: Faster R-CNN: towards real-time object detection with region proposal networks. In: Conference on Neural Information Processing Systems, pp. 91–99 (2015)
6. Redmon, J., Divvala, S., Girshick, R., et al.: You only look once: unified, real time object detection. In: Computer Vision and Pattern Recognition (CVPR), pp. 779–788. IEEE, Las Vegas (2016)
7. Liu, W., Anguelov, D., Erhan, D., Szegedy, C., Reed, S., Fu, C.-Y., Berg, A.C.: SSD: single shot MultiBox detector. In: Leibe, B., Matas, J., Sebe, N., Welling, M. (eds.) ECCV 2016. LNCS, vol. 9905, pp. 21–37. Springer, Cham (2016). https://doi.org/10.1007/978-3-319-46448-0_2
8. He, K., Zhang, X., Ren, S., et al.: Deep residual learning for image recognition. In: Proceedings of the IEEE Conference on Computer Vision and Pattern Recognition, pp. 770–778 (2016)
9. Hu, J., Shen, L., Sun, G.: Squeeze-and-excitation networks. In: Proceedings of the IEEE Conference on Computer Vision and Pattern Recognition, pp. 7132–7141 (2018)
10. Woo, S., Park, J., Lee, J.-Y., Kweon, I.S.: CBAM: Convolutional Block Attention Module. In: Ferrari, V., Hebert, M., Sminchisescu, C., Weiss, Y. (eds.) ECCV 2018. LNCS, vol. 11211, pp. 3–19. Springer, Cham (2018). https://doi.org/10.1007/978-3-030-01234-2_1
11. As, A., Fa, A., Ar, A., et al.: Aerial imagery pile burn detection using deep learning: The FLAME dataset. Comput. Networks (2021)

Research on Intrusion Detection Based on Convolutional Neural Network

Min Sun, Min Gao(✉), and Ni Liu

School of Computer and Information Technology, Shanxi University, Taiyuan 030006, China
932944929@qq.com

Abstract. This paper proposes an intrusion detection model based on a convolutional neural network. First, a one-dimensional convolutional neural network structure is used to speed up the convergence of the model and prevent overfitting. Then, using the method based on SMOTE-GMM, the sample data is equalized by the method of combined sampling. Finally, a wrapped recursive feature addition algorithm is introduced to select the feature subset that makes the model detection effect the best. This paper uses the UNSW-NB15 and CICIDS2017 data sets to verify the effect of the model. Experimental results show that the intrusion detection model proposed in this paper has been improved in various evaluation indicators in the binary-classification and multi-classification. The model proposed in this paper can meet the real-time detection requirements of current complex networks.

Keywords: Intrusion detection · Depth learning · Smote · Convolutional neural network

1 Introduction

With the advent of the era of big data and the vigorous development of new network technologies, network security issues have become increasingly severe. As a real-time monitoring system for network traffic transmission, intrusion detection system identifies abnormal traffic by analyzing the characteristics of network connection data, thereby protecting the network environment from attacks and intrusions [1]. Traditional intrusion detection methods usually use rule matching to detect intrusion behaviors, which cannot effectively extract characteristic information in data traffic, and cannot meet the needs of today's network environment in terms of generalization ability, false alarm rate and detection efficiency. Therefore, introducing new technologies into the field of intrusion detection has important research value.

With the rise of deep learning technology in recent years, it has been widely used in various fields. Convolutional neural network [2] is one of the representative algorithms of deep learning. It can learn independently and effectively extract data feature information by using a multilayer neural network structure. Applying it to intrusion detection can improve the system's data feature analysis and generalization capabilities.

Q. Liang et al. (Eds.): Artificial Intelligence in China, LNEE 854, pp. 473–480, 2022.
https://doi.org/10.1007/978-981-16-9423-3_59

Therefore, this paper proposes an intrusion detection model based on convolutional neural network. First, for the dimensional characteristics of the data set, a one-dimensional convolutional neural network structure is used to speed up the convergence speed of the model and prevent overfitting. Then, for the problem of sample imbalance, based on the SMOTE-GMM method, the sample data is balanced by the combined sampling method, and then combined with the convolutional neural network to build the model, thereby improving the binary-classification effect of the model and the ability to detect categories with a small total number of samples under multi-classification. Finally, for the efficiency of model training and detection, a wrapped recursive feature addition algorithm is introduced, a feature recursive addition algorithm based on a greedy search strategy, combined with the convolutional neural network model, and finally the feature subset that makes the model's detection effect the best is selected, so that the model can improve the detection efficiency while maintaining the detection effect as much as possible, and reduce the consumption of computing resources.

This paper uses UNSW-NB15 and CICIDS2017 datasets to verify the effect of the model. The experimental results show that the model achieves a detection rate of 99.74% in the binary-classification on the UNSW-NB15 dataset. On the UNSW-NB15 and CICIDS2017 datasets, the detection rates of this model on multi-classification reached 96.54% and 99.85%, respectively. By comparing with other five imbalance processing methods and two classification algorithms, the model can effectively improve the intrusion detection capability and is superior to the current intrusion detection methods.

2 Proposed Solution

The architecture of the intrusion detection model proposed in this paper is shown in Fig. 1. The model is composed of three important modules: data preprocessing, imbalance processing and classification decision module. The data preprocessing module is responsible for performing operations on the original data to make the data more conducive to the prediction of the model. Using wrapped recursive feature addition algorithm in feature selection. We proposed a method, SMOTE-GMM in imbalance processing. Finally, in the classification decision module, we use a six-layer 1D-CNN model.

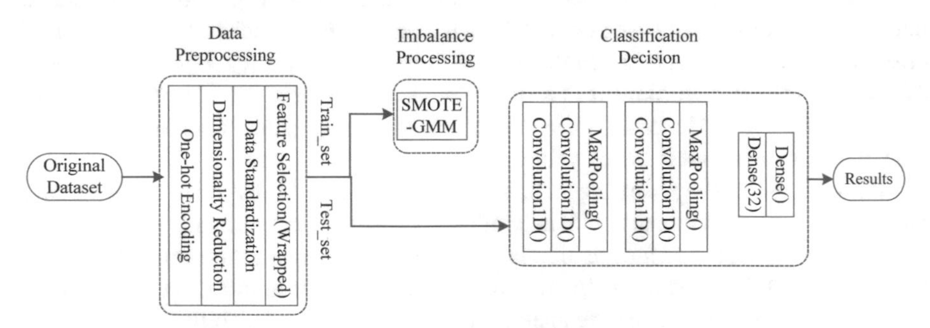

Fig. 1. A schematic diagram of the intrusion detection model

2.1 Datasets' Description

UNSW-NB15 dataset [3] was collected and distributed by the cybersecurity research group at the Australian Centre for Cyber Security. This dataset contains a total of 2,540,044 network traffic samples, involving nine attack categories. Each sample has 49 features, two of which are class label features. This dataset has serious class imbalances, in which normal traffic accounts for 87.35% of the entire dataset, and all attack traffic accounts for only 12.65%.

The CICIDS2017 dataset [4] was collected and compiled by the Canadian Cyber Security Institute with the help of the B-Profifile system [5] at the end of 2017. The dataset contains 2,830,473 network traffic samples, including one benign class and 14 attack categories, with benign traffic accounting for 80.30% and attack traffic accounting for 19.70%. The dataset extracts 84 features from the generated network traffic, of which the last column is the multiclass label.

2.2 Data Preprocessing

One-Hot Encoding. It is used to quantify the nominal features in the UNSW-NB15 dataset that cannot be processed by machine learning algorithms. After applying one-hot encoding, the feature dimension of the UNSW-NB15 dataset changes from 47 to 208. It is used in class label numeralization of two datasets.

Dimensionality Reduction. We drop redundant and meaningless features in the dataset. The feature dimensions of the UNSW-NB15 dataset and the CICIDS2017 dataset are reduced to 202 [6] and 77 [7], respectively.

Data Standardization. We standardize all remaining features and normalize them to a Gaussian distribution with a mean of 0 and a variance of 1.

Feature Selection. Wrapped feature selection mainly uses the recursive feature elimination (Recursive Feature Elimination) method, which uses a base model (leaner) for multiple rounds of training, removes several features after each round of training, then performs the next round of training based on the new feature set. The base model used is the convolutional neural network model sampled by SMOTE-GMM, and the performance of the learner is evaluated by the detection rate. This paper adopts the opposite method and uses the recursive feature addition algorithm. Essentially, recursive feature elimination and addition algorithms are the same idea. The recursive feature adds a search strategy based on greed. Its principle is to initialize an empty feature set first, and then continuously add new features to it. If the added new feature can improve the model effect, keep the feature, otherwise discard it. Because of this recursive method, new feature sets need to be continuously used to train the model, a lot of calculations are required.

2.3 Class Imbalance Processing

The number of abnormal samples in the intrusion detection dataset is inherently small. We propose a method SMOTE-GMM that resample all categories of samples to a uniform number of instance $I_{Resample}$ (hereinafter referred to as I) [7].

$$I_{Resample} = int\left(\frac{N}{C}\right) \tag{1}$$

where N is the total number of samples in the training set, and C is the number of classes.

The method uses SMOTE to oversample classes with less than I to I. SMOTE is a classic oversampling method proposed by Chawla et al. in 2002 [8]. SMOTE increases the quantity of minority class samples by “synthesizing” minority class samples. The “synthesis” is to generate a sample that does not exist in the original dataset and avoid overfitting in the process of building a classification model.

For classes with more samples than I, we use a GMM-based clustering method to undersample the majority class samples to I. GMM is a parameterized probability distribution model that represents a linear combination of multiple Gaussian distribution functions. Assuming that all samples come from multiple Gaussian distributions with different parameters, and samples that belong to the same distribution are divided into the same cluster, GMM returns the probability that sample x belongs to different clusters according to Eq. (2):

$$P(x|\theta) = \sum_{k=1}^{K} \alpha_k \Phi(x|\theta_k) \tag{2}$$

SMOTE-GMM not only avoids the excessive time and space cost caused by using oversampling alone, but also prevents random undersampling from losing important samples. It significantly improves the detection rate of minority classes.

2.4 CNN Modeling

We use 1D convolution. The convolution of a signal sequence $x_1, x_2, \ldots$ at time t can be expressed by:

$$y_t = f\left(\sum_{k=1}^{m} (\omega_k \otimes x_{t-k+1}) + b_t\right) \tag{3}$$

The pooling layer needs to undersample the features to reduce the network complexity and avoid overfitting.We use Max-Pooling [9], as shown in Eq. (4).

$$Y_i = \max_{i \in \Re} x_i \tag{4}$$

We use a six-layer 1D-CNN suitable for NIDS. The data is simply mapped into two-dimensional information as input to the network. The first four layers are convolutional layers. The number of convolution kernels in the four 1D convolution layers is 32–32-64–64. Every two convolutional layers are followed by a Max-Pooling layer and a Dropout layer with a parameter of 0.2. The Max-Pooling layer undersamples the parameters of the convolution layer by two times. The Dropout layer prevents overfitting. Multiple stacked convolutional layers have fewer parameters and more nonlinear transformations than a large-size convolutional layer, and can provide stronger feature learning capabilities [10]. The last two layers are dense layers, which integrate the previously learned local features into global features, of which the fifth layer has 32 neural units. The sixth layer is the output layer, which is mainly used for classification prediction. The optimization algorithm uses "Nadam", the learning rate is 0.008, and the loss function uses "categorical_crossentropy".

3 Experimental Results and Analysis

3.1 Evaluation Indicator

The evaluation index of the network intrusion detection model mainly includes these indicators: accuracy (ACC), detection rate (DR), false alarm rate (FAR), Recall, Precision and F_1 score. For each attack type, we consider the samples as positive ones and the others as negative ones. ACC is defined as the percentage of correctly classified ones among all samples. Recall, which is essentially DR, refers to the ratio of positive samples that are correctly detected. FAR is defined as the ratio of negative samples that are wrongly judged as positive ones. Precision refers to how many of the samples that the model judges to be positive are truly positive samples. F_1score is the harmonic average of Precision and Recall.

3.2 Binary Classification Experiment Results

In the binary classification experiment, to demonstrate the effectiveness of SMOTE-GMM, we compare four different class imbalanced processing techniques. Also, to evaluate the effectiveness of the proposed CNN model, we compare two machine learning classification algorithms, namely, Random Forest (RF) and Multi-Layer Perceptron (MLP).

Table 1 shows the binary classification results on the UNSW-NB15 dataset, where the bold part is the optimal result on a certain index. We observe that the best classification results are obtained by using the SMOTE-GMM-CNN model. The classification effect of 1D-CNN is better than RF and MLP. In terms of the overall performance, no matter which classification algorithm is used, such as RF, MLP, and CNN, SMOTE-GMM consistently achieves the best results. The combined performance of undersampling and oversampling is better than using oversampling alone.

Table 1. Performance evaluation of the proposed model in binary classification on the UNSW-NB15 dataset (%)

Model	Imbalanced Process	ACC	DR	FAR	Precision	F_1score
RF	ROS	98.67	99.99	1.52	90.49	95.01
	SMOTE	98.68	99.99	1.52	90.53	95.02
	RUS + SMOTE	98.66	99.87	1.51	90.52	94.97
	K-means + SMOTE	98.68	99.99	1.51	90.53	95.03
	SMOTE-GMM	**98.68**	**99.99**	**1.51**	**90.54**	**95.03**
MLP	ROS	98.66	**99.86**	1.51	90.52	94.96
	SMOTE	98.67	99.80	1.50	90.61	94.98
	RUS + SMOTE	98.65	99.82	1.52	90.49	94.93
	K-means + SMOTE	98.71	99.84	1.46	90.84	95.13
	SMOTE-GMM	**98.74**	99.82	**1.42**	**91.08**	**95.25**
CNN	ROS	98.68	**99.99**	1.52	90.53	95.02
	SMOTE	98.78	99.93	1.39	91.22	95.38
	RUS + SMOTE	98.78	99.91	1.39	91.25	95.39
	K-means + SMOTE	98.76	99.97	1.41	91.10	95.33
	SMOTE-GMM	**98.82**	99.74	**1.31**	**91.66**	**95.53**

3.3 Multi Classification Experiment Results

In the multi classification experiments, we use the same comparative experiment with the binary classification scenario. The 10-class classification results on the UNSW-NB15 dataset of CNN are shown in Fig. 2, The 15-class classification results on the CICIDS2017 dataset of CNN are shown in Fig. 3.

The experimental results show that the CNN model after class imbalance processing significantly improves the detection rate of attack classes. With the CNN classification algorithm, the oversampling method alone is better than the combination of RUS or K-means undersampling and SMOTE, but is not as good as the combination of GMM-based clustering undersampling and SMOTE.

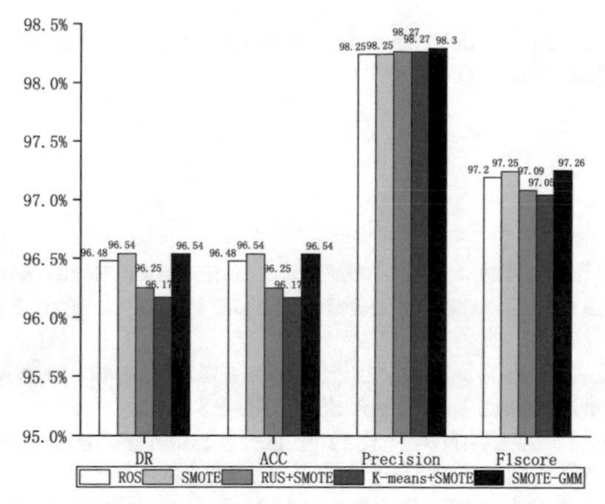

Fig. 2. Performance evaluation of the model in 10-class classification on the UNSW-NB15 dataset using CNN

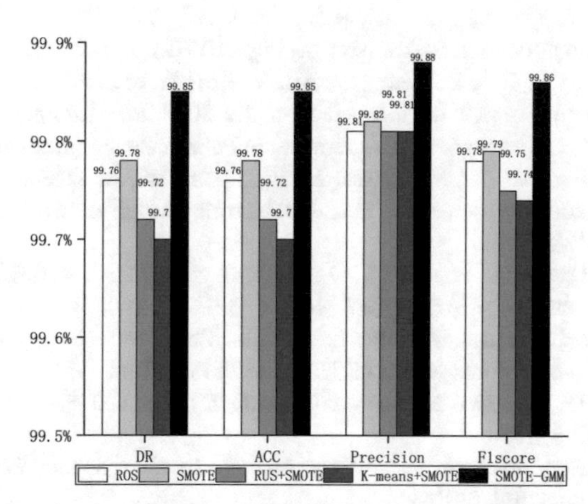

Fig. 3. Performance evaluation of the model in 15-class classification on the CICIDS2017 dataset using CNN

4 Conclusion

The intrusion detection model based on convolutional neural network proposed in this paper introduces a wrapped recursive feature addition algorithm, and uses a combined sampling method to balance the sample data, which can effectively improve the intrusion detection capability and is superior to the current intrusion detection methods.

But there are still shortcomings. In future work, we will try to combine two intrusion detection methods, anomaly and misuse. Intrusion detection based on misuse is more convenient and accurate, but has a high rate of false negatives, while intrusion detection based on anomalies can identify unknown new attacks, but has a high rate of false positives. Under the premise of ensuring system performance, hybrid intrusion detection methods will be one of the future research directions.

Acknowledgements. The work described in this paper is supported by Shanxi Natural Science Foundation (No. 201701D121052).

References

1. Zong, W., Chow, Y.-W., Susilo, W.: Interactive three-dimensional visualization of network intrusion detection data for machine learning. Future Gener. Comput. Syst.- Int. J. Esci. **102**, 292–306 (2020)
2. Liu, Y., Wang, C., Zhang, Y., Yuan, J.: Multiscale Convolutional CNN model for network intrusion detection. Comput. Eng. Appl. **55**(3), 90–95 (2019)
3. Moustafa, N., Slay, J.: UNSW-NB15: a comprehensive data set for network intrusion detection systems (UNSW-NB15 Network Data Set). IEEE (2015)
4. Sharafaldin, I., Lashkari, A.H., Ghorbani, A.A. (eds.) Toward Generating a New Intrusion Detection Dataset and Intrusion Traffic Characterization. International Conference on Information Systems Security and Privacy (2018)
5. Sharafaldin, I., Gharib, A., Lashkari, A.H., Ghorbani, A.A.: Towards a reliable intrusion detection benchmark dataset. Softw. Networking. **2017**(1), 177–200 (2017)
6. Zhang, H., Wu, C.Q., Gao, S., Wang, Z., Xu, Y., Liu, Y., et al.: An effective deep learning based scheme for network intrusion detection. In: 2018 24th International Conference on Pattern Recognition. International Conference on Pattern Recognition, pp. 682–687 (2018)
7. Abdulhammed, R., Musafer, H., Alessa, A., Faezipour, M., Abuzneid, A.: Features dimensionality reduction approaches for machine learning based network intrusion detection. Electronics **8**(3), 322 (2019)
8. Chawla, N.V., Bowyer, K.W., Hall, L.O., Kegelmeyer, W.P.: SMOTE: Synthetic minority over-sampling technique. J. Artif. Intell. Res. **16**, 321–357 (2002)
9. Tran, N.N., Sarker, R., Hu, J.: An approach for host-based intrusion detection system design using convolutional neural network. In: Hu, J., Khalil, I., Tari, Z., Wen, S. (eds.) MONAMI 2017. LNICSSITE, vol. 235, pp. 116–126. Springer, Cham (2018). https://doi.org/10.1007/978-3-319-90775-8_10
10. Simonyan, K., Zisserman, A.: Very deep convolutional networks for large-scale image recognition. Comput. Sci. (2014)

A Robust and Efficient SLAM System in Dynamic Environment Based on Deep Features

Bin Wang[1], Shaoming Wang[1], Lin Ma[1(✉)], and Danyang Qin[2]

[1] School of Electronics and Information Engineering, Harbin Institute of Technology, Harbin, China
malin@hit.edu.cn

[2] School of Electronics Engineering, Heilongjiang University, Harbin, China

Abstract. In the field of mobile robots, positioning and mapping is one of the most basic problems. A robust and efficient Synchronous Localization and Mapping (SLAM) system is essential for autonomous movement of robots. However, due to the complexity and time-varying nature of the real environment, the positioning and mapping effects will be greatly reduced due to scene changes. At the same time, because of its importance in pattern recognition, deep learning has a relatively mature theoretical foundation and practical framework for feature extraction. In this paper, we propose a visual SLAM system based on deep features in dynamic scenes, which combines mature convolutional neural networks (CNNs) HF-Net into an existing SLAM system. First, use HF-Net to detect the input image, and give local descriptors and global descriptors of the image. Then, these descriptors are used by different modules of the SLAM system. Because the features are not obtained by hand, they are very robust to scene changes. In loop closure detection, a distributed bag-of-words (DBoW) is used to form a vocabulary table, and local and global features are all considered at the same time, so the performance is more reliable. The results show that the entire system has lower trajectory error and higher accuracy on the evaluation data set.

Keywords: VSLAM · HF-Net · DBoW · Nonlinear optimization

1 Introduction

With the application and development of probabilistic robots [1], SLAM problems have gradually gained attention and development over the last decades, and related research fields have made great progress [2]. The core of the SLAM problem is to use the feature matching relationship between adjacent key frames to track feature points. After solving and optimization, the camera pose and trajectory are obtained. The essence is still a data association problem. The frontend of a SLAM system can be divided into direct method and feature point method. The direct method uses the pixel information of the image to estimate the camera motion and uses pixel gradients instead of corner points for feature matching, which is very suitable for use in scenes with sparse texture and missing

Q. Liang et al. (Eds.): Artificial Intelligence in China, LNEE 854, pp. 481–489, 2022.
https://doi.org/10.1007/978-981-16-9423-3_60

features, such as RTAB-MAP [3] and DSO [4]. However, the direct method is based on the assumption of constant illumination, so it is relatively affected by illumination, and the effect of loop closure detection and re-localization after feature loss is not good, and related applications are limited. The feature point method uses feature point matching to track road signs, which is not sensitive to illumination changes, and the system has a certain degree of invariance. Related research is extensive and in-depth, such as ORB-SLAM [5] and VINS-Mono [6]. However, the feature point method assumes that the environment is static or only slight movement exists. In the actual application environment, dynamic objects are inevitable. When the scene and viewpoint change due to the surrounding dynamic environment, the feature point method, taking ORB-SLAM2 [7] as an example, will greatly reduce the positioning and mapping effect.

With the development of deep learning, Convolutional Neural Networks (CNNs) have been continuously used in computer vision because of their powerful capabilities in pattern recognition. Since the performance of deep learning in image classification, recognition, object detection, image segmentation and other fields is much higher than that of traditional artificially designed algorithms, its combination with SLAM can effectively improve visual odometry and scene recognition due to hand-designed features. The application limitations brought about, potentially improve the robustness and efficiency of the system. The SLAM scheme based on deep learning has good invariance to lighting and can work under harsh lighting conditions. At the same time, it can identify and extract moving objects in the environment and perform SLAM modeling in a dynamic environment at the same time. Although most deep learning applications use CNNs to process image region-level features (such as semantics), there are also studies focusing on image pixel-level features [8, 9].

At present, the main research hotspots of SALM system include Geometric SLAM, Semantic/Deep SLAM, Multi-Landmarks/Object SLAM, Sensor Fusion, Dynamic SLAM, Mapping and Optimization. Some closely related topics for SLAM include visual re-localization and loop closure detection (LCD). Generally speaking, research in the SLAM field usually needs to extract the global features of each image. This can be achieved through the distributed bag of words (DBoW [10]) method, or through an end-to-end CNN inference network, such as NetVLAD [11].

In this paper, we propose a robust and efficient visual SLAM system, which aims to improve the accuracy and efficiency of visual odometry and relocation by using depth features in complex scenes such as dynamic environments. First, use the HF-Net [11] to obtain local features, local descriptors and global descriptors from each frame of image. Later, the local features are applied to links such as position tracking, local mapping and loop detection, while the global features help to relocate and realize the functions of the entire system, as shown in Fig. 1. The contributions of this article are as follows:

- Proposes a robust and efficient complete SLAM system with functions such as closed-loop detection, global optimization and map reconstruction.
- Comprehensive use of local features and global features, a robust positioning method that combines matching and global features for image retrieval is proposed. Compared with the traditional BoW method, it has higher efficiency and higher accuracy.

- Introduce HF-Net network to extract image depth features, which are applied to loop detection based on global and local features, and the visual vocabulary can be trained offline.

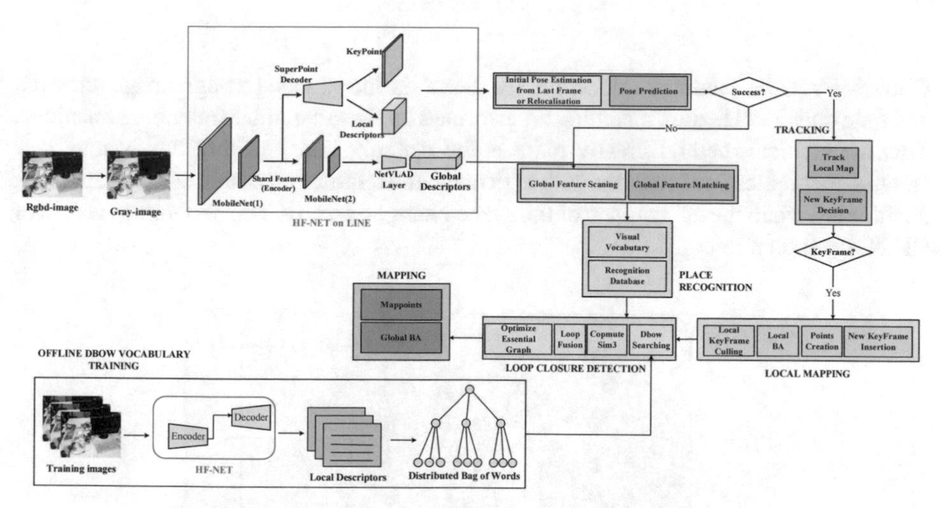

Fig. 1. System framework of the proposed visual SLAM system

2 System Models

2.1 Visual SLAM

The VSLAM system framework consists of the following four parts: frontend tracking, backend optimization, loop detection and trajectory reconstruction, as shown in Fig. 2. The frontend tracking module consists of sensors and visual odometry (VO), which is used to estimate the camera inter-camera state and at the location on the map point; the backend optimization module is responsible for receiving the positional postage information measured at the visual mileage tester and calculates the maximum post-probability estimate; the loop detection module is responsible for judging whether the object is go back to the original position and returning the loop closed correction estimation to eliminate the cumulative error, get the global consistency trajectory and mapping; the map reconstruction module is responsible for building a map adapted to the task requirements according to the camera position and image [12].

2.2 HF-Net

HF-Net is a hierarchical positioning algorithm based on a CNN that can calculate local features and image descriptors and global features at the same time to achieve accurate pose estimation. The hierarchical positioning algorithm is a coarse-to-fine positioning

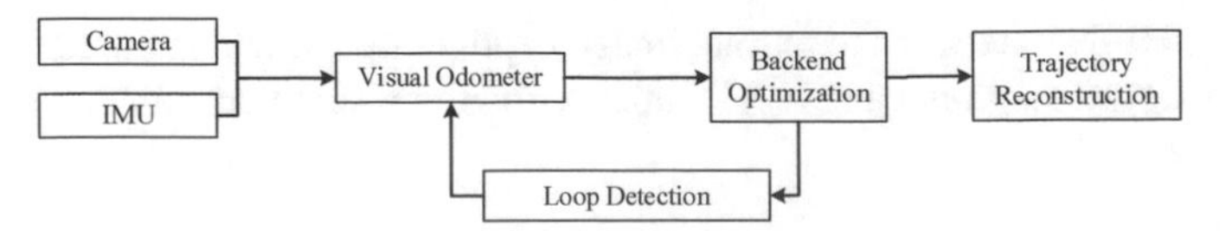

Fig. 2. VSLAM framework

method. First, the candidate "matching locations" is determined through image retrieval, and then only local feature matching is performed on these candidate matching locations. This hierarchical method greatly reduces the running time, making the system very suitable for application scenarios with high real-time requirements, and can realize robust positioning when the appearance of the scene changes greatly. The network structure of HF-Net is shown in Fig. 3.

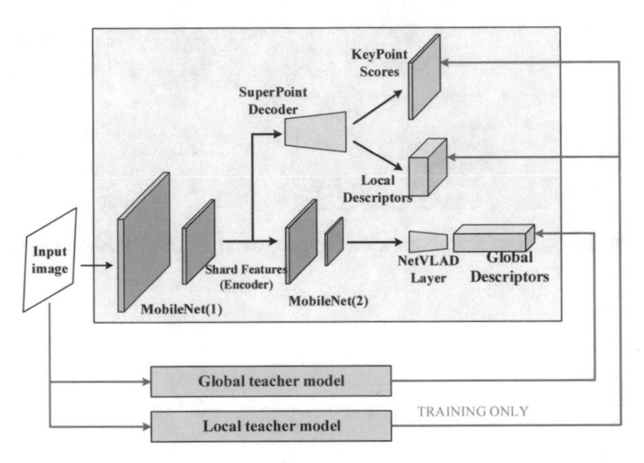

Fig. 3. HF-Net system

As shown in Fig. 3, the HF-Net network consists of a common coding network and three "head networks". The three "head networks" can respectively output: global image descriptors, feature point detection response maps, and feature point descriptors. The coding network is constructed by a MobileNet; the global image descriptor is output by the NetVLAD layer, the SuperPoint decoder is used to realize the extraction of feature points and the calculation of the descriptor. HF-Net can use a network to complete three tasks: generate global image descriptors, detect feature points, and generate feature point descriptors.

3 Experiments and Results

3.1 Datasets

TUM RGB-D is the mostly used SLAM benchmark in literature. It provides a variety of data sequences with precise ground-truth trajectories. At the same time, it provides a lot of evaluation tools. In this section, we mainly use TUM RGB-D to evaluate localization accuracy, while use EuRoc to reload map. They are both collected with a mobile robot. The former includes many dynamic objects like pedestrians and vehicles. Besides, the sunlight, wind and viewpoint changes can pose challenges to our algorithms.

3.2 Feature Points and Feature Matching

In Fig. 4, there shows a result of extracting ORB feature points with OpenCV, in which different colored circles represent different feature points. According to statistics, a total of 79 feature points were extracted. Based on this, feature matching can be realized between adjacent frames.

Fig. 4. The extracted feature points

As can be seen in Fig. 4, feature points always appear in the position with rich texture information in the image, including various corner points, boundary points and places with parallax. Based on the feature points extracted in Fig. 4, the force matching algorithm was used to carry out ORB feature matching. The results are shown in Fig. 5:

Fig. 5. ORB feature matching

Figure 5 shows the running results of the feature matching algorithm. As you can see, unfiltered matches have a large number of false matches. After a filter, the number of matches is reduced considerably, but most of the matches are correct. Here, the selection is based on the fact that the Hamming distance is less than twice the minimum distance, which is an empirical method in engineering without theoretical demonstration. Although the correct matches can be screened out in the figure above, there is no guarantee that all other matches will be correct.

3.3 Full System Evaluation

This paper builds a SLAM system based on the Ubuntu16.04 operating system, and selects the "fr3/walking" series data sequence under the dynamic target classification in the TUM data set to evaluate the performance of the SLAM system in a dynamic environment. The TUM data set is a SLAM measurement data set provided by the Computer Vision Experimental Group of the Technical University of Munich, Germany. It is collected by a handheld Kinect device. The collected trajectory is moving back and forth in the x, y, and z directions, and there are two people back and forth in front of the camera. walk. The results of dynamic point detection during SLAM operation shows in Fig. 6.

Fig. 6. Schematic diagram of dynamic point detection results

Figure 6 shows one person standing and one person sitting on a chair. The green points and squares in the figure represent the feature points and their neighborhoods corresponding to the local map, the blue represents the feature points only corresponding to the reference frame, and the red is the detected dynamic point. It can be seen that the dynamic information in the image is the pedestrian, and the chair that shakes when the pedestrian sits down. The dynamic point detection algorithm proposed in this paper, which combines semantic information and multi-view geometry, is basically reliable in judging dynamic points. Figure 7 shows the trajectory of the entire sequence, including the trajectory estimated by the algorithm in this paper and the ORB-SLAM2 algorithm, and the real trajectory groundtruth.

In practical engineering, the difference between the estimated trajectory and the real trajectory of an algorithm is often needed to evaluate the accuracy of the algorithm. The actual trajectory is usually obtained by some higher precision system, while the estimated trajectory is calculated by the algorithm to be evaluated. To calculate the error of two trajectories, consider an estimated trajectory $\mathbf{T}_{\text{esti},i}$ and a real trajectory $\mathbf{T}_{\text{gt},i}$, of which $i = 1, \cdots, N$. Then some error indicators can be defined to describe the difference between them, the most common one is absolute trajectory error, such as:

$$\text{ATE}_{\text{all}} = \sqrt{\frac{1}{N}\left(\sum_{i=1}^{N}\left\|\log\left(\boldsymbol{T}_{\text{gt},i}^{-1}\boldsymbol{T}_{\text{esti},i}\right)\right\|_2^2\right)} \tag{1}$$

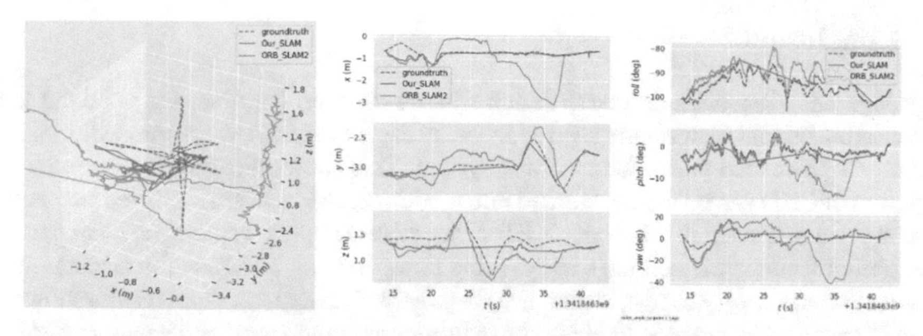

Fig. 7. Schematic diagram of the overall trajectory of the evaluation sequence

This is actually the root-mean-square error (RMSE) of the Lie algebra of each position pose, which can describe the rotation and translation errors of the two trajectories. Meanwhile, only the translation error can be considered, so the absolute translation error can be defined as:

$$\mathrm{ATE}_{\mathrm{trans}} = \sqrt{\frac{1}{N}\left(\sum_{i=1}^{N}\left\|\mathrm{trans}\left(\boldsymbol{T}_{\mathrm{gt},i}^{-1}\boldsymbol{T}_{\mathrm{esti},i}\right)\right\|_2^2\right)} \tag{2}$$

Where trans refers to the shifted part of the variable inside the brackets. From the point of view of the whole trajectory, after the rotation error, the subsequent trajectory will also have the error in the translation, so the two indexes are applicable in practice.

Table 1. Ate RMSE results on Tum RGB-D

Sequence	ORB-SLAM2	Our SLAM
fr3_walking_xyz	0.849426	0.226403
fr3_walking_static	0.344096	0.116807
fr3_walking_half	0.644414	0.248307
fr3_sitting_xyz	0.008726	0.006849
fr3_sitting_static	0.020846	0.015306

In order to compare our SLAM system to ORB-SLAM2, we use the scenes of office, cafe and market, as shown in Table 1. Compared to ORB-SLAM2, our SLAM has more correct pose estimation and re-localization, which can evaluate that our built visual-vocabulary method is more efficient and reasonable. For example, compared to ORB-SLAM2, our SLAM tracks more trajectory and ATE RMSE is far lower. The accuracy of localization is also evaluated with TUM RGB-D. It can be seen that our SLAM has a strong capability to resist dynamic environments.

4 Conclusion

In this paper, after obtaining deep features based on deep learning, the obtained local features and global features are used to act on all aspects of SLAM system, which significantly improves the performance of the system. This shows that the feature extraction function of CNN can be well integrated into modern SLAM systems. The feature extraction algorithm chosen in this paper is HF-Net, because it can calculate the local features and global features of the image at the same time, and can be located in a hierarchical manner, which is very robust, accurate and efficient. The experimental results show that SLAM system combined with HF-Net has faster operating speed and smaller operating error, and the depth feature has a significant effect on the improvement of system performance. Two aspects can be considered for further research in the future. One is to further research a deep learning framework that is more compatible with SLAM system, and the other direction is to research a more convenient and efficient SLAM system. In the final analysis, the pursuit of the unity of system reliability and effectiveness is the future direction of development.

Acknowledgment. This paper is supported by national nature science foundation of China (41861134010, 61971162) and National Aeronautical Foundation of China (2020Z066015002).

References

1. Thrun, S., Burgard, W., Fox, D.: Probabilistic Robotics, 1st edn. China Machine Press, Beijing (2020)
2. Cadena, C., et al.: Past, present, and future of simultaneous localization and mapping: toward the robust-perception age. IEEE Trans. Robot. **32**(6), 1309–1332 (2016)
3. Labbé, M., Michaud, F.: RTAB-Map as an open-source lidar and visual simultaneous localization and mapping library for large-scale and long-term online operation. J. Field Robot. **36**(2), 416–446 (2019)
4. Engel, J., Koltun, V., Cremers, D.: Direct sparse odometry. IEEE Trans. Pattern Anal. Mach. Intell. **40**(3), 611–625 (2017)
5. Mur-Artal, R., Montiel, J.M.M., Tardos, J.D.: Orb-slam: aversatile and accurate monocular slam system. IEEE Trans. Robot. **31**(5), 1147–1163 (2015)
6. Qin, T., Li, P., Shen, S.: VINS-Mono: A robust and versatile monocular visual-inertial state estimator. IEEE Trans. Robot. **34**(4), 1004–1020 (2018)
7. Mur-Artal, R., Tardos, J.D.: ORB-SLAM2: an open-source SLAM system for monocular, stereo, and RGB-D cameras. IEEE Trans. Robot. **32**(5), 1–8 (2017)
8. DeTone, D., Malisiewicz, T., Rabinovich, A.: SuperPoint: self-supervised interest point detection and description. In: IEEE Conference on Computer Vision and Pattern Recognition Workshops, pp. 224–236 (2018)
9. Dusmanu, M., Rocco, I., Pajdla, T., Pollefeys, M., Sivic, J., Torii, A., Sattler, T.: D2-Net: a trainable CNN for joint detection and description of local features. IEEE Robot. Autom. Lett. **4**(4), 3505–3512 (2019)
10. Galvez-López, D., Juan, D.: Bags of binary words for fast place recognition in image sequences. IEEE Trans. Robot. **28**, 1188–1197 (2015)
11. Arandjelovic, R., Gronat, P., Torii, A., Pajdla, T., Sivic, J.: NetVLAD: CNN architecture for weakly supervised place recogni-tion. IEEE Conference on Computer Vision and Pattern Recognition, pp. 5297–5307 (2016)

12. Sarlin, P.-E., Cadena, C., Siegwart, R., Dymczyk, M.: From coarse to fine: robust hierarchical localization at large scale. In: 2019 IEEE/CVF Conference on Computer Vision and Pattern Recognition (CVPR), Long Beach, CA, USA (2019)
13. Qin, T., Peiliang, L., Shen, S.: VINS-Mono: a robust and versatile monocular visual-inertial state estimator. IEEE Trans. Robot. **99**(5), 1–17 (2017)

Multi-view 3D Reconstruction Based on View Selection Network

Bocong Sun and Yongping Xie(✉)

School of Information and Communication Engineering, Dalian University of Technology, Dalian, Liaoning, China
xieyp@dlut.edu.cn

Abstract. In this article, we propose an end-to-end 3D reconstruction algorithm that uses a double-layer attention mechanism for accurate and complete point clouds reconstruction, which can expand the input scale and select a suitable reference view in a larger range, to obtain more detailed pixel information. At the same time, the robustness and integrity of the network can be improved. In addition, we have added a dual-channel attention mechanism to determine the accurate three-dimensional coordinates of pixels and restore more accurate pixel information. The experimental results show that our method has a significant improvement on the DTU data set.

Keywords: Three-dimensional reconstruction · Multi image stereo reconstruction · Attention module

1 Introduction

Multi-view stereo matching reconstruction is a problem that has been widely studied in the field of 3D reconstruction in recent years, and it is also the core problem of 3D reconstruction. In recent years, the research on convolutional neural networks has successfully aroused people's interest in improving the quality of 3D reconstruction. Based on deep learning methods, theoretically, convolutional networks can introduce global semantics more effectively, so they can effectively restore such as the complex texture of diffuse or specular reflection areas, and the deep learning algorithms in recent years have also shown good results. In fact, the stereo matching problem is very suitable for CNN to solve, because all input images are pre-corrected. No need to consider the camera parameters, it can be directly sent to the network. Although many current MVS algorithms have achieved relatively good results from/of the DTU data set [9, 10, 21], many of them have major limitations. These methods learn and infer stereo to obtain information that is difficult to deal with matching ambiguities, and there is a number of inputs images. Due to the limitations, there is no reasonable screening ability.

2 Related Work

2.1 Multi-view Stereo Reconstruction

Multi-view stereo reconstruction (MVS), MVS algorithm can be divided into the following three types: 1) generating point cloud reconstruction [5, 14], 2) generating voxel

Q. Liang et al. (Eds.): Artificial Intelligence in China, LNEE 854, pp. 490–497, 2022.
https://doi.org/10.1007/978-981-16-9423-3_61

reconstruction [11–13, 18], 3) indirect reconstruction based on the depth map [2, 6, 17], based on point cloud propagation is Sequential propagation, relying on the propagation strategy to make the points more and more dense [5, 14], it is difficult to completely deal with, while voxel reconstruction is to transform the space into different regular grids, and judge whether the voxels are attached to the surface. This method is problematic in contrast to spatial discrete errors and high memory usage, depth map reconstruction is the most flexible. Depth map reconstruction reduces the dimension of complex three-dimensional problems of a two-dimensional plane. It is solved by studying the main view and some of its neighbors. Since the depth map represents the characteristics of pixels, the depth map can also be easily restored to a point cloud [16]. In recent years, Efficient reconstruction algorithms are based on the reconstruction of depth maps, and generate high-quality point clouds through view selection, propagation strategies, and multi-scale aggregation [2, 6].

2.2 Matching Method Based on Deep Learning

Recently, neural networks have achieved good results of many computer vision tasks, and there are also applications of three-dimensional reconstruction, such as MVSNet [21], which pioneered the MVS process through deep learning through view matching and differentiable homography transformation. First of all, the camera parameters are encoded into the network at the same time, and an end-to-end training method is realized.

In addition, YaoYao, the author of RMVS [22] and MVS, made subsequent improvements to his MVSNet network, mainly replacing 3D convolution with a timing network to reduce memory consumption. P-MVSNet [15] applies patch-wise in traditional algorithms to the network to learn the confidence in isotropic matching within the cost. In particular, those methods follow the philosophy that the feature amount contributions to different view images are equal, and the heterogeneous image capture characteristics are ignored due to different lighting, camera geometric parameters, and scene content variability. Cascade MVSNet [7] changes the original model to hierarchical, using different depth intervals and depth intervals, so that more data can be trained at one time. Fast-MVSNet [23] uses sparse Cost Volume and Gauss-Newton Layer to increase the speed of the network. UCSNet [4] obtains a more accurate depth map by automatically adjusting the depth interval. D2HC-MVSNet [20] uses LSTM to improve the network and proposes a new detection method.

3 Method

3.1 Overview

- We use the CBAM [19] dual-channel attention mechanism to focus our target of the object to be reconstructed, not on the unrelated background or shadow.
- We add an attention mechanism before generating the feature cost body to increase the number of input images, with the goal of finding views that have a greater impact on the main view.

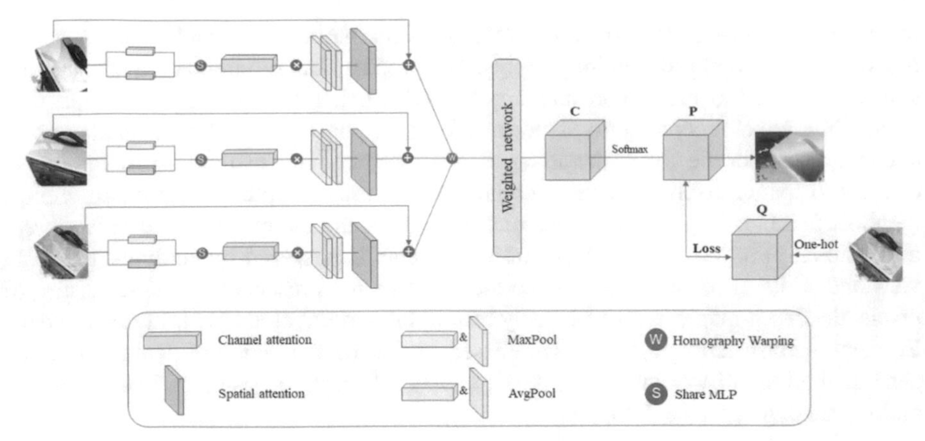

Fig. 1. Our network architecture. After the input image is first sent to the channel attention module, the obtained feature map is sent to the spatial attention module, superimposed on the initial feature map to generate the final feature map, and then the obtained Cost Volume is sent to the homograph change selection network of the view. According to different weights, the final cost body is obtained, and the cross entropy of the probability problem generated after Loss passes the soft-max probability body and the ground truth one-hot.

3.2 Dual-Channel Attention Mechanism

Convolutional Block Attention Module (CBAM) [19] represents the attention mechanism module of the convolution module. It is an attention mechanism module that combines space and channel. Compared with the attention mechanism that only pays attention to the channel, it can achieve better results.

In the process of 3D reconstruction, obtaining the pixel information about the pixel is the key to the texture restoration process, and the traditional channel attention mechanism can effectively infer the pixel information on the pixel, but at the same time, it only pays attention to the pixel information and ignores it. Spatial information will lead to a decrease in overall accuracy, so the dual-channel attention mechanism can obtain accurate pixel information and accurate spatial position coordinates.

First, the perspective map that has passed the homography transformation is sent to the global max pooling and global average pooling based on width and height to obtain two 1 × 1 × C feature maps, and then they are sent to a two-layered In the neural network, the number of neurons in the first layer is C/r (r is the reduction rate), the activation function is Relu, and the number of neurons in the second layer is C. This two-layer neural network is shared. Then, the MLP output features are subjected to an element-wise addition operation, and then the sigmoid activation operation is performed to generate the final channel attention feature, namely M_c. Finally, M_c and the input features map F are subjected to an element-wise multiplication operation to generate the input features required by the Spatial attention module.

$$M_c(F_i) = \sigma(MLP(AvgPooling(F_i)) + MLP(Maxpooling(F_i))) \tag{1}$$

Then input the feature maps through CAM, first do a channel-based global max pooling and global average pooling to obtain two H × W × 1 feature maps, and then perform to concat operations on these two feature maps based on the channel (channel Splicing). Then after a 7 × 7 convolution (7 × 7 is better than 3 × 3) operation, the dimension is reduced to 1 channel, that is, H × W × 1. After sigmoid, the spatial attention feature, namely M_s, is generated. Finally, the feature and the input feature of the module are multiplied to obtain the final generated feature.

$$M_s(F_i) = \sigma(f^{7\times7}([AvgPooling(F_i);\ MaxPooling(F_i)])) \tag{2}$$

3.3 View Selection Network

We believe that each feature map from different perspectives will be distorted to the reference camera plane through homography matrix transformation to construct Cost Volume. Generally speaking, in previous MVSNet papers, all views are considered to give equal contributions, but size of the input view N is limited. N = 3 in MVSNet [21] and N = 5 in Cascade [7]. However, each view should not be given the same consideration due to different lighting, camera position, occlusion order, image content, etc. The occluded near reference view can provide more accurate geometric and photometric information than the partially occluded far view, so we can select the view on an attention network that shares weights.

We believe that the feature cost of different views is raised as $V_{\mathrm{i}=0\ldots N-\mathrm{i}}$, The standard Cost Volume is $C_{\mathrm{i}=0\ldots N-\mathrm{i}}$, then

$$\mathrm{v}_{i,d,w,h}{}' = v_{i,d,w,h} - v_{0,d,w,h} \tag{3}$$

$$\mathrm{c}_{\mathrm{d,w,h}} = \frac{\sum_{i=1}^{N-1} (1 + w_{h,w}) \otimes v_{i,d,w,h}{}'}{N-1} \tag{4}$$

Where i is the reference view, h and w are the width and height of the view, and d is the depth. $w_{h,w}$ represent a 2D weighted attention map, used to filter out different reference views. The view selection network is composed of several 2D convolutions and ResNet, among them, each v_i' is compressed into a two-dimensional feature for input. The two-dimensional feature map is obtained by adopting the cascadc of maximum pooling and average pooling.

$$w_{h,w} = ChooseNet(f_{h,w}) \tag{5}$$

$$f_{h,w} = CONCAT(\max\_\ \mathrm{pooling}\left\|\mathrm{v}_{\mathrm{i,d,w,h}}{}'\right\|_1, average\_pooling\left\|\mathrm{v}_{\mathrm{i,d,w,h}}{}'\right\|_1) \tag{6}$$

3.4 Loss Function

Most MVS papers use the softmax operation for depth output. This method can be interpreted as the expected value along the depth direction. It is feasible for each reference view to have the same contribution to the final cost volume. But when the reference view contributions are different, cross entropy is a more effective way.

$$Loss = \sum_{P} (\sum_{i=1}^{D} -P(i, p) \cdot \log Q(i, P)) \quad (7)$$

Where p is the spatial image coordinates, and $P(i, p)$ is a voxel in the probability volume P. Q is the real binary occupancy volume of the ground, which is generated by the one-hot encoding of the ground truth depth map. $Q(i, p)$ is the voxel corresponding to $P(i, p)$.

4 Experiment

4.1 Experiment Details

We train the overall network on the DTU training set, which consists of 124 different indoor scenes, which are scanned by fixed camera trajectories under 7 different care conditions. According to the conventional practice, we train the model in training set, test part of the test model, and use the same depth map provided by MVSNet [21]. During training, set the input picture to $W \times H = 640 \times 512$, and the number of input images is $N = 7$. Sampling from the depth hypothesis from 425 mm to 935 mm with the depth plane number 192, we implemented the network through Pytorch and used the Adam method for end-to-end training, lasting 16 epochs, and the initial learning rate was 0.001.

In the testing phase, the number of views with $N = 9$ and the view with $D = 192$ are used for evaluation. We use the original image size (1600×1184) on the DTU dataset for testing.

Fig. 2. The middle picture represents MVSNet, and the right is our result. The fine texture area and diffuse reflection area are shown in the picture. It can be observed that our reconstruction effect and completeness are better than MVSNet.

4.2 Experiment Result

We use the test part of the DTU data [1] set to test the model. The results of some depth maps are shown in Fig. 2. The specific data of accuracy and completeness are shown in Fig. 1. The numerical calculation of accuracy and completeness uses DTU data [1]. The official MATLAB code provided by the set finally estimates the final reconstruction quality through the Overall score. Compared with some previous methods, we have made a big improvement in completeness, and the accuracy rate has also been improved compared with some methods. Obtained more delicate texture features, the improvement of integrity mainly depends on the selection network and spatial attention mechanism, which solves the problem that the correct pixels appear in the wrong spatial position. At the same time, the view selection network still obtains more the value of the reference view gets more intuitive pixel features, thereby improving the accuracy. Finally got a complete and smooth point cloud.

4.3 Ablation Experiment

In this section, we provide ablation experiments to analyze the strength of the network. In order to eliminate the influence of other factors, all experiments use the same consistency detection parameters.

We measure the reconstruction quality by the indicators on the validation set, select 18 types of picture groups based on the DTU data [1] set, and set the training view to W = 640, H = 512, D = 256.

During training time, we use a different number of input views. Because the view selection network can adapt to the input of any number of views, to obtain more accurate information, in the testing phase, a different number of views are also selected for input. When the training view N = 5 is compared with N = 3, there is a gap in accuracy. Not big, but there is a more obvious improvement in completeness. Keep the training view unchanged. When the test view changes, the final index gap is not large, which proves that the view selection network we proposed can be better enhanced Close to the effective information in the view, remove the bad information (Table 1).

Table 1. Comparison of our algorithm with previous algorithm results

Method	Acc. (mm)	Comp. (mm)	Overall. (mm)
Ours	*0.377*	*0.346*	*0.363*
MVSNet	0.396	0.527	0.462
R-MVSNet	0.385	0.459	0.422
Point-MVSNet	0.361	0.421	0.391
Cascade	0.325	0.385	0.355

Compared with previous algorithms, our algorithm is better than R-MVSNet [22] and Point-MVSNet [3] in accuracy, but it is inferior to Cascade [7], but in terms of completeness, the error index on the verification set has dropped significantly (Table 2).

Table 2. The impact of different input and output on our model

Train Num	Test Num	Acc. (mm)	Comp. (mm)	Overall. (mm)
N = 3	N = 2	0.415	0.407	0.413
N = 3	N = 3	0.380	0.374	0.377
N = 3	N = 5	0.381	0.365	0.372
N = 3	N = 7	0.380	0.361	0.370
N = 5	N = 7	0.381	0.359	0.369
N = 7	N = 7	0.377	0.346	0.363

5 Conclusion

We have made improvements to the poor part of the 3D reconstruction results based on the DTU data set. Through analysis, we found that most of the images with relatively general reconstruction effects are caused by excessive diffuse or specular reflection areas. The network input is not enough to obtain complete information in these areas, so we propose to try to select the network to increase the input scale and select more information content views from it. At the same time, the dual-channel attention mechanism is used to improve the accuracy and the reconstruction result respectively Completion.

In addition, we believe that most of the current work is based on depth maps, or to get a more complete and accurate depth map, and in the process of fusing the depth maps, there will be a lot of details lost, so I think that next The focus can be placed on the direct fusion of the point cloud, and the depth map can be encoded in the map structure to obtain richer adjacent point information, thereby improving the point cloud.

References

1. Aanæs, H., Jensen, R.R., Vogiatzis, G., Tola, E., Dahl, A.B.: Large-scale data for multiple-view stereopsis. IJCV **120**(2), 153–168 (2016)
2. Campbell, N.D.F., Vogiatzis, G., Hernández, C., Cipolla, R.: Using multiple hypotheses to improve depth-maps for multi-view stereo. In: Forsyth, D., Torr, P., Zisserman, A. (eds.) ECCV 2008. LNCS, vol. 5302, pp. 766–779. Springer, Heidelberg (2008). https://doi.org/10.1007/978-3-540-88682-2_58
3. Chen, R., Han, S., Xu, J., Su, H.: Point-based multi-view stereo network. In: ICCV (2019)
4. Cheng, S., Xu, Z., Zhu, S., et al.: Deep Stereo using Adaptive Thin Volume Representation with Uncertainty Awareness (2019)
5. Furukawa, Y., Ponce, J.: Accurate, dense, and robust multiview stereopsis. IEEE Trans. Pattern Anal. Mach. Intell. (TPAMI) (2010)
6. Galliani, S., Lasinger, K., Schindler, K.: Massively parallel multiview stereopsis by surface normal diffusion. International Conference on Computer Vision (ICCV) (2015)
7. Gu, X., Fan, Z., Zhu, S., et al.: Cascade cost volume for high-resolution multi-view stereo and stereo matching. In: 2020 IEEE/CVF Conference on Computer Vision and Pattern Recognition (CVPR). IEEE (2020)
8. Honari, S., Molchanov, P., Tyree, S., Vincent, P., Pal, C., Kautz, J.: Improving landmark localization with semi-supervised learning. In: CVPR (2018)

9. Huang, P.H., Matzen, K., Kopf, J., Ahuja, N., Huang, J.B.: DeepMVS: learning multi-view stereopsis. In: CVPR (2018)
10. Im, S., Jeon, H.G., Lin, S., Kweon, I.S.: DPSNet: end-to-end deep plane sweep stereo. In: ICLR (2019)
11. Ji, M., Gall, J., Zheng, H., Liu, Y., Fang, L.: SurfaceNet: an end-to-end 3D neural network for multiview stereopsis. In: International Conference on Computer Vision (ICCV) (2017)
12. Kar, A., Häne, C., Malik, J.: Learning a multi-view stereo machine. In: Advances in Neural Information Processing Systems (NIPS) (2017)
13. Kutulakos, K.N., Seitz, S.M.: A theory of shape by space carving. Int. J. Comput. Vis. (IJCV) (2000)
14. Lhuillier, M., Quan, L.: A quasi-dense approach to surface reconstruction from uncalibrated images. IEEE Trans. Pattern Anal. Mach. Intell. (TPAMI) (2005)
15. Luo, K., Guan, T., Ju, L., Huang, H., Luo, Y.: P-MVSNet: learning patch-wise matching confidence aggregation for multi-view stereo. In: ICCV (2019)
16. Merrell, P., et al.: Real-time visibility-based fusion of depth maps. In: International Conference on Computer Vision (ICCV) (2007)
17. Schönberger, J.L., Zheng, E., Frahm, J.-M., Pollefeys, M.: Pixelwise view selection for unstructured multi-view stereo. In: Leibe, B., Matas, J., Sebe, N., Welling, M. (eds.) ECCV 2016. LNCS, vol. 9907, pp. 501–518. Springer, Cham (2016). https://doi.org/10.1007/978-3-319-46487-9_31
18. Seitz, S.M., Dyer, C.R.: Photorealistic scene reconstruction by voxel coloring. Int. J. Comput. Vis. (IJCV) (1999)
19. Woo, S., Park, J., Lee, J.-Y., Kweon, I.S.: CBAM: convolutional block attention module. In: Ferrari, V., Hebert, M., Sminchisescu, C., Weiss, Y. (eds.) ECCV 2018. LNCS, vol. 11211, pp. 3–19. Springer, Cham (2018). https://doi.org/10.1007/978-3-030-01234-2_1
20. Yan, J., Wei, Z., Yi, H., et al.: Dense Hybrid Recurrent Multi view Stereo Net with Dynamic Consistency Checking (2020)
21. Yao, Y., Luo, Z., Li, S., Fang, T., Quan, L.: MVSNet: depth inference for unstructured multi-view stereo. In: Ferrari, V., Hebert, M., Sminchisescu, C., Weiss, Y. (eds.) ECCV 2018. LNCS, vol. 11212, pp. 785–801. Springer, Cham (2018). https://doi.org/10.1007/978-3-030-01237-3_47
22. Yao, Y., Luo, Z., Li, S., Shen, T., Fang, T., Quan, L.: Recurrent MVSNet for highresolution multi-view stereo depth inference. In: CVPR (2019)
23. Yu, Z., Gao, S.: Fast-MVSNet: sparse-to-dense multi-view stereo with learned propagation and Gauss-Newton refinement. In: 2020 IEEE/CVF Conference on Computer Vision and Pattern Recognition (CVPR). IEEE (2020)

The Sentiment Analysis and Sentiment Orientation Prediction for Hotel Based on BERT-BiLSTM Model

Yu Du, Yongchong Wang(✉), Kaiyue Wei, and Jingjing Jia

Business School, Beijing Language and Culture University, No. 15, Xueyuan Road, Haidian District, Beijing 100083, China
hamel112020@163.com

Abstract. With development of social network, online comments have influenced various areas especially for the hotel industry. However, how to utilize text resource and even guide business decision with it becomes a problem to be solved. This paper proposes a compound model which combines BERT with BiLSTM respectively as upstream and downstream task to be trained. Then, classification results of the star hotel online reviews predicted by BERT-BiLSTM model are used to construct Sentiment Index. Finally, it can be concluded that the sentiment orientation is relevant with sales and it makes sense to predict sentiment orientation and sales. The contribution of this work is that we predict hotel sales based on the relationship between Sentiment Index and sales.

Keywords: Sentiment analysis · Star hotel sales · BERT-BiLSTM model · Sentiment Index · Sentiment orientation prediction

1 Introduction

The progress of social network in our society has been refining in pace with the development of information technology. An increasing number of people can even book hotel online in advance before the trip begins. The service, in addition, has been experienced followed by comments written on the web pages of product or service. Sentiment of customers' online reviews promotes the improvement of product and service provided by the companies, especially for the hospitality industry. How to wisely utilize the sentiment expressed in the comments posted on the web is still a hot spot problem. In order to realize the purpose, there have been various models to mine the viewpoints. We call it sentiment analysis.

In recent years, due to better performance of deep learning in NLP, researchers pay more attention to it. The popular deep learning algorithms include Convolutional Neural Network (CNN) [1], Recurrent Neural Network (RNN) [2], Long Short Term Memory (LSTM), etc. Xiao et al. [3] came up with the Bidirectional LSTM to solve the disadvantage of LSTM. Devlin J et al. [4] presented a genuine deep bidirectional language model based on Transformer architecture, namely Bidirectional Encoder Representations from Transformer (BERT).

Q. Liang et al. (Eds.): Artificial Intelligence in China, LNEE 854, pp. 498–505, 2022.
https://doi.org/10.1007/978-981-16-9423-3_62

These online comments extremely affect future customer's behavior, which manifests the high correlation between the sentiment of online sentiment and product sales. Li et al. [5] studied the effect of online comments and analyzed how the numerical and textual comments influence product sales. Sonneir et al. [6] believed positive comments promote the product sales. Reimer et al. [7] believed the relationship between the comments and sales depends on reliability of comments to a great extent. Bollen et al. [8] analyzed the measurement of collective sentiment value and how the value affect Dow Jones Industrial Average and predicted it.

In this paper, we propose a BERT-BiLSTM based sentiment analysis model. Then, on the basis of sentiment polarity, the experiment is performed on the open source hotel reviews. The results show that the model achieves higher accuracy than single BERT and BiLSTM model. Then, it adopts the model to classify the comment data set of star hotel. According to the relationship between the growth rate of sentiment and sales, it proves that it makes sense to predict sentiment orientation.

The contribution of this work is that we predict hotel sales based on the relationship between Sentiment Index and sales after the sentiment polarities are classified by BERT-BiLSTM model that we combine.

2 Methodology

2.1 BERT

BERT proposes two pre-training tasks where the process is learned without supervision on Transformer encoder which is the distinguished features of the model based: Masked Language Modeling (MLM) and Next Sentence Prediction (NSP) [4]. The MLM adopts the idea of Denoising Auto Encoder (DAE). The NSP is designed to represent the deep-lying features about the consistency of sentences for BERT model.

The core part of BERT pre-training model is Transformer consisting of several units that are made up of three modules: position encoding, self-Attention Mechanism, Feed Forward Network.

The paramount module of the Transformer is self-Attention mechanism which multiplies the word vector with three matrices to obtain three new vectors. The formulas are as follows:

$$\mathrm{Q} = \mathrm{Linear}\left(X_{embdedding}\right) = X_{embdedding} W_Q \tag{1}$$

$$\mathrm{K} = \mathrm{Linear}\left(X_{embdedding}\right) = X_{embdedding} W_K \tag{2}$$

$$\mathrm{V} = \mathrm{Linear}\left(X_{embdedding}\right) = X_{embdedding} W_V \tag{3}$$

Finally, the output vectors are calculated by the formula:

$$X_{attention} = softmax\left(\frac{QK^T}{\sqrt{d_k}}\right)V \tag{4}$$

The sub-layer in the modules is connected by Residual Connection and Layer Normalization of standard normal distribution as shown in following equations respectively:

$$X_{RC} = X_{embedding} + X_{attention} \tag{5}$$

$$\mathrm{LayerNorm(x)} = \alpha \times \frac{x_{ij} - \mu_i}{\sqrt{\sigma^2 + \varepsilon}} + \beta \tag{6}$$

where μ_i is the mean value taken by matrix row and σ^2 is variance taken by matrix row.

The Feed Forward Network is a fully connected layer with two layers, the activation function of the first layer is Relu, and there is no activation function in the second. It can be shown as:

$$\mathrm{FFN(Z)} = \max(0,\ ZW_1 + b_1)W_2 + b_2 \tag{7}$$

2.2 BiLSTM

There are three gate structures in the LSTM model: forget gate, input gate and output gate [9]. The function of forget gate is to determine which information should be saved or kept. The output value of this gate is between 0 and 1. The formula is shown:

$$f_t = \sigma\left(W_f * \left[h_{t-1}, x_t\right] + b_f\right) \tag{8}$$

Input gate consists of two parts: the previous hidden state information and the current input information. The calculation methods are as follow:

$$i_t = \sigma\left(W_i * \left[h_{t-1}, x_t\right] + b_i\right) \tag{9}$$

$$\mathrm{C_t} = tanh\left(W_{\mathrm{C}} * \left[h_{t-1}, x_t\right] + b_{\mathrm{C}}\right) \tag{10}$$

The output gate is used to determine the value of the next hidden state. The hidden state contains information from the previous input. Its specific formulas are as follow:

$$o_t = \sigma\left(W_o * \left[h_{t-1}, x_t\right] + b_o\right) \tag{11}$$

$$h_t = o_t * \tan\left(\mathrm{C}_t\right) \tag{12}$$

where the $\mathrm{C_t}$ refers to the current cell state.

The BiLSTM consists of a forward and reverse LSTM stacking on the model. It combines the results of the two layers corresponding to every moment. The formulas are as follow:

$$h_t = LSTM\left(x_t, h_{t-1}\right) \tag{13}$$

$$h_t' = LSTM\left(x_t, h_{t-1}'\right) \tag{14}$$

$$o_t = wh_t + vh_t' + b_t \tag{15}$$

2.3 BERT-BiLSTM

As mentioned above, the BERT pre-training model can extract more textual features due to the presence of unique masked method and Transformer mechanism. Likewise, the BiLSTM model is specialized in dealing with contextual information which needs to detect its deep lexical characters. Therefore, this paper sets this two models as upstream task and downstream task respectively. The overall structure of the compound model is as follow in Fig. 1.

The input data go through the BERT pre-training model described earlier and then the output ***C*** including the [CLS] is taken by the last layer as the input of the BiLSTM model to realize downstream task. Finally, the output P^* is classified by the softmax function which is probability calculation. The formulas are shown in following equations:

$$p(\mathrm{y|S}) = softmax\left(W^{(S)}P^* + b^{(S)}\right) \tag{16}$$

$$softmax_i = \frac{e^i}{\sum_j e^j} \tag{17}$$

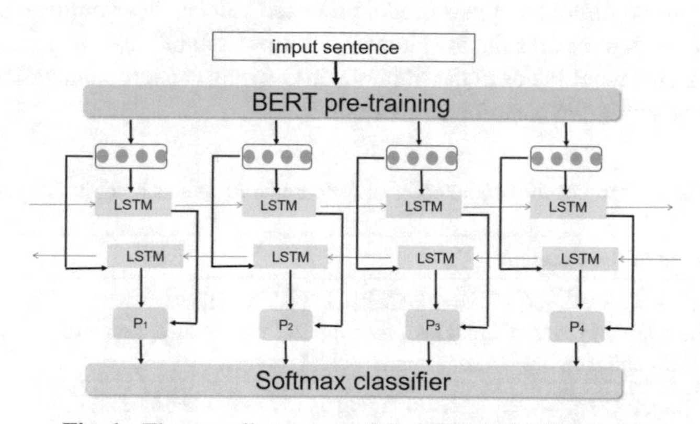

Fig. 1. The overall structure of the BERT-BiLSTM model

3 Experiment and Results

3.1 Data Set

In this paper, about 16 thousand open source Chinese hotel reviews which include 8000 positive data and 8000 negative data are collected to be fed into the training model. Parts of the data are shown in Table 1.

Then, the training data are segmented into three subsets which are training set, validation set and test set respectively in the proportion of 6:2:2.

Finally, all of the subsets are used to train and test the BERT-BiLSTM model.

Table 1. Parts of the open source hotel review samples

No	The content of the comments (English meanings)	Label
01	据说是太仓最好的酒店, 作为一个5星级的酒店让我很失望...(It is said that the best hotel in Taicang, but as a 5star hotel, I was very disappointed with it ...)	0
02	我们是5月1日通过携程网入住的, 条件是太差了, 根本达不到四星级的标准...(We checked in through Ctrip.com on May 1st, and the hotel's condition was so bad that we couldn't believe it reaches the four-star standard...)	0
03	酒店位置很好, 就在市中心广场边上, 出行非常方便...(The location of the hotel is very good, right next to the square in the city center, and it is very convenient to travel...)	1
...	...	...

After the Bert-BiLSTM model is trained, we crawl over 50 thousand starred hotel reviews in Guiyang on Ctrip.com which is a Chinese leading internet hotel reserve service platform from January 2017 to January 2020 which avoid the impact of Covid-19 on the data. There are still 38594 online comments remained by removing invalid data.

Based on the sentiment analysis model proposed before, these online comments are classified and parts of the results are shown as follow in Table 2.

In addition, the whole sales of the star hotel in Guiyang can be calculated by gathering the information about the customers' check-in time.

Table 2. Parts of the predicted label for comments of star hotel in Guiyang

No	The content of the comments	Predicted Label
01	早餐品种丰富, 补菜及时, 带着小朋友出门, 酒店安排的亲子房...(There are varieties of breakfast. And the staff timely supplements dishes, takes children out, arranges parent-child room...)	1
02	常住的酒店, 服务不错, 培训慢慢好起来了, 工作人员见到都会微笑打招呼...(The service is good, the staff will smile and greet to us...)	1
03	相对五星级酒店而言, 早餐品种太少...(Compared with five-star hotels, the menu is a little simple...)	0
...	...	...

3.2 The Results of Training BERT-BiLSTM Model

In order to show the performance of the compound model we propose, this paper compares it with original BERT and BiLSTM model separately, and the results are shown in Table 3.

From Table 3, we can conclude that BERT and BERT-BiLSTM get higher Accuracy than the single BiLSTM. The BERT-BiLSTM is slightly better than base BERT model,

Table 3. Comparison of evaluation indicators of three models

Model	Accuracy	Precision	Recall	F1
BiLSTM	0.8835	0.8994	0.8901	0.8947
BERT	0.9599	0.9592	0.9574	0.9583
BERT-BiLSTM	0.9652	0.9636	0.9643	0.9639

which proves the combined model has more outstanding performance to obtain more semantic features and extract textual information. We can use BERT-BiLSTM to classify the comment dataset of star hotel in Guiyang City, Guizhou Province.

3.3 The Prediction of Sentiment Orientation

3.3.1 Constructing Sentiment Index

Frank and Antweiler [10] proposed bullish index to compute the information sentiment orientation on the stock message board. Based on the idea of this index, this paper proposes a new sentiment orientation index to detect the public sentiment orientation to the hotel product and service. The formula is as follow:

$$\text{Index} = \frac{S^{neg}}{S^{POS}} \tag{18}$$

where S^{neg} is the number of negative comments during a given period, and S^{pos} is the number of positive reviews. If Index > 0, the sentiment orientation reflected by online comments is negative.

Finally, this paper calculates the growth rate of whole star hotel sales in Guiyang and sentiment orientation index we propose before, we find that there is strong relationship between them. The part of result is shown in Fig. 2. Therefore, it can be concluded that predicting the sentiment orientation is actually meaningful.

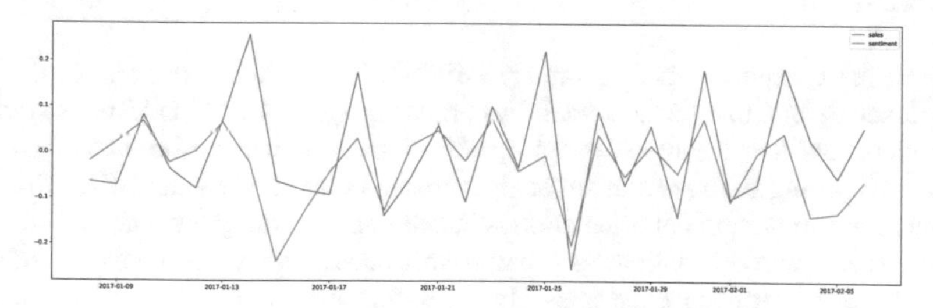

Fig. 2. The growth rate of sales and Sentiment Index

3.3.2 The Result of Predicting Sentiment Orientation

The paper has demonstrated the LSTM model earlier, we use this model to predict sentiment orientation which is in line with the trend of sales. The model is trained by the calculated sentiment index to determine the optimal time window so as to improve the performance. The data are divided into training set and test set with 9:1 scale. The model adopts MAE (mean_absolute_error) and Adam optimizer to train the data. The batch size is 32, Epoch is 20, and Dropout is 0.2.

In this section, the evaluation indicators include RMSE (root-mean-square error), MAE.

The growth rate of Sentiment Index during T period is identified as the target value, the model selects the growth rate from T-N to T-1 as the feature. According to the RMSE, MAE, the results of the prediction are as follow in Table 4.

The experiment results show that the whole result of prediction is optimal when the N is 5. When the number of N improves, the forecasting performance shows the trend goes up first and down then, which also demonstrates that the growth rate is also affected by workday and weekends.

Table 4. The predicting results in diverse time window

N	RMSE	MAE
1	0.892	0.704
2	0.841	0.660
3	0.828	0.650
4	0.825	0.647
5	0.783	0.619
6	0.823	0.654
7	0.822	0.658

4 Conclusion

Based on the compound model, namely BERT-BiLSTM, the paper adopts the open source hotel reviews to train the model. By comparing single BERT and BiLSTM model, the results show that the model we propose achieves higher performance than other models. According to the sentiment analysis results classified by the BERT-BiLSTM, the paper finds the sentiment orientation is in keeping with the growth rate of sales, and predicting sentiment orientation has actual significance. Finally, the prediction indicates that weekends will influence the sales. However, this method doesn't consider the other features to predict sales due to the lack of other variables. In future work, we will consider how to combine other non-sentiment variables and financial metrics to predict sales.

Acknowledgments. This research project is supported by Science Foundation of Beijing Language and Culture University (supported by "the Fundamental Research Funds for the Central Universities") (Approval number: 20YBB04 and 18PT02).

References

1. Santos, D., Nogueira, C., Gatti, M.: Deep convolutional neural networks for sentiment analysis of short texts. In: Proceedings of COLING 2014: Technical Papers, pp. 69–78 (2014)
2. Tang, D.Y., Qin, B., Liu, T.: Document modeling with gated recurrent neural network for sentiment classification. In: Proceedings of the 2015 Conference on Empirical Methods in Natural Language Processing, pp. 1422–1432 (2015)
3. Xiao, Z., Liang, P.: Chinese sentiment analysis using bidirectional LSTM with word embedding. In: Sun, X., Liu, A., Chao, H.-C., Bertino, E. (eds.) ICCCS 2016. LNCS, vol. 10040, pp. 601–610. Springer, Cham (2016). https://doi.org/10.1007/978-3-319-48674-1_53
4. Devlin, J., Chang, M.W., Lee, K., et al.: BERT: pre-training of deep bidirectional transformers for language understanding. In: Proceedings of the 2019 Conference of the North American Chapter of the Association for Computational Linguistics: Human Language Technologies, pp. 4171–4186 (2019)
5. Li, X., Wu, C., Mai, F.: The effect of online reviews on product sales: a joint sentiment-topic analysis. Inf. Manag. **56**, 172–184 (2019)
6. Sonnier, G.P., Mcalister, L., Rutz, O.J.A.: Dynamic model of the effect of online communications on firm sales. Mark. Sci. **30**, 702–716 (2011)
7. Reimer, T., Benkenstein, M.: When good WOM hurts and bad WOM gains: the effect of untrustworthy online reviews. J. Bus. Res. **69**, 5993–6001 (2016)
8. Bollen, J., Mao, H., Zeng, X.: Twitter mood predicts the stock market. J. Comput. Sci. **2**, 1–8 (2011)
9. Liu, G., Guo, J.B.: Bidirectional LSTM with attention mechanism and convolutional layer for text classification. Future Gener. Comput. Syst. 297–309 (2019)
10. Antweiler, W., Frank, M.Z.: Is all that talk just noise? The information content of internet stock message boards. J. Finance **59**, 1259–1294 (2004)

Large Pretrained Models on Multimodal Sentiment Analysis

Yunfeng Song, Xiaochao Fan(✉), Yong Yang, Ge Ren, and Weiming Pan

Xinjiang Normal University, Ürümqi, China
fxc1982@mail.dlut.edu.cn

Abstract. Previous researches on multimodal sentiment analysis mainly focused on the design of hand-crafted features and fusion approaches. Manually extracted features are fixed and cannot be fine-tuned. The choice of extraction methods also requires prior knowledge. With the development of Bert and GPT models, we can see the great potential of unsupervised learning in NLP. In this paper, we use large-scale pretrained text and audio models. We use Albert (text) and PANNs (audio) to train the model from the raw text and audio in an end-to-end way. In addition, we also perform modal fusion on the two pretrained models from different domains. The experimental results show that the accuracy and F1-score of CMU-MOSI dataset are up to 84.98 and 85.08 respectively, outperforming the current SOTA by 1.07 and 3.91.

Keywords: Multimodal · Sentiment analysis · ALBERT · Audio pattern recognition · Attention mechanism

1 Introduction

Sentiment analysis is the process of analyzing, processing, inducing and reasoning subjective texts with emotional color. It can be used to evaluate the emotional state of the speaker or to convey the emotional effect. Using the ability of sentiment analysis, we can automatically judge the positive and negative tendencies of natural language text with subjective description and give the corresponding results. It is widely used in comment analysis and decision-making, e-commerce comment classification and public opinion monitoring [1].

A modality is called a source of information. Modality refers to the way people receive information. People have hearing, vision, smell, touch and other perceptual ways to understand things. In recent years, the development of social networks allows for the rapid growth of multimedia data, such as images, text, sound, video and so on. More and more users are inclined to use multiple forms of media, such as text plus images, text plus songs, text and videos, to express their attitudes and emotions. Traditional single modality, such as text, relies on words, phrases and their semantic associations. It is not enough to identify complex emotional information, and multimodal text can often provide a more vivid description, convey more accurate and rich emotional information, and show

Q. Liang et al. (Eds.): Artificial Intelligence in China, LNEE 854, pp. 506–513, 2022.
https://doi.org/10.1007/978-981-16-9423-3_63

the information that may be hidden in the text, Therefore, multimodal sentiment analysis not only has important theoretical significance but also contains a great social value. Multimodal learning is composed of information from different modalities, generally including two or more modalities. It aims to jointly represent the data of different modalities, capture the internal relationship between different modalities, and achieve the mutual transformation of information of each modality. Even in the case of some missing modality, it can fill in the missing information in the transmission process. Multimodal deep learning brings great opportunities and challenges to machine learning [2].

With the popularity of BERT, [3] and GPT [4], we can see the great potential of unsupervised learning in NLP. Thus in this paper, we use large-scale pretrained text and audio models to sentiment classification problems. For text, we use Albert, an efficient variant of BERT which reduces a great number of model parameters. For audio, we use PANNs, a large-scale neural network trained on AudioSet [5]. Furthermore, previous works on multimodal learning are mainly based on manually extracted features. Once they are extracted, they are fixed and cannot be further fine-tuned, thus leading to sub-optimal performances. So in this paper, We directly train our model on raw text and waveform in an end-to-end way. We compare modality contribution on two modalities and find that textual information contributes the most. By adding acoustic signals, the fusion model achieves the best performance on the CMU-MOSI dataset.

2 Related Work

2.1 Multi-modal Sentiment Analysis

According to the fusion method used, we can divide multi-modal models into two categories: The first category is models used to model single-modality sequence data, such as RNN, GRU, and LSTM. Early fusion is the simple concatenation of the obtained different modality features and regards them as single modalities. This type of model is unable to model the modalities interaction. Late fusion votes after the output of each sub-model. Since the fusion process of late fusion has nothing to do with features, the error of the sub-model is usually also irrelevant [6,7]. Hybrid fusion combines the early and late fusion methods which combine the advantages of the two methods but increases the structural complexity [8,9]. Poria et al. [10] proposed the BC-LSTM (Bi-directional Contextual Long Short-Term Memory) model, which uses a bidirectional LSTM to capture global context information. Chen et al. [11] proposed GME-LSTM, which combines LSTM with gating mechanism and attention mechanism to perform modal fusion at the word level. However, the above research methods ignore the fusion between single-modal internal information and multi-modal information.

The others are specifically designed fusion models. Zadeh et al. [12] proposed Tensor Fusion Network for multi-modal sentiment analysis, using a multi-view gated memory module to synchronize multi-modal sequences. In multi-view sequence modeling, there are interactions within views and interactions between views. Zadeh et al. [13] proposed MFN (Memory Fusion Network) that continues to model these two interactions on a time scale. Graph-MFN (Graph Memory Fusion Network) [14] is a dynamic fusion graph built on top of MFN. Graph-MFN models each modality and changes its structure according to the importance of each modality and selects a suitable fusion map.

3 Proposed Method

Figure 1 demonstrated the overall architecture of our proposed multimodal fusion models. It mainly has two sub-networks where a pretrained ALBERT [15] is used to deal with textual information and a pretrained audio neural network PANNs [16] which is used to handle raw audio waveform. As stated before, we did not manually extract hand-crafted features. The whole network is trained end to end. The output of ALBERT and PANNs will be concatenated followed by the modality fusion module where modality interaction is learned. Finally, after the fusion process, the model gave the result of sentiment classification.

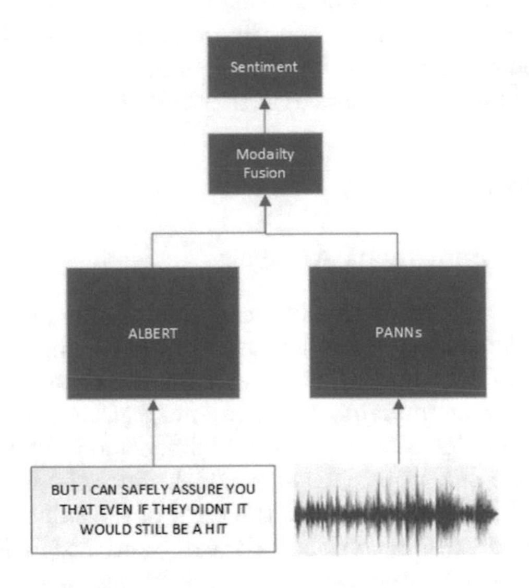

Fig. 1. Proposed model architecture

3.1 Text

ALBERT [15] is a variant of BERT [3]. It mainly made three improvements to Bert, which reduce the overall parameters, speed up the training, and increases the model effect. Bert provides two kinds of pretraining tasks when it is proposed. One is the masking language model. It randomly masks a certain proportion of input markers and then predicts the pretraining tasks of those obscured markers. The other is to predict whether the second sentence is the next sentence of the first sentence. But in practice, the second task was proved not to benefit the model, mainly because it was too simple. Therefore, in Albert, the task is changed to predict the order of sentences to increase the ability of the model to language understanding.

The second is factorized embedding parameterization. ALBERT uses factorization to reduce the parameters of embedding space, that is, mapping the one-hot vector of a word to a low dimensional (E) space, and then mapping it back to a high dimensional (H) space. The model parameter is reduced to $O(V \times E + E \times H)$ from $O(V \times H)$. When $E < H$, the number of parameters decreases a lot.

The third is sharing weights. Previous studies only share the full connection layer or the attention layer, while Albert shares all the weights. From the experimental results, the cost of sharing all the weights is acceptable. At the same time, sharing weights brings training difficulty, which makes the model more robust.

3.2 Audio

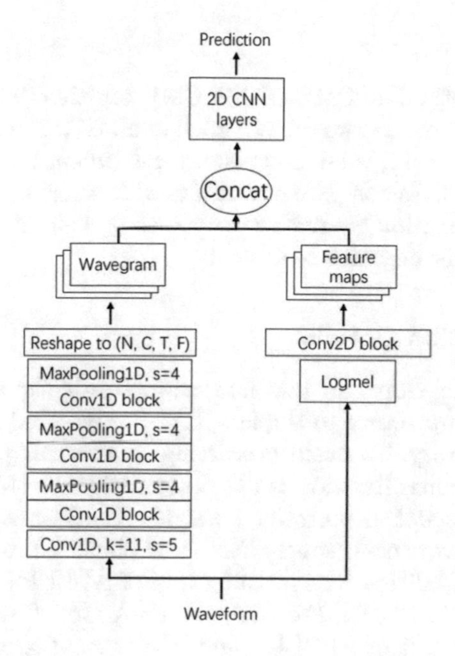

Fig. 2. Architecture of Wavegram-Logmel-CNN

Audio pattern recognition is an important research topic in sentiment analysis. The tone, the stress on certain words are vital clues to identify the speaker's true emotion. The speaker's true intention might be opposite to the words uttered, especially when people are being sarcastic. So we believe adding acoustic information would help to improve the accuracy of sentiment classification.

As we know, ALBERT is a pretrained model leveraging self-supervised learning. So for audio, we also used a pretrained model PANNs [16] (Large-Scale Pretrained Audio Neural Networks for Audio Pattern Recognition). The network architecture of PANNs is shown in Fig. 2. The basic component of PANNs is Convolutional Neural Network (CNN) combined with the pooling function. Kong et al. [16] proposed some variants of PANNs using different configurations and architecture. We chose Wavegram-Logmel CNN which utilizes the information from both time-domain waveforms and log mel spectrograms. The raw waveform of the audio event is one-dimensional. Previous work done on raw waveform has not outperformed the systems trained with mel spectrogram. Since they are unable to capture the pitch shifting in the audio event. To combine the advantage of waveform and log mel spectrogram, Wavegram-Logmel CNN simultaneously processes them. Figure 2 represents the model architecture of Wavegram-Logmel-CNN.[1]

4 Experiment

4.1 Dataset

We evaluated our method on CMU-MOSI: CMU Multimodal Opinion level Sentiment Intensity which was proposed by Zadeh et al. [17]. There are 1284, 229, and 686 utterances in the train, valid, and test set. Sentiment intensity is defined with a scale from -3 to $+3$. To compare our results with other existing approaches, we report binary classification results where values ≥ 0 signify positive sentiments and values < 0 signify negative sentiments.

4.2 Implementation Details

We trained our model from raw text and raw acoustic signals. For text, we pad and truncated each utterance to the length of 32. We used Albert-base-v2 from Transformers repository. To avoid overfitting, the learning rate of ALBERT is set to 0.000005. The loss function is mean squared error (MSE).

For audio, we used the pretrained weights Cnn14-mAP = 0.431.pth[2]. The learning rate of Wavegram-Logmel-CNN is 0.00005. We use different learning rate for different modalities. The learning rate of ALBERT is ten times smaller than Wavegram-Logmel-CNN. For the joint T+A model, we apply feature concatenation to the output of ALBERT and Wavegram-Logmel-CNN followed by full connected layer. The full connected network acts the role of feature fusion module.

---

[1] The figure was take from the original paper.

[2] https://zenodo.org/record/3987831#.YPVAX-gzb-h.

4.3 Baselines

We compare our approach with following classic multi-modal analysis models.

1) **TFN** (Tensor Fusion Network) was proposed by Zadeh et al. [12] that models intra-modality and inter-modality dynamics end-to-end.

2) **GME-LSTM(A)** was proposed by Chen et al. [11] that is able to perform modality fusion at the word level.

3) Georgiou et al. [18] proposed a hierarchical fusion scheme. Two bidirectional Long-Short-Term-Memory networks (BiLSTM), followed by multiple fully connected layers, are trained in order to extract feature representations for each of the textual and audio modalities.

4) Ghosal et al. [19] proposed a recurrent neural network based multi-modal attention framework that leverages the contextual information for utterance-level sentiment prediction.

5) Sun et al. [20] proposed a model that learns multi-modal embeddings from text, audio where individual features derived from the three views are combined into a multi-modal embedding using Deep Canonical Correlation Analysis.

6) Kumar et al. [21] proposed an approach to improve the multimodal sentiment analysis using self attention to capture long term context and gating mechanism to selectively learn cross attended features.

4.4 Results and Discussion

Table 1. The experimental results on CMU-MOSI. T: Text; A: Audio; T+A: Text+ Audio.

CMU-MOSI		
Approach	Accuracy	F1-score
Zadeh et al. [12]	77.1	79.1
Chen et al. [11]	76.5	73.4
Georgiou et al. [18]	76.9	76.9
Ghosal et al. [19]	82.31	80.69
Sun et al. [20]	80.6	80.57
Kumar et al. [21]	83.91	81.17
Proposed Approach (T)	84.55	84.52
Proposed Approach (A)	57.72	57.98
Proposed Approach (T+A)	**84.98**	**85.08**

The experimental results are shown in Table 1. Previous state-of-the-art results on CMU-MOSI was the approach proposed by Kumar et al. [21] which had 83.91 accuracy score and 81.17 F1-score. Our proposed approach T which is a fine-tuned ALBERT reaches an 84.55 accuracy score and 84.52 F1-score. We can see that although it is only trained on text without other modalities, it outperformed all other multimodal fusion models. The proposed A model performed poorly compared with other models. We posit that although PANNs is suitable for audio pattern recognition, it failed to understand the linguistic meaning of acoustic signals. The good performance of T model also proved that. The fusion model T+A had the best accuracy score of 84.98 and F1-score 85.08. By adding the acoustic signals, the performance of T model raised 0.43 on accuracy and 0.56 on F1-score, respectively. The experimental results show that language is the leading factor in sentiment analysis. With the help of acoustic information and modality fusion, the performance is improved.

5 Conclusion

Multimodal Learning is a challenging research area that requires knowledge from different modality domains. Furthermore, the efficient way to model each modality and their interaction remain an open problem. Tremendous research works have been done on different aspects of deep learning (i.e., image, language, audio). Although their works are done separately, we can still learn from their ideas. In this paper, we apply large pretrained language and audio models to the task of sentiment classification. The experiment was conducted on the CMU-MOSI dataset. The experimental results showed that our fine-tuned model outperformed SOTA by 1.07 on accuracy score and 3.91 on F1-score score. We show that language is the primary modality for sentiment classification. By adding acoustic signals, the performance is further improved.

References

1. Huddar, M.G., Sannakki, S.S., Rajpurohit, V.S.: A survey of computational approaches and challenges in multimodal sentiment analysis. Int. J. Comput. Sci. Eng. **7**(1), 876–883 (2019)
2. Soleymani, M., Garcia, D., Jou, B., et al.: A survey of multimodal sentiment analysis. Image Vis. Comput. **65**, 3–14 (2017)
3. Devlin, J., Chang, M.W., Lee, K., et al.: BERT: pre-training of deep bidirectional transformers for language understanding. arXiv preprint arXiv:1810.04805 (2018)
4. Radford, A., Wu, J., Child, R., et al.: Language models are unsupervised multitask learners. OpenAI Blog **1**(8), 9 (2019)
5. Gemmeke, J.F., Ellis, D.P.W., Freedman, D., et al.: Audio set: an ontology and human-labeled dataset for audio events. In: 2017 IEEE International Conference on Acoustics, Speech and Signal Processing (ICASSP), pp. 776–780. IEEE (2017)
6. Snoek, C.G.M., Worring, M., Smeulders, A.W.M.: Early versus late fusion in semantic video analysis. In: Proceedings of the 13th Annual ACM International Conference on Multimedia - MULTIMEDIA 2005, p. 399. ACM Press, Hilton (2005)
7. Vielzeuf, V., Pateux, S., Jurie, F.: Temporal Multimodal Fusion for Video Emotion Classification in the Wild. arXiv:1709.07200 [cs] (2017)

8. Wu, H., Mao, J., Zhang, Y., et al.: Unified visual-semantic embeddings: bridging vision and language with structured meaning representations. In: 2019 IEEE/CVF Conference on Computer Vision and Pattern Recognition (CVPR), pp. 6602–6611 (2019)
9. Andreas, J., Rohrbach, M., Darrell, T., et al.: Learning to compose neural networks for question answering. In: Proceedings of the 2016 Conference of the North American Chapter of the Association for Computational Linguistics: Human Language Technologies, pp. 1545–1554. Association for Computational Linguistics, San Diego (2016)
10. Poria, S., Cambria, E., Hazarika, D., et al.: Context-dependent sentiment analysis in user-generated videos. In: Proceedings of the 55th Annual Meeting of the Association for Computational Linguistics (Volume 1: Long Papers), pp. 873–883. Association for Computational Linguistics, Vancouver (2017)
11. Chen, M., Wang, S., Liang, P.P., Baltrusaitis, T., Zadeh, A., Morency, L.P.: Multimodal sentiment analysis with word-level fusion and reinforcement learning. In: Proceedings of the 19th ACM International Conference on Multimodal Interaction, ICMI, pp. 163–171 (2017)
12. Zadeh, A., Chen, M., Poria, S., Cambria, E., Morency, L.P.: Tensor fusion network for multimodal sentiment analysis. In: Proceedings of the 2017 Conference on Empirical Methods in Natural Language Processing, EMNLP, pp. 1103–1114 (2017)
13. Zadeh, A., Liang, P.P., Mazumder, N., et al.: Memory fusion network for multi-view sequential learning. In: Proceedings of the AAAI Conference on Artificial Intelligence, vol. 32, no. 1 (2018)
14. Bagher Zadeh, A., Liang, P.P., Poria, S., et al.: Multimodal language analysis in the wild: CMU-MOSEI dataset and interpretable dynamic fusion graph. In: Proceedings of the 56th Annual Meeting of the Association for Computational Linguistics (Volume 1: Long Papers), pp. 2236–2246. Association for Computational Linguistics, Melbourne (2018)
15. Lan, Z., Chen, M., Goodman, S., et al.: ALBERT: a lite BERT for self-supervised learning of language representations. arXiv preprint arXiv:1909.11942 (2019)
16. Kong, Q., Cao, Y., Iqbal, T., et al.: PANNs: large-scale pretrained audio neural networks for audio pattern recognition. IEEE/ACM Trans. Audio Speech Lang. Process. **28**, 2880–2894 (2020)
17. Zadeh, A., Zellers, R., Pincus, E., Morency, L.P.: MOSI: multimodal corpus of sentiment intensity and subjectivity analysis in online opinion videos. CoRR, vol. abs/1606.06259 (2016)
18. Georgiou, E., Papaioannou, C., Potamianos, A.: Deep hierarchical fusion with application in sentiment analysis. In: Proceedings of the Interspeech 2019, pp. 1646–1650 (2019)
19. Ghosal, D., Akhtar, M.S., Chauhan, D, Poria, S., Ekbal, A., Bhattacharyya, P.: Contextual inter-modal attention for multi-modal sentiment analysis. In: Proceedings of the 2018 Conference on Empirical Methods in Natural Language Processing, pp. 3454–3466 (2018)
20. Sun, Z., Sarma, P.K., Sethares, W., Bucy, E.P.: Multimodal sentiment analysis using deep canonical correlation analysis. In: Proceedings of the Interspeech 2019, pp. 1323–1327 (2019)
21. Kumar, A., Vepa, J.: Gated mechanism for attention based multi modal sentiment analysis. In: ICASSP 2020-2020 IEEE International Conference on Acoustics, Speech and Signal Processing (ICASSP), pp. 4477–4481. IEEE (2020)

3D Point Cloud Mapping Based on Intensity Feature

Xiangyu Zheng, Haiyun Gan(✉), Xinchao Liu, Weiwen Lin, and Peng Tang

School of Automotive and Transportation, Tianjin University of Technology and Education, Tianjin 300000, China

ganhaiyun_com@sina.com

Abstract. A 3D LIDAR simultaneous localization and mapping scheme with intensity information corrected point clouds is proposed for the traditional simultaneous localization and mapping (slam) methods. Four modules are included in the algorithm: data processing, feature extraction, odometry and mapping. The improved scheme is tested and analysed against the currently popular LOAM and LEGO-LOAM schemes using real measurement data. The results show that the proposed method changes the existing pre-processing method without any increase in pre-processing time. The test results show that the optimization algorithm outperforms the LOAM scheme while meeting the system's real-time requirements, but is inferior the LEGO-LOAM scheme.

Keywords: Mapping · Feature extraction · Intensity information · LOAM · LEGO-LOAM

1 Introduction

With the increasing proportion of robots in the market, occupying a larger share, riding an important role in service, household and industry, the new energy industry is also getting hotter and hotter under the guidance of national policies, and intelligent vehicles have become mainstream with it, which is a complex system integrating precise positioning, environment perception, decision control and other functions. Accurate access to the position and attitude information of the vehicle provides important guarantees for real-time control, decision - making and path planning of unmanned vehicles [1].

SLAM technology realizes the localization of an unmanned vehicle in an unknown environment and the simultaneous establishment of a map of the surrounding environment. Depending on sensor types, the current slam technologies are mainly divided into laser slam and visual slam. Fusing multiple mainstream sensor data through filters to avoid problems such as low performance and poor robustness of single sensors [2]. SLAM based on 3D liar does not require GNSS signal and can directly past spatial 3D information, which is not affected by light transformation, does not have scale drift phenomenon, and does not require pre-arranged scenes and can fuse multiple sensors [3]. With the development of graph optimization methods, the current mainstream 3D simultaneous localization and mapping methods mainly apply graph optimization, but

Q. Liang et al. (Eds.): Artificial Intelligence in China, LNEE 854, pp. 514–521, 2022.
https://doi.org/10.1007/978-981-16-9423-3_64

filtering method also plays a very important role. Traditional algorithms rely only on geometric information to minimize point cloud differences, and intensity information is often ignored, resulting in many limitations in positional estimation, low output positional frequency, and gradually increasing cumulative error. It cannot meet real-time fast localization and navigation. In contrast, the use of intensity and geometric information for localization estimation compensates for the lack of state estimation by geometric information alone and can provide more reliable position and attitude information.

Most existing lidar SLAM efforts focus on the ensemble information of the environment. Nelson proposed BLAM-SLAM, which obtains a better map by using the GICP [4] algorithm to calculate the initial transformation of the point cloud that obtains the nearest proximity point of the map corresponding to the current frame, and then executes the GICP matching algorithm to perform the exact transformation. Zhang proposed the simultaneous localization and mapping scheme (LOAM) [5], which is based on line surface features for point cloud feature extraction, and combines LIDAR odometer output and point cloud sub-map for feature matching, but without data pre-processing loop detection and back-end optimization; Shan proposed the LEGO-LOAM scheme [6], which introduces closed-loop detection of nearest pro search based on historical position, data processing, and back-end nonlinear optimization based on incremental smoothing on top of LOAM. The dynamic target leads to a complex mapping process and large positioning errors. Based on the traditional surface-based mapping methods, semantic information is added to solve the problem that dynamic objects influence localization [7]. Wang proposed SLAM method based on point cloud segmentation, which uses IMU for motion compensation, and then apply a full convolutional neural network on the point cloud to detect possible moving targets and use the traceless Kalman filter for motion target tracking and dynamic and static target differentiation [8]. In contrast to point-line surface features, abstract semantic features reflect thc 3D scene and provide correct feature matching. The Segmatch is proposed to eliminate false category labels, which uses a diffuse filling method. To improve the accuracy of localization by constructing a semantically constrained ICP model [9]. In this paper, a 3D laser SLAM algorithm, which is a method of combination of line and surface features, uses intensity information to correct the point cloud in the pre-processing stage, which is proposed for optimizing the LIDAR odometer poses (Fig. 1).

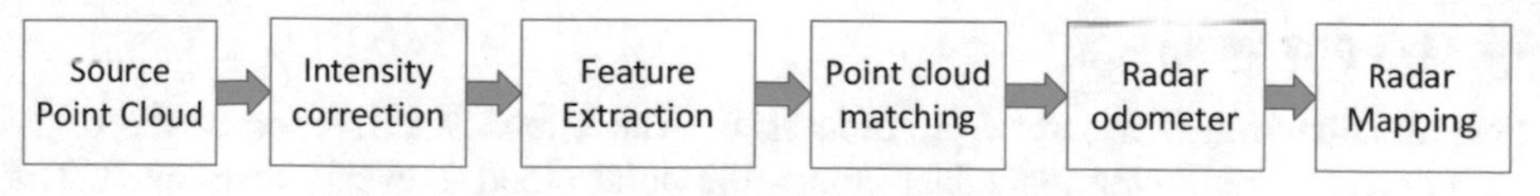

Fig. 1. The overall flow chart of this paper

2 Algorithm Theory and Implementation

2.1 Intensity Correction

LIDAR emits a laser beam and measures the arrival time and the energy of the reflected signal. The position of the object in the sensor coordinates, using the emission angle and the distance (arrival time), is determined. The intensity value, using the ratio between the received energy of the reflected signal and the transmitted laser power, is determined. The physical principle of the received power p_r can be determined as:

$$\mathrm{Pr} = \frac{\mathrm{PeD}_{\mathrm{r}}^{2}\rho}{4R^{2}}\eta\mathrm{sys}\eta atm\cos\alpha \tag{1-1}$$

Where P_e is the power of the emitted laser beam, η_{atm} is the atmospheric emission factor, Dr is the receiver aperture, η_{sys} is the system emission factor, and ρ is the material reflectance of the object, α is the angle of incidence between the object surface and the laser beam. The measured distance R and angle of incidence α are external parameters. η_{atm} and η_{sys} are constant parameters. Therefore, the intensity measurement I equation:

$$I = \frac{P_{\mathrm{r}}}{P_{\mathrm{e}}} = \frac{D_{\mathrm{r}}^{2}\rho}{4R^{2}}\eta\mathrm{sys}\eta\mathrm{atm}\mathrm{cos}\alpha = \eta\mathrm{all}\frac{\rho\mathrm{cos}\alpha}{\mathrm{R}^{2}} \tag{1-2}$$

Where η_{all} is a constant. Therefore, the surface reflectivity is only related to the angle of incidence and the measured called distance R by:

$$\rho \propto \frac{IR^{2}}{\mathrm{cos}\alpha} \tag{1-3}$$

For lidar scanning, this distance can be easily measured. Therefore, the angle of incidence can be estimated because of the local normal. For each point, the two closest points p_1 and p_2 are searched in terms of the local surface normal n can be determined as:

$$\mathrm{n} = \frac{(\mathrm{p} - \mathrm{p}_1) \times (\mathrm{p} - \mathrm{p}_2)}{|\mathrm{p} - \mathrm{p}_1| \times |\mathrm{p} - \mathrm{p}_2|} \tag{1-4}$$

Then the angle of incidence versus the normal is obtained as:

$$\cos\theta = \frac{\mathrm{p}^{\mathrm{T}} \cdot \mathrm{n}}{|\mathrm{p}|} \tag{1-5}$$

2.2 Pre-processing

Feature extraction is a part of pre-processing. The intensity correction of the original points and removing redundant points, the point cloud is evenly distributed. The retained features for large buildings and trees are retained. The ground point cloud and non-ground point cloud split, which can obtain line feature points and surface feature points for promoting improved speed and accuracy of feature extraction. The only points considered to be candidates are clusters containing more than 35 points clouds. Then, the corresponding line feature points and surface feature points are extracted from the segmented point cloud again. The calibrated intensity information is derived from the geometric information P and intensity information I. The calibrated intensity information is applied to the incident angle. In addition, because low intensity values lead to a lower signal-to-noise ratio (SNR), so one of the methods that can improve the ranging accuracy is to include intensity information. Using the line surface feature extraction method, the local smoothness of the laser point is defined as:

$$c = \frac{1}{|S| \cdot ||X^L_{(k,i)}||} || \sum_{j \in S, j \neq i} \left(X^L_{(k,i)} - X^L_{(k,j)} \right) || \tag{1-6}$$

By the characteristic curvature, c is the curvature value, S is the smoothness, *L* indicates that it is in laser coordinates. The laser beam is distinguished and those with high curvature and high smoothness are used as edge points. Conversely, those with low curvature and low smoothness are used as flat points. The curvature is set at a threshold to determine the high and low (Fig. 2).

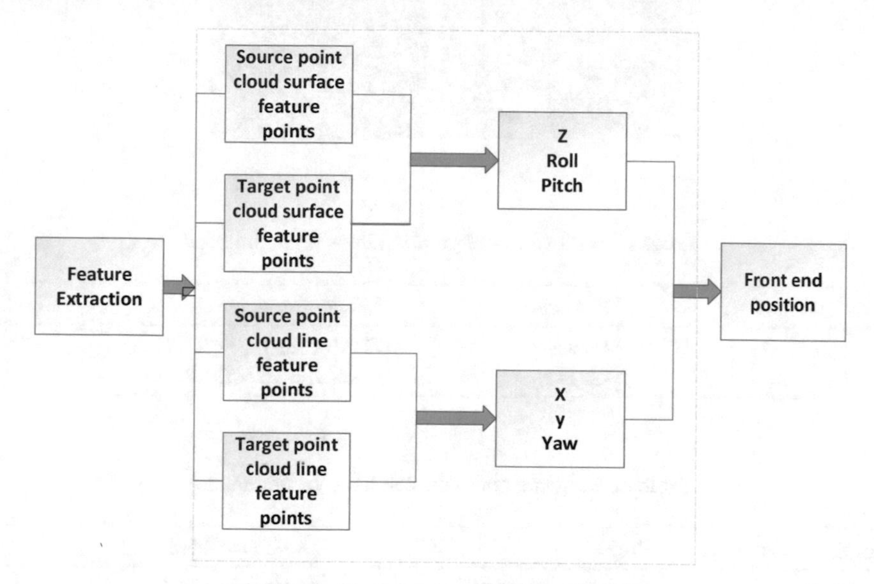

Fig. 2. Feature extraction flow chart

2.3 Front-End Odometry

With the odometry, the positional estimation of each frame of the point cloud can be obtained. The structured features of the infrared system are calculated to obtain the positional relationship of adjacent frames using aberration correction and scan matching parallel algorithms. Because the vehicle-mounted LIDAR works in a rotational scanning mode and the LIDAR works while the vehicle is moving, resulting in each laser point frame not being in a local coordinate system. This problem can be improved by aberration correction.

The extracted plane points are recorded as H_{k+1}, the point cloud of the K_{th} scan is known as P_k, the point cloud of the k_+1th scan is known as $P_{k+1,}$ the extracted edge points are recorded as E_k, the extracted plane points are recorded as H_k, the extracted edge points are recorded as E_{k+1}, the transformation relationship between E_k, H_k and P_{k+1}, E_{k+1}, the transformation between the point clouds is obtained, and the corresponding attitude is obtained. transformation information. Among multiple attitude solving methods, the Gauss-Newton method is used for calculations (Tables 1 and 2).

3 Experiment and Result Analysis

3.1 Software and Hardware System

Table 1. Hardware configuration Item Version/model

Configuration	Projects	Version/Model
Hardware	Laptop LIDAR	i7, 16 G memory, 2.6 GHz Velodyne-16 LIDAR

Table 2. Software configuration Item Version/model

Configuration	Projects	Version/Model
Software	Operating System	Ubuntu 18.04
	Robot Operating System	ROS MELIDOIC
	Opening Source Solutions	LOAM, Lego-LOAM
	Point cloud software	Unity

3.2 Experimental Results

The three laser odometer times without IMU fusion were evaluated using data collected from the actual campus and obtained through auto ware acquisition. The experimental environment is a dynamic, closed-loop campus road, shown in the real scenario figure(a), where the three schemes are objectively evaluated for map building times without closed-loop detection, and the building effects are compared qualitatively and quantitatively.

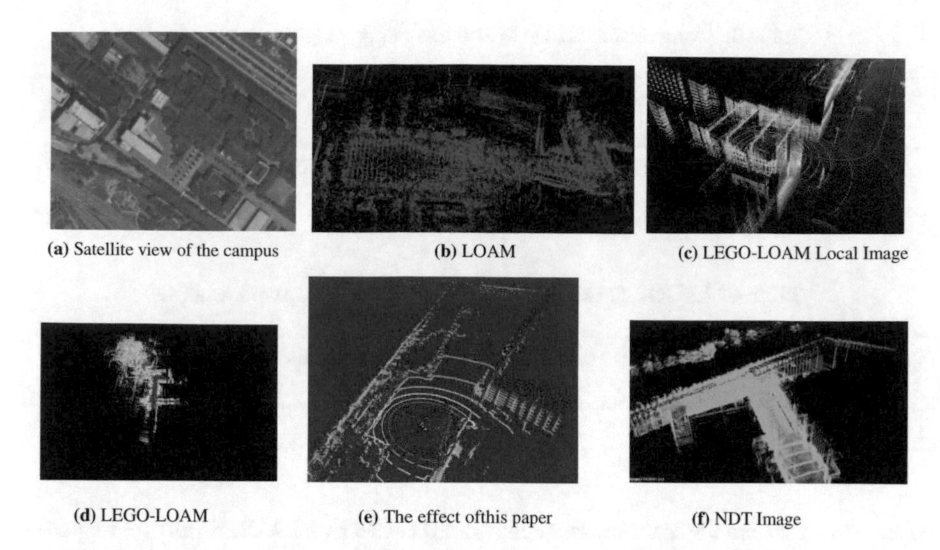

(a) Satellite view of the campus (b) LOAM (c) LEGO-LOAM Local Image

(d) LEGO-LOAM (e) The effect ofthis paper (f) NDT Image

Since the LOAM(b) scheme does not maintain a global map in the laser building module, it leads to truncation effects in the ROS rviz real-time display of laser odometer trajectories and map building, and dynamic obstacles affect the feature point extraction.

The Lego-LOAM (c d) solution and the improved solution perform screening point processing in the pre-processing stage to remove unreliable feature points and prevent unstable feature points from affecting the point cloud matching accuracy, so the map building effect is complete and clear. However, the back-end factor optimization of the Lego-LOAM scheme does not integrate the inertial guidance factor to jointly constrain the positional information. The feature point cloud extracted by the Lego-LOAM scheme is sparse, which affects the point cloud matching accuracy during frame map optimization, although it achieves light weight and improves the real-time performance of the system.

The number of point clouds based on the auto ware using 3D-NDT to build a map(f) is extremely large, which has certain requirements on the computational volume of subsequent navigation and positioning. Successive multi-sensor fusion 3D offline building algorithms are continuously improved to ensure accuracy while reducing the number of point clouds and improving efficiency. The 3D-NDT method is the basis for the development of the other algorithm (Tables 3 and 4).

3.3 Runtime Performance Evaluation

From the experimental data, the LOAM scheme, the Lego- LOAM and the improved algorithm scheme run part of the thread time as shown in the table. In this paper, we add intensity correction and loam algorithm comparison analysis in the preprocessing section, the data processing time is similar. The LOAM scheme and the improved algorithm are second to the LEGO-LOAM algorithm. Because the Lego-LOAM scheme implements lightweight by extracting sparse feature point clouds, and using two-step optimized L-M optimization to solve the bit pose. The computation time of all three

Table 3. Comparison of LOAM and this paper's algorithm time (ms)

Method	Mapping processing	Total mapping	Methods	Data processing	Total mapping
LOAM	51.03	97821.09	Program of this paper	50.37	97597.17

Table 4. LEGO-LOAM time for each thread at a given moment (ms)

Scheme	Frame Matching	Data Fusion	Optimization Time	Total Time
Lego-LOAM	18.82	11.94	32.45	73.21

schemes is less than the scanning frequency of 10 Hz set by LIDAR. So all three schemes can ensure the real-time performance of their respective systems.

4 Conclusion

In this paper, we propose a mapping scheme using intensity information and geometric information for pre-processing in outdoor dynamic road environment to address the problems of low odometry accuracy and lack of robustness of existing laser odometry methods which only rely on geometric information to extract feature points. Data pre-processing, feature extraction, laser odometry and mapping construction are the main modules of this paper. Experimental tests are conducted using actual data collected from real vehicles. The proposed scheme is analyzed qualitatively and quantitatively against the most popular LOAM and Lego-LOAM schemes for mapping. The running time of each module in the three schemes is evaluated using the measured data. However, in terms of improving the accuracy of laser odometry, the removal of dynamic obstacles in the road environment, such as pedestrians and vehicles, during feature extraction needs further improvement. Future in-depth research can be conducted in the following directions.

a) In terms of data fusion framework. In order to improve the frequency of matching each frame with the local map, the data fusion framework can be optimized using factor graphs, of which the core part can be optimized for methods such as L-M, setting the radius of the trust region, and further elaborating the expansion multiplier to set the threshold, in order to achieve nonlinear optimization purposes.
b) In dynamic obstacle processing. Combining deep learning methods for point cloud detection and using methods such as filters for point cloud differentiation to improve the accuracy of map building and localization.
c) In terms of improving the accuracy of positional estimation. Optimize the closed-loop detection and fuse the laser odometer with the true value data of multiple sensors

available in the market according to the key point poses, thus reducing the relative cumulative drift error.

Acknowledgments. This work was supported in part by the Tianjin Artificial Intelligence Project of China (No. 18ZXAQSF00090), the Tianjin University Discipline Leading Talent Training Program of China (No. SSW181030105), and the Tianjin Artificial Intelligence Project of China (No. 2020YJSZXS3 2).

References

1. Wei, S.F., Pang, F., Liu, Z.B.: A review of LIDAR-based simultaneous localization and map construction methods. Comput. Appl. Res. **37**(2), 327–332 (2020)
2. Li, W.-H.: Lightweight multi-sensor fusion SLAM system based on ESKF and graph optimization. Sci. Technol. Innov. **4**(08), 15–18 (2021)
3. Ye, L.J., Zhou, Z.F., Wang, L.D., Zheng, Y.: Cache pool building based on GNSS/INS with LiDAR. Laser Infrared **50**(03), 300–304 (2020)
4. Vlaminck, M., Luong, H., Philips, W.: Surface-based GICP. In: 2018 15th Conference on Computer and Robot Vision (CRV), pp. 262–268 (2018)
5. Zhang, J., Singh, S.: Low drift and real time lidar odometry and mapping. Auton. Robot **41**(2), 401–416 (2017)
6. Shan, T.X., Brendan, E.: LeGO-LOAM: lightweight and ground optimized LiDAR odometry and mapping on variable terrain. In: Proceedings of IEEE/RSJ International Conference on Intelligent Robots and Systems, pp. 4758–4765. IEEE Press, Piscataway (2018)
7. Chen, X., Milioto, A., Palazzolo, E., Giguère, P., Behley, J., Stachniss, C.: SuMa++: efficient LiDAR-based Semantic SLAM. In: 2019 IEEE/RSJ International Conference on Intelligent Robots and Systems (IROS), pp. 4530–4537 (2019). https://doi.org/10.1109/IROS40897.2019.8967704
8. Wang, Z.L., Li, W.Y.: Motion target tracking based on point cloud segmentation with SLAM method. Robotics **43**(02), 177–192 (2021)
9. Dube, R., Dugas, D., Stumm, E., et al.: SegMatch: segment based place recognition in 3D point clouds. In: IEEE International Conference on Robotics and Automation, pp. 5266–5272 (2017)

Customer Stickiness Evaluation Model Research Based on Machine Learning

Yuejia Sun(✉)

No. 32 Xuanwumen West Street, Xicheng District, Beijing, China

Abstract. With the rapid development of communication technology, the mobile market is becoming saturated. How to improve customer stickiness has become a concern for telecom operators. An important work to improve customer stickiness is to construct an effective customer stickiness evaluation model. Firstly, a customer stickiness evaluation variable system is designed from three levels of basic attributes, behavior and psychology. Secondly, this paper proposes a targeted feature extraction method based on customer grouping, and adopts WOE method to perform feature encoding, and then uses IV value and multicollinearity elimination method to perform feature selection. Finally, this paper builds a logistic regression scorecard model, and then introduces several integrated machine learning models for comparison. The experimental results show that the logistic regression scorecard model can be selected to solve the problem of customer stickiness evaluation due to its better market interpretability and acceptable data predictability.

Keywords: Customer stickiness evaluation · AHP · Customer segmentation · WOE encoding · IV feature selection · Logistic regression scorecard model · Integrated machine learning model

1 Introduction

As the mobile market becomes saturated, how to improve customer stickiness has become one of the core issues that telecom operators are concerned about. An important work to improve customer stickiness is to construct an effective customer stickiness evaluation model. Based on the big data accumulated by telecom operators over the years, the customer stickiness evaluation system can be built by machine learning method. Through the customer stickiness evaluation model, the customer stickiness score can be calculated, so as to assist the positioning of segmented target customers. By formulating personalized stickiness improvement strategies for different customers, it can provide strong support for enterprises to achieve refined operations.

Domestic and foreign scholars have carried out research on customer stickiness evaluation from influencing factors and evaluation methods. Anis Allagui [1] proposed a conceptual model of user stickiness, and believed that the four major elements of core service, support service, product design, and customization affect user satisfaction, which in turn affects user stickiness. Ellen Reid Smith [2] put forward five measurement factors of user stickiness, which are product availability, discount, purchase process,

Q. Liang et al. (Eds.): Artificial Intelligence in China, LNEE 854, pp. 522–527, 2022.
https://doi.org/10.1007/978-981-16-9423-3_65

service support and communication relationship. He Qing from China Unicom [3] quantified the impact of various stability factors on the customer performance, scored their contribution to stability, and established a quantification of customer stability model. Ji Li [4] refined the influencing factors of customer loyalty through a sample survey of individual customers of China Mobile, and constructed a customer loyalty driving model. In summary, most researches on customer stickiness focus on influencing factors and quantitative evaluation methods based on traditional statistical methods. Based on previous studies, this paper aims to construct a stickiness evaluation model through machine learning methods.

2 Customer Stickiness Evaluation Model

This paper designs the model based on both the logistic regression scorecard model and the integrated machine learning model, and finally selects the suitable model from two factors: market interpretability and data predictability.

2.1 Definition of High and Low Stickiness

Customer stickiness refers to the degree of reliance and re-consumption expectation formed by the combination of customer loyalty, trust, and benign experience to the brand or product. The analytic hierarchy process is used to extract high and low stickiness customers. First, key pre-rating variables are designed, including stability variables (network age, off-grid status), value variables (ARPU), and integration variables (family and group integration, contracts, apps). Then, the weights of the key variables are calculated by the analytic hierarchy process, and the pre-rating scores of customers are calculated according to the weights and scoring rules of key variables. Finally, based on the pre-rating scores, typical "high stickiness" customers and "low stickiness" customers are extracted for model training and testing.

2.2 Stickiness Evaluation Variable System

Customer stickiness is not only a reflection of behavior, but also a reflection of psychological. Therefore, this paper constructs an evaluation variable system with three dimensions including basic attributes, behavioral factors, and psychological factors. The dependent variable is the label of "high stickiness" or "low stickiness" in observation period (month T), and the independent variables are the states of the customer in performance period (month T-1 to T-6). In this way, an evaluation system is constructed with 88 variables (Table 1).

2.3 Feature Extraction

Before feature extraction, customers should be grouped according to whether they are actively off-grid or passively off-grid. The two types of customers can be judged from whether there is a sinking period before going off the grid. The effective feature extraction

Table 1. Stickiness evaluation variable system

Dimensions	Variables
Basic Attributes	Age, Network age, Occupation type, Education level, Primary and secondary cards, Credit score, etc.
Behavioral Factors	ARPU, DOU, MOU, Caller duration, Contract period, Broadband usage, V network usage, Audio-visual apps usage, Arrears, Package overage, Competitor broadband usage, etc.
Psychological Factors	Satisfaction score, After-use score, Number of complaints, Brand event participation times, etc.

time period for actively off-grid customers is from T-1 to T-3, and that for passively off-grid customers is from T-4 to T-6.

The extracted features include conventional variables and derivative variables. Conventional variables are mainly designed for status variables and take the value of the past one month. Derivative variables are mainly designed for numerical variables and calculated in combination with the past 3 months. The calculation types include total value, average value, difference value, variance, standard deviation, dispersion coefficient, fluctuation value, etc.

2.4 Feature Encoding and Selection

Feature Encoding: The variables are binned based on WOE (weight of evidence) value. The WOE calculation method is shown as formula 1, where G_i is the number of good customers in the i-th bin, G_T is the number of all good customers, B_i is the number of bad customers in the i-th bin, B_T is the number of all bad customers. What WOE actually expresses is the difference between the proportion of good customers in current bins that account for all good customers and the proportion of bad customers in current bins that account for all bad customers. The principles of binning are that the negative sample difference of each bin is as large as possible; the total number of bins is about 2–10; the number of samples in each bin is not less than 5% of the total number of samples. According to the binning results, WOE coding is performed on each bin, that is, the concentration attribute of the predicted category is used as the encoding value.

$$WOE_i = log\left(\frac{p_{i1}}{p_{i0}}\right) = log\left(\frac{G_i/G_T}{B_i/B_T}\right) \tag{1}$$

Feature Selection: The feature Selection is performed based on the IV value (information value). The IV value calculation method is shown as formula 2. The level of the IV value indicates the strength of the predictive ability of the independent variable. Therefore, variables with IV values greater than a certain threshold (usually 0.02) are selected for training the model.

$$IV = \sum_i \left(\frac{G_i}{G_T} - \frac{B_i}{B_T}\right) * WOE_i \tag{2}$$

Multicollinearity Elimination: The VIF (variance inflation factor) is used to measure the multicollinearity of the variables. The VIF calculation method is shown as formula 3, where R^2 is the multiple determination coefficient of regression on other independent variables when x_i is regarded as the dependent variable. The larger the VIF, the higher the correlation between this variable and the other variables. Therefore, by deleting variables whose VIF is greater than a certain threshold (usually 10), multicollinearity can be weakened or eliminated.

$$VIF = 1/(1 - R^2) \tag{3}$$

2.5 Logistic Regression Scorecard Model

Firstly, the logistic regression classification model is constructed to obtain the classification probability, which is preferred because of its simplicity, parallelization, and strong interpretability. The weight of each variable is calculated through logistic regression model training, as formula 4. The probability of high stickiness is calculated through probability mapping, as formula 5. In order to prevent overfitting and avoid the weight of a single variable being too high, L2 regularization method is adopted. The L2 optimal parameters are obtained through cross-validation and grid search.

$$z = w_1 * x_1 + w_2 * x_2 + \ldots + w_n * x_n + b \tag{4}$$

$$p = \frac{1}{1 + e^{-z}} \tag{5}$$

Secondly, the score mapping model is designed to map from probability to score, using the ODD (good to bad sample ratio) method, as formula 6, where B means that for every increase of N score, the ODD increases by 1 time.

$$s = A + B * log\frac{p}{1-p},\ B = \frac{N}{\log(2)} \tag{6}$$

Thirdly, for good explanatory of business operations, the score is further mapped to a percentile system. Construct the X vectors with the highest and lowest scores of all vectors in the feature space and get the $max(s)$ and $min(s)$, and then perform percentile score mapping as formula 7.

$$s_{100} = \frac{s - min(s)}{max(s) - min(s)} * 100 \tag{7}$$

Finally, the scorecard generation model is designed to obtain the adding score for each bin of each variable. By mapping function derivation, for percentile scoring, the score change caused by the change of bins is shown as formula 8.

$$\Delta s_{100} = \frac{100}{\max(s) - \min(s)} * \Delta s = \frac{100 * B * w_i}{\max(s) - \min(s)} * \Delta x_i \tag{8}$$

2.6 Integrated Machine Learning Models

Integrated machine learning models are introduced for comparison, including random forest in the bagging methods, adaboost and GBDT in the boosting methods. This paper compares the effects of the above models with the logistic regression model, and select the suitable model for the stickiness evaluation.

3 Experimental Results and Analysis

3.1 Logistic Regression Scorecard Model Experiment Results

The experimental results of logistic regression on the test data set are as follows: the accuracy of the model is 92.7%, the recall rate for low-stickiness customers is 71.2%, the F1 score is 0.77, and the KS value is 0.73. Taking into account the needs of market regulation, the adding scores of each variable are slightly adjusted based on the calculation results of the logistic regression scorecard model to obtain the final scorecard result, which is shown in Table 2.

Table 2. Logistic regression scorecard model results

Variables	Bins	Scores	Variables	Bins	Scores
Network age	<2	0	Family V network	no	0
	[2, 5]	2		yes	12
	>5	4	Group V Network	no	0
ARPU difference in recent 3 months	<−5	−5		yes	18
	[−5, 5]	0	Audiovisual app member	no	0
	>5	3		yes	4
Double down in recent 3 months	No	−1	Email or cloud disk login times	0	0
	Yes	5		[1, 3]	2
Package overage	[0, 20]	0		>3	5
	>20	−5	Satisfaction score	0	0
Contract period	0	0		(0, 50]	3
	[1, 6]	2		>50	6
	>6	5	…	…	…
Broadband order quantity	0	0	Network quality complaints	0	0
	1	5		1	−2
	>1	8		>1	−4

3.2 Integrated Machine Learning Experiment Results

The comparative experimental results of the integrated machine learning and the logistic regression are shown in Table 3.

Table 3. Comparative experimental results

Model	Accuracy	Recall for low stickiness	F1	KS
Random Forest	93.5%	72.9%	0.79	0.77
Adaboost	94.2%	79.0%	0.83	0.81
GBDT	93.9%	69.6%	0.80	0.83
Logistic Regression	92.7%	71.2%	0.77	0.73

4 Conclusion and Summary

The characteristics of logistic regression model and integrated machine learning model are as follows:

- Logistic regression scorecard model can convert probabilities into score values for each bin of each variable. This model requires high data quality and achieves slightly lower accuracy, but is easy to explain from a business perspective.
- Integrated machine learning model can only convert the probability to a total score and output the importance of each variable. This model can achieve slightly higher accuracy, but is more difficult to explain from a business perspective.

It can be seen from the above experimental results that the accuracy of the logistic regression and the integrated machine learning model has a gap but relatively close, and the gap value is within the acceptable range. Therefore, considering both market interpretability and data predictability, the logistic regression scorecard model is suggested to be selected for the customer stickiness evaluation.

References

1. Allagui, A., Temessek, A.: Testing an E-loyalty conceptual framework. J. E-Bus. **4**(6), 1–6 (2004)
2. Smith, E.R.: Seven steps to building e-loyalty. Med. Mark. Media **6**(3), 94–102 (2001)
3. He, Q., Chen, J.: Yidong kehu wendingdu pinggu moxing ji yingyong [Mobile customer stability evaluation model and application]. C-Enterp. Manag. **2**, 72–73 (2017)
4. Li, J.: Zhongguo yidong guke zhongchengdu ji yingxiang yinsu yanjiu [Research on the customer loyalty and its influential factors for China Mobile]. Nanjing University of Posts and Telecommunications, Nanjing (2015)

Matching Based Intelligent Edge Resource Allocation

Peng Qin[1,2](✉), Miao Wang[1,2], Xiongwen Zhao[1,2], Jiayan Liu[1,2], Shuo Wang[1,2], Haoting He[1,2], and Yuanbo Xie[3]

[1] State Key Laboratory of Alternate Electrical Power System With Renewable Energy Sources, School of Electrical and Electronic Engineering, North China Electric Power University, Beijing 102206, China
qinpeng@ncepu.edu.cn

[2] Hebei Key Laboratory of Power Internet of Things Technology, North China Electric Power University, Baoding 071003, China

[3] School of Control and Computer Engineering, North China Electric Power University, Beijing 102206, China

Abstract. With the rapid development of the Internet of Things, multi-access edge computing (MEC) has emerged as a key technique and intelligent edge resource allocation has become one of the important research issues in MEC. In this paper, we utilize advantages of matching game to integrate users' needs and the position of edge computing nodes in IoT networks. The proposed GS-based two-sided matching approach simulates the competition and negotiation relationship between different user sets in IoT edge nodes, and provides feasible options based on the preference between different entities. Simulations show that it not only solves the problem of edge resource allocation in a semi-distributed manner, but also effectively improves SP's revenue.

Keywords: Matching theory · Intelligent edge computing · Resource allocation

1 Introduction

With the rapid development of the Internet of Things (IoT), multi-access edge computing (MEC) has emerged as a key technique. By allocating computing and storage resource at the edge of networks, it deals with large amounts of data generated by terminals to respond to the request of delay-sensitive applications [1–3]. The concept of edge computing is driven by actual demand, since cloud computing cannot meet the differentiated application needs. Thus, computing is delegated to the edge to supplement the deficiencies of cloud computing. There was no strict definition of edge computing at the beginning. Literature [2] defines edge computing as follows. It is a type of computing anywhere between the data source and the cloud. This is a relatively broad definition of edge computing and it focuses on defining the position of the edge. In order to improve the efficiency of computing resource and optimize system performance, the issue of edge resource allocation has received widespread attention. The geographical dispersion and

Q. Liang et al. (Eds.): Artificial Intelligence in China, LNEE 854, pp. 528–536, 2022.
https://doi.org/10.1007/978-981-16-9423-3_66

heterogeneity of edge computing resource, as well as energy consumption, cost, and stability, all increase the complexity of resource allocation.

Artificial intelligence (AI), as a revolutionary technology, has been widely used in image processing, natural language recognition, electronic game and other fields [4, 5]. In order to solve the problems and challenges in edge resource allocation, academia begin to focus on AI. AI has strong learning and prediction ability, good nonlinear fitting and adaptive characteristics. It is good at mining complex features in high-redundancy data and conducting efficient mass data processing [6].

Intelligent edge computing proposes a new paradigm: each edge device of the IoT is equipped with data collection, analysis, calculation, communication, and the most important intelligence. The new intelligent edge computing draws on advantages of cloud computing, utilizes the cloud to carry out large-scale security configuration, deployment and management, and can intelligently allocate network resource according to device types and application scenarios.

Edge computing has become a hot research topic since it was proposed, and it has important applications in fields of the Internet of vehicles, industrial production, and smart cities. Paper [7] proposed the concept of intelligent robot factory and solutions for operation, data fusion, communication and interaction based on edge computing. It proved the intelligent robot factory had higher efficiency by experiments. Paper [8] addressed the problem of edge-assisted content distribution in autonomous driving service, and proposed a two-level edge computing architecture, which significantly reduced resource utilization and improved service quality by content caching at the wireless edge and content sharing in vehicular networks. Intelligent edge resource allocation is one of the important research issues of MEC. A global resource allocation strategy was proposed in [9]. A dynamic edge resource scheduling strategy based on deep reinforcement learning was developed in [10]. [11] studied the resource allocation strategy of the device-to-device (D2D) assisted edge computing system with hybrid energy harvesting. Mobile users could supplement the extra energy of nearby users when the energy consumed by environmental radio frequency source was about to run out, which improved user satisfaction.

However, as far as we know, there is no framework for resource allocation in MEC by jointly considering the needs of users in intelligent edge resource allocation and their position with edge computing nodes in IoT edge computing network. Therefore, by proposing the GS-based two-sided matching algorithm, this article has the following three advantages: Firstly, a GS-based matching model is proposed. In the intelligent edge resource allocation system model, it could be applied on HAPSs, satellites, ground base stations, etc. Secondly, we utilize advantages of matching to integrate user requirements and the position of edge computing nodes in the IoT edge computing network. Matching games can overcome the limitations of game theory and stochastic optimization [12–14]. It can simulate the competition and negotiation between different user sets at the edge node, and solve the problem in a semi-distributed manner. It provides feasible solutions based on the preferences and mutually beneficial relationships between different entity sets, while taking into account the local preferences of all entities. Finally, extensive simulations demonstrate that our approach has high stability and low complexity.

2 System Model

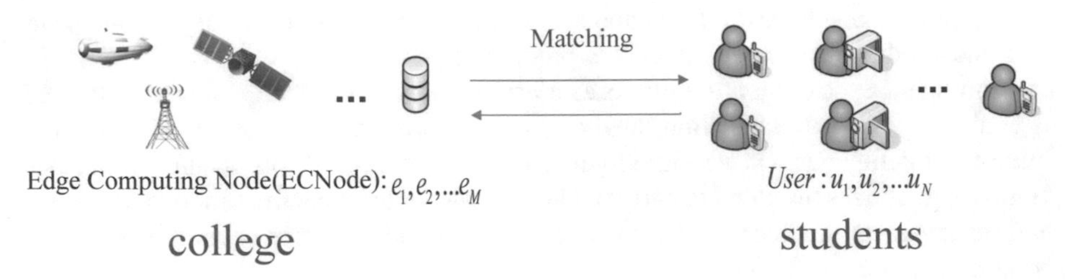

Fig. 1. Intelligent edge resource allocation system model.

We consider an intelligent edge resource allocation system model as shown in Fig. 1, with a set of M edge computing nodes (ECNodes) $E = \{e_1, e_2 \ldots e_M\}$ and a set of subscribed users $U = \{u_1, u_2 \ldots u_N\}$, where N is the number of all users who subscribe from service providers (SPs). The two-sided matching between *e* and *u* can be represented by $\mathcal{M} \subseteq$ E × U. We call *e* and *u* as matching agents. Each edge computing node can be shared among multiple users and is limited by its maximum capacity. In other words, the matching between the edge computing node *E* and the user *U* is many-to-one matching. Note that in intelligent edge resource allocation, SPs manage edge resource and provide to users based on their quotations. Users who need more edge service will provide higher price to obtain edge resource and such users are more favored by SPs. We define performance indicators from the following two perspectives: user experience and SP's revenue.

2.1 User Experience

One of the most important aspects of edge IoT service is user experience, which is also the concern of SPs. In order to improve user satisfaction, we consider the QoS of e_j, which is measured by the key indicator network delay t_i between e_j and u_i, since it determines the distance between user u_i and ECNode e_j. It determines data transmission delay, and hence, the quality of edge computing service.

We set the distance from the farthest IoT terminal to the edge computing node e_j plus the distance from the ECNode e_j to the user u_i as $d_{i,j}$. For simplicity, we assume that the network resources from the ECNode e_j to the user are always sufficient. The cost of network delay t_i generally follows the distance from the IoT terminal to the ECNode e_j plus the distance from ECNode e_j to the user u_i, i.e.,

$$t_i = \rho d_{i,j}, \ \forall i \in \{1, 2, \ldots, N\}, \ \forall j \in \{1, 2, \ldots, M\} \tag{1}$$

where ρ is a scalar. We define the users' maximum service delay tolerance as t_{max}, thus.

$$t_i \leq t_{max}, \tag{2}$$

Since users in the network need to pay SPs for service, according to literature [13], users' utility can be expressed as the revenue obtained from the workload data minus both the cost of network delay and payment to the SPs. It can be expressed as.

$$UT_i^u = \sum_{j=1}^{M} g_{i,j}\xi_{i,j}\lambda_i - \beta_i t_i - \varepsilon_i O_i, \forall i \in \{1, 2, \ldots, N\}, \ \forall j \in \{1, 2, \ldots, M\} \tag{3}$$

where $\xi_{i,j}$ represents the revenue that user u_i can obtain from e_j by purchasing a unit edge computing resource block (CRB) and $\xi_{i,j}\lambda_i$ represents the total revenue of each u_i, since we assume that the workload of each user u_i follows the Poisson arrival process with arrival rate of λ_i. β_i and ε_i are weight factors indicating the importance of service delay and payment of u_i in the utility function, respectively. We introduce the binary variable $g_{i,j}$. If user u_i purchases service from SPs and SPs offload the task to the edge computing node e_j, the node will eventually provide edge computing service for u_i, which is equal to 1, and otherwise 0. The detailed definition is as follows

$$g_{i,j} = \begin{cases} 1, & if\ u_i\ purchases\ service\ from\ SPs,\ and\ SPs\ offload\ to\ e_j \\ 0, & otherwise \end{cases} \tag{4}$$

2.2 Service Provider's Revenue

The revenue that SPs receive from users is another factor we use to measure system performance. Mandatory income is an incentive for SPs to provide better services to their subscribers. We define the revenue of SP R_{SP} as the sum of the price provided by matching users minus the sum of the cost paid for computing service, which can be represented as

$$R_{SP} = \sum_{i=1}^{N} O_i - \sum_{j=1}^{M} Q_M = \sum_{i=1}^{N} q_i c - \sum_{j=1}^{M} Q_M, \ \forall i \in \{1, 2, \ldots, N\} \ \forall j \in \{1, 2, \ldots, M\} \tag{5}$$

where O_i is revenue obtained by the SP from user u_i, based on the expected number of computing resource blocks (CRB) q_i of u_i, c is the price for unit CRB service, and Q_M is the price SP paid to the matched edge nodes providing edge computing resource and service.

3 Problem Formulation

The objective of this paper is to maximize the total revenue for SPs while maintaining user experience. SPs offload service to ECNodes due to users' QoS, because the more edge service a user buy or the higher price users pay, the more SPs and edge nodes prefer the user. In this regard, we formulate the optimization problem of intelligent edge resource allocation as.

$$\mathrm{P1}: \max R_{SP}, \tag{6}$$

$$\text{s.t. } \mathcal{N}(\mathcal{M}, u_i) \leq 1, \ \forall i \in \{1, 2, \ldots, N\} \tag{7}$$

$$\mathcal{N}(\mathcal{M}, e_j) \leq N_e, \ \forall j \in \{1, 2, \ldots, M\} \tag{8}$$

$$g_{i,j} \in \{0, 1\}, \tag{9}$$

$$BP(u_i, e_j) = 0 \tag{10}$$

where $\mathcal{N}(\mathcal{M}, \mathrm{x})$ represents the number of partners in the matching $\mathcal{M}$. (6) represents the system goal to obtain the maximum SP's revenue. (7) and (8) denote the constraints of the maximum capacity of user and edge node respectively. (9) indicates that $g_{i,j}$ is a binary variable. (10) ensures that there exists no blocking pair in matching $\mathcal{M}$. The blocking pair is defined in Definition 1.

Definition 1. In a two-sided matching algorithm, $(u_i, e_j) \notin \mathcal{M}$ is a blocking pair, if it satisfies.

(1) u_i prefers e_j to their current matching partner $\mathcal{M}(\mathrm{u_i})$.
(2) e_j prefers u_i to at least one of its current matching partners $\mathcal{M}(\mathrm{e_j})$.

4 Algorithm Analysis

4.1 Two-*Sided* *M*atching

Matching game was born in the social economy and has wide range of applications in marketing and social life. Each agent in the matching game has a preference to the other set, and the goal is to find a stable matching. Basically, according to the number of agents, the matching game can be divided into two-sided matching and three-sided matching. In a two-sided matching game, it can be further divided into three categories, namely one-to-one matching, many-to-one matching and many-to-many matching [12].

4.2 GS-Based Two-Sided Matching Algorithm for Intelligent Edge Resource Allocation

Algorithm GS-based Two-sided Matching Algorithm for Intelligent Edge Resource Allocation

Input: U, E
Output: Two-sided stable matching $\mathcal{M}$
1: Initialization and construct the preference lists PL_i^u and PL_j^e;
2: Set $\mathcal{M} = \emptyset, flag = 1$;
3: **while** $flag == 1$ **do**
4: Set $flag = 0$;
5: **for** each $u_i \in \mathcal{U}$ **do**
6: **if** $PL_{i,j}^u \neq \emptyset$ **then**
7: Choose the best ECNode $e_j \in PL_i^u$ as $\mathcal{M}(u_i)$;
8: **if** $\mathcal{N}(\mathcal{M}, u_i) == N_e$ **then**
9: Choose the worst matched user $u_i{}'$ in $\mathcal{M}(e_j)$;
10: **if** $u_i \succ PL_j^e(u_i{}')$ **then**
11: Swap u_i and $u_i{}'$ in $\mathcal{M}(e_j)$
12: **else**
13: $PL_i^u = PL_i^u \setminus e_j$
14: **end if**
15: Set flag = 1.
16: **end if**
17: $\mathcal{M} = \mathcal{M} \cup (u_i, e_j)$;
18: **end if**
19: **end for**
20: **end while**
21: Output Two-sided stable matching $\mathcal{M}$

From two-sided matching to GS-based two-sided matching, two constraints are added. First, each ECNode's preference list PL_j^e comes from a master preference list, which means that all users are sorted in a strict order and PL_j^e is obtained all or part from the master list. The master list sorts users in descending order according to their offer O_i. Users who require more edge service will offer higher price and are more favored by ECNodes. In our case, all ECNodes create the same preference list as.

$$PL_j^e = O_i, \ \forall i \in \{1, 2, \ldots, N\}, \ \forall j \in \{1, 2, \ldots, M\} \tag{11}$$

At the same time, user u_i ranks the acceptable edge computing nodes according to the service quality of e_j, which is measured by $d_{i,j}$. Therefore, users indirectly select edge nodes based on the desired CRBs. We denote each user's preference list as.

$$PL_i^u = -t_i = -\rho d_{i,j}, \ \forall i \in \{1, 2, \ldots, N\}, \ \forall j \in \{1, 2, \ldots, M\} \tag{12}$$

The GS-based two-sided matching algorithm for intelligent edge resource allocation proposed in this paper starts from user u_i. It selects the favorite ECNode e_j in its preference list PL_i^u. If the selected ECNode still has capacity, the pair (u_i, e_j) is added to $\mathcal{M}$ directly. If the ECNode e_j has no capacity ($\mathcal{N}(\mathcal{M}, \mathrm{u_i}) == \mathrm{N_e}$), u_i will be compared with the user $\mathcal{M}(\mathrm{e_j})$ currently matched by e_j. If u_i is better than the worst matched user $\mathcal{M}(\mathrm{e_j})$, then u_i and $\mathcal{M}(\mathrm{e_j})$ are exchanged. When there is no blocking pair or the preference list of all users is empty, the GS-based Two-sided matching algorithm will stop, and a stable matching $\mathcal{M}$ is obtained.

4.3 Stability

Considering the previously mentioned matching $\mathcal{M} \subseteq \mathrm{E} \times \mathrm{U}$, where M is a set of two-tuples from $E \times U$, each agent wants to match the other as a pair, rather than staying with the currently matched partner in $\mathcal{M}$. A matching $\mathcal{M}$ is said to be stable if there exists no blocking pair [15].

Theorem 1: The GS-based two-sided matching algorithm will stop after a limited number of steps and output stable matching result.

Proof: Due to space limitation, detailed proof is omitted here. Similar proof can be referred in [16].

5 Simulation Results

In this section, we compare the proposed GS-based two-sided matching algorithm with the optimal matching approach and random approach using MATLAB. We consider the intelligent edge computing networks with a radius of 1 km, which consists of M = 40 edge nodes and N ∈ [50,210] users.

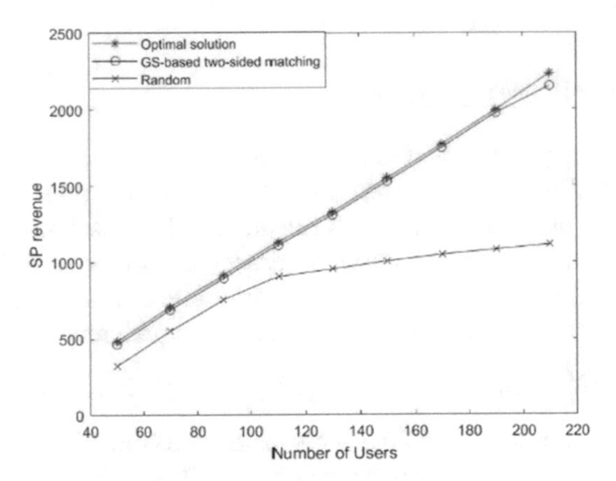

Fig. 2. SP's revenue versus number of users.

Figure 2 shows SP's revenue versus number of users. It can be seen that SP's revenue of our approach is far better than the random algorithm and is approximately optimal. For example, when the number of users is 130, SP's revenue obtained by our approach is 98.4% of the optimal algorithm but 136% of the random method.

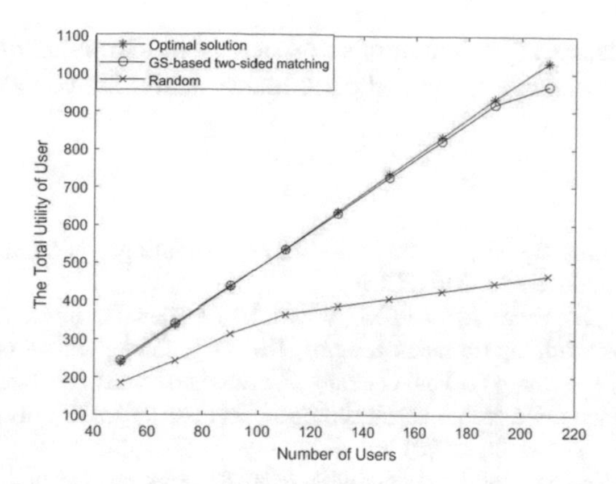

Fig. 3. The total utility of users versus users' number.

Figure 3 shows the total utility of users versus users' number. It can be seen that with more users' requirement satisfied, the total revenue of users increases. Our proposed approach is near-optimal but is significantly better than the random method. For example, when the number of users is 150, the total user utility obtained by our algorithm is 98.2% of the optimal algorithm but 179% of the random approach.

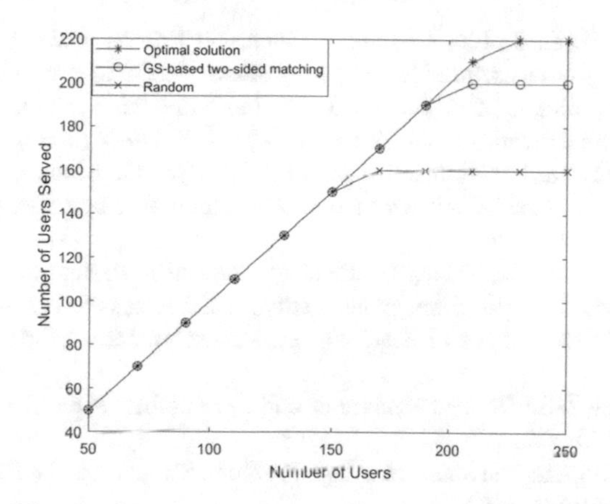

Fig. 4. Number of users served versus users' number.

Figure 4 illustrates the cardinality of the matching result, which indicates the number of users served. It can be seen that the both the optimal algorithm and the proposed approach can serve all users until the number is close to 200. This also indicates the superior performance of our method.

6 Conclusion

In this paper, we propose a matching based intelligent edge resource allocation problem and develop a GS-based two-sided matching algorithm to solve it, which always produces stable matching results in limited number of steps. Simulation results fully show that our approach can achieve near-optimal performance with relative low complexity.

References

1. Shi, W., Cao, J., Zhang, Q., Li, Y., Li, X.: Edge computing: vision and challenges. IEEE Internet Things J. **3**(5), 637–646 (2016)
2. Georgakopoulos, D., Prem, J., Fazia, M., Villari, M., Ranjan, R.: Internet of Things and edge cloud computing roadmap for manufacturing. IEEE Cloud Comput. **3**, 66–73 (2016)
3. Jiang, F., Wang, K., Dong, L., Pan, C., Yang, K.: Stacked autoencoder-based deep reinforcement learning for online resource scheduling in large-scale MEC networks. IEEE Internet Things J. **7**(10), 9278–9290 (2020)
4. Lin, C., Lu, J., Wang, G., Zhou, J.: Graininess-aware deep feature learning for robust pedestrian detection. IEEE Trans. Image Process. 3820–3834 (2020)
5. Zhu, S., Cao, R., Yu, K.: Dual learning for semi-supervised natural language understanding. IEEE/ACM Trans. Audio Speech Lang. Process. **28**, 1936–1947 (2020)
6. Li, J., Gao, H., Lv, H.: Deep reinforcement learning based computation offloading and resource allocation for MEC. In: Proceedings of 2018 IEEE Wireless Communications and Networking Conference (WCNC), pp. 1–6 (2018)
7. Hu, L., Miao, Y., Wu, G., Hassan, M., Humar, I.: iRobot-Factory: an intelligent robot factory based on cognitive manufacturing and edge computing. Future Gener. Comput. Syst. **90**, 569–577 (2018)
8. Yuan, Q., Zhou, H., Li, J., Liu, Z., Yang, F., Shen, X.: Toward efficient content delivery for automated driving services: an edge computing solution. IEEE Netw. **32**(1), 80–86 (2018)
9. Jiang, Y., Ge, H., Wan, C., Fan, B., Yan, J.: Pricing-based edge caching resource allocation in fog radio access networks. Intell. Converg. Netw. **1**, 221–233 (2020)
10. Nath, S., Wu, J.: Deep reinforcement learning for dynamic computation offloading and resource allocation in cache-assisted mobile edge computing systems. Int. Converg. Netw. **1**(2), 181–198 (2020)
11. Chen, J., Zhao, Y., Xu, Z., Zheng, H.: Resource allocation strategy for D2D-assisted edge computing system with hybrid energy harvesting. IEEE Access **8**, 192643–192658 (2020)
12. Manlove, D.F.: Algorithmics of Matching Under Preferences. World Scientific, London (2013)
13. Gale, D., Shapley, L.S.: College admissions and the stability of marriage. Am. Math. Mon. **69**(1), 9–15 (1962)
14. Gusfield, D., Irving, R.: The Stable Marriage Problem: Structure and Algorithms, p. 240. The MIT Press, Cambridge (1989)
15. Cui, L., Jia, W.: Cyclic stable matching for three-sided networking services. Comput. Netw. **57**(1), 351–363 (2013)
16. Qin, P., Fu, Y., Feng, X., Zhao, X., Wang, S., Zhou, Z.: Energy efficient resource allocation for parked cars based cellular-v2v heterogeneous networks. IEEE Internet Things J. (2021). https://doi.org/10.1109/JIOT.2021.3094903

Research on Intrusion Detection Method Based on Deep Convolutional Neural Network

Min Sun, Ni Liu(✉), and Min Gao

School of Computer and Information Technology, Shanxi University, Taiyuan 030006, China
2416728826@qq.com

Abstract. In order to solve the problem of insufficient detection ability of massive network intrusion data and a few attack samples, a deep convolutional neural network model based on Resnet (Residual Network) was proposed for intrusion detection. In the data preprocessing stage, Borderline-SMOTE algorithm is firstly used to artificially synthesize a few attack samples and add them to the data set. In order to meet the input data format of the model, the principal component analysis method combined with bicubic interpolation algorithm is proposed to transform the sample data into a 32*32 gray image matrix form. Finally, the network model is used for classification. NSL-KDD data set is used in the experiment. The results show that compared with the traditional shallow convolutional neural network based on Lenet5, deep belief network algorithm and stacked sparse denoising autoencoder network algorithm, while improving the detection ability of a few types of attack samples, the overall detection rate of network intrusion is improved and the false positives rate is reduced.

Keywords: Network security · Deep convolutional neural nctwork · NSL-KDD data set

1 Introduction

With the rapid development of Internet technology, intrusion detection is another important research direction in addition to firewall technology to ensure network security [1]. In the face of increasingly complex and intelligent network attacks, intrusion detection system monitors and filters network behavior by collecting network traffic information of some key points of the computer system, log files and other data, which can initiate alarm to the computer system in time, so as to achieve the role of protecting network security.

In the face of massive network intrusion data sets, intrusion detection technology based on deep learning is becoming more and more mature. Agarap et al. [2] used gate control unit construction (GRU) to construct the application of cyclic neural network in intrusion detection. Compared with the traditional machine algorithm, the detection accuracy is significantly improved. Alom et al. [3] proposed to apply deep belief network to intrusion detection, In literature [4], Xian Min et al. proposed a deep belief network based on multi-layer limited Boltzmann machine to reduce the dimension of

Q. Liang et al. (Eds.): Artificial Intelligence in China, LNEE 854, pp. 537–544, 2022.
https://doi.org/10.1007/978-981-16-9423-3_67

network intrusion data and classify it with support vector machine, which can improve the detection speed and accuracy of intrusion detection. In literature [5], Javaid et al. proposed a combination algorithm based on deep sparse autocoding network (SAE) and BP in order to reduce data dimensions and remove redundant information. However, such algorithms mostly choose the appropriate number of network layers and neurons based on experience. If they are not set properly, the detection rate will fluctuate within a large range. The convolutional layer of Convolutional Neural Network (CNN) is characterized by parameter sharing and sparse connection, relative to the standard neural network to learn less parameters, and CNN is good at mining the local characteristics of data, as a kind of machine learning model in the case of supervised learning is widely used in computer vision applications, such as image classification, target detection, etc. At present, some researchers have used CNN to extract more abstract and complex features of network intrusion data. In literature [6], Yihan Xiao et al. used PCA (principal component analysis) dimensionality reduction algorithm to remove redundant and irrelevant features and converted the original network intrusion data into a gray image format to build an intrusion detection model with Lenet-5. Solve the problem of high false positive rate and poor generalization ability. In image classification, the performance of convolutional neural network also shows an upward trend as the number of layers of convolutional neural network increases appropriately. Based on this idea, this study builds a deep convolutional neural network model based on RESNET to carry out classification and recognition of network intrusion data, In order to meet the data format of deep convolutional neural network, PCA(principal component analysis) combined with bicubic interpolation algorithm is proposed to transform network traffic data, use PCA combined with double three interpolation algorithm for network traffic data transformation, and in light of the characteristics of network intrusion data sample categories imbalances, By using the improved SMOTE algorithm to expand the sample of a few classes, the detection accuracy of a few attack classes can be significantly improved.

2 Intrusion Detection Model Based on ResNet

The intrusion detection model framework based on RESNET deep convolutional neural network consists of two parts, namely data preprocessing and model training, as shown in Fig. 1.

2.1 Data Processing

A Few Attack Class Samples were Expanded. Due to the uneven data distribution of NSL-KDD dataset, the detection rate of U2R and R2L attack types is low. SMOTE algorithm [7] through the way of interpolation by synthetic new samples of a few categories so as to achieve the aim of improve the unbalanced distribution of data sets, its core idea is to calculate a few categories of each sample point p and k neighbor sample of Euclidean distance, random interpolation between neighboring samples, achieve the purpose of expansion of the minority class samples.

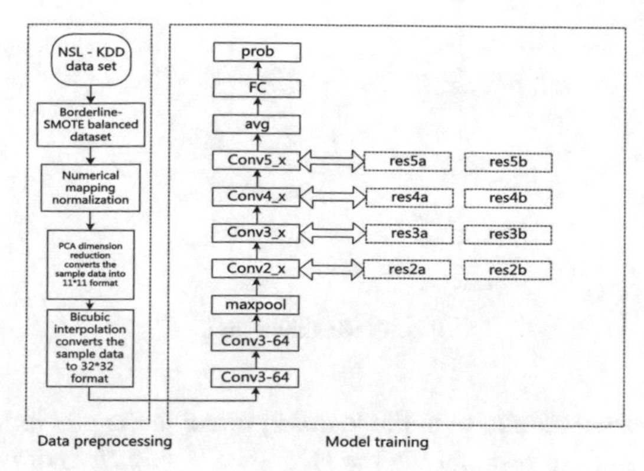

Fig. 1. Intrusion detection model framework

Borderline-SMOTE [8] algorithm is an improvement of SMOTE algorithm. Borderline-SMOTE algorithm focuses on the interpolation expansion of boundary sample advance performance, making the distribution of the synthesized samples of a few classes more reasonable.

Data Format Conversion. In order to meet the data format of convolutional neural network, network intrusion data needs to be transformed. After numerical mapping and normalization of the sample data, each sample data changes from the original 41-dimensional feature to 122-dimensional feature, but the 122-dimensional feature vector cannot be transformed into an N * N matrix form. In order to map the 122-dimensional feature vector to the matrix form of 11*11, PCA algorithm is needed to reduce the dimension.

After PCA dimension reduction operation, the 121-dimensional feature vector is transformed into an 11*11 feature matrix. Since the experimental intrusion detection model needs to go through multiple subsampling operations, the size of the feature matrix is doubled after each subsampling operation, and the 11*11 feature matrix cannot meet the input requirements of the model. The 11*11 eigenmatrix needs to be expanded to the size of 32*32 using bicubic interpolation algorithm.

2.2 Model Design

2.2.1 Residual Unit

Although the BN algorithm solves the problems of gradient disappearance and gradient explosion to a large extent, researchers such as Kaiming He have found through a lot of experiments that as the number of network layers is superimposed, deep neural networks will have performance that is not caused by overfitting. The problem of falling. Based on the performance degradation problem, the residual unit can be used to construct a deep residual neural network to solve it. In the literature [9], the residual unit structure is shown in Fig. 2:

The residual unit "basic_block" in the figure is composed of two 3*3 convolutional layers. Assuming that the input data is x, σ represents the Relu function, and H(x)

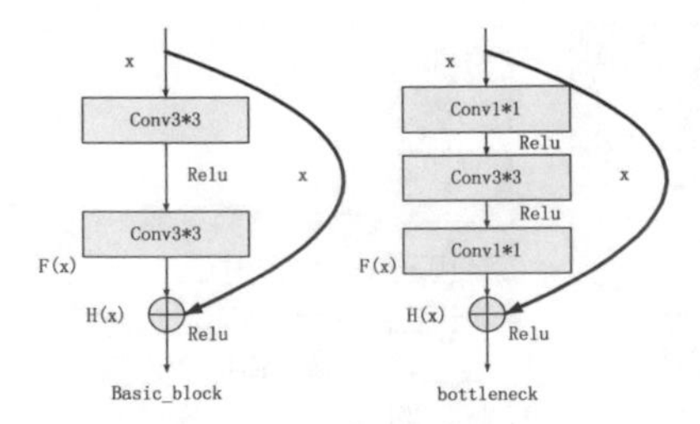

Fig. 2. Residual unit

represents the expected output, then the learning target is the previous complete output H(x) Becomes the current residual F(x) = H(x) − x, x skips two or more layers so that the subsequent layers can directly learn the residuals and perform identity mapping:

$$F(x) = W_2\sigma(W_1x) \tag{1}$$

$$H(x) = F(x, \{w_1, w_2\}) + x \tag{2}$$

Compared with "basic_block", this paper uses another kind of residual unit, "bottleneck", which consists of two 1*1 convolution kernels and a 3*3 convolution kernels. Compared with "basic_block", it can increase the number of network layers and requires fewer parameters to train.

2.2.2 Model Structure

The model in this paper adopts the ResnetV2 proposed by He et al. [10]. In ResNetV2, the residual unit uses the structure of "pre-activation", and the activation function also adopts the identity mapping, and changes the traditional "CONV-BN-RELU" structure. The more easily optimized "Bn-ReLu-Conv" structure is adopted. The configuration parameters of the deep convolutional neural network model based on ResNet are shown in Table 1:

The initial input data size is 32*32*1. Based on the idea of the VGG model [11], two 3*3 small convolution kernels are used instead of the large convolution kernel of the ResNet model 7*7 in the literature [9] to reduce the parameters At the same time, due to the increase of the nonlinear activation layer, the model discrimination ability is improved. The "size" in the table represents the output size after passing through the corresponding layer, and the Relu function is used as the activation function. After multiple convolutional group operations, a global average pooling layer [12] with stronger generalization performance and fewer training parameters is used instead of the fully connected layer, and finally classified by the SOFTMAX activation function.

3 Experiments and Results

3.1 NSL-*K*DD Data Set

The experiment uses an improved version of the KDD99 network intrusion data set, the NSL-KDD data set [13], compared with the KDD99 data set, the NSL-KDD data

Table 1. Model structure settings

Layer	Size	Content
Conv1		3*3, 64, stride = 1
Conv2	32*32	3*3, 64, stride = 1
Maxpool	32*32	2*2, stride = 1, same
Conv3_x	16*16	$\begin{bmatrix} 1*164 \\ 3*364 \\ 1*1256 \end{bmatrix} *2$
Conv4_x	8*8	$\begin{bmatrix} 1*1128 \\ 3*3128 \\ 1*1512 \end{bmatrix} *2$
Conv5_x	4*4	$\begin{bmatrix} 1*1256 \\ 3*3256 \\ 1*11024 \end{bmatrix} *2$
Conv6_x	2*2	$\begin{bmatrix} 1*1512 \\ 3*3512 \\ 1*12048 \end{bmatrix} *2$
Gap	1*1	-
FC	5	-

set deletes a large number of redundant records, so that the classifier will not be biased more frequently record of. The NSL-KDD data set divides the attack types into 4 major categories, The sample distribution of NSL-KDD is shown in Table 2.

Table 2. NSL-KDD sample distribution

Category	Normal	DOS	Probe	R2L	U2R	Total
Train+	67343	45927	11656	995	52	125973
Test+	9711	7458	2421	2754	200	22544

3.2 The Evaluation Index

The evaluation indicators of the experiment mainly include classification accuracy (A), detection rate (DR), and false alarm rate (RFP):

$$A = \frac{TP + TN}{TP + TN + FP + FN} \tag{3}$$

$$DR = \frac{TP}{FN + TP} \tag{4}$$

$$RFP = \frac{FP}{TN + FP} \tag{5}$$

TP indicates that the model classification result is attack type data and the classification result is correct; TN indicates that the model classification result is normal data, that is, the label is "Normal" and the classification result is correct; FP indicates that the model classification result is attack data but the classification result is incorrect; FN indicates the model classification result It is normal data but the classification result is wrong.

3.3 Experimental Results and Analysis

The ResNet-based intrusion detection model uses the batch gradient descent method to adjust the weights of back propagation, and uses the Adam algorithm for optimization. The initial learning rate is set to 0.01, the batch size is set to 256, and the number of iterations (epoch) is set to 40. The experiment uses python3.7 programming language, tensorflow2.0 deep learning framework, etc. for coding, tensorboard1.14.0 visualization tool observes and tracks loss, acc and other scalars, and after training, prints the confusion matrix of the test set, and calculates A, DR, RFP and other indicators, and compare other models.

3.3.1 Global Average Pooling Layer Instead of Fully Connected Layer

After multiple convolutional group operations, use GAP (Global Average Pooling Layer) instead of FC (Fully Connected Layer). It can be found that the use of GAP makes the training parameters less, and all the spatial information is summarized, avoiding the problem of overfitting. Compared with using FC Reduce training time. The calculation of ACC and training time (unit: minutes) takes the average of the last 5 epochs, which is known from Table 3.

Table 3. GAP and FC performance comparison

Layer	A	Time
GAP	78.78%	79
FC	78.54%	83

3.3.2 Influence of Borderline-SMOTE Algorithm on Detection Performance

The detection results are shown in Fig. 3. After the Borderline-SMOTE algorithm was used to expand the training data set, the detection accuracy of Probe attack class decreased slightly, but the accuracy of R2L and U2R attack class was significantly improved.

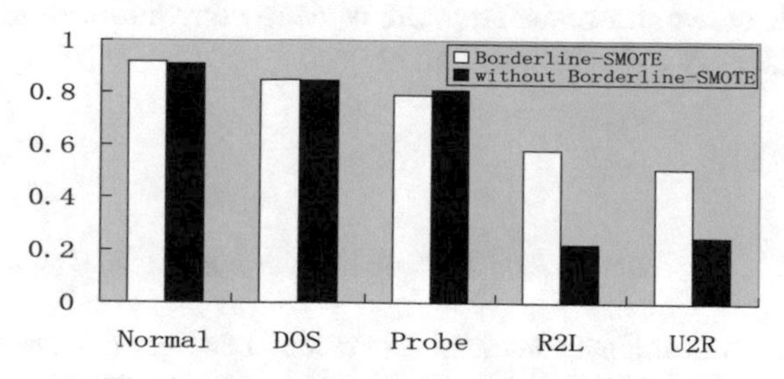

Fig. 3. Comparison of experimental results

3.3.3 Comparison of Existing Algorithms

By using the confidence network (DBN), deep sparse self-coding network (SAE) and PCA-Lenet5 training model, the number of pre-training iterations and fine-tuning iterations is set to 100, which is compared with the intrusion detection model based on RESNET in this paper. Without using Borderline-SMOTE algorithm to process the data set, the detection rate increased by 5.8% on average, and the false positive rate decreased by 0.6% on average. After using Borderline-SMOTE, the detection rate increased by 12.6% on average, and the false positive rate decreased by 1.5% on average. The results are shown in Table 4.

Table 4. Comparative experimental results

Algorithm	A	DR	RFP
DBN	75.36	66.21	5.04
SAE	76.54	66.77	5.06
PCA-LeNet5	76.47	67.01	4.92
Without Borderline-SMOTE-ResNet	78.78	72.49	4.94
Borderline-SMOTE-ResNet	80.51	80.78	4.84

4 Conclusion

This paper proposes to build an intrusion detection model of deep convolutional neural network based on residual units. Firstly, Borderline-SMOTE algorithm is used to process unbalanced data sets to improve the accuracy of a few attacks. Secondly, residual units are used to solve the problem of deep network performance degradation. The detection accuracy is further improved. Since the NSL-KDD data set is old and cannot accurately reflect the current network environment, the next research work can use the standard data set covering the current latest network attacks for research. Secondly, based on the idea of multi-scale convolution, the structure of deep convolutional neural network can be broadened to identify network attacks.

5 References

1. Qing, S., Jiang, J., Ma, H., et al.: Review of intrusion detection techniques. J. Commun. **25**(7), 19–28 (2004)
2. Agarap, A.F.M.: A neural network architecture combining gated recurrent unit (GRU) and support vector machine (SVM) for intrusion detection in network traffic data. In: Proceedings of the 2018 10th International Conference on Machine Learning and Computing, USA, pp. 26–30. ACM Press (2018)
3. Alom, M.Z., Bontupalli, V.R., Taha, T.M.: Intrusion detection using deep belief networks. In: 2015 National Aerospace and Electronics Conference (NAECON), USA, pp. 339–344. IEEE Press (2015)
4. Zhang, K., Xian, M.: Research on hybrid intrusion detection model based on DBN and TSVM. Comput. Appl. Softw. **35**(5), 319–323 (2018)
5. Javaid, A., Niyaz, Q., Sun, W., et al.: A deep learning approach for network intrusion detection system. In: Proceedings of the 9th EAI International Conference on Bio-inspired Information and Communications Technologies (formerly BIONETICS), pp. 21–26. ICST (Institute for Computer Sciences, Social-Informatics and Telecommunications Engineering) (2016)
6. Xiao, Y., Xing, C., Zhang, T., et al.: An intrusion detection model based on feature reduction and convolutional neural networks. IEEE Access **7**, 42210–42219 (2019)
7. Chawla, N.V., Bowyer, K.W., Hall, L.O., et al.: SMOTE: synthetic minority over-sampling technique. J. Artif. Intell. Res. **16**, 321–357 (2002)
8. Yang, Y., Lu, C., Xu, G.: An improved borderline-SMOTE method for unbalanced data sets. J. Fudan Univ.: Nat. Sci. (5), 537–544 (2017)
9. He, K., Zhang, X., Ren, S., et al.: Deep residual learning for image recognition. In: Proceedings of the IEEE Conference on Computer Vision and Pattern Recognition, pp. 770–778 (2016)
10. He, K., Zhang, X., Ren, S., et al.: Identity mappings in deep residual networks. In: Leibe, B., Matas, J., Sebe, N., Welling, M. (eds.) Computer Vision, vol. 9908, pp. 630–645. Springer, Cham (2016). https://doi.org/10.1007/978-3-319-46493-0_38
11. Simonyan, K., Zisserman, A.: Very deep convolutional networks for large-scale image recognition. arXiv preprint arXiv:1409.1556 (2014)
12. Lin, M., Chen, Q., Yan, S.: Network in network. arXiv preprint arXiv:1312.4400 (2013)
13. Canadian Institute for Cybersecurity. NSL-KDD dataset [EB/OL], 18 August 2017

Asynchronous Multi-sensor Data Fusion with Updating Bias Correction

Ping Han[1,2], Zhibo Ni[1,2], Wanwei Wang[1,2], Xiaoguang Lu[1,2], and Zhe Zhang[1,2(✉)]

[1] College of Electronic Information and Automation, Civil Aviation University of China, Tianjin 300300, China
z-zhang@cauc.edu.cn

[2] Tianjin Key Laboratory, Civil Aviation University of China, Tianjin 300300, China

Abstract. Multi-sensor data fusion is capable of providing more reliable and accurate estimations of targets than that from a single sensor. However, the performance of multi-sensor fusion is greatly affected by updating intervals, especially in asynchronous scenarios. A method correcting updating bias for asynchronous multi-sensor data fusion is proposed. Linear resampling for updating interval is applied in time domain for each individual surveillance source, then the state estimation of a target from a single source is given by Kalman filtering. The fused state estimation of each target then can be obtained under the condition of the optimized minimum of the system error covariance. Simulation results show that the proposed method is feasibility and effectiveness for asynchronous multi-sensor data fusion problem.

Keywords: Multi-sensor data fusion · Variable updating interval · ADS-B · Beidou RDSS

1 Introduction

Multi-sensor data fusion is a kind of technology to integrate the data from multiple surveillance sources, and derive a comprehensive estimation of them [1, 2]. Uncertainty of data obtained by each sensor then can be effectively eliminated. The data fusion framework prevent system from vulnerability since the final fusion results do not solely depend on the measurements from any single sensor. Moreover, in the multi-sensor fusion system, multiple sensors detect one target at same time, which makes the detected data redundant and complementary. Global Flight Tracking (GFT) is airlines obtaining the 4D position (longitude, latitude, altitude, and timestamp) of a flight during its operation everywhere on this planet [3]. One of the candidates for global flight tracking is the ADS-B (Automatic Dependent Surveillance - Broadcast). In order to improve the reliability for tracking a flight that use the ADS-B only as its data source, an airline can also introduce Beidou RDSS into its ground tracking system and fuse data from these two independent tracking sources.

Multi-source data fusion is widely applied in various fields such as integrated navigation systems, target tracking, and etc. [4–7], however, there are still some difficulties

Q. Liang et al. (Eds.): Artificial Intelligence in China, LNEE 854, pp. 545–553, 2022.
https://doi.org/10.1007/978-981-16-9423-3_68

need to be addressed in real world applications, such as data registration, fusion model matching and so on [8]. Gan Q used distributed fusion method with two methods for track fusion [9]. One was simply merging the observation data of multi-sensor; the other was merging the multi-sensor data based on the least mean square error estimation. Chong C used simple convex combination and decorrelated information for distributed fusion [10]. And Chang K proposed information matrix filtering fusion method [11]. However, they did not take the optimization of system and observation noise in track fusion into consideration. Given that, Sun S proposed a two-layer fusion structure [12]. Firstly, the cross-covariance matrix between any two sensors was determined by using mesh parallel structure, and then track fusion was carried out. In addition, the optimal adaptive information fusion Kalman multi-step prediction was proposed by Sun S [13]. Based on the linear least square weight, the information was combined to find the sensor and the related noise, and then the data is fused. At the same time, Hu proposed the application of two-tier fusion structure [14], which was based on the method proposed by Qi [15]. The cross-covariance matrix between the pseudo-measurement noise is calculated by constructing the state-independent pseudo-measurement of the sensor deviation. Under the premise of not ignoring the correlation between process noise and pseudo-measurement noise, the global deviation estimation based on Kalman filter is obtained. As previously discussed, both synchronous and asynchronous sensors default that the updating time of data is uniform. Therefore, Yan presented an asynchronous multi-rate multi-sensor state fusion estimation algorithm which was different from the lowest rate fusion, the use of multiscale system theory [16]. In comparison to conventional fusion systems, it successfully removes irregular updating periods, communication delay, random observation and other negative factors, instead, directly performed data fusion, while the prerequisite is the stable minimum updating period. Unfortunately, the sensor is not stable enough to obtain data.

This paper offers a practical solution to asynchronous multi-sensor data fusion with updating bias, by providing an algorithm for improving the reliability and stability of the obtained data by correcting updating bias. Theoretically, the estimate is proven to be linear unbiased, and simulation results show that the proposed method is feasibility and effectiveness for asynchronous multi-sensor data fusion scenarios.

This paper is organized as follows: in Sect. 2 the system modeling is presented. Section 3 presents updating bias correction. Then, Sect. 4 provides the State fusion estimation algorithm. Finally, Sect. 6 demonstrates the simulation results.

2 System Modeling

For track fusion, there are two structures [17]. One is sensor-to-sensor fusion structure and the other is sensor-to-system fusion structure.

The Fig. 1 shows the sensor-to-sensor fusion framework. The upper and lower circles represent track nodes of two different sensor. From the left to right in the figure represents the direction of time. The fusion track is generated by the fusion of the tracks from different sensors at the fusion node.

The Fig. 2 shows the sensor-to- system fusion framework. Whenever the fusion node receives a track point, the fusion algorithm will fuse the information into the system track to obtain the current system track state estimate and form the system track.

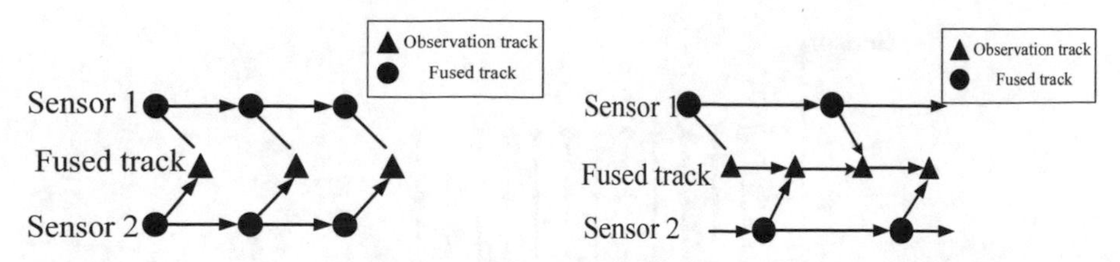

Fig. 1. Sensor-to-sensor fusion structure **Fig. 2.** Sensor-to-system fusion structure

This article will use a sensor-to-sensor framework, because the acquired track updating time is relatively fixed. In addition, the flight motion model of the carrier aircraft is relatively stable, and there will be no sudden changes in speed or heading.

Based on the sensor-to-sensor fusion structure, this paper proposed an updating bias correction fusion system model. Firstly, correct the acquired track which has updating bias. Namely, Linear resampling for updating interval is applied in time domain for each individual surveillance source. When the updating time of acquired track is different from the standard updating time, the updating position should be extrapolated or interpolated to the standard updating time. Additionally, the track with large updating interval should be interpolated, in doing so, the track has corresponding position information at fusion node; subsequently, in order to make the acquired track smoother, Kalman prediction and updating formulas are adopted to filter the track; the final step is to put the smoothed track into the sensor-to-sensor fusion system for track fusion. In this way, the fused state estimation of the target can be obtained.

3 Updating Bias Correction

As is known to all, the updating between different sensors can be asynchronous, it is beyond dispute that sensor as it shows in Fig. 3. Sensor 3 updates data at each updating point; Sensor 2 updates data at every two updating points; and Sensor 1 updates data at every three updating points. developing an algorithm which can fuse the measurements observed by all the sensors and generating the optimal estimation are the objective of this paper.

Assuming that the time updating interval of the sensor is known, the third point position of the target can be obtained by extrapolating the point trace of the first two intervals of the target. The position information which is not at the updating point should be calculated to the updating point, so that a smoother track can be obtained when the track is Kalman filtered. In this case, the motion of the target can only be the first order model, that is, the target is in the state of uniform linear motion. And it should be noted that the model is an ideal model without considering noise or interference [18].

Suppose the first measurement of the target is $\mathbf{Z}_a$, the second measurement is $\mathbf{Z}_{a+1}$, the position coordinate points are (x_a, y_a) and (x_{a+1}, y_{a+1}), respectively. According to the object movement equation, the predicted or extrapolated value of the third point is $\mathbf{Z}_{i+2}$, Its coordinates are:

$$\begin{cases} x_{a+2} = x_{a+1} + V_x(a+1) \times T \\ y_{a+2} = y_{a+1} + V_y(a+1) \times T \end{cases} \quad (1)$$

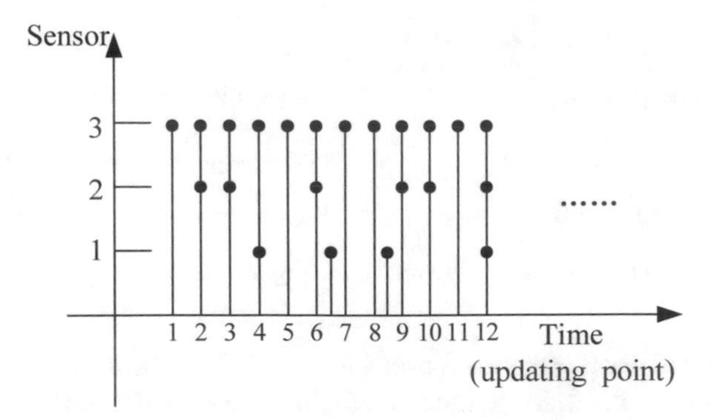

Fig. 3. Sketch map of multi-sensor with updating bias

where T represents the time updating interval of the sensor; $V_x(a+1)$ denotes the first measurement speed of the target on the X axis; $V_y(a+1)$ denotes the second measurement speed of the target on the Y axis. The velocities of the target are:

$$\begin{cases} V_x(a+1) = (x_{a+1} - x_a)/T \\ V_y(a+1) = (y_{a+1} - y_a)/T \end{cases} \tag{2}$$

So, under the condition of known updating bias, the position information obtained by extrapolation method is:

$$\begin{cases} x_a^- = x_{a-1}^- + (x_a - x_{a-1}^-) \times \frac{T}{T+\Delta t_a} \\ y_a^- = y_{a-1}^- + (y_a - y_{a-1}^-) \times \frac{T}{T+\Delta t_a} \end{cases} \tag{3}$$

where x_a^- represents the position information after correction; x_a denotes the position information before correction; Δt_a denotes the time updating bias at the point k.

4 State Fusion Algorithm

The aim of fusion estimation is to find optimal estimation $\hat{\mathbf{x}}$ and error covariance matrix $\mathbf{P}$. In the sensor-to-sensor fusion structure, the tracks should come from two different sensors [17, 19].

Supposing that there are two tracks observed by sensors i and j, which have state estimation $\hat{\mathbf{x}}_i$, $\hat{\mathbf{x}}_j$, error covariance $\mathbf{P}_i$, $\mathbf{P}_j$ and cross covariance matrix $\mathbf{P}_{ij} = \mathbf{P}_{ji}^T$ respectively. The corresponding state estimations error which are independent are given by:

$$\begin{cases} \tilde{\mathbf{x}}_i = \mathbf{x} - \hat{\mathbf{x}}_i \\ \tilde{\mathbf{x}}_j = \mathbf{x} - \hat{\mathbf{x}}_j \end{cases} \tag{4}$$

The static linear estimation equation of the posterior mean $\hat{\mathbf{x}}$ calculated by the prior mean $\overline{\mathbf{x}}$ and the measured value $\mathbf{Z}$ can be written as:

$$\hat{\mathbf{x}} = \mathbf{x} + \mathbf{P}_{XZ}\mathbf{P}_{ZZ}^{-1}(\mathbf{Z} - \overline{\mathbf{Z}}) \tag{5}$$

Moreover, if the information obtained by sensor i is regarded as observation data D_i and the time parameter is omitted, the measurement $\hat{\mathbf{x}}_j = \mathbf{x} - \overline{\mathbf{x}}_j$ can represent as

D_j. Then it can be proved $\mathbf{P}_{XZ} = \mathbf{P}_i, \mathbf{P}_{ZZ} = \mathbf{P}_i + \mathbf{P}_j$. By doing so, formula (5) can be specifically expressed as:

$$\hat{\mathbf{x}} = \hat{\mathbf{x}}_i + \mathbf{P}_i(\mathbf{P}_i + \mathbf{P}_j)^{-1}(\hat{\mathbf{x}}_j - \hat{\mathbf{x}}_i) \tag{6}$$

When the cross covariance between the two estimates can be ignored, namely, $\mathbf{P}_{ij} = \mathbf{P}_{ji} \approx 0$. In this situation, the track fusion algorithm can be given by the following formula.

State estimation of systems:

$$\hat{\mathbf{x}} = \mathbf{P}_j(\mathbf{P}_i + \mathbf{P}_j)^{-1}\hat{\mathbf{x}}_i + \mathbf{P}_i(\mathbf{P}_i + \mathbf{P}_j)^{-1}\hat{\mathbf{x}}_j = \mathbf{P}(\mathbf{P}_i^{-1}\hat{\mathbf{x}}_i + \mathbf{P}_j^{-1}\hat{\mathbf{x}}_j)^{-1} \tag{7}$$

System error covariance:

$$\mathbf{P} = \mathbf{P}_i(\mathbf{P}_i + \mathbf{P}_j)^{-1}\mathbf{P}_j = (\mathbf{P}_i^{-1} + \mathbf{P}_j^{-1})^{-1} \tag{8}$$

This simple convex combination algorithm has been used extensively because of its simple implementation. It is optimal if the cross-covariance of state estimation of two tracks can be ignored. Otherwise, it is only sub-optimal and **P** is not the actual error covariance. In addition, it can be seen from (4) and (5) that if the fusion system is composed of N sensors, it is easy to extend it to the general form.

State estimation:

$$\hat{\mathbf{x}} = \mathbf{P}(\mathbf{P}_1^{-1}\hat{\mathbf{x}}_1 + \mathbf{P}_2^{-1}\hat{\mathbf{x}}_2 + \cdots + \mathbf{P}_n^{-1}\hat{\mathbf{x}}_n) = \mathbf{P}\sum_{i=1}^{n}\mathbf{P}_n^{-1}\hat{\mathbf{x}}_n \tag{9}$$

Estimated weight of each sensor:

$$\mathbf{W}_k^i = \mathbf{P}\mathbf{P}_i^{-1} \tag{10}$$

error covariance:

$$\mathbf{P} = (\mathbf{P}_1^{-1} + \mathbf{P}_2^{-1} + \cdots + \mathbf{P}_n^{-1})^{-1} = (\sum_{i=1}^{n}\mathbf{P}_i^{-1})^{-1} \tag{11}$$

It can be seen from the formula (9) that the error covariance of the observation track needs to be used to obtain the weight of the fusion during track fusion. Therefore, this paper used Kalman filtering to obtain the error covariance of the track [20].

Consider a dynamic target tracked by N sensors. Target dynamics and sensors are showed by following models:

$$\mathbf{x}(s, k+1) = \mathbf{A}(s, k)\mathbf{x}(s, k) + \mathbf{w}(s, k) \tag{12}$$

$$\mathbf{z}(s, k) - \mathbf{C}(s, k)\mathbf{x}(s, k) + \mathbf{v}(s, k) \qquad s = 1, 2, \ldots, N \tag{13}$$

where k represents the discrete-time index; $\mathbf{x}(s, k)$ is the k-th state vector for sensor s; $\mathbf{z}(s, k)$ is the k-th measurement observed by sensor s; $\mathbf{A}(s, k)$ is the systems transition matrix; $\mathbf{C}(s, k)$ is the design matrix; $\mathbf{w}(s, k)$ and $\mathbf{v}(s, k)$ are the zero-mean white Gaussian noises with covariance matrices $Q(k)$ and $R_s(k)$, respectively. In addition, it is assumed that $\mathbf{w}(s, k)$ and $\mathbf{v}(s, k)$ are independent of each other.

The Kalman filter provides a linear unbiased. The Kalman filter prediction model can be described by the following equations:

$$\hat{\mathbf{x}}(s, k \mid k-1) - \mathbf{A}(s, k)\hat{\mathbf{x}}(s, k-1 \mid k-1) \tag{14}$$

$$\mathbf{P}(s, k \mid k-1) = \mathbf{A}(s, k)\mathbf{P}(s, k-1 \mid k-1)\mathbf{A}^T(s, k) + Q(s, k) \tag{15}$$

The Kalman filter estimate model can be described as:

$$\mathbf{K}(s, k) = \mathbf{P}(s, k \mid k-1)\mathbf{C}^T(s, k)[\mathbf{C}(s, k)\mathbf{P}(s, k \mid k-1)\mathbf{C}^T(s, k) + R(k)]^{-1} \tag{16}$$

$$\hat{\mathbf{x}}(s, k \mid k) = \hat{\mathbf{x}}(s, k \mid k-1) + \mathbf{K}(s, k)[\mathbf{z}(s, k) - \mathbf{C}(s, k)\hat{\mathbf{x}}(s, k \mid k-1)] \tag{17}$$

$$\mathbf{P}(s, k \mid k) = [I - \mathbf{K}(s, k)\mathbf{C}(s, k)]\mathbf{P}(s, k \mid k-1) \tag{18}$$

where $\hat{\mathbf{x}}(s, k|k)$ represents the estimate of the state-vector for sensor s observation, $\mathbf{P}(s, k|k)$ is the state estimate covariance matrix under sensor s observation, and $\mathbf{K}(s, k)$ is the corresponding Kalman gain matrix.

5 Simulation Studies

There are 15 operations in providing both ADS-B and Beidou RDSS 4D tracking data for demonstrating the proposed method. The nominal updating interval of Beidou RDSS for operation tracking 4D messages is 60 s, which the ruminal updating interval ADS-B position reports is 15 s. The standard deviation of the interval for the Beidou RDSS dataset is 5 s and the standard deviation of ADS-B position message interval is 13 s.

CV model was used in the experiment. uniform velocity was set to 220 m/s. The dt in the systems transition matrix is the time updating period. The system state vector is $\mathbf{x} = [x, v_x]$, note that x represents the position of the target at updating time k and v_x represents the velocity of the target at updating time k. $\mathbf{z}(s, k)$ is the measurement observed by two sensors, which observe the velocity and the position. $\mathbf{v}(s, k)$ is the zero-mean Gaussian noise with variances $R_s(k)$ which is 5. $\mathbf{w}(t)$ is the zero-mean Gaussian noise with variances $Q(k)$ which is 100. the covariance matrix of initial state estimation $\mathbf{P}$ is $\begin{bmatrix} 1 & 0 \\ 0 & 1 \end{bmatrix}$. Moreover $\mathbf{C}(1) = [1 \;\; 0]$, $\mathbf{C}(2) = [1 \;\; 0]$.

In this experiment, two sensors were used to observe the same target. The observation time is 2910 s, and the fusion interval is 30 s. Our goal is to have the estimation $\hat{\mathbf{X}}$ by fusing the data from two sensors. And the root mean square error is calculated by the following formula:

$$RMSE = \sqrt{\frac{\sum_{i=1}^{n} \left(X_{estimation,i} - X_{model,i}\right)^2}{n}} \tag{19}$$

Where n represents the numbers of all points in a track, $X_{estimation,i}$ represents the estimated value of the i-th point on track, and $X_{model,i}$ is the model value of the i-th point on track (Figs. 4 and 5).

Table 1. RMSE of different methods (The Monte Carlo simulation method repeatedly 500 times)

Tracking mode	Sensor **1**	Sensor **2**	The proposed method	The method in [16]
RMSE with correcting updating bias	4.913 m	4.9783 m	4.365 m	428.217 m
RMSE with updating bias	4.943 m	5.194 m	883.613 m	716.488 m

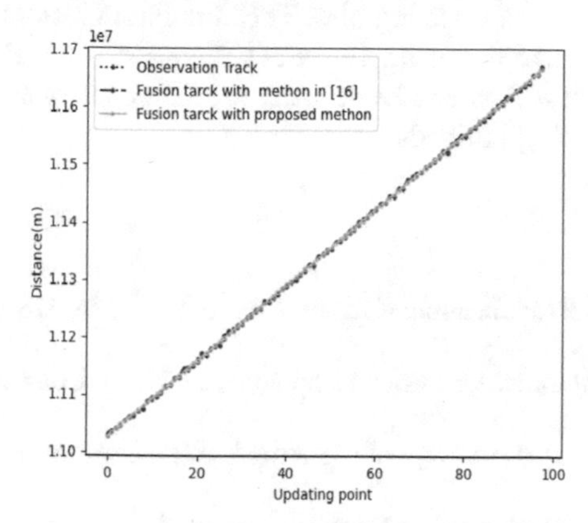

Fig. 4. Simulated track (overall)

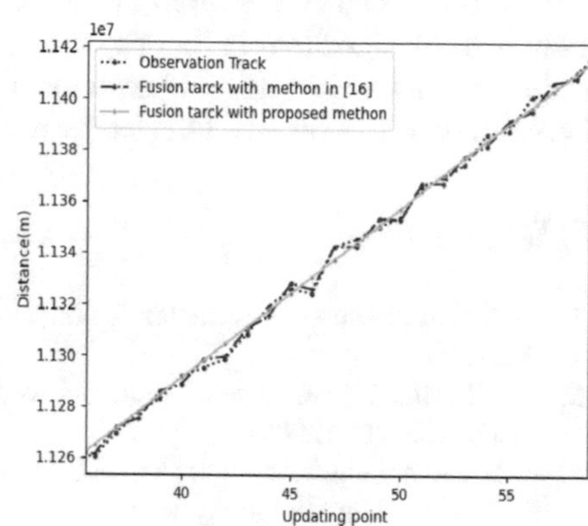

Fig. 5. Simulated track (part)

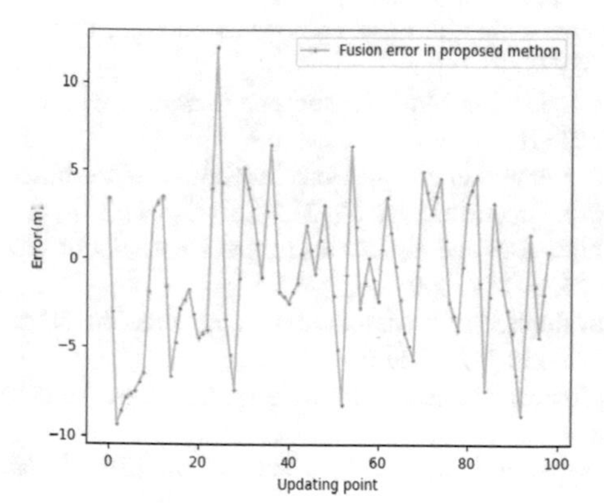

Fig. 6. Fusion error this paper

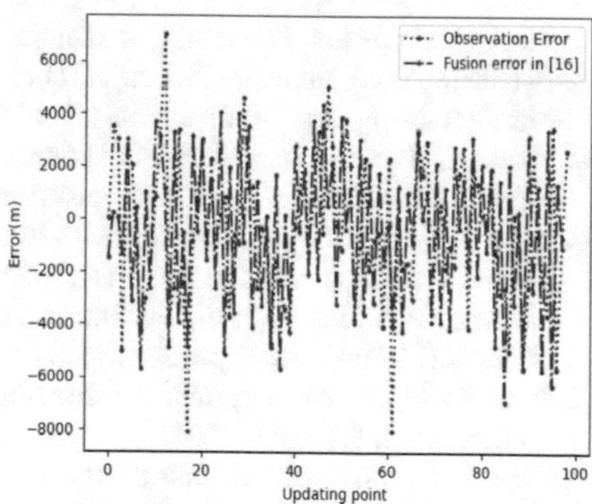

Fig. 7. Observation error and fusion erroring this paper [16].

From the Table 1, we find that the performance of multi-sensor fusion is greatly affected by updating intervals. The value of the RMSE will increase if the updating intervals have biases. And comparing the Fig. 6, 7 and last two columns of Table 1, we can see that the RMSE with correcting updating bias created by fusing sensors 1 and 2 by using of the proposed algorithm are smaller than those by using of Yan's [16]. It shows that the proposed algorithm is more effectual than Yan's.

6 Conclusions

This paper presented a general while critical view of multi-sensor data fusion. If an asynchronous multi-sensor state fusion within the updating bias, the reliability and stability of track fusion can be improved by correcting the updating bias. The simulation results show that the algorithm in this paper can make the estimation track closer to the real value. Therefore, it is an urgent need to witness further advances in the emergence of an algorithm which cannot be affected by updating intervals.

References

1. Li, S.: Multi-sensor information fusion and its application: a survey. Control Decis. **16**, 518–522 (2001)
2. Jia, H., Shi, L.: Advance on multi-sensor information fusion technologies. Chin. J. Constr. Mach. 227–232 (2009)
3. International Civil Aviation Organization (2017). Concept of Operations - Global Aeronautical Distress and Safety System
4. Tan, W., Lu, B., Zheng, L., Ma, H., Hu, W.: Research on ship track fusion based on genetic particle swarm optimization. J. Chongqing Jiaotong Univ. (Nat. Sci. Ed.) **29**, 828–831 (2010)
5. Wu, S., Zheng, Z.: Situation assessment of ship network threat based on multi-source data fusion. Ship Sci. Technol. 163–165 (2021)
6. Zhu, W., Li, R., Zhang, X.: The application of multi-source data fusion to blast target location. J. Projectiles Rockets Missiles Guidance. **31**, 203–206 (2011)
7. Duan, H., Zhang, B., Yang, C.: A spatial bias estimation and compensation algorithm for target track fusion. Electron. Opt. Control 25–28 (2021)
8. Khaleghi, B., Khamis, A., Karray, F.O., Razavi, S.N.: Multi-sensor data fusion: a review of the state-of-the-art. Inf. fusion **14**, 28–44 (2013)
9. Gan, Q., Harris, C.: Comparison of two measurement fusion methods for Kalman-filter-based multi-sensor data fusion. IEEE Trans. Aerosp. Electron. Syst. **37**, 273–280 (2001)
10. Chong, C., Mori, S., Barker, W., et al.: Architectures and algorithms for track association and fusion. IEEE Aerosp. Electron. Syst. Mag. **15**, 5–13 (2000)
11. Chang, K.C., Tian, Z., et al.: Performance evaluation of track fusion with information matrix filter. IEEE Trans. Aerosp. Electron. Syst. **38**, 455–466 (2002)
12. Sun, S.: Multi-sensor optimal information fusion Kalman filters with applications. Aerosp. Sci. Technol. **8**, 57–62 (2004)
13. Sun, S.: Optimal and self-tuning information fusion Kalman multi-step predictor. IEEE Trans. Aerosp. Electron. Syst. **43**, 418–427 (2007)
14. Hu, Y.Y., Zhou, D.H.: Bias fusion estimation for multi-target tracking systems with multiple asynchronous sensors. Aerosp. Sci. Technol. **27**, 95–104 (2013)
15. Qi, Y.Q., Jing, Z.L., Hu, S.Q.: General solution for asynchronous sensors bias estimation. In: Proceedings of the 11th International Conference on Information Fusion, pp 258–264 (2008)

16. Yan, L.P., Liu, B.S., Zhou, D.H.: The modeling and estimation of asynchronous multi-rate multisensory dynamic systems. Aerosp. Sci. Technol. **10**, 63–71 (2006)
17. Yang, W.: Multi-sensor Data Fusion and its Application. Xidian University Press, Xi'an (2004)
18. Yan, C.: Research on time-alignment in error-correction. In: Electronic Information Warfare Technology, pp 13–17 (2003)
19. Chong, C.Y., Mori, S., Chang, K.C., et al.: Architectures and algorithms for track association and fusion. IEEE Trans. Aerosp. Electron. Syst. **15**, 5–13 (2000)
20. Peng, D.C.: Basic principle and application of Kalman filter. In: Software Guide. pp 32–34 (2009)

Research and Practice of Video Recognition Based on Deep Learning

Jie Ren[1], Heping Shi[2], and Jihua Cao[1](✉)

[1] School of Electronic Engineering, Tianjin University of Technology and Education, Tianjin 300222, China
caojihua@sina.com

[2] College of Automobiles and Transportation, Tianjin University of Technology and Education, Tianjin 300222, China

Absrtact. This paper used tensorflow and keras library to build a deep learning environment. Designed and established a deep 3D convolutional network model and Long Short-Term Memory network model, using UCF-101 dataset known category videos as training samples to train the network. Some videos in the dataset were used as test samples to verify the recognition performance of the network model and realize classification. Finally, Tensorboard was used to visually analyze the network training process. The experimental results show that the model has better video recognition performance.

Keywords: Deep 3D convolutional network · Long short-term memory · Tensorboard · Video recognition

1 Introduction

In recent years, researchers have proposed many research directions and methods in the field of video recognition [1–3]. Traditional video analysis methods mostly use HOG [4] and SIFT [5] to extract the apparent features of static information when extracting features in action classification, encode the features, and then input the video features into a support vector machine (SVM) to obtain Action category. However, it is difficult to have better results for behavior recognition in complex environments.

Literature [6, 7] proposed to use 3D convolutional neural network (3D CNN) to extract video content features while extracting video time features, which is a classic model in video classification and behavior recognition. Obviously it is better than the traditional network model, but the video needs to be segmented, and the segmented segment must be used as the input of the 3D CNN. Later, Tran et al. proposed a deep 3D CNN model (C3D). Compared with the traditional network model, the C3D model has a deeper network structure and can take a complete video frame as input without relying on any preprocessing, so it is more suitable for large-scale data sets [8].

Based on the establishment of a deep learning development environment, this paper establishes a C3D and LSTM network model, inputs the training video into the C3D network to extract the features of the video segment, and trains the C3D network. Then

Q. Liang et al. (Eds.): Artificial Intelligence in China, LNEE 854, pp. 554–560, 2022.
https://doi.org/10.1007/978-981-16-9423-3_69

input the features of the video clips extracted by C3D into the LSTM network to extract the overall features of the video, and complete the recognition and classification of human actions.

2 Related Theories

2.1 3D Convolutional Neural Network Structure

Compared with 2D convolutional neural networks, the use of 3D CNN can better capture the feature information of time and space dimensions in the video. 3D CNN forms a cube by stacking multiple consecutive frames, and then uses a 3D convolution kernel in the cube. Each feature map in the convolutional layer is connected to multiple adjacent consecutive frames in the previous layer to capture motion information. The formula of three-dimensional convolution as follow:

$$V_{ij}^{xyz} = \tanh(b_{ij} + \Sigma_m \Sigma_{p=0}^{P_i-1} \Sigma_{q=0}^{Q_i-1} \Sigma_{r=0}^{R_i-1} W_{ijm}^{pqr} V_{(i-1)m}^{(x+p)(y+q)(z+r)} \tag{1}$$

W_{ijm}^{pqr} represents the weight of the position (p, q, r) in the m-th picture of the j-th feature block in the i-th laye;.R_i represents the size in the time dimension.

2.2 Long and Short-Term Memory Neural Network Structure (LSTM)

LSTM (Long Short-Term Memory), a long and short-term memory network can effectively solve the problems of gradient disappearance and gradient explosion that occur when recurrent neural networks process too long time series. The core of this processor is the three doors: forget gate, Input gate, output gate. The definition of LSTM neuron is as follows:

$$i_t = \sigma(W_{xi}x_t + W_{hi}h_{t-1} + b_i) \tag{2}$$

$$f_t = \sigma(W_{xf}x_t + W_{hf}h_{t-1} + b_f) \tag{3}$$

$$o_t = \sigma(W_{x0}x_t + W_{h0}h_{t-1} + b_o) \tag{4}$$

$$g_t = \sigma(W_{xg}x_t + W_{hg}h_{t-1} + b_g) \tag{5}$$

$$C_t = f_t \cdot C_{t-1} + i_t \cdot g_t) \tag{6}$$

$$h_t = o_t \phi(C_t) \tag{7}$$

3 Model Architecture

3.1 Overall Framework

This article used C3D and LSTM network to classify the actions in the video. It mainly includes the following four steps: ① Set up a TensorFlow environment in Anaconda software, build the neural network model with python language and tensorflow library functions in the built environment. ② Input the video into the C3D network, extract the features of the video segment, and adjust the parameters of the network to obtain the trained C3D model. ③ Input the features of the video clips extracted by the C3D network model into the LSTM network, further extract the overall features of the video, and realize the recognition and classification of human actions. ④ Use Tensorboard to visually analyze the network training process. The system block diagram is as follows:

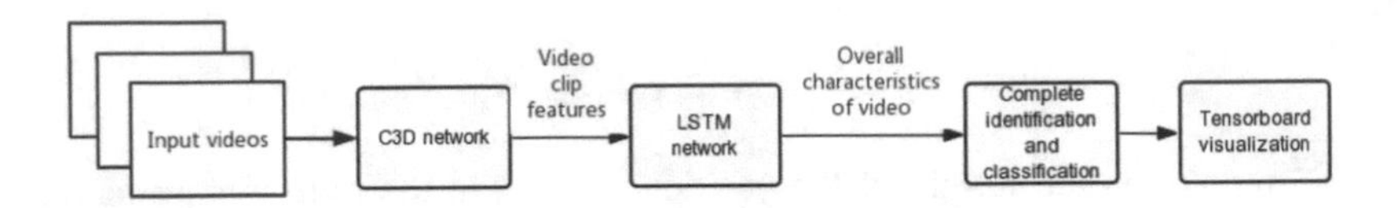

3.2 Fusion Model

The C3D network in this topic includes 8 convolution, 5 Max-pooling, 2 fully connected layers, followed by a softmax output layer. The number of convolution kernels of the 8 convolutional layers are 64, 128, 256, 256, 512, 512, 512, 512, respectively. In the convolutional layer, all 3D convolution kernels is $3 \times 3 \times 3$ with stride 1 in both spatial and temporal dimensions. The pooling layer uses a 3-dimensional pooling operation and all pooling kernel are $2 \times 2 \times 2$ with stride 2, except for pool1 is $1 \times 2 \times 2$ with stride 1, preventing the image from shrinking too fast. The 3D convolutional neural network as follows:

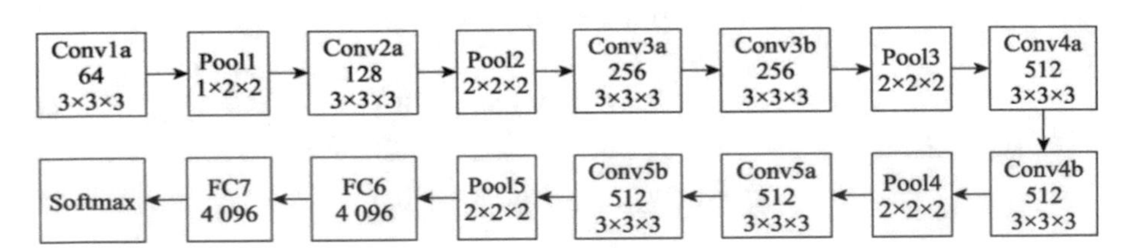

After the fifth pooling layer are two fully connected layers, both fully connected layers have 4096 outputs. Finally, the video sequence features are input into LSTM for training. The LSTM network consists of two layers of LSTM. In this paper, the hidden layer nodes are set to 128. The fully connected layer and the softmax layer classify the human actions in the result. In the training phase of the fully connected layer of the network, Dropout is set to 0.5, which can prevent the model from overfitting. The activation function uses the Relu function, which is 1 in the forward part, which can avoid problems such as gradient explosion.

4 Experimental Test

4.1 Dataset

This article uses UCF-101 dataset, UCF-101 as a special dataset for behavior recognition challenge THUMOS. It contains 13,320 short videos of different lengths (27 h in total), involving 101 human behavior categories and includes five types of actions.The number of each video frame is different, about 150 frames, and the resolution of each video is 320 × 240. Use the three official data set division methods to divide the training set and test set. In this paper, the first division method is adopted, there are 9,996 videos in the training set and 3324 videos in the test set (Fig. 1).

Fig. 1. Some video clips of UCF-101 data set

4.2 Experimental Parameter Settings

During the training process, Adam is used for feedback adjustment. This function is essentially RMSprop with a momentum term. The principle is to adjust the learning rate of each parameter. In this topic, we set the learning rate to 10^{-4}, and the learning rate decay value after each update is set to 10^{-6}.In each iteration of each model, add a Droopout value of 0.5. In the training process, since the total number of training set samples is 9,996, batch_size = 5 in this topic, so steps_per_epoch is 1999, and the number of data iteration rounds epochs is set to 10. Similarly, the total number of samples in the validation set is 3324, so validation_steps is 664 (Table 1).

4.3 Experimental Results

The network models are all carried out on the 64-bit win7 system. First, build the software environment, create the tensorflow environment, and build the C3D and LSTM network models in the created environment. Then input video data of known categories to train and test the network model, so that the network model can recognize the human behavior in the video. Finally, the training process and results are visualized through Tensorboard.

Table 1. Global parameter setting

Parameter	Value
Activation function	RULE
Learning rate	10^{-4}
Dropout	0.5
Batch-size	5
Data iteration rounds epochs	10
Step-per-epoch	1999
Validation-steps	664

After 27 rounds of training, the visualization results are shown in Fig. 2 and Fig. 3. Figure 2 is the change curve of the accuracy rate on the training set and the test set during the training process. It can be seen that the accuracy rate is on the rise.The accuracy rate on the training set increased from 2.19% to 90.62%, and on the test set increased from 4.24% to 83.22%. Figure 3 is the change curve of the loss on the training set and the test set during the training process. It can be seen that the loss is showing a downward trend.The loss on the training set dropped from 4.590 to 0.318 and on the test set dropped from 4.452 to 0.8184. This is only the result of 27 rounds of training. If the number of training rounds increases, the experimental data will eventually stabilize. It can be seen that the network model proposed in this paper achieves a good recognition performance for video recognition.

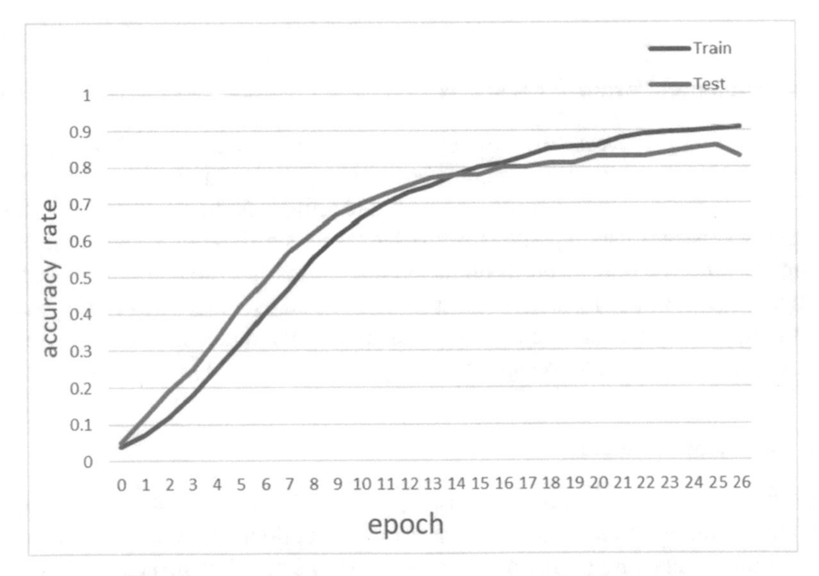

Fig. 2. The accuracy rate on the training set and the test set

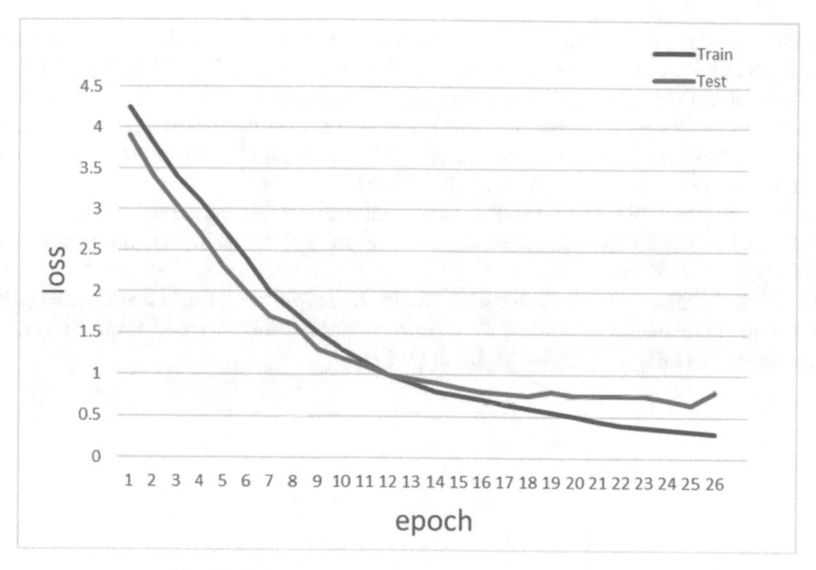

Fig. 3. The loss on the training set and the test set

5 Conclusion

In this paper, the process of building the C3D network model and the LSTM network model is studied in detail, and the two models are combined for video recognition. From the convolutional layer, the pooling layer, the fully connected layer to the softmax output layer, improvements have been made and the settings of various parameters have been performed using the Tensorflow deep learning open source framework for simulation verification. The experimental results were obtained and analyzed.The experimental test data shows that the C3D and LSTM fusion model proposed in this topic has good recognition performance for the classification of human actions in videos. It shows that the research of this subject has a certain practicability.

References

1. Zhu, F., Shao, L., Xie, J., et al.: From handcrafted to learned representations for human action recognition: a survey. Image Vis. Comput. **55**, 42–52 (2016)
2. Najafi, A., Hasanlou, M., Akbari, V.: Land cover changes detection in polarimetric SAR data using algebra, similarity and distance based methods. Int. Arch. Photogram. Remote Sens. Spat. Inf. Sci. **42**, 195–200 (2017)
3. Dhulekar, P., Gandhe, S.T., Chitte, H., et al.: Human action recognition: an overview. In: Satapathy, S., Bhateja, V., Joshi, A. (eds.) Proceedings of the International Conference on Data Engineering and Communication Technology, vol. 468, pp. 481–488. Springer, Singapore (2017). https://doi.org/10.1007/978-981-10-1675-2_48
4. Gupta, V., Singh, J.P.: Study and analysis of back-propagation approach in artificial neural network using HOG descriptor for real-time object classification. In: Ray, K., Sharma, T., Rawat, S., Saini, R., Bandyopadhyay, A. (eds.) Soft Computing: Theories and Applications, vol. 742, pp. 45–52. Springer, Singapore (2019). https://doi.org/10.1007/978-981-13-0589-4_5

5. Rashid, M., Khan, M.A., Sharif, M., et al.: Object detection and classification: a joint selection and fusion strategy of deep convolutional neural network and SIFT point features. Multimedia Tools Appl. **78**(12), 15751–15777 (2019). https://doi.org/10.1007/s11042-018-7031-0
6. Liu, L., Hu, F., Zhao, J.: Action recognition based on features fusion and 3D convolutional neural networks. In:2016 9th International Symposium on Computational Intelligence and Design (ISCID), vol. 1, pp. 178–181. IEEE (2016)
7. Xu, Z., Vilaplana, V., Morros, J.R.: Action tube extraction based 3D-CNN for RGB-D action recognition. In: 2018 International Conference on Content-Based Multimedia Indexing (CBMI), pp. 1–6. IEEE (2018)
8. Li, C., Sun, S., Min, X., et al.: End-to-end learning of deep convolutional neural network for 3D human action recognition. In: 2017 IEEE International Conference on Multimedia and Expo Workshops (ICMEW), pp. 609–612. IEEE (2017)

Adaptive Multi-threshold Image Segmentation Technology Based on Intelligent Algorithm

Man Wang[1,2], Baoju Zhang[1,2](✉), and Cuiping Zhang[1,2]

[1] Tianjin Key Laboratory of Wireless Mobile Communication and Power Transmission, Tianjin Normal University, Tianjin, China
wdxyzbj@163.com

[2] College of Electronic and Communication Engineering, Tianjin Normal University, Tianjin, China

Abstract. Image segmentation technology is to assign a label to each pixel in the image. Accurate image segmentation technology can play a key role in various deep learning tasks such as medical analysis and satellite image object detection. The choice of threshold in image segmentation will directly affect the effect of segmentation. The traditional OTSU algorithm has a poor traversal effect, and the segmentation effect is not obvious for images with concentrated histogram distribution. Based on the optimization principle proposed by the intelligent algorithm, this paper treats each pixel value in the picture as an individual, which is a good way to bypass the disadvantages of local optimization. It can complete automatic global optimization according to the original situation of the image to determine the threshold. Complete the image segmentation, and can retain more internal details of the image.

Keywords: Intelligent algorithm · Image segmentation technology · Adaptive multi-threshold

1 Introduction

In the field of vision, image segmentation often refers to assigning a label task to each pixel in the image. It can also be regarded as a dense predicate task, which classifies each pixel in the image. The precise location of image segmentation can play a key role in various deep learning tasks such as medical analysis, satellite image object detection, and iris detection.

In 1979, the maximum between-class variance method was proposed by Japanese scholar Otsu N [1]. It uses the between-class variance between them according to the gray-scale characteristics of the image to divide the image into two parts, the foreground and the background, and the most The optimal threshold is the key to the entire segmentation technique. Most of the subsequent algorithms focus on the research of the optimal threshold. The determination of the threshold can also be divided into multi-threshold segmentation and single-threshold segmentation according to the characteristics of the

Q. Liang et al. (Eds.): Artificial Intelligence in China, LNEE 854, pp. 561–567, 2022.
https://doi.org/10.1007/978-981-16-9423-3_70

image itself and subsequent work requirements. Then, Cheng J proposed a segmentation method based on histogram [2], which aimed at Otsu's need to traverse the inherent defects of the entire pixel gray scale, through the histogram pre-processing and contour tracking, so as to quickly find the optimal threshold to achieve image segmentation. Fu L exported the Otsu method multiple times [3]. Obtains several simple calculation formulas similar to Otsu. Its automatic threshold can be well applied to real-time image analysis systems. Hua C et al. proposed a fast algorithm for image segmentation based on two-dimensional entropy threshold [4], and reduced the complexity of the traditional two-dimensional threshold method from $O(W^2S^2)$ to $O(W^{2/3}S^{2/3})$. Chen Z et al. proposed an adaptive threshold segmentation method based on wavelet [5], using wavelet analysis to obtain the optimal threshold, and got a better segmentation effect.

Based on the "intelligent algorithm" inspired by the laws of natural evolution, the best individual combination is used to form the next generation of optimization principles. Inspired by this, Holland JH simulates the natural selection and genetics mechanism of Darwin's biological evolution theory and proposes a way to simulate the natural evolution process Genetic Algorithm (GA) for searching the optimal solution [6]. Simon D proposed a Biogeography-Based Optimizer (BBO) based on the mathematics of biogeography as the basis for the development of new fields [7], and solved optimization problems by discussing natural biogeography and its mathematics. Based on the "intelligent algorithm" of the physical method, which imitates the optimization principle proposed by the physical rules of the universe, Kirkpatrick S and others will combine the optimization process with statistical thermodynamics and propose the Simulated Annealing (SA) [8, 9], by simulating solids find the optimal solution for the change of the internal energy of the object during the annealing process. Hatamlou A combined the black hole phenomenon in the universe to propose a Black Hole (BH) [10], through the process of black hole search and swallow new stars to propose an optimization algorithm to solve the clustering problem. Based on the "intelligent algorithm" of animal group behavior, the optimization principle proposed by simulating the social behavior of animal groups in nature. Kennedy J and Eberhart R proposed Particle Swarm Optimization (PSO) [11] based on the social behavior of birds in flocks, which uses particles (candidate solutions) to search for their own route at the same time when looking for the best position in the air. As best position to obtain optimal solution. Mirjalili S and Lewis A used the special hunting social behavior of humpback whales to propose the Whale Optimization Algorithm (Whale Optimization Algorithm, WOA) [12], which achieved the optimal solution by simulating the humpback whale bubble net foraging and attacking with spirals.

This paper will use the optimization ideas of intelligent optimization algorithms and the principle of finding the optimal solution, combined with the difficult problem of adaptive multi-threshold determination in image segmentation technology, and propose an adaptive multi-threshold image segmentation algorithm based on intelligent algorithms. Use PSO and WOA algorithms to solve the combined optimization problem with the smallest objective function value, find the optimal solution of multiple thresholds, and achieve better image segmentation. The output result is compared with the image segmentation result of MSRA10K Salient Object Database [13], which can retain more details and do a good job of preprocessing for subsequent image classification.

2 Intelligent Algorithm

Intelligent algorithm is a combination of random algorithm and local search algorithm that uses a group. By defining certain group behavior and individual behavior, the group has population diversity and behavior orientation, and has good flexibility and wide range. Features such as flexibility, bypassing local optimality, etc.

2.1 PSO Algorithm

The PSO algorithm realizes the global search by adjusting the optimal individual extremum corresponding to the current individual particle and the determination of the optimal candidate individual in the entire particle swarm. Through multiple iterations, the velocity value and position information of the particles are updated, and the optimal solution is determined when the termination criterion is met (the maximum number of iterations is reached).

Suppose the target image A is the target search space, N pixel values represent a community composed of N particles, and the i particle can be represented by a vector X_i:

$$X_i = (X_{i1}, X_{i2}, X_{i3}, \cdots, X_{im}), i = 1, 2, 3, \cdots, N \tag{1}$$

The flight speed of the i particle and the searched optimal individual extremum (individual optimal solution):

$$V_i = (V_{i1}, V_{i2}, V_{i3}, \cdots, V_{im}), i = 1, 2, 3, \cdots, N \tag{2}$$

$$P_i = (P_{i1}, P_{i2}, P_{i3}, \cdots, P_{im}), i = 1, 2, 3, \cdots, N \tag{3}$$

The global optimal individual extreme value of the particle in the entire particle swarm (the optimal solution particle):

$$G_{best} = (P_1, P_2, P_3, \cdots, P_N) \tag{4}$$

When the speed of each particle does not meet the termination condition, it is continuously updated to find the optimal solution:

$$v_i^{k+1} = \omega \cdot v_i^k + c_1 \cdot a \cdot (P_i^k - x_i^k) + c_2 \cdot a \cdot (G_{best\,i}^k - x_i^k) \tag{5}$$

$$x_i^{k+1} = x_i^k + v_i^{k+1} \tag{6}$$

Where ω is the inertia factor (non-negative number), the global optimization performance and local optimization performance can be adjusted by adjusting $\omega.a$ is a random value from 0 to 1. c_1 is the individual learning factor of each particle, c_2 is the social learning particle of each particle. P_i^k is the individual extreme value of the i particle at the k iteration. $G_{best\,i}^k$ is the optimal individual particle of all particles in the k iteration. x_i^k, x_i^{k+1} are the location information up-dated each time.

By calculating the size of the fitness value of different pixels in the picture, find out the pixel position with the smallest fitness value of different pixel values, and globally determine the fixed number of multi-thresholds that best meet the requirements, so as to determine the optimal solution of the whole image, and realize adaptive multi-threshold image segmentation. This method uses a random search method to achieve global optimization and make the picture segmentation more refined.

2.2 WOA Algorithm

The WOA algorithm updates the whale's search position by assuming that the current optimal candidate solution is the target solution (close to the optimal solution) or randomly choosing a position to force the whale to deviate from the prey, thereby realizing a global search. When the termination criterion is met (the maximum number of iterations is reached) When, the WOA algorithm is terminated.

Calculate the distance $\vec{D}$ between the whale position (x, y) and the prey (x*, y*). When $|\vec{A}| > 1$, the position of a humpback whale is randomly selected for position update $X(t+1)$, and when $|\vec{A}| < 1$ selects the current position for position update $X(t+1)$.

$$\vec{D} = \left|\vec{C}\vec{X}^*(t) - \vec{X}(t)\right| \tag{7}$$

$$\vec{X}(t+1) = \vec{X}^*(t) - \vec{A}\vec{D} \tag{8}$$

$$\vec{A} = 2\vec{a} \cdot \vec{r}_1 - \vec{a},\ \vec{C} = 2\vec{r}_2,\ \vec{a} = 2 - 2t/T_{\max} \tag{9}$$

Where t is the number of iterations, $\vec{r}_1$ and $\vec{r}_2$ are random vectors from 0 to 1, $\vec{a}$ is a vector value that randomly decreases linearly from 2 to 0, and $X^*(t)$ is the best whale position at the current position.

During the hunting behavior, the whales will hunt along a spiral path in a narrowing circle at the same time. Therefore, the two methods of shrinking envelopment and spiral update are adopted, and the switching between the two modes is completed by adjusting the possibility coefficient k.

$$\vec{X}(t+1) = \begin{cases} \vec{X}^*(t) - \vec{A}\vec{D} & k < 0.5 \\ \vec{D}^i \cdot e^{bl} \cdot \cos(2\pi l)\vec{X}^*(t) & k > 0.5 \end{cases} \tag{10}$$

$$\vec{D}^i = \left|\vec{X}^*(t) - \vec{X}(t)\right| \tag{11}$$

$\vec{D}$ is the distance from the i search agent to the target, b is the parameter of the logarithmic spiral, and l is a random vector obeying the uniform distribution of $[-1, 1]$.

By calculating the size of the fitness value of different pixels in the picture, the pixel position with the smallest fitness value is found as the optimal position as the optimal solution according to the requirements. And the optimal solution is used as the optimal threshold of the adaptive threshold function to perform image multi-threshold segmentation. Compared with the traditional adaptive optimal value, the local optimal limit can be jumped out and the image segmentation can be more refined.

3 Experimental Results

The MSRA10K Salient Object Database, which originally provides salient object annotation in terms of bounding boxes provided by 3–9 users, is widely used in salient object detection and segmentation community. Although an invaluable resource to evaluate saliency detection algorithms, the database with the marked bounding boxes, however, is often too coarse for fine-grained evaluation.

The adaptive multi-threshold image segmentation technology of the intelligent algorithm using the intelligent algorithm can select the required optimal threshold number according to the distribution of the image pixel value according to the demand. While ensuring the edge contour of the image, it retains the internal details of the image, making the image label more detailed, providing good image preprocessing and detail highlighting for subsequent image classification work and similarity comparison work.

This article takes three thresholds in three dimensions as an example for experimental verification. The adaptive multi-threshold of the MSRA10K Salient Object Database No. 77 image is 72, 144, 197 through the PSO intelligent algorithm, as shown in Fig. 1(c), the adaptive multi-threshold of the same image is 68, 146, 191 through the WOA intelligent algorithm, as shown in Fig. 1(d). The adaptive multi-threshold of the No. 23934 image is 51, 104, 179 through the PSO intelligent algorithm, as shown in Fig. 2(c), the adaptive multi-threshold of the same image is 78, 124, 183 through the WOA intelligent algorithm, as shown in Fig. 2(d). The adaptive multi-threshold of the No. 54071 image is 72, 107, 158 through the PSO intelligent algorithm, as shown in Fig. 3(c), the adaptive multi-threshold of the same image is 93, 147, 173 through the WOA intelligent algorithm, as shown in Fig. 3(d).

By assigning color blocks within different thresholds, a segmentation map with preserved details can be obtained. At present, the algorithm has not performed any enhancement, filtering and other operations on the original image, nor has it performed RGB values for different regions. The specific display results of the two intelligent algorithms and the comparison results with MSRA10K Salient Object Database are shown in the Fig. 1 below.

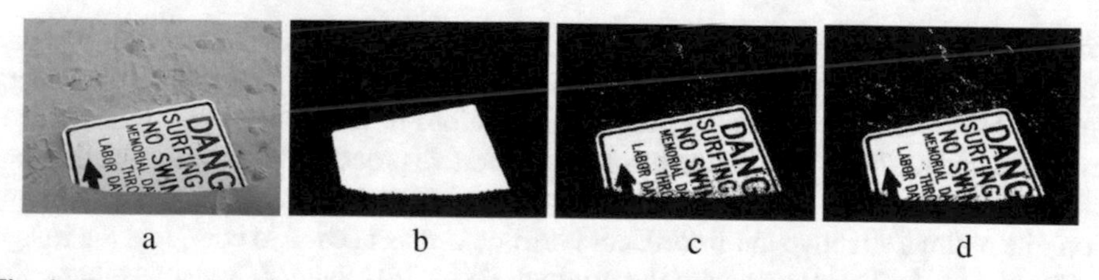

Fig. 1. From left to right: (a) MSRA10K original image number 77 (b) MSRA10K saliency label segmentation diagram (c) PSO algorithm adaptive multi-threshold image segmentation diagram (d) WOA algorithm adaptive multi-threshold image segmentation map

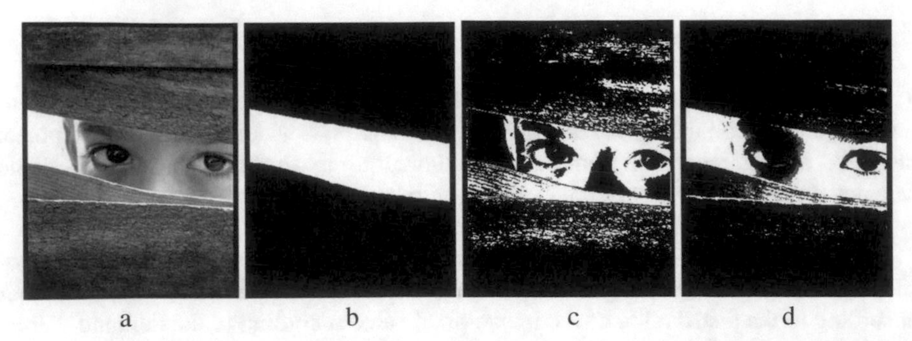

Fig. 2. From left to right: (a) MSRA10K original image number 23934 (b) MSRA10K saliency label segmentation diagram (c) PSO algorithm adaptive multi-threshold image segmentation diagram (d) WOA algorithm adaptive multi-threshold image segmentation map

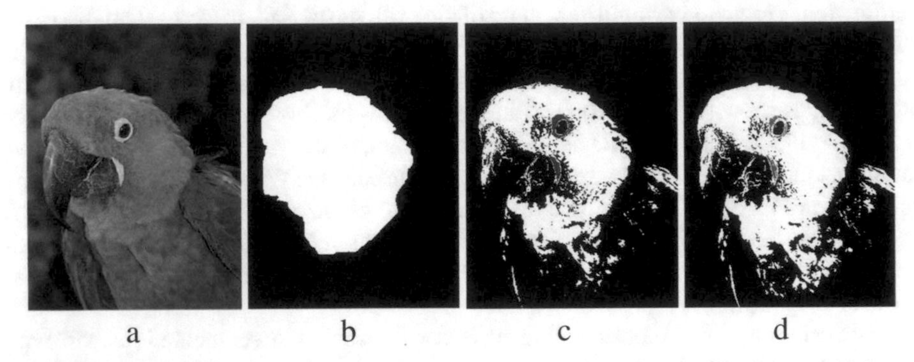

Fig. 3. From left to right: (a) MSRA10K original image number 54071 (b) MSRA10K saliency label segmentation diagram (c) PSO algorithm adaptive multi-threshold image segmentation diagram (d) WOA algorithm adaptive multi-threshold image segmentation map

4 Analysis of Experimental Result

Through intelligent algorithms (this article takes PSO and WOA as examples), you can find the optimal value in the current environment in each cycle by setting the maximum number of iterations according to the pixel conditions of different positions in the picture. And update the optimal value through an iterative process to determine the optimal value within the pixel. According to the requirements of the required output dimension, the optimal adaptive multi-threshold is allocated as needed to complete the image segmentation. And compared with the original MSRA10K Salient Object Database, the segmented image retains the overall external contours while retaining the internal details.

The original image of this article has not yet undergone basic image pre-processing operations such as noise reduction and enhancement, which can maximize the preservation of image details and hide non-detailed parts to reduce the impact of redundant information during subsequent image classification and add image tag tasks. Highlight the attributes and categories of the pictures to the utmost extent, and provide good pre-processing support for subsequent picture classification and similarity comparison work.

5 Conclusion

Intelligent algorithm-based adaptive multi-threshold image segmentation technology based on intelligent algorithm, the use of intelligent algorithm has the characteristics of random search, global optimization, independent of gradient, and non-linear and non-directive. We can adaptively select the optimal segmentation point according to the picture situation, and different areas can be filled and assigned according to the requirements. The increase of the picture noise reduction, enhancement pre-processing operations, the effect will be better segmentation (this operation is not performed in this article). After processing by intelligent algorithms, the segmented image with added details can well retain the image details required for classification and similarity comparison, and occlude the non-detailed parts, providing good technical support for subsequent work.

References

1. Otsu, N.: A threshold selection method from gray-level histograms. IEEE Trans. Syst. Man Cybern. **9**(1), 62–66 (1979)
2. Cheng, J.: Split method based on histogram. J. Huazhong Univ. Sci. Technol. **27**(1), 84–86 (1999)
3. Fu, L.: Threshold selection method based on image gap metric. J. Comput. Res. Dev. **38**(5), 563–567 (2001)
4. Hua, C., Fan, J., Gao, C., Wu, L.: Image segmentation based on 2D entropic thresholding and its fast algorithm. Pattern Ident. Artif. Intell. **13**(1), 42–46 (2000)
5. Chen, Z., Tao, Y., Chen, X.: Wavelet-based adaptive thresholding method for image segmentation. Opt. Eng. **40**(5), 868–874 (2001)
6. Holland, J.H.: Genetic algorithms. Sci. Am. **267**, 66–72 (1992)
7. Simon, D.: Biogeography-based optimizer. IEEE Trans. Evol. Comput. **12**(6), 702–713 (2008)
8. Kirkpatrick, S., Gelatt, C.D., Jr., Vecchi, M.P.: Optimization by simulated annealing. Science **220**, 671–680 (1983)
9. Černý, V.: Thermodynamical approach to the traveling salesman problem: an efficient simulation algorithm. J. Optim. Theor. Appl. **45**, 41–45 (1985). https://doi.org/10.1007/BF00940812
10. Hatamlou, A.: Black hole: A new heuristic optimization approach for data clustering. Inf. Sci. **222**, 175–184 (2013)
11. Kennedy, J., Eberhart, R.: Particle swarm optimization. In: Proceedings of ICNN 1995 - International Conference on Neural Networks, vol. 1944, pp. 1942–1948 (1995)
12. Mirjalili, S., Lewis, A.: The whale optimization algorithm. Adv. Eng. Softw. **95**, 51–67 (2016)
13. Cheng, M.M., Mitra, N.J., Huang, X., Torr, P.H.S., Hu, S.M.: Global contrast based salient region detection. IEEE Trans. Pattern Anal. Mach. Intell. **37**(3), 569–582 (2015)

End-to-end Transmission Link Optimal Selection Method based on Electromagnetic Strength Distribution

Lin Ma[1(✉)], Jiyue Chen[1], Xuedong Wang[1], and Zhongwang Zhang[2]

[1] School of Electronics and Information Engineering, Harbin Institute of Technology, Harbin, China
malin@hit.edu.cn

[2] China National Aeronautical Radio Electronics Research Institute, Shanghai, China

Abstract. In recent years, the communication demand for mobile ad hoc networks has become stronger and stronger. It can not only be widely used in mobile networking in mountainous areas, but also in anti-terrorism, stability maintenance, and field rescue in urban buildings. This paper proposes a route planning algorithm based on electromagnetic distribution, which obtains environmental electromagnetic parameters through parabolic equation, and then realizes end-to-end link adaptive transmission by using branch and bound method. This article makes full use of the electromagnetic environment in space, and under the constraint of the minimum number of hops, the data transmission delay of a single link from the source node to the destination node is minimized, thereby effectively improving the link communication quality.

Keywords: Electromagnetic calculation · Route planning · Transmission optimization

1 Introduction

The mobile ad hoc network is a complex distributed network system composed of several wireless mobile nodes. Due to the large depth of the unknown semi-enclosed space, the complicated spatial structure, and the serious multipath phenomenon, these characteristics have seriously affected the communication quality of the mobile ad hoc network. In this complex environment, it is necessary to effectively analyze the channel environment and plan the path of the mobile node to achieve the optimization of the forwarding path. In addition, according to the communication environment in which the mobile node is located, the number and location of relay nodes need to be reasonably set to achieve the overall optimal network throughput. Compared with the traditional mobile ad hoc network communication, the unknown semi-enclosed space environment puts forward higher requirements for the selection of transmission routing strategy and the setting of relay nodes.

There are many methods for solving electromagnetic problems. The method of moments is a classic method for solving electromagnetic fields in the frequency domain

Q. Liang et al. (Eds.): Artificial Intelligence in China, LNEE 854, pp. 568–576, 2022.
https://doi.org/10.1007/978-981-16-9423-3_71

[1, 2]. Electromagnetic problems in structures with arbitrary shapes can be solved by the method of moments. The frequency domain finite element method [3] is also a classic method for solving electromagnetic fields in the frequency domain. The time domain method mainly includes time domain difference method and time domain finite element method [4]. In addition, there are some high-frequency methods, such as the ray tracing method [5]. It is currently mostly used in the analysis of problems between buildings. The parabolic equation method (PE method) is a forward full-wave analysis method, which can quickly solve the problem of radio wave propagation in a large area, and can accurately describe the complex atmospheric structure and the electromagnetic characteristics of the complex surface, and can accurately predict radio waves in complex environments Propagation characteristics. This article mainly uses parabolic equations to realize electromagnetic calculations in complex environments.

This paper studies the relationship between the link adaptive rate and the number of node hops. This process studies the optimal node path selection under the constraints of the minimum number of hops and the minimum transmission delay, which is an integer programming process. The algorithm framework for solving general linear integer programming problems is mainly branch and bound algorithm, and branch-cut algorithm combined with cutting plane technology [6]. Various extensions of the branch and bound method are studied in the literature [7–11].

In this paper, the branch and bound method is integrated into the algorithm of end-to-end optimal link selection. According to the electromagnetic results calculated by the parabolic equation, a routing planning algorithm based on electromagnetic distribution is obtained. The rest of this article is organized as follows: The second section will build system models and introduce them. The third section will introduce the parabolic equation and transmission optimization model which were based on the principles introduced in the second section. The fourth section will provide implementation and performance analysis. Finally draw a conclusion.

2 System Model

2.1 System Framework

In this paper, the parabolic equation model is used for electromagnetic calculation, the optimization model of the optimal relay is established, and the solution is based on the branch and bound method. The researched system block diagram is shown in Fig 1.

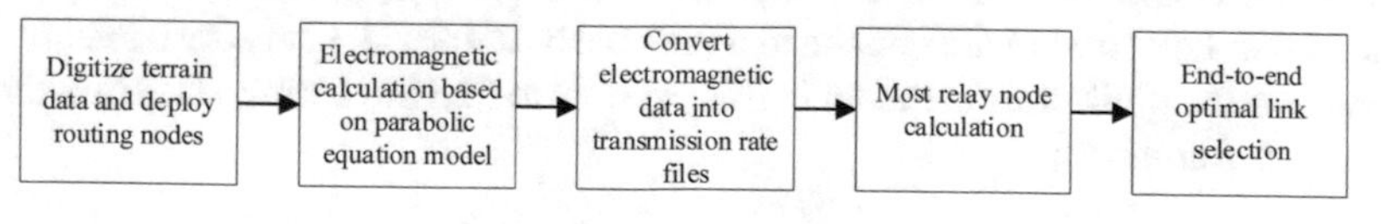

Fig. 1. System block diagram

In the research process, firstly we convert the target terrain into a digital map and set the location parameters of known routing nodes. Then calculate the electromagnetic distribution of the existing terrain according to the parabolic equation model to obtain the field strength at each position and the path loss between each node. Then set the speed files corresponding to different path losses, and convert the results obtained from the electromagnetic calculations. Finally, an optimization model and constraints are established according to the optimization goal, and the branch and bound method is used to solve and calculate the optimal relay node to obtain the end-to-end optimal link.

2.2 Parabolic Equation Model

The Parabolic Equation (PE) model is a formula derived from the wave equation and belongs to the deterministic propagation model. The numerical method for solving PE is mainly the Split-step Fourier Transform method (SSFT). In each step of solving the parabolic equation, the SSFT algorithm combines the boundary conditions to perform the Fourier transform operation on the parabolic equation, and then multiplies it with the refractive index term to obtain the final solution. When the SSFT algorithm is used, its iteration step is almost independent of wavelength, which can accurately and quickly solve the problem of radio wave propagation in a large area and complex environment. This paper uses the SSFT algorithm to solve the parabolic equation.

2.3 Transmission Optimization Model

Assuming there are N routing nodes in the transmission network, the network model is shown in Fig 2.

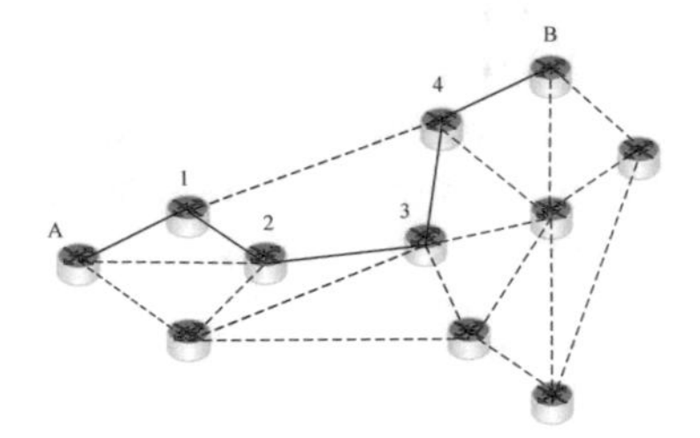

Fig. 2. Transmission optimization diagram

As shown in Fig 2, each node in the network is connected by a dotted line (part not shown), indicating that data can be transmitted to each other. The solid line between node A and node B represents the transmission link. After transmission optimization, the optimal transmission rate is obtained after the hops1, 2, 3, and 4, but the shortest distance node 1 and node 4 are hopped to obtain the best link. The purpose of establishing the transmission optimization model in this paper is to find the optimal hop link through dynamic programming.

3 The Proposed Method

3.1 Calculate the Parabolic Equation

Suppose the time-harmonic factor of the electromagnetic field is $e^{-i\omega t}$ and the scalar ψ represents any field component. In the process of radio wave propagation, ψ satisfies the following two-dimensional scalar wave equation:

$$\frac{\partial^2 \psi}{\partial x^2} + \frac{\partial^2 \psi}{\partial z^2} + k_0^2 n^2 \psi = 0 \quad (1)$$

where k_0 is the propagation constant in vacuum, $n = \sqrt{\varepsilon_r}$ is the refractive index of the medium. Assuming that in the process of radio wave propagation, the refractive index n hardly changes with the distance x, that is, $\partial n^2 / \partial x = 0$, we get the wave function as:

$$\left[\left(\frac{\partial}{\partial x} + jk_0(1-Q) \right) \left(\frac{\partial}{\partial x} jk_0(1+Q) \right) \right] u = 0 \quad (2)$$

where Q is called a pseudo-differential operator, and

$$Q = \sqrt{\frac{1}{k_0} \frac{\partial^2}{\partial z^2} + n^2(x, z)} \quad (3)$$

In the Cartesian coordinate system, using different approximation methods for Q will result in different PE forms. SSFT is a numerical algorithm for solving PE in the frequency domain. After calculation, the SSFT solution of the parabolic equation can be obtained as:

$$u(x, z) = e^{ik_0(n-2)x} \Im^{-1} \left[e^{i\Delta x \sqrt{k_0^2 - p^2}} U(x_0, p) \right] \quad (4)$$

where $U(x, p) = \Im[u(x, z)] = \int_{-\infty}^{+\infty} u(x, z) e^{-ipz} dz$ represents the Fourier transform of u.

3.2 Establish a Transmission Optimization Model

Assuming that there are N nodes in the network, according to the different field strengths and path losses between the nodes, set M files with different transmission rates, assuming that the data packet length is L, as shown in Fig 3.

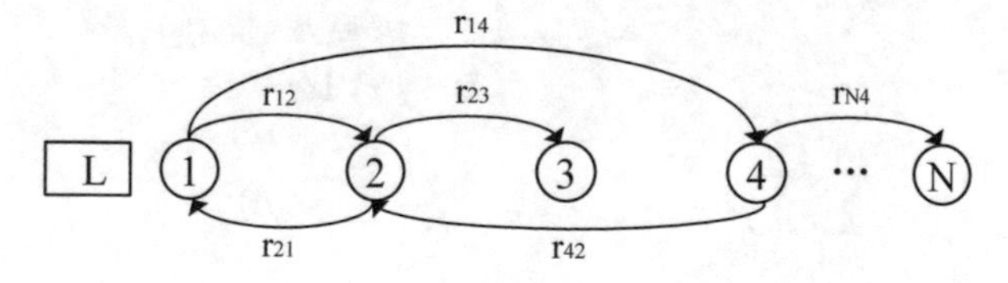

Fig. 3. Transmission optimization model

In Fig. 3, define $r_{i,j}$ as the transmission rate of data from node i to node j, where the value of $r_{i,j}$ can be obtained by looking up the table after the electromagnetic calculation result interval conversion, as shown in Table 1.

Table 1. Transmission rate value

$r_{i,j}$	M_1	M_2	M_3	M_4	M_5
RSS	[rss_1, rss_2]	[rss_2, rss_3]	[rss_3, rss_4]	[rss_4, rss_5]	[rss_5, rss_6]

Among them, Table 1 only shows part of the values, and the specific value of M is set according to actual needs. Combined with the electromagnetic calculation results, when the received signal strength RSS of a node in the network is within a set interval value, it is converted into the corresponding transmission rate file.

Let the cost function $t_{i,j}$ be the transmission delay corresponding to the data packet transmission from node i to node j. The goal of this research is to maximize the average transmission rate from node 1 to node N under the constraint of the minimum number of hops. The mathematical problem is modeled as follows:

$$
\begin{aligned}
&P1: \qquad R* = \max \frac{L}{\sum_{i=1}^{N}\sum_{j=1}^{N} x_{i,j}t_{i,j}} \\
&\text{s.t.} \quad t_{i,j} = \frac{L}{r_{i,j}}, \qquad (a) \\
&i, j, k \in \{1, 2, \cdots N\}, \qquad (b) \\
&x_{i,j} \in \{0, 1\}, \qquad (c) \\
&\sum_{i=1}^{N} x_{i,j} \leq 1, \forall j, \qquad (d) \\
&\sum_{k=1}^{N} x_{j,k} \leq 1, \forall j, \qquad (e) \\
&\sum_{i=1}^{N}\sum_{j=1}^{N} x_{i,j} \leq m, \forall i, j \qquad (f) \\
&\sum_{i=1}^{N} x_{i,j} - \sum_{k=1}^{N} x_{j,k} = \begin{cases} -1 & if j = 1 \\ 1 & if j = N \\ 0 & j \neq 1 or N \end{cases} \qquad (g) \\
&\sum_{i=1}^{N}\sum_{j=1}^{N} t_{i,j}x_{i,j} \leq T \qquad (h)
\end{aligned}
\tag{5}
$$

Among them, the optimization goal is the maximum average transmission rate of data through all nodes, $x_{i,j}$ is the communication link establishment factor. When $x_{i,j} = 1$, it means that the communication link for data transmission from node i to node j is established. Otherwise, when $x_{i,j} = 0$, the communication link from node i to node j is not established. L represents the size of the data packet, and m represents the maximum hop value. The cost function $t_{i,j}$ is the transmission delay corresponding to the data packet transmission from node i to node j. The cost function $t_{i,j}$ can be obtained by formula (5). Considering the directionality of signal transmission, only in the case of $1 \leq j - i \leq M$, $r_{i,j}$ exists. In other cases, a link cannot be established between node i and node j, $r_{i,j}$=0. The above optimization problem is a 0-1 integer programming problem. This problem can be solved by a classic branch and bound algorithm. Different branch variables and sub-problems are selected for branching, and the optimal solution is calculated.

4 Implementation and Performance Analysis

4.1 Experiment Environment

In the simulation experiment, the solution of the parabolic equation was tested first. Assuming that the transmitting frequency of the antenna is 300 MHz, within a range of 50 km, the parabolic equation model is solved, and the obtained field strength varies with the height and propagation distance. Then, it is assumed here that there are $N = 20$ nodes in the network, the maximum number of hops is $m = 4$, the data packet size is 1, the transmission rate is set to 5, and the node transmission power is set to 20 dBm. Table 2 shows the correspondence between signal path loss and transmission rate between nodes.

Table 2. Path loss and transmission rate conversion table

$r_{i,j}$ (Mbps)	$M_1 = 32$	$M_2 = 16$	$M_3 = 8$	$M_4 = 4$	$M_5 = 2$	$M = 0$
Pr (dBm)	$[-50, 20]$	$[-80, -50]$	$[-110, -80]$	$[-140, -170]$	$[-170, -140]$	$(\infty, -170]$

Randomly use trigonometric functions to simulate a mountainous terrain map, as shown in Fig 4a), as shown in the terrain contour map. The number in the figure is the height of the corresponding terrain. Choose any 20 nodes in the terrain, as shown in Fig 4b), where every four nodes are adjacent groups, distinguished by different colors.

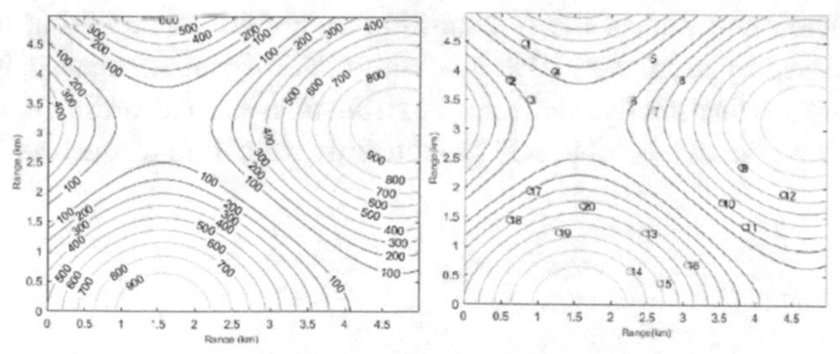

a) Establishment of topographic map b) Select routing node

Fig. 4. Random topographic map

Set the transmitting antenna as an omnidirectional antenna, the erection height is increased by 3 m on the basis of the height of the transmitting node, the transmitting frequency is 500 Mhz. The positions of the selected 20 nodes are used as the transmitting nodes, and the electromagnetic model of the terrain area is obtained by simulation, so that the path loss of each transmitting node corresponding to the remaining 19 receiving node positions is obtained.

4.2 Experiment Results

According to the above parameter settings, the simulation results are shown in Fig 5.

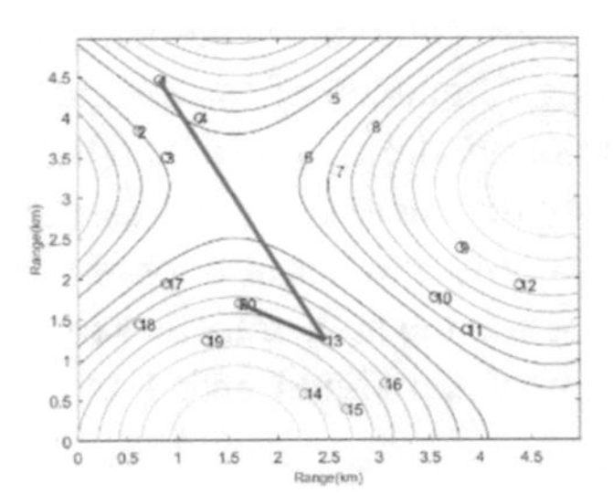

Fig. 5. Example of optimal link selection (transmission path 1-13-20)

In Fig. 5, the signal is transmitted by node 1 and the signal is received by node 20. The entire link is completed by node 13 during the transmission process. The minimum transmission delay obtained is 0.1875 (1/Mbps) and the maximum average transmission rate is 5.3333 (Mbps). Changing the transmitting node and receiving node. If there is no path that meets the constraints including the maximum transmission delay and the maximum number of hops, the transmission will fail. In order to prove the validity and practicability of the above research results, some of the best links for transmission were selected and compared with the transmission rate of the link composed of its neighboring nodes. The results are shown in Table 3.

Through comparison, the transmission rate of the best link obtained by the algorithm of this study is always higher than the other two adjacent links, and also higher than the direct transmission rate of the transmitting node and the receiving node, and can be stably increased by more than 10%.In addition, the signal attenuation between some nodes is too large to be directly connected for transmission. According to the algorithm, the optimal jump plan can be obtained to realize the selection of the best link.

Table 3. Transmission rate comparison (Mbps)

Best link	Optimal transmission rate	Adjacent link 1	Transmission rate	Adjacent link 2	Transmission rate
1-13-15	5.33	1-4-15	1.33	1-20-15	1.33
1-13-20	5.33	1-19-20	1.60	1-3-20	1.33
2-13-14	6.40	2-3-14	3.20	2-20-14	1.00
3-8-5	3.20	3-7-5	1.33	3-4-5	1.78
3-4-17-10	4.00	3-4-10	1.78	3-17-10	1.60
3-4-16	3.20	3-13-16	1.60	3-20-16	1.33
3-13-20	5.33	3-17-20	1.00	3-19-20	1.60
4-13-14	6.40	4-20-14	1.00	4-3-14	1.78
4-13-15	5.33	4-13-14-15	1.52	4-20-15	1.00
5-7-12	8.00	5-8-12	1.00	5-9-12	1.60
6-4-1	8.00	6-3-1	1.78	6-2-1	1.89
6-7-12	10.67	6-9-12	1.78	6-8-12	1.78

5 Conclusion

This paper uses the parabolic equation to establish the electromagnetic field distribution map of the mountain, and proposes a wireless ad hoc network end-to-end link adaptive transmission method based on the branch and bound method. Based on the relationship between the link adaptive rate and the number of node hops, this paper establishes a single link from the source node to the destination node to minimize the data transmission delay optimization model under the constraint of the minimum number of hops, and sets reasonable settings based on the electromagnetic distribution. The conversion interval between electromagnetic data and speed provides parameters for solving the optimization model. Finally, the branch and bound method is used to solve the model, which maximizes the end-to-end rate, thereby increasing the throughput of the entire network.

Acknowledgment. This paper is supported by national nature scinece foundation of China (61971162, 41861134010) and National Aeronautical Foundation of China (2020Z066015002).

References

1. Jiao, D., Ergin, A., Shanker, B., et al.: A fast higher-order time-domain finite element boundary integral method for 3-D electromagnetic scattering analysis. IEEE Trans. Antennas Propag. **50**(9), 1192–1202 (2002)
2. Jin, J.M., Riley, D.J.: Finite Element Analysis of Antennas and Arrays. Wiley, Hoboken (2008)

3. Jin, J.M.: The finite element method in electromagnetics, 3rd edn. Wiley-IEEE Press (2014)
4. Cohen, G., Pernet, S.: Scientific Computation: Finite Element and Discontinuous Galerkin Methods for Transient Wave Equations. Springer, Dordrecht (2017). https://doi.org/10.1007/978-94-017-7761-2
5. James, G.L.: Geometrical Theory of Diffraction for Electromagnetic Waves. Peregrinus, Chicago (1976)
6. Junger, M., Liebling, T., Naddef, D., et al.: 50 Years of Integer Programming 1959–2008: From the Early Years to the State-of-the-Art. Springer, Berlin (2010). https://doi.org/10.1007/978-3-540-68279-0
7. Little, J.D.C., Murty, K.G., Karel, C., et al.: An algorithm for the traveling salesman problem. Oper. Res. **11**(6), 972–989 (1963)
8. Lawler, E.L., Wood, D.E.: Branch-and-bound methods: a survey. Oper. Res. **14**(4), 699–719 (1966)
9. Horst, R.: A general class of branch-and-bound methods in global optimization with some new approaches for concave minimization. J. Optim. Theory Appl. **51**(2), 271–291 (1986). https://doi.org/10.1007/BF00939825
10. Nakariyakul, S.: A comparative study of suboptimal branch and bound algorithms. Inf. Sci/ **278**, 545–554 (2014)
11. Morrison, D.R., Jacobson, S.H., Sauppe, J.J., et al.: Branch-and-bound algorithms: a survey of recent advances in searching, branching, and pruning. Discret. Optim. **19**, 79–102 (2016)

Visual Target Measurement Method in Unknown Environment Based on Stereo SLAM

Lin Ma[1(✉)], Jiayi Zheng[1], Shizeng Guo[1], and Zhongwang Zhang[2]

[1] School of Electronics and Information Engineering, Harbin Institute of Technology, Harbin, China
malin@hit.edu.cn
[2] China National Aeronautical Radio Electronics Research Institute, Shanghai, China

Abstract. With the development of UAV technology and deep learning, the demand for 3D information acquisition of unknown environment is increasingly strong. The 3D information of unknown environment can be widely used in unmanned driving, augmented reality, disasters rescue and other scenes. Using computer vision to estimate the 3D information of the target in the scene has become the mainstream method. Because the single image lacks depth information, it is usually necessary to know the size of a reference object in the scene, or to shoot continuously by using SLAM algorithm to obtain the 3D information. The stereo vision can solve the shortcomings of monocular vision. It uses two cameras to shoot and uses the spatial geometric relationship to recover the 3D information. This paper uses stereo SLAM to track and map, and extracts feature points to calculate its 3D coordinates of the feature points. After a series of image processing operations, the feature points are used to estimate the size of target object. Finally, the accurate measurement of road width and tree height is realized.

Keywords: 3D measurement · Stereo vision · Straight line fitting · Edge detection

1 Introduction

Recently, with the development of UAV technology and deep learning, the demand for 3D information acquisition of unknown environment is increasingly strong. The 3D information of unknown environment can be widely used in unmanned driving, augmented reality, disasters rescue and other scenes. The methods for obtaining 3D information of unknown environment can be roughly divided into two categories, namely, active measurement and passive measurement. Passive measurement is to use the reflection of surrounding light on the object during the measurement process, such as using a camera to measure. Active measurement is to illuminate the measured object actively in the measurement process, such as structured light technology, laser measurement and so on. Among them, the most commonly used is monocular vision and stereo vision.

Q. Liang et al. (Eds.): Artificial Intelligence in China, LNEE 854, pp. 577–585, 2022.
https://doi.org/10.1007/978-981-16-9423-3_72

The monocular vision method uses a camera to take a single photo. However, due to the lack of depth information in a single image, it is usually necessary to know the size of an object in the image as a reference to estimate the size of target object [1–3] uses this method to estimate the size of an object in an unknown scene or the height of the camera itself [4, 5] look for 2D properties such as vanishing lines and vanishing points, to establish the 3D size relations among objects.

When the monocular camera shoots continuously, SLAM can be used to obtain 3D information. MonoSLAM [6] is the first SLAM system that runs in real time. MonoSLAM uses the EKF algorithm to mathematically model the system, but the linearization of the SLAM problem by the EKF algorithm will result in the inability to obtain the global optimal value, which limits the accuracy. PTAM [7] uses nonlinear optimization to deal with the SLAM problem, which can effectively reduce the nonlinear cumulative error and improve the accuracy.

The stereo vision can solve the problem of scale uncertainty in the monocular vision. The stereo vision uses two cameras to take two pictures at the same time, and then uses the spatial geometric relationship to determine the 3D coordinates of a pixel. Based on this principle, the size of object can be measured. [8] uses the principle to continuously collect Martian surface image pairs with the same camera at two different positions, and then solves the camera pose between two different positions, so as to recover the 3D coordinates information of target objects.

Stereo SLAM includes RTAB-MAP [9] and ORB-SLAM2 [10], etc. ORB-SLAM2 is based on ORB-SLAM [11], and is the current mainstream SLAM solution, which supports monocular, stereo and RGBD devices.

Based on the above research, it can be seen that using stereo vision to measure the size of object has high accuracy and real-time performance. In this paper, ORB-SLAM2 is selected for real-time tracking and mapping, and then a series of image processing operations are used to estimate the 3D information of target object. The tree height and the road width in unknown environments can be measured finally.

2 System Model

2.1 Stereo SLAM

Stereo SLAM adopts ORB-SLAM2. ORB-SLAM2 extracts ORB feature points. Because the extraction speed of ORB feature points is fast, the camera pose can be calculated in real time without GPU acceleration, and it has high robustness and strong adaptability to the environment. The ORB-SLAM2 system consists of the following three parallel threads:

1. Tracking

   This part mainly completes the extraction of ORB feature points and inter-frame camera pose estimation, and then determines new key frame.
2. Local mapping

   This part mainly completes the local map construction. Including the processing of new key frame and using Local BA method for local mapping.

3. Loop closure

   This part mainly completes loop detection and loop correction. Loop detection is realized by DBoW algorithm, and then calculate SE3 to obtain the similar transformation. Loop correction includes loop fusion and graph optimization. Finally, perform global BA and update the map.

   The ORB-SLAM2 system diagram is shown in Fig. 1.

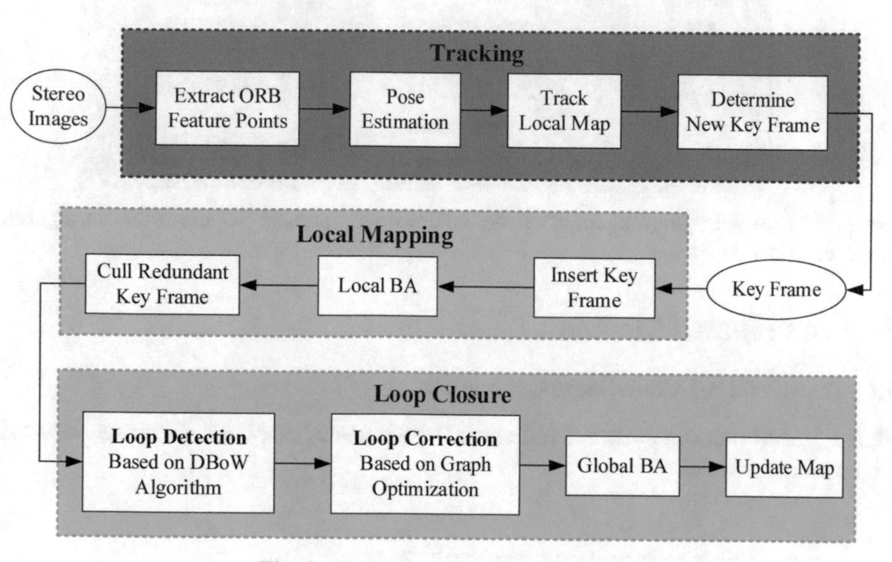

Fig. 1. Stereo SLAM system diagram

2.2 3D Measurement Based on Stereo SLAM

Figure 2 shows the overall system diagram. The system can be divided into two parts:

1. Stereo SLAM and depth calculation

   This part mainly performs stereo SLAM to obtain the camera's motion trajectory and the depth of the object, and extract the ORB feature points. Then using feature points to transform from 2D to 3D to obtain the 3D coordinates of a point in space.
2. 3D measurement of unknown environment

   This part is based on stereo SLAM. After a series of image processing operations, using the feature points obtained by stereo SLAM can obtain the contour and size information of the target object. The height of tree and the width of road in unknown environments can be measured finally.

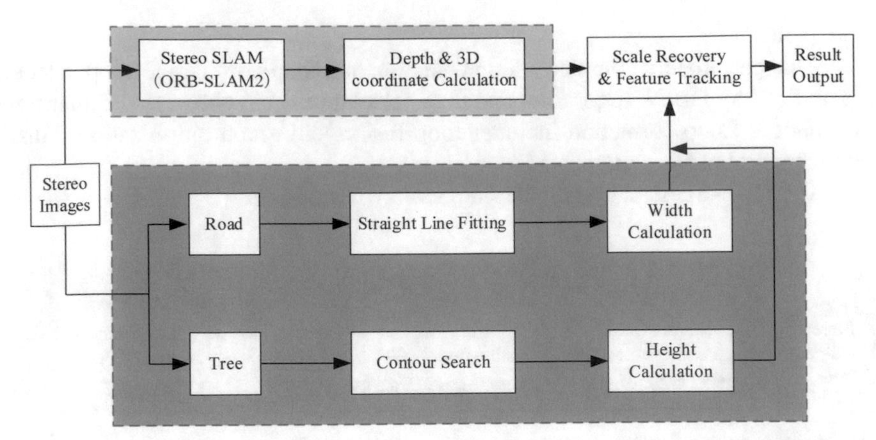

Fig. 2. The overall system diagram of the unknown environment 3D measurement algorithm proposed in this paper

3 The Proposed Method

3.1 Depth and 3D Coordinate Calculation

Stereo SLAM can obtain the depth and 3D coordinates of pixels. Figure 3 shows the stereo camera model.

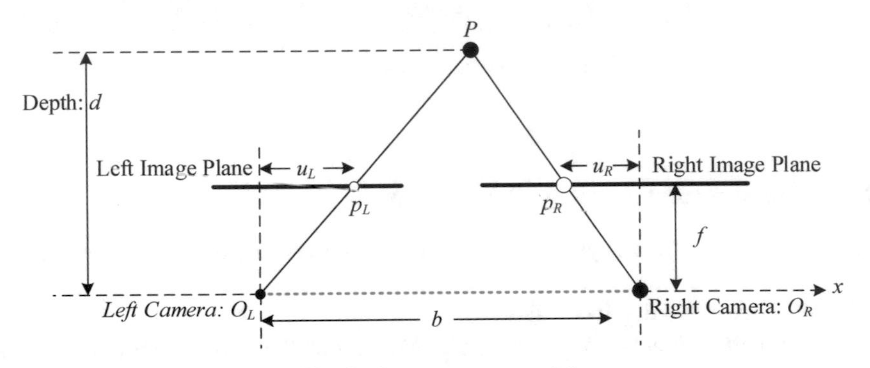

Fig. 3. Stereo camera model

From the geometric relationship in Fig. 3, we can get:

$$\frac{d-f}{d} = \frac{b-u_L+u_R}{b} \tag{1}$$

where d is the depth, f is the focus of camera, b is the baseline distance of the two camera, u_L and u_R are the pixel coordinate values of a point in the left camera and the right camera respectively. The difference between the two pixels is called parallax which is denoted as Δu:

$$\Delta u = u_L - u_R \tag{2}$$

The depth can be calculated from (1) and (2):

$$d = \frac{bf}{\Delta u} \tag{3}$$

According to the pinhole model of camera, the coordinate relationship between space point P and the image pixel is:

$$\begin{cases} s_L p_L = KP \\ s_R p_R = K(RP + t) \end{cases} \tag{4}$$

where s_L and s_R are the scales of the left and right cameras which are unknown quantities, K is the internal parameter matrix, R and t are the relative pose between the left and right cameras. K, R and t can be obtained by camera calibration. The 3D coordinates of the point P in space can be calculated by using (4).

3.2 Road Width Calculation

The steps for calculating the road width are as follows:

1. Image binaryzation and point set division
   First, the image is binarized. Then only keep the road area and remove other irrelevant areas. For the road area, search for the left and right borders. The leftmost pixel in each row of the road area is divided into the left boundary, which is recorded as the set L, and the rightmost pixel is divided into the right boundary, which is recorded as the set R.
2. Straight line fitting
   The least square method is used to fit straight lines to the left boundary set L and the right boundary set R respectively. Supposing the form of the fitted straight line is $y = \varphi(x) = kx + b$. The least squares method requires the smallest sum of squared errors:

$$e^2 = \sum_{i=1}^{m} (\varphi(x_i) - y_i)^2 = \sum_{i=1}^{m} ((kx_i + b) - y_i)^2 \tag{5}$$

So the parameters k and b can be calculated by using this principle:

$$\begin{bmatrix} \sum_{i=1}^{m} x_i^2 & \sum_{i=1}^{m} x_i \\ \sum_{i=1}^{m} x_i & m \end{bmatrix} \begin{bmatrix} k \\ b \end{bmatrix} = \begin{bmatrix} \sum_{i=1}^{m} x_i y_i \\ \sum_{i=1}^{m} y_i \end{bmatrix} \tag{6}$$

3. Width calculation

   In the same row of pixels, find the two feature points closest to the left and right straight lines, and calculate the distance. The distance is the estimated value of the road width. It's worth noting that, in order to improve the accuracy, it is necessary to avoid using far points as much as possible.

3.3 Tree Height Calculation

The steps for calculating the tree height are as follows:

1. Image binaryzation

   First, the image is binarized. Then only keep the tree area and remove other irrelevant areas.
2. Edge detection

   Edge detection is based on Canny operator. Canny operator belongs to the first-order differential filter. It has three advantages: multi-peak response, optimal zero-crossing point positioning and maximum SNR. The Canny operator is an optimization operator with strong filtering and diversified detection stages. Before processing the Canny operator, it is necessary to use a Gaussian smoothing filter to smooth the image to remove noise. By calculating the gradient difference, the neighborhood or local intensity value can be completed, highlighting the significant change of the enhanced edge.
3. Mathematical morphology processing

   Mathematical morphology processing includes dilation and erosion. They can be used to extract the boundary which are used here to make the extracted edge more clear and obvious.
4. Find the feature points to calculate tree height

   Calculate the smallest rectangle that can enclose the outline of the tree, and find the highest feature point inside the rectangle. Then find the ground feature point in the entire image, and use the 3D coordinates of these two feature points to calculate the height of the tree.

4 Implementation and Performance Analysis

This paper uses the open source dataset cityscapes to conduct experiments. This dataset contains 1100 stereo pictures. It's often used for computer vision. The simulation environment is Ubuntu16.04. Figure 4 shows the camera trajectory obtained by stereo SLAM.

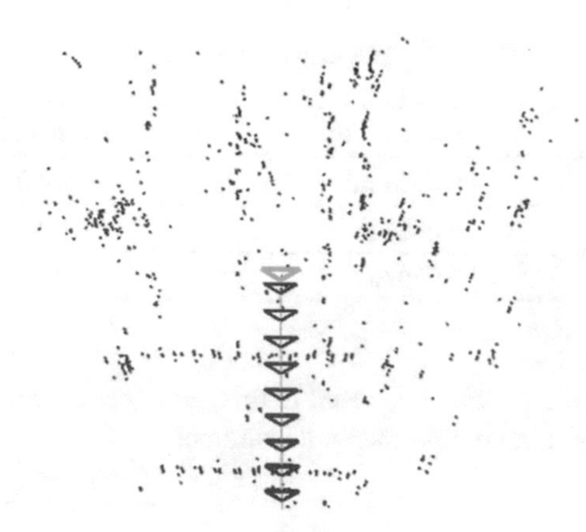

Fig. 4. The trajectory of camera obtained by stereo SLAM

Figure 5 shows the result of ORB feature points extracted by stereo SLAM and the calculation results of road width and tree height in the image.

Fig. 5. The calculation results of road width and tree height

The three pictures selected in Fig. 5 have different lane numbers. The measurement results are shown in Table 1.

Table 1. The measurement results

Lanes number	Road width	Tree height
3	6.0624 m	9.7390 m
3	6.0624 m	5.7967 m
2	4.0567 m	6.0016 m

It can be seen that the method proposed in this paper can effectively obtain the road width and the tree height in an unknown environment.

5 Conclusion

This paper proposes a 3D measurement method of unknown environment based on stereo SLAM. Using stereo SLAM to solve the real-time motion trajectory of the camera and the 3D coordinates of the feature points. Based on the stereo SLAM, a series of image processing operations are used to estimate the size of target object. The experiment show that the stereo SLAM can accurately estimate the camera's motion and calculate the 3D coordinates of the feature points in real time, and the whole system can accurately obtain the road width and the tree height.

Acknowledgment. This paper is supported by national nature scinece foundation of China (41861134010, 61971162) and National Aeronautical Foundation of China (2020Z066015002).

References

1. Deutscher, J., Isard, M., MacCormick, J.: Automatic camera calibration from a single Manhattan image. In: Heyden, A., Sparr, G., Nielsen, M., Johansen, P. (eds.) ECCV 2002. LNCS, vol. 2353, pp. 175–188. Springer, Heidelberg (2002). https://doi.org/10.1007/3-540-47979-1_12
2. Hoiem, D., Efros, A.A., Hebert, M.: Putting objects in perspective. Int. J. Comput. Vision **80**(1), 3–15 (2006). https://doi.org/10.1007/s11263-008-0137-5
3. Lalonde, J.F., Hoiem, D., Efros, A.A., et al.: Photo clip art. ACM Trans. Graph. **26**(3), 3 (2007)
4. Andalo, F.A., Taubin, G., Goldenstein, S.: Efficient height measurements in single images based on the detection of vanishing points. Comput. Vis. Image Underst. **138**, 51–60 (2015)
5. Criminisi, A., Reid, I., Zisserman, A.: Single view metrology. In: Seventh IEEE International Conference on Computer Vision, vol. 1, pp. 434–441 (2002)
6. Davison, A.J., Reid, I.D., Molton, N.D., et al.: Monoslam: real-time single camera slam. IEEE Trans. Pattern Anal. Mach. Intell. **29**(6), 1052–1067 (2007)
7. Klein, G., Murray, D.: Parallel tracking and mapping for small AR workspaces. In: IEEE & ACM International Symposium on Mixed & Augmented Reality. ACM (2008)
8. Di, K., Peng, M.: Wide baseline mapping for mars rovers. Photogramm. Eng. Remote. Sens. **77**(6), 609–618 (2011)

9. Labbe, M., Michaud, F.: Online global loop closure detection for large-scale multi-session graph-based SLAM. In: 2014 IEEE/RSJ International Conference on Intelligent Robots and Systems. IEEE (2014)
10. Mur-Artal, R., Tardos, J.D.: ORB-SLAM2: an open-source slam system for monocular, stereo and RGB-D cameras. IEEE Trans. Rob. **33**(5), 1255–1262 (2017)
11. Mur-Artal, R., Montiel, J.M.M., Tardos, J.D.: ORB-SLAM: a versatile and accurate monocular slam system. IEEE Trans. Rob. **31**(5), 1147–1163 (2015)

Resource Allocation for Full-Duplex V2V Communications Based on Deep Reinforcement Learning

Keshan Zheng[1], Liang Han[1,2(✉)], and Yupeng Li[1,2]

[1] College of Electronic and Communication Engineering, Tianjin Normal University, Tianjin 300387, China
hanliang@tjnu.edu.cn

[2] Tianjin Key Laboratory of Wireless Mobile Communications and Power Transmission, Tianjin Normal University, Tianjin 300387, China

Abstract. In this paper, we investigate the application of full-duplex technology in vehicular networks and develop a novel decentralized resource allocation mechanism for full-duplex vehicle-to-vehicle (V2V) communications based on deep reinforcement learning. According to the mechanism, a V2V link is regarded as an agent, and everything beyond the particular V2V link is regarded as the environment. Each agent can learn the information of the environment independently and meet the requirements of V2V delay constraints while minimizing interference in the vehicle-to-infrastructure (V2I) communication process. From the simulation results, the overall capacity of the full-duplex is higher than that of the half-duplex.

Keywords: Full-duplex V2V · Resource allocation · Deep reinforcement learning

1 Introduction

Efficient traffic control and management is a fundamental and urgent challenge in our ever-growing cities. However, the increase in vehicles and the insufficient communications between neighboring vehicles during the driving status has caused too frequent traffic accidents and congestion of urban roads. Vehicle-to-vehicle (V2V) communication has been developed to solve this problem [1–3].

Spectrum resources are becoming increasingly scarce with the increase in the number of vehicles connected to the network. To improve the spectrum efficiency under the situation of an increasing shortage of spectrum resources, we use reinforcement learning (RL) to solve the problem of resource allocation in vehicle communication.

In the previous works, the research of V2V communication was mostly based on the transmission mode of half-duplex, which uses the spectrum insufficiently. To address this problem, we develop the full-duplex [4–6] transmission mode in V2V communication. The full-duplex transmission mode uses the same frequency band for simultaneous transmission of information. Compared with the half-duplex transmission mode, the full-duplex transmission mode has the characteristics of high-frequency band utilization and

Q. Liang et al. (Eds.): Artificial Intelligence in China, LNEE 854, pp. 586–592, 2022.
https://doi.org/10.1007/978-981-16-9423-3_73

short communication delay. The innovation of this paper is to use the deep reinforcement learning [7–9] method to find the optimal sub-band and power level for transmission for each vehicle to be communicated, to realize full-duplex V2V communication, thereby realizing the improvement of frequency band utilization.

2 System Model

As shown in Fig. 1, we consider the area covered by a base station (BS), with $\mathcal{M} = \{1, 2, \cdots, M\}$ cellular users (CUEs). Cellular users access the BS through V2I links to communicate with the BS. The communication resources they use are allocated by the BS. Now there are $\mathcal{K} = \{1, 2, \cdots, K\}$ vehicle users (VUEs) in the area covered by the same BS, A vehicle performs V2V full-duplex communication with neighboring vehicles.

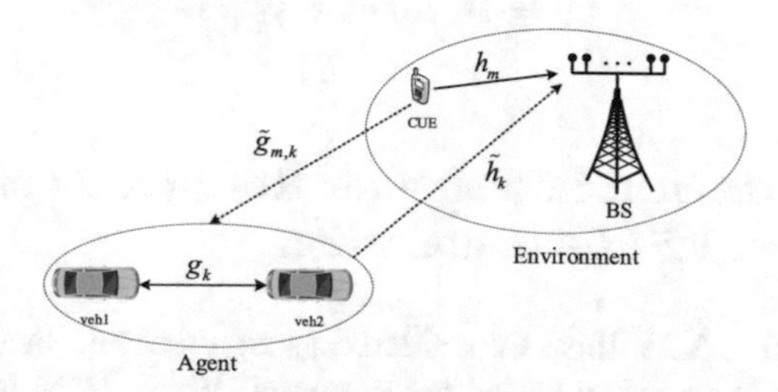

Fig. 1. System model

To improve the utilization of spectrum resources and minimize the interference to cellular users, we assume that the V2V link reuses the spectrum resources of the vehicle-to-infrastructure (V2I) uplink. As a result, their communication links will interfere with each other, which will affect the communication rate. We consider that the gain of the transmit and receive channels of full-duplex communication is the same, for $m \in \mathcal{M}$, the SINR of the mth CUE can be expressed as:

$$\gamma_m^c = \frac{P_m^c h_m}{\sigma^2 + \sum\limits_{k \in \mathcal{K}} \rho_k^m \cdot 2P_k^v \tilde{h}_k} \tag{1}$$

where P_m^c and P_k^v denote the transmission powers of the mth CUE and the kth VUE, h_m is the power gain of the channel corresponding to the mth CUE, $\tilde{h}_k$ is the interference power gain of the kth VUE for V2I, ρ_k^m is the spectrum allocation indicator with $\rho_k^m = 1$ if the kth VUE reuses the spectrum of the mth CUE and $\rho_k^m = 0$ otherwise, and σ^2 is the noise power. The capacity of the mth CUE is:

$$C_m^c = W \log(1 + \gamma_m^c) \tag{2}$$

Assuming that the kth VUE and the k'^{th} VUE are in full-duplex communication. Since the distance of V2V communication is relatively short, then the SINR at the kth VUE can be expressed as:

$$\gamma_{k,k'}^v = \frac{P_{k'}^v g_{k',k}}{\sigma^2 + \beta_k P_k^v + G_d + G_c} \tag{3}$$

where: $G_c = \sum_{m \in \mathcal{M}} \rho_k^m P_m^c g_{m,k}, G_d = \sum_{m \in \mathcal{M}} \sum_{k'',k''' \in \mathcal{K}, k'' \neq k'''} \rho_k^m \rho_{k''}^m \left(P_{k''}^v g_{k'',k'''} + P_{k'''}^v g_{k''',k''} \right)$ denote the interference power introduced by sharing the spectrum with V2I and the interference power introduced by sharing the spectrum with other V2V respectively, $g_{m,k}$ is the interference power gain of the mth CUE, $g_{k'',k'''}$ and $g_{k''',k''}$ are the interference power gain of the k''^{th} and k'''^{th} VUE. $\beta_k P_k^v$ denote the interference power after self-interference suppression, and σ^2 is the noise power. $P_{k'}^v$ denote the transmission power of k'^{th} and the transmission, and $\beta_k, \beta_{k'}$ denote the self-interference cancellation coefficient of the kth VUE and the self-interference cancellation coefficient of the k'^{th} VUE, respectively. Hence the full-duplex capacity of the kth VUE can be expressed as:

$$C_k^v = 2W \log\left(1 + \gamma_{k,k'}^v\right) \tag{4}$$

3 Deep Reinforcement Learning for Resource Allocation of Full-Duplex V2V Communication

In our system model, a V2V link is considered as an agent and everything beyond the particular V2V link is regarded as the environment. When V2V link communication is to be carried out, since the BS does not have information about the V2V link, the selection of spectrum and power during V2V link communication is made based on their current observations. We make the environment and the agent interact with each other continuously to obtain rewards and maximize the expected cumulative discounted rewards. In this process, the state of the environment observable by each V2V link consists of several parts: instantaneous channel information of the V2V link: G_t^m, link interference at the previous moment: I_{t-1}, power from V2V to BS: P_t, the selection of the neighbor sub-band: N_{t-1}, the remaining load of the vehicle:L_t, the remaining time after the transfer is completed:U_t. As shown in Fig. 2, at each time t, the agent observes a state, s_t, from the state space,S, and accordingly takes an action, a_t, from the action space, A, selecting sub-band and transmission power. Based on the actions taken by the agents, the environment transits to a new state, s_{t+1}, and each agent receives a reward, r_t, from the environment. Assuming that after the end of the interaction, the rewards received in each round in this process are as follows: $r_t, r_{t+1}, r_{t+2}, ..., r_n$. we expect that the sum of all rewards from time t to the end is:

$$R_t = r_t + \mu r_{t+1} + \mu^2 r_{t+2} + ... + \mu^{n-t} r_n \tag{5}$$

where μ is called the discount factor, and we expect that the value R_t can be maximized. A simple transformation of the above formula is:

$$R_t = r_t + \mu(r_{t+1} + \mu r_{t+2} + ...) = r_t + \mu R_{t+1} \tag{6}$$

This equation reflects the relationship between the sum of rewards at two adjacent moments. In our model, when the agent interacts with the environment, we express the

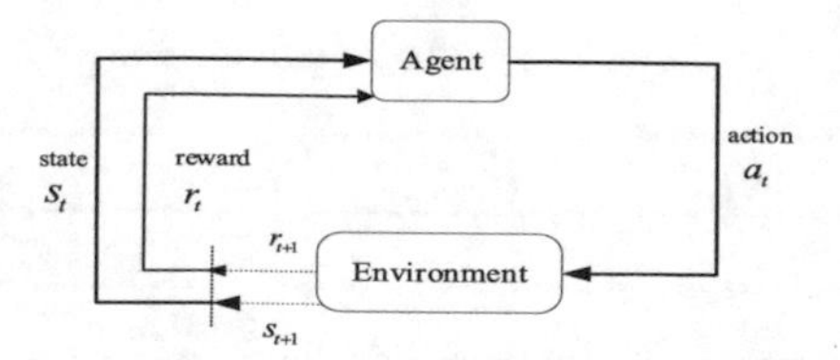

Fig. 2. Reinforcement learning process diagram

reward from the environment to the agent in the following form:

$$r_t = \lambda_c \sum_{m \in M} C_m^c + \lambda_d \sum_{k \in K} C_k^v - \lambda_p (T_0 - U_t) \tag{7}$$

where T_0 is the maximum tolerable latency, and λ_c, λ_d and λ_p are denote weights. The reward is determined by the capacities of the V2I and V2V links and the latency constraints of the corresponding V2V link. the largest sum of discount rewards is:

$$R_t = \sum_{n=0}^{\infty} \mu^n r_{t+n} \tag{8}$$

where $\mu \in (0, 1)$ is the discount factor. Therefore, we need to train the agent to make it act based on a certain strategy π, so that the final R_t is the largest. In reinforcement learning, the action-value function is defined as follows:

$$Q_\pi(s_t, a_t) = E[R_t | s_t, a_t] \tag{9}$$

It is called the value-function, whose calculated value is denoted as q. Since the result also depends on the strategy π, further maximize both sides of Eq. (9):

$$Q^*(s_t, a_t) = \max_\pi Q_\pi(s_t, a_t) \tag{10}$$

Thus, the final Q value will be independent of the influence of strategy π. Similarly, the above transformation is also performed on the Eq. (8) to obtain:

$$Q(s_t, a_t) = r_t + \mu \cdot Q(s_{t+1}, a_{t+1}) \tag{11}$$

The value of q can be calculated by using the equation and the real-time rewards obtained in each round. We use a neural network, DQN, to simulate a function to implicitly calculate Q each round. The neural network can be trained using the TD algorithm [10]. The r_t can be used as the supervision signal when training the neural network.

4 Simulation Results

The simulation environment is as follows: a BS is located in the center of Manhattan streets simulated by 3GPP TR 3.855 and covers the entire street area. Vehicles are randomly generated in lanes around the street according to the spatial Poisson distribution,

Table 1. Simulation parameters

Parameter	Value
Carrier frequency	2 GHz
Bandwidth per channel	1.5 MHz
BS antenna height	25 m
BS antenna gain	8 dBi
BS receiver noise figure	5 dB
Vehicle antenna height	1.5 m
Vehicle antenna gain	3 dBi
Vehicle receiver noise figure	9 dB
Vehicle speed	36 km/h
Neighbor distance threshold	150 m
Number of lanes	3 in each direction (12 in total)
Latency constraints for V2V link T_0	100 ms
V2V transmit power level list	[23, 20, 10, 5] dBm
Noise power σ^2	−114 dBm
$[\lambda_c, \lambda_d, \lambda_p]$	[0.1, 0.9, 1]
SINR threshold in broadcast	1 dB
Self-interference cancellation coefficient β	−120 dB
Activation function	ReLu
Learning rate	0.01 (decreases exponentially)

and each vehicle communicates with the four closest vehicles. The detailed parameters can be found in Table 1.

In Fig. 3, we plot the relationship between the number of vehicles and the sum of V2I link capacity. As can be seen from Fig. 3, with the increase of the vehicles, the number of V2V links increases, as a result, the interference to the V2I link grows, therefore the V2I capacity drops. However, the full-duplex V2V communication obtained by the reinforcement learning method has better performance to mitigate the interference of V2V links to the V2I communications.

In Fig. 4, different V2V links have a different probability of selecting transmission power based on the different remaining time limits. Generally, we think that when the remaining transmission time is small, the link should choose high power to transmit the data as soon as possible, but according to the simulation results, when the remaining transmission time is extremely short, the number of links with high power transmission is reduced. Since the remaining transmission time is too short, and the transmission cannot be completed even if the power is increased. Therefore, by saving power and reducing the interference to V2I, greater rewards can be obtained. This shows that DQN can learn some hidden information.

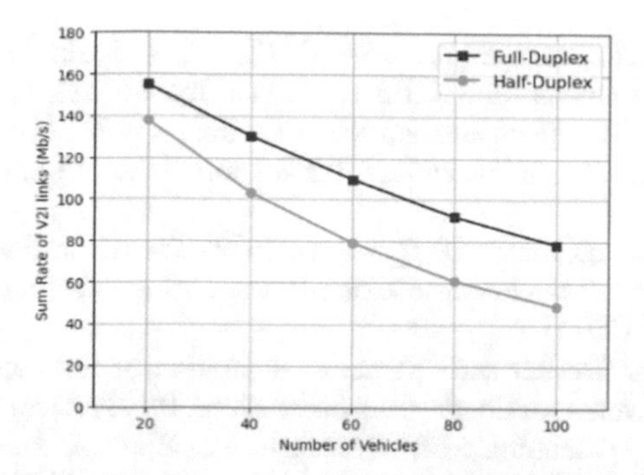

Fig. 3. Mean rate versus the number of vehicles

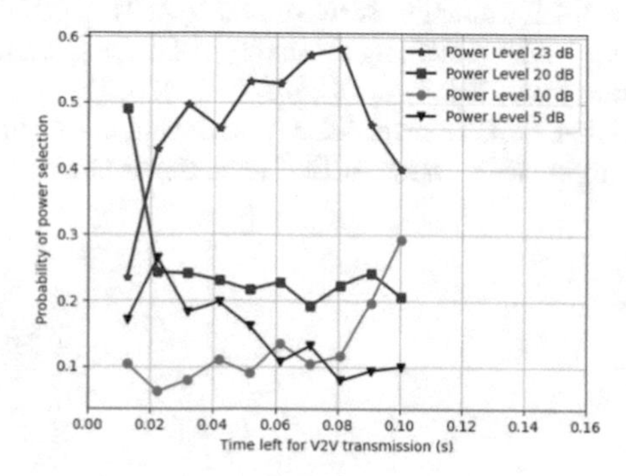

Fig. 4. The probability of power level selection with the remaining time for transmission

5 Conclusion

In this paper, we proposed a way to use deep reinforcement learning to solve the resource allocation problem in full-duplex V2V communication. According to the characteristics of reinforcement learning, each agent can learn the information of the environment independently meanwhile which can meet the requirements of V2V delay constraints while minimizing interference in the V2I communication process. Eventually, the purpose of improving spectrum utilization can be achieved.

Acknowledgment. This work was supported by the National Natural Science Foundation of China (61701345, 61901301), Natural Science Foundation of Tianjin (18JCZDJC31900, 18JCQNJC70900), and Tianjin Education Commission Scientific Research Plan (2017KJ121).

References

1. Chen, S., et al.: Vehicle-to-everything (v2x) services supported by LTE-based systems and 5G. IEEE Commun. Stand. Mag. **1**(2), 70–76 (2017)
2. Liu, C., Chau, K.T., Wu, D., Gao, S.: Opportunities and challenges of vehicle-to-home, vehicle-to-vehicle, and vehicle-to-grid technologies. Proc. IEEE **101**(11), 2409–2427 (2013)

3. Sun, W., Strom, E.G., Brannstrom, F., Sou, K.C., Sui, Y.: Radio resource management for D2D-based V2V communication. IEEE Trans. Veh. Technol. **65**(8), 6636–6650 (2016)
4. Tang, A., Wang, X.: A-duplex: medium access control for efficient coexistence between full-duplex and half-duplex communications. IEEE Trans. Wirel. Commun. **14**(10), 5871–5885 (2015)
5. Huang, X., He, J., Li, Q., Zhang, Q., Qin, J.: Optimal power allocation for multicarrier secure communications in full-duplex decode-and-forward relay networks. IEEE Commun. Lett. **18**(12), 2169–2172 (2014)
6. Tan, L.T., Le, L.B.: Design and optimal configuration of full-duplex MAC protocol for cognitive radio networks considering self-interference. IEEE Access **3**(1), 2715–2729 (2015)
7. Arulkumaran, K., Deisenroth, M.P., Brundage, M., Bharath, A.A.: Deep reinforcement learning: a brief survey. IEEE Signal Process. Mag. **34**(6), 26–38 (2017)
8. Luong, N.C., et al.: Applications of deep reinforcement learning in communications and networking: a survey. IEEE Commun. Surv. Tutor. **21**(4), 3133–3174 (2019)
9. Mao, Q., Hu, F., Hao, Q.: Deep learning for intelligent wireless networks: a comprehensive survey. IEEE Commun. Surv. Tutor. **20**(4), 2595–2621 (2018)
10. Keerthisinghe, C., Verbic, G., Chapman, A.C.: A fast technique for smart home management: ADP with temporal difference learning. IEEE Trans. Smart Grid **9**(4), 3291–3303 (2018)

Macloed Estimation Algorithm for Real-Time Measurement of Heart Rate with CW Doppler Radar

Zi-Kai Yang[1], Heping Shi[2(✉)], and Junhua Gu[1]

[1] The School of Artificial Intelligence, Hebei University of Technology, Tianjin 300401, China
zikai_yang@tju.edu.cn, jhgu@hebut.edu.cn

[2] The School of Automobile and Transportation, Tianjin University of Technology and Education (TUTE), Tianjin 300222, China
shiheping@tju.edu.cn

Abstract. Real-time monitoring of the heart rate (HR) based on a non-contact continuous-wave (CW) Doppler radar is necessary for the diagnosis and prevention of cardiovascular diseases. To achieve it, the window length should be shortened, which will result in the spectrum leakage and degrade the measurement accuracy of HR. In this paper, the Macloed estimation algorithm is proposed to realize the real-time detection of HR. Compared with the traditional discrete Fourier transform (DFT) method, simulation results show that the proposed method can degrade the measurement error of HR to 1.75%, which verifies the effectiveness of the proposed method.

Keywords: Non-contact heart rate (HR) detection · Continuous-wave (CW) Doppler radar · Macloed estimation

1 Introduction

Nowadays, cardiovascular diseases have become the main cause of human death [1]. To diagnose and prevent the cardiovascular diseases, measurement and long-term monitoring of human heart rate (HR) are essential [2]. In the past few decades, many contact sensors, such as electrocardiogram (ECG) sensors [3] and ultrasonic cardiogram (UCG) sensors [4], have been invented to monitor the HR by connecting to the chest, wrist, or fingertips [5]. However, they will make people uncomfortable under long-term monitoring [6]; In addition, they are difficult to be applied to some special populations, such as burn patients with damaged skin and mental patients with sensitive skin [7]. Fortunately, non-contact sensors can overcome these drawbacks well [8, 9].

Among many non-contact sensors, due to insensitive to light and temperature, continuous-wave (CW) Doppler radar is the most popular one, which can detect micro motions caused by human physiological movements through the phase modulation effect [10, 11]. With the CW Doppler radar, the average HR in a long time window can be accurately extracted when the measurement time exceeds 10 s [12, 13]. However, compared with the average HR, instantaneous HR, i.e., the HR is monitored in real-time,

Q. Liang et al. (Eds.): Artificial Intelligence in China, LNEE 854, pp. 593–599, 2022.
https://doi.org/10.1007/978-981-16-9423-3_74

is more useful for the diagnosis and prevention of cardio-vascular diseases [14]. As a result, it is necessary for the CW Doppler radar to achieve the real-time monitoring of HR.

To monitor the HR in real-time, a time window length of less than 5 s is necessary [15, 16]. However, when using the discrete Fourier transform (DFT) method, which is the most common method in extracting HR [17–20], the 5 s time window length is too short to obtain sufficient spectrum resolution [14]. To overcome this obstacle, various studies have been proposed on real-time monitoring of the HR using the CW Doppler radar. The time-window-variation technique combined with the fast Fourier transform and wavelet transform were proposed in [15] and [14], respectively. A polyphase-based discrete cosine transform method was proposed in [21]. Although these methods mentioned above can improve the extraction accuracy of HR, they still exhibit significant errors (greater than 3.4%). According to the Ref. [9], it is necessary to reduce the measurement error of HR to less than 2%.

In this paper, the Macloed estimation algorithm is introduced to achieve the real-time monitoring of the HR with high accuracy. In order to verify the effectiveness of the proposed method, simulations are performed and the results show that the measurement error of HR can be reduced to less than 2%.

2 CW Doppler Radar Sensor

Figure 1 shows the block diagram of the CW Doppler radar system for HR monitoring. The transmitted signal of the radar can be expressed as

$$T(t) = \cos\left[2\pi ft + \phi(t)\right] \tag{1}$$

where f is the carrier frequency, t is the elapsed time, and $\phi(t)$ is the phase noise of the transmitter. Subsequently, the phase of $T(t)$ is modulated by the human physiological motion $x(t)$, which can be expressed as

$$\begin{aligned} x(t) &= x_h(t) + x_r(t) \\ &= m_h \sin \omega_h t + m_r \sin \omega_r t \end{aligned} \tag{2}$$

where $x_h(t)$ and $x_r(t)$ are the chest-wall displacement caused by heartbeat and respiration, respectively. Specifically, m_h and m_r are the amplitude caused by heartbeat and respiration, respectively; ω_h and ω_b are the angular frequency caused by heartbeat and respiration, respectively. Assume the radar is positioned at an initial distance of d_0 facing a subject. Neglecting the amplitude variation, the reflected signal captured by the receiver can be expressed as

$$R(t) = \cos[2\pi ft - \frac{4\pi x_h(t)}{\lambda} - \frac{4\pi x_r(t)}{\lambda} - \frac{4\pi d_0}{\lambda} + \phi(t - \frac{2d_0}{c})] \tag{3}$$

where c is the velocity of radio wave (the propagation velocity of light), $\lambda = c/f$ is the wavelength of carrier. Then, the $R(t)$ is mixed with the $L(t)$, which is derived from the same source as the transmitted signal. As a result, the resulting baseband signal can be expressed as

$$\begin{aligned} B_I(t) &= \cos(\frac{4\pi x_h(t)}{\lambda} + \frac{4\pi x_r(t)}{\lambda} + \frac{4\pi d_0}{\lambda} + \Delta\phi) \\ B_Q(t) &= \sin(\frac{4\pi x_h(t)}{\lambda} + \frac{4\pi x_r(t)}{\lambda} + \frac{4\pi d_0}{\lambda} + \Delta\phi) \end{aligned} \tag{4}$$

where $\Delta\phi = \phi(t) - \phi(t - 2d_0/c)$ is the residual phase noise, which can be neglected according to range-correlation theory [22]. To solve the null point problem encountered in the process of analyzing the baseband spectrum, the B_I(t) and B_Q(t) are combined as a complex signal, which can be expressed as [17]

$$B(t) = B_I(t) + jB_Q(t) \tag{5}$$

where j is the square root of -1. After sampling, the digitized $B[n]$ is generated as the output of the analog-to-digital conversion (ADC), for the real-time extraction of HR in digital signal processing (DSP), where the Macloed estimation algorithm proposed in this study will be used.

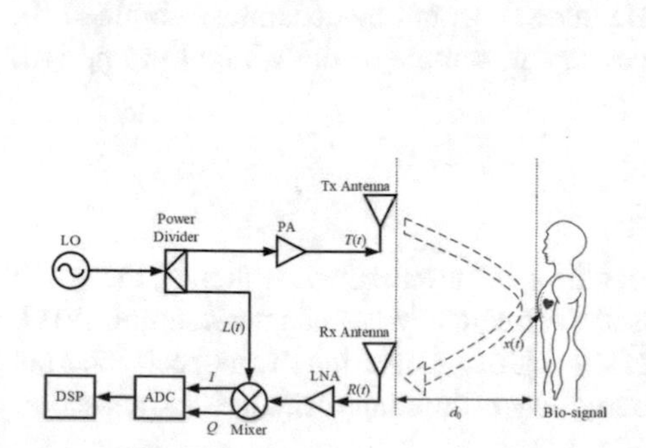

Fig. 1. The CW Doppler radar system

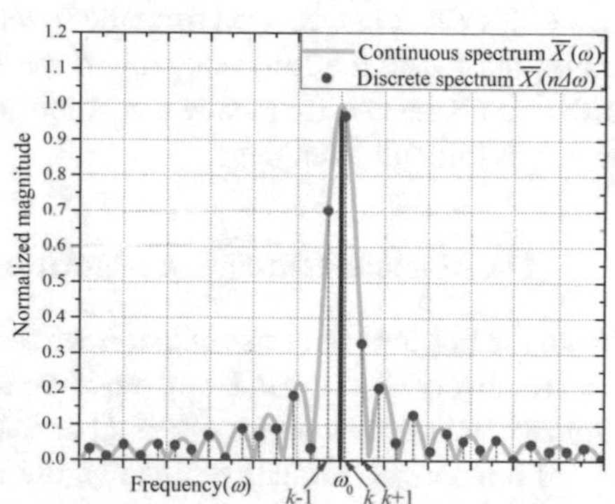

Fig. 2. The spectrum of $x(t)$, $\overline{x}(t)$ and $\overline{x}[n]$

3 Spectrum Leakage and the Macloed Estimation Algorithm

3.1 Spectrum Leakage

In order to facilitate the description of the spectrum leakage phenomenon, it is assumed that the tested signal $x(t)$ is a single-frequency signal, which is expressed as

$$x(t) = A_0 e^{j\omega_0 t} \tag{6}$$

where A_0 and ω_0 are the amplitude and angular frequency of the signal, respectively. As a result, the Fourier transform of $x(t)$ is

$$X(\omega) = A \cdot 2\pi \cdot \delta(\omega - \omega_0) \tag{7}$$

where $\delta(\omega)$ is the Dirac Delta function. It can be seen from the Eq. (7) that for the signal $x(t)$, there is only one line in the spectrogram, which is located at ω_0, as shown by the red line in Fig. 2. When the time duration is T, the signal $x(t)$ can be considered to be truncated by a rectangular window $w_T(t)$, which is expressed as

$$w_T(t) = \begin{cases} 1; \ 0 < t \le T \\ 0; \ \text{otherwise} \end{cases} \tag{8}$$

Therefore, the truncated signal $\overline{x}(t)$ can be expressed as

$$\overline{x}(t) = x(t) \cdot w_T(t) \tag{9}$$

Neglecting the change of phase, the spectrum of $\overline{x}(t)$ is shown by the cyan line in Fig. 2. Compared with the spectrum of signal $x(t)$, it can be seen that the spectrum of $\overline{x}(t)$ is no longer a single line and the leakage has arisen.

Then, the signal $\overline{x}(t)$ is sampled as

$$\overline{x}[n] = \overline{x}(n\Delta t), \ n = 0, 1, 2, \ldots, N - 1 \tag{10}$$

where $N = T / \Delta t$. The spectrum of $\overline{x}[n]$ is shown by the magenta points in Fig. 2, where $\Delta\omega = 2\pi / T$. It can be seen from the spectrum of $\overline{x}[n]$ that since the ω_0 is not an integer multiple of $\Delta\omega$, the frequency ω_0 of the signal $x(t)$ cannot be accurately obtained. In order to obtain the frequency ω_0, it is necessary to sample in the whole time period, which is difficult to achieve.

3.2 The Macloed Estimation Algorithm

From the Fig. 2, it can be seen that if we select the k as the extracted frequency, there will be a relatively large error. However, if we used the frequency-domain peak sample $\overline{X}[k]$, and two adjacent samples $\overline{X}[k-1]$ and $\overline{X}[k+1]$, the estimation of the peak location could be more accurate. Specifically, the frequency estimation formula is expressed as [23]

$$k_{peak} = k + \left(\sqrt{1 + 8d^2} - 1\right)/(4d) \tag{11}$$

where

$$d = \frac{\text{Re}\left[\overline{X}[k-1]\overline{X}[k] - \overline{X}[k+1]\overline{X}[k]\right]}{\text{Re}\left[2\left|\overline{X}[k]\right|^2 + \overline{X}[k-1]\overline{X}[k] + \overline{X}[k+1]\overline{X}[k]\right]} \tag{12}$$

4 Results

To verify the effectiveness of the Macloed estimation algorithm and compare its performance with the conventional DFT approach, we numerically simulated the frequency estimation scattered by a 3-s baseband signal $B(t)$ in the presence of noises. For relaxed people, the HR varies within 1–3 Hz, while the respiration rate typically varies in a lower range of 0.1–0.3 Hz [5]; Meanwhile, the body movement caused by heartbeat is less than 2 mm and by respiration is typically less than 1 cm [16]. As a result, we assume the frequency and amplitude of respiration are 0.21 Hz and 1 mm, respectively. The amplitude of heartbeat is 0.4 mm and the frequency of HR varies from 1.4 Hz to 1.7 Hz. In all cases, Gaussian noises were added to the scattered fields so that the SNRs of the received signals are both 20 dB [24]. The extraction results of HR are shown in Fig. 3. It is obvious that the accuracy of using Macloed estimation algorithm to extract the HR is higher than that of the traditional DFT method. In order to further demonstrate the effectiveness of the proposed method in this paper, the HR varies from 1 Hz to 2.2 Hz and the absolute relative measurement errors of HR is shown in Fig. 4, where the absolute relative measurement error is expressed as

$$error = \left| \frac{f_{\text{meas}} - f_{\text{ref}}}{f_{\text{ref}}} \right| \times 100\% \tag{13}$$

where f_{meas} is the measured HR, f_{ref} is the reference HR. Obviously, the HR errors measured using the Macloed estimation algorithm are lower than that of traditional DFT method. Specifically, the average measurement error of HR is shown in Table 1.

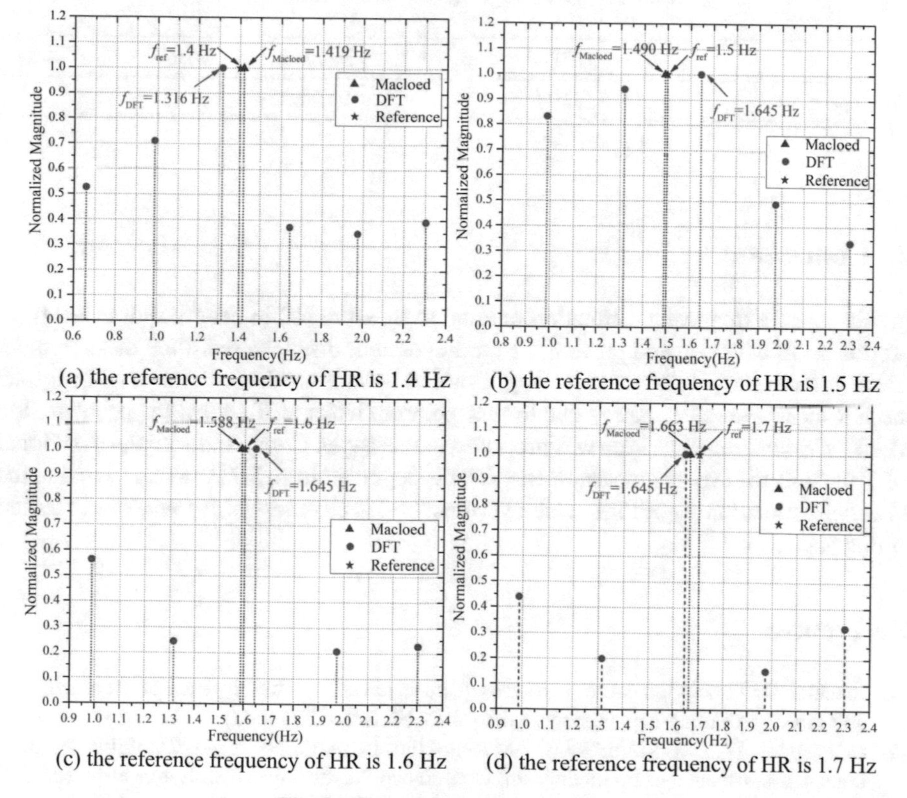

(a) the reference frequency of HR is 1.4 Hz
(b) the reference frequency of HR is 1.5 Hz
(c) the reference frequency of HR is 1.6 Hz
(d) the reference frequency of HR is 1.7 Hz

Fig. 3. The extraction results of HR

Compared to the traditional DFT method, the Macloed estimation algorithm reduces the average measurement error of HR from 5.71% to 1.75%, which verifies the effectiveness of the Macloed estimation algorithm.

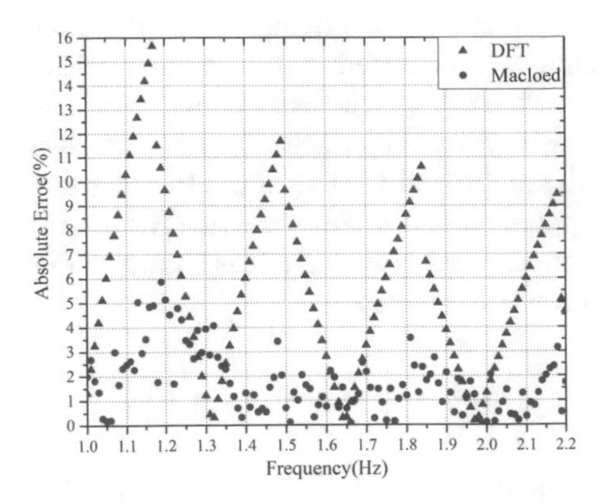

Fig. 4. The absolute relative measurement errors of HR

Table 1. The average measurement error of HR

	DFT	Macloed estimation
Average errors	5.71%	1.75%

5 Conclusion

In this paper, a frequency estimation method to measure HR in near real-time has been proposed. To diagnose and prevent the cardiovascular diseases, real-time measurement of HR is necessary, so the window length should be shortened. However, it will degrade the HR measurement accuracy due to the spectrum leakage. To solve the problem, the Macloed estimation algorithm was proposed in this paper. Compared with the traditional DFT method, the average measurement error is degraded from 5.71% to 1.75% using the Macloed estimation algorithm, which verifies the effectiveness of the proposed method in this paper.

References

1. Stuckler, D.: Population causes and consequences of leading chronic diseases: a comparative analysis of prevailing explanations. Milbank Q. **86**, 273–326 (2008)
2. Massagram, W., Boric-Lubecke, O., Macchiarulo, L., et al.: Heart rate variability monitoring and assessment system on chip. In: Conference Proceedings IEEE Engineering Medical Biology Society (2005). https://doi.org/10.1109/IEMBS.2005.1616214

3. Alghatrif, M., Lindsay, J.: A brief review: history to understand fundamentals of electrocardiography. J. Commun. Hosp. Intern. Med. Perspect. **2**, 14383 (2012)
4. Anavekar, N.S., Gerson, D., Skali, H., et al.: Two-dimensional assessment of right ventricular function: an echocardiographic–MRI correlative study. Echocardiography **24**, 452–456 (2007)
5. Ye, C., Toyoda, K., Ohtsuki, T.: A stochastic gradient approach for robust heartbeat detection with Doppler radar using time-window-variation technique. IEEE Trans. Biomed. Eng. **66**, 1730–1741 (2019)
6. Yang, Z.-K., Zhao, S., Huang, X.-D., et al.: Accurate Doppler radar-based heart rate measurement using matched filter. IEICE Electron Exp. **17**, 1–6 (2020)
7. Nosrati, M., Tavassolian, N.: Accurate Doppler radar-based cardiopulmonary sensing using chest-wall acceleration. IEEE J. Electromagn. RF Microw. Med. Biol. **3**, 41–47 (2019)
8. Yang, Z.-K., Liu, W.-K., Zhao, S., et al.: A concurrent dual-band radar sensor for vital sign tracking and short-range positioning. Freq **74**, 369–376 (2020)
9. Petrovic, V.L., Jankovic, M.M., Lupsic, A.V., et al.: High-accuracy real-time monitoring of heart rate variability using 24 GHz continuous-wave Doppler radar. IEEE Access **7**, 74721–74733 (2019)
10. Ye, C., Ohtsuki, T.: Spectral Viterbi algorithm for contactless wide-range heart rate estimation with deep clustering. IEEE Trans. Microw. Theory Tech. **69**, 2629–2641 (2021)
11. Yang, Z.-K., Shi, H., Zhao, S., et al.: Vital sign detection during large-scale and fast body movements based on an adaptive noise cancellation algorithm using a single Doppler radar sensor. Sensors **20**, 4183 (2020)
12. Droitcour, A.D., Boric-Lubecke, O., Lubecke, V.M., et al.: Range correlation and I/Q performance benefits in single-chip silicon Doppler radars for noncontact cardiopulmonary monitoring. IEEE Trans. Microw. Theory Tech. **52**, 838–848 (2004)
13. Tu, J., Lin, J.: Respiration harmonics cancellation for accurate heart rate measurement in non-contact vital sign detection. In: 2013 IEEE MTT-S International, vol. 1, p. 3 (2013)
14. Li, M., Lin, J.: Wavelet-transform-based data-length-variation technique for fast heart rate detection using 5.8-GHz CW Doppler radar. IEEE Trans. Microw. Theory Technol. **66**, 568–576 (2018)
15. Tu, J., Lin, J.: Fast acquisition of heart rate in noncontact vital sign radar measurement using time-window-variation technique. IEEE Trans. Instrum. Meas. **65**, 112–122 (2016)
16. Yang, Z.-K., Shi, H., Zhao, S., et al.: Fast heart rate extraction using CW Doppler radar with interpolated discrete Fourier transform algorithm. AIP Adv **10**, 075113 (2020)
17. Li, C., Lin, J.: Random body movement cancellation in Doppler radar vital sign detection. IEEE Trans. Microw. Theory Tech. **56**, 3143–3152 (2008)
18. Girbau, D., Lazaro, A., RamosÁ, et al.: Remote sensing of vital signs using a Doppler radar and diversity to overcome null detection. IEEE Sens. J. **12**, 512–518 (2012)
19. Liu, T., Hsu, M., Tsai, Z.: High ranging accuracy and wide detection range interferometry based on frequency-sweeping technique with vital sign sensing function. IEEE Trans. Microw Theory Tech. **66**, 4242–4251 (2018)
20. Tang, D., Wang, J., Hu, W., et al.: A DC-coupled high dynamic range biomedical radar sensor with fast-settling analog DC offset cancelation. IEEE Trans. Instrum. Meas. **68**, 1441–1450 (2019)
21. Park, J., et al.: Polyphase-basis discrete cosine transform for real-time measurement of heart rate with CW Doppler radar. IEEE Trans. Microw. Theory Tech. **66**, 1644–1659 (2018)
22. Budge, M.C., Burt, M.P.: Range correlation effects on phase and amplitude noise. In: Proceedings IEEE Southeastcon, p. 5 (1993)
23. Macleod, M.D.: Fast nearly ML estimation of the parameters of real or complex single tones or resolved multiple tones. IEEE Trans. Signal Process. **46**, 141–148 (1998)
24. Dong, S., et al.: Doppler cardiogram: a remote detection of human heart activities. IEEE Trans. Microw. Theory Tech. **68**, 1132–1141 (2020)

Keywords Clustering for the Interview Texts Based on Kmeans Algorithm

Xu Gao[1], Xiaoming Ding[1](✉), Wei Wang[1], Guangming Wang[2], Yueyuan Kang[2], and Shaofang Wang[2]

[1] College of Artificial Intelligence, Tianjin Normal University, Tianjin 300387, China
xmding@tjnu.edu.cn

[2] Faculty of Education, Tianjin Normal University, Tianjin 300387, China

Abstract. We generally use the extraction of keywords to study the factors affecting the growth of excellent teachers. However, an automatic research algorithm is needed to let us understand more clearly when there are many keywords and more complex which aspect has greater influence on teachers. Therefore, this paper proposes an unsupervised learning clustering method based on keyword extraction. First, segments the text and use the Word2Vec tool to train the word vector of the segmentation results; secondly, uses the TF-IDF algorithm to extract the keywords of the text, and extracts five from each document; finally, uses the extracted keywords as the clustering sample, through the Kmeans algorithm obtains the clustering result and manually marks the clustering results to evaluate the clustering effect.

Keywords: TF-IDF · Word2Vec · Kmeans · Clustering

1 Introduction

As our country pays more and more attention to education, more outstanding teachers have become the objects of attention of parents, schools and the society, and more interview articles of outstanding teachers have appeared to provide more material for the growth of fledgling junior teachers. Therefore, how to mine and analyze these texts has important significance and value to research.

We use the word vector training of the Word2Vec framework [1]. Word2Vec is an open source word vector toolkit released by Google in 2013. The algorithm theory of this project refers to the neural network language model designed by Bengio in 2003 [2]. This neural network model uses two nonlinear transformations, there are many network parameters, and the process of training is slow, so it is not suitable for large corpus. The team of Mikolov simplified it and implemented the Word2Vec word vector model [3].

Keyword clustering is an unsupervised research method in natural language processing. Common unsupervised clustering methods include: hierarchical methods, grid-based methods, model-based methods and classification methods.

The clustering method based on the hierarchical method [4] has agglomerative hierarchical clustering that first treats each object as a cluster, and then merges these atomic

Q. Liang et al. (Eds.): Artificial Intelligence in China, LNEE 854, pp. 600–606, 2022.
https://doi.org/10.1007/978-981-16-9423-3_75

clusters into larger and larger clusters, until all objects are in a cluster. However, the computational complexity of this method is too high, and the algorithm does not need to pre-determine the number of clusters, but which level of clustering as the clustering effect we need to choose requires us to complete it in accordance with the actual objective conditions and experience.

Network-based clustering methods include SOM clustering. The SOM neural network was proposed by Professor Kohonen [5], a neural network expert. The algorithm assumes that there are some topological structures or sequences in the input object, which can achieve dimension reduction mapping from the input space (n-dimensional) to the output plane (2-dimensional), Its mapping has the nature of maintaining topological characteristics, and has a strong theoretical connection with the actual brain processing. However, the processing time is too long, and further research is needed to adapt it to large databases.

The topic model-based clustering method [6] can reduce the dimension of the vector space from high-dimensional feature word vectors to low-dimensional semantic topic vectors, which can effectively solve the problem of too high dimension of feature word vectors, lack of semantics, etc. But there are probabilistic latent semantic analysis and the influence of ignoring low-frequency terms on the clustering effect.

For the above problems, this paper proposes a Kmeans clustering algorithm based on classification method. The k-means algorithm [7] takes K as a parameter, divides n objects into K clusters, so that the clusters have a higher degree of similarity, but the similarity between clusters is lower. The word vector model of Word2Vec is to improve the semantic loss problem when extracting keywords by the TF-IDF algorithm. The word vector is trained according to the upper and lower n words of a word in the word segmentation result, and the result of each word segmentation is obtained, and then extract the word vector clustering of the candidate keywords we need.

2 Training of Word Vector Model

This paper proposes a semantic-based word vector training, using the Word2Vec tool, the Skip-gram model to train the word vectors of the word segmentation results [8]. Word2Vec considers the context information of the current word, and the learned word vector contains rich semantic and grammatical relationships. We selected 22 interviews for outstanding mathematics teachers for research. Specific steps are as follows:

- Text preprocessing.
  - Word segmentation, stop word filtering.
  - According to the TF-IDF algorithm, the word with the highest TF-IDF value of each text is extracted.
  - Filter the similar or identical words in the extracted words according to the word similarity.
- Training word vector. According to the word segmentation result, train and save the word vector of each word segmentation before filtering stopwords.

- Extract the word vectors of the keywords extracted from the first step as the clustering corpus.

The parameters used when training the word vector model are as follows (Table 1):

Table 1. Description table of word vector training parameter

Parameters	Description
sentences	Word segmentation result, namely word vector training corpus
window	Indicates the maximum possible distance between the current word and the predicted word, the setting value is 5
min_count	Represents the minimum number of occurrences, set to 1
workers	Indicates the number of threads used when training word vectors
size	The dimension of the word vector for training, the default is 100
n_clusters	Number of cluster categories, artificially designated as 5

3 Kmeans Keywords Clustering

3.1 Kmeans Clustering Algorithm

Kmeans algorithm first randomly selects K objects, each object initially represents the average or center of a cluster, and then for each remaining object, according to its distance from the center of each cluster, it is assigned to the nearest cluster. Finally, the average value of each cluster is recalculated. This process is repeated until the criterion function converges. Usually, the squared error criterion [9] is used, which is defined as formula (1):

$$E = \sum\nolimits_{i=1}^{k} \sum\nolimits_{p \subset c_i} |p - m_i|^2 \quad (1)$$

E refers to the sum of the square errors of all objects in the database, p is a point in the space, representing a given data object, and m_i is the average value of the cluster c_i.

3.2 Experimental Process

Input: the number of clusters K, and the database D containing n data objects
Output: K clusters meeting the minimum standard of variance

- Reduce the 100-dimensional data to 2 dimensions, build a two-dimensional word vector dictionary and visualize the data with a scatter diagram, as shown in Fig. 1.
- Set the value of K, randomly select K from n data objects as initial clustering centers, this article sets K = 7.

- Allocate clusters for each data, calculate the similarity between each data and the cluster center, and assign it to the most similar cluster.

Update the center of each cluster according to the data points in the cluster. Repeat until convergence.

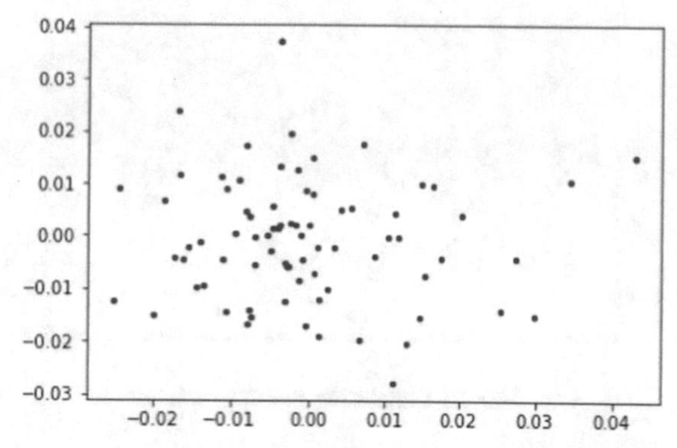

Fig. 1. Scatter plot of keyword word vectors

The clustering results are as follows in Table 2:

Table 2. Clustering results

Category	Keywords
The first category	Love; master; lecture; delve into; practice; special teacher; mathematical knowledge; professor; guide; lesson plan; sense of accomplishment; spirit; Math teacher; diligent; assessment; assessment; atmosphere; family education; work hard; research ability; reflection; accompany; life-long learning
The second category	Profession; classroom; excellence; young teacher; faculty and researcher; professional knowledge; idea; game; premium; enthusiasm
The third category	Ability; knowledge
The fourth category	Excellent teacher; principal; school; awareness; course; teacher ethics; management; profession
The fifth category	Sense of responsibility; organizational management ability; innovative mind; teach and educate people; individual; return; classroom teaching; early education; bole; self-cultivation; surroundings; understand mathematics; platform
The sixth category	Ability; knowledge
The seventh category	Values; patient; certificate; attitude; classmate; self conscious; teacher occupation; teacher; career planning; competition; welfare

We visualize the clustering results, use different colors or shapes to represent different clusters, and get Fig. 2:

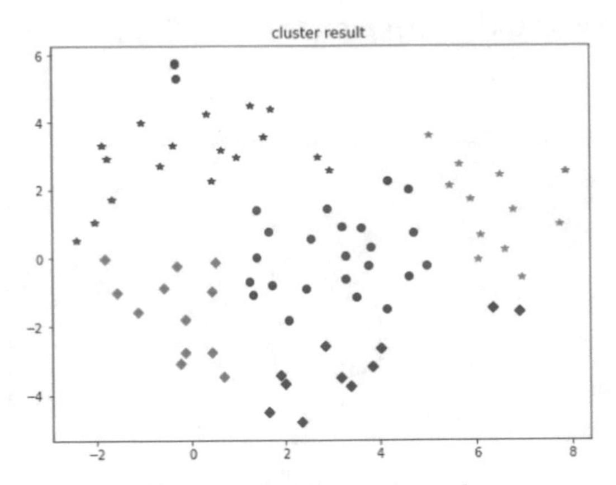

Fig. 2. Clustering result graph

We tried to take different K values, and finally found that when K = 7, we best met our requirements for this clustering. Although some words are not accurately located in the cluster where they should be, they still have great reference values. For example, in the first category, some words such as love, spirit, diligence, hard work, scientific research ability, lifelong learning, etc. indicate that in order to become an excellent teacher, the teacher needs to have certain abilities and qualities. And some words such as the master, the super teacher, the professor, and the mathematics teachers also explain that in the process of growing into an excellent teacher, they cannot do without the encouragement and help of people around them; in the fifth category, sense of responsibility, organization and pipeline ability, teaching and educating, innovative awareness, reward, self-cultivation, etc. The word means that if a teacher wants to grow into an excellent teacher, he must have some abilities and qualities to set an example for the students and achieve success in teaching; in the seventh category, a teacher's values, attitude, patience, self-awareness, and career planning represent that an excellent teacher should have firm belief, a clear understanding of himself, and be full of enthusiasm for the teaching profession, so that he or she can go further and further.

3.3 Evaluation Index

Clustering itself is an unsupervised learning process, but the clustering effect evaluation can be done by manually labeling the effect of the clustering, and then the accuracy rate (Accuracy), recall rate (Recall), F1 value (H-mean value) are calculated to evaluate how good the clustering effect is.

- Accuracy: It is the proportion of all predicted correct values in the total. The formula is (2):

$$\text{Accuracy} = \frac{TP+FN}{TP+TN+FP+FN} \tag{2}$$

- Recall rate: recall rate, that is, the proportion of the correct prediction that is positive to all the actual positive. The formula is (3):

$$Recall = \frac{TP}{TP+FN} \tag{3}$$

- F1 value: arithmetic mean, geometric mean of the first day, the bigger the better. The formula is (4):

$$F1 = \frac{2TP}{2TP+FP+FN} \tag{4}$$

In the above formula, *TP* (True Positive) stands for predicting positive classes as positive classes; *FP* (False Positive) stands for predicting negative classes as positive classes; *TN* (True Negative) stands for predicting negative classes as negative classes, *FN* (False Negative) represents the prediction of a positive class as a negative class.

Our keyword clustering is divided into seven clusters, and the confusion matrix is shown in the following Table 3, which is used to show the corresponding results of the real and predicted labels of the classification model. The seven evaluation standard values finally obtained by the seven clusters are shown in the following Table 4. Most of the obtained percentages are above 80%, which proves that our analysis results have a certain degree of accuracy.

Table 3. Confusion matrix

Confusion matrix	1lei	2lei	3lei	4lei	5lei	6lei	7lei	Pass rate (%)
1lei	18	2	0	0	1	1	1	78.26
2lei	0	8	2	0	0	0	0	80.00
3lei	0	1	9	1	0	1	0	75.00
4lei	0	0	0	7	0	0	0	100
5lei	0	0	1	2	10	0	0	76.92
6lei	0	0	0	0	0	3	0	100
7lei	0	2	0	0	0	0	9	81.82

Table 4. List of evaluation standard values

Index (%)	1lei	2lei	3lei	4lei	5lei	6lei	7lei
Accuracy	93.67	91.14	92.41	96.20	94.94	97.47	96.20
Sensitivity (Recall)	78.26	80.00	75.00	100	76.92	100	81.82
F1-Score	87.80	69.57	75.00	82.35	83.33	75.00	85.71

4 Conclusions

This thesis is mainly keywords extraction based on the TF-IDF algorithm for the results of 22 excellent mathematics teacher interview texts, then training word vector, and finally achieving the clustering of keywords according to the Kmeans clustering algorithm which is meaningful for teachers and society. Teachers can constantly make themselves better based on such a result, and our society will have more excellent teachers and the education will be better and better.

Acknowledgement. The work was supported by the Doctoral Foundation of Tianjin Normal University (52XB2004), the Natural Science Foundation of China (62001328) and TJNU "Artificial Intelligence + Education" United Foundation.

References

1. Zhao, X.-P., Huang, Z.-Y.: A short text clustering algorithm combining TF-IDF method and word vector. Electron. Design Eng. **28**(443, 21), 11–15 (2020)
2. Mikolov, T., Chen, K., Corrado, G., et al. Efficient estimation of word representations in vector space. Comput. Sci. (2013)
3. Bengio, Y., Ducharme, R., Vincent, P., et al.: A neural probabilistic language model. J. Mach. Learn. Res. (2003)
4. Li, X.-R., Xia, Y.: Interest point recommendation algorithm based on similarity fusion and dynamic prediction Comput. Eng. Appl. **54**(905, 10), 110–114+217 (2018)
5. Le, Q.V., Mikolov, T.: Distributed representations of sentences and documents. In: Proceedings of the 31st International Conference on Machine Learning, pp. 1188–1196 (2014)
6. Yuan, H.-C.: Research on the spectral clustering method of the complex network of the adjacent surface of the three-dimensional model. Shandong Normal University (2014)
7. Lin, J.-H., Zhou, Y.-M.: Analysis of the evolution of news comment topics combined with word vectors and clustering algorithms. Comput. Eng. Sci. **38**(11), 2368–2374 (2016)
8. Li, X.-M.: Research on internet personal consumption loan pricing based on clustering algorithm. Shanghai University of Finance and Economics (2014)
9. Mao, Y.-X., Qiu, Z.-X.: Research on information technology document clustering based on Word2Vec model and K-means algorithm. China Inf. Technol. Educ. (008), 99–101 (2020)

Research on the Application of Artificial Intelligence in Brain Functional Imaging

Zongyi Han(✉) and Libo Qiao

Factual of Psychology, Tianjin Normal University, Tianjin 300387, China
psy-hzy@mail.tjnu.edu.cn

Abstract. Magnetic resonance imaging (MRI) can directly reflect the functional activity pattern of the brain, which helps to better understand the rules of human thinking and activity patterns. It has become an important means of brain science research and clinical examination. In recent years, the rise of artificial intelligence (AI) method can simulate the thinking habits of the human brain and extract more invisible information than the traditional statistical analysis methods. How to better integrate AI technology with brain medical imaging and promote the diagnosis and treatment of brain diseases has become the focus and hotspot of future research. Therefore, this paper reviews the application and development of AI in brain function imaging in the past three years.

Keywords: Artificial intelligence · Magnetic resonance imaging · Brain science · Applications

1 Introduction

Brain is the central organ of all physiological activities, thinking and emotions in the human body, and has an extremely complex structure and function. It brings new experiences such as learning, seeing, remembering, hearing, perceiving, and understanding to humans. At the same time, it also threatens human health due to brain atrophy, Parkinson's disease, Alzheimer's disease (AD), brain tumors and other diseases. The US Centers for Disease Control and Prevention once pointed out that if brain diseases can be correctly diagnosed and cured, the life span of patients may be extended by 10–20 years.

With the rapid development of modern medical equipment like magnetic resonance imaging, positron emission tomography (PET), electroencephalogram (EEG), etc. [1], the visualization of brain structure and the diagnosis of brain diseases have gradually become possible. Among them, MRI technology has the advantages of higher imaging contrast of body soft tissues, higher spatial resolution, and no radiation damage to patients. It has slowly become an important tool for scientific research and disease diagnosis [2].

In recent years, artificial intelligence has gradually become one of the hot methods of brain science [3]. This technology simulates the thinking learning mode of the human cerebrum. Through the analysis and induction of known subjects, new knowledge or a certain model can be formed. And the AI algorithm makes realistic judgments and

Q. Liang et al. (Eds.): Artificial Intelligence in China, LNEE 854, pp. 607–613, 2022.
https://doi.org/10.1007/978-981-16-9423-3_76

predictions about the unknown. This paper reviews the research situation of AI in brain function images at domestic and abroad in the past three years. From the three aspects of existing applications, existing problems, and development trends, summarize the research status and point out the direction of future expansion.

2 Application of Artificial Intelligence in Brain Functional Imaging

MRI technology drives the rapid development of brain science, accompanied by the exponential growth of various data. The information contained therein can be analyzed by AI algorithm, which is mainly used in image acquisition or reconstruction, automatic image segmentation or recognition, disease recognition or classification, and prognosis judgment. It is shown in Fig. 1.

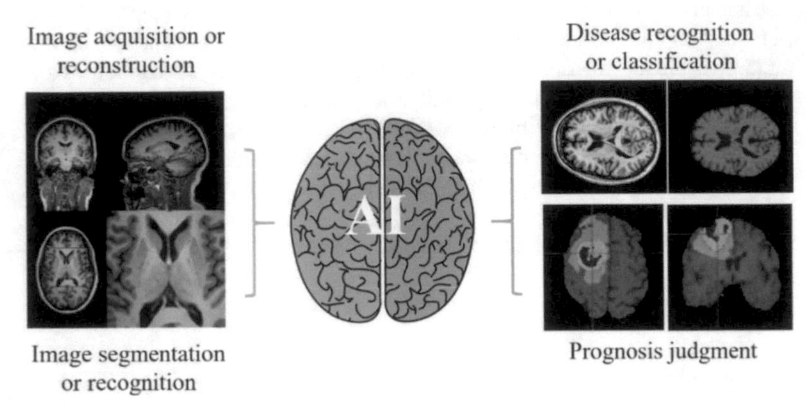

Fig. 1. Common applications of AI in MRI

2.1 Brain Functional Image Acquisition and Rapid Reconstruction

Brain function imaging is non-radiation and non-invasiveness and has become one of the most commonly used auxiliary techniques for brain research and diagnosis. However, long scanning and closed environment can easily lead to claustrophobia and indirectly increase medical costs. How to use new technologies to shorten the time required for MRI examinations has gradually become a problem that scientists urgently need to solve.

Souza et al. [4] first performed a reconstruction method using convolutional neural networks (CNN) for spatial data and images respectively. The spatial data is regarded as an image, and the unsampled points are zero filled (ZF). The k-space data is reconstructed by CNN, and then the inverse discrete fourier transform (IDFT) is performed. The ascendancy of this approach is that there are no complex transformations, and a large number of network parameters can be shared.

Huang Li [5] designed and implemented a three-dimensional magnetic resonance image (3D-MR) super-resolution reconstruction algorithm applying S3D-RBDN network based on 3D separable convolution. Through residual learning and dense connection, the hierarchical features in the original low-resolution image are fully utilized. Increase the useful information of the image, thereby improving the quality of the reconstructed image.

Han [6] proposed unsupervised medical anomaly detection generative adversarial network (MADGAN). Training on axial MRI slices of three healthy brains to reconstruct and detect brain abnormalities at different stages. Then use 1133 healthy T1-weighted brains MRI scans and 135 healthy contrast-enhanced T1 brains MRI scans to form two different data sets, which are respectively for detecting AD, mild cognitive impairment (MCI), brain metastasis and various brain diseases.

Xu Zhenhao [7] put forward an MRI reconstruction algorithm centered on generative confrontation networks. The U-net network acts as the generator, and the residual structure is used as the coding part to alleviate network degradation. The context information of different scales is merged at distinct expansion rates of the hole convolution, and the sigmoid function is used to complete the feature classification. Meanwhile, the idea of integrated learning is integrated into the algorithm to further improve the reconstruction effect. Compared with mainstream reconstruction networks, this model ranks first in the test sets with sampling rates of 10%, 20%, 30%, and 50%.

2.2 Automatic Segmentation and Recognition of Brain Function Images

The subject of brain segmentation is to divide the images into different regions according to a given standard. Because of its great medical and social significance, it has attracted many domestic and foreign scholars. At present, the segmentation processing of brain imaging depends largely on the experience of anatomy experts. This method is not only time-consuming and labor-intensive, but the results are more subjective. The introduction of AI can effectively improve the accuracy of segmentation and make automatic segmentation possible.

Sathish [8] proposed a network model of supervised learning due to deep learning. Diffusion-weighted imaging (DWI) and perfusion-weighted imaging (PWI) sequence diagrams of ischemic stroke lesions were segmented. Cross-entropy is used to record the segment loss, and the loss is aggregated by three discriminators. With triple cross-validation, the Dice coefficient on the dataset is 82%.

Xiong Zijin [9] designed a novel convolutional neural network based on thalamic MRI images called residual dense U network (RDU-Net). Through the introduction of DenseNet connection, Res-Net residual learning and InceptionNet bottleneck design in the algorithm. Improve the feature extraction ability of each layer network, reduce the difficulty of algorithm training, and solve the shortcomings of dense connections that require high video memory.

Zhang [10] improved the convolutional neural network under the characteristics of brain tumors. Three modes of distributed dense connection (DDC) cross-skip, skip-1, and skip-2 are proposed. The features of CNN are enhanced by constructing tunnels between the key layers of the network, and then the DDU-Nets algorithm is integrated and used for the segmentation of brain tumor MRI images. After the evaluation of the BraTS2019 dataset, the effectiveness of the method is proved, and the calculation cost is greatly decreased.

Zhou Hang [11] studied an adaptive segmentation algorithm for cerebral vascular infarction disease. By way of the classic Fuzzy C-Means (FCM) algorithm, the idea of quantum particle swarm and space transformation is further introduced. The local gray-scale statistical method is added, and then the Cereburum FCM algorithm is researched out. The experimental accuracy rate on the dataset is 86%, and the performance far exceeds the existing FCM and its improved algorithm.

2.3 Classification, Recognition and Diagnosis of Brain Diseases

The AI model can also realize the diagnosis and identification of diseases by extracting the characteristics of the brain image.

Zhou Cong [12] collected structural magnetic resonance imaging (sMRI) and diffusion tensor imaging (DTI) data of 48 patients with obsessive-compulsive disorder (OCD) and 45 healthy controls. Preprocessing the sMRI data under the voxel-based morphological (VBM) analysis method to obtain the whole brain gray matter volume (GMV) and white matter volume (WMV). Use FMRIB Software Library (FSL) based on Tract-based spatial statistics (TBSS) to compare the two groups to get the fractional anisotropy (FA) and the mean diffusivity (MD). According to the above GMV, WMV, FA and MD, support vector machines (SVM) are used to carry out machine learning classification research on OCD patients and healthy people, to determine the optimal indicators for identifying OCD and the 10 brain regions that are most meaningful for classification.

Liu [13] developed a novel 3D CNN architecture that can identify patients with AD, patients with mild cognitive impairment, and healthy controls. The model consists of the convolutional layer, normalization layer, activation layer and maximum pooling layer. On the ANDI dataset, the test accuracy improves by about 14% compared to existing models.

Hua Minghui [14] compared the GMV structural index and functional connectivity (FC) differences between 97 patients with chronic schizophrenia and 105 healthy controls. K-means cluster analysis method was used to explore the heterogeneity of schizophrenia based on the altered pattern of the whole brain voxel level GMV. This result provides a certain pathophysiological basis for the clinical classification and precise treatment of schizophrenia.

Mostafaie et al. [15] proposed an MRI image detection method for brain glioma, which mainly includes imaging selection, preprocessing, lesion segmentation, tumor feature map extraction, decision-making comparison, and preselection box output. Dice scores on both high-grade gliomas (HGG) and low-grade gliomas (LGG) images were 0.73. The method can not only fully detect the range of the tumor, but also classify it according to the brain tumor features.

2.4 Disease Prognosis Assessment and Risk Stratification

Contemporary people pay further attention to physical health, and the prediction of brain diseases, the examination of diagnosis and treatment effects, and the assessment of diseases have gradually been put on the agenda. The emergence of AI technology provides more and more complex variables for this research.

Zhang Yue [16] calculated nine brain imaging features involving fractional anisotropy, mean diffusion, axial diffusion rate and radial diffusion rate through the white matter parcellation map. Individual predictive studies on the positive and negative symptom scales of patients with schizophrenia (SZ) were compared by utilizing five regression algorithms, including least squares regression, ridge regression, LASSO regression, linear support vector regression and elastic network regularization linear regression. In the MRI brain map, the frontal and temporal lobes contribute the most to the prediction.

Suri et al. [17] combined the MRI imaging symptoms of myocardial injury, hypoxia, plaque rupture, arrhythmia, venous thromboembolism, coronary thrombosis, encephalitis, ischemia, inflammation and lung injury, adopting machine learning and Deep learning technology calculates the probability of COVID-19 infection and classifies the severity of its infection.

Qu Taiping [18] adopted a new end-to-end 3D connected convolutional neural network architecture. It has a layer jump connection function, which can effectively use the context information in the T1 weighted space of the brain MRI image. Apply CMI-HBN, ABIDE and ADHD200 three public children and adolescent brain MRI datasets to conduct experiments to estimate the age of their brains.

3 Current Unsolved Problems and Future Trends

AI technology to assist in the research of brain conditions is still in the initial trial stage. In practical applications, there are the following problems that need to be solved in the future:

(1) The quality of feature extraction needs to be improved. The internal structure of the brain is very complicated. Even if its MRI images are manually labeled, it is easy to define deviations between various tissues. The ambiguity of AI feature extraction is a difficulty that demands to be overcome. The classification effect of morphological attributes is not ideal in post-processing, mainly because the feature extraction process of different brain regions in the preliminary stage is not done well. Meanwhile, artificial intervention is also minimized as much as possible to realize fully automatic image feature set construction or selection.

(2) Randomness generated by the instrument. For brain MRI images, the same case may have different morphological responses under different medical machines. Although there have been gradually more research results in the field of MRI at domestic and abroad in recent years. However, different manufacturers and different models of MRI machines from different scanning parameters also have a very large impact on AI imaging. Therefore, it is necessary to establish a variety of evaluation

systems from various aspects such as machine models and scanning parameters to match suitable detection standards for each brain area.

(3) The data samples are relatively small. AI algorithms usually require huge datasets for training to ensure the accuracy and stability of the model. At present, there are fewer brain-related datasets and also exist errors in the labeling between different data sets, which makes it difficult to evaluate with a unified standard. Academicians often need to collect by themselves, which further increases the workload of scientific research. Therefore, it is very necessary to apply multimodal MRI data to fully reflect the image characteristics of the brain.

(4) The segmentation algorithm and network structure require to be optimized. The timeliness and accuracy of the existing traditional and improved brain science algorithms are not very rational. New algorithms are developed to enable them to recognize various types of brain science. At the same time, the calculation cost is cut down, and the algorithm complexity is reduced and optimized. Furthermore, improve the artificial neural network architecture to achieve a balance between obtaining deeper information of the image and reducing the time complexity.

4 Conclusion

With the acceleration of the aging of the population and the increasing involution of study and work, young people have become a generation under pressure. This has led to the frequent occurrence of various brain diseases, which has escalated into a public health issue that attracted much attention. Among them, image acquisition or reconstruction, automatic image segmentation or recognition, disease recognition or classification, and prognosis judgment play a crucial role in the research. Although the current research of AI in brain functional imaging is still in the initial stage, the continuous breakthrough of the above problems, can help to further explore brain science research and effectively improve the diagnosis and treatment efficiency of brain diseases.

Acknowledgement. This paper is supported by Youth Research Project of Tianjin Normal University (52XQ2101).

References

1. Pizurica, A., Wink, A.M., Vansteenkiste, E., et al.: A review of wavelet denoising in MRI and ultrasound brain imaging. Curr. Med. Imaging Rev. **2**(2), 247–260 (2006)
2. Wang, Q.: Research on Segmentation of Deep Brain Structures in MRI. Beijing University of Technology (2015)
3. Amin, J., Sharifa, M.: Brain tumor detection using statistical and machine learning method. Comput. Methods Programs Biomed. **177**, 69–79 (2019)
4. Souza, R., Frayne, R.: A hybrid frequency-domain/image-domain deep network for magnetic resonance image reconstruction. In: IEEE Patterns and Images, pp. 257–264 (2019)
5. Li, H.: 3D MR Image Super Resolution Reconstruction Algorithm Based on Deep Learning. University of Electronic Science and Technology of China (2020)

6. Han, C., Rundo, L., Murao, K.: MADGAN: unsupervised medical anomaly detection GAN using multiple adjacent brain MRI slice reconstruction. BMC Bioinform. **22**, 1–20 (2021)
7. Xu, Z., Shen, X., Li, X.: Magnetic resonance images reconstruction based on generative adversarial network. Comput. Eng. Appl. (2021)
8. Sathish, R., Rajan, R., Vupputuri, A., Ghosh, N., Sheet, D.: Adversarially trained convolutional neural networks for semantic segmentation of ischaemic stroke lesion using multisequence magnetic resonance imaging. In: IEEE Engineering in Medicine and Biology Society, pp. 1010–1013 (2019)
9. Xiong, Z.: Research on Thalamic Segmentation Based on Convolutional Neural Network. Southeast University (2019)
10. Zhang, H., Li, J., Shen, M., Wang, Y., Yang, G.-Z.: DDU-Nets: distributed dense model for 3D MRI brain tumor segmentation. In: Crimi, A., Bakas, S. (eds.) BrainLes 2019. LNCS, vol. 11993, pp. 208–217. Springer, Cham (2020). https://doi.org/10.1007/978-3-030-46643-5_20
11. Zhou, H.: Research on the Recognition and Segmentation Algorithm of Vascular Infarction in MRI Cerebral Images. Harbin Engineering University (2020)
12. Zhou, C.: Altered Structural Networks of Obsessive-compulsive Disorder and Machine Learning Analysis with Multiple Neuroimaging Indices. Kunming Medical University (2020)
13. Liu, S., Yadav, C.: On the design of convolutional neural networks for automatic detection of Alzheimer's disease. In: Machine Learning for Health Workshop, pp. 184–201 (2020)
14. Hua, M.: Investigation of Brain Abnormalities and Neurobiological Subtypes in Schizophrenia Using MRI and Machine Learning Techniques. Tianjin Medical University (2020)
15. Mostafaie, F., Teimouri, R.: Region of interest identification for brain tumors in magnetic resonance images. In: 2020 28th Iranian Conference on Electrical Engineering (ICEE), pp. 1–5 (2020)
16. Zhang, Y.: Automatic Classification and Individualized Prediction of Schizophrenia Studied with Brain Imaging Features Using Multi-modal MRI. South China University of Technology (2019)
17. Suri, J.S., et al.: COVID-19 pathways for brain and heart injury in comorbidity patients: a role of medical imaging and artificial intelligence-based COVID severity classification: a review. Comput. Biol. Med. (2020)
18. Qu, T.: Research on Prediction Method for Children and Adolescents' Brain Age Based on Deep Learning. Jilin University (2020)

Integrated Navigation for UAV Precise Flight Management with No-Fly Zones

Shangwen Yang[1], Yuxin Hu[1], and Rui Sun[2](✉)

[1] State Key Laboratory of Air Traffic Management System and Technology, Nanjing 210007, China
[2] College of Civil Aviation, Nanjing University of Aeronautics and Astronautics, Nanjing 211106, China
rui.sun@nuaa.edu.cn

Abstract. The use of Unmanned Aerial Vehicles (UAVs) has increased significantly in recent decades. Nevertheless, the abuse of the UAVs raises concern in regards to regulations and public safety. One effective method to solve this problem is that using a government or manufacture owned platform to conduct real-time UAV monitoring. In particular, the performance of the on-board navigation sensor is essential for the monitoring system. In this paper, we have proposed a sensor integration-based method for the UAV precise flight management with no-fly-zones. It is demonstrated that the designed system could perform an 3D accuracy of 1 m with 200 Hz output rate from the field test results and therefore could meet the requirement of the precise Air Traffic Management (ATM) with no-fly-zones.

Keywords: UAV · Integrated navigation · No-fly-zone

1 Introduction

Unmanned aerial vehicle (UAV) generally refers to the aircraft without pilots. It was first used in military by the U.S. Department of defense in the 1970s. In recent years, with the development of relevant critical technologies, the reduction of production costs and the improvement of flight safety, the application fields of civil UAV are becoming wider, and the commercial application benefits are becoming higher [1, 2]. According to the estimate of the Federal Aviation Administration (FAA), there will be about 7 million UAVs in the United States in 2020; By 2022, the number of commercial UAVs operating in US airspace will reach 450000 (US FAA, 2018). At present, the UAV control has been improved from the early manual operation on the ground to carrying a micro flight control computer on the UAV. At the same time, it integrates the global navigation satellite system (GNSS), inertial measurement unit (IMU) and ground control station or controller to conduct semi-automatic or full-automatic navigation flight through the wireless communication system, The United States FAA calls this flight system unmanned aerial system (UAS).

Q. Liang et al. (Eds.): Artificial Intelligence in China, LNEE 854, pp. 614–621, 2022.
https://doi.org/10.1007/978-981-16-9423-3_77

The world began to enter the era of UAVs, with explosive growth in military, commerce, agriculture, exploration and shooting, tourism and leisure. Due to the fuzzy boundary between the official use of UAV and flight remote control toys, and individuals can also own UAV, it poses a great threat to the safety of civil aviation. According to the data, in the second half of 2015 alone, FAA received reports from airlines, private aircraft pilots and air traffic controllers that UAVs approached the flight path up to 175 times, and there were many cases in which UAVs and civil aircraft might collide only a few seconds or a few meters away. The takeoff and landing section of the busiest airport in the United States has the most cases, which poses a great threat to the safety of public life.

As a new transport tool, UAV has flexibility, but it may also affect public security and even pose a threat to national security due to improper use or malicious use. National security and public security are the premise of all, and the competent authorities must give consideration to national security and the protection of legitimate activities at the same time. The biggest problem with drones is that terrorists may transform them into new weapons. How to manage UAVs and maintain national and public safety is an urgent problem to be solved. Looking at the management of unmanned aerial vehicles in various countries, it seems that they have stagnated, and have not formulated a set of laws focusing on unmanned aerial vehicles. Most of them are the addition and deletion of provisions in aviation laws and regulations, thus specifying some relatively simple norms. European countries have slightly different details on the specifications. Germany sets an upper limit of 25 kg for the weight of UAVs, while Britain regards 20 kg UAVs as general aircraft. In order to unify the management scheme, the European Aviation Safety Agency (EASA) mentioned in the a-npa 2015-10 declaration that EASA has been instructed by the EU to formulate UAV supervision rules, which divides the aircraft into three levels. The lowest model aircraft can only be operated within the operator's sight, and the flight altitude is limited to less than 150 m. FAA issued a relatively complete UAV management draft in early February 2015, which mentioned that commercial small UAVs under 25 kg should not fly at an altitude of more than 150 m, speed no more than 100 mph, and need UAV business license.

FAA has launched an online map and updated the content of the b4ufly mobile app to facilitate users to view the location and relevant information of the UAV no fly zone.

The domestic UAV no fly zones are listed as follows:

1. Military and civil airports;
2. Over government agencies, over military units;
3. Facilities with strategic status (such as large reservoirs, hydropower stations, etc., flight is prohibited);
4. No flights are allowed over prisons, detention centers, detention centers, drug treatment centers and other places of supervision.

The geo system launched by Dajiang company will dynamically cover all kinds of restricted flight areas around the world, and flight users will obtain relevant restricted information in real time, including but not limited to airports, temporarily restricted flight areas caused by some emergencies (such as forest fires, large-scale activities, etc.), and some permanently prohibited flight areas (such as prisons, nuclear factories, etc.).

In addition, users may also receive flight alerts in some areas, such as wildlife reserves, densely populated towns and other areas that allow flight. The above areas that cannot fly completely freely are collectively referred to as restricted areas, including alert areas, enhanced alert areas, authorized areas, height restricted areas, no fly areas, etc.

2 Objectives of the Research

The existing no fly zones for UAVs are set by limiting the UAV operation area according to specific control measures and marking it on the digital map. The navigation system carried by UAVs is used for management through real-time control or post-processing tracking analysis, Therefore, the key to the precise control of UAV no fly zone lies in the accuracy of the map used to delimit the no-fly zone and the accuracy of UAV navigation system. Taking Google Maps (including images) as an example, its two-dimensional absolute positioning accuracy is about 5 m. At present, the navigation systems carried by general commercial UAVs mostly use GNSS real-time single point positioning mode to calculate the trajectory. The plane positioning accuracy of this kind of system is about 5–10 m and the height accuracy is about 10–15 m. The above GNSS based navigation system will cause the interruption of positioning solution or large error under the specific UAV operating environment, such as high-speed rise or turn, and the signal is blocked or reflected by urban high-rise buildings and trees. Therefore, it is not conducive to the accurate control of UAV no fly zone [3]. At the same time, the height accuracy of GNSS real-time single point positioning mode solution does not seem to meet the requirements of UAV height limiting operation, as shown in Fig. 1.

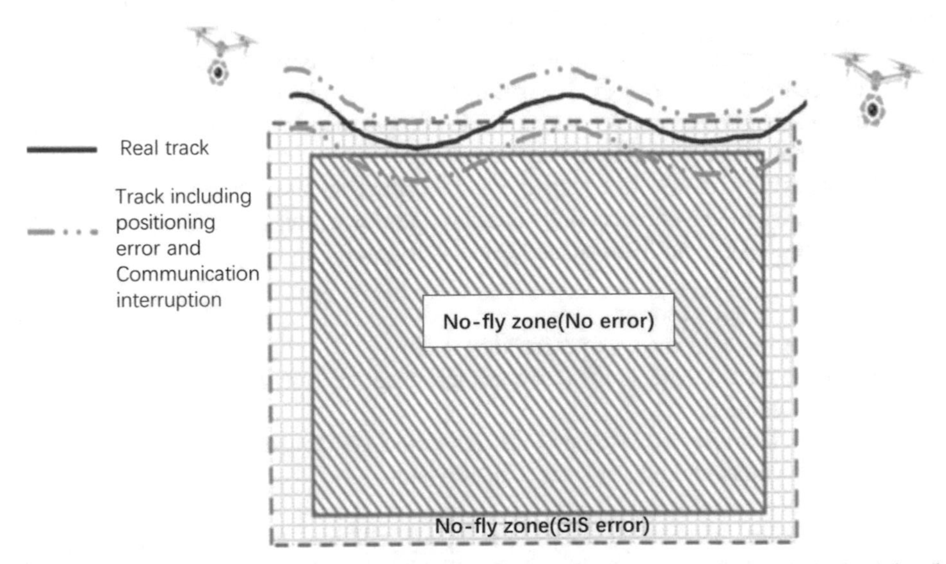

Fig. 1. The influence of the precision of map and navigation system on the control of restricted navigation area

Therefore, when delimiting the UAV no-fly zone, the relevant authorities must consider the accuracy limitations of the map and the UAV navigation system to set up a buffer zone, so as to avoid subsequent law enforcement disputes due to the insufficient accuracy of the map and navigation system. For example, the width of the buffer zone extending outward from the original no fly zone needs to be at least 8–10 m; In terms of height, the width of the buffer zone extending downward from the original no fly zone needs to be at least 10–12 m. When the coordinates calculated in real time by the UAV navigation system fall into the buffer zone extended by the no way zone, the flight control software shall immediately correct the aircraft trajectory to fly out of the no way zone or land immediately, and notify the operator to recover the aircraft as soon as possible, as shown in Fig. 2. In conclusion, continuous and high-precision trajectory is very important for the control of UAV restricted area. Therefore, this study proposes that it is a proper strategy to develop the control architecture of UAV restricted area by using integrated navigation.

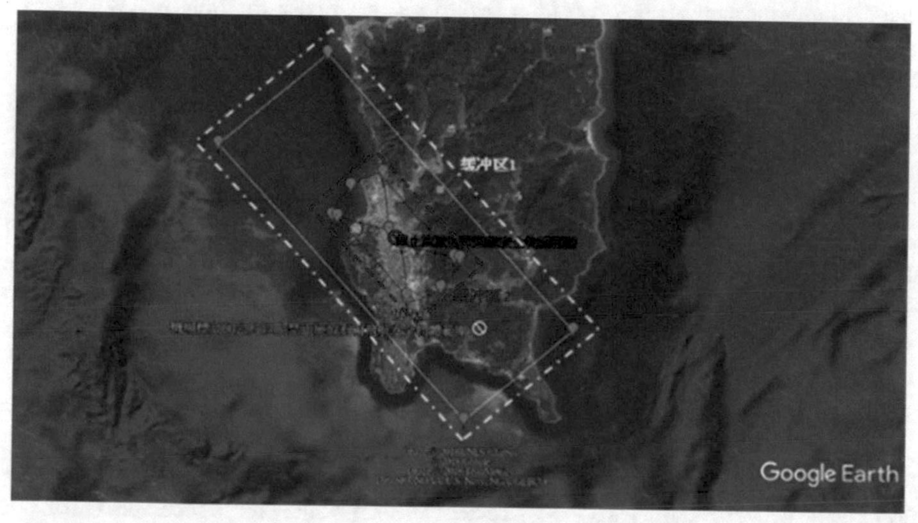

Fig. 2. An example of a buffer zone for a no-fly zone

3 Design of UAV Control Architecture Based on Integrated Navigation

At present, UAVs usually carry multi-frequency observation measurements provided by China's Beidou, the United States GPS, the EU Galileo and Russia's GLONASS satellite navigation systems, so as to obtain high-precision real-time single point positioning accuracy. However, multi GNSS system still cannot meet the application requirements of complex environment. For example, GNSS will cause whole cycle jump and satellite lock loss under the condition of high dynamics of UAV; It will also be affected by signal blocking in areas passing through urban areas, underpasses, tunnels, forests and so on. Inertial navigation system (INS) has the function of autonomous navigation and can obtain the position and attitude information of UAV in real time. At present, it is also widely used in UAV navigation, but its estimation error will accumulate over time. Through GNSS and INS integrated navigation, the estimation, correction and fusion of GPS and INS data can be realized, and the two systems can learn from each other to solve the problem of unable positioning caused by GPS signal locking and the drift caused by INS long-time accumulated error [4–8], and effectively improve the output frequency of the system, so as to improve the accuracy and reliability of UAV navigation, Finally, it meets the needs of UAV no flight zone management and control.

In the development of INS/GNSS Integrated Positioning and orientation algorithm, the Kalman filter theory plays a very important role in the optimal estimation data processing theory, which provides an important theoretical basis for integrated navigation. Kalman filter (KF) has been widely used in navigation, control and other related fields since Dr. Rudolph E. Kalman published his famous paper on the recursive solution of discrete data linear filter in 1960 [9]. At present, UAV navigation mostly adopts nonlinear filtering. Extended Kalman filter (EKF) realizes linear solution by Taylor series expansion of nonlinear function. At present, it is a widely used algorithm in Integrated Navigation Engineering.

In integrated navigation, in addition to the errors of navigation solutions (9 elements in total of 3 positions, 3 velocities and 3 attitudes), the errors of sensors (3 acceleration deviations, 3 gyro drift, 3 accelerometer scale factors and 3 gyro scale factors) are usually considered to form an error state vector with 21 parameters. In this paper, INS and GNSS are regarded as two independent systems by using loosely-combination architecture. INS uses navigation equation; GNSS has an independent Kalman filter to find the derivative solution. Finally, the extended Kalman filter and post-processing smoother are used to obtain the best integrated solution, as shown in Fig. 3.

4 Experiment and Analysis

Figure 3 shows the integrated navigation system used in this experiment. IMU is a low-order tactical level system MMQ-G (gyro drift 100°/h), while GNSS receiver is UBLOX EVK-6T. The flight test area is selected in the area with less traffic and pedestrians and good air permeability. The trajectory accuracy of the integrated navigation system under the condition of UAV dynamic flight is evaluated. In order to verify the accuracy of the combined system, another ground road test is carried out. The test vehicle is equipped with a high-level INS/GPS integrated system (SPAN-CPT) as the reference system for this experiment (gyro drift 1°/h). The system has tactical level IMU and GPS dual frequency carrier phase observation, and uses the smooth solution of differential solution as the reference solution with inertial Explorer; The data processing of the system to be tested is based on the software developed by this study. The difference between MMQ-G inertial navigation observation and single frequency L1 carrier phase observation is used to calculate the real-time filter solution, and compared with the reference solution. Because the UAV cannot carry the reference system due to the load limit, the positioning accuracy of the test system is tested in the on-board environment.

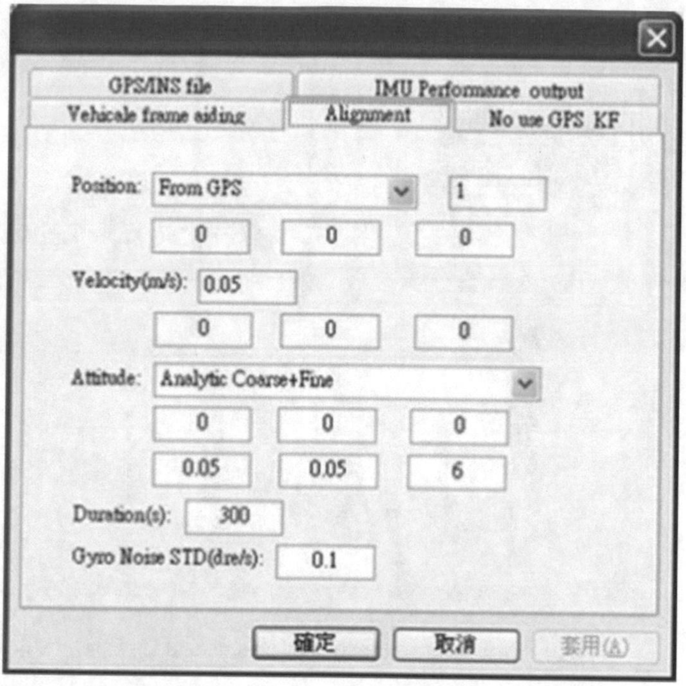

Fig. 3. The loosely-coupled architecture diagram and self-developed software used in this study

The experimental length is 1 h. The experimental content is to simulate the flight situation of the aircraft, so it includes straight line and 180° turning. Figure 4 shows the experimental track, and Fig. 5 shows the three-axis positioning error of the combined module in East, North and up. The results show that the RMS values of three-axis positioning errors are 0.77, 0.66 and 0.83M respectively. Figure 6 shows the trajectory of the dynamic test of the test system according to the flight altitude of 600 m and 300 m. The data processing strategy is the same as that of the on-board test. The measured results show that the integrated navigation system proposed in this study can continuously provide 1 m 3D positioning accuracy (sampling rate is 200 Hz) in a dynamic environment, which is suitable for the control application of UAV no-fly zone in the future. Of course, the map accuracy is also a very important issue for UAV air traffic control. If the map accuracy is also improved to the sub meter level, the buffer area can be reduced in a large area to effectively reduce the disputes when law enforcement agencies judge violations.

In terms of the actual use in the future, the control and management application of real-time UAV no-fly zone may be closed or disturbed artificially in practical operations. Therefore, this paper suggests that for UAVs with a certain load or more, in addition to report the flight plan in advance to apply for the flight permit, it is also very important to upload the trajectory diagram of UAV afterwards for the review of relevant departments.

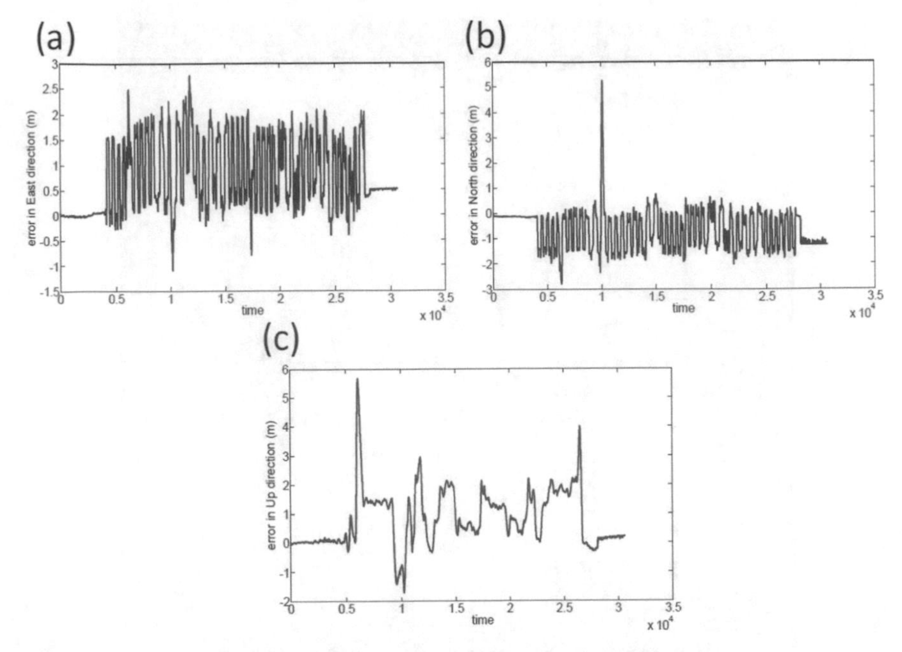

Fig. 4. Positioning errors of East, North, and Up axis

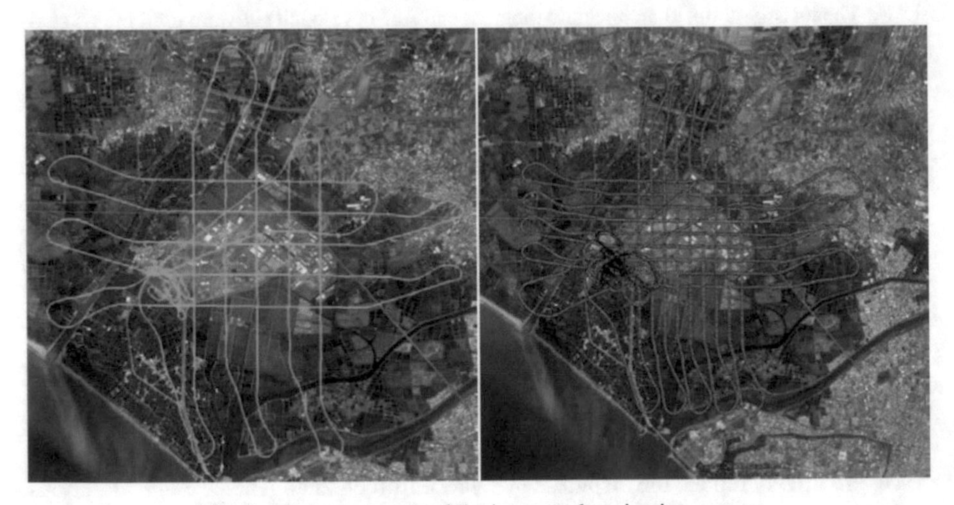

Fig. 5. Flight test track of the integrated navigation system

5 Conclusion

In this paper, an accurate positioning method of UAV Based on integrated navigation is designed and applied to the accurate control of UAV no-fly zone, and better estimation accuracy is obtained in comparison with high-precision reference trajectory. Experiments show that the designed algorithm has good positioning accuracy and can meet the needs of accurate control of UAV, regardless of whether the UAV is moving in a straight line or curve.

References

1. Colomina, I., Molina, P.: Unmanned aerial systems for photogrammetry and remote sensing: a review. ISPRS J. Photogramm. Remote Sens. **92**(2), 79–97 (2014)
2. Birnbaum, Z., Dolgikh, A., Skormin, V., et al.: Unmanned aerial vehicle security using behavioral profiling. In: Proceedings of International Conference on Unmanned Aircraft Systems, pp. 1310–1319. IEEE (2015)
3. Macgougan, G., Lachapelle, G., Klukas, R., et al.: Performance analysis of a stand-alone high-sensitivity receiver. GPS Solutions **6**(3), 179–195 (2002)
4. Tian, S., Xue, S., He, Y.: Application and development of satellite navigation technology. Command Inf. Syst. Technol. (02), 50–55 (2014)
5. Soloviev, A., Van Graas, F.: Use of deeply integrated GPS/INS architecture and laser scanners for the identification of multipath reflections in urban environments. IEEE J. Sel. Top. Signal Process. **3**(5), 786–797 (2009)
6. Milanes, V., Naranjo, J.E., Gonzalez, C., et al.: Autonomous vehicle based in cooperative GPS and inertial systems. Robotica **26**(5), 627–633 (2008)
7. Sun, R., Han, K., Hu, J., et al.: Integrated solution for anomalous driving detection based on BeiDou/GPS/IMU measurements. Transp. Res. Part C **69**, 193–207 (2016)
8. Wang, R., Liu, J., Xiong, Z., et al.: Double-layer fusion algorithm for EGIbased system. Aircr. Eng. Aerosp. Technol. **85**(4), 258–266 (2013)
9. Kalman, R.E.: A new approach to linear filtering and prediction problems. J. Basic Eng. Trans. **82**, 35–45 (1960)
10. Wang, J.: Research of algorithm for mobile node localization in mine based on Kalman filtering. J. Electron. Measur. Instrum. **27**(2), 120–126 (2013)
11. Mao, Y., Yang, Y., Hu, Y.: Research into a multi-variate surveillance data fusion processing algorithm. Sensors (Peterborough, NH) **4975**(2019), 1–12 (2019)

LFMCW Milimeter Radar Detection

Xinyu Wang[1,2], Yan Xu[1,2], Ming Li[1,2(✉)], Jiawei Han[1,2], and Li Li[3(✉)]

[1] Tianjin Key Laboratory of Wireless Mobile Communications and Power Transmission, Tianjin Normal University, Tianjin 300387, China
mliece@163.com

[2] College of Electronic and Communication Engineering, Tianjin Normal University, Tianjin 300387, China

[3] School of Economics and Management, Tianjin University of Science and Technology, Tianjin 300222, China
498718213@qq.com

Abstract. The so-called millimeter wave is a segment of radio waves, located in the microwave and far infrared wave overlapping wavelength range, so it has the characteristics of two spectrums. Millimeter-wave radar can detect distant targets in both day and night, and has the characteristics of being less affected by the environment and higher resolution in ranging, velocity measurement and Angle measurement. With the research of intelligent detection, the requirement of radar is getting higher and higher, and the requirement of ranging accuracy and resolution is more and more stringent. The LFM pulse signal has a large time-bandwidth product, which can detect longer-distance targets and meet the requirements. Pulse compression is usually required for LFM echo signal to get specific information of the target more directly and accurately, remove redundant interference, better distinguish multiple aliased echoes, and get target distance and speed information. The common pulse compression methods are dechirp method and matched filter method. This paper mainly introduces the pulse compression technology of linear frequency-modulated continuous wave radar, which can get the distance and velocity information of the target.

Keywords: Milimeter radar · Radar speed measurement · Radar ranging · Dechirp pulse compression · Matched filtering

1 Introduction

Radar (Radio detection and ranging), which we can think of as the “eyes” and “ears”. It uses radio methods to detect and find targets, and determine their direction, distance and speed of movement. In the early days, radar was mostly used for military [1, 2]. Early warning can only achieve the detection of the target distance. But with the in-depth research of radars in various countries, its function became increasingly powerful. At present, radar can well achieve the measurement of distance, angle, speed and shape of the target.

X. Wang and Y. Xu—Contribute equally to this work.

Q. Liang et al. (Eds.): Artificial Intelligence in China, LNEE 854, pp. 622–628, 2022.
https://doi.org/10.1007/978-981-16-9423-3_78

Radar has a wide range of uses, and is generally divided into two categories: military and civilian. Military radar refers to the radar used in the military field, such as air intelligence radar, maritime alert radar, airborne early warning radar, over-the-horizon radar, ballistic missile early warning radar, etc. Radar used in the non-military field is civilian radar. It has been widely used in people's daily life. The positioning of mobile phones and bicycles makes it easier to find and plan routes in our daily lives. Environmental monitoring and weather forecasts convenient for us to have a certain understanding of the environment in advance and prepare for response. Car reversing radar facilitates us to conduct objects in blind areas of vision Judgment. Automatic doors in shopping malls reduce people's physical labor. The detection of patients' vital signs and the diagnosis of diseases in hospitals are also inseparable from sensing [3].

The driverless technology that has become so popular in recent years also makes extensive use of radar detection technology, which can be described as "no radar, no intelligence". Millimeter radar is the most common application at present, and it is also the best choice for accuracy and stability. Regardless of the influence of humidity, temperature and light, millimeter-wave can work normally. No matter the influence of humidity, temperature and light, millimeter wave can work normally. Radar also has the following advantages [1–3]:

(1) High resolution, small, low cost, portability.
(2) Ability to work in complex environments with little influence from external environmental factors.
(3) It can accurately analyze and record the location, number and specific information of the target.
(4) High target positioning accuracy, high resolution, wide signal band and good immunity to electromagnetic interference.

2 LFMCW Radar Pulse Compression

Radar can be used to measure the distance of the target. The propagation of electromagnetic waves in a homogeneous medium can be approximated as a straight line with the speed of light ($c = 3 \times 10^8$ m/s). The emitted electromagnetic wave will be reflected back when it hits the target in this direction and will be received by the receiver. Therefore, the distance of the target (the relative distance between the target and the radar) can be judged by the time delay of the transmit and the received signal.

Since the signal transmission to reception is a round-trip process, the time delay is assumed to be t_r, then the time taken by the radar to the target is $0.5t_r$, Suppose the distance is 'R', and 'c' is the speed of light, consequently: $R = 0.5ct_r$ (Fig. 1).

According to the working principle of radar, it is known that the detection capability of millimeter-wave radar is related to its own average transmitting power. And only by continuously improving the transmitting power of the radar system can the detection capability of the radar system be better improved. We generally use the method of increasing the pulse width of the transmitter to improve the average power of the system, but the increase of the pulse width will cause the decrease of the range resolution of the radar system. So, there is a contradiction between improving the measurement accuracy

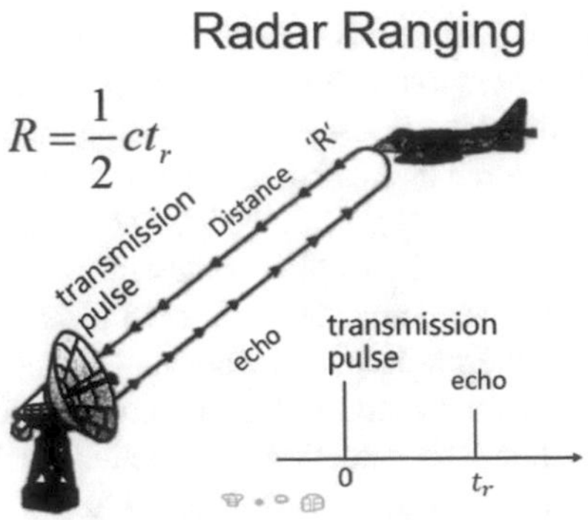

Fig. 1. Priciple of Radar ranging

and resolving power of radar system. At this time, we need to introduce pulse compression technology. Transmitting a large time and wide bandwidth signal at the transmitting terminal to improve the speed measurement accuracy and speed resolution of the signal. Compressing the wide pulse signal into narrow pulse at the receiving terminal to improve the distance measurement accuracy and distance resolution of the radar to the target. This method effectively improves the detection and resolution capabilities of the radar system.

At present, the most widely used signal in the pulse compression technology is the LFM (linear frequency modulation) pulse signal, which is a common large-time and wide-band product signal, and that is insensitive to the Doppler shift of the echo signal. The graph of the LFM signal is shown in the following figure. Figure 2(a) shows the LFM pulse signal waveform and Fig. 2(b) shows the envelope of the signal, range is A and width is T, Fig. 2(c) shows the linear change rule of carrier frequency. It changes according to the slope $(f_2 - f_1)/T$ during the time T.

2.1 Dechirp Method

In order to solve the contradiction between the acting distance and spatial resolution faced by the traditional single frequency pulse, the pulse compression technology is generally adopted: the transmitting pulse with relatively wide width and low peak power makes the signal have enough energy to ensure the acting distance; After receiving, we compress the low peak width pulse into the narrow peak value pulse, so as to avoid pulse overlap and improve the spatial resolution [4–6] (Fig. 3).

Linear frequency modulated continuous wave signal originates from the transmitter antenna:

$$S(t) = u(t)e^{j2\pi f_0 t} = Arect\frac{t}{T}e^{j2\pi\left(f_0 t + \frac{1}{2}\mu t^2\right)}, \tag{1}$$

where, μ = B/T is the frequency modulation slope, B is the bandwidth, T is the frequency modulation time; f_0 is carrier frequency; A is the signal amplitude. The echo signal can be expressed by

$$S_1(t) = A_1 rect\frac{t-\tau}{T}e^{j2\pi\left(f_0(t-\tau) + \frac{1}{2}\mu(t-\tau)^2 + f_d t\right)}, \tag{2}$$

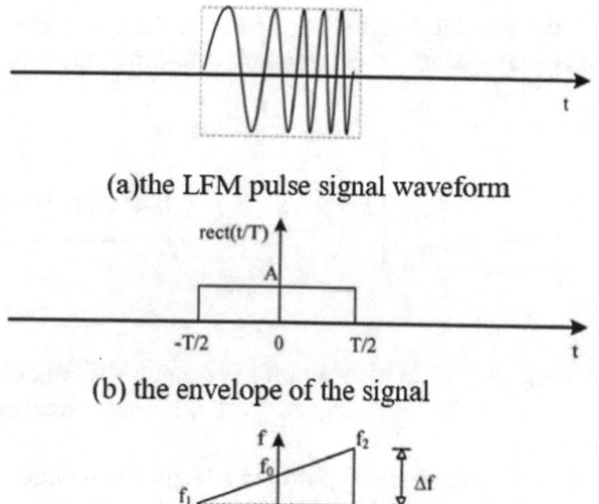

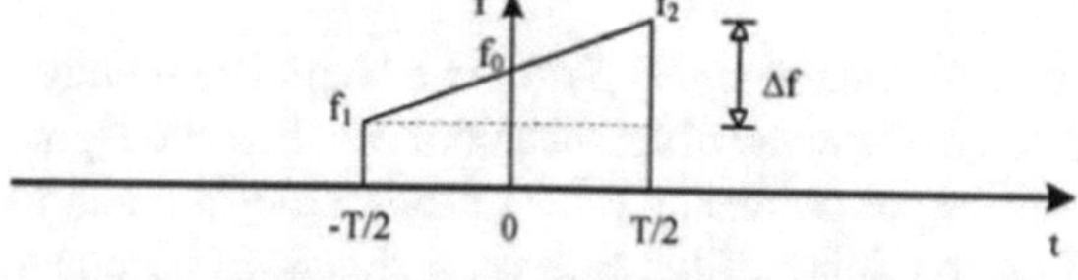

(c) the linear change rule of carrier frequency

Fig. 2. LFM pulse signals

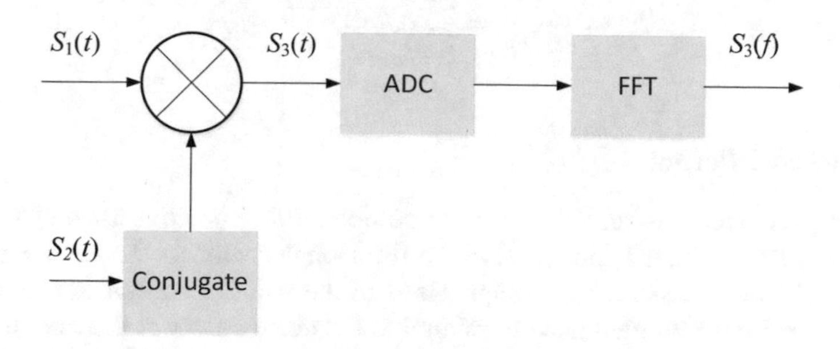

Fig. 3. Dechirp method

where $\tau = (2R)/c$ is the signal delay, R is the target distance, c is the vacuum speed of light; A_1 is the echo signal amplitude, including attenuation factor,fd is the Doppler frequency (difference between transmitting frequency and receiving frequency).

The reference signal is generated by a local oscillator signal with appropriate delay, wherein the amount of delay is usually estimated from the bandwidth signal ranging result (same FM slope as the input signal). Eventually, the reference signal has the form

$$S_2(t) = A_2 rect\frac{t - \tau_0}{T} e^{j2\pi\left(f_0(t-\tau_0)+\frac{1}{2}\mu(t-\tau_0)^2\right)}, \tag{3}$$

where τ_0 is the relative transmission delay of the reference signal; A_2 is the amplitude of the reference signal.

In this method, the reference signal S_2(t) with the same carrier frequency as the transmitting signal is beat against the echo signal. It follows that

$$S_3(t) = S_1(t)S_2^*(t)$$
$$= A_1A_2 rect\frac{t-\tau}{T} rect\frac{t-\tau_0}{T} e^{j\left[2\pi\left(\underbrace{\mu(\tau_0-\tau)}_{f_b}+f_d\right)t+\underbrace{2\pi f_0(\tau_0-\tau)+\pi\mu\left(\tau^2-\tau_0^2\right)}_{\phi}\right]} \tag{4}$$

If the Doppler frequency f_d is not taken into account, the beat difference signal is a single frequency signal with frequency f_b, which is much smaller than the carrier frequency f0 of the signal.

It can be seen that the mixed signal S_3 (t) is a single-frequency signal, and its frequency value is related to the delay difference between the echo signal and the reference signal. Therefore, the output frequency value of the single-frequency signal can be controlled by controlling the delay of the reference signal. Subsequently, the expression of beat signal in frequency domain is able to be obtained by merely using the FFT. Eventually, resolution in range and time can be derived with the following forms.

$$\Delta t = \frac{1}{\mu N T_S} = \frac{1}{B}, \Delta R = \frac{c\Delta t}{2} = \frac{c}{2B} \tag{5}$$

2.2 Matched Filtering

The LFM pulse signal is suitable for pulse compression. The advantage of this signal is that the matched filter is not sensitive to the Doppler shift of the echo signal. The maximum signal-to-noise ratio is emphasized in the radar maximum range equation, so the matched filter is often used to minimize the noise power at the output of radar receiver. There are two approaches to pulse compression matched filtering based on time domain and frequency domain. The essence of the two methods is the same, and the frequency domain FFT method is mostly used in the simulation. The following figure shows the simulation model of matched filter [7–10] (Fig. 4).

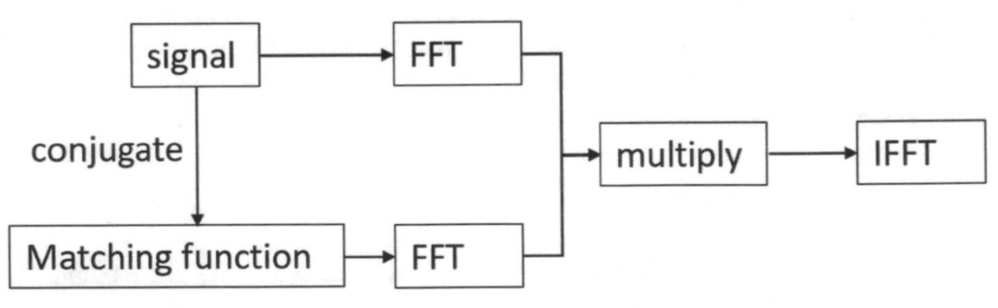

Fig. 4. Matched filtering method

Objects with different distances in the same direction can be clearly distinguished by matching filtering. The expression of the LFM signal is:

$$s(t) = \exp(j\pi f_0 t + j\pi k^2 t^2) \tag{6}$$

where the starting frequency is f_0, FM slope is 'k', ($k = B/T$). Let the pulse width of the transmitted signal be 'T', bandwidth is 'B', sampling frequency is f_s, One period of the received echo signal is $t = T + \frac{2R_{\max}}{c}$, number of Samples is '$N$', sampling frequency is $f_s = \frac{N}{t}$, The echo signal can be obtained as:

$$s_r(t) = \exp(j\pi k t^2), (0 \le t \le T) \tag{7}$$

Using the frequency domain method to find the matching function, the frequency domain can directly take the complex conjugate, and then the matching function is transformed by FFT. Multiplied by the FFT variable frequency domain of the echo function to obtain the pulse compression of the echo signal and match the pulse. The matching pulse compression method eliminates the frequency change faster part, can more clearly obtain the target.

IFFT transformation is performed on the echo signal at this time. In this process, FFT is circular convolution, and there may be cases that need to fill in zero. The following figure shows the echo function image obtained after pulse compression processing. It can be clearly seen that the distances and waveforms of the three targets have been distinguished.

3 Conclusions

This article briefly describes the principle of radar ranging in theory, and introduces the purpose and method of pulse compression. The two pulse compression methods (dechirp method and matched filter method) of linear frequency modulation signals are mathematically deduced and analyzed. At present, the pulse compression technology is mostly used in the processing of radar echo, so that we can get the required information more conveniently and intuitively when processing the echo. The LFM pulse signal has a large time-bandwidth product, can detect longer-distance targets, has good anti-interference ability, and is currently the most practical signal for pulse compression technology. Pulse compression technology can increase the detection range of targets and maintain a high resolution when the transmission power is limited. This is an important means for radar multi-target resolution and anti-jamming. Pulse compression techniques are widely used in the application of radar systems.

Acknowledgement. The author acknowledges the support from the National Natural Science Foundation of China (Grant 61805176, Grant 62001327), and the Philosophy and Social Science Youth Foundation of Tianjin (No. 17JCQNJC14800).

References

1. Will, C., et al.: Human target detection, tracking, and classification using 24-GHz FMCW radar. IEEE Sens. J. **19**(17), 7283–7299 (2019)
2. Panpan, H.: Research on frequency domain pulse compression of LFM signal. In: Journal of Physics: Conference Series, vol. 1550, no. 4. IOP Publishing (2020)
3. Li, X., Ma, X.: Joint Doppler shift and time delay estimation by deconvolution of generalized matched filter. EURASIP J. Adv. Signal Process. **2021**(1), 1–12 (2021). https://doi.org/10.1186/s13634-021-00741-7
4. Wang, Q., Song, Y., Xu, D.: Millimeter-wave radar and video fusion vehicle detection based on adaptive Kalman filtering. In: Journal of Physics: Conference Series, vol. 1966, no. 1. IOP Publishing (2021)
5. Levy, C., Pinchas, M., Pinhasi, Y.: Coherent integration loss due to nonstationary phase noise in high-resolution millimeter-wave radars. Remote Sens. **13**(9), 1755 (2021)
6. Wang, Z., et al.: Research of target detection and classification techniques using millimeter-wave radar and vision sensors. Remote Sens. **13**(6), 1064 (2021)
7. Farnett, E.C., Stevens, G.H., Skolnik, M.: Pulse compression radar. Radar Handb. **2**, 10–11 (1990)
8. Ramp, H.O., Wingrove, E.R.: Principles of pulse compression. IRE Trans. Mil. Electron. **2**, 109–116 (1961)
9. Burrascano, P., et al.: Pulse compression in nondestructive testing applications: reduction of near sidelobes exploiting reactance transformation. IEEE Trans. Circuits Syst. I Regul. Pap. **66**(5), 1886–1896 (2018)
10. Lu, Z., et al.: Study of terahertz LFMCW imaging radar with Hilbert transform receiver. Electron. Lett. **50**(7), 549–550 (2014)

Radar Signal Sorting Based on Intra-pulse Features

Mingyue Ma, Jiaqi Zhen(✉), and Weijian Qi

College of Electronic Engineering, Heilongjiang University, Harbin 15008, China
zhenjiaqi@hlju.edu.cn

Abstract. The current radar signal sorting methods are exposed to the problems of poor sorting effect and easy interference in the increasingly complex electromagnetic environment. A new sorting method has recourse to the characteristics of complexity and spectral resemblance. Primarily, the received pulse needs to be normalized and denoised, then the resemblance coefficient, box dimension and sparseness are extracted from the preprocessed signals in the frequency domain. And the three sets of features are combined into a joint feature vector, and the clustering algorithm is used to achieve the classification of 8 kinds of modulation signals. Finally, the experimental simulation shows the characteristic parameters have splendid inter-class separation, with the signal to noise ratio (SNR) of 0 dB and a minimum sorting accuracy of 85% among different modulation types of signals, this approach has considerable engineering value.

Keyword: Radar signal sorting · Box dimension · Resemblance coefficient · Complexity

1 Introduction

Radar signal sorting plays a considerable role in electronic countermeasures (ECM) system. The basic radar signal sorting method is to judge which radar each pulse belongs to in the dense overlapping pulse flow by using the regularity of the signal. The selection of parameters occupies the primary position in the whole radar signal sorting. Radar signal sorting approaches are mostly based on various intercepted parameters. Among them, the sorting approaches depended on pulse description word (PDW) and pulse repetition interval (PRI) are the most widely adopted [1, 2].

With the conditions of electronic warfare on the battlefield become changeable, various new radar systems with changing laws continue to appear, the traditional pulse description word sorting method and pulse repetition interval sorting method can not meet the actual needs of modern electronic warfare. As the fingerprint feature of radar signal, intra-pulse feature can extract many useful features that can realize radar signal sorting [3]. Compared with other conventional parameters, the subtle characteristics in the pulse are more universal, stable, unique and measurable. For decades, numerous academic experts have extracted the intra-pulse characteristics of radar signals as signal sorting parameters, and fused these parameters into various algorithms and achieved good sorting results [4].

Q. Liang et al. (Eds.): Artificial Intelligence in China, LNEE 854, pp. 629–637, 2022.
https://doi.org/10.1007/978-981-16-9423-3_79

Some of methods proposed by some scholars are mainly for several specific modulation signals, or they are susceptible to noise interference. Since extracting complexity features in time domain is susceptible to complex electromagnetic environment, here a novel approaches depended on the frequency domain is presented. Radar signals are normalized and denoised after frequency domain conversion. Then the box dimension of the signal is extracted to reflect the geometric scale information of the signal. To further illustrate the complexity characteristics, the sparseness of distribution is introduced into the processing of radar signals, and the relevant threshold is deduced to obtain sparseness. The box dimension and the sparseness distribution characteristics in the pre-processed frequency domain are extracted as joint as the complexity features. Then, the resemblance coefficient is calculated in the frequency domain and combined with the complexity characteristics, and the sorting is completed by kernel fuzzy c-means clusterin (KFCM). The excellent property of the presented method is verified by simulation results on 8 kinds of radar signals.

2 Signal Pre-processing

In general, different modulation characteristics can be obtained in different transformation domains. However, the influence of noise varies in different transformation domains. The modulation style of the signal can be directly represented on the waveform of the signal. However, in the current complex electromagnetic environment, the direct extraction of time domain features is susceptible to noise. Therefore, it is a good choice to convert the intercepted radiation source signal into frequency domain and extract its characteristics in frequency domain. In this paper, after the intercepted radar radiation source signal is transformed by fast Fourier transform (FFT) and normalized by energy, the radar signal sequence $\{x(k), k = 1, 2, \cdots, N\}$ is obtained. The pre-processed sequence is used to extract the frequency domain complexity feature, and the box dimension and sparseness feature parameters are extracted.

Since the effective signal energy is concentrated in a narrow band range, the noise energy is uniformly distributed throughout the band (in this case, the noise is all Gaussian white noise). Therefore, denoising must be done. The denoising signal sequence is $f(k)$ and the denoising equation is

$$f(k) = \begin{cases} x(k) - \frac{1}{N}\sum_{k=1}^{N} x(k), & x(k) > \frac{1}{N}\sum_{k=1}^{N} x(k) \\ 0, & x(k) \le \frac{1}{N}\sum_{k=1}^{N} x(k) \end{cases} \quad k = 1, 2, \cdots, N \tag{1}$$

Here is a signal model with a bandwidth of 20 MHz and an initial frequency of 100 MHz. Its modulation mode is linear frequency modulation. Here SNR is set to 15 dB, the above model is used for denoising operation. Figure 1 and Fig. 2 represent the spectrum before and after denoising. By comparing the two figures, they show the obvious effect of noise suppression in the whole spectrum, which will be beneficial to the subsequent feature extraction.

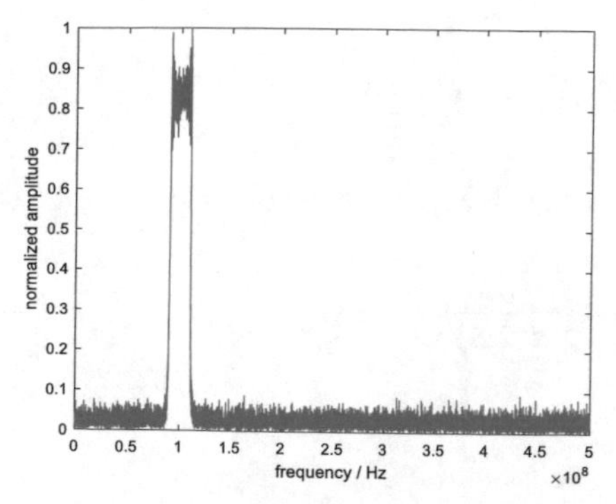

Fig. 1. Spectrum before noise removal

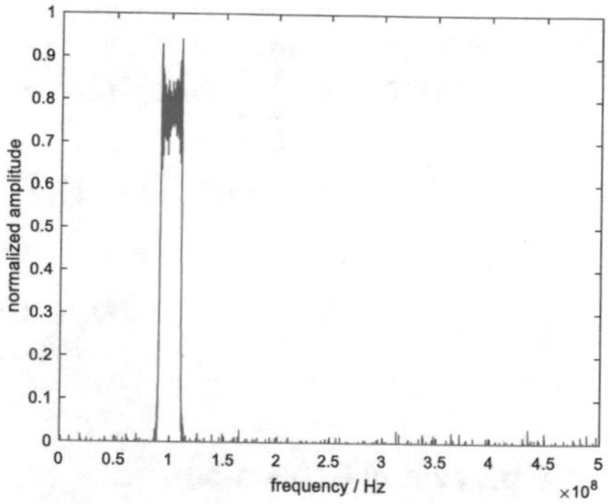

Fig. 2. Spectrum after noise removal

3 Feature Extraction

Many classical intra-pulse characteristics of radar signals extraction methods have been proposed [5]. For example, phase difference method, time domain autocorrelation method, modulation domain analysis method, wavelet transformation method, and so on. The algorithm proposed in this section combines the frequency domain extraction complexity and resemblance coefficient features to obtain higher noise immunity and better separation.

3.1 Box Dimension Extraction

The signal sequence after pretreatment of different modulation types of radar radiation source signals has different complexity characteristics. Fractal characteristics are a kind of complexity characteristics. The fractal properties of sequence can quantitatively describe its complexity and irregularity. The box dimension parameter in fractal theory can accurately describe the geometric scale of the series.

Suppose (F, d) is a measurement space and $Q\ (f, \lambda)$ is used to represent a closed sphere with a spherical radius of λ and a spherical center of f. Where λ is a constant and its range of values is $[0, \infty)$. Assuming P is a non-empty subaggregate of F, for each λ, $Y(P, \lambda)$ represents the minimum amount of closures that cover P, that is

$$Y(P,\lambda) = \left\{ N : P \subset \bigcup_{k=1}^{N} Q(f_k, \lambda) \right\} \tag{2}$$

Where $f_1, f_2, \cdots, f_N$ is a distinct point of F.

Set P is a bicompact set, non-negative real number. If there is

$$D_f = \lim_{\varepsilon \to 0} \frac{\ln Y(P,\lambda)}{\ln(1/\lambda)} \tag{3}$$

D_f represents the box dimension of set P. The calculation equations of box dimension of the pre-processed signal sequence are

$$d(\Delta) = \sum_{k=1}^{N} |f(k) - f(k+1)| \tag{4}$$

$$d(2\Delta) = \sum_{k=1}^{N/2} (\max\{f(2k-1), f(2k), f(2k+1)\} - \min\{f(2k-1), f(2k), f(2k+1)\})|f(k) - f(k+1)| \quad (5)$$

$$D_f = 1 + \log_2 \frac{d(\Delta)}{d(2\Delta)} \quad (6)$$

3.2 Sparseness Extraction

The frequency domain scale data of the pre-processed signal sequence can be represented by the box dimension. In order to fully reflect the complexity characteristic of the sequence, the sparseness distribution characteristic of the sequence is also required. Here it is referred to the classification of unknown complex radar signals.

Let $G_N = \{g_1, g_2, \cdots g_N\}$ be a sequence in one-dimensional (0, 1) space containing Y elements '1'. In order to describe the density characteristics at different positions in sequence G_N, a window function of length N_0 is constructed to slide from left to right on sequence G_N, and the '1' elements in the first window are Y_k, then the sparseness D_k in this window is

$$D_k = \left| \frac{Y_k}{Y} - \frac{k}{\lfloor N/N_0 \rfloor} \right|, \quad k = 1, 2, \cdots, \lfloor N/N_0 \rfloor \quad (7)$$

The mean sparseness μ of sequence G_N is

$$\mu = \frac{1}{\lfloor N/N_0 \rfloor} \sum_{k=1}^{\lfloor N/N_0 \rfloor} D_k, \quad k = 1, 2, \cdots, \lfloor N/N_0 \rfloor \quad (8)$$

The mean sparseness reflects the overall sparseness of '1' element in the sequence. The pre-processed signal sequence is not a sequence in one-dimensional (0, 1) space. Therefore, it needs to be treated with 0–1, and the processing equation is

$$g_k = \begin{cases} 1, f(k) \geq \xi \\ 0, f(k) < \xi \end{cases} \quad k = 1, 2, \cdots, N \quad (9)$$

After 0–1 processing of $f(k)$, g_k is obtained, and it is a sequence in one-dimensional space (0, 1). Rearrange this signal sequence according to its size to obtain $g'_k = \{g'_1, g'_2, \cdots g'_N\}$, then the threshold ξ is

$$\xi = \begin{cases} g'_{\frac{N+1}{2}}, & N \text{ is odd} \\ \frac{1}{2}(g'_{\frac{N}{2}} + g'_{\frac{N+1}{2}}), & N \text{ is even} \end{cases} \quad (10)$$

After the threshold ξ is determined, for the signal sequence $f(k)$, it should be transformed according to Eq. (9), and then the average sparseness μ is obtained according to Eqs. (7) and (8).

3.3 Resemblance Coefficient Feature Extraction

Radar radiation source signals with different modulation modes have different spectrum shapes, and the change of the spectrum contains the information of the frequency, phase and amplitude of the signal [6]. The feature of resemblance coefficient is designed to judge diverse shape of the spectrum. The resemblance coefficient feature is essentially a feature mapped to the feature space, which can represent the degree of difference between two functions or two discrete sequences.

Defines the resemblance coefficient of one-dimensional continuous positive function $f_1(x)$ and $f_2(x)$ as

$$C_r = \frac{\int f_1(x)f_2(x)dx}{\sqrt{\int f_1^2(x)dx} \cdot \sqrt{\int f_2^2(x)dx}.} \tag{11}$$

C_r denotes the degree of difference in the trend of change between the two functions. It is not difficult to see that when $f_1(x)$ and $f_2(x)$ are orthogonal to each other, $C_r = 0$;When $f_1(x)$ and $f_2(x)$ are exactly the same or the same scale, $C_r = 1$; The C_r of the two signals decreases positively with their trend and thus with the difference in outline.

After discrete processing of $f_1(x)$ and $f_2(x)$, the resemblance coefficient of these two discrete signals is

$$C_{rc} = \frac{\sum F_1(k)F_2(k)}{\sqrt{\sum F_1^2(k)} \cdot \sqrt{\sum F_2^2(k)}} \tag{12}$$

Where $\{F_1(k)\}_{k=1}^N$ and $\{F_2(k)\}_{k=1}^N$ are the discrete signals of $f_1(x)$ and $f_2(x)$ respectively. Thc resemblance coefficient of the discrete signal has the same characteristics as that of the continuous signal.

In this paper, the resemblance coefficient of the signal spectrum are clearly calculated by using the triangular non-orthogonal signal sequence as the projected signal, and its sequence expression is

$$T(k) = \begin{cases} 2k \cdot mx/N, & 1 \le k \le N/2 \\ 2mx - 2k \cdot mx/N, & N/2 \le k \le N \end{cases} \tag{13}$$

Where m represents the maximum signal amplitude. Because the ability distribution of triangular signals is concentrated, the energy distribution of various modulation signals can be better reflected by projecting the radar emitter signal to the triangle.

4 Radar Signal Sorting Process

To sum up, the sorting steps for the received radar radiation source signal are as follows:

Step 1: Transform the radar radiation source signal sequence in frequency domain and normalize the energy, and perform denoising pretreatment according to Eq. (1);
Step 2: Calculate box dimension and sparseness of pre-processed signal sequence according to Eqs. (6) and (8) respectively;

Step 3: Extract the resemblance coefficient feature of its spectrum from Eq. (12);
Step 4: Take box dimension, sparseness and resemblance coefficient as the feature parameters of signal sorting, and complete the final sorting based on kernel fuzzy c-means clusterin (KFCM).

5 Simulation Experiment and Analysis

In this paper, the effectiveness of the sorting approach based on complexity and resemblance coefficient is proved by Matlab simulation experiment feature extraction and simulation experiments are carried out for 8 types of typical intra-pulse modulation signals. These 8 types of signals are continuous wave signal (CW), LFM signal, frequency shift keyed signal (FSK), binary phase shift keyed signal (BPSK), quadrature phase shift keyed signal (QPSK), linear frequency modulation and binary phase shift keyed signal (LFM-BPSK), frequency shift keyed and quadrature phase shift keyed signal (FSK-QPSK) and nonlinear frequency modulation signal (NLFM). Among them, the frequency of both FSK and FSK-BPSK are set to 25 MHz and 45 MHz and the rest of the samples are set to 35 MHz. In addition, this paper selects sinusoidal frequency modulation signal as NLFM signal. At 0–20 dB, each type of signal generates 100 signals for subsequent experiments.

Firstly, the generated 8 types of signals are converted to frequency domain by FFT, and then denoised after energy normalization. The box dimension, sparseness and resemblance coefficient features of 8 different modulation types of radar radiation source signals are calculated according to the equations. In the corresponding SNR, the average value of each feature is obtained. Figure 3, Fig. 4 and Fig. 5 show the average value of radar signal sparseness features, box dimension features and resemblance coefficient features under different SNR ranges. In these figures, legends 1 to 8 represent CW, LFM, FSK, LFM-BPSK, BPSK, QPSK, FSK-BPSK and NLFM respectively. It can be seen that the complexity characteristics of the frequency domain of various modulation signals are obviously independent, and there is an excellent degree of inter-class separation between various modulation signals. In addition, the fluctuation of these characteristic parameters is small and has strong anti-interference. This verifies the rationality of the algorithm, so these features can be applied to signal sorting.

In this paper, KFCM algorithm is used to realize the clustering and sorting of 8 forms of radar emitter signals. Before that, three eigenvectors with excellent inter-class separation have been obtained. Now the three component joint eigenvectors are sorted by clustering. In this simulation experiment, the Gaussian radial basis kernel is selected as the kernel function, and the initial number of clusters is set as 2, and the superior limit of categories is set as 8. The iterated algebra is set to 60, and the stop condition is $\varepsilon \leq 0.001$. The simulation results are listed in Table 1.

According to the simulation results, when the SNR is above 10 dB, the accuracy of 8 kinds of signal sorting is 100%. With the decrease of SNR, the sorting accuracy of radar signals also decreases. When the SNR is 0 dB, the sorting accuracy of other kinds of signals except CW and NLFM signals decreases obviously, because the characteristic overlap probability of this part of modulated signals further increases with the decrease of

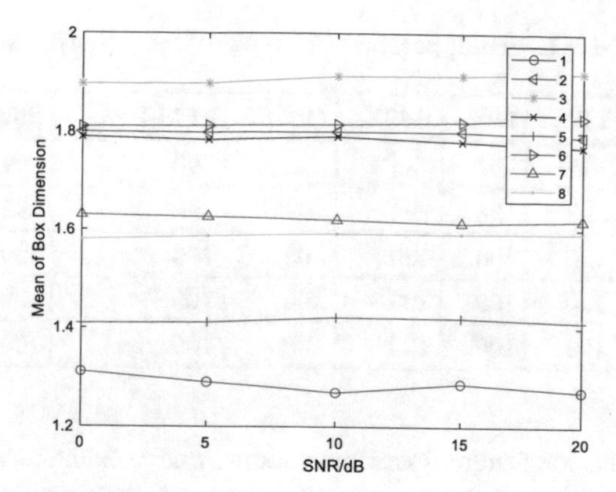

Fig. 3. Mean of box dimension under different SNR

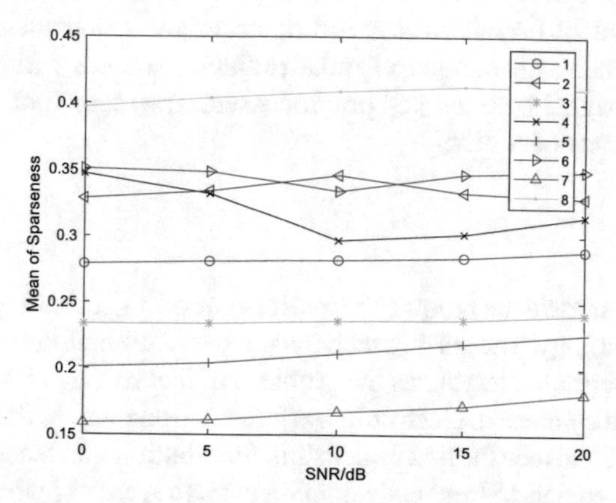

Fig. 4. Mean of sparseness under different SNR

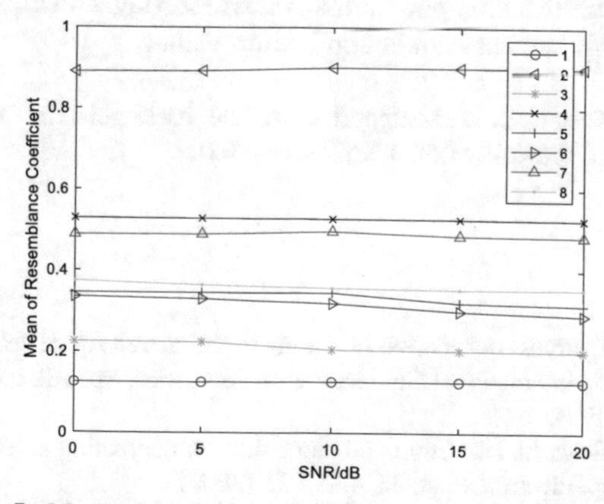

Fig. 5. Mean of resemblance coefficient under different SNR

Table 1. Sorting accuracy of 8 types of radar signals (%)

SNR (dB)	CW	LFM	FSK	BPSK	QPSK	LFM-BPSK	FSK-BPSK	NLFM
0	100	95	92	86	85	90	89	100
5	100	98	98	94	93	99	95	100
10	100	100	100	100	100	100	100	100
15	100	100	100	100	100	100	100	100
20	100	100	100	100	100	100	100	100

SNR. The box dimension feature, sparseness feature and resemblance coefficient feature extracted in this paper can compensate each other and overcome the disadvantage of single feature. Even in the case of low SNR, 8 kinds of modulation signals can maintain relatively good degree of separation between classes. At this time, the minimum sorting accuracy is 85%, and the overall average sorting accuracy is more than 92%. In the actual battlefield environment, the number of radar radiation sources will be greatly reduced after the receiving signal is sorted and preprocessed. The method of the paper will also obtain higher accuracy of sorting.

6 Conclusion

Due to the progress of radar technique, the traditional sorting technology is challenged by the new radar technology. The sorting technology based on traditional pulse description word and pulse repetition interval has the problem of low sorting accuracy, and it is very easy to be disturbed by noise in electronic warfare environment. In this paper, the signals are denoised at first, and then the box dimension, distribution sparseness and resemblance coefficient of radar signals in frequency domain are extracted as feature parameters. The features extracted in this paper have good inter-class separation and noise resistance. Finally, eight different modulation types of signals are classified through KFCM. The simulation results show that this approach can realize sorting with a excellent recognition rate for these 8 signals and has certain application value.

Acknowledgments. This work was supported by the basic scientific research projects of Heilongjiang Provincial University (2020-KYYWF-1005).

References

1. Nelson, D.: Special purpose correlation functions for improved signal detection and parameter estimation. In: IEEE International Conference on Acoustics, Speech, and Signal Processing, vol. 4, pp. 73–76 (1993)
2. Nishiguchi, K., Kobayashi, M.: Improved algorithm for estimating pulse repetition intervals. IEEE Trans. Aerosp. Electron. Syst. **36**, 407–421 (2000)
3. Jun, H., Minghao, H., Yixin, Y., et al.: A novel method for sorting radar emitter signal based on the ambiguity function. Aerosp. Electron. Warfare **25**, 35–38 (2009)

4. Guo, Q., Nan, P., Zhang, X., et al.: Recognition of radar emitter signals based on SVD and AF main ridge slice. J. Commun. Netw. **17**, 491–498 (2015)
5. Guo, Q., Teng, L., Qi, L., et al.: A novel radar signals sorting method-based trajectory features. IEEE Access **7**, 171235–171245 (2019)
6. Wang, S., Gao, C., Zhang, Q., et al.: Research and experiment of radar signal support vector clustering sorting based on feature extraction and feature selection. IEEE Access **8**, 93322–93334 (2020)

Resource Allocation for Full-Duplex D2D Communications Based on Deep Learning

Xueqiang Ren[1] and Liang Han[1,2(✉)]

[1] College of Electronic and Communication Engineering, Tianjin Normal University, Tianjin 300387, China

[2] Tianjin Key Laboratory of Wireless Mobile Communications and Power Transmission, Tianjin Normal University, Tianjin 300387, China

hanliang@tjnu.edu.cn

Abstract. As a branch of artificial intelligence, deep learning has begun to be explored and applied to the field of wireless communication systems. This paper proposes a deep neural network (DNN) model suitable for resource allocation of full-duplex device-to-device (D2D) communications. This deep neural network is an unsupervised learning model. It uses channel information generated by a random algorithm as the input and the optimal power allocation strategy as the output of the neural network. Since the opposite of spectral efficiency is added to the loss function, the model can maximize spectral efficiency through gradient descent. Compared with the traditional exhaustive search algorithm, the trained model will greatly shorten the running time of the program. This is especially obvious when the number of D2D users is large enough and the area is large enough.

Keywords: DNN · Full-duplex · Unsupervised learning · D2D communications · Spectrum efficiency

1 Introduction

With the development of mobile communication technology, the number of wireless devices in the world is rapidly increasing. However, the spectrum resources that can be used for wireless mobile communication are limited. As the number of these smart devices grows, the available spectrum resources become less and less. Therefore, how to improve the utilization rate of spectrum resources has become an urgent problem.

To improve the utilization rate of spectrum resources and solve the problem of shortage of spectrum resources, full-duplex communications systems and device-to-device (D2D) communications systems have been extensively studied in recent years. The reason of full-duplex communications systems gained significant attention is the potential that can further improve or even double the capacity of conventional half-duplex systems. The benefits of full-duplex communications systems are brought by allowing the downlink and uplink channels

Q. Liang et al. (Eds.): Artificial Intelligence in China, LNEE 854, pp. 638–645, 2022.
https://doi.org/10.1007/978-981-16-9423-3_80

to function at the same time and frequency [1]. Though the gains of full-duplex systems can be easily foreseen, practical implementations of such full-duplex systems pose many challenges and a lot of technical problems still need to be solved. The crucial barrier in implementing full-duplex systems resides in the self-interference (SI) from the transmit antennas to receive antennas at a wireless transceiver. More explicitly, the radiated power of the downlink channel interferes with its own desired received signals in the uplink channel. Clearly, the performance of full-duplex systems depends on the capability of SI cancellation at the transceiver [2]. D2D communications have been envisioned as an allied technology of 5G wireless systems for providing services that include live data and video sharing. A D2D communication technique opens new horizons of device centric communications, i.e., exploiting direct D2D links instead of relying solely on cellular links. Offloading traffic from traditional network-centric entities to D2D network enables low computational complexity at the base station (BS) besides increasing the network capacity [3].

In recent years, many breakthroughs in SI cancellation techniques have been reported, e.g., in [4,5]. Especially, these studies demonstrate the feasibility of full-duplex communications systems. The amalgamation of full-duplex D2D communications networks and cellular communications networks has provided an interesting direction for further research. The D2D links could enhance cellular reuse and improve the utilization rate of spectrum resources [6]. However, the interference issues in an environment involving cellular network and D2D network could impair system performance. How to assist D2D pairs in resources allocation according to the required quality-of-service (QoS) has become especially important [3].

In previous studies, the calculation formulas were often too complicated to derive the formula solution when calculating the D2D resource allocation. The common approach is to use an exhaustive search algorithm for approximation to achieve the purpose of calculating the allocated power. However, calculating the numerical solution is still sometimes not applicable. The traditional iterative method will consume a lot of time and even be powerless when the number of users is large. The traditional solution is using a computer to run an exhaustive algorithm to find the optimal power allocation strategy. Due to the rise of machine learning and neural networks in recent years, more and more people have begun to actively explore the use of deep learning (DL) neural networks to solve D2D power allocation problems. DL is a technique based on deep neural networks (DNNs), which relates to the activity of neurons in the brain, and which has seen a recent surge in interest. The current enthusiasm for DL is mainly due to its significant advantages in terms of its better performance than conventional schemes [7]. Researchers have recently begun to apply DL in many aspects of wireless communications systems.

In this paper, we propose a deep neural network model suitable for resource allocation of full-duplex D2D communications. This deep neural network is an unsupervised learning model. It uses channel information generated by a random algorithm as the input and the optimal power allocation strategy as the output of the neural network. In this model, a custom loss function is used to calculate the loss of the entire model. The Adam algorithm is used in gradient descent. Since the opposite of spectral efficiency is added to the loss function, the model can maximize spectral efficiency through gradient descent. By comparing with the exhaustive algorithm, it is confirmed that the accuracy of the model can meet the requirements.

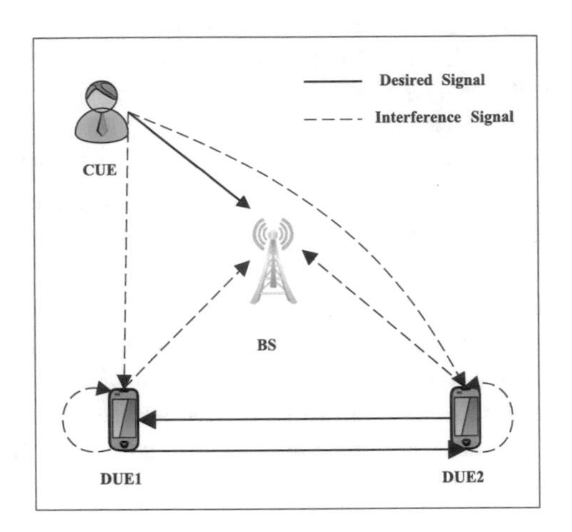

Fig. 1. System model of full-duplex D2D communications.

2 System Model

We consider a multi-channel cellular system which allows direct communication among D2D communication users. We assume that the number of channels is K, and for each channel, one cellular user equipment (CUE) transmits data to the BS, i.e., uplink transmission [8]. The number of D2D user equipments (DUEs) is set to N, where each DUE is not only a transmitter but also a receiver. The channel gain between the i-th transmitter and the j-th receiver for channel k, which comprises both the distance-related channel gain (i.e., path loss) and multipath fading, is denoted as $h^k_{i,j}$. We use last two indexes to denote CUE and BS, e.g., $h^k_{c,b}$ is the channel gain between the CUE and BS for channel k.

In the proposed communication system, the transmission power of the i-th device is denoted as P_i. The useful signal power transmitted by the i-th device received by the j-th device denotes $P_{i,j}$. The receiving equipments include the BS and all D2D equipments. Among them, $i = \mathrm{c}$ indicates that the transmitting device is a CUE, and $j = \mathrm{b}$ indicates that the receiving device is a BS. P^k_i denotes the transmit power of the i-th D2D device on the k-th channel. N_0 represents the spectral density of Gaussian white noise.

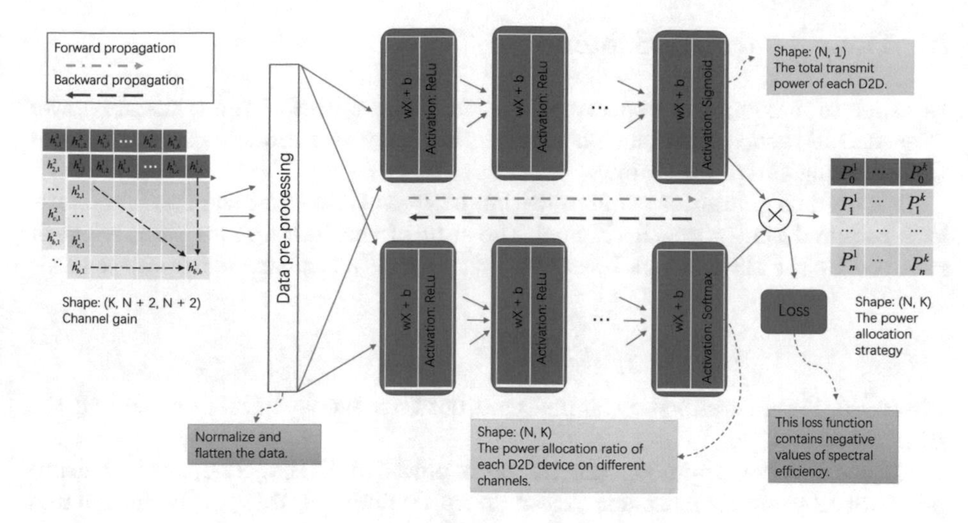

Fig. 2. Deep Neural Network Model.

2.1 Full-Duplex D2D Communication System Model

We assume that users are randomly distributed over an area D × D. There are multiple cellular users communicating with the BS, and only the uplink is considered. At the same time, there are multiple groups of D2D devices communicating. In Fig. 1, the two device users are marked as DUE1 and DUE2, and they use full-duplex communications mode. It should be noticed that due to the shortage of spectrum resources, the spectrum used in the communication between D2D devices in this paper is the spectrum of the uplink of the communication between the CUE and the BS. For example, if there are K CUEs communicating with the BS, then the D2D communication devices can use K channels for communications.

The BS will not only receive useful signals from cellular users but also interference signals from all D2D devices that use this channel for communications. In addition to the useful signal of the same group DUE, every D2D user can also receive the interference signals of all other DUEs and all CUEs in this area. At the same time, each device can also receive its SI caused by the transmitted signal.

2.2 Deep Neural Network Model

As shown in Fig. 2, we have built a network structure similar to [8] to solve the problem of full-duplex D2D communications. In the input layer, the input is a tensor of the channel gains of all devices in the entire communication system. It is more suitable to be the input of the neural network through data preprocessing (normalization, flatten). The neural network is mainly composed of two parts, each one is a four-layer network structure, and they multiplied to be the output. The output result of this network is a matrix of N rows and K columns for each sample.

3 Deep Learning Scheme

In order to maximize the spectrum efficiency and optimize the transmit power of each D2D device under such a communication system model, we need to meet the following three constraints.

We need to guarantee communication between CUEs and the BS. Therefore, it is required that on each channel, the sum of the interference signals power received by the BS must be less than a threshold I_{th}. It can be formulated as

$$\sum_{i=1}^{n} P_i |h_{i,b}|^2 < I_{th}, \tag{1}$$

where $|h_{i,b}|^2$ represents the channel gain between the i-th D2D device and the BS.

There is a constraint on the transmit power of DUEs. The total transmit power of all channels must not exceed an upper limit (P_T). It can be formulated as

$$P_i{}^k \geq 0, \sum_{k \leq K} P_i{}^k \leq P_T. \tag{2}$$

We need to ensure the QoS of D2D devices. We use spectrum efficiency to measure the performance of the model. In this model, P_M indicates the transmit power of the CUE. $\prod$ represents the set of D2D devices. $\left|h_{i,j}^k\right|^2$ represents the channel gain between the i-th device and the j-th device. σ_{SI}^2 represents the ratio of average SI power before and after the cancellation process in full-duplex D2D communications. The spectrum efficiency of the group i is $SE_i = SE_{oi} + SE_{ei}$. The formula for calculating SE_{oi} and SE_{ei} can be written as follows:

$$SE_{oi} = \sum_{k \leq K} \log_2 \left(1 + \frac{\left|h_{i,i+1}^k\right|^2 P_i^k}{\sum_{j \in \prod \setminus \{i\}} \left|h_{j,i+1}^k\right|^2 P_j^k + P_{i+1}^k \sigma_{SI}^2 + P_M \left|h_{0,i+1}^k\right|^2 + N_0 W} \right) (i = 1, 3, 5 \ldots), \tag{3}$$

$$SE_{ei} = \sum_{k \leq K} \log_2 \left(1 + \frac{\left|h_{i+1,i}^k\right|^2 P_{i+1}^k}{\sum_{j \in \prod \setminus \{i+1\}} \left|h_{j,i}^k\right|^2 P_j^k + P_i^k \sigma_{SI}^2 + P_M \left|h_{0,i}^k\right|^2 + N_0 W} \right) (i = 2, 4, 6 \ldots). \tag{4}$$

An unsupervised learning method is used in this DNN model. The determination of the loss function does not depend on the existing label. We use a customized loss function calculation method to optimize the parameters of this network model. Since the output of the neural network may not be able to meet the minimum requirements of communication, we considered these output results in the loss function to train the neural network model to get better performance and better output results. The loss function can be written as

$$loss = -\lambda_0 \sum SE_i + \lambda_1 \sum \tanh\left([R_T - SE_i]^+\right) + \lambda_2 \sum \tanh\left(\left[I_{CUE}^k - I_{th}\right]^+\right), \tag{5}$$

where λ_0, λ_1 and λ_2 are the controlling parameters. R_T is the minimum requirement to guarantee the spectrum efficiency of QoS. I_{CUE}^k is the interference power caused to the communication between the i-th CUE and the BS at the channel k.

In order to avoid overfitting when training the network model as far as possible, we have added Dropout solutions to the network. We uses the Adam algorithm to update parameters when doing backpropagation in this paper.

4 Performance Valuation

In this section, we investigate the performance of the proposed scheme. The comparison with the exhaustive algorithm and the random algorithm proves that the accuracy of the model proposed in this paper is very close to the exhaustive algorithm. In addition to the exhaustive algorithm, we also compared a random allocation scheme. As shown in Fig. 3, the scheme and the exhaustive algorithm in this paper are significantly better than the random algorithm.

We set a communication scene which ranges to be a square area and the size of area $D = 30$ m. The BS is located in the center of the square area and the positions of other devices are randomly distributed in this square. We set the maximum transmit power of the D2D device (P_T) and the transmit power of the cellular user to be 20 dBm. The bandwidth and the noise spectral density are set to 10 MHz and −174 dBm/Hz, respectively. Furthermore, $I_{th} = -50$ dBm, $R_T = 3$ bps/Hz. In addition, the DUEs and CUEs are randomly distributed over an area in which the maximum distance between the transmitter and receiver of the same transmit pair is set at 15 m. An independent and identically distributed circularly symmetric complex Gaussian random variable is used for multipath fading, with zero mean and unit variance [8].

In Fig. 3, we compared the performance of three algorithms in maximizing spectral efficiency under the different sizes of area. The simulation results show that the neural network model trained in this paper is very close to the best performance.

In Fig. 4, we compared the spectral efficiency of the full-duplex with the half-duplex D2D communications model under different sizes of area and different σ_{SI}^2 (−100 dB, −90 dB). The result shows that when the residual SI is relatively small, the spectrum efficiency of the full-duplex communications mode is significantly higher than that of the half-duplex communications mode. Under this simulation data, half-duplex communications mode should be selected when the SI is too large to meet the requirements of full-duplex communications. Full-duplex communications mode should be selected for communications systems.

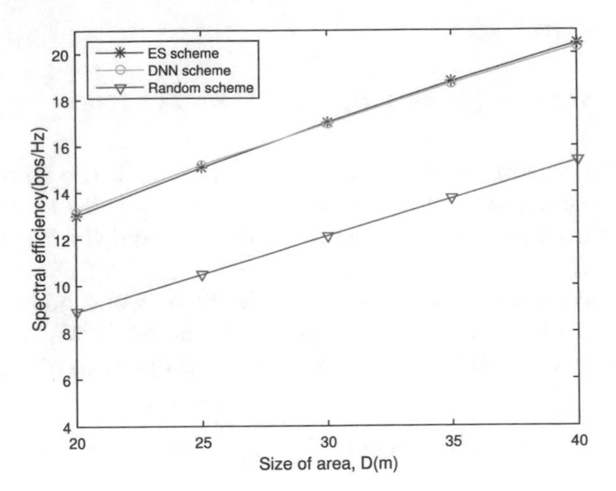

Fig. 3. Comparison of spectrum efficiency of full-duplex D2D communications under different schemes.

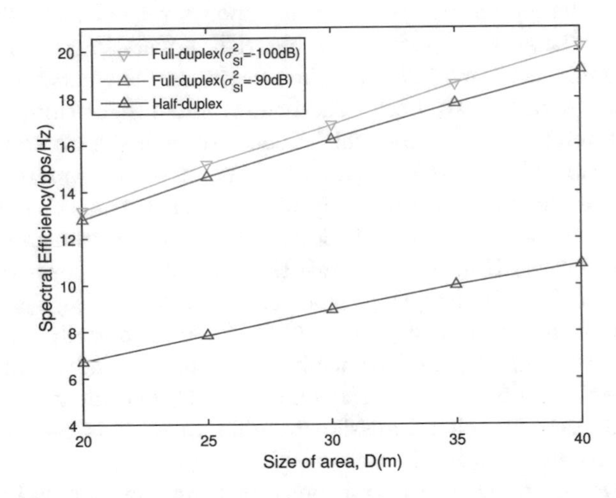

Fig. 4. Half-duplex and full-duplex D2D communications spectrum efficiency comparison under DNN algorithms.

5 Conclusion

This paper proposed a neural network model suitable for the full-duplex D2D communications mode and successfully applies the DL algorithm to the field of wireless mobile communication. DL can solve the problems of excessive iterations and time-consuming that traditional exhaustive algorithms cannot solve. This full-duplex DL algorithm has good performance as shown in Sect. 4. Using a trained model, it can quickly calculate the transmission power of devices without iterations. By comparing the spectrum efficiency with the half-duplex communication mode, the problem of selecting the duplex communication mode in a specific scenario is solved.

Acknowledgment. This work was supported by Tianjin Education Commission Scientific Research Plan (2017KJ121).

References

1. Nguyen, D., Tran, L.N., Pirinen, P., et al.: On the spectral efficiency of full-duplex small cell wireless systems. IEEE Trans. Wirel. Commun. **13**(9), 4896–4910 (2014)
2. Duarte, M., Dick, C., Sabharwal, A.: Experiment-driven characterization of full-duplex wireless systems. IEEE Trans. Wirel. Commun. **11**(12), 4296–4307 (2012)
3. Ansari, R.I., Chrysostomou, C., Hassan, S.A., et al.: 5G D2D networks: techniques, challenges, and future prospects. IEEE Syst. J. **12**(4), 3970–3984 (2017)
4. Sim, M.S., Chung, M.K., Kim, D., et al.: Nonlinear self-interference cancellation for full-duplex radios: from link-level and system-level performance perspectives. IEEE Commun. Mag. **55**(9), 158–167 (2017)
5. Zhang, Z., Chai, X., Long, K., et al.: Full duplex techniques for 5G networks: self-interference cancellation, protocol design, and relay selection. IEEE Commun. Mag. **53**(5), 128–137 (2015)
6. Li, J., Song, J.B., Han, Z.: Network connectivity optimization for device-to-device wireless system with femtocells. IEEE Trans. Veh. Technol. **62**(7), 3098–3109 (2013)
7. LeCun, Y., Bengio, Y., Hinton, G.: Deep learning. Nature **521**(7553), 436–444 (2015)
8. Lee, W., Jo, O., Kim, M.: Intelligent resource allocation in wireless communications systems. IEEE Commun. Mag. **58**(1), 100–105 (2020)

Application and Design of Capacitive Hand-Painted Screen in Psychotherapy of Painting Art

Chaoran Bi[1](✉), Hai Wang[2], Weiyue Cao[1], Liang Liu[3], and Yu Liu[4]

[1] College of Media Design, Tianjin Modern Vocational Technology College, Tianjin 300350, China
806281351@qq.com

[2] Teaching Center for Experimental Electronic Information, College of Electronic Information and Optical Engineering, Nankai University, Tianjin 300350, China

[3] College of Art, Tianjin University of Technology and Education, Tianjin 300222, China

[4] College of Art and Design, Tianjin University of Technology, Tianjin 300384, China

Abstract. Due to the rapid development of human society and the transition from agricultural society to industrial society, people's material life is constantly enriched, but industrialization also brings a tense pace of life, and the sense of alienation between people brings endless stream of social problems. Art psychotherapy is regarded by contemporary people as one of the important ways to persuade and solve psychological problems. Painting psychotherapy is also one of the important forms of art psychotherapy. This therapy can not only accurately and realistically complete the subconscious inner expression and spiritual release of patients, but also reduce the work pressure of psychologists. This design is an innovative upgrade and expansion of the traditional painting treatment mode.

Keywords: Art therapy · Drawing tablet · Interface design · Natural human-computer interaction

1 Prospect and Current Situation

Art therapy, also known as art psychotherapy, first appeared in the late 1940s. People's paintings reflect their true and unique self. Through the observation of art, they can be used as an important basis for diagnosis and evaluation in psychotherapy. In psychotherapy, artistic activities are often used to relieve patients' physical and mental disorders [1, 2].

Drawing plate, also known as index plate, is a general term for electronic plate drawing. A computer input device is generally composed of capacitive touch screens, which determines the position where the digital board is touched, showing the drawing track and image of the operator [3].

Q. Liang et al. (Eds.): Artificial Intelligence in China, LNEE 854, pp. 646–653, 2022.
https://doi.org/10.1007/978-981-16-9423-3_81

2 Treatment Mechanism of Traditional Painting Therapy

Traditional painting therapy, in short, is the connection between art therapists, patients and works. The initial art therapy originated from the study of psychotic artists in the early 20th century. In 1940, Margaret naumburg established the model of using artistic expression as treatment, encouraging the use of various artistic materials in psychotherapy to deal with personal inner fears and contradictions. So far, art psychotherapy has become a basic psychotherapy. Contemporary French psychologist g. H. laquet divides painting therapy into "objective realism" and "mental realism". It is believed that when a mental patient is ill, his consciousness is often inconsistent with the object, which turns the reality in his mind into a non-objective reality. However, this "spiritual realism" is the biggest feature of the gifted painting of mental patients. Jung, the founder of analytical psychology, also attaches great importance to the positive effect of painting therapy on mental patients. He stressed: "drawing what we feel in our hearts and drawing what we see in front of us are two different arts." According to clinical experience, Jung believes that painting as a tool to express subconscious experience is more direct than language, and it is an effective means of "psychotherapy" for mental patients. Jung's method of more direct expression of inner experience has been systematized by other physicians and psychologists and has developed into a "dynamically oriented art therapy". In the West now, the implementation of "art therapy" has changed the previous definition of the word "therapy", that is, therapy is not only physical, but also ideological and spiritual.

2.1 Operation Process of Cooperation Between Art Therapist and Patient

First, the art therapist provides drawing tools for patients and establishes contact through language. During this period, sometime may be invested, because psychological counselors need to imperceptibly help patients relax, remove all psychological defenses, and introduce art therapy at the right time [4].

After painting, the content in the picture can be classified as "personification" and saved digitally. The art therapist should communicate with the patient and let the author talk about the painting content and creative feelings. Especially abstract art [5]. Shown in Fig. 1.

2.2 Equipment, Tools and Problems Required by Traditional Painting Therapy

Traditional painting needs a certain space and painting tools. If it is color painting, it needs brushes, pigment consumables and paper. If it is oil painting or watercolor pigment, the operator also needs detergent and water washing to change the color, which is more complex and will cause environmental pollution and resource consumption. Patients without painting foundation lack experience in mastering the properties of pigments and color reduction. Too complex operation will virtually increase the action cost of painters and affect the treatment effect. The hand-painted flat panel does not occupy a lot of space and is safe. The famous painter Van Gogh swallowed a can of his favorite yellow pigment when he was suffering from mental illness. In the later stage of painting preservation, the preservation and archiving of a large number of patients' paintings is also a problem of occupying space [6, 7].

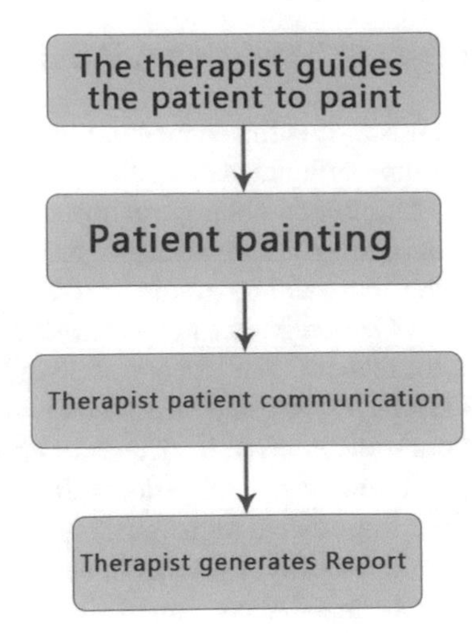

Fig. 1. Operation process of cooperation between art therapist and patient

3 Conception and Feasibility of Intelligent Hand-Painted Tablet in Art Psychotherapy

In the domestic psychological counseling and treatment industry, the psychological counseling room and related personnel resources are also relatively short. The traditional painting art materials and tools occupy a large space and consume human and material resources. If the weak artificial intelligence electronic equipment can replace the traditional painting and produce treatment and interaction, it can reduce the consumption of human and material resources and improve the efficiency of art therapy.

3.1 Improvement of Hardware Data Storage and Calculation

As we all know today, the space occupied by electronic storage devices is far less than that in our real life. The amount of information stored in high-density disk is huge. At present, the capacity of mainstream hard disk can reach 500 g – 2 TB, which can store a large amount of data information, store a large number of patients' creative works for a longer time to search and search is more convenient, more systematic and effective observation of patients' treatment process and effect, and record patients' data, including demographic data The records and information of personal growth history, mental state, physical state and social function can be saved in large quantities, making future treatment plans.

The data processing resource library can provide users with different art materials and tools in a more organized way and show their effects. Users can find the way they want to draw more conveniently and accurately, save the time of therapists and treatment, make the treatment more consistent and the effect more remarkable [8].

3.2 Expression Function of Capacitive Touch Screen in Painting Therapy

The operation of hand-painted flat panel is very simple. Besides, it does not occupy a lot of space, and it is safe. In case of misoperation, you can use the "regret key" to return upward, so as to avoid some unalterable defects in traditional painting.

The capacitive touch screen can be equipped with sound effect to enhance the texture of painting strokes, optimize picture lines, increase the operation pleasure and fun of painters. It greatly reduces the risk of painting failure and improve the terapeutic effect of artistic creation.

The capacitive touch screen can be equipped with a front camera to record the micro expression records of the painter's creative process without the painter's knowledge, and the data can be used for later analysis. At the same time, close shooting can also capture the synchronous emotional fluctuation response of painters in different stages of painting in the process of creation, and can repeatedly analyze and confirm the crux, avoiding the short board that is easy to miss in traditional therapy.

The electromagnetic pressure sensing control and rate recording of capacitive touch screen can record the synchronous emotional response of painters at different stages of drawing, and analyze their emotions through stroke pressure. Later data observation is used to better help therapists analyze patients, specify treatment planning process, design records and feedback for user behavior.

4 Operation Design and Implementation Path of Capacitive Touch Screen

4.1 Design of Usage Rules of Capacitive Touch Screen

As for the capacitive touch hand-painted screen, the capacitive technology touch panel works by using the current induction of the human body. The capacitor screen is the four layers composite glass screen. The inner surface and interlayer of the glass screen are coated with a layer of nano indium tin metal oxide respectively. The outermost layer is only 0.0015 mm thick silica glass protective layer. The interlayer ITO coating is used as the working surface, Four electrodes are led out from the four corners, and the inner ITO is used as the screen layer to ensure the working environment. The capacitive touch screen only needs touch without too much pressure, so it has little physical requirements for users. It is suitable for most people, including some users with physical disabilities. At the same time, the capacitive technology is mature, with high precision, fast response, wear resistance, long service life, no correction, and high-definition imaging without drift, low maintenance cost for users [9].

Hand painted screen painting art therapy enhances the therapeutic effect through the theoretical basis of traditional painting psychotherapy and the drawing software of electromagnetic induction hand-painted screen. This design is based on the original expressive art therapy research theory, through the self-expression of patients' artistic creation, psychological counselors' analysis, diagnosis and guidance for treatment [10, 11].

4.2 Interactive Mode Design of Hand-Painted Screen Painting Art Therapy

Hand painted screen painting art therapy relies on the practice of image creation and treatment belief, and interaction is an important expression in treatment. As for image creation and painting, visual feedback focuses on suspension. The interaction between people and objects is visual feedback. Compared with traditional painting, hand-painted screen painting art therapy can provide users with visual, auditory and tactile feedback. Through timely feedback, users can understand that the current program is running normally and can input safely to complete timely expression, giving it a sense of smoothness to control everything and bringing users a good interactive experience.

4.3 Operation Process Design of Hand-Painted Screen Painting Art Therapy

The therapist introduces the operation mode, and the patient adjusts the equipment and learns to use the operation under the guidance of the consultant. In the hand drawn system, enter the drawing interface matching with the patient according to the operation tips and step instructions. After relaxation, devote yourself consciously to painting therapy and create painting. The therapist should monitor the patient at this stage using relevant equipment.

After the final completion of the work, the therapist shall communicate, discuss and interact with the patient. Let the patient explain the meaning of the content of the picture created by himself. The hard disk computing device shall give a rough evaluation result of the player's paintings according to the existing database. Before the formal discussion, the consultant should also guide the players to observe their works as a whole for a short time, and fine tune them. According to the observation of the whole process and the rough evaluation results of background data, provide more information for the consultant. Finally, according to the degree of control of the information, the therapist can decide the guided treatment intervention in time. Shown in Fig. 2.

4.4 Operation Interface and Environment Design of Hand-Painted Screen Painting Art Therapy

The first thing to contact when entering the game is the guidance interface. The focus of interface design should highlight the details of interactive design and visual feedback. The design of good design guidance interface is directly related to the player's good impression and acceptance of hand-painted screen painting art therapy, and then related to the treatment effect.

4.4.1 Visual Design of Guiding Interface of Hand-Painted Screen Painting Art Therapy

According to the variety of different interface design styles can be provided in the painting game. For different players, the player himself or the therapist can choose the interface style, that is suitable for the patient's visual characteristics, tool size and color brightness.

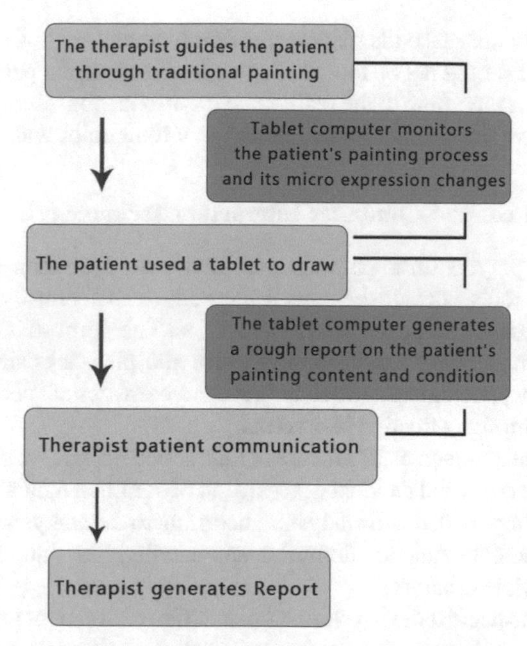

Fig. 2. Operation process design of hand-painted screen painting art therapy

4.4.2 Function Design of Guidance Interface

The menu process should not be too complex. The animation special effects in the interface can clearly attract the user's attention and attention to the important details in the interface, so that the user will not get lost in the process, as a result of wasting time, increasing action cost, generating resistance, and even giving up the treatment operation [12].

The inevitable task operation requires multi-step operation across multiple pages. You can use the wizard control to guide the user to complete the multi-step operation. The design follows the user's daily life and behavior habits, from left to right, from top to bottom or the opposite reading habits. At the same time, we can use significant prompts to guide the user's operation when necessary. For example, motion guidance is a good way, because the human subconscious likes to capture moving objects [13]. This kind of motion is not only the movement of position, but can attract the player's visual center through changes in size, flicker or transparency. At the same time, there are also prompts to tell the user how many steps are needed, which stage to go to now, and how much remains to be completed. To sum up, we should guide the players on the interactive interface, try to be concise, and feed back information through visual elements to enable users to enter the painting operation better and faster [14, 15].

4.4.3 Studio Scene Space Design

Considering the space design based on the traditional psychological counseling room, its purpose is to make the counselor feel warm, safe and casual. The studio should not be too large and should not make the players feel empty, resulting in a sense of insecurity. At the same time, it should not be too small, which will make the painter feel depressed and claustrophobic. According to the reality, the traditional consulting room can be used as a reference to give the human eye the feeling of a light color walls of 15 m^2.

4.5 Utilization of Human-Computer Interaction Technology

During the game, players draw pictures with their hands or with an electronic pen. Different painting tools contact the screen to produce different dynamic effects and sound effects of strokes, which are fed back to users. The purpose is to make users feel happy quickly through sharp and accurate texture and pleasant painting sound effect, understanding the psychological and behavioral characteristics of "people", to strengthen and expand them through effective interaction.

According to the common basic process of interaction design, we should first conduct data research on the types and incidence proportion of common mental diseases, and then formulate the preliminary demand analysis. According to the analysis results, convey the meaning of requirements, function definition and detailed rules, and formulate function overview and function structure.

The tablet can also assist the psychotherapist in the treatment of patients. The picture generated after the patient's painting gives a rough evaluation of the patient's picture by retrieving the information in the database, the information meaning corresponding to the concrete performance and abstract performance images drawn by the patient, so that the therapist can understand the patient and formulate the corresponding treatment plan. For example, today's AI painting Xiaobing can draw the corresponding picture through some theme keywords. On the contrary, it can also draw up the corresponding theme and picture interpretation according to the picture. The same object may have several meanings, so the interpretation of a work may be different, which requires the doctor to cooperate with the patient's multiple communication and different communication means to obtain more information according to the prompts of the machine.

4.6 Hardware Equipment

The common tablet computers on the market include IOS operating system and iPad series tablet computers produced by apple. Domestic tablet computers with Android system are also more widely used, such as Lenovo tablet Xiaoxin pad, Lenovo M10 plus Huawei matepad, etc. They can be used for screen hand-painted treatment, and can be equipped with capacitive stylus if necessary.

5 Summary

Based on the study of art psychotherapy, this paper uses modern and contemporary hand-painted capacitive screen tablet computer technology to upgrade art therapy, so that to enhance the effect of painting art therapy. The possibility of the combination of the two is found, which makes it not only a game of art psychotherapy, but also an evaluation mechanism. Through the discussion of the psychotherapy mechanism of traditional painting art and the expression form of virtual imaging technology, this paper formulates the elements of hand-painted screen art psychotherapy, such as content, interactive behavior, role composition, game task, and analyzes the feasibility of the technical route. It is improved on the basis of traditional art therapy. Its purpose and significance let patients better release their consciousness and emotion, so as to achieve cure as a result.

References

1. Piper, A.M., Lazar, A.: Co-design in health: what can we learn from art therapy? Interactions **25**(3), 70–73 (2018)
2. Liu, Y.C., Chang, C.L.: The application of virtual reality technology in art therapy: a case of tilt brush. In: 1st IEEE International Conference on Knowledge Innovation and Invention (ICKII) (2018)
3. Fan, L., Yu, C., Shi, Y.: Guided social sharing of emotions through drawing art therapy: generation of deep emotional expression and helpful emotional responses. In: The Seventh International Symposium of Chinese CHI (2019)
4. Ranjgar, B., Azar, M.K., Sadeghi-Niaraki, A., et al.: A novel method for emotion extraction from paintings based on Luscher's psychological color test: case study Iranian-Islamic paintings. IEEE Access **7**(1), 120857–120871 (2019)
5. Brahnam, S., Brooks, A.L.: Two innovative healthcare technologies at the intersection of serious games, alternative realities, and play therapy. Stud. Health Technol. Inform. **207**, 153–162 (2014)
6. Wu, J.J., Lin, W.S., Shi, G.M., et al.: Visual orientation selectivity based structure description. IEEE Trans. Image Process. Publ. IEEE Signal Process. Soc. **24**(11), 4602–4613 (2015)
7. Zhang, W.R., Pandurangi, A.K., Peace, K.E.: YinYang dynamic neurobiological modeling and diagnostic analysis of major depressive and bipolar disorders. IEEE Trans. Biomed. Eng. **54**(10), 1729–1739 (2007)
8. Iyengar, A.K.: Enhanced clients for data stores and cloud services. IEEE Trans. Knowl. Data Eng. **31**(10), 1969–1983 (2018)
9. Lcl, A., Mgo, B.: Touch gesture performed by children under 3 years old when drawing and coloring on a tablet. Int. J. Hum. Comput. Stud. **124**, 1–12 (2019)
10. Kim, J., Nakamura, T., Kikuchi, H., et al.: Covariation of depressive mood and spontaneous physical activity in major depressive disorder: toward continuous monitoring of depressive mood. IEEE Biomed. Health Inform. **19**(4), 1347–1355 (2015)
11. Noble, S.L.: Control theoretic scheduling of psychotherapy and pharmacotherapy for the treatment of post-traumatic stress disorder[J]. IET Control Theory Appl. **8**(13), 1196–1206 (2014)
12. Soper, D.S.: Informational social influence, belief perseverance, and conservatism bias in web interface design evaluations. IEEE Access **8**, 218765–218776 (2020)
13. Miraz, M.H., Excell, P.S., Ali, M.: Culturally Inclusive Adaptive User Interface (CIAUI) framework: exploration of plasticity of user interface design. Int. J. Inf. Technol. Decis. Mak. **20**(1), 199–224 (2020)
14. Wu, H., Li, G., et al.: Correction to: Innovation and improvement of visual communication design of mobile app based on social network interaction interface design. Multimed. Tools Appl. **79**(1), 17–18 (2020)
15. Yi, M., Wang, Y., Tian, X., et al.: User experience of the mobile terminal customization system: the influence of interface design and educational background on personalized customization. Sensors **21**(7), 2428 (2021)

Research on Improved Collaborative Filtering Recommendation Algorithm Based on User Behavior

Yong Ai[1,2](✉), Zixian Li[1,2], Chong Sun[1,2], and Jun Tie[1,2]

[1] College of Computer Science, South-Central University for Nationalities, Wuhan, China
aiy_scuec@qq.com

[2] Hubei Provincial Engineering Research Center for Intelligent Management of Manufacturing Enterprises, Wuhan, China

Abstract. Recommendation algorithms have been widely used in many fields. This paper studies an improved collaborative filtering recommendation algorithm based on user behavior to improve the accuracy. In this paper, user behavior is used as the influencing factor of item collaborative filtering recommendation and user collaborative filtering recommendation. Two improved collaborative filtering recommendation algorithms are proposed. The experimental results show that the improved collaborative filtering recommendation algorithm has higher recommendation accuracy for the recommended topic.

Keywords: Collaborative filtering · Recommendation algorithm · Similarity matrix · User behavior

1 Introduction

Recommendation algorithms have been applied in many fields, such as shopping, learning, travel and so on [1]. With the rise of e-commerce, it opens up a vast space for the development of recommendation algorithms, which brings great commercial value. Nowadays, the widespread use of mobile Internet provides a large amount of mobile information data [2]. It brings new application fields for recommendation algorithms. The successful application of recommendation algorithm makes people pay more attention to the practicability, security and robustness. This has further promoted the research of distributed computing and offline recommendation algorithms [3]. The rise of artificial intelligence and deep learning provides some new directions for the development of recommendation algorithms. Future recommendation algorithms will be more intelligent and personalized [4]. The application field will be more extensive.

At present, collaborative filtering recommendation is still the most commonly used recommendation method in applications. Later, researchers also proposed some recommendation algorithms based on this, including content-based recommendation algorithms, knowledge-based recommendation algorithms and so on.

Q. Liang et al. (Eds.): Artificial Intelligence in China, LNEE 854, pp. 654–661, 2022.
https://doi.org/10.1007/978-981-16-9423-3_82

Content-based recommendation algorithm was first used in information retrieval system. The content-based algorithm recommend similar items to the items user has liked [5]. This algorithm tags the user's preferences, which simplifies the recommendation.

Compared with the content-based algorithm, the knowledge-based recommendation algorithm pays more attention to the items to be recommended. Instead of focusing on the user's preference for the item. This algorithm pays more attention to the direct contact between the user and the knowledge content of the object. Users actively put forward the content they want to be recommended [6].

This paper proposes two improved collaborative filtering recommendation algorithms which are based on item and user. The algorithms introduce the user behavior to improve the accuracy of the recommendation, which could improve the accuracy of recommendations.

2 Related Works

2.1 Content-Based Recommendation Algorithm

Content-based recommendation algorithms were first used in information retrieval systems. The algorithm will recommend similar items to the items user has liked. From the perspective of users' interests and preferences, it is considered that users with the same tag have higher similarity. Content with the same tag will be more popular with similar users. The algorithm can be expressed as the following three steps.

1) Item Representation [7]: Labeling the representation of an item. A representative feature vector is extracted from the content of the items. Then the items can be represented by this vector.
2) Profile Lcarning [8]: Analyze and learn the user's feature vector, and calculate the user's interest feature vector. From this, the preference characteristics of the user are obtained. In specific applications, different keywords will be weighted to improve the accuracy and effectiveness of the recommendation.
3) Recommendation Generation [9]: Calculate and analyze the feature vector of the user and the feature vector of the content to be recommended. The relevance between the user and the content to be recommended is obtained. According to the size of the similarity, the top N contents are recommended to the user.

2.2 Knowledge-Based Recommendation Algorithm

Content-based recommendation algorithms are not suitable for all situations. There are some contents or items that are not suitable to be converted into specific feature vectors. For example, cell phone, real estate and other contents that do not have a high degree of generality. At this time, content-based recommendation algorithms can not recommend reliable content to users.

In order to solve this kind of problem, knowledge-based recommendation algorithm is proposed [10]. This algorithm pays more attention to the recommended item itself, rather than the user's preference for items. Knowledge-based recommendation algorithm achieves more precise recommendation through user's active interaction. It includes the following two ways.

1) Limit the range: The user actively gives some range for the expected recommendation result. These ranges are transformed into user knowledge sources according to the set rules of the algorithm. The user knowledge source will serve as a rule for later searching and matching, and the items to be recommended must be matched according to such rules.
2) Limiting the case: The way of limiting the case is not to use the requirements put forward by the user as a matching rule, but as a reference standard. That is, the attributes proposed by the user are given different weights. By virtue of this weight, that correlation between the item with similar attributes and the instance is calculate. According to the different distribution of weights, similar items with high correlation are recommended.

2.3 Collaborative Filtering Recommendation Algorithm

The core idea of collaborative filtering algorithm is to find some similarity through the behavior of the group [11]. The similarity is used to make decisions and recommendations for the user. This algorithm is not based on the single category feedback of the item or the similarity of the content. The core of collaborative filtering is how to calculate the similarity between items and users. Collaborative filtering algorithms fall into the following two categories.

1) **Item Collaboration Filter**

   The item collaboration filter algorithm does not calculate the similarity between items and analyze the user's hobby behavior from the perspective of items [12]. Instead, it starts from the user's behavior, and the degree of coincidence of user's behavior determines the similarity between items. The recommended steps are divided into two steps.

   a) Calculating item similarity matrix

$$S_{ij} = \frac{|u(i) \cap u(j)|}{\sqrt{|u(i)| \cup |u(j)|}} \tag{1}$$

   Where $u(i)$ and $u(j)$ represent the set of users who have acted on item i and j. $|u(i) \cap u(j)|$ represents the degree of coincidence of items i and j. The denominator portion is expressed as the product of users generated for items i and j.

   b) Recommend by the item similarity matrix and the user's past behavior

$$P_{uj} = \sum\nolimits_{i \in N(u) \cap s(j,k)} S_{ij} * r_{ui} \tag{2}$$

   Where P_{uj} represents the user's level of interest in item j. r_{ui} represents the user's evaluation of the behavior of the item i. $N(u)$ represents the set of items that the user has acted on. $s(j, k)$ represents a set of k items similar to item i, where j belongs to an element of the set.

2) **User Collaboration Filter**

User collaborative filter algorithm starts from the overall behavior of the target user to find users with similar behavior. And recommend items of interest to similar users to the target user [13]. This algorithm is also divided into two steps.

a) Calculate the coincidence degree between users

$$S_{uv} = \frac{|N(u) \cap N(v)|}{\sqrt{|N(u)||N(v)|}} \tag{3}$$

Where $N(u)$ and $N(v)$ represent the set of user behaviors that user u and v have generated for themselves. The numerator represents the degree of overlap between two users.

b) Recommending target users according to user coincidence degree

$$P_{ui} = \sum\nolimits_{v \in s(u,k) \cap u(i)} S_{uv} r_{vi} \tag{4}$$

Where r_{vi} represents the degree of behavior of user v toward item i.

Formula 4 expresses that item i is recommended to user u by similar user v. Then user v is an element of the user set $s(u, k)$ similar to user u. And the item i that user v has acted on has not been generated in the user action of user u.

3 Improved Algorithm Based on User Behavior

In this paper, the user behavior is introduced into the item collaborative filtering algorithm and the user collaborative filtering algorithm. The collaborative filtering algorithm is improved to increase the progress of recommendation.

3.1 Improved Item Collaborative Filtering Algorithm

In the actual application of recommender system, users will have a variety of user behaviors for multiple items. So for such users, their click operation on the item obviously can not fully show their interest. In contrast, some users only have user behavior for certain types of items. This would further illustrate their interest. Therefore, the number of user behaviors will provide a strong contribution to the recommendation algorithm. The similarity calculation formula is improved as Formula 5.

$$S_{ij} = \frac{\sum_{u \in u(i) \cap u(j)} \frac{1}{\log(1+|N(u)|)}}{\sqrt{|u(i)||u(j)|}} \tag{5}$$

Where $N(u)$ is the number of actions user u has taken. That is to say, the more behavioral items a user u produces, the less it contributes to the size of the molecule. In this way, the contribution value of multi-behavior users to the similarity is effectively reduced.

The improved item collaborative filtering algorithm is shown in Algorithm 1. The algorithm consists of three main steps. (I) Calculate the similarity between the items according to Eq. 5. (II) calculating a list of items to be recommended according to the similarity of the items and the historical behavior of the target user. And (III) recommend that top K high-score items as the recommend objects according to the list of the items to be recommend.

Algorithm 1: Improved item collaborative filtering algorithm

Input: User history behavior set H. Item content information M, Target user

Output: Top K items to be recommended

```
1:  for each user in H :
2:          Get an inverted set of items for the user as User_Click
3:  end for
4:  for each item in M :
5:          Get the information set of the item as Item_Form
6:  end for
7:  for each item in User_Click :
8:          The similarity between items is calculated using Formula 5 and
    sorted in descending order, which is expressed as Item_Sim_Form
9:  end for
10: for target user in User_Click & Item_Sim_Form & Item_Form :
            According to the items that the target user is interested in, a
    descending order set Recom_List of the items to be recommended is
    generated.
11: end for
12: Return top K items of the Recom_List as recommend items.
```

3.2 Improved User Collaborative Filtering Algorithm

In the actual recommendation system, there will be such a situation that the target user and other users have only one item with common behavior. And this item does not have the functionality to represent the user's interests. That is to say, the user group generally has a user behavior on this item. In this case, it is unreasonable to equate the contribution of such items to the similarity of users as representative items. This will affect the practicality of the recommendation results. Therefore, for such popular items, their contribution to user similarity should be reduced. The modified formula is shown in Formula 6.

$$S_{uv} = \frac{\sum_{i \in N(u) \cap N(v)} \frac{1}{\log(1+|u(i)|)}}{\sqrt{|N(u)||N(v)|}} \tag{6}$$

Where $u(i)$ represents the set of users that generated the user action for item i. It can be seen from Formula 6 that the more users generate behaviors, the smaller the contribution of item i to the similarity between user u and user v. Therefore, the more representative the feature is, the more the similarity contribution is.

Improved user collaborative filtering algorithm shows as Algorithm 2. In the data preprocessing stage, the user's behavior set on the item is obtained. In the process of recommendation, the items of interest of similar users are no longer calculated as item similarity, but are directly used as recommended items.

Algorithm 2: Improved user collaborative filtering algorithm

Input: User history behavior set H. Item content information M, Target user

Output: Top K items to be recommended

1: for each user in H :
2: Get a set of behaviors of this user on each item as User_Clicktoitem
3: end for
4: for each item in M :
5: Get the information set of the item as Item_Form
6: end for
7: for each user in User_Clicktoitem :
8: for each item in User_Clicktoitem :
9: The similarity between items is calculated using Formula 6 and sorted in descending order, which is expressed as User_Sim_Form
10: end for
11: end for
12: for target user in Item_Form & User_Sim_Form & User_Clicktoitem: Acquiring the top i items of interest of the similar user as recommended items for the target user as Recom_List.
13: end for
14: Return top K items of the Recom_List as recommend items.

---

4 Experiment

The data set is derived from http://grouplens.org/datasets/movielens/. In the data set, user behavior consists of user ID, topic ID, rating, and rating time. A total of 671 users and 100005 valid user actions. Topic content consists of topic ID, name, and content. A total of 9126 non-consecutive ID topics.

4.1 Experiment on Improved Item Collaborative Filtering Algorithm

The above data set was experimented with Algorithm 1. The experimental results are compared with the original algorithm. Because the two algorithms have different recommendations for the same user. Therefore, this item takes the same recommended topics and compares them with the first seven. The recommended results are shown in Fig. 1.

The recommendation results show that the improved algorithm is better than the original algorithm for the same topic recommended by the same user. This shows that the introduction of user behavior can improve the accuracy of topic recommendation.

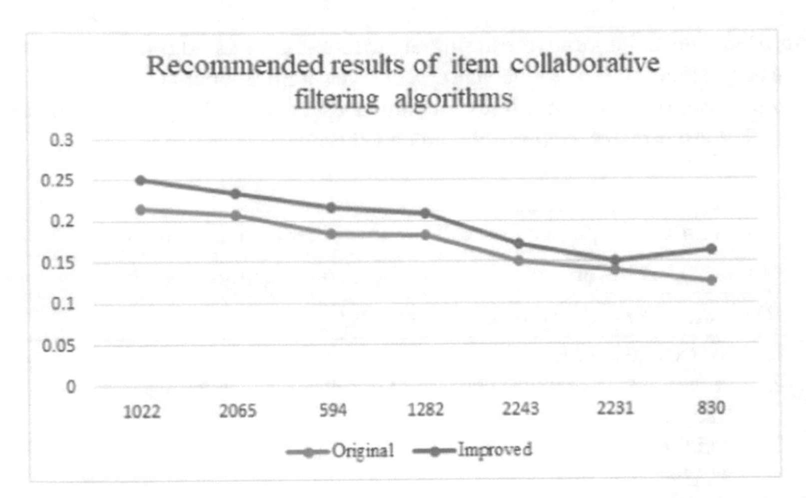

Fig. 1. Recommended results of original and improved item collaborative filtering algorithms

4.2 Experiment on Improved User Collaborative Filtering Algorithm

The above data set was experimented with Algorithm 2. The experimental results are compared with the original user collaborative filtering algorithm. The recommended results are shown in Fig. 2.

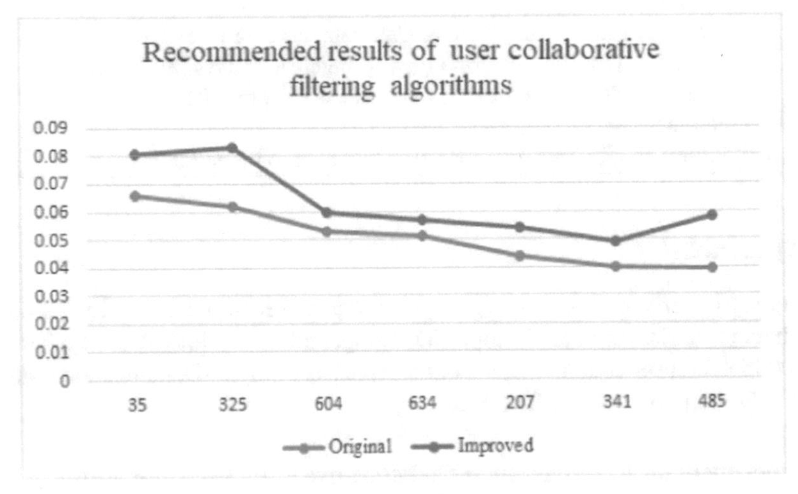

Fig. 2. Recommended results of original and improved user collaborative filtering algorithms

The recommendation results show that the improved algorithm is better than the original algorithm for the same topic recommended by the same user. This shows that the introduction of user behavior can improve the accuracy of topic recommendation.

5 Conclusion

In this paper, user behavior is introduced into collaborative filtering recommendation algorithm as an influencing factor. Two improved item-based and user-based collaborative recommendation algorithms are proposed. The two algorithms are tested on the data set. The experimental results show that the recommendation method with user behavior as the influencing factor has a certain improvement in user similarity accuracy.

Acknowledgments. This work is supported by the Major Projects of Technological Innovation in Hubei Province, China (No. 2019ABA101).

References

1. Sandvig, J.J., Mobasher, B., Burke, R.D.: A survey of collaborative recommendation and the robustness of model-based algorithms. IEEE Data Eng. Bull. **31**(2), 3–13 (2008)
2. Hsu, C., Lu, H., Hsu, H., et al.: Adoption of the mobile internet: an empirical study of multimedia message service (MMS). Omega-Int. J. Manage. Sci. (2007)
3. De Myttenaere, A., Golden, B., Grand, B.L., et al.: Study of a bias in the offline evaluation of a recommendation algorithm. arXiv: Information Retrieval (2015)
4. Li, H., Li, H., Zhang, S., et al.: Intelligent learning system based on personalized recommendation technology. Neural Comput. Appl. **31**(9), 4455–4462 (2019)
5. Lops, P., de Gemmis, M., Semeraro, G.: Content-based recommender systems: state of the art and trends. In: Ricci, F., Rokach, L., Shapira, B., Kantor, P.B. (eds.) Recommender Systems Handbook, pp. 73–105. Springer, Boston, MA (2011). https://doi.org/10.1007/978-0-387-85820-3_3
6. Tarus, J.K., Niu, Z., Mustafa, G.: Knowledge-based recommendation: a review of ontology-based recommender systems for e-learning. Artif. Intell. Rev. **50**(1), 21–48 (2018)
7. Maples-Keller, J.L., Williamson, R.L., Sleep, C.E., et al.: Using item response theory to develop a 60-item representation of the NEO PI–R using the International Personality Item Pool: development of the IPIP–NEO–60. J. Pers. Assess. **101**(1), 4–15 (2019)
8. Libon, D.J., Bondi, M.W., Price, C.C., et al.: Verbal serial list learning in mild cognitive impairment: a profile analysis of interference, forgetting, and errors. J. Int. Neuropsychol. Soc. **17**(5), 905–914 (2011)
9. Sarwar, B., Karypis, G., Konstan, J., et al.: Analysis of recommendation algorithms for e-commerce. In: Proceedings of the 2nd ACM Conference on Electronic Commerce, pp. 158–167 (2000)
10. Rositch, A.F., Atnafou, R., Krakow, M., et al.: A community-based qualitative assessment of knowledge, barriers, and promoters of communicating about family cancer history among African-Americans. Health Commun. **34**(10), 1192–1201 (2019)
11. Gong, S.: A collaborative filtering recommendation algorithm based on user clustering and item clustering. J. Softw. **5**(7), 745–752 (2010)
12. Wei, S., Ye, N., Zhang, S., et al.: Item-based collaborative filtering recommendation algorithm combining item category with interestingness measure. In: 2012 International Conference on Computer Science and Service System, pp. 2038–2041. IEEE (2012)
13. Zhao, Z.D., Shang, M.S.: User-based collaborative-filtering recommendation algorithms on Hadoop. In: 2010 Third International Conference on Knowledge Discovery and Data Mining, pp. 478–481. IEEE (2010)

Application of Pattern Recognition in Airport Target Conflict Alert

Yan Liu(✉)

State Key Laboratory of Air Traffic Management System and Technology, Nanjing 210007, China
16108456@qq.com

Abstract. The operational safety of aircraft and vehicles at the airport currently mainly relies on the ATC and drivers, and there is a risk of collision caused by negligence of personnel. At present, control voice and video surveillance are important information that can be obtained from mode information, but they are not fully integrated into the control automation system application. Through the intelligent pattern recognition technology, the control voice and video monitoring information are recognized and processed, and the control system is integrated into the control system to accurately determine the operation trend of the airport target, and promptly give warnings, which has important practical significance. In response to this need, this paper studies structured command generation based on control voice, three-dimensional relationship calculation based on target semantics, future trajectory prediction and target conflict warning methods, to achieve knowledge-incorporated flight conflict detection between airport targets, and automate air traffic control System application method is designed to provide support for airport operation safety guarantee.

Keywords: Pattern recognition · Trajectory prediction · Airport target conflict

1 The Introduction

Flight safety accidents happened at home and abroad in recent years on signs and accidents, according to statistics, in flight safety accident happened in the past, why the machine and equipment about 23%, other factors (1.5%), whereas human factors accounted for 75.5%, in the human factors, the unit reason accounted for 35%, controllers for 25%, the unit and controller together account for nearly 40%. In particular, in the recent "Hongqiao incident", two flights almost collided on the runway of the airport. According to the Civil Aviation Administration of China (CAAC), the vertical distance between the two planes was as short as 19 m, and the wingtip distance was 13 m. This is a serious accident sign caused by the tower controller forgetting the dynamics of the aircraft and violating the working standards.

Data and real instances is obvious, "human error" has the potential to affect aviation safety, with the high-speed development of air transportation, the air traffic control work put forward higher requirements, the controller as an important participant of aviation

Q. Liang et al. (Eds.): Artificial Intelligence in China, LNEE 854, pp. 662–669, 2022.
https://doi.org/10.1007/978-981-16-9423-3_83

operations, its control behavior through sliding, take-off, cruise, landing the whole flight process, are also important factors affecting the flight safety. In the above events, if the automatic at C system can collect the voice information of the controller, timely find the wrong or forgotten commands sent by the controller, predict the possible collision events, and provide the controller with auxiliary alarm warning, the accident chain will inevitably be interrupted and the occurrence of accident symptoms will be avoided.

There is also some research in this area abroad. Honeywell is working on voice recognition technology for pilots in noisy environments to verify the correctness of repeated commands [1].

2 Control Speech to Generate Structured Commands

The transformation of control speech into structured control command includes two steps: one is to convert the control speech into text form through speech recognition; the other is to structure the control command in text form through natural language processing. Speech recognition includes acoustic model and language model, and controlled speech can be converted into controlled text by deep neural network, while natural language processing includes part of speech analysis, syntactic analysis and semantic analysis [2]. The steps for the structured extraction of textual control instructions are shown in Fig. 1.

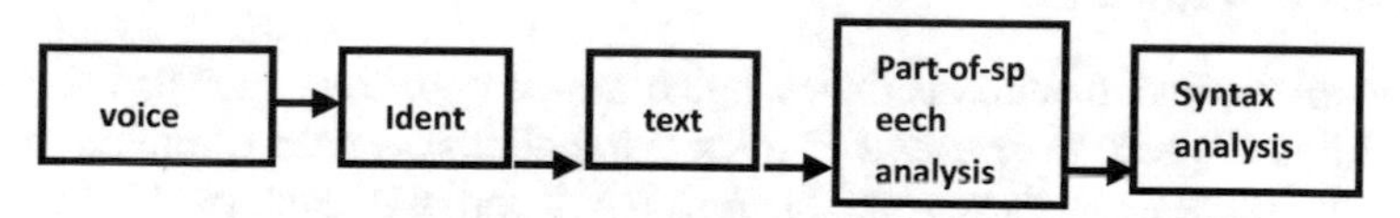

Fig. 1. Extraction process of structured control instruction

There are four steps to extract structured control commands from controlled speech. First, speech recognition is used to convert the controlled speech into the corresponding control commands in text format. Second, part of speech analysis, including Chinese word segmentation and part of speech tagging, converts the textual control instruction into word sequence, and tagging the corresponding part of speech for each word; Third, parsing, analyzing the order and part of speech of each word in the word sequence, and generating dependency syntax tree; The fourth is semantic analysis, through the semantic analysis of the results of syntactic analysis to find the predicate-argument relationship, and fill in the structured template [3].

3 Object Extraction and Position Relationship in Image

Image recognition technology is relatively mature, this article is not in this, this paper presents a three-dimensional space based on target semantic monocular vision relationship calculation method, combining with the target category, semantic information and exterior contour segmentation knowledge, and with the aid of automatic camera calibration based image plane and three-dimensional space coordinate mapping, process.

1) Select a typical target with known size in the monitoring scene, combine the camera imaging geometric transformation relationship, realize the automatic camera calibration, and establish the coordinate system mapping relationship between two-dimensional image plane and three-dimensional space;
2) With the target category information, the object appearance contour knowledge is extracted from the target knowledge base to provide data support for automatic camera calibration, attitude estimation and depth estimation; Furthermore, by learning the local details of the target image, a deep neural network based target pose estimation model is constructed to obtain the target pose information.

In video surveillance, information such as location relationship, attribute, flight number and tail number of the target can be obtained.

4 Future Trajectory Projections

Based on the voice and video recognition information, the flight number, command action and other information can be extracted and correlated with the target track in the system, and the target's future motion trajectory can be inferred by direct extrapolation method or combining with the ATC data.

4.1 Direct Extrapolation

If the control command action is the change of motion position, such as the change of course, altitude, speed, the method of track extrapolation is used to predict the future trajectory. If the command is "CES0234 turn left 45°", adjust the course 45° according to the current speed and the current heading, and predict the trajectory. If it is consistent with the planned trajectory, the motion trajectory can be inferred by integrating the real-time motion state with the planned trajectory [4]. The specific algorithm is as follows:

Consider the motion of the object in two dimensions. According to the two-dimensional target kinematics equation:

$$\begin{cases} \dot{x}(t) = v(t)\cos\varphi(t) \\ \dot{y}(t) = v(t)\sin\varphi(t) \\ \dot{v}(t) = a_q(t) \\ \dot{\varphi}(t) = a_f(t)/v(t) \end{cases} \tag{1}$$

(x, y) is the position of the target; $v(t)$ is the tangential velocity of the target; $\varphi(t)$ is the Angle at which the target course changes; $a_q(t)$ and $a_f(t)$ are the tangential acceleration and normal acceleration of the target motion respectively [5]. For aircraft, except in specific stages (such as takeoff, climb, etc.), the target mainly moves in a uniform straight line and a uniform turn. Therefore, further assumptions $a_q(t) = 0$ and $a_f(t) = 0$ constants are made, so that the target motion includes the following two special forms:

(1) When, the target moves in a straight line. $a_f(t) = 0$
(2) At that time, the target moves at a uniform curve. $a_f(t) \neq 0$

If, represents the turning angular velocity of the target, then the extrapolation model and model parameters that are directly extrapolated to the target are dependent on, thus forming the following direct extrapolation method structure, $\omega = \dot{\varphi}(t)\omega\omega$.

(1) estimator ω

To estimate the target's turning angular velocity: ω

$$\omega_{k+1} = e^{-T/\tau_\omega}\omega_k + w_{\omega,k} \tag{2}$$

Where, ω_k and ω_{k+1} are respectively the turning angular velocity at time k and time $k+1$; T is the sampling interval; τ_ω is the time-dependent constant of angular velocity; $w_{\omega,k}$ and is white noise. Formula (2) is solved by the least square method according to the observed input.

(2) Extrapolation model

The target extrapolation model is:

$$X_{k+1} = A(\omega)X_k + Bw_k \tag{3}$$

$X_k = \left[x_k, v_{xk}, y_k, v_{yk}\right]^T$ x_k, v_{xk}, y_k, v_{yk} is the axial position, axial velocity, y and w_k is white noise. The noise matrix is shown in Eq. (4).

$$B = \begin{bmatrix} T^2/2 \; T \; 0 & 0 \\ 0 & 0 \; T^2/2 \; T \end{bmatrix}^T \tag{4}$$

The transfer matrix is $A(\omega)$ obtained by discretization of the two-dimensional target kinematics equation. The rotation depends ω on the angular velocity of the turning and has two forms:

At the time: $\omega = 0$

$$A(\omega) = \begin{bmatrix} 1 \; T \; 0 \; 0 \\ 0 \; 1 \; 0 \; 0 \\ 0 \; 0 \; 1 \; T \\ 0 \; 0 \; 0 \; 1 \end{bmatrix} \tag{5}$$

At the time: $\omega \neq 0$

$$A(\omega) = \begin{bmatrix} 1 \; \sin \omega^T/_\omega & 0 \; \cos \omega T - {}^1/_\omega \\ 0 \; \cos \omega T & 0 - \sin \omega T \\ 0 \; 1 - \cos \omega^T/_\omega & 1 \; \sin \omega^T/_\omega \\ 0 \; \sin \omega T & 0 \; \cos \omega T \end{bmatrix} \tag{6}$$

Through matrix calculation, the coordinates and target velocity of the extrapolated trajectory at time K can be obtained as x_k, v_{xk}, y_k, v_{yk}.

4.2 Combined with Basic Data Method of ATC

If the control instruction includes airport runway, runway holding point, taxiway, take-off procedure, landing procedure, parking area, airway route, corridor entrance point, navigation point and other information, use this information to predict the future operation trajectory. If the command is "CES0234 can take off", the trajectory needs to be inferred based on basic data such as runway position and length, departure procedures and the principle of energy conservation. The following is an example of future trajectory prediction during take-off and taxiing.

The vertical profile mainly represents the longitudinal motion of the aircraft. During the flight, the external forces on the aircraft mainly include engine thrust, lift, resistance, gravity and ground friction during take-off and taxiing [6]. The aircraft performance library provides performance parameters such as thrust for each type of aircraft at each flight stage. Newton's second law and the principle of energy conservation are adopted to list the model equation:

$$\begin{cases} m\frac{dv}{dt} = F\cos(\alpha + \varphi_p) - D - mg\sin\theta \\ mv\frac{d\theta}{dt} = F\sin(\alpha + \varphi_p) + L - mg\cos\theta \end{cases} \tag{7}$$

$$(F - D)v_{TAS} = mg\frac{dh}{dt} + mv_{TAS}\frac{dv_{TAS}}{dt} \tag{8}$$

5 Object Conflict Detection

In combination with the current position of the aircraft or vehicle, flight parameters, predicted trajectory and the interval standard adopted, the conflict judgment between target A and all other targets is carried out. First, the horizontal conflict is judged. When the horizontal conflict exists, the vertical conflict is detected. The position of target A in the future period of time (for example, 5 s) is obtained, and the position of target A in the future 5 s is speculated. Whether the horizontal positions of targets meet the interval standard is judged. When the horizontal interval violates, the vertical interval is judged, and then the position relationship between targets is judged with the interval of 5 s until the end of the prediction time [7]. As long as the horizontal and vertical interval standards are violated once, a conflict is considered. The specific calculation steps are as follows.

The tuple (x, y, V, h, t) is used to represent the state of the target at time T, where (x, y) is the geographic projection coordinate of the target at the current time T, v is the target's velocity at time t, and h is the target's height at time t [8]. Set the current state of target A as (X1, y1, v1, H1, T1), and that of target B as (x2, Y2, v2, H2, T1). Adopt the aforementioned direct extrapolation method to extrapolate target A and target B to maintain the speed and heading for the next 5 s, respectively. After the extrapolation, the target states are (x3, Y3, V1, H3, T1 + 5) and (X4, Y4, V2, H4, T1 + 5). The extrapolated distances of target A and target B relative to the current time are denoted as S1 and S2 respectively. The horizontal interval standard is set as D and the vertical interval standard is set as G.

Step 1: Judge $|h1 - h2| \leq K$ whether the vertical spacing standard is violated, and no;
Step 2: Judge whether the horizontal interval standard is violated, as shown in Fig. 2.

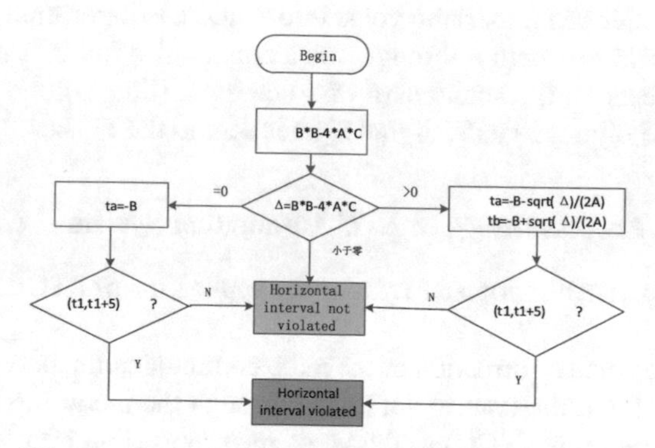

Fig. 2. Whether the horizontal interval violates the judgment method

Follow the following procedure to determine whether the horizontal interval is violated:

(1) First, variables A, B and C are calculated, as shown in formulas (9), (10) and (11)

$$A = \left(\frac{x3 - x1}{s1} \cdot v1 - \frac{x4 - x2}{s2} \cdot v2\right)^2 + \left(\frac{y3 - y1}{s1} \cdot v1 - \frac{y4 - y2}{s2} \cdot v2\right)^2 \tag{9}$$

$$\begin{aligned} B = 2 \cdot &\left(\frac{x3 - x1}{s1} \cdot v1 - \frac{x4 - x2}{s2} \cdot v2\right) \cdot \left(x1 - x2 - \frac{x3 - x1}{s1} \cdot v1 \cdot t1 + \frac{x4 - x2}{s2} \cdot v2 \cdot t1\right) \\ +2 \cdot &\left(\frac{y3 - y1}{s1} \cdot v1 - \frac{y4 - y2}{s2} \cdot v2\right) \cdot \left(y1 - y2 - \frac{y3 - y1}{s1} \cdot v1 \cdot t1 + \frac{y4 - y2}{s2} \cdot v2 \cdot t1\right) \end{aligned} \tag{10}$$

$$\begin{aligned} C = &\left(x1 - x2 - \frac{x3 - x1}{s1} \cdot v1 \cdot t1 + \frac{x4 - x2}{s2} \cdot v2 \cdot t1\right)^2 \\ &+\left(y1 - y2 - \frac{y3 - y1}{s1} \cdot v1 \cdot t1 + \frac{y4 - y2}{s2} \cdot v2 \cdot t1\right)^2 - d^2 \end{aligned} \tag{11}$$

(2) the symbol of calculation $\Delta = B^2 - 4AC$ and judgment Δ;
(3) Let $\Delta = 0$, ta = −B, determine whether it is within the interval $(t1, t1 + 5)$. If yes, it means that the horizontal interval is violated; otherwise, it means that the horizontal interval is not violated;
(4) let $\Delta > 0$, $ta = -B - \sqrt{\Delta}/(2A)$, $tb = -B + \sqrt{\Delta}/(2A)$, judge ta or tb whether or is within the interval $(t1, t1 + 5)$. If yes, it means that the horizontal interval is violated; otherwise, it means that the horizontal interval is not violated;
(5) $\Delta < 0$, indicating that the horizontal interval is not violated.

6 Design of System Application Method

6.1 Control Voice Collection Methods

Acquisition controller and pilot radio voice information, to determine whether a control instruction, a control instruction through voice communication unit record output port input to control seats on the sound card of voice input (line in), the voice sound card analog signals into digital signals, digital signals sent to the PC bus.

6.2 Application Process Design of ATC Automation System

The following steps can be adopted to realize the application of ATC automation system.

Step 1: Collect the voice information of the radio communication between the controller and the pilot, and input the control command through the recording output port of the voice communication unit (VCU) to the voice input port (line in) of the sound card at the control seat. The sound card converts the analog signal into digital signal, and the digital signal is sent to the PC bus;
Step 2. Convert the digital signals into text information by using the collected digital signals and speech recognition software, and send the text information of all control seats to the control system server in a unified manner;
Step 3. Use the generated text information to form structured instructions according to the control instruction generation method in Sect. 1;
Step 4. According to the extracted control instructions and the existing target information in the server of the control system, related targets are associated, and the running trajectory of the target to be processed from the current to the future VSP time (such as 3 min) is predicted to form the forecast trajectory.

7 Conclusion

The application of controller's control speech recognition to target conflict detection is the embodiment of intelligent air traffic management. However, if the current research content is really applied to control duty, it needs to improve the accuracy of control speech recognition, including the recognition rate in the combination of Chinese and English and noisy environment, which is the key to the application of the system. In addition, the recognition of voice control has other applications, such as the direct generation of system commands by voice control commands, reducing the input burden of controllers, and the rationality judgment of controllers' voice commands, which have a wide application prospect.

Acknowledgment. This work was financially supported by National Key Research and Development Program of China (2020YFB1600100).

References

1. Liu, W., Hu, J., Yuan, W.: Research on voice command recognition technology of standard language (English) for land and air communication. Comput. Sci. (7), 131–137 (2013)
2. Yu, J., Zhang, R.: Speaker recognition based on MFCC and LPCC. Comput. Eng. Des. (05), 1189–1191 (2009)
3. Design and research of voice application system for AIR traffic control simulator. Nanjing University of Aeronautics and Astronautics, Nanjing (2002)
4. Peng, Y., Hu, M., Zhang, Y.: J. Traffic Transp. Eng. **3**(1), 3–8 (2005)
5. Wang, C., Guo, J., Shen, Z.: J. Southwest Jiaotong Univ. **44**(2), 295–300 (2009)
6. Tang, X., Han, Y., Han, S.: Flight path prediction of aircraft 4D based on hybrid system model. J. Nanjing Univ. Aeronaut. Astronaut. **44**(1), 105–112 (2012)
7. Tang, X., An, H., Wang, C.: Aircraft ground taxi guidance method for conflict avoidance. J. Southwest Jiaotong Univ. **46**(6), 1032–1039 (2011)
8. Zhang, J., Wang, J., Zhang, J,, et al.: Conflict resolution strategy of sliding. Command Inf. Syst. Technol. **3**(1), 59–63 (2012)

Airspace Conflict Detection Method Based on Subdivision Grid

Zhiqi Liu[1,2(✉)], Ying Nan[1,2(✉)], and Yi Yang[2(✉)]

[1] Nanjing University of Aeronautics and Astronautics, Nanjing 211106, China
zhiqi_liu@foxmail.com, nanying@nuaa.edu.cn
[2] State Key Laboratory of Air Traffic Management System and Technology, Nanjing 211106, China
nuaa_yang@nuaa.edu.cn

Abstract. Based on the meshing method, a representation and conflict detection method for the traditional airspace is designed. The method performs grid representation and conflict detection for the traditional geometric airspace. Firstly, the global airspace is divided and gridded, and all grids are coded and expressed digitally; Next, a grid coded multiway tree is established by using the latitude, longitude and height codes of the airspace grid to represent the traditional geometric airspace; Finally, the airspace conflict detection method based on grid is constructed. According to the airspace grid multiway tree, the conflict airspace grid is obtained and the conflict area is represented by the conflict airspace grid. Simulation results show that this method can improve the efficiency of conflict detection compared with the traditional conflict detection algorithm.

Keywords: Airspace · Earth tessellation grid · Conflict detection · Grid encoding · Multi-tree

1 Introduction

The status of air traffic management system has become important with modern air transportation industry, but the pressure it faces is also increasing. Although the airspace conflict detection function plays an increasingly important role in the current air traffic control system, its performance needs to be improved.

The research on airspace conflict detection began in the 1940s–1950s. Many scholars around the world have proposed a variety of relevant models and algorithms. At present, geometric floating-point calculation is used the most widely, that is, to cross and judge whether there is an airspace conflict through the edges of airspace required by each airspace use plan. Although this method can accurately calculate the conflicts and the range of conflicts among airspace use plans, there are some problems for large-scale airspace conflict detection, high algorithm complexity, long calculation time, low efficiency and so on [2–4, 6, 9, 12, 17].

In this paper, the airspace to be detected is meshed based on subdivision grid. Then the multi-tree containing the airspace grid representation information is used to detect the conflicts among airspace use plans. The simulation results show that this method can improve the detection efficiency and shorten the calculation time of conflict detection.

Q. Liang et al. (Eds.): Artificial Intelligence in China, LNEE 854, pp. 670–677, 2022.
https://doi.org/10.1007/978-981-16-9423-3_84

2 Grids Generation in Geometric Airspace

Earth tessellation grid (ETG) is a kind of earth fitting grid that can be subdivided infinitely without changing the shape. When subdivided finely enough, it can achieve the purpose of simulating the geometric earth. And the discreteness, hierarchy and global continuity of ETG just meet the requirements of computers for data discretization processing [1, 11]. The military grid reference system (MGRS) was proposed by the U.S. military according to the European grid map in the 1940s. The system contain 10 km level, kilometer level even meter level grids around the world, which make it can establish a corresponding relationship between longitude and latitude coordinates and grid coordinates to simplify position reporting and coordination among soldiers [12]; Global area reference system (GARS) was proposed by the U.S. Geospatial Intelligence Agency in 2006. GARS grids are divided into three levels of different sizes by using the equal longitude and latitude grid division method, which are mainly used to meet the representation and description of geographical location in joint operations [5, 7, 10, 14].

GeoSOT-3D earth subdivision framework proposed by Professor Chengqi Cheng's team from Peking University has attracted the attention of scholars around the world because of excellent characteristics such as globality, hierarchy and multi-scale. The framework has also been applied in the expression of spatial objects [18]. Airspace grid model has also been applied in some air traffic control fields, such as airspace scheduling, flow control, meteorological impact prediction, UAV formation and other research fields [8, 15, 16], but there is no relevant research on airspace conflict detection and determination.

The first step is global airspace grid modeling. The earth surface is projected to a rectangular plane by using the isometric projection of the positive axis cylinder. After the projection, the longitude and latitude lines form two groups of parallel lines perpendicular to each other. Then the airspace is divide. The longitude and latitude plane is divided level by level in the longitude and latitude direction according to the 64 equal divisions of 8 * 8 specification, forming the plane grids of mutually inclusive and gapless at each level, which the highest level is level 10. Similarly, in the altitude direction, the airspace is divided level by level according to the 8 equal divisions, which the highest level is level 9. Besides, the altitude of the first level in the altitude direction is not divided. Finally, the plane grids and altitude grids are combined to form a gridded airspace structure, as shown in Fig. 1 below.

Fig. 1. Schematic diagram of partial global grids division

The second step is coding digitally for airspace grid. The coding consists of longitude and latitude direction and altitude direction coding. Double digit octal coding is used for grids in the longitude and latitude direction. The longitude and latitude codes from level 1 to level 10 are obtained by nested codes from low to high by level (that is, grids size are from large to small). And similarly, the altitude direction is also coded from low to high level along the altitude but do not code for the first level of altitude, which octal is also used to obtain the altitude codes of level 2 to 10. Finally, combine the longitude and latitude codes with the altitude codes according to the corresponding levels. Then the three-dimensional codes of airspace grids are formed. The first two levels coding methods of this process are shown in Fig. 2.

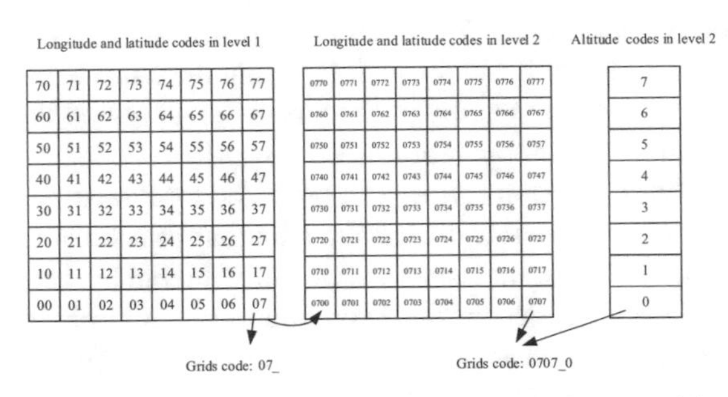

Fig. 2. Longitude, latitude and altitude grids coding from level 1 to level 2

3 Conflict Detection Using Codes Multi-tree

Now the airspace grid code multi-tree can be constructed.

In the first step, calculate the grid in level 1 according to the coordinates of arbitrary point on the bottom of the airspace, where the point should be in this grid. Then expand grids in the longitude and latitude direction and obtain all the grids in level 1 where they intersect the bottom of the airspace. And at the same time, the longitude and latitude codes of these grids in level 1 can be generated. Then a multi-tree can be constructed, which the ID of the airspace is the root node while all the longitude and latitude codes in level 1 are sub nodes.

The second step is to decompose the grids in level 1 to the next level and obtain the grids in level 2 where they intersect the bottom of the airspace, while the longitude and latitude codes of these grids in level 2 can be generated. Then expand these grids in the altitude direction and similarly generate the altitude codes in level 2. Decompose the airspace layer by layer until the target level is reached (for example, if the target level is the 10th level of the longitude and latitude direction grids in the airspace grid, the altitude grids should be also decomposed to the 10th level).

Finally, combine the longitude, latitude and altitude codes from low to high levels to form each sub nodes. And all these sub nodes should be inserted into the corresponding node in the multi-tree. In this way, a multi-tree structure of airspace grid codes is formed, which can represent the airspace. The multi-tree structure formed by all the subdivision process is shown in Fig. 3.

As shown in the figure, because altitude grids in this level needn't be divided, the codes in level 1 only contain codes of longitude and latitude grids, but the codes in the level 2 or higher level are all composed of longitude and latitude codes and altitude codes. Grid code xy2 in the first node of the level 2 signify that this code represent the longitude and latitude direction grids in this level. And followed by the code xy2 is a digital code of the longitude and latitude direction grids. Similarly, the code z1 represents the level 1 code in the altitude direction and followed by the code z1 is a digital code of the altitude direction grids. As for codes of higher-level grids, they are obtained in the same way. Then the corresponding codes are also written into the deeper sub nodes of the multi-tree according to the above rules. Finally the airspace can be completely represented as the multi-tree structure in Fig. 3.

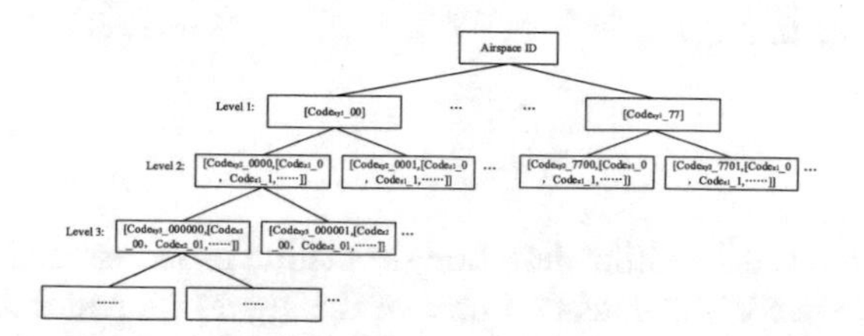

Fig. 3. Grid code multi-tree of geometric airspace

Fig. 4. Multi-tree conflict detection

Since the airspace has been represented by grids and a multi-tree composed of grid codes, the conflict detection has the following two characteristics.

1. The actual conflict space of airspace can be represented by the same grid codes in multiple airspace use plans, which is obviously the airspace grids set represented by these codes;
2. Although the grids can be divided into deferent levels and size spans, for airspace with conflicts, if a grid at a certain level is occupied by multi airspace use plans, some of all the low-level grids contained in the grid must also be occupied by multi plans inevitably.

Above characteristics are shown in Fig. 4. The root level (represented by rectangular) of the multi-tree signifys the airspace to be detected, the hollow node indicates that there are grids in conflict with other airspace use plans at this level. And conversely the solid nodes have no conflicts with other airspace use plans. According to these features, it is obvious that all nodes in the deep nodes under a solid node are also solid nodes. Therefore, traverse from the low level nodes to the high level and stop traversing for the grids without conflict (solid nodes), which obviously can save unnecessary traversal steps in conflict detection.

By applying the above principles, the process will be described in detail.

General airspace conflict detection needs to consider both time conflicts and space conflicts. Firstly, the conflict detection algorithm detects the time overlapping of the two airspace use plans to determine the time conflict. Obviously, if there is no coincidence in occupation time, there is no possibility of conflict between the two airspace even if there is airspace intersection and that signifys there are no airspace conflicts. Then carry out space exclusion detection for the two airspace use plans to be detected, which needs to compare the longitude, latitude and altitude boundaries of the two pieces of airspace. If there is a dimension without overlapping section in the three dimensions of longitude, latitude and altitude, there are no conflict space between the two pieces of airspace.

Then the conflict space is calculated by using the grid coded multi-tree structure. In level 1, obtain the longitude and latitude codes of the two pieces of airspace to be detected and find the intersection. Respectively obtain the next level longitude and latitude codes of under the first level intersection grids in the multi-tree. Then calculate the longitude and latitude codes for the next level intersection grids. Similarly, the codes of intersection grids can be obtained layer by layer until the highest level. In this way, the altitude code of each intersection grid is also obtained step by step. If the longitude, latitude and altitude codes of intersection grids are combined, the set of grids represented by codes can signify the conflict space of airspace.

4 Simulation Analysis

By using different detection methods, the conflict detection simulation is performed for different airspace in the same airspace. The detection time of the algorithm under the conditions of different number of airspace occupancies and different number of conflict space is separately analyzed. And the analysis verified the computational efficiency of the conflict detection algorithm based on grids.

Firstly, the experimental conditions and scenes are constructed. If a series of polygonal space occupations with large scale and random location, shape are established in the airspace, it is obvious that there will probably be conflicts between them.

In the experiment a large number of random airspace are generated repeatedly, and the total number of space occupations is respectively set to 100, 150, 200,..., 500. Each time the traditional method and the method in this paper are used for conflict detection. After the detection, the airspace conflict detection information is analyzed as well as the time consumed by each method.

Examples are as follows. When the total number of airspace is 100, the airspace conflicts output by the traditional airspace conflict detection algorithm is shown in Fig. 7, and the airspace conflict information is output in the overlapping area in the figure. Similarly, the conflict airspace output by the airspace conflict detection algorithm based on subdivision grids is shown in Fig. 8.

From the comparison between Fig. 5 and Fig. 6, the results of the two conflict detection algorithms are consistent in detecting the same airspace conflict. Obviously, the new method can detect airspace conflict equally effectively.

For the time consumed, comparing the detection method in this paper with traditional method, it can be found that the calculation time of this method is shortened, as shown in Table 1. In the table, C_1 represents the number of airspace occupancies. C_2 represents the number of airspace conflicts. T_1 represents the calculation time of traditional method and T_2 represents the calculation time of new method. R signifys saving ratio of the calculation time.

Figure 7 shows the change of detection time both of the traditional airspace conflict detection algorithm and the algorithm in this paper. As can be seen from the figure, with

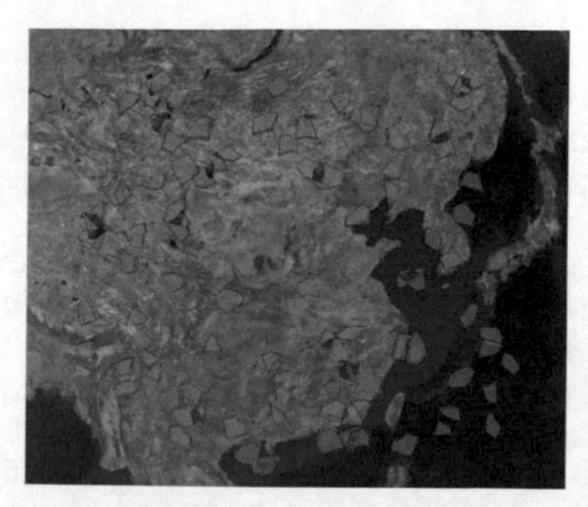

Fig. 5. Results of traditional airspace conflict detection method

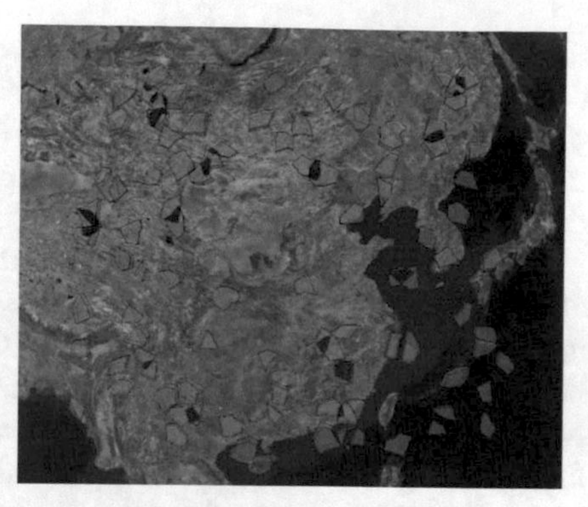

Fig. 6. Results of conflict detection method based on subdivision grid

Table 1. Comparison result of conflict calculation time of different methods

C_1	C_2	T_1 (s)	T_2 (s)	R
100	27	0.12	0.07	71.43%
150	66	0.21	0.08	61.90%
200	110	0.26	0.1	61.54%
250	173	0.43	0.16	62.79%
300	220	0.54	0.21	61.11%
350	238	0.67	0.28	58.21%
400	422	0.78	0.39	50.00%
450	417	1.09	0.52	52.29%
500	451	1.31	0.62	52.67%

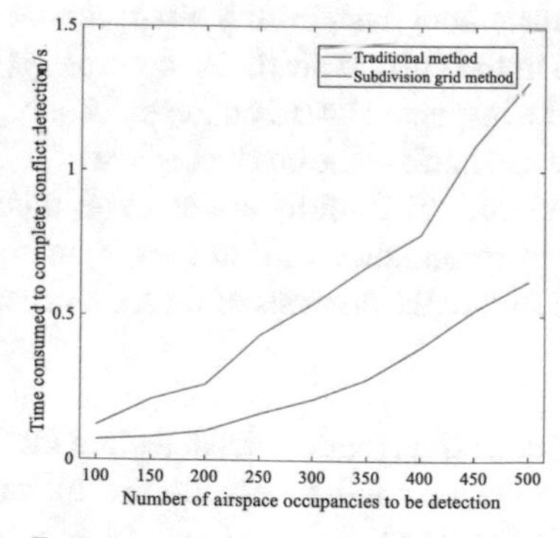

Fig. 7. Result of detection time consumed for two methods

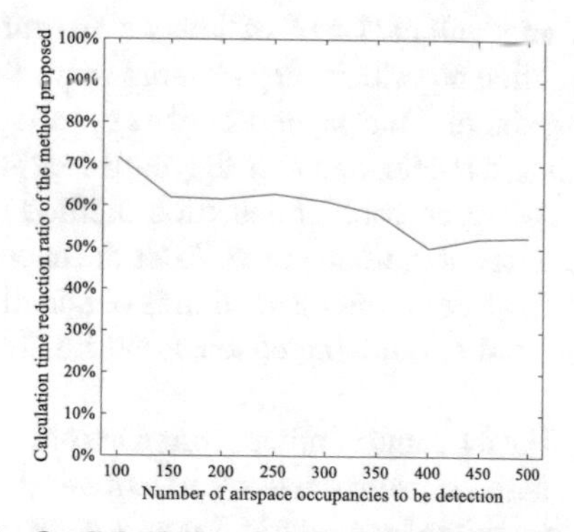

Fig. 8. Calculation time shorten by the method based on subdivision grid

the increase of the number of airspace occupancies in each experiment, the calculation time of the two methods increases. However, the time consumed of the algorithm in this paper is significantly less than that of the traditional algorithm. With the increase of the total number of airspace, the speed of the calculation time is increasing significantly less than that of the traditional airspace conflict detection algorithm.

Figure 8 shows the characteristic of the saving ratio of the calculation time for the algorithm in this paper compared with the traditional conflict detection algorithm. With the total number of airspace occupancies increasing, although the number of airspace to be detected in each experiment is increasing, the saving ratio of the calculation time can be maintained at no less than 50%, comparing with the traditional method. That means the efficiency of the conflict detection method based on subdivision grids is significantly higher than that of the traditional detection algorithm.

From the simulation analysis results above, the airspace conflict detection algorithm based on earth subdivision grids proposed in this paper can effectively detect the conflicts in the space. Compared to traditional airspace conflict detection algorithm, this method can shorten the original calculation time by more than 50% under the premise of calculating the same results. Thus the method greatly improves the detection efficiency of a large number of airspace conflicts.

As for the underlying reasons, when the airspace researched has been determined, the grids generation method and its scale corresponding to the airspace are accordingly determined. Therefore, the multi-tree traversal scale and upper limit of the new method can be determined, which will not change with the increase of the total number of detection airspace occupancies. However for the traditional method, the detection method is based on the geometric parameters of each airspace occupancy. Thus the actual calculation process for the traditional method will become complex with the increase of the total airspace occupancies number in the region. As mentioned above, the new detection method avoids the impact of the number and scale of airspace occupancies on the detection process.

5 Conclusion

As an significant part of today's air traffic management system, the airspace conflict detection algorithm plays a more important role in the current air traffic control system than before. This paper describes the airspace to be inspected based on the earth subdivision grid model, and uses the multi-tree structure to calculate the conflict space of it. The new airspace conflict detection method proposed and the traditional detection method are tested at the same time. After the experiment the calculation time of the two methods for airspace conflict detection is obtained. According to the analysis of the experimental results, the following conclusions are clear.

1. For the same number of airspace occupancies detecting, the conflict detection method based on subdivision grid can get the same airspace conflict detection result as the traditional geometric floating-point calculation method.
2. Using the traditional method and the grid based detection method proposed to detect the same conflict airspace, it can be found that the calculation time of the new method

can be reduced by more than 50% compared with the traditional method. Therefore, this method is superior to the traditional method in computational performance.

Fund Support. Supported by the State Key Laboratory of Air Traffic Management System and Technology, No: SKLATM202002.

References

1. Zhou, C., Yang, O., Ting, M.: Progresses of geographical grid systems researches. Prog. Geogr. **28**(05), 657 (2009). (Chinese)
2. Mei, D., Wu, W.H., Xu, J.Y., et al.: Flight conflicts detection and elusion in modern air combat. Electron. Opt. Control **14**(06), 131 (2007). (Chinese)
3. Cui, D.G., Cheng, P., Geng, R.: Conflict probability analysis of automatic air traffic control. J. Tsinghua Univ. (Sci. Technol.) **40**(11), 119 (2000). (Chinese)
4. Jianping, W., Zili, Z., Hua, W.: Study on model of determining tactic airspace conflict. In: Proceedings of 14th Chinese Conference on System Simulation Technology & Application (CCSSTA'2012), vol. 14, p. 210 (2012). (Chinese)
5. Ling, C.: The Detection of Airport Runways Based on Bayesian Classification in PolSAR Image. Civil Aviation University of China, Tianjin (2016).(Chinese)
6. Miao, S., Cheng, C., Zhai, W., et al.: A low-altitude flight conflict detection algorithm based on a multilevel grid spatiotemporal index. ISPRS Int. J. Geo-Inf. **8**(6), 289 (2019)
7. Nan, L., Jianjun, Z.: Error analysis and quality control of spatial data. Sci. Technol. Inf. **15**(15), 23 (2017). (Chinese)
8. Ping, C.: The Application of Geographic Information System on Air Traffic Control System. South China University of Technology, Guangzhou (2010).(Chinese)
9. Prandini, M., Hu, J., Lygeros, J., et al.: A probabilistic approach to aircraft conflict detection. IEEE Trans. Intell. Transp. Syst. **1**(4), 199 (2000)
10. Xiaojuan, S.: Application of new surveying and mapping technology in cadastral survey. Huabei Nat. Resour. **22**(1), 118 (2018). (Chinese)
11. Zhao, X., Ben, J., Sun, W., et al.: Overview of the research progress in the earth tessellation grid. Acta Geodaetica et Cartographica Sinica **45**(S1), 1 (2016). (Chinese)
12. Yang, Y., Mao, Y., Xie, R., Hu, Y., Nan, Y.: A novel optimal route planning algorithm for searching on the sea. Aeronaut. J. **125**(1288), 1064 (2021)
13. Yi, D., Yuan, C.: Research on military geographic reference system. Electron. Qual. **33** (11), 79 (Chinese)
14. Yongke, Y.: Accuracy Assessment of Large Scale Land Over Datasets. Nanjing University, Nanjing (2014).(Chinese)
15. Yongwen, Z.: Introduction to Airspace Management. Science Press, Beijing (2018).(Chinese)
16. Zhu, Y., Chen, Z., Pu, F., et al.: Development of digital airspace system. Strat. Study CAE **23**(03), 135 (2021). (Chinese)
17. Chen, Z.: Theory and Method of Airspace Management, Science Press, Beijing (2011). (Chinese)
18. Li, Z., Chen, C., Li, S.: Research on spatial objects fast visualization based on GeoSOT-3D. J. Geo-Inf. Sci. **17**(07), 810 (2015). (Chinese)

Passive Positioning and Filtering Technology Using Relative Height Measurement Information

Lai-Tian Cao[1,2](✉), Kai-Wei Chen[1,2], Peng Sun[1,2], Xiao-Min Qiang[1,2], Yi-Ming Zhang[3], and Li-Jia Zhang[1,2]

[1] Beijing Aerospace Automatic Control Institute, Beijing 100854, China
ccclllttt@163.com

[2] National Key Laboratory of Science and Technology on Aerospace Intelligent Control, Beijing 100854, China

[3] Beijing Institute of Computer Technology and Application, Beijing 100854, China

Abstract. Under the condition of obtaining the passive measurement angle information and the relative height information of the positioning system and the target, using the least squares algorithm to locate the radiation source target, and then using the Kalman filter to filter and predict the position data after the least square estimation, so as to make the best estimation of the motion state of the radiation source target.

Keywords: Least squares · Kalman filter · Passive location

1 Introduction

On the basis of receiving the signal of the radiation source target and obtaining the relevant positioning parameters, the positioning system uses the continuous measurement of multiple observation parameters with complex data processing methods, combined with the trajectory data of the positioning system, the location of radiation source target can be accurately estimated. This is how the passive positioning system realizes the positioning process of the space target [1].

The traditional passive positioning method has been widely used due to its strong anti-interference ability and mature technology, but its positioning error is relatively large, so it is necessary to take corresponding measures to reduce its positioning error and improve positioning accuracy [2]. This paper proposes a passive positioning and filtering technology, which uses passive measurement angle information and the relative height measurement information of the positioning system and the target, Single-point positioning of the radiation source target, then using the Kalman filter method to make the best estimation of the movement state of the radiation source target, thereby further reducing the positioning error and improving the positioning accuracy.

Q. Liang et al. (Eds.): Artificial Intelligence in China, LNEE 854, pp. 678–683, 2022.
https://doi.org/10.1007/978-981-16-9423-3_85

2 Estimation of the Position of the Radiation Source Target

The passive positioning system can obtain the relative height information of the positioning system and the target by installing a height measuring device. On this basis, the passive detector in the passive positioning system can be used to obtain information parameters such as the azimuth angle and pitch angle of the radiation source target required for positioning. These parameters all correspond to a positioning surface. The target positioning point can be obtained through multiple measurements at different observation positions [3].

The schematic diagram of the target detection by the positioning system is shown in the Fig. 1 below.

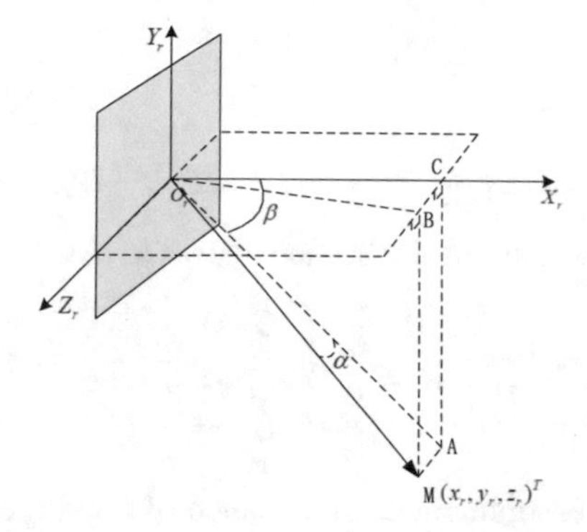

Fig. 1. Schematic diagram of target detection by positioning system

In the above figure, the $X_rO_rY_r$ plane is the passive detector antenna plane, O_r is the center point of the passive detector antenna, the target position of the radiation source is M, and the coordinate is $(x_r, y_r, z_r)^T$, $\overrightarrow{O_rB}$ is the projection of $\overrightarrow{O_rM}$ on the $X_rO_rZ_r$ plane, $\overrightarrow{CA}$ is the projection of $\overrightarrow{BM}$ on the $X_rO_rY_r$ plane. $\angle AO_rC = \beta$, $\angle MO_rA = \alpha$, β and α are defined as the pitch angle and azimuth angle of the radiation source target. The relative height (y_r) of the positioning system and the target can be measured.

Due to the error in the measurement of the positioning system, the single-frame target least squares positioning method is used to estimate the position of the radiation source target.

Obtain the observation equation:

$$\begin{cases} \beta = \arctan \frac{y_r}{x_r} + v_\beta \\ \alpha = -\arctan \frac{z_r}{\sqrt{x_r^2 + y_r^2}} + v_\alpha \\ y = y_r + v_y \end{cases} \quad (1)$$

In the formula, β, α and y are the pitch angle, azimuth angle of the radiation source target and the relative height measurement value of the positioning system and the target, v_β, v_α and v_y are the pitch angle, azimuth angle of the radiation source target and the relative height of the positioning system and the target measurement noise, respectively.

Further denoted as:

$$Z_0 = HX + V \tag{2}$$

In the formula, observation vector $Z_0 = \left[\beta\ \alpha\ y\right]^T, X = \left[x_r\ y_r\ z_r\right]^T, V = \left[v_\beta\ v_\alpha\ v_y\right]^T$.

Measurement matrix $H = \left[H_{\alpha\beta}^{2\times3}\ H_y^{1\times3}\right]^T$, $H_{\alpha\beta}^{2\times3}$ is a non-linear function for $(x_r, y_r, z_r)^T$, it can be linearized by Taylor series expansion, and the linearization expressions (H) can be obtained from retaining the first two terms:

$$H_{\alpha\beta}^{2\times3} = \frac{\partial(\beta,\alpha)}{\partial(x_r, y_r, z_r)} = \begin{bmatrix} -\frac{y_r}{x_r^2+y_r^2} & \frac{x_r}{x_r^2+y_r^2} & 0 \\ \frac{x_r \cdot z_r}{(x_r^2+y_r^2+z_r^2)\cdot\sqrt{x_r^2+y_r^2}} & \frac{y_r \cdot z_r}{(x_r^2+y_r^2+z_r^2)\cdot\sqrt{x_r^2+y_r^2}} & -\frac{\sqrt{x_r^2+y_r^2}}{x_r^2+y_r^2+z_r^2} \end{bmatrix} \tag{3}$$

$$H_y^{1\times3} = \left[0\ 1\ 0\right] \tag{4}$$

The recurrence equation:

$$\hat{X}_n = \hat{X}_{n-1} + (H^T R^{-1} H)^{-1} H^T R^{-1} (Z_0 - \hat{Z}_0) \tag{5}$$

In the formula, the measurement variance matrix R:

$$R = \begin{bmatrix} \sigma_\beta^2 & 0 & 0 \\ 0 & \sigma_\alpha^2 & 0 \\ 0 & 0 & \sigma_y^2 \end{bmatrix} \tag{6}$$

σ_β^2, σ_α^2 and σ_y^2 are the measurement error variance of the target pitch angle, the target azimuth angle, and the relative height between the positioning system and the target.

The error covariance matrix P_n:

$$P_n = (H^T R^{-1} H)^{-1} \tag{7}$$

Predicted value of measurement vector:

$$\hat{Z}_0 = \left[\hat{\beta}\ \hat{\alpha}\ \hat{y}\right]^T \tag{8}$$

And has the following relationship:

$$\begin{cases} \hat{\beta} = \arctan\frac{\hat{y}_r}{\hat{x}_r} \\ \hat{\alpha} = -\arctan\frac{\hat{z}_r}{\sqrt{\hat{x}_r^2+\hat{y}_r^2}} \\ \hat{y} = \hat{y}_r \end{cases} \tag{9}$$

In the formula, $\hat{x}_r$, $\hat{y}_r$ and $\hat{z}_r$ are the estimated position of the radiation source target. $\hat{\beta}$ is the estimated value of the pitch angle of the radiation source target, $\hat{\alpha}$ is the estimated value of the azimuth angle of the radiation source target, $\hat{y}$ is the estimated value of the relative height between the positioning system and the target.

Under the condition that the initial recursion value ($\hat{X}_0$) of the state variable is determined, according to formula (5), the best estimate of the target position in a single frame can be obtained by the iterative method. Since the position of positioning system and radiation source target change in real time, the target position is estimated according to the above-mentioned method in each data processing cycle.

3 Passive Positioning Filter Processing

Kalman filter is used to filter and predict the position data after the least squares estimation, and make the best estimation of the motion state of the radiation source target, thereby improving the positioning accuracy.

3.1 Filter State Equation

The filter state equation adopts a 6-dimensional state equation:

$$X_k = \Phi_{k,k-1} X_{k-1} + GW \tag{10}$$

The target filter state vector is the estimated value of the target position and velocity:

$$X_k = \left[\hat{x}_r, \hat{y}_r, \hat{z}_r, \hat{v}_{xr}, \hat{v}_{yr}, \hat{v}_{zr}\right]^T \tag{11}$$

Input control matrix G:

$$G = \begin{bmatrix} 0\,0\,0 \\ 0\,0\,0 \\ 0\,0\,0 \\ 1\,0\,0 \\ 0\,1\,0 \\ 0\,0\,1 \end{bmatrix} \tag{12}$$

W is the noise matrix.

The state transition matrix $\Phi_{k,k-1}$:

$$\Phi_{k,k-1} = \begin{bmatrix} 1\,0\,0 & \Delta T & 0 & 0 \\ 0\,1\,0 & 0 & \Delta T & 0 \\ 0\,0\,1 & 0 & 0 & \Delta T \\ 0\,0\,0 & 1 & 0 & 0 \\ 0\,0\,0 & 0 & 1 & 0 \\ 0\,0\,0 & 0 & 0 & 1 \end{bmatrix} \tag{13}$$

ΔT is the time difference between the two filtering processes.

3.2 Measurement Equation

The measurement equation is as follows:

$$\begin{cases} \beta = \arctan \frac{\hat{y}_r}{\hat{x}_r} + v_\beta \\ \alpha = -\arctan \frac{\hat{z}_r}{\sqrt{\hat{x}_r^2 + \hat{y}_r^2}} + v_\alpha \\ y = \hat{y}_r + v_y \end{cases} \tag{14}$$

In the formula, v_β, v_α and v_y are the measurement noise of the target pitch angle, the target azimuth angle and the relative height of the positioning system and the target.

The measurement equation is a nonlinear equation, and the linearized range equation can be obtained by sampling processing:

$$y_k = H_k x_k + v_k \tag{15}$$

And has the following relationship:

$$H_k = \begin{bmatrix} H_{\alpha\beta}^{2\times 3} & 0^{2\times 6} \\ H_h^{1\times 3} & 0^{1\times 6} \end{bmatrix} \tag{16}$$

$$H_{\alpha\beta} = \frac{\partial(\beta, \alpha)}{\partial(x_r, y_r, z_r)} = \begin{bmatrix} -\frac{\hat{y}_r}{\hat{x}_r^2+\hat{y}_r^2} & \frac{\hat{x}_r}{\hat{x}_r^2+\hat{y}_r^2} & 0 \\ \frac{\hat{x}_r\cdot\hat{z}_r}{(\hat{x}_r^2+\hat{y}_r^2+\hat{z}_r^2)\cdot\sqrt{\hat{x}_r^2+\hat{y}_r^2}} & \frac{\hat{y}_r\cdot\hat{z}_r}{(\hat{x}_r^2+\hat{y}_r^2+\hat{z}_r^2)\cdot\sqrt{\hat{x}_r^2+\hat{y}_r^2}} & -\frac{\sqrt{\hat{x}_r^2+\hat{y}_r^2}}{\hat{x}_r^2+\hat{y}_r^2+\hat{z}_r^2} \end{bmatrix} \tag{17}$$

$$H_h^{1\times 3} = \begin{bmatrix} 0 & 1 & 0 \end{bmatrix} \tag{18}$$

3.3 Sequential Kalman Filter Equation

The Kalman filter equation [4] is as follows:

$$\begin{cases} \hat{x}_{k,k-1} = \Phi_{k,k-1}\hat{x}_{k-1} \\ P_{k,k-1} = \Phi_{k,k-1}P_{k-1}\Phi_{k,k-1}^T + Q_{k-1} \\ K_k = P_{k,k-1}H_k^T(H_k P_{k,k-1}H_k^T + R_k)^{-1} \\ P_k = (I - K_k H_k)P_{k,k-1} \\ \hat{x}_k = \hat{x}_{k,k-1} + K_k(y_k - \hat{y}_{k,k-1}) \end{cases} \tag{19}$$

In the formula, $P_{k,k-1}$ is the covariance matrix of estimation error, K_k is the matrix of filter gain, and R_k is the matrix of measurement noise.

Under the condition that $\hat{x}(0)$ and $P(0)$ are determined, the state estimate ($\hat{X}_k$) can be calculated recursively based on the measured value (y_k). $\hat{x}(0)$ can take the average value of the position data after the previous M least squares estimation.

4 Simulation Results and Conclusions

The main simulation conditions are selected as follows: the target pitch angle and azimuth angle measurement error are both 1.5° (1 σ), the height measurement error of the positioning system and target is 10 m (1 σ), the initial value of the P array is $\boldsymbol{P}(0) = diag\left(1500^2, 10^2, 200^2, 20^2, 0.1^2, 0.1^2\right)$, the filter period is $\Delta T = 2\,s$, and 1000 Monte Carlo simulations are performed. The simulation results are as follows (Fig. 2):

In the above figure, Δx, Δy and Δz represent the difference between the estimated target position and the true target position, Δvx, Δvy and Δvz represent the difference between the estimated target velocity and the true target velocity. From the simulation results, the positioning method has a better correction effect on the target position error, and the following conclusions can be drawn:

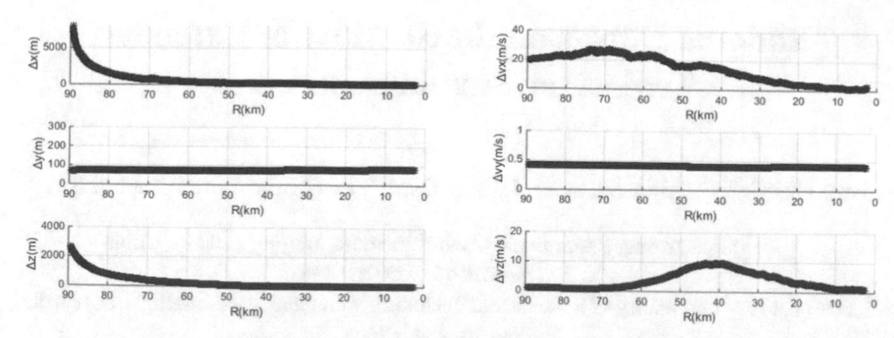

Fig. 2. Position and velocity estimation accuracy

The correction result of the algorithm is real-time. Under the condition of determining the trajectory of the positioning system, the target position error and velocity error converge rapidly. In the last stage, the target position error in 3 directions all converged to less than 100 m, and the velocity error in 3 directions converged to less than 0.5 m/s. The simulation results show that the passive positioning and filtering technology using relative height measurement information can make the best estimation of the position and motion state of the radiation source target and improve the positioning accuracy.

References

1. Qiu, H.: Analysis of passive location technology and influencing factors. Master's thesis of Xidian University, p. 1 (2014)
2. Liu, X.: Research on airborne passive location technology and tracking algorithm. Doctoral Dissertation of Harbin Engineering University, p. 3 (2012)
3. Sun, Z.K., Guo, F.C., Feng, D.W.: Single station passive location and tracking technology, p. 1. National Defense Industry Press (2008)
4. Qin, Y.Y., Zhang, H.Y., Wang, S.H.: Kalman filtering and integrated navigation principle, p. 6. Northwestern Polytechnical University Press (2015)

Radome Effects on the Stability of Extended Trajectory Shaping Guidance

Kaiwei Chen[1,2](✉), Lijia Zhang[1,2], Laitian Cao[1,2], Peng Sun[1,2], and Xuehui Shao[1,2]

[1] Beijing Aerospace Automatic Control Institute, Beijing 100854, China
kwillchan@outlook.com

[2] National Key Laboratory of Science and Technology on Aerospace Intelligent Control, Beijing 100854, China

Abstract. Effects of radome refraction error slope of radar seeker on the stability of extended trajectory shaping guidance (ETSG) are discussed in this paper. The formation mechanism of radome error was analyzed. The radome error parasitic loop was established. The guidance system model with radome error parasitic loop was constructed based on ETSG to analyze impacts of radome error parasitic loop on stability of the guidance system. The variation laws of stable region of ETSG with radome slope and guidance parameters under different system order and time-to-go are investigated. Research results demonstrate that the stable region in negative feedback of radome error parasitic loop is easy to be unstable under higher system order, and the maximum tolerance of ETSG to radome slope under positive feedback of parasitic loop decreases continuously as the missile approaches to the target.

Keywords: Radome error · Parasitic loop · Extended trajectory shaping guidance · Stable region

1 Introduction

Radome is an essential protector for radar homing missiles flying in the atmosphere. It can protect radar antenna of seeker from airflow disturbance and impact during the missile flight. Moreover, the streamlined design of radome can decrease flight resistance of missile [1, 2]. However, the refraction of radar beam by the radome and distortion of radiation pattern caused by shape and material of radome [3, 4] can cause electrical boresight error of antenna, thus producing radome refraction error angle. In consequence, the attitude disturbance of missile will be coupled into the seeker and the radome error parasitic loop will be formed in the missile guidance and control system. This radome error parasitic loop decreases stability of the guidance system.

Many researches have been conducted studying effects of radome error on the missile guidance. Garnell [5] pointed out firstly that radome slope is against stability of guidance and control system. Nesline and Zarchan [6–8] studied impacts of radome error on the performance of the missile guidance system. They concluded that radome slope

Q. Liang et al. (Eds.): Artificial Intelligence in China, LNEE 854, pp. 684–690, 2022.
https://doi.org/10.1007/978-981-16-9423-3_86

deteriorates the stability of the guidance system and increase the miss distance of the missile, especially in high airspace.

With respect to effects of radome slope on the guidance system of the missile, researchers mainly focus on the traditional proportional navigation guidance (PNG). There are few studies concerning effects of radome slope on other types of guidance systems. Extended trajectory shaping guidance (ETSG) is a multi-constraint optimal guidance law with terminal positon, impact angle and terminal acceleration constraints. Compared to PNG, it has time-variant and multi-parameter characteristics due to the involvement of impact angle constraint and time-to-go. Effects of radome slope on ETSG are more complicated and more significant. Hence, studying effects of radome slope on the performance of ETSG is essential to engineering application of multi-constraint optimal guidance and the overall design of corresponding guidance system.

Based on above problems, effects of radome slope of radar seeker on the stability of ETSG are discussed here. This paper is structured as follows: Sect. 2 introduces ETSG law and the formation mechanism of radome error, and the radome error parasitic loop model is developed. Section 3 sets up ETSG system with radome error parasitic loop. Under the assumption of frozen time, the variation laws of stable region of ETSG with radome slope and guidance parameters under different system order and time-to-go are investigated. Section 4 concludes this paper.

2 Problem Formulation

2.1 The Extended Trajectory Shaping Guidance

Without considering the target maneuver, the expression of ETSG is [9]:

$$a_c = N_1 V_r \dot{q} + N_2 V_r \frac{q - q_F}{t_{go}} \tag{1}$$

where a_c is the acceleration command, V_r is closing velocity, q is line-of-sight (LOS) angle, $\dot{q}$ is LOS rate, q_F is the expected impact angle, t_{go} is time-to-go, N_1 and N_2 are weight coefficients that are determined by the guidance factor n^*:

$$\begin{aligned} N_1 &= 2(n^* + 2) \\ N_2 &= (n^* + 1)(n^* + 2) \end{aligned} \tag{2}$$

2.2 Modeling of Radome Error

Different from air dielectric, radome has a unique dielectric constant and cone shape. It refracts electromagnetic wave at a certain incident angle with its longitudinal axis, thus causes angular deviation of antenna electrical boresight from the real target. Therefore, the radome refraction angle Δq is formed (Fig. 1).

For different gimbal angle of the seeker ϕ_y, the refraction angle Δq varies with look angle through the radome and properties of radome. Therefore, there's positive or negative refraction error, accompanied with value uncertainty. When the refraction error

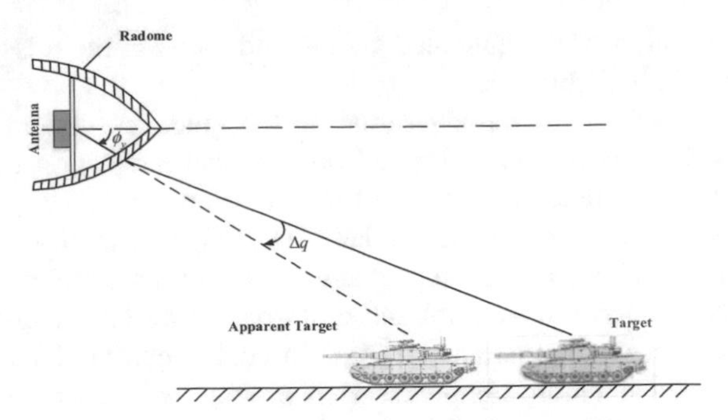

Fig. 1. Radome refraction physics

changes with ϕ_y, such nonlinear relationship can be described by the radome refraction angle curve: $\Delta q = f(\phi_y)$.

The slope of $\Delta q = f(\phi_y)$ is called radome error slope (R):

$$R = \frac{d\Delta q}{d\phi_y} = \frac{df(\phi_y)}{d\phi_y} \tag{3}$$

Under normal conditions, Δq is very small (generally $<1°$) and has negligible influences on the guidance system. However, the variation rate of Δq, that is, radome slope, can change guidance parameters and performance, thus influencing the terminal guidance precision of the missile and even cause pre-instability of the guidance system. Although the radome slope varies during a flight, it is often convenient to think of the slope as a constant.

2.3 Modeling of the Radome Error Parasitic Loop

Based on above analysis, the existence of radome affects seeker measurement on real target position. The basic missile-target geometry under the existence of radome error is shown in Fig. 2.

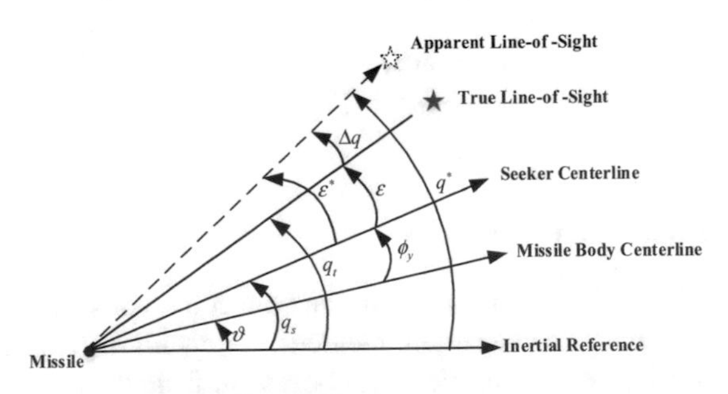

Fig. 2. Basic geometry

In Fig. 2, ϑ is attitude angle of the missile body, q_s is the inertial angle to the seeker centerline, q_t is true LOS angle, q^* is apparent LOS angle caused by radome error, ε is the included angle between seeker centerline and true LOS, ε^* is the error angle between seeker centerline and apparent LOS.

According to angular relationships in Fig. 2, the q^* can be expressed as:

$$q^* = q_t + (q_s - \vartheta)R \tag{4}$$

Given stable seeker tracking, ε is generally a small parameter. So, it can believe approximately that $q_s \approx q_t$. Then, (4) can be rewritten as:

$$q^* \cong q_t + (q_t - \vartheta)R = q_t(1 + R) - \vartheta R \tag{5}$$

Generally speaking, $R \ll 1$. Hence, (5) can be simplified into:

$$q^* = q_t - \vartheta R \tag{6}$$

It can be seen from (6) that due to the existence of radome error, missile attitude will be fed back to the seeker, generating a measurement error of q_t. Consequently, this measurement error is delivered from radar seeker to the noise filter and then to the flight control system, followed by a wrong guidance command being sent. The missile responds to wrong command, which causes extra missile attitude and thereby produces new radome error. In this way, a radome error parasitic loop is formed in the guidance loop. The radome error parasitic loop model based on ETSG is shown in Fig. 3.

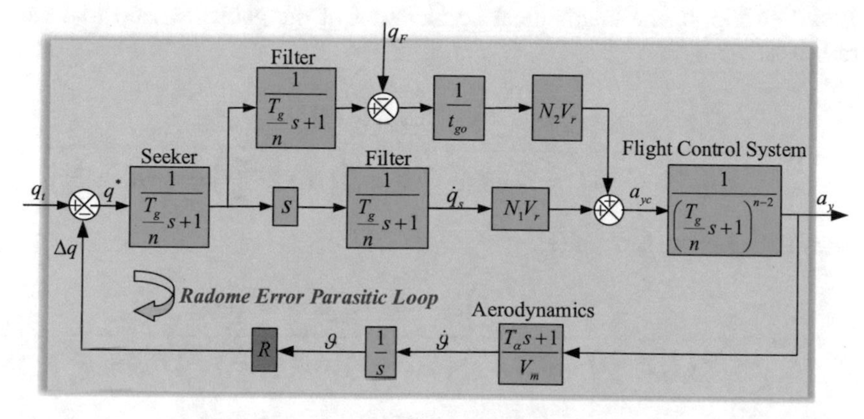

Fig. 3. The radome error parasitic loop model

In Fig. 3, T_g is guidance time constant, T_α is missile turning rate time constant, V_m is missile velocity, and a_y is missile longitudinal acceleration. For the purpose of universality, total dynamic order of the seeker, noise filter and flight control system is set to n. Most studies take $n \leq 5$.

It can be seen from Fig. 3 that whether the parasitic loop is positive feedback or negative feedback depends on R. The radome error parasitic loop generates negative feedback when $R > 0$, while it generates positive feedback when $R < 0$. According to definition of radome slope, its sign is uncertain, indicating that both positive and negative feedbacks of radome error parasitic loop exist during the flight of missile.

3 Effects of Radome Slope on the Stability of ETSG

In engineering, it can be assumed that the guidance system is a slow time-varying system, and the stability of the system is approximately determined by the Routh-Hurwitz criterion, which was used to analyze the system stability in this paper. The guidance loop model with radome error parasitic loop is shown in Fig. 4.

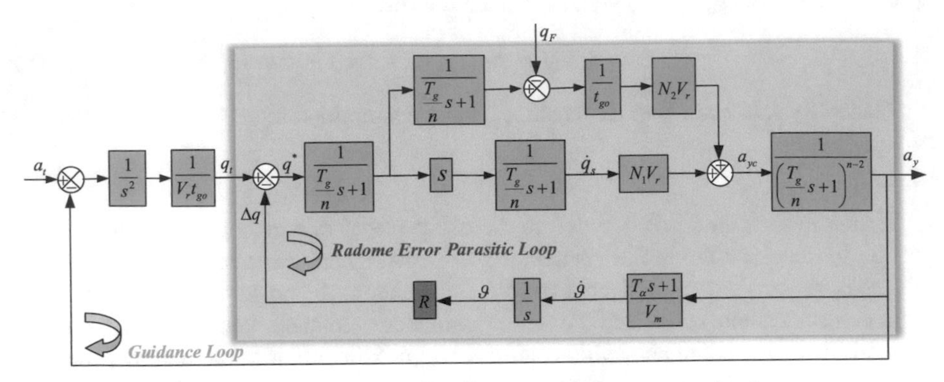

Fig. 4. The guidance loop model with radome error parasitic loop

Based on Fig. 4, the longitudinal acceleration of the guidance loop (a_y) can be expressed as:

$$a_y(s) = \frac{\left(N_1 s + \frac{N_2}{t_{go}}\right)a_t - N_2 V_r \left(\frac{T_g}{n}s + 1\right)^2 s^2 q_F}{s^2 t_{go}\left(\frac{T_g}{n}s + 1\right)^n + s t_{go} R \frac{V_r}{V_m}(T_\alpha s + 1)\left(N_1 s + \frac{N_2}{t_{go}}\right) + \left(N_1 s + \frac{N_2}{t_{go}}\right)} \tag{7}$$

The dimensionless form of (7) is:

$$a_y(\bar{s}) = \frac{\left(N_1 \bar{s} + \frac{N_2}{\bar{t}_{go}}\right)a_t - N_2 V_r \left(\frac{1}{n}\bar{s} + 1\right)^2 \frac{\bar{s}^2}{T_g} q_F}{\bar{s}^2 \bar{t}_{go}\left(\frac{1}{n}\bar{s} + 1\right)^n + \bar{s}\bar{t}_{go} R \frac{V_r}{V_m}(\bar{T}_\alpha \bar{s} + 1)\left(N_1 \bar{s} + \frac{N_2}{\bar{t}_{go}}\right) + \left(N_1 \bar{s} + \frac{N_2}{\bar{t}_{go}}\right)} \tag{8}$$

It can be seen from (8) that stability of ETSG with radome error parasitic loop is unrelated to the input. The characteristic equation of (8) is:

$$F(\bar{s}) = \bar{s}^2 \bar{t}_{go}\left(\frac{1}{n}\bar{s} + 1\right)^n + \bar{s}\bar{t}_{go} R \frac{V_r}{V_m}(\bar{T}_\alpha \bar{s} + 1)\left(N_1 \bar{s} + \frac{N_2}{\bar{t}_{go}}\right) + \left(N_1 \bar{s} + \frac{N_2}{\bar{t}_{go}}\right) \tag{9}$$

It determines the stability of ETSG system. Equation (9) reveals that stability of ETSG is determined by $\bar{t}_{go}$, $\bar{T}_\alpha$, R, n and n^* together. Given fixed $\bar{t}_{go}$, effects of radome slope on the stability of ETSG can be concluded according to the Routh-Hurwitz criterion.

The relation curves between stable region of ETSG with radome error parasitic loop and $\bar{T}_\alpha$ and R when $n = 4$ or 5 are shown in Fig. 5.

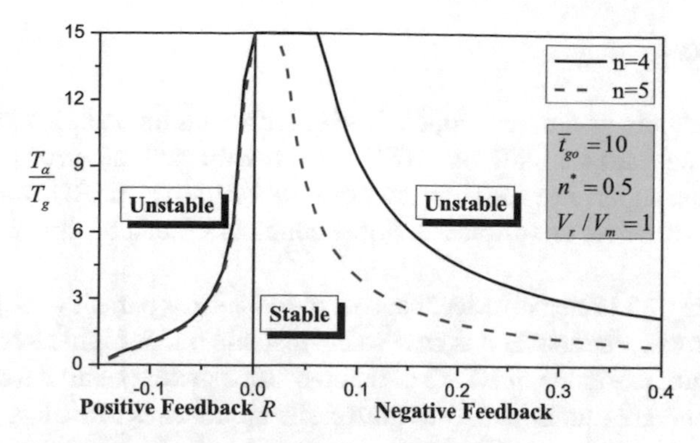

Fig. 5. Stable region of ETSG under different system order

It can be seen from Fig. 5 that when $R > 0$ (negative feedback), the stable region of ETSG is larger than that when $R < 0$ (positive feedback). The stable region decreases with the increase of $\overline{T}_\alpha$. When $R > 0$, the stable region of ETSG is easy to be unstable under higher system order. When $R < 0$, the stable region of ETSG is basically unaffected by n.

The relation curves between stable region of ETSG with radome error parasitic loop and $\overline{T}_\alpha$ and R when $\bar{t}_{go} = 5$, $\bar{t}_{go} = 10$ and $\bar{t}_{go} = \infty$ are shown in Fig. 6. The stable region of ETSG system when $\bar{t}_{go} = \infty$ is the stable region of radome error parasitic loop.

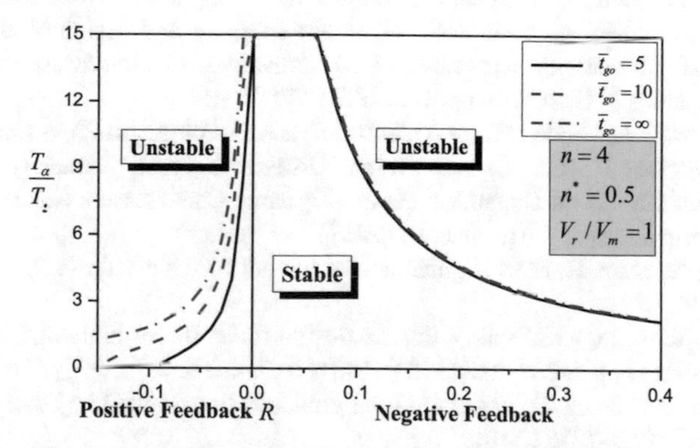

Fig. 6. Stable region of ETSG under different time-to-go

Figure 6 shows that when $R < 0$, the stable region of ETSG is positively correlated with dimensionless $\bar{t}_{go}$. In other words, the maximum tolerance of ETSG to radome slope under positive feedback of radome error parasitic loop decreases continuously as the missile approaches to the target, thus making the guidance loop lose stability before the parasitic loop. This puts forward higher requirements on controlling radome slope. When $R > 0$, the stable region of ETSG remains basically the same with the change of $\bar{t}_{go}$.

4 Conclusion

To investigate effects of radome slope of radar seeker on the stability of the extended trajectory shaping guidance law, an ETSG system with radome error parasitic loop is established in this paper. The variation laws of stable region of ETSG with radome slope and guidance parameters are disclosed. Some conclusions can be drawn as following:

1. The stable region in negative feedback of radome error parasitic loop is larger than that in positive feedback, and is easy to be unstable under higher system order.
2. The maximum tolerance of ETSG to radome slope under positive feedback of parasitic loop decreases continuously as the missile approaches to the target, thus making the guidance loop lose stability before the parasitic loop.

These conclusions can provide theoretical references for radome slope index constraint and guidance system design of different radar homing missiles.

References

1. Zarchan, P.: Tactical and Strategic Missile Guidance. American Institute of Aeronautics and Astronautics (2012)
2. Siouris, G.M.: Aerodynamic Forces and Coefficients. Missile Guidance and Control Systems, pp. 53–154 (2004)
3. Arhip, M., Gavrila, G.: The influence of radome on radar antennas system. In: Communications (COMM). In: 2012 9th International Conference on 2012, pp. 151–154. IEEE (2012)
4. Moreno, J., Fernandez, M., Somolinos, A., et al.: Analysis and design of antenna radomes. In: 2013 IEEE International Conference on Microwaves, Communications, Antennas and Electronics Systems (COMCAS), pp. 1–5. IEEE (2013)
5. Garnell, P., East, D.J.: Guided Weapon Control Systems. Pergamon Press New York (1980)
6. Nesline, F., Zarchan, P.: Radome induced miss distance in aerodynamically controlled homing missiles. In: 17th Fluid Dynamics, Plasma Dynamics, and Lasers Conference. American Institute of Aeronautics and Astronautics (1984)
7. Nesline, F.W., Zarchan, P.: Missile guidance for low-altitude air defense. J. Guidance Control **2**, 283–289 (1979)
8. Nesline, F.W., Zarchan, P.: Missile guidance design tradeoffs for high-altitude air defense. J. Guidance Control Dyn. **6**, 207–212 (1983). ISSN 0731-5090
9. Wang, H., Lin, D., Cheng, Z., et al.: Optimal guidance of extended trajectory shaping. Chin. J. Aeronaut. **27**, 1259–1272 (2014)

Design of Multi-target Tracking Algorithm Based on Multi-sensor Fusion

Zhangchi Song[1,2(✉)], Xiaomin Qiang[1,2], Peng Sun[1,2], Laitian Cao[1,2], Lijia Zhang[1,2], and Baisheng Yang[1,2]

[1] Beijing Aerospace Automatic Control Institute, Beijing 100854, China
984358097@qq.com

[2] National Key Laboratory of Science and Technology on Aerospace Intelligent Control, Beijing 100854, China

Abstract. In the modern battlefield environment, the traditional single sensor often cannot give full play to its optimal performance. The advantages of different sensors are used for combined detection and tracking to obtain all-round multi-dimensional information of the target. Through multi-sensor fusion processing, interference is eliminated, and the target state estimation is obtained to achieve precise positioning of the target.

Keywords: Multi-sensor fusion · Multi-target tracking · Track management

1 Introduction

With the development of modern science and technology, the environment of modern warfare is becoming more and more complicated. The precision strikes of missile weapons are faced with complex natural environment interference, electronic countermeasures and optoelectronic countermeasures. In the new generation of combat systems, relying on a single sensor to provide information cannot meet combat needs. Multiple sensors must be used for coordinated detection and identification, including microwave, millimeter wave, television, infrared, laser, electronic support measures (ESM), and electronic intelligence. Multi-sensor integration including various active and passive detectors covering wide frequency bands such as ELINT to provide a variety of observation data, using the advantages of different sensors for combined detection and tracking, to obtain all-round multi-dimensional information of the target, Through multi-sensor fusion processing, interference is eliminated, target state estimation is obtained, and precise target positioning is achieved.

2 Multi-target Tracking Algorithm Based on Multi-sensor Fusion

2.1 Multi-sensor Fusion System Model

In real application scenarios, a single sensor is limited by a complex environment, and often cannot exert its optimal performance. By fusing the measurement information of

Q. Liang et al. (Eds.): Artificial Intelligence in China, LNEE 854, pp. 691–697, 2022.
https://doi.org/10.1007/978-981-16-9423-3_87

multiple sensors, the performance is complemented. For example, the typical sensor active radar and passive radar, through information fusion, can not only give full play to the active radar homing, have range information, not be affected by the control of the target radiation source, and have the performance advantages of all-weather detection, but also can give play to the covert detection of passive radar homing. The advantages of being free from enemy electronic interference and long operating distance [1].

In the multi-sensor fusion system, according to the structure and function of the system, the whole is divided into a centralized fusion system and a distributed fusion system. The centralized fusion system realizes the use of all sensor measurement information, but the communication overhead is relatively high, and the data is interconnected. It is more difficult, and requires the system to have large-capacity processing capabilities, and the computational burden is heavy. In this paper, the storage capacity required by the distributed fusion system is smaller and the fusion speed is faster. This article uses a distributed fusion system [2, 3].

In the distributed fusion system, the local tracking system first completes the point-track tracking, prediction and track association of the observation data to form their respective local trajectories, and then transmits the processing results to the fusion center for fusion to generate accurate system navigation trace. Since the fusion center is no longer the original data, the distributed system can only achieve locally optimal or global sub-optimal fusion performance, but the burden on the fusion center is greatly reduced [4]. This type of system not only has local independent tracking capabilities, but also has global monitoring and evaluation features.

The distributed fusion system is shown in Fig. 1 below.

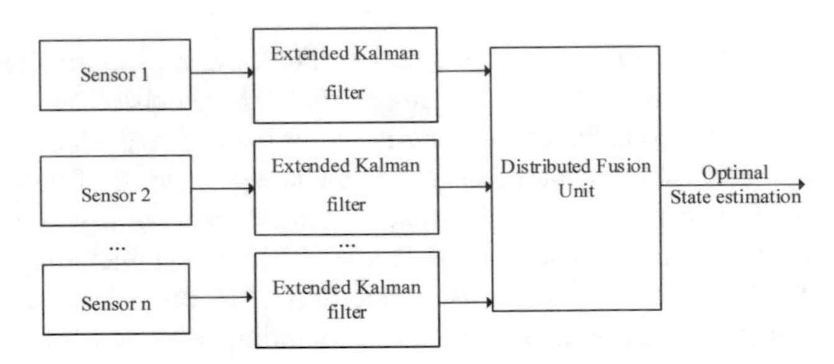

Fig. 1. Distributed fusion system

2.2 Track Management Algorithm for Multi-target Tracking

Due to the complex external environment and the limitations of the sensor itself, the measurement information sources include: tracked targets, false alarms and other targets, and there are problems such as crossover between target tracks. Therefore, the initiation, confirmation, association, and cancellation of the target track have become key issues in the application of target tracking engineering [5].

The steps of the multi-target tracking track management algorithm used in this paper are as follows:

(1) Potential track generation: tentatively search for possible target trajectories under the condition of uncertain whether there is a target in the detection scene. Here, gates are used to filter the measurement of two adjacent time frames. If there are two adjacent time frames the measured data is in the wave gate, then the track formed by the two measurements can be used as the potential track.
(2) Track initiation: using M/N logic method to identify possible tracks through prediction and related gates through multiple hypotheses to achieve track initiation.
(3) Track correlation: The empirical JPDA algorithm is used to realize the track probability interconnection. The potential track is generated from the measurement with the correlation probability of adjacent data frames higher than the predetermined gate, and the track parameters are initialized according to the selected adjacent frame measurement values. The candidate echo (the output of the tracking gate rule) is compared with the known target track, and finally the measurement with the highest correlation probability is selected to pair with the track. For the measurement j and the track t, the probability of interconnection:

$$\beta_{jt} = \frac{\mathbf{G}_{jt}}{\mathbf{S}_t + \mathbf{S}_j - \mathbf{G}_{jt} + \mathbf{B}} \quad (1)$$

$$\mathbf{N}_{jt}[\mathbf{v}_j(k)] = \frac{1}{2\pi|\mathbf{S}(k)|^{1/2}} \exp\{-\frac{1}{2}\mathbf{v}_j'(k)\mathbf{S}^{-1}(k)\mathbf{v}_j(k)\} \quad (2)$$

$$\mathbf{G}_{jt} = \mathbf{N}_{jt}[\mathbf{v}_j(k)], \quad \mathbf{S}_t = \sum_{j=1}^{m} \mathbf{G}_{jt}, \quad \mathbf{S}_j = \sum_{t=1}^{T} \mathbf{G}_{jt} \quad (3)$$

In the formula: B is a clutter density related constant, $\mathbf{G}_{jt}$ is the effective likelihood function of the interconnection between the measurement j and the target track i, $\mathbf{S}_t$ is the sum of all $\mathbf{G}_{jt}$ of a certain target, $\mathbf{S}_j$ is the sum of all $\mathbf{G}_{jt}$ for a certain measurement, **v** is innovation, **S** is the covariance of innovation [1].
(4) Track cancellation: In this paper, the sequence probability ratio test algorithm is used to realize the track termination. The sequence probability ratio test algorithm determines whether the track is terminated or not by calculating test statistical variables and comparing them with the cancellation threshold [6].

$$a_1 = \ln \frac{P_D/(1 - P_D)}{P_F/(1 - P_F)} \quad (4)$$

$$a_2 = \ln \frac{1 - P_D}{1 - P_F} \quad (5)$$

The measurement equation is as follows: In the formula, P_D and P_F are the track detection probability and false alarm probability.

$$ST(k) = ma_1 \quad (6)$$

In the formula, m is the number of detections. Define the track cancellation threshold at time k as:

$$T_c(k) = \ln c1 + ka2 \quad (7)$$

Where $c_1 = \beta/(1-\alpha)$, α and β are the predetermined allowable error probability, α is the probability of missed withdrawal (when the track should be cancelled but the judgement of the track is not cancelled), β is the probability of false withdrawal (when there is a real track but it is judged as the track withdrawn) probability. In this way, the tracking termination decision logic can be expressed as follows:

$$\textit{Cancel the track}: ST(k) < T_c(k) \tag{8}$$

$$\textit{Keep the track}: ST(k) \geq T_c(k) \tag{9}$$

That is, at time k, if the gate of a certain track falls into the measurement point trace, the statistic $ST(k)$ increases by a_1. If there is no measurement value in the gate of the track, $ST(k)$ remains unchanged, and the track cancellation threshold $T_c(k)$ increases by a_2 every time. When the statistic $ST(k)$ is lower than the threshold $T_c(k)$, the algorithm decides that the track is cancelled.

2.3 Multi-target Tracking Algorithm Based on Multi-sensor Fusion

A multi-target tracking algorithm based on multi-sensor fusion is designed, as shown in Fig. 2.

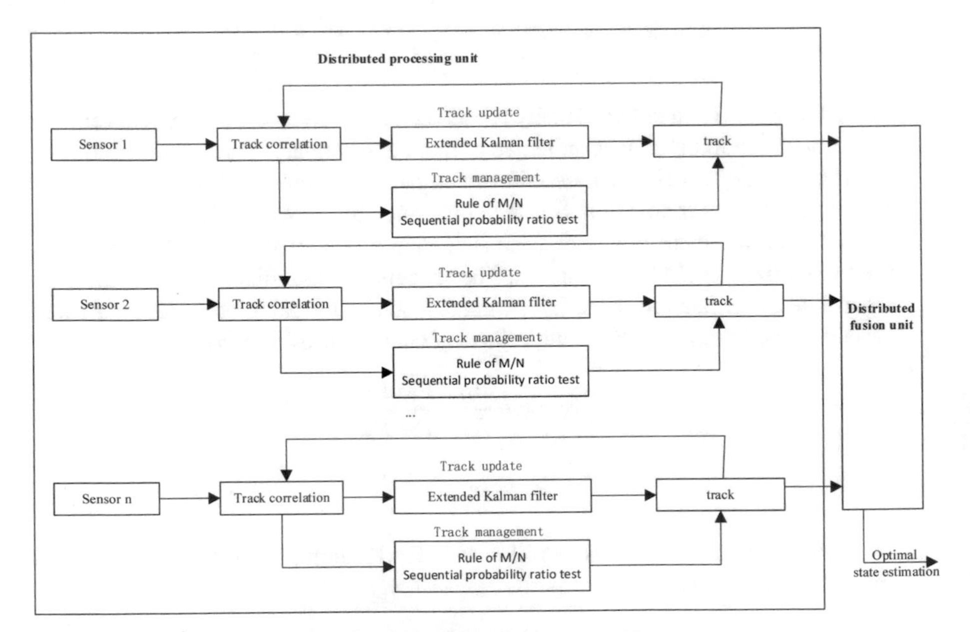

Fig. 2. Distributed Fusion algorithm

The steps of multi-target tracking algorithm based on multi-sensor fusion are as follows:

(1) Use the measured values output by the sensors in two adjacent frames to generate potential tracks;
(2) Use the track correlation algorithm described in Sect. 2.2 to perform track correlation calculations between the sensor output measurement in current frame and the existing track, and perform track start and track termination operations;
(3) Apply the extended Kalman filter algorithm to update the track;
(4) Finally, the track tracked by each distributed processing unit is sent to the distributed fusion unit, and the optimal state estimation is output.

3 Simulation Results

The multi-target tracking algorithm based on multi-sensor fusion is simulated, and the simulation scenarios are as follows: There are 3 sensors involved in the fusion. The scene contains 3 targets, and all targets move in a straight line at a uniform speed. The initial state is:

$$\begin{cases} x_1 = [x_1, \dot{x}_1, y_1, \dot{y}_1, z_1, \dot{z}_1] = [0\,\text{m}, 20\,\text{m/s}, 0\,\text{m}, 0\,\text{m/s}, 0\,\text{m}, \text{m/s}] \\ x_2 = [x_2, \dot{x}_2, y_2, \dot{y}_2, z_2, \dot{z}_2] = [50\,\text{m}, 10\,\text{m/s}, 0\,\text{m}, 0\,\text{m/s}, 100\,\text{m}, 0\,\text{m/s}] \\ x_3 = [x_3, \dot{x}_3, y_3, \dot{y}_3, z_3, \dot{z}_3] = [0\,\text{m}, 10\,\text{m/s}, \text{m}, 0\,\text{m/s}, -100\,\text{m}, 0\,\text{m/s}] \end{cases} \tag{10}$$

The sampling interval is T = 0.02 s. The probability of detection, the clutter density is. The number of Monte Carlo simulations is 500 times.

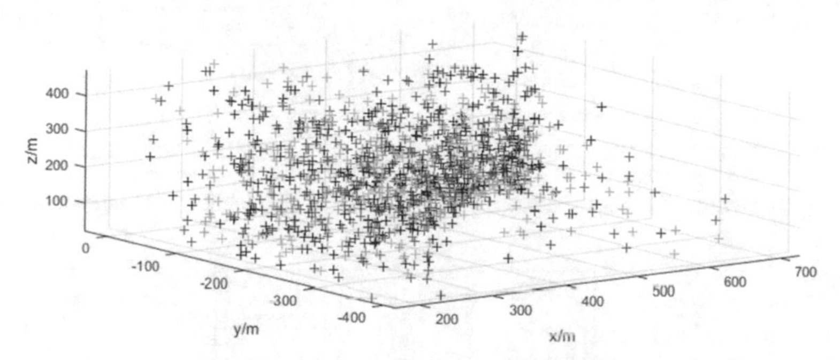

Fig. 3. Multi-target tracking scene diagram

Figure 3 above shows the multi-target tracking scene diagram of a random Monte Carlo test, and Fig. 4 shows the multi-target tracking track output after the track management processing and 3 sensor fusion filtering. It can be seen from the figure that through the track management processing, the false target information in the environment is eliminated, and the stable tracking of multiple targets is achieved.

Figure 5 indicates the multi-target tracking accuracy in the X, Y and Z directions. It can be seen that the three target tracks in the observation environment have better convergence speed and convergence accuracy.

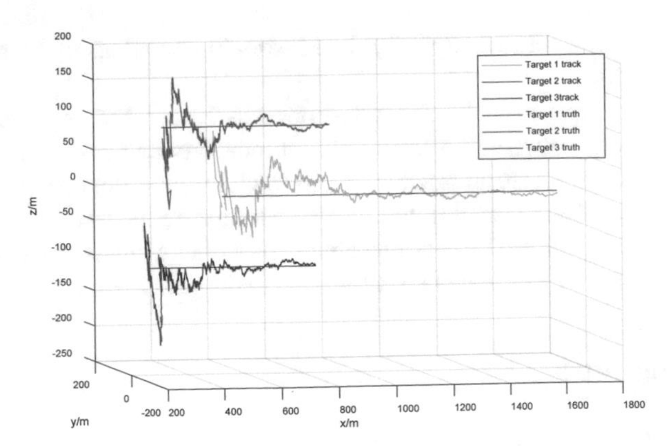

Fig. 4. Multi-target tracking track diagram

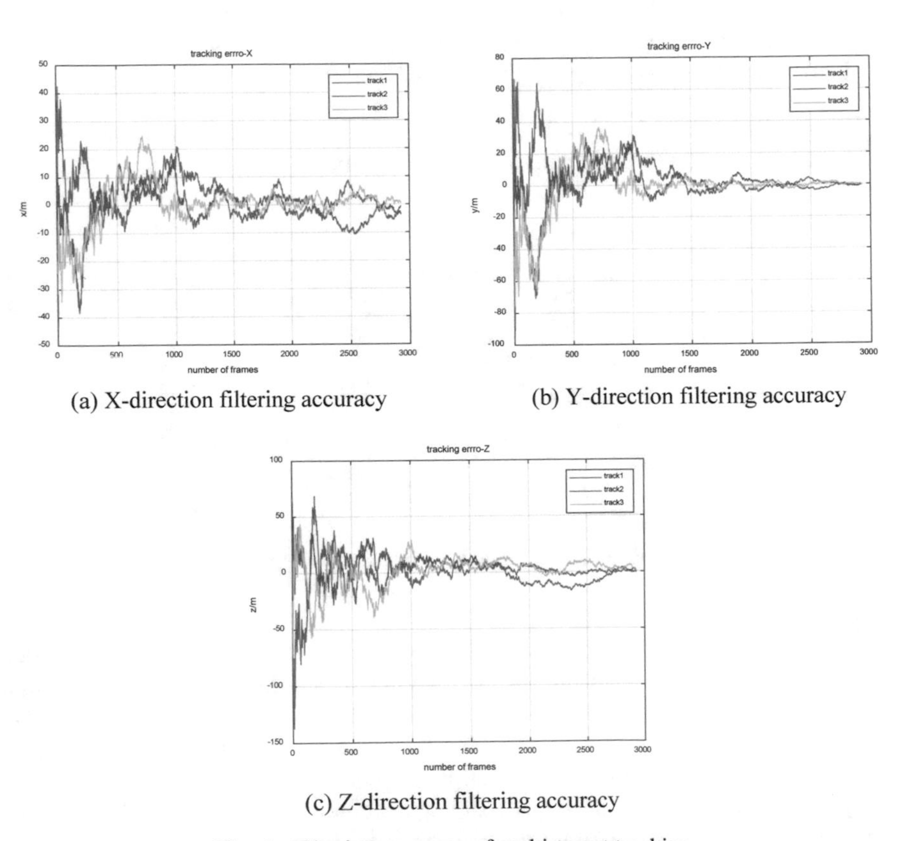

(a) X-direction filtering accuracy

(b) Y-direction filtering accuracy

(c) Z-direction filtering accuracy

Fig. 5. Filtering accuracy of multi-target tracking

4 Conclusion

In this paper, a Multi-target tracking algorithm based on multi-sensor fusion is designed for the limitation of single sensor performance in complex interference environment. Through information fusion and track management, the algorithm makes full use of the advantages of different sensors to achieve stable target tracking in complex background environment.

References

1. He, Y., Wang, G.: Radar Data Processing with Application. Publishing House of Electronics Industry (2000)
2. Kaity, S., Jana, B., Gupta, P., et al.: Multi-sensor Tracking Simulator Design and Its Challenges (2021)
3. Ehrman, L.M.: Multitarget, Multisensor Tracking. IET Digital Library (2012)
4. Wang, Z.: Research on Multi-sensor Cooperative Target Tracking Method. Xidian University (2011)
5. Zhu, J., Meng, X.: Track correlation technology and application of multi-sensor target tracking. J. Electron. Measur. Instrum. **17**(2), 51–55 (2003)
6. Zhu, J.: Research on multi-objective Fast tracking data association Algorithm. Harbin Engineering University (2019)

Edge-Cloud Synergy Based Construction of Multi-task and Lifelong Learning Knowledge

Hailong Zhang[1(✉)], Hanwu Luo[2], Wenzhen Li[2], and Ruofeng Qin[2]

[1] Wuhan NARI Limited Liability Company, State Grid Electric Power Research Institute, Wuhan, People's Republic of China
8278799@qq.com

[2] East Inner Mongolia Electric Power Co., Ltd., Ordos, People's Republic of China

Abstract. Deep learning based algorithms have achieved good results in many applications. However, most of the traditional machine learning methods appear to have catastrophic oblivion problem. The reason for this is that the model can't persist the knowledge it learned so it can not recognize the difference between the knowledge and the old one. Thus, constructing knowledge base become a common option to solve this problem. Nevertheless, the previous models still face huge challenges of small samples, uneven distribution and oblivion. In this paper, we come up with a new model of multi-task and lifelong learning knowledge base which is based on edge-cloud synergy. The experiment shows our model can handle the problems above better than the previous work.

Keywords: Lifelong learning · Knowledge · Edge-cloud synergy · Multi-task

1 Introduction

1.1 Background

With the developing of deep learning methods, we have achieved great work in many computer vision tasks. However, when it's applied to edge-cloud synergy, it appears to have many problems such as catastrophic oblivion. The knowledge it learned can't be persistent so the new model will replace the old one completely and the differences between them will be ignored [1, 2]. As a result, the persistence of knowledge becomes crucial.

The knowledge of the deep learning procedure obtained is mostly contained in its model and the training of model usually cost a lot of time. So, the opening and sharing of models become a common sense in industry. The existing knowledge base is mostly retrieved by the key words of the model and then picked out manually. Since every model in the knowledge base has its own application scenarios while these information is not provided when retrieving, it's pretty hard to pick up a proper model for a new task in a new environment.

In the same time, the new project often faces a severe lack of samples [1]. It's hard to use the traditional manual transfer learning method to transfer the model and thus leads to a large predict error.

Q. Liang et al. (Eds.): Artificial Intelligence in China, LNEE 854, pp. 698–703, 2022.
https://doi.org/10.1007/978-981-16-9423-3_88

And the edge devices are also short of computation and storage resources. So we need to obtain the knowledge from the base directly or indirectly.

1.2 Previous Work

The knowledge base is a rule set which contains the facts and data that the rules connect [1]. The existing model knowledge base like MLFlow and TensorFlow have an architecture like Fig. 1. Users can search the models by its name and control the version of these models [3, 4]. But this kind of search can't get the meta-data (e.g. the application scenarios). The knowledge base can't store and update the samples and it's hard to merge the models when doing incremental maintenance.

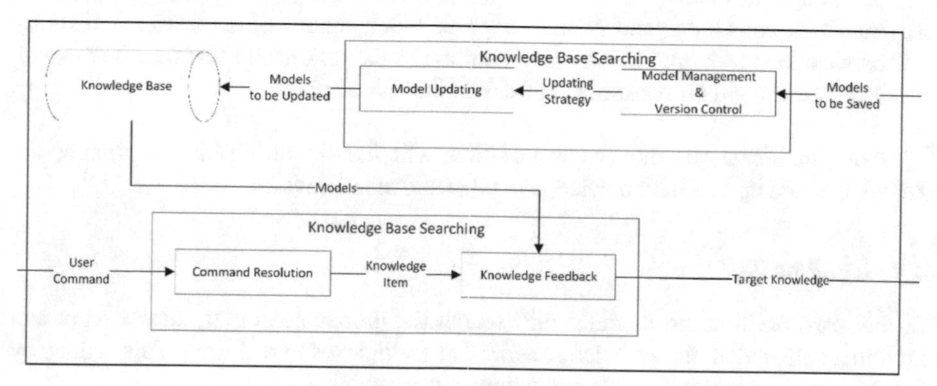

Fig. 1. Basic architecture of MLFlow and TensorFlow

1.3 Our Work

In this paper, we first come up with a basic architecture of multi-task knowledge base search engine in the edge-cloud synergy environment, which, contains 4 modules.

Based on our architecture, we also propose 4 key algorithms to search in and maintenance the knowledge base, which are multi-task knowledge and index extraction, multi-task search and sort, multi-task knowledge updating decision-making and multi-task index updating.

2 Construction Algorithm of Multi-task and Lifelong Learning Knowledge

2.1 Overview of Algorithm

In this section, we'll introduce our algorithm used in the knowledge base. The algorithms can be divided to 2 major part: the initialization part and the runtime part.

In the initialization part, we'll introduce the initialization method of multi-task knowledge and the synchronization method of edge-cloud knowledge.

In the runtime part, we'll introduce the searching algorithm of our knowledge base and the updating methods used to renew the knowledge and indexes in the base.

2.2 Initialization Part

In this part, the knowledge base will initialize itself by means of multi-task knowledge and the corresponding extractor. After initialization, the extractor and the indexes will be downloaded and stored by edge devices using our synchronization module.

The initializer mainly works on 3 blocks:

1) Inner-task knowledge and its indexes. This block contains a sample sets and its attributes. The indexes will be constructed based on the task attributes.
2) Task knowledge and its indexes. This block contains multiple tasks. Each task includes its samples, models and attributes. The indexes will be constructed based on the task attributes.
3) Inter-task knowledge and its indexes. This block contains the transfer probability between one task and the others in addition to the task group that task belongs to. The indexes will be constructed based on the task attribute.

After initialization, the synchronization will handle the collisions among the knowledge and update the knowledge in edge and cloud devices.

2.3 Runtime Part

In this part, the runtime modules will search the proper models by user's input and incrementally update the knowledge as well as the indexes in real time. This is done by 2 submodule: searching module and maintenance module.

2.3.1 Searching Module

The searching module is designed to search for knowledge depending on runtime data and the target type of knowledge that desired. This module use hierarchical extractor to gain the knowledge and the index from different layers. The resolver, extractor and sorter can work together to accomplish this task.

To achieve this, we first need to resolve the input command. The input command can be something like a dictionary below (Table 1):

Table 1. Example of input command

Command ID	User	User address	Knowledge type	Runtime data ID
01	User A	192.162.x.x	Task	03

The output of resolution will be a list of triples like {Attribute, Model, Samples}. If the search engine can't find a relative result, it'll push the input into candidate knowledge cache as data to be filled [6].

The extraction of output is done by an index extractor. The extractor can obtain inner-task knowledge, task knowledge and inter-task knowledge.

For inner-task knowledge, the goal is to get its samples, models and indexes; For task knowledge, our mission is to obtain its attributes, models and samples; For inter-task knowledge, the extractor will first get the task index and then obtain the relationship between the target task and the others by asking the sorter.

In order to find out the relationship between tasks, one possible way is to calculate the similarity between 2 tasks. There're many possible factors that can cause a difference between 2 tasks such as their training samples, the models and the application scenarios [5]. Here, we propose an algorithm based on decision tree [3]. The main idea is to construct a decision tree for every task and then compare the tree of tasks instead of the task itself. The similarity of trees reveals the similarity of tasks.

For every task, we first use the dataset of the task to predict some results and compare to its ground truth. According to the comparison result (True or False), the tree is divided to 2 branch. The comparison is done based on a set of division rules and some pre-defined parameters which makes the "purity" (such as entropy or gini) of the nodes as higher as possible (Fig. 2).

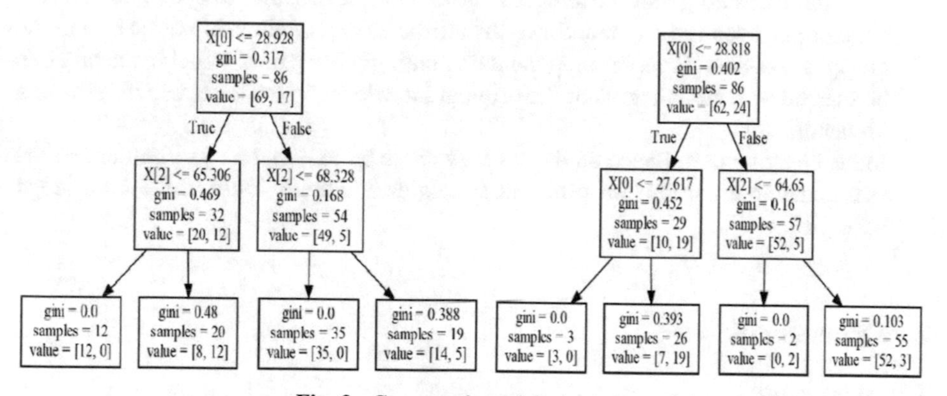

Fig. 2. Construction of decision tree

The similarity algorithm of 2 trees is described below:

Algorithm 1. Calculate similarity of trees

Input:	2 decision trees – clf1 , clf2
Output:	real number N
1	- Get the division rule set of every leaf nodes
2	- Calculate the segment points of every dimension
3	- For every area A corresponding to the segment point:
4	Calculate the confidence coefficient of clf1 and clf2
5	Calculate the distance D of clf1 and clf2 in A
6	- Output D / Na , where Na is the total area count

After obtaining the similarity between tasks, we can output a relevant list that contains the relative knowledge in descending order by the similarity.

2.3.2 Maintenance Module

This module mainly focus on merging the candidate knowledge in the candidate cache with the annotated knowledge to form the complete knowledge [8]. Also it'll determine to use some strategies to hierarchically update the knowledge and indexes in the base. Here in the module there're 4 sub parts to accomplish the missions above.

1) Async merging part. Sometimes a few samples will lack of annotations so they can't be directly pushed to our base. The async merging part will put these samples into cache and wait until the notations arrive [6].
2) Policy determination part. In this part, the submodule will give an update policy depending on the candidate knowledge. The updating strategy is mainly measured by the differences between the candidate knowledge and the existed knowledge in the base. Specifically, we measured their similarity using the formula in Sect. 2.3.1, their confidence coefficient, their amount of samples (the more samples it has the higher c-c it should have and their stability when tested on different datasets.

   After measuring their differences, we can decide whether to update the knowledge or just keep the old one. Since the attributes and relationship of tasks and the group of tasks itself support incremental updating [6], the new model in the task can be merged or accumulated into the original knowledge in the base, which leads to a higher efficiency.
3) Index updating part. Based on the strategy made by part 2), this part implement the real updating algorithm for different strategies and apply them to the knowledge base [5].

3 Experiment

3.1 Background

The data used in experiment come from the structure data of coal blending optimization. Our goal is to predict the CSR (Coke Strength after Reaction). As time goes by, we have observed that the precision for new samples will gradually goes down so we need to increment the existing knowledge in order to fit the new scenery.

3.2 Conditions and Index

We first construct the initial knowledge base using 86 samples. Then we randomly pick up 34 samples and let them arrive in order as the candidate incremental samples. When there is a new sample arrived, we test our searching algorithm using 53 samples. The index are error_rate and pccs listed below. Suppose that the ground truth of the i-th sample is y_{test-i}, and its prediction is $y_{prediction-i}$. The vector $\overrightarrow{t}$ and $\overrightarrow{p}$ refer to the set of GT and the set of prediction. They can be calculated by the formula below:

$$\overrightarrow{t} = (y_{test-1}, y_{test-2} \cdots y_{test-n}) \tag{1}$$

$$\overrightarrow{t} = (y_{test-1}, y_{test-2} \cdots y_{test-n}) \tag{2}$$

$$\text{error rate} = \frac{100}{n}\sum_{i=1}^{n} |(y_{test-i} - y_{prediction-i})|/y_{test-i} \tag{3}$$

$$\text{pccs} = \frac{E[(\vec{t} - \mu_{\vec{t}})(\vec{p} - \mu_{\vec{p}})]}{\sigma_{\vec{t}}\sigma_{\vec{p}}} \tag{4}$$

4 Conclusion

In this paper, we propose a new knowledge base architecture to solve a series of problems while distributing machine learning models in scenarios of edge-cloud synergy.

The experiment shows that our knowledge base can search the knowledge in a high efficiency and achieve pretty high precision of prediction. Meanwhile, the knowledge base can accumulate the samples the further lower reduce the error rate in runtime, which may contribute to the problems of machine learning in edge devices [8].

In the future, we may focus on the problems of privacy protection. Such problems can occur when there're lots of renters using our knowledge base. And these problems can appear in company of federation learning and distributed lifelong learning, which leads to a more complicated systematic challenge waiting to be overcome.

Acknowledgments. This work was Funded by the State Grid Science and Technology Project (Rese arch on Key Technologies of Intelligent Image Preprocessing and Visual Perception of Transmission and Transformation Equipment).

References

1. Loster, M.: Knowledge base construction with machine learning methods. Doctoral thesis of University Potsdam, 13 April 2021
2. Jiang, Z., Chi, C., Zhan, Y.: Research on medical question answering system based on knowledge graph. IEEE Access **9**, 21094–21101 (2021)
3. Zhang, H.: Neural network-based tree translation for knowledge base construction. IEEE Access **9**, 38706–38717 (2021). https://doi.org/10.1109/ACCESS.2021.3063234
4. Hamdi, G., Omri, M.N., Benferhat, S., Bouraoui, Z., Papini, O.: Query answering DL-lite knowledge bases from hidden datasets. Ann. Math. Artif. Intell. **89**, 1–29 (2020). https://doi.org/10.1007/s10472-020-09714-2
5. Chen, Y., Li, C., Gong, L., et al.: A deep neural network compression algorithm based on knowledge transfer for edge devices. Comput. Commun. **163**, 186–197 (2020)
6. Li, G., Dong, M., Yang, L.T., et al.: Preserving edge knowledge sharing among IoT services: a blockchain-based approach. IEEE Trans. Emerg. Top. Comput. Intell. **4**(5), 653–665 (2020). https://doi.org/10.1109/TETCI.2019.2952587
7. Hai, N., Gong, D., Liu, S.: Ontology knowledge base combined with Bayesian networks for integrated corridor risk warning. Comput. Commun. **174**(1), 190–204 (2021)
8. Qiu, G., Zhang, S.-H.: Research on data optimization method of software knowledge base operation and maintenance based on cloud computing. In: Liu, S., Xia, L. (eds.) ADHIP 2020. LNICSSITE, vol. 347, pp. 229–238. Springer, Cham (2021). https://doi.org/10.1007/978-3-030-67871-5_21

A Federated Learning System for Target Recognition over 5G MEC Networks

Haiyang Wang[1], Mingyu Zhang[2(✉)], Minggang Liu[1], Jiangtao Hong[2], Bo Zhang[1], and Zhilong Zhang[2]

[1] Shandong Electric Power Engineering Consulting Institute Co., Ltd., Jinan 50100, Shandong, China

[2] School of Information and Communication Engineering, Beijing University of Posts and Telecommunications, Beijing 100876, China

mingyu_zhang@bupt.edu.cn

Abstract. In the scenario of electricity substation and other construction sites, it is necessary to monitor workers' wearing of helmets through image recognition technologies for safety production. However, fast data training and target recognition usually require high communication and computation capacities. To relieve the burden of networks and data centers caused by large amount of data uploading and processing, this paper combines the 5G MEC network with the federated learning framework. An adaptive compression mode selection mechanism is proposed to efficiently transmit training models over time-variant bandwidth, and a federated learning system is designed for real-time target recognition. The accuracy of our proposed model is 6.73% higher than that of a traditional non-cooperative method. In terms of communication delay, the training delay is reduced by 24.5% compared with centralized processing method.

Keywords: Target recognition · 5G network · Federated learning · Adaptive lossless compression

1 Introduction

In recent years, with the development of wireless networks and the increase of smart devices, artificial intelligence (AI) technologies have been widely used to replace manual tedious and dangerous construction operations. Image recognition, as one of the most popular AI technologies, has gained considerable attentions due to its advantages in quick deployment and high performance. However, image recognition task usually requires high-cost data acquisition and marking, which may cause a series of problems in data processing and privacy security. For example, in the scenario of electricity substation, the amount of data collected in a single site is limited. Moreover, the collected data set usually contains sensitive information such as professional equipment. Transmitting private data across stations over public networks may lead to the risk of privacy disclosure. In this situation, a target detection model cannot be well trained by using a single data set, and may fail to meet the requirement of high recognition accuracy.

Q. Liang et al. (Eds.): Artificial Intelligence in China, LNEE 854, pp. 704–711, 2022.
https://doi.org/10.1007/978-981-16-9423-3_89

As a general distributed learning framework, federated learning (FL) is expected to overcome the above problems. At the same time, 5G mobile edge computing (MEC) enables the edge of networks with data processing capability, and provides a platform for deploying federated learning algorithms. The integration of 5G MEC and federated learning has been widely concerned. For example, Zeng R [1] proposes an incentive mechanism for MEC edge nodes with low cost to participate in the learning process, which eventually improves the performance of federated learning. Zhu Z [2] formulates an age-sensitive MEC model which combines the edge federated mode with the multi-agent actor-critic reinforcement learning. However, the abovementioned works take no consideration of non-ideal communications [3]. In addition, Liu KH [4] studies fine-grained offloading for multi-access edge computing in 5G. Ma X [5] and Zhu Z [6] propose power allocation schemes to reduce time cost. Moreover, with respect to image recognition technology, the spatial pyramid pool network (SPP-Net) proposed in [7] addresses the issue that R-CNN can only input fixed size pictures. Fast R-CNN [8] combines ROI technology with R-CNN. However, most of the existing works have not considered the influence of time-varying bandwidth of wireless channels on training speed.

Motivated by the above concern, we consider the scenario of FL over 5G MEC networks, and propose an image recognition system to monitor the wearing of personnel safety equipment in electricity substation and other construction sites. Experimental results show that the accuracy and the delay performance can be improved by adopting our proposed method.

2 System Design

To monitor the wearing of personnel safety equipment in electricity substation and other construction sites, a target recognition system is proposed based on the FL framework and 5G MEC network.

As is shown in Fig. 1, task execution is mainly divided into four processes: local model updating, local parameters compression and uploading, global model updating and global model distribution. The aim is to detect whether the workers in a construction site are wearing safety helmets. Firstly, the data set is divided into two parts and distributed to different clients for subsequent federated learning and training. Then, each client uses YOLOv5 [9] as local safety helmet detection model. The model parameters are transmitted between the server and clients through PUSCH and PDSCH in 5G networks. In the process of transmission, the compression coding level is adaptively determined according to the available bandwidth. The client updates the model parameters locally and sends them to the server. After the server collects the model parameters of all participants, it uses the federal weighted average to aggregate the model parameters, and sends the updated model back to the clients. The training process continues until the model converges or reaches a specified number of training rounds. In the following, four main modules of the task execution process will be discussed in details.

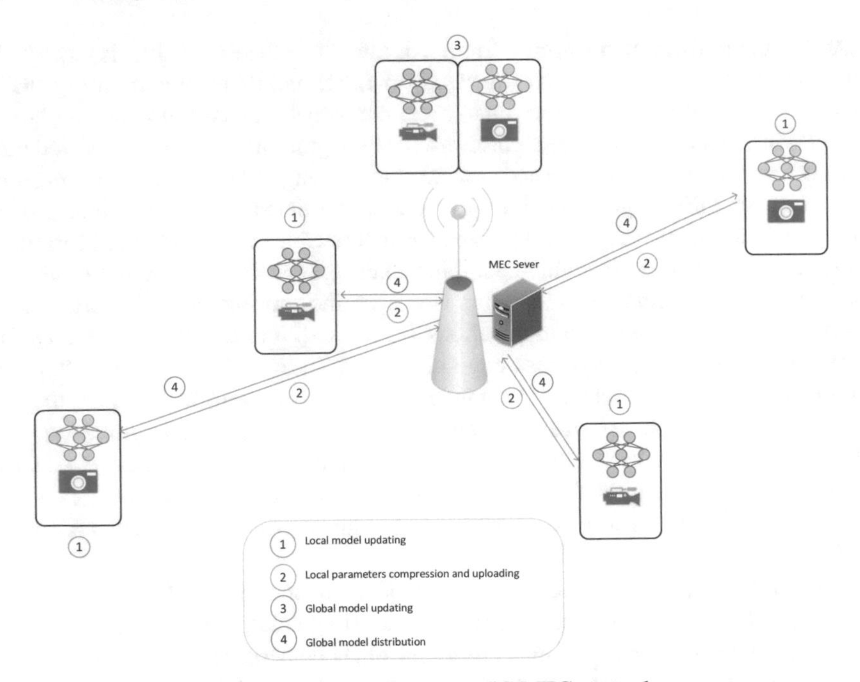

Fig. 1. Structure of FL over 5G MEC networks

2.1 Adaptive Compression Module for Data Transmission

In this paper, an adaptive lossless compression level switching mechanism is proposed. The current 5G transmission rate is detected before each round of parameter transmission, and the lossless compression coding level is selected adaptively according to the current available bandwidth. Choose the level according to the standard of the shortest time:

$$T_{all} = T(D, P) + \frac{DP}{R}, \tag{1}$$

where T_{all} is the total time cost, D represents the size of the model parameters, P denotes the compression rate, and R represents the bandwidth. $\frac{DP}{R}$ is the transmission time cost. $T(D, P)$ denotes the processing time for data compression. In this paper, Zlib module is adopted for lossless compression of model parameters. There are 4 different levels. Level 0 refers to no compression mode, and levels from 1 to 3 refer to compressed mode. In particular, level 1 has the shortest compression time and the lowest compression ratio; in level 2, both the compression time and the compression ratio are moderate; level 3 refers to the longest compression time and the highest compression ratio. Moreover, the 4 levels correspond to 4 transmission rates (i.e. level 0: original, level 1: fast, level 2: ordinary, and level 3: slow). In our experiment, the compression time, compression size and compression rate are shown in Table 1.

According to experimental results, if the network bandwidth is less than 2.64 Mbps, compression level 3 is suggested; if the transmission rate is greater than 2.64 Mbps and less than 34.72 Mbps, level 2 is selected; when the network speed is greater than

Table 1. Compression parameters

Compression level	Compression time (s)	The size after compression (MB)	Compressibility (%)
0	0	55.3	100
1	1.17	35.5	64.2
2	1.63	33.5	60.6
3	4.91	32.4	58.6

34.72 Mbps and less than 135.36 Mbps, level 1 is adopted; otherwise, uncompressed mode (i.e. level 0) achieves lowest delay. Before each round of parameter transmission, the available bandwidth is estimated, and the compression level is adaptively adjusted. For example, when the transmission rate drops to 1 Mbps, the compression mode automatically switches to level 3, which can reduce the burden of communication and the delay of parameter transmission. Thereby, the total time cost is reduced accordingly.

In practice, the compression time and compression rate are calculated before the first transmission of the model parameters. The thresholds shown in Table 1 are calculated, and the compression level is adaptively selected to reduce the total delay cost. Compared with the scheme with fixed compression rate, our proposed adaptive mechanism can be more flexible and easier to satisfy the transmission constrains in different scenarios. In face of the fluctuation of wireless channel conditions, our proposed method can effectively improve the training efficiency of federated learning. Notice that lossless compression that we adopted completely preserves the model information, and does not affect the accuracy of the model.

2.2 Federated Learning Module

The main process of federated learning can be seen in Fig. 1. The whole process mainly includes local model updating, local model compression and uploading, global model averaging and global model distribution. The specific process is as follows:

- Each client starts the local training procedure based on the its own data set.
- After the training is procedure completed, the model parameters are sent to the server. The data to be uploaded are compressed immediately based on the adaptive level switching mechanism proposed in Sect. 2.1.
- The server updates the global model with a weighted average of the number of datasets owned by the participants, which is given by

$$\overline{w}_{t+1} = \sum\nolimits_{k=1}^{K} \frac{n_k}{n} w_t^{(k)}, \tag{2}$$

  where t is the round index, w represents the model parameters, K is the number of participants, and n denotes the amount of data that participants have uploaded.
- Finally, the server sends the updated global model to each client, and a client carries out a new round of training based on the local model and the newly received data. This

process is repeated until the algorithm converges. The stop condition of the iteration is all local model parameters are unified with global model parameters, or the number of iterations reaches a specified threshold.

2.3 Data Processing Module

The data processing module mainly deals with the images of the data set. Firstly, the image of the dataset is annotated, including the size of the image, the category name and location information of the target. Then, the original label format is converted to the format required by YOLOv5, including normalizing coordinates, indicating the category number and other operations.

2.4 Target Detection Module

The target detection algorithm adopted in our system is YOLOv5 based on Ultralytics. The number of parameters in YOLOv5 is relatively small, so that the burden on computing and communication is affordable for edge computing devices. The input layer of YOLOv5 adopts Mosaic data enhancement method to expand the data set, and the backbone network uses Focus and CSP structures. The output of YOLOv5 is the same as YOLOv3.

3 Experimental Results

In this paper, Intel (R) Xeon (R) CPU E5 - 2609 v4 @ 1.70 GHz is used for federated learning and training with two edge devices which are equipped with display cards NVIDIA GTX1080Ti. The operating system is Ubuntu 16.02, and the framework is PyTorch. The network bandwidth in our experiments is randomly distributed within a range from 1 Mbps to 150 Mbps.

In this paper, three safety helmet detection methods are trained as comparison baselines. The first one is our proposed method which contains two federal participants. The federal aggregation is carried out after every 5 rounds of local training. Each participant has a certain number of security hat data sets. The main time cost is concentrated on the model parameter transmission. The second baseline is the non-cooperative method, where the model of each edge device is trained separately. Only one single data set is used for training and the training time is relatively short. The third baseline is the centralized method, which collects all original data sets from two participants. Due to the fixed local computing capacity, the training time is doubled compared with the non-cooperative method.

An open-source safety helmets dataset, Safety helmets (hardhat) wearing detection dataset (SHWD), is adopted in this paper. The dataset contains 7581 images including 9044 people who wear safety helmets and 111514 who do not wear safety helmets.

Target detection not only needs to correctly classify the detected target bounding box, but also needs to determine whether the predicted box is consistent with the actual bounding box, and whether the target to be detected is correct. The positioning performance depends on the Intersection over Union (IoU). If IoU ≥ 0.5, it is considered as a correct detection. The accuracy and recall can be calculated according to IoU and confidence threshold, further calculate the average precision (AP), and finally calculate the average value of each category of AP. The notations mAP@0.5 and mAP@0.5:0.95 mean the mAP with IoU threshold 0.5 and the average mAP with IoU thresholds from 0.5 to 0.95 in steps of 0.05, respectively. The comparison of different performance metrics changing with training rounds of the three methods are shown in Fig. 2. The horizontal axis is the training cycle, and the vertical axis is the accuracy rate, the recall rate, mAP@0.5, mAP@0.5:0.95, respectively.

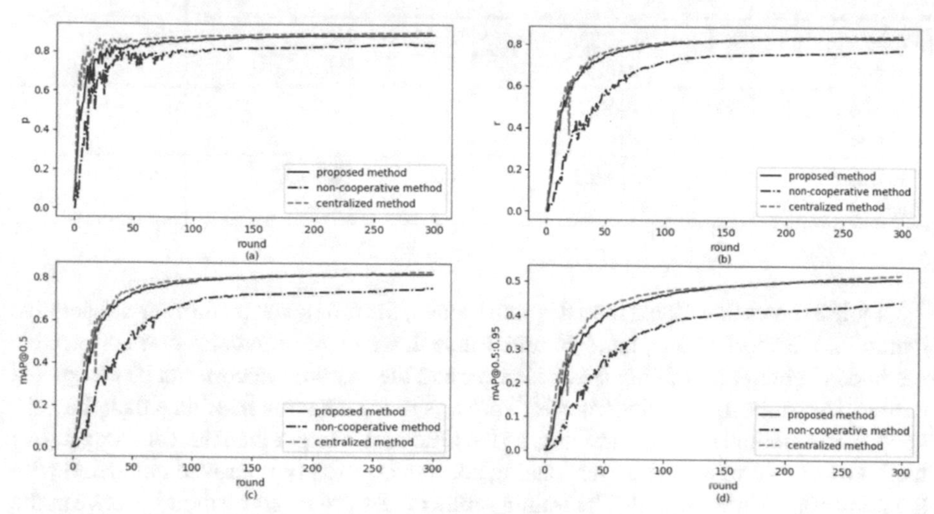

Fig. 2. The comparison of each evaluation index in the different rounds of three methods. (a) is precision. (b) is recall. (c) is mAP@0.5. (d) is mAP@0.5:0.95

From Fig. 2, we can observe the results after 300 rounds of training. Compared with the fluctuation at the beginning of training, the accuracy of the model starts to converge at about 100 rounds. The precision, recall and mAP of our proposed method and the centralized method are significantly better than those of the non-cooperative method. After 300 rounds of training, the mAP@0.5:0.95 of proposed method is 50.8%, the mAP@0.5:0.95 of non-cooperative method is 44.0%, and the mAP@0.5:0.95 of centralized method is 52.1%. The mAP@0.5:0.95 of our proposed method is 6.73% higher than that of the non-cooperative method, and 1.29% lower than that of the centralized method.

Compared with the non-cooperative method, since the proposed method can obtain the model parameters of other participants besides the local ones, it can obtain more information of the cap feature than the non-cooperative method, which makes the generalization ability of the model better. Therefore, the proposed method outperforms the non-cooperative method in accuracy performance. Since the data are not exposed to other participants in the training process of our proposed method, and only lightweight training parameters such as training gradient are transmitted to the data center for average, the accuracy will be reduced compared with the centralized method. However, these performance losses are traded off for data security and privacy protection, as well as the speed of training.

In addition, the training time, accuracy, recall, mAP@0.5 and mAP@0.5:0.95 of the three methods are shown in the Table 2.

Table 2. Training performance comparisons of 3 methods

Methods	Time (min)	Precision (%)	Recall (%)	mAP@0.5 (%)	mAP@0.5: 0.95 (%)
Proposed method	564	88.9	83.4	81.9	50.8
Non-cooperative method	366	83.5	77.6	75.3	44.0
Centralized method	747	89.9	84.5	82.9	52.1

Table 2 shows that after a model is well trained, the time spent on the non-cooperative method is the shortest one, i.e., 366 min. Since the amount of data for non-cooperative method is lighter, the training speed is faster and the training process can be completed earlier. The centralized method needs to process twice the amount of data than the non-cooperative method in model training, so the time cost is larger than the non-cooperative method, i.e., 747 min. The corresponding accuracy is greatly improved compared with the non-cooperative method. The training time of our proposed method is between the non-cooperative method and the centralized method, which is 564 min. The training time is reduced by 24.5% compared with that of the centralized method.

In terms of recognition accuracy, all the 4 performance metrics of our proposed method are very close to that of the centralized method. Specifically, after all the methods have completed the training, the precision, recall, mAP@0.5 and mAP@0.5:0.95 of our proposed method is 5.4%, 5.8%, 6.6% and 6.8% higher than that of the non-cooperative method, respectively. Our proposed method greatly reduces the time cost with a guaranteed training accuracy. As a result, the proposed method can obtain a high-performance model within a shorter time, and data security and privacy are well guaranteed compared with the centralized method.

4 Conclusion

In this paper, we propose a federated learning system for real-time target recognition in 5G MEC networks, which not only ensures the privacy of users, but also increases the recognition accuracy. In this system, an adaptive lossless compression level switching mechanism is adopted, which can effectively improve the training efficiency over time-varying wireless channels. Experimental results show that compared with the non-cooperative method and the centralized method, our proposed method has 6.73% and 24.5% improvements in accuracy and delay reduction, respectively.

Acknowledgement. This work is supported by National Natural Science Foundation of China (No. 61801051, No. 61971069), and the Open Project of a Laboratory under Grant No. 2017XXAQ08.

References

1. Zeng, R., et al.: FMore: an incentive scheme of multi-dimensional auction for federated learning in MEC. In: 2020 IEEE 40th International Conference on Distributed Computing Systems (ICDCS), pp. 278–288. IEEE (2020)
2. Zhu, Z., et al.: Federated multi-agent actor-critic learning for age sensitive mobile edge computing. IEEE Internet of Things J. **9**, 1053–1067 (2021)
3. McMahan, B., et al.: Communication-efficient learning of deep networks from decentralized data. In: Artificial Intelligence and Statistics, pp. 1273–1282. PMLR (2017)
4. Liu, K.H., et al.: Fine-grained offloading for multi-access edge computing with actor-critic federated learning. In: 2021 IEEE Wireless Communications and Networking Conference (WCNC), pp. 1–6. IEEE (2021)
5. Ma, X., et al.: Scheduling policy and power allocation for federated learning in NOMA based MEC. In: GLOBECOM 2020–2020 IEEE Global Communications Conference, pp. 1–7. IEEE (2020)
6. Zhu, Z., et al.: An edge federated MARL approach for timeliness maintenance in MEC collaboration. In: 2021 IEEE International Conference on Communications Workshops (ICC Workshops), pp. 1–6. IEEE (2021)
7. He, K., et al.: Spatial pyramid pooling in deep convolutional networks for visual recognition. IEEE Trans. Pattern Anal. Mach. Intell. **37**(9), 1904–1916 (2015)
8. Girshick, R.: Fast R-CNN. In: Proceedings of the IEEE International Conference on Computer Vision, pp. 1440–1448 (2015)
9. Zhou, F., et al.: Safety helmet detection based on YOLOv5. In: 2021 IEEE International Conference on Power Electronics, Computer Applications (ICPECA), pp. 6–11. IEEE (2021)

Author Index

Q. Liang et al. (Eds.): Artificial Intelligence in China, LNEE 854, pp. 713–716, 2022.
https://doi.org/10.1007/978-981-16-9423-3

Printed in the United States
by Baker & Taylor Publisher Services